Historical & Cultural Astronomy

The Historical & Cultural Astronomy series includes high-level monographs and edited volumes covering a broad range of subjects in the history of astronomy, including interdisciplinary contributions from historians, sociologists, horologists, archaeologists, and other humanities fields. The authors are distinguished specialists in their fields of expertise. Each title is carefully supervised and aims to provide an in-depth understanding by offering detailed research. Rather than focusing on the scientific findings alone, these volumes explain the context of astronomical and space science progress from the pre-modern world to the future. The interdisciplinary Historical & Cultural Astronomy series offers a home for books addressing astronomical progress from a humanities perspective, encompassing the influence of religion, politics, social movements, and more on the growth of astronomical knowledge over the centuries.

The Historical & Cultural Astronomy Series Editors are: Wayne Orchiston, Marc Rothenberg, and Cliff Cunningham.

W. M. Goss • Claire Hooker • Ronald D. Ekers

Joe Pawsey and the Founding of Australian Radio Astronomy

Early Discoveries, from the Sun to the Cosmos

 Springer

W. M. Goss
National Radio Astronomy Observatory
Socorro, NM, USA

Claire Hooker
Sydney Health Ethics, Sydney
School of Public Health
Sydney, NSW, Australia

Ronald D. Ekers
Australia Telescope National
Facility, CSIRO
Epping, NSW, Australia

This work was supported by National Radio Astronomy Observ. CSIRO ATNF.

ISSN 2509-310X   ISSN 2509-3118   (electronic)
Historical & Cultural Astronomy
ISBN 978-3-031-07915-3   ISBN 978-3-031-07916-0   (eBook)
https://doi.org/10.1007/978-3-031-07916-0

Cover figure: Fritz Goro photo of Pawsey "adjusting" the 21 cm Potts Hill grating array at the edge of the Potts Hill reservoir in March 1951. Credit: Getty Images, Fritz Goro, The Life Picture Collection, licensed by Getty November 2021 / Licence organised by Shutterstock, Inc., New York, NY (5 Nov 2021) Original, Premium Editorial All Media Cover

This Springer imprint is published by the registered company Springer Nature Switzerland AG
The registered company address is: Gewerbestrasse 11, 6330 Cham, Switzerland

Joe Pawsey on the catwalk at the top of the linefeed support tower of the Vermilion River Radio Observatory, University of Illinois, September 1961

For: Hastings and Elizabeth (Liz) Pawsey,
who inspired and accompanied our interest in
J.L. Pawsey throughout the last decade

Foreword

Radio astronomy, once a scientific *wunderkind*, is now almost a century old. It is hard to teleport ourselves back into the mid-twentieth century when astronomy was synonymous with tubes holding glass mirrors and lenses. But into this world came celestial observations with radio receivers and arrays of oddly shaped antennas, leading to fundamental contributions to both astronomy and physics. And during succeeding decades observers expanded their view ever wider, eventually using all other parts of the electromagnetic spectrum (gamma rays, X-rays, ultraviolet, infrared), cosmic ray particles, neutrinos, and, just in recent years, even gravitational waves. But it all began with radio astronomy.

Radio astronomy is now old enough to have spawned many studies using tools such as intellectual history, sociology of research groups, national and institutional histories, and interdisciplinary interactions. One important genre, however, has received little attention—biography. In fact, setting aside *auto*biographies, there have been only *two* substantial biographies of key radio astronomers![1] Yet the development of any science can only be weakly understood without close examination of the personalities and motivations of the individuals who actually *did* it. For the story of early radio astronomy, any consideration of persons needing a biography easily points to one man, Joseph Lade Pawsey. Thus, it is especially welcome that Miller Goss, Claire Hooker, and Ron Ekers have devoted more than a decade of study and collaboration to produce this comprehensive biography of Pawsey, the most important figure responsible for the phenomenal success and world leadership of early Australian radio astronomy.

We learn of Pawsey's rural roots in Victoria, Australia, his academic success culminating in a PhD in Physics from Cambridge University in 1934, his early radio research in England, and then his return home at the start of World War II in order to join the newly created Radiophysics Laboratory (RPL) in Sydney, devoted to top-secret development of radar for myriad uses on the ground, in the air, and at

---

[1] (1) Robertson, P. (2017) and (2) Goss and McGee (2009).

sea. At war's end in 1945 Pawsey became the leader of a small group of physicists and engineers that investigated recently discovered radio wave bursts from the Sun. This soon led to an important new understanding of the solar atmosphere, as well as to development of analysis techniques that have proved fundamental to radio astronomy ever since. Over the next 15 years as his group grew and matured, it arguably became the most distinguished Australian research group in the world *in any area of science*. And its only peer in radio astronomy was in England. Eventually, Pawsey's technical guidance and personal mentorship led to many outstanding researchers outgrowing RPL and establishing their own groups in universities at home and abroad.[2] For reasons described by the authors in detail, Pawsey himself also decided at this time to leave RPL in order to lead a new national radio observatory in the USA. But before he could begin in 1962 he was diagnosed with a brain tumour and soon died at the age of 54.

While following Pawsey's life and career, the reader learns not only about his own research, but also about other major developments in radio astronomy swirling around him. Several long overseas tours allowed attendance at conferences, visits to major facilities, and consultation with foreign radio researchers. In particular, we learn of sometimes bitter battles at conferences over the nature and number of radio sources in the universe, the reliability of data gathered from different radio telescope designs, and the validity of any subsequent cosmological inferences. Ironically, Pawsey and RPL's Bernard Mills fought especially with Martin Ryle's group in the Cavendish Laboratory, Cambridge, precisely where Pawsey had earned his PhD 25 years earlier. Pawsey also co-authored the definitive textbook of radio astronomy for its time (Pawsey & Bracewell, 1955, *Radio Astronomy*).

To produce this volume the authors have scoured the archives of institutions around the world, as well as those of Pawsey's family and colleagues. The authors' varied backgrounds have meant that all aspects of Pawsey's life have been exhaustively covered. Goss began his deep dive into the history of Australian radio astronomy with a biography of Ruby Payne-Scott, an important early colleague of Pawsey's and the first woman in the world to make radio astronomical measurements.[3] Both Goss and Ekers have had distinguished careers in radio astronomy around the world, as well as having been observatory directors—Ekers in fact was formerly director of a descendant of RPL, the Australia Telescope National Facility. Hooker is a historian and sociologist of science and medicine who also studied the career of Payne-Scott as part of her doctoral dissertation, and since then has investigated many aspects of the history of Australian science and medicine.

In summary, historians of astronomy and of technology will welcome this much-needed biography. They will find that the authors have convincingly made the case

---

[2] Prominent among those who left RPL in the period 1955–1965 were Ron Bracewell (to Stanford University), Frank Kerr (to University of Maryland), John Bolton (to Caltech for 6 years, then returning), and Bernard Mills and Wilbur Christiansen (both to University of Sydney).

[3] Goss and McGee (2009), *Under the Radar: The First Woman in Radio Astronomy, Ruby Payne-Scott*. A less technical and shorter version by Goss (2013) is *Making Waves: The Story of Ruby Payne-Scott, Australian Pioneer Radio Astronomer*.

for the importance, even centrality, of Joe Pawsey in shaping the exciting development of early radio astronomy, a field that in turn revolutionised astronomy as a whole.

University of Washington,   Woodruff T. Sullivan III
Seattle, WA, USA

Preface

J.L. Pawsey and a New Understanding of Early Radio Astronomy

Why did we write a book about Joseph Lade Pawsey? What lessons does his life provide in the second decade of the twenty-first century? How did an only child from a farm family of modest means, in rural Victoria in Australia, become one of the leading radio astronomers by the mid-twentieth century? How did a new, sparsely populated nation, with very constrained experience in and resources for academic research, largely invent and dominate the new scientific field of radio astronomy in the years that immediately followed the devastations of World War II? How can we best characterise the scientific challenges of the first two decades of radio astronomy, and how do we gauge the conditions for success?

These are the questions that underpin our exploration of the life of this remarkable scientist, science leader, and father. The new post-war radio astronomers opened the second window on the universe. Before 1945, optical astronomers could study the universe in a narrow band of wavelengths from about 400 to 700 nm, hardly a factor of two. The mid-twentieth century radio astronomers had an impressive wavelength range of a factor of 10,000 as they studied the new universe at wavelengths of 1 mm to 10 m. The new window resulted in a vastly expanded view and many new discoveries; five Nobel Prizes (to eight recipients) were to be awarded. Uncharacteristically for a small nation with a still emerging sense of national identity, Australians were to play a major role in these new discoveries. That this came about at all and the way in which it came about is a story that centres on Pawsey.

Who Was J.L. Pawsey?

Anyone familiar with the bare-bones story of the beginnings of radio astronomy knows something of Joe Pawsey by proxy, simply because he led the radio astronomy group in Sydney at the Radiophysics Laboratory (RPL) CSIRO, one of the three original research groups that established this new field of science between them. But Pawsey is less well known than the leaders of the other two groups, Nobel Prize winner (Sir) Martin Ryle (Cavendish Laboratory, Cambridge) and (Sir) Bernard Lovell (Manchester). That he is less well known is due to several factors, among them: his early death; his lack of interest in public attention; his practice of declining authorship on most of the group's papers, despite making extensive contributions to many; his transition from active research to science leadership during the 1950s; and the general tendency for histories of physics to be centred on the USA, UK, and Europe.

Accidents of history placed Pawsey, a fine radio physicist with a speciality in antenna design, in a position to pursue curiosity-driven research in a field (astronomy) in which he was totally naïve. In his own words (from an Australian Broadcasting Company television interview on 11 November 1960), he reflected on how radio astronomy had begun:[4]

> I had done very little astronomy, or no astronomy, until the end of the war ... I'd been working some years in England on the early development of television and then on radar in Australia and we were interested in finding really interesting things for research. Now there was a complete mystery evident to us in those days. A man from America in the early 1930s had discovered radio waves coming from the Milky Way and this seemed to be worth trying to find out something about. Then the thing was complemented by the parallel discovery of finding radio waves from the sun.

Something "really interesting" had been stumbled upon indeed! This story, the story of the beginnings of radio astronomy, has been retold many times. But its unexpectedness, its audacities, and its achievements remain fresh: how a small number of radio "boffins",[5] knowing nothing whatsoever about astronomy, collected a pile of Army surplus radar equipment, held it together with curses and soldering irons, and used it to make a rapid series of transformational discoveries that changed human understandings of the universe within two short decades.

Looking back, Pawsey occasionally marvelled at where a career of "following his nose" had led him. Growing up in a family of very modest means, he credited his mother's fierce dedication to his education with the academic achievements that gained him a scholarship and access to a university education, at that time an advantage out of reach for most Australians. His skills then placed him as a leader

---

[4] Australian Broadcasting Corporation, television programme HORIZONS, 1960, Nov 11 interview with Joseph Pawsey and Ron Giovanelli by Moderator George Baker. NRAO ONLINE.56 has a partial transcript of the interview.

[5] Boffin was a British slang term common in World War II for someone engaged in technical or scientific research.

in Sydney's radar research group (at the coyly named Radiophysics Laboratory) and thus, when the war ended, in charge of a group of first-class radio scientists and engineers. Knowing of Karl Janksy's 1933 observations of radio emission from the Milky Way, and of wartime observations of solar radio emission, Pawsey placed investigation of "cosmic noise" on a list of possible projects for investigation, as the years of peace began. And the rest, as the saying goes, is history.

Pawsey's distinctive leadership style of supporting and fostering the independence of promising young scientists generated an environment perfectly suited to discovery science. Within a few years, Ruby Payne-Scott, John Bolton, Bernard Mills, W.F. "Chris" Christiansen, Frank Kerr, and Paul Wild had all become internationally known for their various lines of investigation, which slowly developed into fully fledged research programmes. Meanwhile, Pawsey's immediate superior, E.G. "Taffy" Bowen, drove funding for these programmes, in part through leveraging his international network of major scientific philanthropists.

By the end of the 1950s, Pawsey could agree, in a letter to his mother, "As you say I have achieved eminence in my profession."[6] But there had been difficulties, too. Some discoveries (e.g. the HI line), which could quite easily have been made at RPL, were missed. A rupture took place with the most successful of the original research group, John Bolton, who left to start a radio astronomy research programme at Caltech, USA, in 1955. Worse, by 1960, Pawsey's and Bowen's increasingly clashing visions for RPL culminated in the complete fracture of Pawsey's research group. Pawsey was faced with the need to adapt to a new era of radio astronomy, with increasing numbers of research groups globally, and moving away from specialised instruments towards much bigger general-purpose research facilities. At the end of 1961, he accepted the Directorship of the new, complex, and politically challenging National Radio Astronomy Observatory, USA,[7] only to die a year later, at the age of 54, of brain cancer.

But before we begin to tell this story, it will be helpful to the reader to learn something of our approach to this book, which is less a biography than a means of developing new insights into the early history of radio astronomy in Australia by considering it from Pawsey's perspective.

Our Approach: Understanding Science Through History

First, we would like to briefly consider the question of what the purpose of history actually is: to entertain? To create a record of the past? To find and celebrate heroes? To untwist the thinking that led to what we now know to be correct, or incorrect, theories and ideas? The three authors of this book, without realising it, had slightly

---

[6] J.L. Pawsey to his mother, 20 May 1956 (Joe and Lenore Pawsey Collection).

[7] See Kellermann et al. (2020) for a frank description of the organisation that Pawsey was about to take on.

different answers to this question. Historians come in various guises, often classified as (1) antiquarian, (2) practitioner, (3) academic, and (4) public (sometimes termed "popular"). This book combines antiquarian, practitioner, and academic approaches.

Antiquarian historians locate, assemble, and delight in the details of the past, for its own sake. Antiquarian history describes family history, local history, and specialist histories, for example, of railways, or dolls. Practitioner histories are written by insiders: doctors who write the history of medicine, pilots who author aviation histories, and scientists who author the histories of their disciplines. Practitioner histories typically select the events and topics of professional interest for inclusion, and represent these events in ways that reflect what is valued or expected in that professional context. In science, this often means writing about discoveries—and not about ideas that led nowhere. These histories may also be written as a means of identifying lessons that can be used to improve professional practice. In this book, we comment in various places on analogies between past problems in radio astronomy and present ones and draw some possible inferences for present approaches.

Academic historians develop conceptually driven (historians say "theoretically driven") analyses of historical circumstances, aimed at understanding the social structures, systems, and processes that drive change over time. Academic historians are also interested in identifying continuities—many institutions, and ideas, remain intact even under conditions of apparent social turbulence. Academic historians of science have often been interested in how particular ideas came to be suggested, and later accepted. This is sometimes termed "the history of ideas". They may also be interested in how other systems or forces, such as the development of new technologies, influence the development of science.

To illustrate these different approaches, we can consider the example of the ionosphere. A practitioner historian such as Stewart Gillmor might provide an account of how the ionosphere was "discovered" by Sir Edward Appleton, and publish technical details about the experimental design and equipment used. An academic historian such as Chen-Pang Yeang might not use the word "discovery", but instead ask how the idea of a conducting layer in the earth's atmosphere came to be held by many different people. Such an historian might identify that recently invented technologies, such as the coaxial cable, might influence this idea. (Heaviside imagined that an entity rather like a giant coaxial cable encircled the earth, with radio waves reflected within it.) Antiquarian historians might fill in this picture with minutiae (e.g. the type of connector used by Marconi when cobbling his first radio transmission equipment together).

This book draws together antiquarian, practitioner, and academic historical approaches, a triangulation that is both rigorous (by virtue of the (frequent!) critiques and checks that each perspective provides of the others) and exploratory (by virtue of the new insights each perspective opens to the others). Our different approaches, combined with the sheer volume of available records, have, of necessity, resulted in some topics and issues being covered in greater detail and analytic depth than others, and there will still be room in the future for additional studies of this period, as well as of those that succeeded it.

Different Perspectives on History

Story is often imagined as a form of archaeology, with the "truth" waiting to be "uncovered" from the concealing layers of paper piled in archival boxes. But writing a history is much more like building a house than it is like disinterring a fossil skeleton. The past is not just "there" waiting to be "discovered" in the same way a supernova is "there". History always involves actively constructing a story for the reader. The historian must select which events, out of a lifetime's worth, should be selected for inclusion in the story, and how these events should be interpreted. History is always a matter of interpretation, and this is as true of those actually involved as of the historian who comes later. As we comment in several places throughout this book, neuroscientists now recognise that memory is an *active* process of creation and is unreliable as a record of the past. Rather, memories provide extensive information about how someone sees themselves and others.

In this book, our interpretation has been influenced by our shared values. Along with the first generation of radio astronomers, very much including Pawsey, Bolton, and Ryle, we value science pursued using instruments with which the scientist is intimately familiar. The early radio astronomers all possessed a mix of skills in both engineering and physics. Their deep understanding of the instruments they built, maintained, and used both grounded their trust in their data and was inseparable from the development of their theoretical (mathematical) and conceptual thinking. The scientist in this circumstance can be considered analogous to a "master craftsman". A scientist who works in this way enjoys a special and privileged relationship with her or his results. This was deeply valued by the first generation of radio astronomers (see quotes in Sullivan, 2009, p. 450), and we also celebrate it through the writing of this book. Other values underpinning our approach include our value for pure, or "curiosity-driven", science; for congenial and collaborative research environments; and for a disciplinary "ecosystem" that encourages the flourishing of research.

Structure of This Book

We have opted for a presentation that is mostly chronological, with events presented as closely as possible to the time order in which they occurred. However, some narratives—such as the narrative of the imagining, funding, planning, constructing, and finally using the Parkes radio telescope—cover more than a decade. Rather than confuse readers by switching between multiple stories in parallel, we have also devoted single chapters to such narratives that, as a result of following events across several years, do not appear in strict time sequence.

Part I, Childhood (Chaps. 1–3), not only follows Pawsey's early years but sets the scene for his career by discussing how strongly "science" in Australia had been, and still was at that time, influenced by the context of colonialism. We continue this theme in **Part II, Becoming a Scientist** (Chaps. 4–8). We identify how government investment in Australian science, through the creation of a national applied science organisation (the Council for Scientific and Industrial Research, CSIR) and a Radio Research Board, not only provided Pawsey with the opportunity to pursue a research career, but also made trained personnel available for crucial radar research during World War II. In this section, we also discuss the hitherto largely unrecognised intellectual foundations for radio astronomy in ionospheric physics. Certainly Pawsey brought these research concepts and methods to bear in his later work on radar and extra-terrestrial radio observation, as well as to his few years working for EMI, where he also became an expert in antenna design at VHF frequencies (near 40 MHz). In **Part III, WWII** (Chaps. 9 and 10), we provide an overview of how this research experience shaped Pawsey's approach to the practical difficulties of developing radar defensive weapons that could be used in the tropical conditions relevant to the Pacific theatre, very different from the design of those used by the British and US forces in Europe.

We follow the emergence of radio astronomy in **Part IV, Hot Corona** (Chaps. 11–14), which revisits the initial exploratory observations that were made in 1945 and early 1946 to such dramatic effect. Drawing from materials found in the Sally Atkinson archive (see below), we make the impact of the initial series of observations made at Collaroy in late 1945, which established the association between radio bursts and sunspots, more visible. We revisit the use of sea-cliff interferometry, which, in 1946, produced Ruby Payne-Scott's observations of radio emission from a sunspot of such intensity that it had to be from a non-thermal source. We trace the puzzle of the emission mechanism later in Chap. 34.

One key feature of this period was the observation of radio emission from the solar corona: neither of the UK or New Zealand groups had instruments with adequate sensitivity to detect the weak signal from the corona that was always present. We provide detailed insights into how Pawsey and theorist David Martyn together came to identify the million degree corona, the publication of which was made difficult by concerns over status and priority, through which Pawsey's calm and constructive leadership style was particularly evident.

Part V, Connections (Chaps. 15–18), discusses Pawsey's strong orientation towards internationalism and follows his efforts to cultivate a global network of scientific colleagues through which to disseminate the observations and achievements made by the group at RPL. Despite Australia's comparative wealth after the War and the clear early dominance of the RPL group, the Sydney researchers continued to face difficulties related to their postcolonial status and geographical location. Publishing papers and accessing research students, colleagues in optical astronomy, and sufficient theorists all presented challenges, while scientific meetings were almost invariably overseas. In this section, we follow the first of Pawsey's many long journeys overseas to build and maintain networks. During the first of these journeys, Pawsey coined the term "radio astronomy"—a term that also suggested itself to others at that time and swiftly was taken up in use. Pawsey's attempts at establishing international collaborations were not always successful. We particularly discuss challenges in interactions between the Sydney group and Martin Ryle's group at the Cavendish Laboratory, Cambridge.

Part VI, Quiet Leadership (Chaps. 19–26), explores what Pawsey rated as his greatest achievement (and we concur)—the creation of a research environment that enabled a series of brilliant young scientists to establish independent lines of research. This section summarises the major events of the period 1950–1955, including the coup of hosting a major scientific meeting, URSI, in Sydney in 1952 (although ironically Pawsey himself was ill with influenza after the first few days of the congress). In these chapters, the development of "cross"-style telescopes, the missed opportunity of the first detection of the HI line, and early solar research are also presented. We also devote a chapter to the history of the discovery of Sagittarius A and its association with the galactic centre. This chapter demonstrates the inaccuracy of the notion that discoveries (or science in general) are made by single individuals, at a single point in time. Instead, we show how the discovery of the galactic centre was a long process to which many people contributed insights and observations: RPL scientists Jack Piddington and Harry Minnett, who first detected a radio source near the galactic centre at 20 cm; John Bolton, who led four colleagues at RPL to build a hole-in-the-ground dish-shaped antenna in their off-duty hours in order to make more detailed observations of the radio emission from the galactic centre region; Pawsey, who then supported the endeavour and championed its results; and Dick McGee, who made the actual observations of the galactic centre and, together with John Bolton, instantly recognised it as such. This section concludes with chapters exploring Pawsey's brief resumption of ionospheric research, in the context of his increasing shift away from active research and towards providing leadership for RPL. By the mid-1950s, he was accumulating a series of significant international roles and honours, including election to the Royal Society.

Part VII, Towards a Bigger Science (Chaps. 27–32), revisits the decade-long development of the Parkes radio telescope, then referred to as the Giant Radio Telescope or GRT. In this narrative, mostly taking place in the years 1955–1961, the relationship between Pawsey and Bowen grew increasingly strained. We discuss how Pawsey played a key role in facilitating the design and development of the GRT by presenting the most convincing science case for its construction. But it was Bowen who was the entrepreneurial force for the project. In the years as the GRT was designed and constructed—with many challenges—Pawsey's attention was more focused on the research of his proteges Mills, Christiansen, and Wild, and on the establishment of needed international resources, such as the formalisation of galactic coordinates. Bowen became increasingly frustrated with Pawsey's lack of focused drive for completion of the GRT, leading to eventual schism at RPL.

In **Part VIII, The Development of Understanding** (Chaps. 33–37), we turn to an exploration of the history of ideas in Pawsey's era. We set the scene in Chap. 33 with a discussion of Pawsey's own philosophy of science, to the degree that we can ascertain it from his own words in letters to his mother and in correspondence with the scientists at RPL. Here we show how closely Pawsey's view of "primary" and "verification" discoveries closely matches the analysis of the "serendipity pattern" by Merton (which followed that of Nobel laureate Irving Langmuir). These categories provide an apt classification for discoveries made during the first two decades of radio astronomy research internationally.

At this point we depart from our chronological progression to show how some of the concepts developed over a wide time range during the period of Pawsey's activity as a radio astronomer at RPL: 1945–1961. We start with a chapter (Chap. 34) on the development of a theory for radio emission from cosmic sources since this played a key role in the developments of radio astronomy between 1945 and 1955. Progress was hindered by the initial radio star concept which had to be abandoned; further progress was not possible until the synchrotron radiation mechanism was accepted as the explanation for both extragalactic and galactic non-solar emission in the late 1950s. In Chaps. 35 and 36, we untwist and recount the tangled controversy over discrete source counts and their implications for cosmology (discussed in detail by Edge and Mulkay). The importance of radio astronomy for cosmology was overshadowed by the bitter conflicts related to the quality of the surveys and also the dispute between the Cambridge radio astronomers and the "steady-state" cosmology theorists. We make what may be the first in-depth study of the underlying reasons for this dispute. We show how, and why, the wrong interpretations were made by Mills (and supported by Pawsey) using the better-quality observations, and how the correct interpretations were made by the Cambridge group based on their flawed catalogue. Chapter 37 is a description of the history of aperture synthesis, with an emphasis on Pawsey's role in the initial formulation of the Fourier synthesis concept in 1947. We then follow the parallel developments by Ryle's Cambridge group and Pawsey's Sydney group. The two endeavours diverged when the Cambridge group adopted the electronic computer to make the Fourier transforms, an outcome of the serendipitous cross-fertilisation of ideas that resulted from close juxtaposing of the crystallography and radio astronomy groups at the Cavendish in the late 1950s. At this time the Australian effort languished, and it was the successful exploitation of the technique that led to Ryle's Nobel Prize in 1974.

We return to the increasing tension and disagreement between Pawsey and Bowen in **Part IX, Death and Legacy**. By the end of the 1950s, the limitations of Pawsey's "learning organisation" model had been reached. By orchestrating John Bolton's return from Caltech (where he had spent a spectacularly successful 5 years building a new radio astronomy facility at Owens Valley and achieving identifications of many radio sources (Robertson, 2017)) to lead research at the newly built GRT, Bowen effectively engineered a serious fracture at RPL. Mills and Christiansen departed for university posts, and Pawsey was left with less direct control over the group as a whole. In these circumstances, Pawsey finally accepted a repeated offer to take the role of director of the newly formed National Radio Astronomy Observatory in the USA. How he might have handled the challenges of this role remains unknown, as he died later that same year.

We conclude the book with summary reflections on the value, and values, of a man like Pawsey, whose comparative modesty and humility enabled an extraordinary scientific leadership, whose lifetime saw a dazzling succession of discoveries, and whose pleasures and satisfactions were many. Paired with the entrepreneurial flair and enjoyment of power and status of Taffy Bowen, Pawsey's steady scientific stewardship underpinned the extraordinary succession of discoveries and achievements that marked the first 17 years of radio astronomy at RPL.

Pawsey in the History of Radio Astronomy

This book is not a traditional biography; it is not devoted to providing a complete record of Pawsey's life. Instead, it primarily uses the story of Pawsey's life as the prism through which we re-evaluate the early years of radio astronomy. These years have already been the subject of published histories, of course. Woodruff T. "Woody" Sullivan's *Cosmic Noise, A History of Early Radio Astronomy*, a true *magnum opus*, provides a humorous, insightful, and exhaustive recapitulation.

Yet there is room to add many details about the Australian experience, and to explore J.L. Pawsey's particular role more thoroughly. *Cosmic Noise* concludes in 1953. This book extends the story over the ensuing decade, 1953–1962, and provides an Australian complement to a recent history of developments in the USA over these years, *Open Skies: The National Radio Astronomy Observatory and its Impact on US Radio Astronomy,* by Ken Kellermann, Ellen Bouton, and Sierra Brandt (2020).[8]

This book was informed by other historical works on the history of radio physics overseas and on the history of science in Australia. We made extensive use of Boris Schedvin's (1987) history of CSIR (later CSIRO), *Shaping Science and Industry: A History of Australia's Council for Scientific and Industrial Research 1926–1949.* We drew our analysis of "the tyranny of distance" affecting Australian science from papers by Rod Home and Roy Macleod. We were also influenced by histories of ideas, and of instruments and technologies themselves, particularly Louise Brown's (1999) drily witty and insightful *Technical and Military Imperatives: A Radar History of WWII* and Chen-Pang Yeang's (2013) *Probing the Sky With Radio Waves*, which, along with the many papers by Stewart Gillmor, provided us with the history of ionospheric research.

Works specifically about the early years of Australian radio astronomy have been published by one author of this book, W.M. Goss, and by science writer and editor Peter Robertson. We also refer readers to *Explorers of the Southern Sky* by Haynes et al. (2010), a comprehensive non-technical account of all Australian astronomy, including Aboriginal and radio astronomy. Robertson's histories include the Parkes Radio Telescope and, more recently, his biography of John Bolton. These are accessible written accounts, to which *Joe Pawsey and the Founding of Radio Astronomy in Australia* adds more details related to the politics, funding, and construction of the Parkes Radio Telescope.

Besides Robertson's biography of Bolton, a set of biographical studies of the first generation of Australian radio astronomers has been published by Goss, who held a postdoc at RPL and who has interviewed many of the people involved in this history. Goss has written two biographies of Ruby Payne-Scott (*Under the Radar, the First Woman in Radio Astronomer, Ruby Payne-Scott,* by Goss & R. McGee, 2009, and *Making Waves: The story of Ruby Payne-Scott*, 2013). This has certainly established

---

[8] Although the substantial portion of this book was written prior to the publication of *Open Skies*, we have integrated references to *Open Skies* where possible.

Payne-Scott's visibility in the historical record very firmly, and her name and story are now widely known. His subsequent 2017 monograph *Four Pillars of Radio Astronomy, Mills, Christiansen, Wild, Bracewell*, authored with R. Frater and H. Wendt, contains biographical sketches of these four major figures. Thus, all the main figures of the early period of Australian radio astronomy—Payne-Scott, Bolton, Mills, Wild, Bracewell, and Christiansen—have received biographical attention. This book supplies a biographical study of the only figure missing—the man who held the entire enterprise together, J.L. Pawsey.

The wholly naïve reader will be able to follow the events of the early years, solely by reading this book. Having said this, the authors have been anxious to avoid duplication of facts and events that are readily found in the other published works. For example, we only very briefly mention the foundational discoveries by Karl Jansky and Grote Reber in the 1930s, as their stories are prominently presented in many places. We have been selective in our discussions for another simple reason, that is, the wealth of material available. As Sullivan (2009) and others have commented, covering even just the first 17 years of radio astronomy greatly exceeds the capacity of any single book. For these reasons, we frequently refer readers to the published works mentioned above and warmly recommend that the reader enjoy the significant insights contained in these works.

We have attempted to use a retelling of J.L. Pawsey's life in science to bring an Australian perspective on the early years of radio astronomy into sharper focus. This is because, despite the even-handed treatment offered by *Cosmic Noise*, perspectives from places like Australia that are outside the USA and UK (or Europe) tend to be less visible in the history of science.

Ideas in This Book

Perspective and History

The history of science is often presented as a highlight reel of major discoveries and great men, from Newton and Kepler to Einstein and Hubble. This version of history is teleological: it presents the history of science as if science was always progressing inevitably from one discovery to the next. This "highlight reel" reflects what Pawsey's contemporary, sociologist of science Robert K. Merton, termed the "reward system" in science—scientists are rewarded, not by riches, but by recognition for priority and for discovery. A highlight reel version of Pawsey's life would tell us that he was the first to detect the Hot Corona, and draw the reader's attention to his election as Fellow to the Royal Society.

But the highlight reel is not, in fact, the truth about the past; it leaves a lot out of the picture. It is also not an accurate description of how science occurs in reality. Just how distorting the "highlight reel" (academic historians term it the "dominant narrative") is becomes immediately obvious when writing about an *Australian* scientist. There are very few Australian scientists in the highlight reel—their resources were more constrained; their attention was consumed by more urgent and local research questions; they had fewer opportunities for theoretical innovation. Told from an Australian perspective, the history of science *cannot* be a narrative of successive discoveries by a set of great men. Instead, the story must be one dependent on global communication systems, of contingencies and networks, of significant but imperfect leaders, and hence, of *collective* achievement, in which many individuals played roles. (We note at this point that isolation can also have its benefits, such as increased independence and innovation, and we include discussion of these in this book.)

We, as authors, have had to resist the temptation to interpret our protagonist as a kind of scientific hero. Pawsey *was* both deeply and broadly influential in those foundational decades. But what is of more interest to us was to explore *how* he came to be the scientific leader that he became. Some of questions we muse on in this book include: how was it that someone could become an internationally known physicist, when virtually none of the tiny number of Australian physics graduates of his era could even obtain a research post? How much did the different leadership styles of Pawsey and others (Ryle particularly) matter in influencing the number and quality of discoveries in the early years of radio astronomy? What explains the lines of research that the Australian radio astronomy group took up, and why were some overlooked and ignored? Why did some controversies continue quite bitterly, even after the main science questions were settled? Can we untangle the rationales and ideas behind different choices in instrument or research design? What *epistemic values* (i.e. values about knowledge, such as preferences for sensitivity, or reliability, generalisability, or specificity; a greater trust in data obtained from a receiver one had built oneself and knew well, or a greater trust in results that could be explained with mathematical theory) influenced how scientists formed and defended their views or reinterpreted their data?

Pawsey and the Philosophy and Sociology of Science

These are the sorts of questions asked by academic and practitioner historians of science. Pawsey himself, clearly interested in questions about the nature of science, would likely have been interested in them—the history, philosophy, and sociological study of science was just beginning during his lifetime. Radio astronomy has long been of interest as a case study of the emergence of a new field of science. How knowledge develops when all we have available to us is partial and limited evidence, constrained instruments, and erroneous implicit assumptions is enduringly fascinating!

Numerous discussions of the philosophy of science informed this book. By "philosophy of science", we mean "analytic explanations for how science 'works.'" For example, we asked ourselves whether early radio astronomy fitted Popper's model or Kuhn's model of how science works. Sir Karl Popper—who was a contemporary of Pawsey's and held an academic appointment at Canterbury

College, Christchurch, at the same time as Pawsey's colleague and superior Sir Fred White—suggested the counter-intuitive idea that science progresses, not by discovering new things, but by being very good at identifying which ideas are *incorrect* (i.e. through *refutations* rather than through *confirmations*). Did early radio astronomy progress through refutations? The answer was no, not often. (There were some examples of important refutations—e.g. the "radio star model" of the early 1950s was clearly refuted by Mills's evidence that the discrete sources they had detected were extended, and not the size of stars—and some more complex examples—e.g. synchrotron emission was initially refuted by lack of polarisation, but this refutation was incorrect because the prediction was flawed.) Did radio astronomy advance through the process that Thomas Kuhn suggested drove scientific change, in which the accumulation of a sufficient number of "anomalies" (data outliers) makes it impossible to retain a central organising concept (e.g. a geocentric universe), until a revolution in thinking takes place?[9] The answer was also "no", although we did repeatedly see this process occur on a much smaller scale and shorter time period than hypothesised by Kuhn.

Many other theoretical frameworks from the extensive philosophy of science literature were similarly trialled to see how well they explained the events, decisions, and actions in this book. For example, given that doing radio astronomy requires an assemblage of financial and political support, the right scientists and engineers, the needed instruments, computers, libraries and geographical resources, and useful ideas and methods for testing and exploring them, would Ankeny and Leonelli's improvisational jazz-band analogy of the development of "repertoires" in science allow us to better understand choices about "cross" or "dish" telescope designs? (Ankenny & Leonelli, 2016). Given that instrumentation and scientific and engineering considerations were so intertwined in this period, could the co-development of ideas and instruments be best understood as a process of "design thinking"? (Farrell & Hooker, 2008). These and other concepts from the philosophy of science such as black boxing, tacit knowledge, and inscription devices (Sullivan, 2009, p. 12) certainly describe aspects of scientific practice, but none of these provided a compelling explanatory framework for us.

This book therefore independently converges on the same lines of analysis as did Sullivan, and before him, on the major sociologically informed study of the early years of radio astronomy, *Astronomy Transformed*, which was the outcome of a collaboration of Martin Ryle's student David Edge and sociologist of science Michael Mulkay (1976). The key feature of the first 5 years of radio astronomy was relatively naïve observation, in which the first radio astronomers simply discovered what, if anything, they could detect using the surplus radar instruments at their disposal. Often, progress would occur by finding that, as Cambridge radio astronomer Francis Graham Smith remarked, "you've got a good technique and you say, 'What's the most interesting way to apply this technique?'"(Sullivan, 2009, p. 451). Sometimes the absence of preconceived assumptions and ideas left the path

---

[9] https://plato.stanford.edu/entries/thomas-kuhn/ and Kuhn (2012).

open for discovery, but mostly progress then took the form of attempts to modify and design new instruments that could improve the woeful inaccuracy and inconsistencies of early observation.

The intellectual interest of the first 17 years of radio astronomy therefore has more to do with analysing the interactions between instrument use and design, observation, and theoretical and conceptual development, with inferences and modifications flowing in both directions (Sullivan, 2009, p. 449). Chen-Pang Yeang's *Probing the Sky With Radio Waves* (Yeang, 2013) provided us with a model of this sort of analysis, showing how both developments in instrument design and the "paper tools" of (mathematical) theory influenced ionospheric physics before 1930.

Sociologically, our interest was drawn to the effects of different workplace cultures, the impact of different personalities and different styles of leadership, and the contingencies of available resources and support, on various episodes in the way the "technoscience" (as Sullivan, 2009, termed it; see e.g. pp. 449–452) of radio astronomy evolved. Over the longer term of this book and of the field, these become questions about how the structural relationships between physicists and engineers, and the different ways in which transitions to "big science" occurred, have affected what research programmes are pursued, and how successful they have been (Galison et al., 1992; Price, 1986; Sullivan, 2009, pp. 420–423).

J.L. Pawsey: A Quintessential Mertonian Scientist

As it turned out, the approach taken by Robert Merton, a sociologist who first used sociological methods to understand how science works, proved the most useful for understanding Pawsey and the early years of radio astronomy. Merton suggested that science functioned as a result of a unique cluster of values and behavioural norms, which he termed "the ethos of science". These norms were those of communalism (knowledge belongs to the scientific community and the public); universalism (scientific claims are not judged by the class, race, or other social characteristics of the scientist making them); disinterestedness (scientists pursue research for its own sake and not for personal reward); and organised scepticism (scientists subject all claims, including their own, to sceptical scrutiny). Merton considered that science "works" only because of these norms: scepticism is key to finding errors; the communalism of knowledge is what makes it possible to question all knowledge claims; universalism underpins the practice of objectivity; and disinterestedness enables the pursuit of knowledge for its own sake (Cetina, 1991).

Although Merton's analyses of these institutions, such as reward systems in science, have been extensively critiqued since their publication in the 1950s and 1960s, they usefully described much of the interactions and events of Pawsey's professional life. Perhaps this should not have been such a surprise, because Merton (1910–2003) was describing the science of his era—which was, of course, the period of Pawsey's career (though Merton did not examine radio astronomers specifically, to our knowledge). These norms did not necessarily describe the behaviour of scientists as individuals—the relative secrecy that characterised Ryle's group at Cambridge, for instance, offers a good example of what Ivan Mitroff (1974) would later term "counter-norms" (and Merton saw them as institutional, rather than individual, features of science). Yet these norms did indeed characterise Pawsey's own ethos and professional behaviour, and many of the first generation of radio astronomers in general.

Edge and Mulkay's 1976 study of early radio astronomy, *Astronomy Transformed,* also primarily drew on Merton for its analysis. As *Astronomy Transformed* is referred to only in passing in this book, we mention here that it was extensively used in developing our analysis. Based largely on interviews with mostly UK radio astronomers, *Astronomy Transformed* covered roughly the same historical period as *Joe Pawsey and the Founding of Radio Astronomy in Australia.* Its focus was to analyse different patterns of competition and cooperation, differentiation and convergence, between the Cambridge and Jodrell Bank groups (Chubin, 1978).

Edge and Mulkay argued that *social differentiation*, by which they meant the difference between Ryle's group (more internally integrated and leadership-driven) and Lovell's group (autonomous teams, similar to Pawsey's group), drove *scientific* differentiation, that is, different choices in research programs and in the instruments built to pursue them. *Astronomy Transformed* mentioned, but did not equally include, Pawsey's group. This is typical of how studies of groups in the USA and UK have been regarded as defining the "history of science", while studies of groups located elsewhere have been seen as being only of "local" interest. In fact, comparing the Sydney and Cambridge groups would have been a better way to test the proposition that social differentiation drives scientific differentiation: though socially different, the experimental situation in each group was nearly identical!

We were drawn to Merton for another reason—the relevance of his analysis of serendipity in scientific discovery (Merton & Barber, 2004). The early years of radio astronomy were characterised by many serendipitous discoveries. (As demonstrated by Harwit (2019), the rate of new discoveries is highest at the inception of a new field.) Merton first unveiled the concept of the "serendipity pattern" in a talk to the American Sociological Society in March 1946, describing it as a pattern "of observing an unanticipated, anomalous, and strategic datum, which becomes the occasion for developing a new theory or for extending an existing theory".[10] He then wrote a book about the idea with Elinor Barber in 1958—*Travels and Adventures in Serendipity.* However, the book was not published until 2002 (in Italian); it was published in English in 2004, after Merton's death the previous year. The ideas in this book helped the authors characterise some discoveries in Pawsey's era.

With these concepts in mind—norms, priority disputes, competition and cooperation, reward systems, epistemic values, implicit assumptions, and serendipity—we shaped our story of the man who would find himself leading the world's foremost research group in a brand new field of science.

---

[10] James L. Shulman introduction p. XXI in Merton and Barber (2004).

Meet the Authors

The writing of this book is a direct result of the dedication, determination, and inspiration of W.M. (Miller) Goss. His devotion to the history of his field of radio astronomy led him to spend countless hours sorting through archive boxes, tracking down images, and piecing together the missing components of events, ideas, and correspondence. His enthusiasm and willingness to drive the project drew in his co-authors, R.D. (Ron) Ekers, his contemporary and a radio astronomer of considerable repute who himself has long been interested in the philosophy of science, and Claire Hooker, an historian turned sociologist of science, health, and medicine, who first connected with Goss through their mutual interest in Ruby Payne-Scott (see Fig. 1).

The backgrounds of the three authors of this book have clearly influenced their approach to the writing of this biography. Here we offer the reader a glimpse of their stories.

W. M. Goss

In August 1967, 5 years after J.L. Pawsey's death, W. M (Miller) Goss ventured to Australia as a postdoc at CSIRO. He has personal knowledge of many of the people who appear in this book, including all of Pawsey's major proteges: his postdoc

Fig. 1 The authors Ron D. Ekers, Claire Hooker, and W. Miller Goss at the Jansky Very Large Array in New Mexico. Credit: C. Hooker

supervisors were John Bolton and Brian Robinson; he knew Bernie Mills, Chris Christiansen, and Paul Wild; and he spent considerable time in 1974–1977 working with Christiansen, Bob Frater, John O'Sullivan (of WiFi fame), Arthur Watkinson, and others, testing and using the Fleurs Synthesis Telescope. Another Australian, Frank Kerr from the University of Maryland, was a frequent visitor to Australia with contact continuing after Goss moved back to the USA in 1986.[11]

As Goss's knowledge of radio astronomy grew, so did his love for the history of Australian radio astronomy, whetted by the influences of Christiansen, Mills, and Ron Bracewell.[12] Mills stimulated Goss's interest in Joe Pawsey. In 2010, Mills gave a quiet chuckle as he said to Goss: "My point of view on Australian history will prevail; I am the last man standing!"[13] We hope that Mills's warm perspective on Pawsey permeates these pages.

From 1972 to 1974 and then 1977 to 1985, Goss was at the University of Groningen, Kapteyn Astronomical Institute, the Netherlands, where he met J.H. Oort, A. Blaauw, and H.C. van de Hulst, the founders of radio astronomy in the Netherlands. Henk van de Hulst was a remarkable mentor, insisting that conversations take place in Dutch, a valuable learning exercise for Goss, who learned the history of radio astronomy and a new language in parallel; see also ESM 23.5, Van de Hulst, in Chap. 23, Van de Hulst's shared interests in the history of astronomy, for more details of visits and discussions between Goss and Van de Hulst. Van de Hulst described his theory of science concerning discoveries that occur over a long period, even decades.[14] Goss contrasts this paradigm with numerous discoveries in early radio astronomy which occurred over traditional short timescales of months or years.

---

[11] Frank Kerr (1918–2000) was hired by Pawsey from the University of Melbourne in mid-1941 to work on radar at RPL. Kerr was a pioneer in the 21 cm HI research at RPL in the 1950s and 1960s, moving to the University of Maryland in 1966 until his retirement in 1978. Sullivan (1988) in "Frank Kerr and Radio Waves: From Wartime Radar to Interstellar Atoms" has described his noteworthy career.

[12] Ron Bracewell moved to Stanford University in 1955, where he founded a radio astronomy institute. In 1962, he had a short sabbatical at the University of Sydney. His ground-breaking observations at the new Parkes telescope are described in NRAO ONLINE.2 "Bowen's Role in Centaurus A *Nature* 1962 publication" (see Chap. 1, Footnote 1, ESM 1.1, Additional Details). Goss and Libby Goss had a 3-day visit with Ron and Helen Bracewell in January 2007, 7 months before Bracewell's death; in the following years, a substantial fraction of the Bracewell archive was acquired by the NRAO archive.

[13] A surprising claim from such a modest colleague! Mills outlived the other stars of *Four Pillars* by several years. Goss visited him about once every 2 years in the period 1991–2010. However, his other colleagues, Bruce Slee (1924–2016) and John D. Murray (1924–2019), survived him for some years. Murray was a prominent (and honoured) attendee for the book launch of the *Four Pillars of Radio Astronomy* at the University of Sydney in early 2018.

[14] See "Nanohertz Astronomy", Sullivan, 1984, p. 385: a nanohertz is 10^{-9} cycles per second which corresponds to a time span of 10^9 s or 30 years. Discussions with van de Hulst continued in 1987 during his visit to NRAO. See Chap. 23 for ESM 23.5, Van de Hulst, for a description of a remarkable tale told during a trip he took to Chaco Canyon with Goss and Goss's son Andrew.

Ekers was a colleague in Groningen until 1980, when he departed to become the Very Large Array Director at the National Radio Astronomy Observatory.[15] Woody Sullivan was a postdoc in Groningen, and he and Goss engaged in discussions about radio astronomy history on a daily basis at the institute. Goss's close association with Sullivan has continued to the present.

These connections in both environments led Goss to wonder: why and how did early radio astronomy develop in Australia and the Netherlands? What were the ingredients for success? What role did engineering innovation and scientific rationale play in the planning process?

Goss's active involvement in writing the history of radio astronomy began with more than a decade of work researching and writing about Ruby Payne-Scott, the extraordinary early Australian solar radio astronomer. His interest in Payne-Scott was first sparked by a casual remark by John Bolton, who referred to her as "the brightest staff scientist" of those early years.[16]

Goss ultimately wrote two biographies of Payne-Scott. The research and writing gave him detailed insights into the early years of Australian radio astronomy. Mills, Christiansen, Wild, and Bracewell were the subject of his third book on the history of Australian radio astronomy, *Four Pillars of Radio Astronomy: Mills, Christiansen, Wild, Bracewell* published in 2017 (Frater et al., 2017). The immersive level of familiarity and understanding of the people and events of this period required by this endeavour exceeds that provided in general histories of the field. Goss frequently reflects on the long-lasting influence of his own interactions with the major characters of this period. This book captures some of these reflections.

Ronald D. Ekers

Ekers was a summer student at the CSIRO Division of Radiophysics in the (Southern Hemisphere) summer of 1962/1963. Pawsey had died earlier that year. While Ekers never met Pawsey, Ekers did attend radio astronomy lectures given by the scientists who had been part of the Pawsey team, whom we encounter throughout this book.

---

[15] Goss succeeded Ekers as VLA Director in 1988, staying in this position until 2002.

[16] See Goss and McGee from 2009 (*Under the Radar, The First Woman in Radio Astronomy: Ruby Payne-Scott*, p. 262). Bolton was not the only one to describe Payne-Scott in this way. "Chris" Christiansen also described her in these terms to Hooker in the late 1990s. In about 1995, Goss told this story from 1968 to Letty Bolton, John's wife. She immediately laughed as she explained the ultimate irony of John's praise. She explained that John and Ruby had experienced a hostile relation that lasted for years, after beginning at the Dover Heights field station in 1946–1948. (Sharing a common site and equipment was impossible.) Pawsey had attempted in vain to arrange a truce during his overseas trip (Chap. 17) and continued this futile task when he returned to Australia in October 1948.

Ekers has always been interested in the philosophy of science, and when he had responsibility for research management (at VLA 1980–1988 and ATNF 1988–2003), this evolved into an interest in the role of the research environment in the discovery process. In the early 1980s, he attended a talk by Derick de Solla Price at Yale and became impressed with his analysis of the exponential growth of scientific discoveries when a new field of research opens up. In 1983, Ekers and Ken Kellermann organised a meeting on *Serendipitous Discoveries in Radio Astronomy* for the 50th anniversary of Jansky's discovery (Kellermann & Sheets, 1983). This conference became a key to his deeper interest in this field as he started exploring the literature. Ekers was fascinated with the book by Robert K. Merton and Elinor G. Barber which was written in the 1950s, but not published until 2004, *The Travels and Adventures of Serendipity*. This book, and additional research papers by Robert K. Merton, included many examples of serendipitous discoveries in science, but Ekers noted that although none of these examples were taken from developments in radio astronomy, Merton's serendipity pattern was an excellent match to the discoveries in radio astronomy. Merton's predicted pattern was confirmed by later discoveries![17]

Ekers had used radio interferometers and aperture synthesis radio imaging techniques throughout his entire career. Later as he co-authored the paper describing the design of the VLA in 1983 (Napier et al., 1983), he started looking more closely into the history of the development of the aperture synthesis concept. This resulted in a number of talks (including the Hewish lecture[18] in Cambridge in 2010) and plans for a book on this topic. There were strong inconsistencies in the view of whether the aperture synthesis concept arose in the Cambridge or the Sydney group. Thus, Ekers found that it was necessary to read all the papers and notes from this era as well as to have discussions with many of the scientists actively involved at the time. These included Bracewell, Christiansen, and Ryle. Searching for records from the 1940s and 1950s involved discussions with Miller Goss who had already combed the archives (the National Archives of Australia and Churchill Archives Centre) for material from this period. Goss suggested incorporating the early history of aperture synthesis in this book. But, once involved in this project, Ekers's interest grew as he looked into the factors and personalities behind the many radio astronomy discoveries as well as the missed discoveries during this period. The research into the historical developments described in this book provides a real-world example of how scientific progress is made.

---

[17] Halley lecture "Paths to Discovery in Radio Astronomy – Prediction and Serendipity", R.D. Ekers, Oxford 23 May 2007.

[18] R.D. Ekers, 8 June 2010, Cavendish Laboratory, Cambridge, "Paths to Discovery in Radio Astronomy, Prediction and Serendipity".

Claire Hooker

Hooker wrote her PhD thesis about the history of women in Australian science in the mid-1990s. At the time she saw this primarily as a feminist exercise: she intended to use historical examples to refute the now debunked claim that women's under-representation in the mathematical and physical sciences is the inevitable result of sex-linked differences in brain physiology and function (Fine, 2017). Hooker attempted to piece together the story of Payne-Scott from the RPL archives (as much as was possible given her lack of background in physics), and at this point she fortunately intersected with W.M. Goss. Goss continued to involve Hooker in the history of radio astronomy after Hooker's thesis was concluded, e.g., by coediting the autobiography of American pioneer radio astronomer Nan Dieter Conklin, and then invited her participation in this book. Hooker's approach is driven by her sense that truly understanding the scientists of the past requires understanding why they were fascinated by the research topics they pursued, and sharing as much as possible, their pleasure in their work.

We authors have never lost our fascination with understanding the imperfect and contextually determined way by which humans slowly figure out how to understand the universe in which we are located. We have never lost our awe and respect at how a complicated and often very problematic set of human institutions and systems—that is, *science*—can somehow produce ever more reliable ideas, insights, and knowledge. We hope this book conveys this fascination and respect to the reader who follows our untwisting of these many strands of ideas, theory, evidence, design, personalities, organisations, systems, and places, which together produced such a stunning new field and such dramatic discoveries in the two decades after World War II.

Additional Texts and Sources

A book is always the tip of an iceberg of writing; so many interesting investigations could not be fitted into the main work. The reader will find many additions, figures, and comments in the Electronic Supplementary Material ("ESM"), accessible from the respective chapters in the publisher's website.

In addition, the NRAO Archives is hosting a website with supplementary reference texts (70 supplementary reference texts to be found at go.sn.pub/IGY9nU), cited in the text as NRAO ONLINE #. In these texts are the often wonderful stories and original documents for which there was no room in the main book. In some cases, we have placed the full narratives of events for which summaries are provided here (e.g. NRAO ONLINE.38 to NRAO ONLINE.47 provide a detailed description of challenges faced by Pawsey and Bowen in constructing the Giant Radio Telescope (GRT)). They are referred to in individual chapters; each text has the prefix

"NRAO ONLINE". An index for these texts is provided in ESM 1.1, Additional Details (see Footnote 1 in Chap. 1).

A few highlights in the NRAO ONLINE texts: One such is J.A. ("Jim") Roberts's text from July 2000: *Have Gen, Will Travel: Imperfect Images from the Life of a Radio Astronomer*. The text contains a delightful and fact-filled summary of his career at Sydney University (circa 1943–1949 for a BSc and MSc degree), Cambridge University (1949–1952), and CSIRO (from 1952 to 1987). Another highlight is the assemblage, for the first time, of a complete narrative of theorist David Martyn's security disasters during WWII, which created a lasting trauma and underpinned his psychosis in 1956. A third highlight is the surprising story of the three different occasions in three different places, recorded as that when the "first sod" for the GRT was broken. This story was uncovered thanks to the dedicated work of colleague John Sarkissian of CSIRO, Parkes (NRAO ONLINE.48 and ESM 29.1, Three Peg Events; see Chap. 29).

As will become apparent, Pawsey was a dedicated internationalist. In NRAO ONLINE.26, the multi-year connections of Pawsey with colleagues in Canada (starting in 1941) are summarised. Of particular importance, we highlight Pawsey's efforts to support the beginning of radio astronomy in India (NRAO ONLINE.32 and NRAO ONLINE.33). Swarup's association with Pawsey in Sydney in 1953–1954 started the process that led to the remarkable group of Indian radio astronomers founded by Swarup and colleagues in 1963, just after Pawsey's death. An additional text, a paper by W.M. Goss (2014) in the Astronomical Society of India Conference Series, *The Metrewavelengh Sky*, describes the connection of Pawsey to the Tata Institute of Fundamental Research, and his efforts to initiate radio astronomy in India in the early 1960s. The text highlights the roles played by M.K. Vainu Bappu and T.K. ("Kochu") Menon in 1961 at the Berkeley [California] International Astronomical Union. A recent publication by Indira Chowdhury in 2016, *Growing the Tree of Science, Homi Bhabha and the Tata Institute of Fundamental Research*, has also described the events of 1961 in Berkeley (p. 185).

Our Sources

Joe Pawsey and the Founding of Radio Astronomy in Australia draws from a large volume of international, institutional, and corporate primary archival material, supplemented by multiple interviews and extensive discussions with many colleagues and relatives who had some connection to the people and events discussed here.

The authors had access to two extensive, and previously unexplored, privately held, archives. The first was the Sally Atkinson archive, discovered in her house after her death on 13 November 2012. The original Pawsey, Payne-Scott, and McCready data from Collaroy in 1945–1946 (Chap. 12) was in this set. Copies of this material have now been included in the CSIRO Radio Astronomy Image Archive (CRAIA), and the originals will be deposited with the Australian Academy of Science.

Secondly, the Joe and Lenore Pawsey Family Collection in their sons' collections (both Hastings Pawsey, Frenchs Forest, Sydney, NSW, Australia, and the late Stuart Pawsey, Berkeley, California, USA) provided valuable material.[19]

We have also used the vast quantities of material held in the National Archives of Australia (which holds all RPL records), the CSIRO archives in Canberra, the Australian Academy of Science records, records at the National Library in Canberra, the Churchill Archives Centre, and the NRAO archives.

Interviews and discussions (aloud and via email) were carried out over many years starting in 1992 (with John Bolton). Most of the interviews occurred in the years 1998–2018 with Bernie Mills, Ron Bracewell, "Chris" Christiansen, Paul Wild, Peter Hall, Fiona Hall, Rachel Makinson, Elizabeth Hall, Don Yabsley, Harry Minnett, Sally Atkinson, Campbell Wade, Ken Kellermann, Ellen Bouton, Hastings Pawsey, Stuart Pawsey, Elizabeth Pawsey, Jasper Wall, Tim Robishaw, Robert Hayward, Joe Fletcher, Sir Bernard Lovell, Sir Frances Graham Smith, Professor Rodney Davies, Professor Adriaan Blaauw, Sergei Gulyaev, Grahame Fraser, Paul Vanden Bout, Frank Bash, Frank Drake, David Heeschen, David E. Hogg, Don Mathewson, Don Melrose, Don McLean, Summer Ash, Harry Wendt, Brian Svoboda, Craig Anderson, John Murray, John Brooks, Malcolm Sinclair, Gwen-Anne Mansfield and Professor Henk van de Hulst, Margaret Clarke, Rupert Clark, Jet Merkelyn, Letty Bolton, John Whiteoak, and Nick Lomb.

The reader can find further details about our primary sources in ESM 1.1, Additional Details as noted in Footnote 1 of Chap. 1, NRAO ONLINE.1 "Series Note for National Archives of Australia, Knuckey System" and NRAO ONLINE.3 "Schedvin Draft of his unpublished CSIRO History—1983".

The CSIRO Radio Astronomy Image Archive (CRAIA) holds a collection of over 15,000 images that relate to the early history of radio astronomy in Australia. The original images were taken as negatives, photographs, and slides. They have been scanned at high resolution to produce TIFF files. The digital collection is maintained by CSIRO Astronomy and Space Science. The scanned images include people, telescopes, events, observatory sites, engineering, and other technical work and science results.

---

[19] Joe and Lenore Pawsey kept almost all their correspondence and daily documentation from the time they left their parental homes. They also retained their many diaries, rental documents from England, medical history correspondence, notes about their children's schooling, and the like. Of specific interest to this book are the wide-ranging letters written by Joe to his parents covering subjects from the philosophy of "what is science", aspects of religion, premarital sex, what could make a good marriage, the British judicial system, and many others.

University degrees, awards/certificates/medals from school to Joe's death, related newspaper and magazine clippings, letters of sympathy after his death, and the many obituaries are included, as well as an extensive photographic archive.

Following the death of Lenore Pawsey in 1974, the collection was inherited by her son Hastings and it remained almost unresearched till he was contacted by the author Miller Goss. Following the completion of a comprehensive cataloguing by Hastings, it will be offered to the State Library of New South Wales for permanent storage and access for future research.

Low-resolution versions of the scanned images can be accessed through an interactive application, and *high-resolution* digital files can be requested from CSIRO. More details on the image archive and how to use it can be found at https://www.atnf.csiro.au/ImageArchive/index.html.

Socorro, NM, USA	W. M. Goss
Sydney, NSW, Australia	Claire Hooker
Epping, NSW, Australia	Ronald D. Ekers

The original version of the book has been revised. A correction to this book can be found at https://doi.org/10.1007/978-3-031-07916-0_43

Acknowledgements

We thank the Pawsey family for their dedicated support over almost two decades: Hastings and Elizabeth Pawsey in French's Forest and the late Stuart Pawsey (1939–2020) and Glenda Pawsey in Berkeley, California. Liz Pawsey transcribed hundreds of letters from J.L. Pawsey to his mother Margaret during the period he was at Cambridge and then working in the UK in the 1930s.

A very special thank you to our ever patient, adaptive, and encouraging editor Robyn Harrison, for pulling the work together so effectively for us. Thanks also to Ramon Khanna and Christina Fehling of Springer for their support over many years.

W. M. Goss:

The staff of the National Archives of Australia (NAA, in Chester Hill (Sydney) and Canberra) have provided continual assistance and numerous email and telephone discussions during the last 12 years. WMG visited Chester Hill NAA in Sydney eight times in the period 2008–2017. The staff of the National Library of Australia (Canberra) provided helpful advice during two successful visits during these years.

We extend our special thanks to Paul Wood, Brian Scales, Fiona Burn (retired December 2015), and Kerrie Jarvis (retired December 2014) for their continual advice and assistance; often, special intervention was provided by the NAA staff. In 2013, Rob Birtles of CSIRO (Records Advisor, Collections and Archives) located a large collection that had been previously transferred to the NAA, but not incorporated into the C3830 category (the main RPL files). This consignment (C3830/6) was located by Fiona Burn and now appears in the online listing for C 3830. Examples of this "new collection" are the Z1 series (Bowen planning and numerous overseas visits) and Z3 series (the same categories for Pawsey). Numerous important documents were located by Goss starting in July 2014. In addition, valuable assistance from NAA staff has been provided by Melanie Grogan, Edmund Rutlidge, Brendon Fenton, Philip Ball, and Amanda Hardie.

Rob Birtles of CSIRO has provided advice and assistance for more than 12 years, helping us locate many items in the CSIRO and the NAA archive.

In Australia:

For the last two decades, Harry Wendt has provided continued support. He has read many draft texts and given excellent advice based on his vast expertise in the history of Australian radio astronomy.

John Sarkissian (CSIRO CASS, Operations Scientist, Parkes Observatory) provided invaluable help in uncovering many aspects of the history of the GRT—the Giant Radio Telescope (the Parkes telescope). John also provided the link with the Phil Jelbart family, neighbours close to the Parkes telescope. Rodney and Penny Jelbart, son and daughter-in-law of Phil, were generous with their hospitality and additional information about the construction of the GRT and the tree plantation. Penny's parents, David Cooke (1932–2020) and Margaret, were also quite helpful. David had been a prominent engineer at the telescope and Officer-in-Charge of the CSIRO Parkes Observatory from 1988 until his retirement in 1993.

David Nash of the Australian National University provided photographs of the construction and early years of the GRT, many from the George Day family. We thank the Day, Nash, and Helm families for assistance.

Douglas Bock and Nic Svenson of CSIRO CASS (Commonwealth Astronomy and Space Science) gave administrative and financial support for the editing of this book. We also would like to thank our colleague Robert "Bob" H. Frater for his continued advice over the last 45 years. Bob is the main author of *Four Pillars of Radio Astronomy, Mills, Christiansen, Wild, Bracewell* (along with Goss and Wendt) of 2017. This book concerns the four most prominent proteges of Pawsey.[20]

Thank you to: the late Peter G. Hall (1951–2016) and Jeannie Hall, Fiona Hall, Dr Elizabeth Hall, Sue Brian, Jan Christensen, Rita Nash, Martin Y. Smith, Helen Sim, Jessica Chapman, Robert Sault, Gwen-Anne Mansfield, the late John Deane (1949–2020), Nick Lomb, Don Melrose, Alison Muir, Christine van der Leeuw, Edward Bowen and the late Angelika Porrey, John Brooks, Malcolm Sinclair, Robert Batchelor, Anne Green, Crys Mills, Laurel Davidson, Bob Hewitt, Roy MacLeod, the late Letty Bolton, Brian Bolton, Bruce McAdam, Tim Christiansen, Ron Stewart, Robyn Williams (ABC-Sydney), Barnaby Norris, and Lisa Phillipps. The use of the CSIRO Radio Astronomy Image Archive, now online since late 2019, remains an essential ingredient of any research on Australian radio astronomy. The many years of Jessica Chapman's support remains an essential ingredient of our research. We thank Jessica, Barnaby Norris, and Shaun Amy for their major achievement.

These late colleagues from Australia assisted Goss as he began these endeavours in 1997: Dick McGee (co-author of the first Payne-Scott book), Sally Atkinson (who introduced me to the children of Ruby Payne-Scott in 1997), and B.Y. Mills (who provided continual advice up to the time of his death in 2011). The following late colleagues provided continued advice in person and by correspondence: Chris Christiansen, John Murray, O. Bruce Slee, John Bolton, Harry Minnett, Paul Wild, Don Yabsley, Kevin Westfold, Dick Hunstead, and Jim Caswell.

---

[20] Frater and Goss collaborated as authors of the biographical memoir for Christiansen (2011) and also with Harry Wendt for the biographical memoir of Mills (2013).

The authors wish to thank Dr Tom Sear from ADFA for his swift assistance in locating material held by the Australian War Memorial.

All authors owe a debt of gratitude to Jim Roberts for continued advice and criticisms over many years. Thanks to Jim for his making available his thorough autobiography: *Have Gen, Will Travel: Imperfect Images from the Life of a Radio Astronomer*, J. A. Roberts, 2002 (available in NRAO ONLINE.49).

We thank Prof C. Boris Schedvin for the use of his unpublished manuscript concerning the history of the CSIRO from 1983. See NRAO ONLINE.3.

From the USA:

We extend our special thanks to Woody Sullivan for his advice and for the foreword for this book. The following have provided numerous comments through the last few years: Nan Janney, Tania Burchell, Ken Kellermann, Bob Hayward, Stephen White, Summer Ash, Mark McKinnon, Claire Chandler, and Judy Stanley. David E. Hogg also provided a number of items from the archive of his late mother Helen Sawyer Hogg.

Tony Beasley, NRAO Director, assisted in providing financial support for editing tasks.

Also, we thank Patricia Henning (whose scientific advisor was Frank J. Kerr at the University of Maryland, his last PhD student), Assistant Director NRAO, Scientific Support and Research, and in September 2021 Assistant Director for New Mexico Operations, for assistance in funding the photograph licence fees as well as National Archives of Australia copying fees.

We extend our gratitude to the following deceased colleagues in the USA: Ron Bracewell, David Heeschen, Gart Westerhout, Nan Dieter Conklin, Allan Sandage, John Graham, Wal Sargent, Harold Weaver, Maarten Schmidt, Geroge Swenson, and Gordon Stanley.

Goss visited the California Institute of Technology Archives in 2009 and 2010. We thank Charlotte (Shelly) E. Erwin and Loma Karklins for numerous suggestions and assistance.

Thank you to: Shri and Hiromi Kulkarni, Helen Bailey Bayly, Marshall Cohen, Linda and Francois Schweizer, Gloria Lubkin, Mark Bracewell, Campbell Wade, Paul Vanden Bout, Mort Roberts, Ellen Bouton, Lance Utley, Kristy Davis, Martin Harwit, Suzy Gurton, Anthea Coster, Jim Moran, R. Curtis Ellison, Alan Maxwell, Stephen Stanley, Teresa Stanley, Luisa Phelps, Bob Lash, Nicole Gugliucci, Sylvia Kowalski, Cornelia Lang, Rebecca Cummins, Pam Winfrey, Pauline Newman Davies, Ylva M. Pihlstrom, the late Roger A. Kopp, and Richard T. and Shirley F. Hansen.

In the UK:

We thank the following for advice and comments: the late A.C.B. Lovell, F. Graham Smith, the late John Baldwin, the late Roger Griffin, the late Donald Lynden-Bell, the late Rodney Davies, Lewis Ball, David A. Green, Antony Hewish, Wendy Bracewell, Stephen Mullaly, the late Peter A.G. Scheuer, and Simon Mitton. David A. Green and John Baldwin were Goss's hosts at Churchill Archives Centre for visits in 2010 and 2012. In 2012, Jane Scheuer provided a number of key documents from her late husband's archive.

In Canada:

We extend our gratitude for assistance to: Joseph Fletcher, Tim Robishaw, Elizabeth Griffin, Alison Ryle Bird, Paul Bird, Chris Purton, the late Norm Broten, the late Jack Locke, and Victor Gaizauskas of the National Research Council in Ottawa. Tim Robishaw was the host for the exciting workshop on the history of Canadian radio astronomy held in Penticton, BC, in July 2016. A primary motivation for the workshop centred around the work of Richard Jarrell, the author of *The Cold Light of Dawn: A History of Canadian Astronomy*. This book was uncompleted at the time of his death in 2013, and now being completed by a number of Canadian colleagues. Jasper Wall has provided copious insights into radio astronomy history of Australia and the UK.

In New Zealand:

We thank: Sergei Gulyaev, Grahame Fraser, Jordan Alexander, Sandra Coney, Peter Hosking and Mary Harris, Dorothy Cooper, Delwyn Dickey, Sally Greenwood, Peter Greenwood, Beverley Chessum, Gordon Greenwood, Patricia Sallis, and Doreen Richards.

In Chile: Lars A. Nyman

In India:

We acknowledge advice and assistance from: the late Govind Swarup, Lakshmi Saripalli, Indira Chowdhury, Yashwant Gupta, Sushan Konar, and Jayaram Chengalur. Swarup made his comprehensive personal archive available to Goss in December 2013. The TIFR in Mumbai provided a copy of the Swarup interview made by Prof Chowdhury in January 2014. We thank Ms. Oindrila Raychaudhuri of TIFR archives and Prof Mustansir Barma, TIFR Director.

In the Netherlands:

Ekers and Goss profited by numerous conversations in the years 1970–2020 with Leonid I. Gurvits and colleagues who are now deceased: J.H. Oort, H.C. van de Hulst, A. Blaauw, H. van Woerden, and Ulrich Schwarz.

In Russia: Rustam Dagkesamanskii, Sergei Trushkin, Adelina Temirova, and the late Oleg Verkhokanov.

Claire Hooker:

Thank you to my co-authors Miller Goss and Ron Ekers, for this wonderful adventure into the past.

I am grateful to my philosopher of science father Cliff Hooker for many stimulating discussions and consultations about the history and philosophy of science as the book progressed and to my brother Giles Hooker for additional such discussions, along with companionship and photography in New Mexico. I am indebted to my Heads of Department Ian Kerridge and Angus Dawson for supporting this project and to colleagues Tom Sear, Peter Robertson, and Peter Hobbins for historically informed discussion and support.

Any book requires a lot of support for the writer on a personal basis. My thanks are due to the staff at The Bunker café, Springwood, where much of this book was written, and where a warm and relaxed welcome accompanied the excellent coffee and food; to John Barron and Rebecca Glenn for the provision of a wonderful home in which to write during one crucial working period; and above all to my family

(Rogan, Selkie, Ursula and Arthur Jacobson, and the grandparent team of Bernadette Jacobson and Jean and Cliff Hooker) for coping with my absences during its production.

Ronald D. Ekers:

I thank Jay Ekers for tirelessly correcting my grammar and joining our retreats in the Blue Mountains in 2017 and 2018. We all thank Jay for her meticulous efforts in renumbering all the chapter and ESM references when we changed the book structure.

For research into aperture synthesis, I am thankful for the invitation to give the Hewish lecture in Cambridge in 2010 and for discussions at that time with John Baldwin and Elizabeth Waldram. Later useful inputs from were received from Ken Kellermann, Bob Frater, and Martin Harwit.

For the source surveys I am greatly indebted to Jasper Wall for many discussions and frequent incisive inputs. Also, thanks to Summer Ash for her feedback on the structure of the chapter on source surveys.

For the development of theories of radiation, I received feedback from the world expert, Don Melrose. In addition, I had many discussions with Jim Roberts, one of the few remaining scientists who were directly involved in the discussions of the 1950s and 1960s. Leonid Gurvits helped trace activities in the USSR.

Marc Price has provided numerous details about early science carried out with the Parkes dish in the 1960s.

We would like to thank the Australian Academy of Science (AAS) Chief Executive Dr Anna-Maria Arabia for access in 2017 to the restricted access files of F.W.G. White concerning David F. Martyn. The handwritten note by Fred White in 1983 is a key contribution: "Boris Schedvin tells me that information in National Archives shows suspicion of Martyn to be much more serious than I imagined!" The April 1941 report of the Intelligence Section of the General Staff of the Eastern Command (Australian Military Forces, the army) "Subject: Confidential Enquiry in Regard to the Association of Dr David Martyn, PhD ... with Mrs Ella Horne of German Birth" was especially critical of Martyn. In addition, we thank Rachel Armstrong of AAS for her assistance.

I would like to express my admiration to the late Sally Atkinson for her meticulous archive of original Division of Radiophysics Laboratory documents.

I thank my co-author Claire Hooker for introducing me to the early developments in ionospheric physics which provided a key to the early developments in radio astronomy.

Contents

Part I
Childhood

Chapter 1
An Inheritance of Intangibles, 1890s

[N]onconformists believe that beauty comes from within, and are mindful of what Christ had to say about the difficulties a rich man may encounter on attempting to enter the Kingdom of Heaven. Methodists can make money as well as anyone, but they cannot enter "society" without imperiling their principles. McCalman (1993)

Joseph Lade Pawsey was born in the Western Districts of Victoria (Australia) at Ararat on 14 May 1908, the only child of Joseph Andrews Pawsey (27 November 1865–30 June 1943) and Margaret née Lade (27 December 1879–8 August 1969). It was a humble enough beginning for a man who would have a life so different to almost all his countrymen. And yet Joseph Lade Pawsey's individual experiences and unique career were shaped by the general trajectories of his generation. Spending money may have been very limited, and the family's fortunes somewhat fragile, but the young "Lade" was nonetheless born into the growing social prosperity of a fresh new nation, the economic and technological growth of the twentieth century. He was a product, too, of the socially progressive values of the time, strongly reflected in his family (McCalman, 1993 and Bashford & Macintyre, eds., 2013).

Joe's parents were idealistic in their own way, and harboured dreams beyond their material circumstances in their youth—as young people do. Such dreams were a feature of the social mobility that the Industrial Revolution produced in the nineteenth century, and the impetus and courage they gave to immigrant families who sailed half way around the world to their new lives in Australia, were not their least important feature. Despite having married a second wife from a successful shopkeeping family, Lade's great grandfather Robert Pawsey, a bootmaker, took advantage of the sponsored fares offered by Presbyterian Minister Reverend John Dunmore Lang (Baker, 1998) to emigrate to the Colony of Victoria in 1849. Robert

More information about the use of primary sources can be found in ESM 1.1, Additional Details.

Supplementary Information The online version contains supplementary material available at [https://doi.org/10.1007/978-3-031-07916-0_1].

© The Author(s) 2023
W. M. Goss et al., *Joe Pawsey and the Founding of Australian Radio Astronomy*, Historical & Cultural Astronomy, https://doi.org/10.1007/978-3-031-07916-0_1

Pawsey, his second wife Elizabeth, their young sons Charles and Henry, and his eldest son Joseph Josiah, aged 15, who was born to Robert's first wife (who died giving birth to a sibling who died also) and who would become Joseph Lade's grandfather, arrived in Hobson's Bay and the city of Melbourne in February, 1850, and headed out for Geelong.

The Pawseys arrived just before the rushed influx of half a million who came to the colony over the next decade in pursuit of gold. The bootmaker Robert brought more than social aspiration to the Colonies, however, and an inheritance of ideas and values is part of what gave young Lade his start in life. Typical of his generation of migrants, Robert Pawsey was one of the "poor but honest" in whom religious and radical idealism were mixed in Nonconformist religious practice (McCalman, 1993). Nonconformist groups and churches—that is, those who did not conform to the Church of England—abounded in the first half of the nineteenth century, rejecting the governance of high society bishops and seeking a more authentic and direct religion through the practice of good works, direct experiences of conversion, faith and prayer, lay preaching, and collective meetings for worship, confession and mutual spiritual support. Nonconformist Christianity, as biographer of Australia's middle class has put it, is perfect for pioneers (McCalman, 1993): it preached a self-disciplined work ethic focused on temperance, parsimony and virtuous conduct, while granting all participants equal value, power and participation in religion, providing experiences of skill and leadership in the congregation-led meetings and preaching and conveying the glowing inner certainty of a personal relationship with one's saviour.

Robert Pawsey was exactly such a man. His passage on the frigate *Clifton* (the last of the three ships chartered by John Dunmore Lang to bring families to Australia) was likely no accident. Lang was a trenchant Scottish Presbyterian whose subsidy of serious minded Protestant immigrants was explicitly intended to build a population whose morals might reform the immorality in Colonial society that the over-representation of convicts and Catholics (the two being strongly equated in his mind) had produced (Baker, 1998). Shortly after arriving in Australia, Robert Pawsey joined a small independent church group that met in each man's home by turn, to worship and pray and engage in Bible study. He was a man of principle. The family remembers that he left this group when it accepted a gift of a block of land on which to erect a church building from the Victorian government. He either objected stringently to taxation and to profiting indirectly from it, or to the blurring of boundaries between Church and State, both hallmarks of a true Congregationalist. Instead Robert established an independent chapel, the "Ebenezer Independent Chapel" and remained its minister, without salary, until his death in 1891.

Robert's son Joseph Josiah established a hardware store, initially for miners in Ballarat, and another later in Stawell; he also established a general supplies store in the small northwestern Victorian town of Jeparit. Material prosperity, however, is one thing, and happiness another. Like his own mother, Joseph Josiah's first wife died in childbirth and the infant did not long survive her. His second marriage, to the daughter of a similarly prospering family (Elizabeth Andrews) was more successful if that is judged by their 8 children, but in the end, 8 children resulted in only two grandchildren: a granddaughter, and the young "Lade". Joseph Josiah, Lade's grandfather, retired to Melbourne. His general store in Jeparit was eventually bought by Prime Minister Sir Robert Menzies's family, which too had a blend of Methodism and Presbyterianism. Joseph Josiah's son Joseph Andrews, Lade's father, however—his middle name was his mother's maiden name—expended money in pursuit of various ideals, shaped strongly by the Nonconformist churches, then having a profound impact on progressive social movements at the end of the nineteenth century. The Churches were evangelical, given to literal interpretations of the Bible and strict in their moral disciplines, but were also determinedly egalitarian and idealist. They powered the temperance movement, movements for women's education and suffrage, the baby health movement, took reformist attitudes to prostitution and to the imprisoned, and forwarded a host of other "progressive" social and political projects of the era (Catterall, 2016). Joseph Andrews was suffused by such visions of a more perfect social future.

The verging-on-middle-class valued education, and Joseph Andrews Pawsey was a scholastically inclined child, who won numerous school prizes, culminating in Fifth form Dux with top prizes in English and Mathematics at the Establishment Stawell Grammar School. He sat and passed the Matriculation examination, needed for entry into the Victorian Civil Service. It is not known why he did not pursue this career. He may not have had much opportunity, for the Depression of the 1890s had severely curtailed spending and opportunities for many. Perhaps an early marriage prompted him to look for more swiftly lucrative support for his family—or perhaps grief at the death of wife and child in childbirth (the third generation to have this melancholy experience) prompted him to seek a change of location. He opened a general store in Ararat, Victoria, but it was not very successful, perhaps as a result of the predilection that his son Lade could remember, for abandoning work only half completed while absorbed in political and social discussion. Whatever the reason, in 1902 he went gold mining in Wild Dog Gully above the small village of Strath Creek.

In Strath Creek he met 22-year-old Margaret Lade, the daughter of a local dairy farmer, one of ten children. Margaret knew a lot about practical farming and life in the dirt, and she too must have dreamed about an enlarged sphere of existence, one where women had more influence and more opportunities. The year before she met Joseph Pawsey—it was 1901, the first year of the new Commonwealth nation of Australia, redolent as it was with possibilities for social improvement and attainment—Margaret had spent a year at Methodist Ladies' College (MLC) Melbourne, one of the string of "colleges" to which the middle class sent their children (and still do) (McCalman, 1993). MLC offered mature age students various courses, and Margaret learned singing and nursing there. This combination of the practical and the exploratory was quintessential of her, and Joseph Andrews Pawsey maybe seemed a man of a similarly expansive vision to herself. He proposed, offering a world tour as a honeymoon, and she accepted. And when shortly after the goldmining venture failed as was inevitable, and he offered to release her from an engagement that was unlikely ever to fulfil the promise of a world tour after all, she did not decide to be released. They were married in 1906.

The Pawseys bought a 2000-acre large farm, "Kuvindra", near Willaura adjacent to the Grampian mountains in Victoria, in partnership with a brother and a brother-in-law from the Lade side, and on 14 May 1908, Joseph Lade Pawsey was born there. He did not remain there long. Kuvindra is remembered for the impracticality of its management, directed by Joseph Andrews's enthusiasm for change and improvement, the application of theory with little practical experience to temper it. It too failed, and the Pawseys moved to different farms and towns during small Lade's early years, eventually purchasing "Glencoe", a small dairy and sheep farm near Camperdown. This farm was approximately 100 acres—they may have leased another 60 acres—and Margaret's practical experience in cheese making and farm work provided the core stability for its management. The family must have struggled financially nonetheless. Joseph Andrews dabbled in a number of non-farming activities, including being the enthusiastic local agent for Riley and Stoewer cars. Margaret became an unofficial neighbourhood midwife. She attended expectant mothers by horse and cart, and later by model T Ford, becoming expert in car maintenance and repair as she coped with frequent burst tyres and mechanical breakdowns. She returned horses to stalls and put away ploughs when her husband left them in fields in his enthusiasm for political debate. And for all her passionate devotion to her son, she did not have another child. Perhaps experience as a midwife was offputting; decades later she confided in her granddaughter-in-law that she was not willing to face birth again.

If Joseph and Margaret were not particularly wealthy, if they struggled to maintain their foothold in the middle class and did not have the knack of turning their ideas into successful ventures, were *they* happy? Perhaps they were. They certainly lived a life actively engaged in politics and in the progressive issues and concerns of the day. Joseph was a prolific writer of letters to various editors on social, financial, political and philosophical questions. In the mid-1920s, Margaret canvassed heavily in support of women voting in Council and government elections. Margaret went even further. In 1924 she was lobbying for the Women's section of the Victorian Farmer's Union to have baby health centres established in their towns. "This is great national work," she wrote, "for if we take care of the babies and start them out in life with health and sound constitutions, we can look forward to many of our social evils naturally disappearing" (Smart & Quartly, 2015; Sheard, 2017; Lovelace, 2012). Two years later, while her son Lade was exploring Europe on an impressive schoolboy tour, she became President of the Section and advocated for temperance, in particular for preventing girls under 21 from having access to alcohol. In the 1930s, she had advanced to being President of the Women's Country Party in Victoria; by the end of the decade, she was advocating strongly for expanded women's roles in the war.

It seems reasonable to infer that the Pawsey family lived a fairly rich social and intellectual life, one formed by the moral framework of Progressive Nonconformism (even if their religious practice had receded into the background), and studded with principles that undergirded their strong opinions on social and political matters. Although Robert and Joseph Andrews were struggling financially (though not disproportionately to many others at that time), they shared with other upwardly mobile families a sense of the importance of their action in the world, and the strong sense of social duty that arises from moral sensibility. Margaret's hopes for what improving population health might achieve—for example, by building baby health centres—is typical of the optimism of the Progressives of the day, despite the setbacks of the first World War and the Depression. They were a family interested in new ideas and confident that scientific advances would bring improvement. The young Lade's long dead Nonconformist great grandfather (Robert) would have understood how rich an inheritance this was.

Chapter 2
Just a Boy from the Bush, 1908–1925

Friday, 23rd April, was observed at the School as Anzac Day. Lieut General Sir John Monash was present, and gave a splendid address . . . The part played by the Public Schools of the Empire in providing leaders and officers has been most important. The privileges enjoyed by boys at Public Schools brings with it responsibility. This is part of the price that has to be paid ... We should be wicked if we did not learn the lesson of the value of comradeship, loyalty, self-sacrifice, and devotion to duty. We must apply that lesson to the problems of peace. *The Wesley College Chronicle*, May 1920, p. 5.

Of course Lade Pawsey's family were enormously keen on education. Writing much later in his life, Joe[1] particularly credited his mother with a single-minded focus on ensuring he had a good education. In this, Margaret Pawsey was typical of many of her generation in highly prizing education, both for its own sake as well as for the opportunities that it could open up, and who were fierce in their defence of it amid other family economic priorities.

Frequent moving early in life meant that Lade was home-schooled until he was 8 1/2 years old. Limited schooling was not unusual for farm boys (or for city boys with fathers engaged in trade), but family stories have it that "although he did not attend school there seems no doubt that he exhibited an unusually inquiring mind which his parents made every effort to develop" (Lovell, 1964). In the meantime he is likely to have imbibed the practical manual education that country boys of that era commonly acquired, able to fix equipment and build things, the constructive play of life on a farm. His formal schooling began in 1916 at the one-teacher school in the small village of Cobrico, about 2 km from the family farm "Glencoe."

In 1919, Lade entered the Camperdown Higher Elementary School, eight miles from his home. There he showed a very solid academic focus and performance. Available records for these years are sketchy, but some results from 1920, 1921 and

---

[1] We use "Lade" in referring to young Joseph Lade Pawsey. He became known as "Joe" after he left home.

1922 are available.[2] He received a "Merit Certificate" in November 1920 for satisfactorily completing the Course of Study prescribed for High School (first and second years) in the form Regular II B, 2-a. Strikingly, Pawsey achieved 81% in physics while the class average was only 28. At the end of 1922, Lade was said to be ranked second out of 18 pupils. Again, the physics ranking was 89 percentage points while the average of the class was only 49.5%. In his annual assessment, the teacher noted, "If given a chance will have a brilliant future ahead. Has tried to improve the mental and physical aspects of education." (This statement may have been charitable. Lade's children heard stories in later years that he provided a challenge to his teachers with many probing questions.)

With his excellent academic record at Camperdown, Pawsey sat examinations for and received a Government Junior Fellowship, starting in February 1923 at the private school Wesley College, St Kilda Rd., Melbourne[3] where he was a boarder. This was both a necessary and desired step to "finishing" schooling. Even more than in other states in Australia, Victoria was then, and remains, deeply committed to a system of private schooling for the elite. Janet MacCalman reports that 81% of the Victorian-educated male elite from this period came from private schools; the other 19% came from just two government high schools. There were few government high schools available at all, such was the bitter opposition to the notion that the government might provide full secondary schooling. The result was that children like Lade Pawsey, clever but poor, were reliant on scholarships. It was perhaps in keeping with the family Nonconformist background that he was sent to Wesley.

Lade certainly justified whatever family sacrifices it took to send him to Wesley College. Lade obtained a First Class Honours in physics and Third Class Honours in algebra and trigonometry. These resulted in the award of a Government Senior Scholarship and access to the privileges of the elite. At the conclusion of his second year, in December 1924 when he was just 16, Lade applied successfully to participate in a tour of Europe organised by the Young Australia League (YAL). A Progressivist organisation typical of the era, aimed at national improvement as part of the British Empire, the YAL grew out of an association formed in 1908 by John Joseph "Boss" Simons (1882–1948) and Lionel "Pop" Boas (1875–1949), to promote Australian rules football in Western Australian schools. Activities grew rapidly to include literature, debating, band music, sport, hiking and camping—very similar to the Boy Scouts, which were also formed in 1908 by the Imperially-minded General Robert Baden-Powell. Setting its activities mainly for boys, the YAL proudly aimed to become the "largest boys club in the British Empire". An important aspect of the YAL became the promotion of the ideals of "Education Through Travel". In 1909 the League began offering interstate tours; overseas tours followed in 1911, 1914, 1925 and 1929. (https://www.yal.org.au/about/our-history/).

---

[2] Pawsey Family Archive.

[3] Details of Pawsey's performance at Wesley were provided in a letter to Joe's wife Lenore Pawsey on 31 May 1963 (6 months after his death), written by the Headmaster Dr. T. H. Coates.

The YAL Tour of Europe December 1924 to July 1925

With about 10 other boys from Wesley College, Lade was chosen for the 1924–1925 European tour, the first after World War I. They would join nearly 300 other boys from throughout Australia. The 16-year-old Lade kept a thorough journal throughout the trip. The point of the tour was to impress on the band of young men their serious responsibilities as upholders of the ideals of Empire and the tasks of its governance. Prior to the Christmas Eve departure, there was a visit to the Federal Parliament House in Melbourne and an address by the Australian Prime Minister Stanley Bruce (1883–1967, Prime Minister 1923–1929 and later the influential High Commissioner in London from 1933 to 1945). Following this solemn occasion Lade experienced a different sort of colonial fellowship by suffering severe sea sickness often during the one-month voyage. On 28 and 29 December there was only one word in his journal: "SICK".

The length of the voyage in itself was a striking reminder of the distance between Australia and her Imperial centre. Pawsey wrote poignantly of the emotions experienced by the boys as they were about to leave Australia completely, having last docked at Fremantle:

> . . . We left Fremantle after dark. It was an extremely memorable scene. We arrived at the wharf at 8.55 though the boat was advertised to sail at 9 p.m. However, we had plenty of time as it did not leave till about 10. Our last glimpse of Australia was a rather wonderful scene [New Year's Eve]. The hundreds of people on the boat and wharf, white faced in the artificial light, and connected by a multitude of many coloured streamers; behind them the goods sheds with black shadows between, with the steam from an occasional railway engine showing white in the surrounding blackness; and further back still the lights of the town.
>
> Most of the people seemed subdued except a few, like the Eastern States boys whose farewells were over and who tried to show their unconcern with a few rather feeble league songs and calls. Somehow they were not very rousing. For some time the rattle of the winches, and the sight of loads of cargo swinging out of the loaded trucks up into the darkness, and descending into the lifted holds formed a background to the parting scene. Finally a hissing engine crawled out of the darkness and drew the last empty trucks away, the gangways were cleared and the hawsers cast off. As the tug drew the stern away from the wharf our only remaining material connection with Australia was the multitude of streamers. These slowly parted and soon we headed for the open sea under our own power. Our band played "Old Auld Syne" as we slipped out to sea.
>
> I watched the shore till the people disappeared and the lights grew dim. Then just as we turned to go below a strong light was thrown on the side of the ship and a tiny steamer steamed towards our side where the pilot was climbing down a rope ladder. It came close enough to let the pilot step onto its bridge, touched our side gently and steamed off as fast as she had come. She signaled us with her shrill siren, wishing us "bon voyage" I suppose, and we replied with our great booming one. Thus we severed our last connection with Australia, and really commenced our long voyage to the other side of the world.

When the YAL boys reached Waterloo Station in London, they were greeted by 2000 London Cadets, a comparable organisation. The two groups marched through the streets, laying a wreath of eucalyptus leaves on the WWI Cenotaph, Whitehall, and viewing a regiment of Cold Stream Guards and the Royal Irish. "They are like parts of one machine and look splendid," Lade wrote.

As well as the tourist sites of London, the group had a meeting with Edward, the Prince of Wales (who would become King Edward VIII for 11 months in 1936) on 4 February 1925 at Whitehall Palace. Lade, at age 12 years, had seen the Prince in April 1920 in Melbourne:

> We had dinner at camp and then marched to Whitehall Palace to be received by the Prince of Wales. We were on some lawns at the other side from where the mounted Life Guards are on sentry duty. We were drawn up in the shape of three sides of a square with the staff officers in the open side.
>
> The Prince came out of the door shown [Lade apparently included a drawing], the officers were introduced to him and then we all filed past him, he stood where the cross is shown and shook hands.
>
> He seems to me to look a lot older than last time I saw him. He was dressed in plain clothes and a bowler hat. After the shaking hands, he spoke to us for a moment. He simply welcomed us and said that he hoped that we should have as good a time here as he had in Australia.

Later the same day, the YAL group was received by the Archbishop of Canterbury.

On 9 February 1925, the group left for France on the Continent. Their visit included famous WWI battlegrounds where the First Australian Imperial Force had been participants, as well as Paris and a service at Notre Dame:

> It is a great cathedral. It is slightly less sombre than Amiens [cathedral] though similar. It is very bad for hearing as it echoes for a second or so after a voice sounds. The service was very queer; I did not understand a word and I don't think anyone else did except for a few notices given out in French. There is a tablet to the memory of the British who died in the War.

In Monaco, Lade learned a valuable life-lesson:

> We visited the Casino [in Monaco], they let our boys in free or in other words they do not charge anything for the privilege of taking our money. I spent a few more francs for the experience, altogether with Menton [his friend] I only spent about half a crown. I do not like it much. There is a kind of unhealthy excitement which I cannot very well explain. Even playing for small stakes one franc at a time, one gets this and I do not know what it would be like for larger. There is no harm in it as long as you go there to spend a bit of money and fix your limit fairly low. It is a case how—even if one goes there with the idea of making money—you are usually not only unsuccessful but get cleaned right out trying to make up your first losses. Of course it is not hard to win. Three of us went in, I lost about 9f, Garnet about 30f and Eric Sewell won about 30 f. I have probably got more good from it than Eric as I will keep off it more.

On 28 February 1925, the group went to Rome and met King Victor Emmanuel III of Italy. Lade commented that the palace was "very dingy from the outside. Looks like a third class hotel. Inside it is very beautiful. It is far more cosy ... than Versailles." The next day they met with Pope Pius XI (Pontiff from 1922 to 1939):

> We had an audience with the Pope at 1 o'clock. We got there about 1 hour early and had to wait till 2 o'clock ... The Pope came in attended by several guards in a fine uniform of blue with large gold trimmings and other attendants. All in the room knelt on the right knee while he walked round to each. You took his hand and inclined the head towards it instead of kissing it and after he had passed stood up. The floor was of hard stone ...

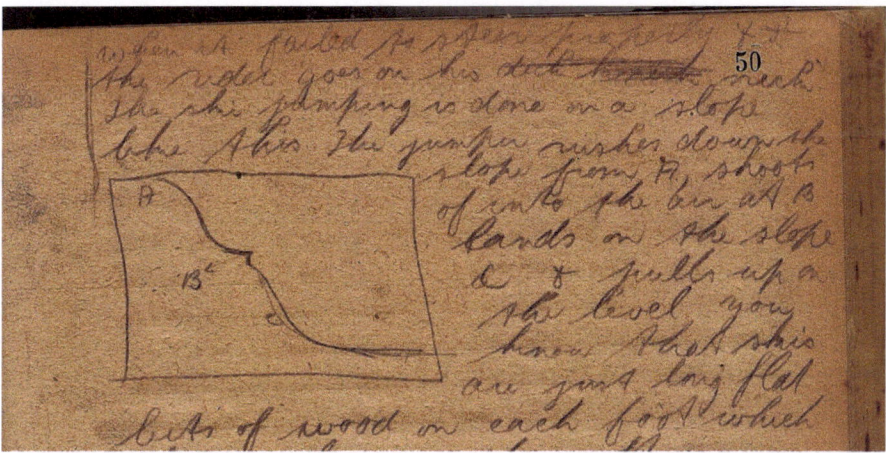

Fig. 2.1 Drawing and text about ski jumping from Lade's diary. Credit: Joe and Lenore Pawsey Family Collection

The tour included Florence, Milan and Switzerland, where of course the Australian boys, with little experience of high mountains, felt impressed, sometimes a little sick and thrilled about snow:

> Caught train at 8.30 for Grindelwald, one of the leading winter sport places. We went a fair distance in an ordinary train then caught a mountain train, with a central cog for pulling, but not the real steep kind which goes at about an angle of 60° with the horizontal. When we got there we had our lunch and then were issued toboggans. The snow was several feet deep. We had great fun going down short slopes. We then walked several miles, mostly up hill to a glacier, we went inside a cave cut in it. The colour of the ice is wonderful … You know that skis are just long flat bits of wood on each foot which slide easily over the soft snow. The jumpers come down the slope at an enormous rate … Most of us were pretty wet about the feet but dry elsewhere. It does not hurt one in the least to fall very hard into moderately deep snow. (See Fig. 2.1).

In March 1925, the group had a personalised meeting with the Belgian King, Albert I (King from 1909 to 1934), where the theme was of gratitude for the actions of the Australian Imperial Forces in Belgium on Flanders Fields in WWI:

> In a large fernery we were lined up along three sides and received by the Royal family of Belgium, the King, Queen, Prince and Princess all came in. First the King greeted the Staff officers, then made a fine speech in English. He said the Belgians would always remember with gratitude and admiration the heroic Australian soldiers. He continued in this strain for some time and said he welcomed us and hoped we had a good time in Belgium. The King, Queen and their two children then walked round the ranks, with Boss [Simons], were introduced to each boy and stopped and had a yarn with nearly every boy. You may judge the time they spent on each when the whole reception took 2 hours and there were 4 of them each talking to different boys. The King and Queen both asked how high I was and one of them how old. The Queen said that they would probably visit Australia, to a group including myself and we said we would be very glad to see them. We were then served some light refreshments and marched off. Altogether I think it was wonderful to be received by a Royal family thus. There is no false pride and exclusiveness about them. I can well understand if

they are always like today why Belgium has a popular King. He is a very capable looking man. The Queen is nice looking and the other two just look ordinary upper class people.

This visit of State was followed by a tour of the Ypres battlefields, the scene of some of the costliest battles for the Australians in WWI, only 8 years earlier. The Australian casualties during the Third Battle of Ypres from 31 July to 10 November 1917 had been about 38,000 killed or wounded. The young men of the YAL were certainly moved as they visited the Australian cemeteries. Lade Pawsey's commentary conveys something of the impact of visiting these still-new battlefields:

> We marched to the municipal buildings where we were welcomed by the burgemaster of Ypres. During the war Ypres was the centre of the fiercest fighting. It was held by the Allies all the time and was consistently shelled so that no buildings escaped. They say that there was not one stone left on another but I have learned that this means simply no complete buildings left. There are many small areas left unrepaired and they are uneven places with bricks, etc. half covered with grass and a few walls standing here and there. The devastation in 1918 must have been terrible . . . After the reception we marched to a military cemetery, an extension of the civil one used mainly for bodies found since the war. It will surprise you to learn that 40 bodies a week are still found in the district. The graves are arranged in rows with about 2 or 3 ft. to each man . . .
>
> Here we found the grave of one of the boys of the YAL who died on active service and laid a wreath on it in respect, [not only for] him but to all the others whose graves we could not visit. We returned to the town for dinner, and after about 1 hr. leave in the town caught the train for Zonnebeke. We walked to a huge cemetery near which the Australians distinguished themselves on 20th September 1917. All along the road, in the ditches, etc. there were shells, bombs, and other remains. They were still unexploded we were told, and consequently dangerous to handle.

The final two months were spent in the UK; the major activity was a tour of Scotland and England. A highlight was Lord Nelson's ship the *Victory* in the historic port of Portsmouth:

> She used to be out in the harbour but the weather played such havoc with her that they brought her in and she now lies on concrete foundations out of the water where she will lie till she ceases to be. They are almost rebuilding her with good new timber but exactly the same as before, as the old was so rotten. She was built about 1780 I think. She was only raised finally from the water the day before we went there. We then went on board and were shown over in companies with a good guide. We saw the spot on deck where Nelson was shot and below in the cockpit where he died. The space between the decks is awfully small especially in the cockpit. It would not pay to be a tall sailor [given his height of 189 cm, he was sensitive to hitting his head] in those days. The ship is built very blunt at both bow and stern. She is just like an old picture of any ship of the line.

Unfortunately for Lade, he fell sick in early April 1925 and, with two of his mates, was left behind in the Winchester hospital when the rest of "the mob" continued their tour. One of his friends had the measles. Vaccination against measles—at that time one of the leading causes of childhood mortality in Australia—would come only after WWII, and Pawsey's unfortunate experience was a reminder of how both commonplace and costly such childhood diseases were. "Horrors," wrote Lade. "Rumour that we are all measles contacts and [must] remain [in Southampton?] for a fortnight. I'll have to spin that doctor fellow some yarn about me never catching measles. I got up for a bit longer and feeling much better." He found one doctor that believed him but the second one did not; he had to stay in isolation.

By 8 May 1925 he was out of quarantine and touring the Highlands:

> We went in private cars for a trip in the Highlands. We were in our host's car. We went first to Kinlock Rannock. The country soon after leaving Perth, became very wild. At first bare hillsides were covered with heather alone, but further on there was a great deal of bare rock. In all the valleys there seems to be a small stream of clear water in a rocky bed ... There is a beautiful waterfall [Fall of Allt Mor] there which comes down a high hillside in a succession of falls and very steep rapids.

The Highlands were followed by Edinburgh,[4] and to the Forth Bridge, 14 km to the northwest. Lade was fascinated by the technical details of this remarkable structure, built in 1890:

> It is really wonderful. A few measurements are length with approaches 1½ miles. Height of railway line above sea level 150 ft. Height of highest point of bridge 361 ft. Length of longest span 1710 ft. Cost in 1880–1890 = £3,500,000. The largest ships go underneath in a wide channel 200 ft. deep. There is a lot of traffic over it.

In West Yorkshire they met the Premier of Western Australia, Philip Collier,[5] and then returned to London. On 28 May 1925, the YAL group went to 10 Downing Street to meet Stanley Baldwin, during his second term as PM.[6] Lade wrote only: "He presented us with the Kings' Colours." On 30 May the group went to Buckingham Palace to meet King George V, who.

> ... said he was glad to see us there on the eve of our departure for home. He had felt the benefit of his trip round the world when he was younger than most of us and thought that this trip would help us when we filled responsible positions in the Empire. He congratulated us on our smart appearance and presented us with a large autographed photo of himself and the Queen to hang in our headquarters at Perth. We then marched off and saluted him as we passed.

Lade Pawsey wrote little in his diary on the way home; the trip was a repeat voyage, his chronic seasickness was unabated. He was continually ill from London to Port Said, continuing his malady most of the voyage to Colombo. Finally, there was relief when the ship docked in Fremantle, Western Australia, on 3 July 1925. A large crowd awaited their return to Australia, greeting them in a form similar to that shown returned soldiers a few years earlier:

> We caught a special train to Perth and at the station were welcomed by a very big crowd which had assembled. We then marched with all our flags through the main streets to the

---

[4]The host family was Mr. and Mrs. Pratt. In 1931, Lizzie Pratt wrote Joe a letter of welcome to Great Britain. They were pleased he was coming to Cambridge: "It shows you have been studying and we feel quite proud of you. When you have a holiday just let us know. We shall be pleased to get a visit from you again, if we are at home." Likely Joe did visit in June 1933 during a visit to Scotland (see Chap. 7).

[5]From 1924 to 1930 and then 1933 to 1936.

[6]The periods were (1) May 1923–January 1924, (2) November1924–June 1929 and (3) June 1935–May 1937.

YAL building. They had a lot of flags across the road and a large banner with "Welcome Home" in large letters. Pop Boas gave a speech of welcome. We then marched to Govt. House where the Governor welcomed us and gave us a good substantial afternoon tea. We returned to the YAL where the mail was given out, had an hour's leave and then marched down to the Palace Hotel . . . We returned to the YAL, at 10.15 or 30 for a dance but had to leave soon. We went to Fremantle in buses and the band came down to play us off. However, the boat did not sail till 4 a.m. and by 2 a.m. almost all the W.A. boys had gone [home].

Fortunately for Lade, the passage across the Australian Bight from 4 to 7 July was smooth and he was not seasick as the ship arrived in Adelaide. The journal ends at this point; likely the arrival in Melbourne was about 8 July 1925,[7] a total journey of just over six months.

Flag waving, military marches and inspections, the voyage "Home", a laudatory lecture on the White Australia Policy, patriotic visits to the carnage of the Great War—this was the world that Joseph Lade Pawsey grew up in. Most people who were children during WWI have spoken of the crushing sense of responsibility they felt for the world that had, in the common narrative, been given to them at an unthinkable cost in ultimate sacrifice. Theirs was a world in which new national pride and identity were sometimes struggled with, but which was firmly rooted in, the grandeur and the vision of the British Empire.

Lade didn't write about it, and such things did not feature in his tour, but this post-war world was in many ways a very progressive world, the world of his mother's imagining: a world in which universal primary schooling was mandated, where kindergartens and baby health centres were being created; where cities were having sewer systems laid (Melbourne's would not be completed until after WWII, and would enjoy a corresponding drop in typhoid fever and other enteric disease outbreaks); one where women could newly vote, attend University and receive degrees in recognition of their scholastic achievements. There were many hopes for a future of social improvement.

We can pause to note, too, how small Lade Pawsey's world was—where entry to an elite high school made it possible to easily meet those who governed: where a schoolboy from the bush could kiss and shake the hands of the Pope, Kings and Kings-to-be, Prime Ministers, Premiers and Generals. The social mobility and access of young colonial men encouraged their aspirations.

Pawsey finished his Wesley College career with honours in late 1925 (age 17), winning a Fred J. Cato Scholarship to begin study at Queen's College of the University of Melbourne, as well as a Senior Government Scholarship.[8] He had very definitely left home, and he chose to assume a new, independently forged identity, being known now to everyone besides his parents, as "Joe". Here he might really begin to imagine what sort of person he might like to be.

---

[7] Uncertain by 2–4 days.

[8] After his return from the YAL tour, the academic year was too advanced to start University. Thus, Lade did the Leaving Honours course again, with two second class honours.

Chapter 3
Becoming a Physicist, 1926–1929

[T]he landscape of colonial science reveals the strong leadership of a few men, frequently dependent on the goodwill and patronage of government; a commitment to empirical utility, as against abstract science; and a close relationship between academic and government science. Then and since, science served Australia as a guide to "moral improvement" and social organisation; as a social elevator for ambitious young men and women; and as a social adhesive for artisans and managers. MacLeod (1988).

In March 1926, Pawsey entered the University of Melbourne, enrolled in a Bachelor of Science.

If we pause to think about it, it seems appropriate for a Young Australian League alumnus to embark on a life of science. To write about the history of science from Australia brings us to a consideration of what Joe had just been treated to: the relations of Empire. In Australia, science was produced by, and in turn produced, the British Empire—one need only consider the rapid developments in navigation (such as longitude), in medicine (understanding of and treatment of that scourge of the British Navy, scurvy), in botany and zoology (developing systems of classifications that made sense of the plants and animals of the New World), and of course in astronomy (for one navigated by the stars, and observing the Transit of Venus was the excuse that took Captain James Cook to the almost unknown southern land) to grasp this fundamental relationship. The problems addressed were the needs of Empire; the resources of Empire funded the curiosity and questions of the savants in Britain and the continent. In this way the social structure of British and European science was built around asymmetries, in which there was a flow of primary data—specimens, calculations—from the colonies back to London, Oxbridge, Edinburgh and Paris, where the scientific aristocrats formed their theories about the processes of geological change, or the origin of species, or the nature of gases, or the structures of matter. This relationship between the "periphery" and the "metropolis" was material, ideological, theoretical and social (Hodge, 2011; Bennett & Hodge, 2011, and MacLeod, 2000).

© The Author(s) 2023
W. M. Goss et al., *Joe Pawsey and the Founding of Australian Radio Astronomy*,
Historical & Cultural Astronomy, https://doi.org/10.1007/978-3-031-07916-0_3

In Australia, where the first English invaders arrived in 1788, the first century of colonial life was mostly a struggle to survive, and such science as was pursued was a scramble to understand the very different natural world—flora, fauna, weather, geology—the colonists faced. Books and journals were expensive and precious resources, and it took months for new scientific publications to arrive from Britain (Moyal, 1976 and MacLeod, 1988). Scientific interest in Australia, however, was lively, and many leading European scientists (Charles Darwin among them) made the long voyage to visit, explore and encourage local scientific activity. Of course one of the leading colonial sciences was astronomy. The southern skies were a necessary and important component of growing understanding of position calculation and meteorology, as well as being of absorbing interest to the new settler, and each Australian colony quickly established an Observatory and government astronomer to go with it; in addition, a number of private observatories were expensively shipped out and built.

But astronomy was not quite the same as physics, and physics itself was a very new field that emerged slowly from studies of electricity and magnetism over the course of the nineteenth century. The first two Universities, in Sydney and Melbourne, were founded in the late 1850s but taught only tiny numbers of students, with science subjects being taught as part of Arts degrees. The first science faculties were established in the late 1880s; the 23-year-old William Bragg came from Cambridge (where experimental physics was not then taught) to be Professor of Mathematics and Experimental Physics at the University of Adelaide in 1886 (https://www.nobelprize.org/prizes/physics/1915/wh-bragg/biographical/); 24-year-old Richard Threlfall came from Cambridge to be Professor of Physics at the University of Sydney in 1888 (Home, 1990); Thomas Rankin Lyle, mature at 29, followed from Dublin, to take the Chair in Natural Philosophy at the University of Melbourne in 1889.

The Professoriate of the late nineteenth and early twentieth centuries were defined by their energy, their singular achievements, and by the intellectual isolation and "tyranny of distance" (Blainey, 1982) that marked so much of life for white Australians, whether immigrants or "Currency" (locally born) lads and lasses. Isolation and the tyranny of distance was particularly acute in physics, and their impacts were keenly felt. William Bragg's experience was a case in point. To teach himself experimental physics, Bragg read a handbook on the ship he sailed on to Adelaide, and then apprenticed himself to a local firm of instrument makers in order to build a laboratory useful to students (or to himself). The formation of the Australian Association for the Advancement of Science in 1888, which allowed Bragg to meet with his counterparts from Sydney and Melbourne, was decisive for those "such as myself," as he told his fiancée Gwendoline, "who are willing to work but who don't know quite where to begin." In particular, exposure to new ideas and direct mentoring from experienced experimentalist Richard Threlfall proved to be the critical stimulus that encouraged Bragg to explore published literature on radioactivity and hesitantly to begin to undertake experiments that might contribute to it. This also was not easy in the Australian colonies. In order to pursue an interest in the exciting new discovery of X-rays, he needed to collaborate with a local chemist,

Samuel Barbour, who could supply a glass discharge tube obtained during a visit to England, and his father-in-law, Charles Todd, the Government Astronomer, who could lend an induction coil and a battery sufficient to produce the required Roentgen rays.

As is well known, Bragg's experimental research program grew in success and confidence, and he began to cultivate research students and colleagues in his turn, including the young John Percival Vissing Madsen, who would later direct the wartime radar research program in which Joe Pawsey played so central a role. Madsen represented the first generation of Australian-born scientists; he graduated from the University of Sydney with a Bachelor of Science and the University Medal in Mathematics in 1900, and Bachelor of Engineering and University Medal in 1901 (Myers, 1986). He then came to the University of Adelaide as Lecturer in Mathematics and Physics where, with and led by Bragg, he conducted experiments in radioactivity and X-rays, leading to a Doctor of Science, the highest degree possible to obtain in Australia until after World War II. Madsen's collaboration with Bragg was unquestionably fundamental in shaping his vision and understanding of what knowledge-driven science could accomplish.

Eventually isolation became too much for Bragg; he wanted colleagues who could understand his research and to partake in the discussion of new ideas and data. In 1909 Bragg returned to Britain, and as is well known, two Nobel prizes quickly followed. Bragg certainly demonstrated that it was *possible* to undertake first class research in physics in Australia. But, as historian of Australian physics Rod Home remarked, it took a Bragg to do it. After his departure, his students were not able to sustain a research programme (Home, 1984 and Home 1990). J.P.V. Madsen, who, right from the beginning, had held a strong interest in practical applications for science, returned to Sydney in the same year that Bragg left, to take up a lectureship in Engineering; and there he found a flair, not for research himself, but for the facilitation and support of research by others. It is in these stories that we can see the relation between "metropolis" and "periphery" at its most stark.

Nonetheless, science was growing slowly in Australia. If Madsen's career was one sign of this, so too was that of Thomas H. Laby, Pawsey's professor at the University of Melbourne. Laby was born less than a year after Madsen (1880), but his father's early death in 1888 left him in straitened economic circumstances, and he was limited to a technical college education. A little coaching enabled him to gain a job as assistant in the chemical laboratory of the NSW Agricultural Department, analysing the chemical composition of fertilisers, and then to becoming Demonstrator in Chemistry at the University of Sydney, just as Madsen left for Adelaide. Laby was a gifted experimentalist, and he began to try a little experimentation on topics drawn from what he could get hold of in the limited scientific literature to which he had access (it took months for the latest publications from scientific journals in Europe to reach Australia, and very few people could afford subscriptions to them).

The fundamental mechanism for nurturing a young scientist in the "Dominions" was a trip "Home", that is, back to England. (Joe's son Hastings recalled that his father-in-law very typically always referred to England as "Home", even though he was a second generation Australian and died in 2004!). In this way, such a young man (a few women began to follow this path in the early twentieth century, too) could learn the latest experimental techniques and the latest ideas, and could form the networks needed to sustain continued work in science after returning. The initial resource that supported a trip "Home" was the 1851 Exhibition Scholarships. These were established with excess funds after the Great Exhibition of 1851 (in London) (Auerbach, 1999) and were intended to develop scientific and technical training in sciences (physics, chemistry, mechanics/engineering) important to British national industries, and a small but growing percentage were made available to students from Dominion/Commonwealth Universities. Indeed such was the demand from outside the UK, that an additional scheme of Overseas Research Scholarships was instituted. While the 1851 Exhibition Scholarships and the Overseas Research Scholarships schemes were Imperial in that they harnessed the best Colonial resources for use in metropolitan science, Rod Home notes that in that period, little distinction was made between Imperial and Colonial interests: what was good for Britain was simply considered to be good for her Empire. To receive a Scholarship, a University judged to be a centre for education in Physics (or Chemistry or mechanics) had to put a candidate's name forward, and the professor in charge often needed to advocate directly to such members of the Scholarship Committee as they had connections with; or else the advocacy needed to come from the Director of the laboratory in the UK, where the candidate hoped to undertake their research.

The impact of the scheme on science might perhaps be judged by the fact that one of the early recipients was Ernest Rutherford, who was awarded an "Ex" in 1895 to study at the Cavendish Laboratory, Cambridge, the step that paved the way to his Nobel prize winning research while at McGill, where he worked out the concept of radioactive half-life, discovered the radioactive element radon, and identified alpha and beta radiation. Like his friend William Bragg, whom he met and encouraged while travelling from New Zealand to the Cavendish, Rutherford eventually desired to return from the periphery to the "centre". In 1919 he became Cambridge Professor and Director of the Cavendish Laboratory himself. In the UK and especially at the Cavendish, he then nurtured the embryonic careers of "Dominion" men.

Rutherford understood that the "centre" was as dependent on the Dominion "periphery" as it was the bestower and owner of scarce intellectual resources. By mentoring Dominion students such as Laby, the Cavendish and other "metropolitan" research centres received a constant influx of extraordinary talent, fresh ideas, different experiences, and in some cases, less constraining education, assumption or experiences than the "Public School" men of the hierarchical UK education system. Rutherford, Bragg and others could thus cultivate a lifelong research network that kept the centre constantly updated about the research springing up in ever widening circles.

Laby's is a beautiful case in point that, as we shall see, had lengthy consequences for the young Joe Pawsey and, in the end, for the shape of radio astronomy. Laby had made a small study of radium occurring in mineral samples at the Department of Mines in Sydney, and he used this as his—successful—claim to be awarded an Ex to study radioactivity at the Cavendish. But at the Cavendish he was shifted to undertaking research with no application or benefit in mind, exploring the total ionisation of various gases by the alpha rays of uranium. When he left Cambridge, first for Wellington, then for Melbourne, he took with him multiple resources that would be crucial for Joe Pawsey and for the development of physics in Australia. One was his networks, in particular, his growing close friendship with Ernest Rutherford, which became so close that both men and their families stayed with the other when visiting the other's country. Another was a view of what an ideal professor in the ideal "Commonwealth" of science, would be: one actively engaged in research, one who advocated, not for an external examination system where students were coached to be good exam-passers, but for the endowment of research to which students could be apprenticed. Laby argued: by having the attractions of a "secure though relatively modest livelihood" and "ample leisure for study", "the nation will contribute its share to the general progress of civilisation … and rising generations will be brought into direct contact with men who are best able to instil into their minds a true conception of the nature and value of knowledge".

Those words—from 1911—might have been written with Joe Pawsey in mind, so perfectly do they describe what he too would value and his way of thinking. We might presume that having Thomas Laby as his professor was inspirational for Joe, who was one of the students to benefit from Laby's perspective and energy.

Joe was having fun, too, in his undergraduate years at the University of Melbourne. Since he became a man whose quiet gravitas was one of his professional qualities, it is pleasurable to see at nearly a century's distance, how much he enjoyed those student years. In 1931, *The Wyvern*, a Queen's College (part of the University of Melbourne) publication, reported:

> In college life, Mr. Pawsey, better known as Joe, was universally admired and respected. From the first he became a prominent figure and identified himself wholeheartedly with the College activities. His enthusiasm in connection with some of the lighter of these has often proved infectious, as those who shared his undergraduate days will remember. Recently he has created a stir and gained a certain amount of notoriety by the introduction of an alleged automobile into College life.

Joe was in a small pool of students. In 1920 there were 13 BSc graduates at Melbourne and about 50 at Sydney. In 1925, reflecting the bump provided by returned servicemen, there were 44 at Melbourne and still about 50 at Sydney. In 1930, under the influence of the Depression, there were but 29 at Melbourne and 48 at Sydney. Thus Joe Pawsey was already in a small, very elite world—one that was often cosy in that period—where having tea with the professor was natural and expected. The number of students who graduated with honours in Physics was extremely tiny (Branagan & Holland, 1985).

Joe's academic abilities shone. In the first year he took four subjects, with First Class Honours in three (mixed maths, natural philosophy [physics], and chemistry) and Second Class in pure maths. He also followed lectures in French and German. He was the Exhibitioner (that is, the student achieving the highest mark in a given subject among all matriculating students) in Mixed Mathematics and won a £60 Queen's College scholarship. The next year (1927), he followed with two First Class Honours (pure maths and physics) and one Second Class Honour (chemistry). He was the Exhibitioner in Pure Mathematics. Again he was awarded a £60 Queen's

College scholarship. In the third year (1928), he again achieved First Class Honours in physics and Second Class Honours in chemistry. He was awarded a First Class Honours BSc in Physics on 13 April 1929.

But what now? The job prospects for graduates who had completed a science degree in 1928, and in particular those who had majored in Physics, were uncertain at best. Graduates in engineering or geology might find work with mining companies or the small but growing number of building, engineering or manufacturing firms operating in Australia. Chemists could find work, as Laby did, in State Agricultural departments or in the nascent pharmaceutical industry or at the Commonwealth Sugar Refineries. But for a physicist, school teaching was the only obvious use for such a degree. The two women who graduated in this field in Sydney—Phyllis Nicol and Ruby Payne-Scott—could not avoid this fate, despite Ruby's obtaining a one-off research role in cancer research for a few years. Laby's daughter Jean—not entirely supported by her father, who held reasonably conservative views on women's roles and abilities—fared only a little better (Goss & McGee, 2009, and Hooker, 2015).

However, while the majority of young men faced the bleak prospects of the Depression, Joe Pawsey discovered that qualifications did open the doors to employment. For a brief period, he held a job in Tasmania working for the Geophysical survey. But swiftly he ventured on a bolder step: into the small, expansive, totally absorbing world of research.

Part II
Becoming a Scientist

Chapter 4
New Opportunities in Australian Science, 1929

Rivett brought to the Council a deep conviction of the importance of theoretical systems analysis, a desire for axiomatic certainty, and a distrust of purely empirical work, a set of beliefs which was to have a major influence on the ethos of CSIR. Schedvin (1987, p. 24).

Joe Pawsey's was the first generation in which an Australian born child could think of growing up to be a scientist, as he was poised to do at the end of his undergraduate years. There was a new sense in Australia that science would be important for a nation growing in independence and confidence, and the modern world was being rapidly and profoundly reshaped by technology. In this chapter we set out the social and intellectual background to Pawsey's Masters and PhD research and introduce the reader to the scientific staff of the Australian Radio Research Board, where Pawsey's Masters was undertaken.

The Developing Independence of Australian Science and the Formation of the CSIR

By the mid-1920s every state capital had a University, and most University science departments included Professors who now managed to pursue *some* research interests, despite the heavy teaching loads created by the influx of returned soldiers and a population with growing interest in education. There was research outside the Universities, too: in state government departments of Mines and Agriculture, in the State Natural History museums, in hospital laboratories, behind chemist shops and for chemical companies.

Of course the institutions, relationships, styles of research and choice of projects in Australian science remained structured by Australia's sense of place within the British Empire, and by Empire ideologies and loyalties, but also by the tension between these and emergent independence and nationalism (MacLeod, 2000;

The original version of the chapter has been revised. A correction to this chapter can be found at https://doi.org/10.1007/978-3-031-07916-0_43

© The Author(s) 2023, corrected publication 2024
W. M. Goss et al., *Joe Pawsey and the Founding of Australian Radio Astronomy*,
Historical & Cultural Astronomy, https://doi.org/10.1007/978-3-031-07916-0_4

MacLeod & Jarrell, 1994; MacLeod & Lewis, 1988; MacLeod, 1980). It is characteristic of science in the Commonwealth that Departmental research programs were
very limited in scope and depth, and yet impressive in their achievements. Most
information about Australian flora, fauna, geology, and climate was yet to be
"discovered" by European scientists. (Tragically, little attention was paid to the
extraordinary systematic knowledge among Aboriginal people, of either the celestial, or the terrestrial environment (Norris, 2016)).

After World War I, scientific research expanded in Australia, and globally: the
pace of commercial and technological change was increasing, dazzling new possibilities—like being able to send voices across vast distances, without wires, the
magic of radio—were reshaping every aspect of work and life. Of course funds were
very scarce—Universities scraped by with small philanthropic scholarships, such as
the one Pawsey would be awarded. Funding only decreased at the time that Pawsey
began his Masters work, due to the global economic Depression. Despite this, a
number of research institutions were constructed in the years after the War, such as
the Commonwealth Serum Laboratories (built to manufacture vaccine against diph-
theria), the Australian Institute of Tropical Medicine (oriented to the fundamentally
racist research questions of whether and how white people could thrive in the
tropics), and the Walter and Eliza Hall Institute (a venture to investigate venom
and infectious diseases, which became a centre for virology under the direction of
future Nobel prize winner Frank Macfarlane Burnet (Brogan, 1990)). It was this
modest growth that enabled a fortunate few Australian children, one of whom was
Pawsey, to become scientists.

Creation of the CSIR, Scene of Most of Pawsey's Career

In 1920 an Institute for Science and Industry was created by an Act of Federal
Parliament and in 1926 the Act was amended to form the Council for Scientific and
Industrial Research (CSIR; later CSIRO). The purpose of the new organisation was
to "initiate and conduct scientific research to assist in the development of the primary
and secondary industries of Australia" (https://www.csiro.au/en/About/History-
achievements/Our-history). This creation of a national laboratory that would undertake major research projects and coordinate scientific research across the country
was a symbol of science becoming a national priority (Schedvin, 1987).[1]

The new CSIR was governed and directed by an Executive Committee of three.
The members of this Committee were selected for their scientific eminence—and for
their eye on business interests (Schedvin, 1987). They comprised George Julius, an
energetic engineer and successful businessman enthusiastic about supporting radio
research, Arnold Richardson, an agricultural scientist and Superintendent of

---

[1]We refer interested readers to excellent histories of CSIRO, which can be found here: https://
csiropedia.csiro.au/

Agriculture to the Victorian government, and above all, ACD Rivett, Professor of Chemistry at the University of Melbourne (and married to chemist Stella Deakin, daughter of Australia's second Prime Minister, Alfred Deakin). As the masterful historian of CSIRO, Boris Schedvin, has argued, this "triumvirate" provided exceptionally balanced leadership for the new organisation through its first two decades (Schedvin, 1987). It was a particular challenge to find a balance between the pressure to undertake applied research that would quickly produce "results" and to support the scientific enterprise in general. Fortunately for Pawsey's career, David Rivett provided consistent support and advocacy for "basic" science throughout his 23 years of service to CSIR (Rivett, 1972).

In fact David Rivett, a generation younger than Pawsey, shared many similarities with him—and indeed the two families were distantly connected, as wartime correspondence between Rivett and Pawsey's mother indicates.[2] Both Rivett and Pawsey were Australian born, and both were from rural backgrounds, with principled, Nonconformist parents and relatives (Rivett's father was a pastor, passionate about equality, pacifism, and social progress through education); both were educated, with the help of scholarships at Wesley College for their high school and matriculation, and then at Queen's College at the University of Melbourne; and both were mentored during their degree in science at the University of Melbourne (BSc 1906, for Rivett) by an enterprising professor. Rivett had a transformative experience of what "pure" research could offer during World War I when he was seconded to research into the factors limiting the production of pure ammonium nitrate, needed for explosives, in Britain.[3] He wrote of his war work:

> any ordinary type of test tube fumbler, if given a handful or two of these materials, could manage by a few hit and miss trials to get some sort of procedure for getting a specimen of ammonium nitrate out of them: but *there was only one possible way of finding how to get the maximum amount of this compound in the purest condition* and that was by going through the whole involved business of getting the complete phase rule model of the highly complex four-component system, with its dozen or more possible phases. Once you got these models ... [y]ou knew you had the one and only best line of procedure: and you realised **what an utterly stupid practice blind empirical stabbing would have been**, since even if it had led to some success, the chances of the highest success being attained were not one in a million. [our emphasis].[4]

Schedvin comments that Rivett brought to the Council a deep conviction of the importance of theoretical systems analysis, a desire for axiomatic certainty, and a

---

[2]Letter from David Rivett to Mrs. Pawsey 21 July 1942, Joe and Lenore Pawsey Family Collection.

[3]Rivett held a Rhodes scholarship to Oxford in 1907 and then worked at the Nobel Institute in Stockholm working under Svante Arrhenius, where his chemical interests moved towards an understanding of equilibria within heterogeneous systems.

[4]Rivett identified the phase variables in the crystallisation of ammonium nitrate (later published as The Phase Rule and the Study of Heterogeneous Equilibria (1923)).

distrust of purely empirical work, which guided his dedication to ensuring the Council supported basic science research.[5]

Rivett's vision for the CSIR was of the recreation of the British and European model of small research teams built around a distinguished scientist, who could exercise great autonomy in setting research goals and methods (Schedvin, 1988). C.B. Schedvin writes that CSIR in these years was profoundly characterised by the scientific norms—or ideals—of community—open sharing of information, individual endeavour, and above all, by commitments to scientific autonomy. The Executive, and in particular Rivett, continued to defend this vision rather remarkably through the pressures of the Depression and beyond.[6]

But the CSIR had no annual appropriation to build such research programs on its own, and pursuing basic research was outside its supposed limited "coordination" role—the actual doing of science was supposed to remain the province of the States. Virtually all the early work of the CSIR was focused on urgent agricultural issues, such as the terrifyingly swift spread of prickly pear cactus in Queensland, or the frequency of rust in wheat crops. Indeed, so strong was this focus that W F Evans calls it an "enigma" that the CSIR should so early have also created the Radio Research Board (Evans, 1973). But wireless was the transformative technology of the decade, and its predecessor, the telegraph, quite literally tied the new nation together (Standage, 1998; Taylor, 1980; Muscio, 1984) (at some cost: the dispossession of Aboriginal people from their Country, and also the life of the great uncle of one of the authors of this book!). 1901, the year the Commonwealth of Australia came into being, was also the year that Guglielmo Marconi first managed to transmit a Morse signal across the Atlantic.[7]

Thus Pawsey's career began in a new national scientific organisation in which one leader, David Rivett, was an active supporter of basic science research—and where there were also strong connections to industry, including the new industry of telecommunications.

Radio: A Technology Transforming Australia

The spread of radio was an early example of the breathtaking speed of technological and social change that marked the twentieth century. The new nation of Australia would need to be able to innovate. This section explains why the Radio Research Board at the CSIR was formed and provides context to understand why Pawsey's Masters research was of importance.

---

[5] Indeed, Rivett's ideal was that half the CSIR budget should be devoted to knowledge-oriented questions.

[6] See Chap. 10.

[7] At least allegedly: and if the 1901 result was dubious, the capability was indeed confirmed the following year.

In 1905 the Marconi company had already started a wireless Morse service for interstate communication in Australia. Amateur wireless transmitters were active in Australia; in 1905 the first legislation regulated such activities, within the Commonwealth Postmaster General (PMG) department. Radiotelegraphy was central to Naval and land communications during WWI. The year before, using Australian and British capital—the Australian government owned 50% of the shares, plus one—Amalgamated Wireless Australasia (AWA) came into existence by buying the rights of the Marconi and Telefunken organisations in Australia. AWA undertook manufacture of radio equipment for a range of customers including merchant shipping and built a necessary research and development arm to support its products. Over the next two decades, the influence of AWA in providing technical innovation and human resources for science, was profound. Its Chairman, Sir Ernest Fisk, acknowledged that it was the second largest wireless organisation in the British Empire. It was, John Madsen pointed out when arguing for the value of including Fisk on the radio research board, "a semi Government department", one that undertook almost all the construction and technical work in Australia and with respect to broadcasting in Sydney and Melbourne specifically (Evans, 1973).

After the war, the commercial broadcasting potential of radio grew exponentially (Jones, 1995; Carty & Griffen-Foley, 2011). In the USA the first advertised service began in 1920; by March 1921 there were 50,000 receiving sets, which grew to 750,000 by May the following year, while 187 new broadcasting stations sprung into being in the same time period. The British Broadcasting Company was formed in May 1922 and by 1925 had issued over a million listeners' licences. In 1923 the British government announced it would erect a long wave transmitter for Empire telegraphy, which was in operation by 1926; in 1924 the first still pictures were successfully sent across the Atlantic by radio.

Radio was transforming Pawsey's world during his formative adolescent and early adult years; but successful commercial radio needed to solve a whole series of challenges to make the technology workable in the many markets now so enthusiastically taking it up. Radio reception was highly variable in geographical range and in quality. Radio operators encountered the phenomena of "skipping" (long distance propagation occurred but signals "skipped" over "dead zones"), fading or "swinging" of signals (variations in the received signal (or in signal attenuation); for example, there were dramatic differences in the distance over which radio signals could be heard between day and night, and "static" or "strays", that is, noise created by electrical disturbances in the atmosphere, also known as "atmospherics", (the topic that Pawsey would investigate for his Masters research). Antenna development itself required (and continues to require) a mix of empirical engineering and mathematical-theoretical research (Gillmor, 1991).

Responding both to the commercial challenges and the strategic potential for Empire-wide communication systems, the British Department of Scientific and Industrial Research (DSIR) —the model on which the CSIR was based—constituted a Radio Research Board in 1920 to "assist in the coordination of radio research work carried out by the fighting services and the Post Office, *and to provide for research work of a fundamental nature in directions where it was lacking and where it would be likely to lead to useful applications.*"

In order to understand Pawsey's Masters and PhD research, and how it provided the "repertoire" (Ankeny, 2019) of ideas, practices, mathematical theory and devices that would underpin wartime radar research and then early radio astronomy (Gillmor, 1991; Sullivan, 2009), we now offer a brief sketch of research into radio communications and the entity that turned out to strongly influence these, the ionosphere. In hindsight, ionospheric research can also reveal—par excellence—how much scientific progress has been driven by a dialectical interaction between science and technology (de Solla Price, 1964). Ionospheric research also involved several other actors—Appleton, Fred White and the new Australian Radio Research Board scientists David Martyn and George Munro, among others—who shaped Pawsey's scientific development enormously, directly or indirectly (Gillmor, 1991). Having set out a sketch of ionospheric research, we will then return to describe the work of the Radio Research Board.

The Creation of the Radio Research Board (CSIR): High Impact in Constrained Circumstances

The case for a Radio Research Board was compelling enough in itself, given the need to improve receivers and broadcast quality and to understand how conditions local to Australia, such as climate and geography, impacted on transmission quality. Local radio broadcasting companies wanted to know what frequencies to use to broadcast to a rural population, and what local conditions of climate and geography would affect the broadcast quality (Gillmor, 1991).

John Madsen and Thomas Laby were among the earliest to see the importance of new radio technologies and the need for research to support their development. Madsen above all is credited with the creation of the Radio Research Board at CSIR. He was able to advocate, network, finagle, hustle and harass a similar Board into being in Australia, bringing together the Chairs, Presidents and leadership of the Wireless Institute of Australia, the Broadcasting Company of Australia (3LO), Farmers Broadcasting Company, Australian National Research Council, HP "Poo-Bah" Brown (Chair of the Postmaster General Department), the Munitions Supply Board, the Department of Defence, and the relevant professors in Melbourne and Adelaide, T H Laby and Kerr Grant (Evans, 1973, and Gillmor, 1991). (Just what a feat it was to constructively manage competing interests and points of difference is worth a pause of appreciation and admiration for John Madsen, and is entertainingly presented in Evans's history of the RRB (Evans, 1973)).

During its first 13 years, the Radio Research Board led a precarious existence, constantly threatened by dire governmental funding cuts during the Depression. Despite this, it generated substantial contributions to science locally and globally. The Board was constituted in order to conduct useable research in six priority areas:

"Field Intensity; Atmospherics; Fading; Distortion and Modulation". However, the training, interests, and connections of the researchers—as well as fact that the physical world impinged directly on radio communications and needed to be understood before various difficulties could be remediated—resulted in the Board making more contributions to "pure" science issues than to patentable improvements to radio communications technology in the years prior to World War II.

From our perspective, it also brought together a remarkable (if small) group of radio researchers whose knowledge and expertise would be available to Pawsey as he embarked on his first significant experience of research in his Masters degree.

David F. Martyn, A.L. Green and G.H. Munro and L.H. Huxley Are Recruited to the Radio Research Board, 1929–1930

Once established, the Radio Research Board found funds—70% from PMG, and the rest from broadcasting companies in Sydney and Melbourne that Laby and Madsen had already been pursuing agreements with—to support 6 research scientists, 3 in Sydney and 3 in Melbourne. High quality researchers with expertise in relevant areas were hard to find. Pawsey was the only Australian candidate appointed; in the absence of students with suitable training, the rest needed to be recruited from Britain.

Four of its six initial research officers were recruited from Britain, via an illustrious selection committee composed of leading British physicists Sir Ernest Rutherford (Chap. 3), Sir Edward Appleton (Chap. 5) and Sir Henry Tizard (Chap. 9).[8] This committee would ensure that the Australian program remained connected to the "centre" of ionospheric research. British scientists offered consistent support to the Australian team. Tizard offered to train the new recruits at Slough; Sir Robert Watson Watt made available at a reduced price, a new cathode ray direction finder which Munro and Huxley brought with them on the voyage from Britain.

The initial four officers recruited to the Radio Research Board were A.L. Green, G.H. Munro (Home, 1995), L.G. Huxley (Crompton, 1991) and D.F. Martyn (Massey, 1971). All arrived intending to pursue the "pure" science questions that had interested them prior to their appointment—they were not jobbing graduates with a narrow interest in solving technical problems for local broadcasters—and all made swift and profound contributions to research exploring the composition and physical features of the ionosphere, at that time the leading issue in radio research.

The pre-eminent researcher among the group was David Forbes Martyn (Piddington & Oliphant, 1971; Home, 2000). Martyn was a Scot who graduated

---

[8]Tizard would lead the British radar research effort in the lead up to, and during, World War II (Chap. 9).

with a PhD from the University of London, and showed considerable talent as both an experimentalist and as a theoretician. He was 23 when he came to Australia in 1929, at first to work in Laby's laboratory (where he doubtless would have met Joe Pawsey) investigating fading of signals from local broadcasting stations, but soon after moving to Sydney to work with Madsen's group. Laby proved touchy and difficult to work with throughout this time; in 1932 and 1933 respectively, George Munro and Thomas Cherry also moved to Sydney to avoid him (Evans, 1973).[9]

Because Martyn's research interests and capacities would have a profound impact on Pawsey in subsequent years, and because his Radio Research Board activities offer a snapshot of "the state of the science" at the time when Pawsey began his career, we sketch a few salient details here. Martyn came to Australia with two proposals of research to put to the Board: one was for what was effectively a Doppler radar—a proposal to study the moon using the reflection of very high frequency waves[10]—and the other to study the ionosphere using an adaptation of Appleton's frequency change technique. Only the second project was approved by the Board, and according to Jack Piddington, it proved to contain a subtle fallacy, prompting Martyn's interest in the pulse echo sounding technique.[11]

At this time Martyn had a gift for activating the otherwise isolated researchers in Australia, and he collaborated with nearly everyone working in the field locally (Evans, 1973, p. 122), showing great flair not only as a theoretician, but in research to solve a spectrum of issues in instrument design. His collaboration with Cherry and later with Green perfected the group's 3-aerial reception system, which he later found to be in advance of any European instrument for studying polarisation and lateral deviation. Thomas Cherry had also recorded some complex wave-length-change fringes in Melbourne (from long distance signals from transmitters in Sydney), and it was Martyn who succeeded in analysing them to reveal a layer between the E and F layers of the ionosphere. With Radio Research Board colleagues George Munro and J.H. "Jack" Piddington, he perfected a "pulse-phase" technique that provided continuous data on the polarisation of reflected radio waves and hence on the dynamics of the layers of the ionosphere from which they were reflected, resolving a debate about which dispersion formula applied to ionospheric reflections. In 1934, his collaboration with Sydney Professor of Physics V.A. Bailey (1895–1964) (Home, 1993) provided a theoretical explanation of the newly discovered "Luxembourg Effect" (cross modulation of radio waves), showing it to be a non-linear effect in the ionosphere. In 1935 he and Green demonstrated that the

---

[9] Interestingly, Madsen wrote to Laby to suggest Pawsey be appointed as Cherry's replacement; but at that time Pawsey was in the middle of his PhD studies in the UK and Laby replied that he was unwilling to wait two years for Pawsey to finish.

[10] A detailed discussion of Martyn's plans for a lunar radar can be found in online supplementary materials, NRAO ONLINE.4.

This was a very prescient suggestion made 20 years in advance of the first doppler radar studies of the moon.

[11] This apparent failure of his research plan would prove useful in the development of radar (Chap. 9 and ESM 9.1, Radar History).

reflection point of radio waves from the ionosphere could move rapidly. In the same year he developed a theorem relating equivalent height and reflection coefficient at oblique incidence to that at vertical incidence. Whilst he was himself rather self-deprecating about this piece of "simple trigonometry" (Evans, 1973, p. 121), the theorem was widely accepted and applied and became known as "Martyn's Theorem".

All of these publications had substantial international impact, as did his 1936 researches with O.O. Pulley on the layers of the upper atmosphere, which yielded several fairly revolutionary assertions, including that above 80 km the temperature rose steadily to values of the order of 1000 C. This paper, communicated to the Royal Society by Rutherford, aroused substantial discussion, and its claims have since been verified by post-war research using rockets and satellites. Martyn's many contributions made him an internationally pre-eminent ionospheric scientist even in those years where only one research trip back to the UK was possible (in 1936). In 1972, Stewart Gillmor surveyed all ionospheric papers published between 1925 to 1960 by 1676 authors. The most cited authors were (1) Appleton, (2) Chapman, (3) Ratcliffe and (4) Martyn (Piddington & Oliphant, 1971; Home, 2000; Evans, 1973, p. 188).

We note here that it was at this time that solar research in Australia began to develop under the stimulus of Madsen, Martyn and others in the Radio Research Board, who were interested in better understanding the sun, since solar radiation produced the structural complexity of the ionosphere. By 1930, Madsen had sought connections to the Mt. Stromlo Solar Observatory, located near Australia's capital of Canberra (some 3.5 hours' drive from Sydney today and considerably more distant in travel time in 1930). A.J. Higgs collaborated with both the Sydney and the Melbourne team, using an atmospherics recorder, a cathode ray direction finder and then building his own manual ionosonde, to explore connections between solar activity and auroral displays, "magnetic storms", and radio fading (Evans, 1973, p. 107).[12] Higgs's pulse-echo recordings became an essential segment of ionospheric research at Sydney. By 1937 the research program at Mt. Stromlo employed two investigators almost full time, and the Radio Research Board was approaching the Observatory authorities to initiate investigations on the spectra of solar eruptive zones and other solar factors that influenced the ionisation of the upper atmosphere, plans that would be disrupted by war. When the Mt. Stromlo Director retired just prior to World War II, Martyn was a strong candidate for his replacement.[13] This research and the connections between Sydney's radio researchers and the Mt. Stromlo astronomers would be very important in the years to come.[14]

As well as indicating the calibre and range of Martyn's research, this very brief precis demonstrates how significantly Australian science could develop as a result of investing in high calibre scientists and training for young scientists. All these

---

[12] See also Chap. 5.

[13] In hindsight, the preference to retain Martyn for radar research instead, would turn out to be an unpredictable tragedy (see Chap. 9 and ESM 9.3, 1941: Difficulties, and NRAO ONLINE.7).

[14] See Chaps. 11, 12, 13 and 14 and 16.

researchers were, or would become, close colleagues of Pawsey. And the Radio Research Board's investment in human resources had other crucial, but largely unforeseen, impacts.

First, Evans's analysis of the history of the Radio Research Board identifies the significant contributions that the Board made to the development of scientific culture in Australia—partly as a result of the frequent interchange of scientific visits to the UK (and the USA) and vice versa; partly because radio researchers often delivered special courses and colloquia to augment teaching at the Universities (Evans, 1973, p. 94); and especially in training postgraduate researchers, with at least 16 scientists gaining postgraduate degrees as a result of support from the Board (Evans, 1973, p. 92). Most went on to distinguished careers.

Secondly, the Radio Research Board was an important supplier of trained personnel to industry. When A.L. Green left the Board to join the staff of the large telecommunications company AWA (Amalgamated Wireless (Australasia)), a formal Board memorandum noted somewhat tartly that it was decided to send the Executive a note "and to point out that the Board has now supplied Messrs AWA Ltd with three Officers" (Evans, 1973, p. 113). W. Baker, G. Builder, H.B. Wood and O. O. Pulley likewise worked in industry during the later 1930s. And of course, all the early Radio Research Board appointees, except Huxley (he returned to the UK when job insecurity was at its peak in 1931), would be key contributors to the wartime radar research program. Perhaps most importantly, the third point to be made is that, had the Board been terminated in the Depression era as was very nearly the case, then Australia would not have had the trained personnel available to pursue the radar research program that provided such crucial defence capability during the war.

We also note that although the Radio Research Board was clearly constituted within the structures of Empire (Egaña & Anduaga, 2009), it held a clear international outlook. Radio communications itself, and Marconi's and other major companies, were supra-national in outlook (Evans, 1973, p. 114)—as, indeed, was meteorological science. There were many close links between the Radio Research Board and Appleton's research group; for example, Radio Research Board scientists O.O. Pulley (1906–1966) and G. Builder (1906–1960)[15] were Australian graduates who spent time working with Appleton before returning to join the ionospheric research team in Sydney, as did Fred White from New Zealand.[16] Jack Piddington interrupted his Walter and Eliza Hall Research Fellowship to gain experience at Cambridge with Appleton. Board scientists were connected to Sir Robert Watson-Watt's program of research in atmospherics,[17] with Munro and others training at Watson-Watt's research station at Slough.

But their connections were not only with the UK. The scientists encouraged correspondence and visitors from the USA, Canada, South Africa and elsewhere.

---

[15] For Pulley's research see https://www.researchgate.net/scientific-contributions/84998060_O_O_Pulley. For Builder, see Home, 1993.

[16] White is discussed extensively in subsequent chapters.

[17] Chapters 5 and 9.

Lloyd Berkner (1905–1967), the prominent entrepreneurial American ionospheric researcher, spent 6 months at the Watheroo Observatory in Western Australia (Evans, 1973, p. 115; Home, 1983) and then some weeks at Sydney and Melbourne, as part of a world tour on behalf of the Carnegie Institution of Washington, and prepared a report jointly with Martyn on collaboration between the Radio Research Board and the Carnegie Institution. This international outlook and especially the US connections, would become important in the post-war years.[18]

In the end, Board scientists published more than 110 papers in the period 1928–1940, developed impressive experience in instrument design and adaptation, built extremely close links with British radio science and an identity in the global networks beginning to develop in radio and meteorology.

Having sketched the organisational environment and scientific colleagues of the institution where Joe Pawsey was to begin his research career, we turn to a brief history of the research questions and methods in ionospheric research (such as the "frequency change" and "pulse-echo" methods of ionospheric sounding), and the role played by significant British researchers Sir Edward Appleton, J.A. Ratcliffe, and Reginald Smith-Rose, to understand the intellectual background to his early career.

---

[18] In Chapters 16 and 17 we discuss the importance of connections between Radiophysics and the Carnegie Institution of Washington, DC. In the funding of the Parkes telescope, the Carnegie Institution of Washington would play a major role (Chap. 27).

Chapter 5
Ionospheric Research, 1895–1935

The triumph of ionic refraction facilitated an even subtler epistemic transformation in radio propagation studies. While the old atmospheric reflection had built predictions and explanations on the geometry and very few material parameters of a featureless, homogeneous Kennelly-Heaviside layer, ionic refraction resorted to a much more structured, heterogeneous, and hence interesting atmosphere. The skip zone and other short-wave irregularities resulted from the upper layer's height, thickness, and electron-density profile or from the geomagnetic rotations of radio waves within it. This assertion had a flip side: the physical characteristics of the ionized atmosphere could account for various wave propagation phenomena, and those phenomena revealed the structure of the upper layer, too. The work of short-wave researchers in the early 1920s prepared the ground for a change of focus from the behaviour of radio waves using the upper atmosphere as an explanatory tool to the properties of the ionized atmospheric layer using radio waves as a probing instrument. Propagation studies were beginning to evolve into atmospheric physics. Yeang (2012, p. 146).

Pawsey's initial foray into research was concerned with an area of applied science –investigations into how "atmospherics" (electrical disturbances in the atmosphere) and ionospheric turbulence affected radio communications. In this chapter we explore the general intellectual background to this science. We return to our account of Pawsey's development in Chap. 6.

This chapter draws from research in the history and philosophy of science, and in particular a recent history of early ionospheric research by Chen-Pang Yeang (2012). We note in footnotes where analogies and connections exist to later events in radio astronomy. In this summary, we discuss the interplay between "pure" and "applied" science, and we explore what a history of science might look like, if it paid particular attention to the instruments that scientists used. We are interested in where scientific ideas come from, and in how some ideas might occur simply because a scientist was familiar with certain sorts of instruments and not others, or because a scientist was primarily interested in improving an aspect of instrument design.

This story is relevant to Pawsey, who, in 1930, was beginning research that could lead to a career in industry as easily as in basic science. It is a story of how immediate, practical problem solving (such as how to obtain clearer reception of

© The Author(s) 2023
W. M. Goss et al., *Joe Pawsey and the Founding of Australian Radio Astronomy*, Historical & Cultural Astronomy, https://doi.org/10.1007/978-3-031-07916-0_5

radio signals) generated broad, conceptual questions, such as how to understand radio wave propagation in a turbulent ionosphere. Conversely, an investigation of the structure or dynamics of the ionosphere, or studies of radio wave propagation, could and did unexpectedly address practical issues in radio communications.[1]

The Beginnings of Radio

Radio communications research began in the mid 1890s: for context, this was not long after the very first Professors of Physics (at Sydney in 1887) and "Natural Philosophy" (at Melbourne in 1889) had arrived in Australia to set up their new Departments, and just 14 years before Pawsey's birth. The pace of change in science can be gauged by considering the difference of a single generation, from Guglielmo Marconi (creator of radio communications) to Joe Pawsey. In this generation, the world moved from a time when "science" was still very much the domain of wealthy "amateurs"—particularly in Australia (Moyal, 1986)—to one where science was conducted by professional scientists in companies and Universities.[2]

Was being able to take more risks an advantage of the amateur era? If so, the origins of radio are a case in point. In 1894, Guglielmo Marconi (1874–1937), the 20-year-old son of minor Italian nobility and educated, as was still common then, at home, spent hours in his room trying to create "wireless telegraphy"; that is trying to send telegraph messages without wires, using the recently discovered "Hertzian" (radio) waves. At that time, natural philosophers (physicists) considered Hertzian waves would be essentially the same as light waves, and physicist Oliver Lodge (1851–1940) (Gregory & Ferguson, 1941; Wilson, 1971) had predicted the maximum transmission distance would be a half mile. But by using a recently invented device, a coherer, which changed resistance when exposed to radio waves, Marconi was able to build a wireless storm alarm, a device that received radio waves generated by lightning, and then transmitted a signal across his attic room to ring a small bell. Later (in 1895–6), outside, and using a grounded receiver and transmitter and a higher monopole antenna, he transmitted radio waves for two miles, and over hills. From there it did not take long for Marconi to begin shipboard experiments—wired telegraphy was of course entirely useless for moving ships at sea—and to

---

[1]Philosophers of science have debated how one might define "pure" and "applied" research ever since the terms were invented in the late nineteenth century. However, what these terms connote in any specific context is usually well understood by the scientists involved. We will discuss how the move to radio astronomy after WWII is an example of Australia accepting the value of pure research; and yet this pure radio astronomy research was somewhat hidden under the applied rain making research program (see Chaps. 16 and 17 and NRAO ONLINE.25). We note that the pursuit of pure research can proceed by designing telescopes using applied research techniques. For example, building the GRT (Chaps. 27 and 29) was applied research; using it was pure research.

[2]There is substantial scholarship discussing amateurs and professionals in science. See e.g. Griffiths (1996) and Meyer (2010).

pursue, not research in physics, but a global radio communications company. In 1901 he famously transmitted a radio signal from Poldhu, Cornwall, in Great Britain, to Newfoundland, Canada (Fleming, 1937).

What mechanism could explain how a signal transmitted in Wales could be received in North America despite the curvature of the earth, which should have blocked the radio wave since it travelled only in a straight line? Between 1902 and 1919, there were two different concepts to explain how transatlantic radio wave propagation could occur: surface diffraction, and atmospheric reflection (later, refraction). The first explained radio wave propagation around the earth as the result of multiple diffraction of radio waves over the edges of cliffs and other terrestrial features. The second explained transatlantic radio wave propagation as a result of being reflected from a hypothesised entity in the upper atmosphere, i.e., the ionosphere. The first (surface diffraction) remained the dominant focus in academic radio research for almost two decades, even though the second (atmospheric reflection) was intuitively accepted by most radio engineers from the very early years.

It is intriguing to consider why this difference in view between academic physicists and radio engineers persisted for such a long time. Following Yeang, we suggest that the reason included factors such as: the constraints of particular instruments; different preferred research styles; and the influence of different people (physicists and engineers) and institutions. These factors would later arise in radar research and radio astronomy.

1902–1925: Surface Diffraction—A Productive Research Program Based on an Incorrect Premise

Why did surface diffraction remain of interest, when it could not explain radio propagation phenomena well known to annoyed radio operators, such as fading, static and diurnal variations in signal? That surface diffraction persisted as a research program reflected, in part, the dominance of mathematical physics among the researchers who worked on it in Cambridge (and elsewhere in the UK), France and Germany.[3] They prized theory, suggests Yeang, not for the breadth of empirical information it could account for, but instead for its "elegance", that is, its logical consistency and accessibility in form.

Researchers in these centres tended to investigate physical problems that could be formally represented, usually in terms of differential equations with boundary conditions that represented the physical circumstances of the problem (in this case transatlantic propagation and antenna directivity). They would then develop various

---

[3] This dominance also provided experimental physicists with constant connection to mathematical theoreticians, the lack of which resource would later hamper early Australian radio astronomy.

mathematical techniques to solve them.[4] Thus, the focus of research soon became a mathematical question: "what was an accurate approximation of the diffracting field's intensity above a large conducting sphere?" (That is, these researchers became less focused on finding a direct answer to the question of how long distance radio wave propagation could be explained.)

The research program that resulted sought proper approximations of the diffracting field's analytical form and debated the legitimacy of these approximations. Due to the lack of available instruments and infrastructure, the surface diffraction theorists had virtually no data on which to test their theories for more than a decade. But contrary to expectation, when empirical data became available and a formula developed to express it (we tell this story below), the formula did *not* resolve the debates over their approximating theories, because the empirical regularity had the wrong wavelength dependence. Mathematical research in surface diffraction continued anyway—new mathematical tools for dealing with approximations of diffraction series or integrals were being developed, of interest for their own sake[5] (Yeang, 2012, p. 106).

1910–1919: The Austin-Cohen Formula: Discarding Anomalous Data

In the early 1910s, the US Navy was able to finance tall transmitting towers and receivers and to equip its ships with radio communications equipment. The Navy then began conducting propagation experiments in order to test how well the equipment worked. As a result, the first empirical data that could be used to test ideas about radio wave propagation became available. Two engineers, Louis Austin and Louis Cohen, developed a formula in 1910–1911 that could serve as a useful approximation for the measured values recorded in these experiments. They measured a well-defined characteristic of transmitters (antenna current) and aimed to represent it through a simple mathematical formula that fit with the framework of surface diffraction, which was the dominant theory of the time. In the process, Austin had to decide what to do with the nighttime data. Given that radio signals often behaved differently between day and night (for instance travelling much farther at night), it was, of course, too variable to fit his calculations. He simply discarded it. He faced a similar question when the formula consistently produced values that were too high for distances of more than 200 km. Austin's decision was

---

[4]Hertz himself had worked within this tradition, and had proposed a theory for his spark gap experiment in which he modelled the spark gap as a tiny radiating dipole source and solved Maxwell's equations under spherical symmetry (Yeang, 2012, p. 21).

[5]And although the problems they addressed mathematically did not necessarily correspond with real physical ones, the program was productive—Yeang comments that the theory of complex series and integrals that prospered from the1930s to the 1950s arose from the diffraction theorists of a generation earlier.

again to discard the anomalous data, by invoking the assumption that the discrepancy resulted from the atmospheric absorption of energy, an assumption that fitted with simple absorption laws elsewhere in physics.

These kinds of judgements, made in order to resolve apparent anomalies, look incorrect in hindsight; they were shortly to be explained by the features of the ionosphere.[6] But at the time the formula, and the models with which it was designed to fit, were convincing, because they were coherent with the knowledge of the day, and they continued to produce useful research. In this case, by discarding anomalous data, Austin and Cohen were able to develop a formula that combined transmitting—antenna current, height of transmitting antenna, height of receiving antenna, and wavelength—with a previous long-distance transmission formula. It seemed to "work". The Austin-Cohen formula was enormously useful to scientists. It provided the only quantitative and empirical basis for understanding long-distance propagation at that time. From that time, researchers focused their concerns on whether their predicted numerical results fitted the formula, rather than examining whether their theories fitted physical intuition, or whether they fit with wireless engineers' knowledge of how their instruments functioned.

The Austin-Cohen formula was compelling to wireless engineers, too, since, as Yeang remarks, what physicists saw as a law for testing mathematical theory, engineers viewed as a reliable design rule: it stipulated the quantitative dependence of incoming signals' strength on antenna height, distance, transmitter power, and wavelength, meaning that engineers could design antennas to provide a minimum signal-level over a given distance. As a result, the Austin-Cohen formula had huge engineering consequences. Because it predicted longer propagating distances at longer wavelengths, the builders of long-range wireless stations lowered their operating frequencies as much as possible and erected giant antenna towers to have their signals reach wider areas.

The irony was that this paradigm[7] became obsolete as soon as it consolidated. The principle reason was that by the end of WWI, radio amateurs and engineers found that short waves (ie 1.7–30 Mhz, or 10-180 m) could *also* propagate over very long distances, and with just moderate transmitting power.

---

[6] Similar misjudgements about anomalous data would occur in radio astronomy. We discuss an example in Chaps. 34 and 35, in which many radio astronomers dismissed data that appeared to point to sources of radio emission outside the galaxy. At that time, the dominant model was that radio emission was generated by "radio stars", all of which were thought to exist within the galaxy.

[7] Yeang refers to this as the "Watson-Austin-Cohen paradigm". It was a paradigm in the sense that it was a particular way to conceptualise long distance radio propagation and could encourage problem solving research within this conceptualisation; but it did not fit other Kuhnian definitions of what a paradigm might be.

Hypothesising an "Ionosphere"

From the early twentieth century, wireless operators were preoccupied with several phenomena that significantly impacted radio communications. One was the fact that the maximum distance a signal could travel varied between night and day time, travelling much farther at night. Another was "static"—clicks, grinding sounds, hissing noises—which often interfered with incoming transmissions. Static was a more serious problem at night, during summer, and in low altitude settings. Radio operators also observed that there was a strong but unexplained association between static and storms and other meteorological events (Yeang, 2012, p. 85), an issue that was to become the topic of Pawsey's Masters research.[8] A third issue was that Marconi's antennas needed to be tilted to generate optimal transmission.

Surface diffraction explained none of these facts. But all of them, together with the phenomenon of long distance radio wave propagation around a curved earth itself, could be explained intuitively by the idea that radio waves were reflected back to earth from the atmosphere. As a result, the concept of a reflecting "layer" gained wide acceptance among wireless engineers early in the twentieth century, even as mathematical physicists were developing theory for surface diffraction.

The Idea of Atmospheric Reflection, 1902

The history of research into this layer, which came to be termed the "ionosphere" after 1930, is repeatedly illustrative of how "pure" scientific investigation of its structure and characteristics shaped, and were shaped, by practical concerns with improving electrical and communications technologies. Even the very concept of the ionosphere had its origins in technological development. The concept of a radio wave reflecting layer in the atmosphere was first published in 1902 by two Britons, separately: physicist Arthur Kennelly (1861–1939) and former telegraph operator Oliver Heaviside (1850–1925). Heaviside was a practical man who invented and patented the coaxial cable. He also taught himself James Clerk Maxwell's 20 equations, and then, to make them available for practical use, simplified them down to the four commonly used today (Buchwald, 1985). He used these equations to predict the existence of an ionised layer in Earth's upper atmosphere (Nahin, 1987). In his model, the earth and atmosphere were conceived rather like a large-scale coaxial cable: a conductor with concentric boundaries of (1) the earth and (2) a hypothesised

---

[8]This provides a link to Karl Jansky, who made the world's first radio astronomy observations. He had built an antenna designed to study the origin of the "noise" on the trans-Atlantic radio communications system. He discovered three sources of noise: nearby storms, distant storms reflected by the ionosphere and noise coming from the Milky Way, discussed further in Chap. 6. See Sullivan (2009) (Chap. 3).

upper atmosphere layer. Radio waves might propagate over long distances by reflecting from these boundaries[9] (Yeang, 2012, p. 88).

Of course much radio wave behaviour did not fit this simple model. William Eccles's (1875–1966) 1912 hypothesised that the hypothetical layer formed when sunlight (radiation) broke apart molecules in the upper atmosphere and produced free ions and massive neutral particles, and that radio waves were *refracted* by these ions rather than simply reflected from a layer (Ratcliffe, 1971). This provided a plausible explanation for diurnal and seasonal variations in signal transmission and for achieving optimum transmission only with tilted aerials.[10]

Investigations into radio wave propagation involving the ionosphere, which became investigations into the characteristics of the ionosphere itself, formed the context for Pawsey's initial research. It is useful to note, that the potential for using radio-wave interference for finding the direction to sources of radio emission was obvious to many who had been involved in ionospheric research.

Direction-Finding Equipment and the Existence of the Ionosphere

The first empirical evidence for the hypothesised layer was found by a man who was both a Cambridge trained theoretical physicist *and* a London-trained engineer, T.L. Eckersley (1886–1959). Pawsey's research style would similarly combine engineering skill with theoretical insight. In order to give the reader a sense of the equipment then in use, the difficulties that scientists and engineers were only just beginning to understand, and of how ideas were connected to tinkering with it, we now briefly recount what Eckersley did.

A popular early wireless direction-finder was the rotating loop antenna, or the "frame aerial" (the Bellini-Tosi system), which determined a radio wave's propagating direction by rotating the vertical receiving loop around the vertical axis until the detected signal strength was minimum. However this, and other early direction-finding systems had many problems, including that they had direction errors of up to 40° and that they became erratic at sunset and fluctuated through the night. These night difficulties persisted despite the steep improvements in antenna loops, goniometers (devices for the precise measurement of angles), rotating mechanisms, and tube amplifiers that were generated by WWI. In the latter part of the war (1916–17), Eckersley, then stationed in the Mediterranean, set out to improve the equipment by demonstrating that that the observed errors were not generated internally, but by

---

[9]Kennelly drew on J.J. Thompson's discovery that low density air could conduct electricity—the more dilute the air, the more the conductivity—to suggest that, supposing air pressure to be proportional to density, at 80 km high, air conductivity would be twenty times that of sea water.

[10]The ground was no longer a relevant boundary condition. (Yeang, 2012, p 145).

interference from waves returning (reflected, refracted or diffracted) from the sky.[11] He designed three experiments to disentangle the mixed polarisation of "ground" and "sky" waves, basing them all on designs for direction-finders. When the results of the three experiments were compared, they showed that sky waves *were* present and *were* the cause of the observed errors; by corollary, they indicated, but did not prove, the hypothesised Keneally-Heaviside layer must be real and could be studied by measuring polarisation. This "pure" research also indicated how direction-finding devices could be improved: by designing them to cope with "sky waves".

Thinking with Equipment: Adapting Direction-Finders to Investigate "Sky Waves"

If Eckersley's experience had demonstrated that searching for sky waves could improve direction-finders, National Radio Laboratory (UK) engineer Reginald Smith-Rose (1894–1980) with assistant R. Barfield found that looking for improvements to direction-finders, could also find evidence for sky-waves (Oatley, 2004). In 1925 (after Appleton's famous confirming experiment, below) they experimented with the Adcock direction-finding system, which located positions using phase-detection instead of signal strength; Pawsey would later use this instrument as well.[12] The new design controlled the direction errors within 1°, which was not only an unthinkable improvement on direction-finding accuracy since Eckersley's wartime work 8 years earlier, but also provided confirmation that the major source of errors in the loop-type direction-finders was interference from sky waves, since these were eliminated in their adapted Adcock system (Yeang, 2012, pp. 212–214).

While this demonstrated additional evidence for the existence of sky waves, it also did not indicate anything about the characteristics of the Kennelly-Heaviside layer, whose existence had been confirmed the year before.

---

[11] The refraction/reflection changed the sense of polarisation of the radio wave, allowing it to be distinguished from the transmitted radio wave's polarisation.

[12] The Adcock used two mutually orthogonal straight rod antenna pairs, where the difference between the two rods in each pair created a phase difference in the radio signals on the rods. Since the two antenna pairs used in the system were highly sensitive to the symmetric condition, Smith-Rose and Barfield found a way to maintain symmetry by lifting the goniometer above the ground, so that each vertical rod was more like a complete Hertzian dipole with two branches, and signals were registered from the midpoints.

Sir Edward Appleton, the Frequency-Change Method and the Magneto-Ionic Theory of the "Ionosphere", 1924

Conclusive evidence for the existence of an ionosphere was famously provided in December 1924 by E.V. (later Sir Edward) Appleton (1892–1965) and his colleague Miles Barnett (1901–1979, originally from New Zealand) (Gabites, 2000). Appleton was eventually awarded a Nobel prize for this work. This research emerged from, and was made possible by, the start and rapid expansion of commercial radio broadcasting in 1922. This meant that powerful continuous broadcasters became available for the first time.[13] Appleton was sponsored by the (newly formed) British Radio Research Board to investigate fading. By this time the idea that fading resulted from interference between ground and sky waves was widely accepted. But there was still no direct experimental evidence for the existence of the Kennelly-Heaviside layer, or for its hypothesised cause (ionisation from solar radiation), or characteristics (for example, that it would show variations in height and ion density, which would in turn cause variations in radio wave propagation).

This direct evidence was supplied by Appleton and Barnett by creating an innovative method of artificial fading. The BBC allowed Appleton and Barnett to use their Bournemouth sender. "The method adopted has been to vary the frequency of the transmitter continuously through a small range and attempt to detect the interference phenomena so produced between the two rays" (Appleton & Barnett, 1925). This method was known thereafter as the "swept frequency" or "Appleton frequency-change" method. With this continuous scan in frequency the difference in distance travelled by the "ground" and "sky" waves, when measured in the number of wavelengths, changes, so the combined signal cycles through periods of cancelation (fading) or reinforcement. When all the frequencies used are combined there is only one delay for which they all reinforce.

This frequency scanning interferometer provided direct evidence for the existence of the ionosphere—and additionally, a direct measurement of its height. As Appleton and Barnett wrote in 1925:

> These effects may be explained in a general way if an atmospheric reflecting layer is postulated which is comparatively ineffective for the waves of this frequency during the daytime but bends them down very markedly at night. According to this view two rays arrived at the receiver at night, one nearly along the ground, which may be called the direct ray and the other returned from the atmosphere, and called the indirect ray ... If we assume the simplest interpretation of these interference phenomenon and regard them as analogous to those of a **Lloyd's mirror fringe system**, [our emphasis] the effects may be viewed as follows ... The experimental observations ... indicate a path difference of order 80 kilometres, which is consistent with a reflecting layer of about 85 kilometres ...

---

[13] https://www.nobelprize.org/uploads/2017/01/appleton-lecture-new.pdf

Thus the 1925 Appleton and Barnett set-up can be viewed as a precursor of the sea-cliff interferometer of 1946 (Chap. 13).[14]

With a scanning monochromatic signal, Appleton's scheme was described by Ratcliffe (1974a, p. 2095):

> The first experiments [of Appleton & Barnett, 1925] were designed to be as simple as possible. A BBC [CW] transmitter, whose frequency could be slowly varied, was used after the end of the normal transmissions at midnight. Reception was at a distance where the ground- and sky-waves were expected to be roughly equal, the receiving apparatus was of the simplest type, and the signal variations were observed on an ordinary table galvanometer. The expected "fringes" were obtained and were counted to give a measure of the virtual height of reflection.

The height was determined using a simple equation based on two or more adjacent frequencies that produced maxima (or minima) in the fringe pattern. The determination of the two wavelengths could be used to derive the virtual height of the reflecting layer.

Ratcliffe (1974a, p. 2096) continued as he described the pioneering results from 1925 as interferometry, succinctly describing the Appleton-Barnett frequency change method in the terms instantly recognisable to radio astronomers:

> ... [T]he strength of the wave received at a distance of about 100 km from a CW transmitter was observed while the frequency was slowly changed. The observed signal fluctuated between maxima and minima as the phase difference between the sky- and ground-waves altered, and, by analogy with similar optical phenomena, the fluctuations were called "fringes". If the "amplitude" of the fringes was to be large, the sky-wave should be roughly equal to the ground-wave.

Appleton and Barnett's empirical verification of the existence of the Kennelly-Heaviside layer—soon termed the "ionosphere", a word invented by Sir Robert Watson-Watt in 1926 and widely taken up from the early 1930s (Gillmor, 1976)—was of course quickly followed by deepening understanding of its formation and properties and consequently, of radio wave propagation phenomena.[15] Its composition of course fluctuated diurnally, and its structure and characteristics, resulting from the electrical characteristics of the movements and collisions of charged particles, and affected by the earth's (fluctuating) magnetic field, quickly turned out to be much more complex than previously considered. The ionosphere turned out to be both layered (Appleton could distinguish, not one, but several "layers") and turbulent. The "magneto-ionic" model of the ionosphere was dominant by the

---

[14]It is interesting to note that Pawsey's colleague Payne-Scott would also later use the swept-frequency technique in interferometry to measure positions of short duration solar bursts (NRAO ONLINE.20 and Goss, 2013, pp. 167–185).

[15]The rapid development of long distance short wave radio communications in the years after World War I contributed to the development of the "magneto-ionic model" of this layer by presenting surprising phenomena in need of explanation—for example, that propagating range varied with wavelength, and that the signal strength of short waves (under 50 m) would become zero after some distance but then rise again ("skipping"). See Yeang (2012, Chap. 6).

mid-1920s, and the "frequency change" method was established as an active experimental method to better understand it.

Connections to Cambridge and London: How the Magneto-Ionic Paradigm Generated a Research Program in Australia, 1929–1939

Appleton's research provided the context for Pawsey's Masters and PhD projects. It was Appleton who found early empirical evidence to connect atmospherics with thunderstorms and other electrically excited weather processes by working with a Cavendish Laboratory researcher who had been conducting cloud-chamber experiments to mimic thunderstorms—an example of the cross fertilisation made possible by the Cavendish's size and breadth. And of course the research of Appleton and his students and colleagues—who included the young J.A. "Jack" Ratcliffe (1902–1987), Pawsey's PhD supervisor—created a new research program in ionospheric studies. Ionospheric physics grew exponentially from 1926 to 1938 with a doubling time of 3.2 years.[16]

This research program strongly influenced the research program at the Radio Research Board in Australia, as discussed in Chap. 4. Twenty-four percent of the Australians started their ionospheric career in the UK. For instance, Radio Research Board scientist A.L. Green had worked with Appleton and Ratcliffe on the polarisation of downcoming radio waves and found them to be elliptically polarised in the left-handed sense. Since the magneto-ionic theory predicted that similar measurements made in the southern hemisphere would show a right-handed polarisation, Green set up in Jervis Bay shortly after his arrival in Australia to verify this prediction experimentally. In 1930 he was able to announce by telegram that (in his own words): "sky waves received from 2BL are approximately circularly polarised, as was the case in England, but that the sense of rotation is on the contrary right-handed ... and it forms the final link in the chain of proof of the Eccles-Larmor-Appleton magneto-ionic theory" (Evans, 1973, p. 170).

We have already mentioned, in Chap. 4, how greatly David Martyn's research during the 1930s was shaped by, and contributed to, the techniques and discoveries of this early period of ionospheric research, including identifying a layer between the E and F layers of the ionosphere (with Cherry), perfecting an adapted "pulse-phase" technique (see below) with Munro and Piddington, and providing the theory of the Luxembourg effect (with Bailey). In the period 1925–1960 Australia was ranked fourth in the world for research output in ionospheric physics, an achievement due significantly to Martyn.

---

[16]From 1947 to 1969 it was still growing but more slowly (5.4 year doubling time).

An American Contribution: The "Pulse-Echo" Method for Ionospheric "Sounding", 1925

In 1925, not long after Appleton and Barnett's frequency-change method was devised,[17] Americans Gregory Breit (1899–1981) and Merle Tuve (1901–1982) generated a mathematically identical "pulse echo" technique. In this method, the researcher sent a train of pulsed waves skyward and used the time difference between the transmitted pulse and its reflected "echo" to both demonstrate the existence of the ionosphere, and to measure its height.[18] The technique turned out to be superior in some respects because all the frequencies in the short duration pulse are present at the same time. No errors can be introduced by changes in the ionosphere during the frequency sweep, and multiple layers are simply seen as multiple pulse echoes. Whereas the frequency-change method required a day for a trained expert to measure the height from an oscillogram, the pulse-echo device provided a compelling, immediate, visual reading for the height.[19] When Breit and Tuve found that the observed ionospheric "height" varied, and that what they measured was a "virtual" rather than real height, new questions about what dynamics and structure existed in the ionosphere came into view.

We note here that despite the enormous success and wide uptake of the pulse-echo method for ionospheric sounding, including a new round of improvements to instrumentation to develop automation (Yeang, 2012, calls this "mechanised objectivity"), and the simple, real-time visual display of signals, no significant ionospheric research program developed in the USA after Breit and Tuve turned their attention to nuclear physics in the late 1920s. But that was not to be the end of the story. Merle Tuve would later take up research in radio astronomy at the Carnegie Institution of Washington—playing a role in the USA that was very similar to Pawsey's in Australia. Indeed, Tuve became a very important colleague, friend and supporter of Pawsey's, as we will see.

---

[17] The frequency scanning method may have fitted more easily in the experience of physicists of the time because the plasma theory, then of interest in physics, was based on propagation as a function of frequency. But as researchers would quickly discovery, it is the spatial structure in the ionosphere that is most important for understanding radio wave propagation, and this structure is more naturally related to the pulse echo delays discussed in this paragraph.

[18] Tuve's Minnesota Professors, John Frayne and William Swann, had tried to develop an echo-pulse device in 1921 using a single antenna that switched between transmitting and receiving functions, which failed because of antenna multiplexing issues (Pawsey would later invent an effective switch for this problem). It is also interesting to note that Breit had originally proposed building a large parabolic reflector that would produce narrow beams at wavelengths of several meters, from which he could obtain directly a sky wave's elevation angle and the layer's height. But Tuve predicted that a parabolic reflector for waves of 50+ m would be required, and not only was that too large to construct, but additionally, Breit had his funding cut. Tuve then proposed using "interrupted continuous waves" instead. (Yeang, 2012, p. 220).

[19] See Yeang (2012). We note that the pulse method would eventually provide the basis for aircraft warning radar which Pawsey developed in WWII.

The story[20] of early radio and ionospheric research lets us understand more about how scientists come to be able to think about or approach research questions—influenced by their workplaces, the norms of how research questions might be designed, their experience with different kinds of instruments, their exposure to other researchers in different areas. This offers us more insight into the intellectual context of ionospheric and radio research that Pawsey was about to enter, and in which he would form his own views on the nature of science.

---

[20] For readers who enjoy connecting the past to our contemporary world, we note that between 1960 and 1970 the development of long distance communications moved to satellites, and ionospheric studies became irrelevant for communications. Ionospheric/magnetospheric research has now moved into the realm of space physics and interplanetary plasma at the interface between the earth's ionosphere and space. The ionosphere poses challenges for modern radio observatories working at low radio frequencies, since it corrupts the radio images. Conversely these corrupted images provide new information on the properties of the ionosphere. A major study by Honours student Cleo Loi at the University of Sydney, made using the MWA and processed at the Pawsey Centre in Perth, has recently revealed new and previously unknown ionospheric structure. (https://theconversation.com/how-an-undergraduate-discovered-tubes-of-plasma-in-the-sky-42810).

Chapter 6
To the Cavendish Laboratory of the University of Cambridge, 1931

From the 1851 Exhibition Scholarship Committee to Pawsey on 1 October 1931:

> [We] approve of your proposal to spend the period of your Scholarship at the University of Cambridge instead of at the University of London as formerly arranged, and after taking all the circumstances into consideration, [we believe] that you have made a wise choice. You, therefore, have the permission of the Commissioners to make whatever arrangements you consider necessary to proceed with your Radio Research at the Cavendish Laboratory in conjunction with Mr. Ratcliffe. (Sir Evelyn Shaw, 1882–1974, Secretary of the 1851 Scholarship Committee).

Pawsey started out in research in the midst of excitement over the possibilities of radio communications and the iteratively developing physical understanding of the ionosphere and of the equipment that might be used to investigate it. During 1926–28 he completed his BSc at the University of Melbourne, Victoria. In 1929 he began a Master's Degree, which was at that time a research-only degree, under the direction of Professor T.H. Laby. He was supported by receiving the M.J. Bartlett Research Scholarship. Presumably this, along with his work as a tutor in Physics at Queens College, provided him with a small, but independent, income. He embarked on a study of "atmospherics"—electrical disturbances in the atmosphere that Appleton, at King's College, London, and others had linked in part with thunderstorm activity—and their impact on radio broadcasting. From January 1930 to August 1931, he carried out observations using a cathode ray direction finder, working with George H. Munro and Lenard Huxley as part of the Australian Radio Research Board (RRB). Pawsey wrote in 1933: "We were able to give strong evidence that all atmospherics originate in lightning flashes, and made measurements of intensity enabling the distance of the thunderstorms to be roughly determined." (Ratcliffe & Pawsey, 1933)

Supplementary Information The online version contains supplementary material available at [https://doi.org/10.1007/978-3-031-07916-0_6].

W. M. Goss et al., *Joe Pawsey and the Founding of Australian Radio Astronomy*, Historical & Cultural Astronomy, https://doi.org/10.1007/978-3-031-07916-0_6

Pawsey was involved in one publication from this period: "Accurate Measurement of the Frequency of the Carrier Waves of Victorian Broadcast Stations", appearing in the 27 March 1930 *Australasian Electrical Times* by J.L Pawsey, W.J. Wark and R. Fallon, all from the Natural Philosophy Laboratory of the University of Melbourne. The purpose of this project was to check the frequency stability of the Victorian AM stations. The method was to take a standard frequency source (from an elinvar tuning fork) and multiply it by a factor of 40 or 41. This tone was then compared with the received carrier of the radio station. Similar to modern AM stations, the transmission band was from roughly 0.5 to 1.6 MHz, with the stations spaced by some tens of KHz. Thus the carrier frequencies of each station were required to have a stability of better than a few KHz. Pawsey and colleagues measured the stability of three Victorian stations. But as Pawsey et al. pointed out: "It should be emphasised, however, that in Australia, with its comparatively few stations, rigid frequency control is not so absolutely necessary as in Europe, where [the density of stations is quite high]." Evans comments that in no small measure as a result of this work, by 1933 understanding of atmospherics was adequate for planning a national broadcasting network (Evans, 1973, p. 88).

As a side note, it is amusing that the radio researchers encountered the same initial suspicion and disengagement from meteorological researchers in the 1930s as radio astronomers would later encounter from optical astronomers in the late 1940s and early 1950s. Investigation of atmospherics had shown that radio instruments could detect the source of an atmospheric from upwards of 1000 miles away. This made the value of radio instruments for weather prediction obvious to radio researchers, but not to meteorologists. The situation was made more difficult by the radio scientists making elementary errors in meteorological analysis (Evans, 1973, p. 102).

The MSc thesis was submitted in March 1931, apparently finished at the end of 1930.[1] In the 116-page thesis "Atmospherics", Pawsey began by defining the problem:

> An "atmospheric" may be defined as a *naturally occurring* variation of the electric and magnetic field of such a nature as to be capable of actuating a radio receiver; and so in the presence of signals, interfering with the reception of such signals. Such variations are propagated in the usual manner, so atmospherics are merely a class of [low frequency] electro-magnetic waves.

He provided a fascinating history of the discovery of atmospherics, starting with the famous Russian physicist Alexander S. Popov (1859–1905) (Smith-Rose, 2021), whose pioneering work occurred in May 1895 when he detected lightning strikes at a distance of 50 km. Pawsey described how early work on the cause of atmospherics (frequency from about 10 kHz to 30 MHz with typical frequencies near 300 kHz) was carried out by Appleton, Watson-Watt and Herd in the years 1922–1926 in the UK. Pawsey summarised the directional evidence of atmospherics which occurred

---

[1] Hastings Pawsey discovered in 2016 that no copy of the MSc thesis could be located at the University of Melbourne; fortunately the Royal Commission for the Exhibition of 1851 in London had a copy in their Pawsey file which was provided to the Pawsey family.

all over the world: "The major sources of atmospherics lie in the region of great thunderstorm activity for the time in question [often tropical afternoons)."

A noteworthy section of the thesis deals with the "treatment of interference [due to atmospherics] by means of Fourier analysis". Pawsey showed that the output of the radio receiver could be represented by a Fourier series or Fourier integral. Given the fundamental early contribution on the use of Fourier analysis for radio astronomy that McCready, Pawsey and Payne-Scott would make in 1946–7 (Chap. 13 and especially Chap. 36), we are intrigued at Pawsey's clear understanding of this property in about 1930, prior to his experiences at Cambridge. Evidently Laby was familiar with the use of Fourier analysis, which appears in a book on geoprospecting that he published with Broughton Edge in 1931 (Laby & Edge, 1931). Perhaps Pawsey had come across this application of Fourier analysis during his short period of research assistant work geoprospecting in Tasmania in 1929.[2] In any case, based on his understanding of Fourier's integral theorem, he gave a number of examples of Fourier pairs: an isolated sine wave, an exponentially damped sine wave and an "infinitely short pulse", later called a delta function.

The thesis was reviewed by David Martyn, presumably in early 1931 (undated):

Since 1920 a rapidly increasing number of original contributions have been made, as the result of the intensive study of the subject carried out largely in Britain, France, America, and Germany. These investigations have been made with the object of discovering the precise electrical nature of atmospheric impulses, their origin, their connection with meteorological influences, and their effects on apparatus commonly used for the reception of wireless signals. The number of original contributions to this branch of science is now very large, but no connected account of the whole subject has hitherto been published, nor have the results of any attempts to analyse critically the data and methods of different observers been made available. Mr. Pawsey has attempted both of these tasks. Commencing with a brief historical survey which serves to present the subject as a whole in perspective, the author proceeds to describe exhaustively work on the intensity and waveform of atmospherics, and on their places of origin. There follows a section on the mode of origin of atmospherics in which all the known possible sources of electrical disturbance are considered, with special emphasis on the lightning flash. A strong case is made out, on several grounds, for the explanation of all individual atmospheric impulses as being due to separate lightning flashes occurring in some part of the terrestrial globe. Finally, there is a comprehensive account, chiefly mathematical in character, of the effect of atmospheric impulses on wireless receivers. The problem naturally possesses great practical importance, but unfortunately the investigations so far carried out have been brought to little practical issue. The author presents the lines of attack which have been employed in attempting a solution. There is a most extensive bibliography in which the author wisely gives weight to the papers which he considers to be of chief importance. This thesis could only be the result of a great amount of diligent and discriminating labour. The author has made a most painstaking survey of his subject in a competently critical manner. The data which he has brought together, and perhaps some of his conclusions, should prove of great use to all types of workers in that field. Mr. Pawsey's practical work is competent, and again he shows much diligence in the reduction and analysis of his data.

---

[2]H. Wendt commented to the authors that it seems that this technique dates back to the 1870s and was used on ship compasses, so it would not be surprising if surveyors were also familiar with it.

Fig. 6.1 J.L. Pawsey, April 1931, awarded MSc with First Class Honours. Credit: Joe and Lenore Pawsey Family Collection

We can perhaps see in this the emergence of a Pawsey "research style": focused, painstaking and meticulous, and attempting always to integrate existing knowledge, and pull all the elements together. There is no doubt that the contribution he had made to the nascent research program at the Radio Research Board was substantial and there was considerable discussion about whether he ought to be recognised as a lead author of the R.R.B. Report No. 5(1) "Atmospherics in Australia" (Evans, 1973, p. 77). The MSc was awarded in April of 1931 with First Class Honours (Fig. 6.1).

We should note that in the same year that Pawsey was awarded his thesis, 1931, Karl Jansky (1905–1950) (Sullivan, 2009), just three years older than Pawsey and an engineer with Bell Telephone, was building a turntable-mounted antenna in order to investigate atmospherics that might affect Bell's plans to wirelessly transmit trans-atlantic telephone calls by reflecting them from the ionosphere. It is a shame that Jansky could not know of or access Joe's useful compendium of information about atmospherics! While Pawsey contemplated his next career steps, Jansky was categorising a year's worth of static into three types: close thunderstorms, distant thunderstorms, and a mysterious additional hiss with a maximum intensity that rose and fell once a day, which turned out to come from the Milky Way. As is now well known, the first observations in "radio astronomy" had been made, only to go largely unrecognised until after WWII (Sullivan, 2009, p. 34). Pawsey could not have seen the Milky Way as Jansky did, because the frequencies he used (around 100 kHz) were below the ionospheric cut off, and hence nothing from outside the ionosphere could be seen.

During his period at the University of Melbourne, Joe Pawsey was also involved in a number of sporting activities. At Queen's College in 1928 he qualified in Jiu-Jitsu and in rifle shooting. "He was awarded a Victoria Rifle Association medal in 1930 and was Vice-Captain of the Melbourne team which visited Adelaide in 1930 and was subsequently Captain." (Lovell, B., 1964) Also, he was the Vice-Captain of the Queen's College second football team.

1931: Award of an 1851 Exhibition Scholarship, Choice Between London and Cambridge[3]

In April 1931, Pawsey began to plan for his future. While we do not know what inner hopes or visions he held, clearly he could not see that future for himself in Australia. Opportunities for scientific work of any kind were very limited; the landscape for work as a physicist—let alone in research as such—was bleak indeed.

1931 was the height of the Great Depression, and Australia was a country with a small population strung out across vast geographical distances. It was a nation of agriculturalists and shopkeepers; it was an economy built on the export of primary materials, not on manufacturing; it was composed of immigrants who had left the centres of the Industrial Revolution. The development of organisational infrastructure for sectors intended in the end to be profitable often still rested with the government, which had to effectively get whole new industries started. State and Federal governments supported limited research in agriculture, geologists could hire their services to mining companies, and scientific medicine was drawing clinician researchers into the laboratories slowly growing in hospitals. And in one sense this activity added up to a quietly maturing Antipodean scientific milieu, with achievements all the more remarkable for being borne from a scarcity of resources.

But physicists? The numbers of positions around the country could be counted on fingers. With the Radio Research Board just beginning, there were no industries to demand them; the sole occasional employer was Amalgamated Wireless Australasia (AWA). The immensely talented Ruby Payne-Scott, who was to become such a key member of Pawsey's research group during and after WWII, found employment for a couple of years as a medical physicist as part of a cancer research project. When

---

[3] See ESM 6.1, Pawsey's letters, for a description of the collection of 189 letters from Joe to his parents from 1931 to the end of 1939 when the Pawsey family departed for Australia at the beginning of WWII.

that concluded, she had no choice but to look (rather anxiously) for work as a school teacher (Goss & McGee, 2009).

It is interesting that Pawsey, then still engaged in research on atmospherics in Laby's group at the University of Melbourne, appears to have explored several possible employment avenues. On 2 April 1931, he wrote a letter to a prominent US electronics company, General Radio Company (founded in 1915 in Cambridge, Massachusetts), hoping to find a position with this firm. J.W. Norton, Chief Engineer, replied on 14 May:

> ... Apparently, you are interested in some temporary arrangement to permit you to become acquainted with American technique. The organisation of our Engineering Department is such that it would be difficult to find an opening of the nature you have in mind. Our engineers are required to acquaint themselves with all phases of our work and, consequently, commercial contacts and an intimate knowledge of the routines and practices of our customers are quite as much a part of their work as is the more fundamental scientific research. Because of this, it is not feasible for us to employ physicists or engineers for technical work alone ...

This remains an issue at the contemporary science/industry interface.

He then sent a series of letters to Cambridge University, partly on the advice of Prof A.M. Wadham, professor of agriculture at the University of Melbourne and a former faculty member of the botany department at Cambridge. (This is a good example of the value of international links—Wadham did not even have to be in the same field to support Pawsey). As we have noted, going to study in the UK was in fact the most predictable path for the most promising young men of science in the interwar period. Most of the Australian professoriate—like Laby and Wadham—were British. They recommended their very best students to return "Home" for the exposure to new ideas and the collegiate company of fellow scientists that they remembered, and they used their carefully sustained networks to nurture these careers. Future Nobel prize winner Frank Macfarlane Burnet (1899–1985),[4] a fellow Melbournian, left for Cambridge just a year before Joe Pawsey.

The biggest barrier, of course, was money, but like many (perhaps the majority) Australian scientists, the crucial vehicle was a substantial scholarship program. By the end of May, Pawsey had submitted his application for an 1851 Exhibition Scholarship; he had been recommended by the University of Melbourne. He wrote, "I hope to continue with radio research, and would presumably be working under the direction of Mr J.A. Ratcliffe ... I hope to arrive in England about the end of September." Ratcliffe had been a student of Sir Edward Appleton's, and remained at the Cavendish when Appleton moved to London.

In an accompanying letter, Pawsey applied to the Cavendish Laboratory with more details of his proposed study programme:

> I wish to carry out radio research at the Cavendish Lab. I understand that facilities for this work are available and that some students are carrying out such work under the direction of Mr. J.A. Ratcliffe, but that apparatus may not be readily available. For this reason Prof Laby

---

[4] See https://www.nobelprize.org/prizes/medicine/1960/summary/

advises that it would be better not to specify a particular problem but to decide this question after consultation with Cambridge authorities . . . I am posting by this mail a letter from Prof Laby (FRS) to Lord Rutherford which deals with this matter.

Pawsey gave a short description of his MSc thesis work on atmospherics: "This work is that on which the Final Honour Exam results and the Dixson Scholarship . . . were awarded and was submitted for the 1851 Exhibition Scholarship." Within a day or so, the news arrived from London that Pawsey had been awarded this scholarship (£280 per annum).

In Cambridge, Priestley sent the papers along to Ratcliffe, asking if Rutherford "can accommodate him in the Cavendish Laboratory, and also if the authorities at Sidney Sussex will take him". At this point, the decisions regarding Pawsey became quite confused due to a misunderstanding of the 1851 Exhibition authorities.

The Royal Commission for the Exhibition of 1851 sent the application and reference letters to Prof Richard Tetley Glazebrook (1854–1935), a prominent physicist who had been the first Director of the National Physical Laboratory in Teddington, London. Glazebrook was the assessor for the 1851 research fellowship, writing a report on Pawsey's application, "on the work of J.L. Pawsey, age 22, [sic, actually 23]".[5]

> The thesis is an admirable piece of work. I endorse the high opinion of it expressed by Mr. Martyn of the Australian Research Board [sic, Radio Research Board]. Personally I know of no such account of Atmospheric Research . . . He clearly has a wide knowledge of what has been done to elucidate the cause of the phenomena and the laws to which they are subject. It is highly desirable that he should be brought into closer contact with the men who are working at the subject here and I recommend him strongly. **I think he would be better advised to work with Professor Appleton rather than at Cambridge** [our emphasis].

Shaw, secretary of the 1851 Scholarship Committee, informed Pawsey in early July 1931[6] that he had been awarded an 1851 Scholarship[7] to go to *London, King's College to work with Edward Appleton*. Lord Rutherford was not pleased to hear that the Cavendish had lost a promising PhD candidate. Pawsey, assuming that the decision was final, wrote Cambridge, withdrawing his application on 9 July 1931. On the same date, Rutherford had a telephone and written conversation with the 1851 Exhibition staff, and Ratcliffe wrote to Raymond Priestley (1886–1974), Secretary to the board of Research Studies, on 10 July 1931:

> I note from [Pawsey's] letter that he wishes to "carry out radio research at the Cavendish Lab", and he mentions my name and says he would like to work with our little band of radio research workers . . . I [just met] . . . Prof Appleton,[8] of King's College, London, and he then told me that Pawsey had been awarded an 1851 Exhibition, but that the authorities had

---

[5]Royal Commission for the Exhibition of 1851 archive, communicated to Hastings Pawsey May 2016.

[6]Cambridge University files for J.L. Pawsey, letter to R.E. Priestley, secretary of Board of Research Studies, 9 July 1931.

[7]Pawsey was in an illustrious group of Dominion (Australian, New Zealand and Indian) recipients: Rutherford, Oliphant, Massey, Laby, Leslie Martin and Bhaha, among others.

[8]Ironically, Appleton had been Ratcliffe's advisor at Cambridge in 1924–1927.

decided that he should go to London and not Cambridge, because there had been so many 1851 Exhibitioners at Cambridge of recent years ... In view of the fact that Pawsey very strongly wishes to come to Cambridge and to work with us, it seems very strange if he has been allocated to London ... I have also seen Lord Rutherford and have ascertained that he would be willing to take Pawsey into the Lab as a research student.

After Priestley wrote Shaw on 14 July 1931, Shaw responded in a defensive manner two days later[9]:

It is quite true that Mr. Pawsey ... applied for permission to conduct his research work at Cambridge, but it is not true that the Commissioners decided that he should go to London "because there had been so many of their Scholars at Cambridge in recent years". The facts are these: Pawsey was awarded a Scholarship, but both the examiners of his papers felt that he would have better facilities for the particular research he proposed at King's College, London, than at Cambridge, and so after consulting with Professor Appleton, I was instructed to make the award to Pawsey conditional upon his going to King's College. I explained this in my telegram to the University. I hope that you will make it clear to everyone that there is not the slightest wish on our part to divert anyone from Cambridge. As you will readily understand, our only wish is to secure the best conditions for our Scholar's studies.

Ratcliffe reported that the committee had assumed that the "obvious place for [Pawsey] to work was with Appleton. They did not know any wireless work was done at Cambridge. They therefore asked Appleton if he would accept Pawsey and he accepted." Rutherford proposed a simple solution: after Pawsey reached England, he could visit both places and make his own decision. On 16 August 1931, two days before the ship (*Oronsay*) departed from Melbourne, Ratcliffe cabled Pawsey, explaining his options. Pawsey must have been relieved, as he told Ratcliffe that he would indeed visit Cambridge.[10] He sought advice from Laby, but Laby simply advised him to ask for more advice on arrival.

Pawsey arrived in the UK during the third week of September 1931 to face the decision of London or Cambridge. Pawsey met Shaw as Laby advised on Friday 25 September to discuss "the merits of my doing a course of research at Cambridge or London."

Pawsey visited Cambridge, being shown around by C.B.O. Mohr, a research student at the Cavendish and fellow graduate of the University of Melbourne physics department.[11]

---

[9] This complex story has been pieced together using the Cambridge academic records of Pawsey and the 1851 Exhibition archive.

[10] A few days after leaving Melbourne, Pawsey received a telegram from Laby in Fremantle that did not provide decisive advice: "See Shaw, 1851 Commissioners, obtain advice English specialists, difficult me decide between London and Cambridge." [sic]

[11] Mohr (1906–1986) had graduated from Melbourne two years earlier than Pawsey. Mohr followed Massey to the Cavendish where they collaborated on nuclear physics problems (e.g. "Anomalous Scattering of Alpha Particles and Long Range Nuclear Forces", Massey and Mohr, 1938). Later in his life he was at the University of Cape Town and the University of Melbourne Physics Department. Massey had been a classmate of Pawsey's at the University of Melbourne.

[Mohr] was able to give me an idea of Cambridge from the students' point of view. I saw Ratcliffe ... when he showed me the work they are doing there and on my return to London I visited Appleton at King's College and went with him to visit **[F.W.G. "Fred"] White**[12] (formerly with Ratcliffe [1930–31]) now demonstrating at King's. We also went to Slough to the Radio Research Station. [our emphasis].

On 30 September 1931, Pawsey outlined the pros and cons of both places to Shaw.[13] He was systematic. His letter laid out categories of consideration: "radio equipment, radio research workers, general research in physics, general conditions of life, duration of the degree and financial considerations". Pawsey liked the fact that the group in Cambridge was much smaller (at present one compared to four at London, leading to more attention from the supervisor). The duration at King's College was usually two years for a PhD compared to three at Cambridge. London would be considerably cheaper.

In the end though, the decision was not so much pragmatic but romantic. The overriding factors were:

(1) Cambridge stands alone in respect to the famous physicists whom one meets (also mathematicians) and, (2) Cambridge alone has the "atmosphere" of university life ... there is not a great deal to choose between the two places, each having special advantages ...Taking these things into consideration I feel that it would be preferable for me to go to Cambridge to do radio research under Ratcliffe.

On 1 October 1931, Shaw wrote Pawsey with the message that the Chair of the Scholarships Committee:

approves of your proposal to spend the period of your Scholarship at the University of Cambridge instead of at the University of London as formerly arranged, and after taking all the circumstances into consideration, he believes that you have made a wise choice. You, therefore, have the permission of the Commissioners to make whatever arrangements you consider necessary to proceed with your Radio Research at the Cavendish Laboratory in conjunction with Mr. Ratcliffe.

On 3 October, Pawsey wrote Shaw pointing out:

I have been up to Cambridge [a second visit on 1 October] and the arrangements for my doing a research for a PhD are all in order, both as regards University and College (Sidney Sussex) authorities. The former is not official as the Board of Studies has not considered it, but I am assured by Priestley that this approval is practically certain. I shall therefore "go up" on [5 October] ... I am taking at Cambridge the proscribed courses leading to the PhD.

On 7 October 1931, Shaw wrote Appleton with an apology:

We did know, when we made his Scholarship conditional upon working with you, that he had already made contact with Cambridge and that Lord Rutherford had agreed to his working with Mr. Ratcliffe. When the Commissioners heard this, they decided that the best thing to do was to let Mr. Pawsey acquaint himself with the conditions of each centre and then be guided by his own inclinations if he should arrive at a definite conclusion one way or the other. He told me that it was extremely difficult for him to choose between the two

---

[12] F.W.G. White (1905–1994) would later join Pawsey at Radiophysics during WWII.

[13] 1851 Exhibition archive, J.L. Pawsey. Pawsey to Shaw 30 September 1931. They had had a personal meeting earlier in the day.

centres available for his research, but that the attractions of university life at Cambridge drew him strongly to Cambridge. On that, I could nothing but advise him to join Mr. Ratcliffe.[14]

Pawsey finally let his family know his status on 7 October 1931[15]:

> I am now more or less settled down. I "came up" on Monday [5 October]. Today I had a look round and Ratcliffe showed me the works in the problem I am starting on. I dined in Hall tonight for the first time. It was not a very unusual business. The only thing of note is the old custom of dining off the bare boards. After, Ratcliffe was to take me to see some of "the works" and asked me up to coffee in his rooms. I met Mrs. Ratcliffe, a charming woman I think … They make a nice pair.

On 16 October 1931, the Degree Committee informed the registrar of the University of Cambridge that J.L. Pawsey had been admitted to a course of research and that Professor Lord Rutherford was the supervisor. Pawsey was informed of this decision on 4 November 1931.

Friends and Student Life: J. L. Pawsey and Frederick H. "Ted" Nicoll from Canada, 1931–1933

In late 1931, Joe Pawsey met Ted Nicoll (1908–2000), a fellow 1851 Exhibition research student at the Cavendish. Nicoll had graduated from the University of Saskatchewan, Canada, and arrived about the same time at Cambridge.

A major source for the interaction of these two colleagues is the 400-page collection of letters from Nicoll to his family in Battleford, Saskatchewan.[16] On page 62 of the collection, Nicoll wrote his family on 15 November 1931 with the first reference to Pawsey:

> Yesterday, Pawsey—a friend of mine from Australia and also in Physics—arranged to get a bike for me from one of his friends. Pawsey had me over for lunch at his place earlier in the week when we decided we would go for the bike ride on Sunday. He got the bike from an Australian friend of his and Pawsey and I had dinner at my place and then left for Ely 16 miles at 1:30. We got there at 3:30 and spent a good deal of time looking over the famous [Norman cathedral from AD 672] there. It is very wonderful indeed … We left Ely at about 04:30 and arrived home at 06:30—well after dark but we had cycle lights with us … The day was misty all day and so damp the moisture was actually condensing but it wasn't at all disagreeable cycling and I thoroughly enjoyed it.

In these first few months that both Pawsey and Nicoll were in Cambridge, they both experienced the remarkable hospitality provided by Lady Frances Ryder (1888–1965, daughter of the fifth Earl of Harrowby), Organiser of the Dominion

---

[14] Appleton was so impressed with the [only] copy of Pawsey's MSc thesis that he borrowed it in July 1931 for "a few weeks". By October 1932, he had still not returned the thesis; after being prodded by the 1851 Exhibition Scholarship staff he returned the copy on 13 October.

[15] Pawsey Family Archive.

[16] Provided by Ted Nicoll's daughter, Patricia Agnew.

Services and Students Hospitality Scheme, which provided assistance to Dominion servicemen in WWI and again in WWII as well as to students in the interwar years. Lady Frances Ryder (CBE) and her friend Miss Celia Macdonald of the Isles provided these services in the 1930s, organising home visits in the UK for Canadian, Australian, New Zealand, South African and American students. At some point, Lady Frances reported that they had card indices of 1600 potentially lonely visitors. By the time of WWII, *The Australian Women's Weekly* of 13 April 1940 reported:

> Go along any afternoon you are in London to 21b Cadogan Gardens and ask for Lady Frances Ryder's rooms, and if you hail from any part of the Empire you are bound to meet someone you know there ... Lady Frances [reported] that her mail amounted to something like 32,000 letters received and answered during the year. If you have a son or daughter who has left a hometown to seek more knowledge at [UK universities] a letter of introduction from the college will assure a welcome always at Lady Frances Ryder's ... As time went on Lady Frances heard of young students who were coming to London, living in boarding houses, and in many cases utterly miserable with loneliness. The tea-parties became more frequent, one friend introduced another, and in the end Lady Frances decided to take the huge suite of rooms which by now are known to so many—including innumerable Australians ...

A Canadian newspaper also had an article about Lady Frances[17]:

> Lady Frances Ryder is so well known among students, especially at Oxford and Cambridge, that the impudents [very rude], grateful though they certainly are, have taken to calling the system "Lady Rydering". It is really a good phrase, however, for pun that it is, it represents innumerable things easily expressed. And often while one is Lady-Rydering it is not unusual to find some other fellow student, who is an acquaintance, Lady-Rydering in another house in the same district. And you have picnics together, and there is great glee.

By early December 1931, Nicoll and Pawsey had numerous invitations for the December Christmas period organised by Lady Frances. For example, on Saturday 5 December, there was "Lady Frances at home" with games and dancing from 08:30 to 10:30 pm. Ted was enthusiastic about this party: "... [N]eedless to say I danced and had an excellent time and met two nice Australian girls (more of them later[18]). I hadn't accepted Lady Frances's invitation for the Sunday night but as I had nothing to do she very nicely asked me to go to it too." The two colleagues also went sightseeing in London during the weekend. The breakneck social schedule continued through Christmas and New Year's, with home visits organised by Lady Frances.

On 28 December, Pawsey moved to the home of two friends of Lady Frances: "My hostesses—the two Misses Bradshaw—are elderly, live in a comfortable house on the outskirts of the town of Retford and are rather nice."

The two friends returned to Cambridge on 8 January 1932 to continue their research studies. In early February, Pawsey and Nicoll decided to move farther outside Cambridge to "get cheaper rooms". (Pawsey was associated with Sidney

---

[17] A quote from an unnamed Canadian newspaper provided by Ted's mother, Mabel Nicoll.

[18] Pawsey and Nicoll took the two Australian girls (the Ainsworth sisters) to a ball organised by Lady Frances the following week. "We danced from 09:30 to 02:30 [am!]." Joe and Ted were to meet them again on 21 December 1931.

Sussex College and Nicoll with Trinity College.) They also would share a sitting room to save expenses. Nicoll wrote to his family on 7 February 1932:

> Pawsey is doing wireless research . . . and he and I are thinking of making some 5 metre sets [60 MHz] for amusement . . . I agree with you, Lenore [Ted Nicoll's sister] that I have probably seen more of some English life than many who live here. I can tell you we are in a marvellous position; being on a scholarship gives you the key to so many things like these and attending Cambridge puts you in such a position that you can meet anyone and consider yourself on their level if not above it. Altogether it gives you a great deal of self-confidence.

During the summer of 1932, William Beare (a fellow Canadian from Toronto, also at Cambridge), Pawsey and Nicoll went on a mountain climbing trip to North Wales, traveling the 700 miles on motorbike, and climbing Mount Snowden (elevation 3560 feet or 1085 metres). Soon afterwards, Nicoll's mother, Mabel, arrived in London (2 July 1932) and Nicoll took her to Cambridge, where she met Pawsey. Nicoll reported to his family: "Pawsey took us out to tea and she [Mabel] also liked him." Mabel returned to London that evening to continue her European tour on the continent.

Later that summer, Pawsey and Nicoll took a 3000-mile tour of Europe, travelling by motorcycles and camping along the way. For Pawsey, some of the route through France, Belgium, Germany, Austria, Switzerland and again France was a repeat of the YAL (Young Australia League) trip of 1925. They were gone for about five weeks. Ted summarised the tour in great detail in a letter from 1 October 1932.

The Flanders battlefields near Ypres were the first destination, a repeat visit for Joe. They visited Sanctuary Wood, a site where many Canadians were killed in April 1916. Ted wrote: "The weapons and instruments of destruction certainly set one thinking about war, why it happens and whether it will happen again and as a result Joe and I have had some excellent [discussions] on the subject though we have not, I fear, solved the world peace problem."

As they visited Germany they were initially impressed: "The Germans are very kind, good-natured and happy-go-lucky, with none of the German reserve . . . We found them easy to talk with and met some amusing people." Then they.

> met a real Hitlerite belonging to the Hitler [storm troops] and traveling also by motorbike. He could speak a bit of English and accompanied us for 50 miles till lunch and we ate together after which he left us. The chief information we got from him was the Nazi motto: "Bread and work for everyman" and their hearty dislike for [World War I] reparations—thinking that they are the source of all their troubles.

A high point of their visit was the Black Forest, where they did some long hikes, including one of 30 miles round trip (Figs. 6.2, 6.3 and 6.4). They visited Munich and the Deutsches Museum, a beer house near the headquarters of the Storm Troopers, and then they went to Innsbruck in Austria. From there, the pair travelled south through the Brenner Pass to Italy where they visited Cortina, Venice and Milan. Heading back north, they saw Geneva and Paris before boarding a cargo ship in Boulogne to sail to Folkstone (UK) and home.

Travelling through London in late September 1932, Pawsey and Nicoll visited the 1851 Exhibition office, meeting Canadian W.J. Henderson (previously from Queen's

Fig. 6.2 Pawsey on a climb in the Black Forest, summer 1932. Credit: Joe and Lenore Pawsey Family Collection

University in Canada), who was to later play a prominent role in the National Research Council of Canada. Henderson was on his way to the Cavendish Laboratory. He would later join Pawsey and Nicoll on numerous motorcycle trips.

For Pawsey, as for so many research students before and after him, this was a time of thoughtful exploration of his appraisal of the world in which he found himself, and of the roles he wanted to play within it. In Chap. 8, we briefly discuss some of his political and social reflections, formed partly from his critical observations of National Socialists in Germany. In a letter to his parents on 14 March 1932, he mused in particular about science. We quote these musings here, both because they capture something of how he approached *doing* science, but more for the values that were coming to define him:

> ... In my last letter I touched on the spirit of science and the effect on the world. I wish to emphasise one point. I defined the scientific method of today as a willingness to question (& investigate systematically) any question which comes up. It may appear as an obvious thing stated thus. I should like to point out that the practical application is the reverse to many or I think most people. One of the most blatant applications is the questioning and re-examination of the old established ideas. These ideas have been upheld by men we look up to, men to whom we do not claim superior intellect. May we then question their

Fig. 6.3 Pawsey on his
motorbike. Credit: Joe and
Lenore Pawsey Family
Collection

Joe Pawsey on his 500 c.c. motorbike. He and Ted camped all over Europe in 1933

conclusions in any given line? The old school of thought held—I feel—that to question one tenent (sic) of the teaching of a great man was to despise him and to reject one tenent (sic) was to reject his whole philosophy. My idea is to accept no man as infallible & to base my conclusions, not in those of greater men than myself but in those cases where it is feasible on my own reasoning from the facts at my disposal.

These points of view are exemplified (a) in the doctrine of the verbal inspiration of the Bible & (b) in the attitude which gives Newton a place as probably the greatest scientist in an age which practically the whole of his conclusions are believed to be accurate. Most people hold views somewhere between the two.

As you will point out it is impossible to personally investigate every question & actually most things must still be accepted "from the accumulated wisdom of the ages" (or ignorance). The point of view however should have, & has, an important secondary

Fig. 6.4 Nazi Storm Troopers. Credit: Joe and Lenore Pawsey Family Collection

effect. It teaches tolerance. Most questions you conscientiously investigate have two sides to them which it is not easy to decide between. Those conclusions which you obtain second hand are almost always one sided. We are told unequivocally that such a thing is right, another wrong, not that a certain person or persons considered the first preferable to the second. This intellectually tolerant attitude possesses the characteristic that there is no great driving force. It does not play on the emotions without which life is flat. Thus one gets neither the saint nor the Inquisition & the "Thirty Years War" (of scenes from European History).

Religions seem to me to take their inspiration from a play on the emotions. Emotion and the satisfaction of instincts are the driving forces of religion—love— the fear of death—etc.

Science has had amazing successes in the conquest of the physical world through a reasoning questioning process. It fundamentally mistrusts emotional conclusions—the only argument for this is that different people agree on the former & usually disagree with the latter. The driving forces are comparatively weak—the curiosity—analogous to my mind to the artistic spirit, & a thirst for power. Curiosity the former, though less potent one would imagine, is more fruitful.

There is thus a clash between religion & science. Religion says unequivocally "Do this!" & gives no reasons. Science says, here are the reasons but does not say "Do this!" with any great power. They should to my mind urge the doing of the same thing.[19]

There are two possibilities. One is that you are bored stiff—the other is that you are very interested. If so state which. Of course this is just a collection of platitudes which have been said thousands of times before. But so have thousands of these things.

---

[19] Apparently, Joe was not "devout". On 4 January 1933, Joe reported to his parents that Mrs. Harford, a Lady Frances Ryder hostess for New Years, was "very devout which may be a strain on me".

In practical terms, the time had arrived when Pawsey wished to consider what kind of scientist he would be. In February 1933, Nicoll reported to his family that Pawsey was already looking for a job in the wireless industry. He wrote: "[Pawsey's] chances of getting the [position] are pretty good and that makes my chance for an extension [of my 1851 Exhibition] still better," since he and Pawsey were possibly competitors for this additional funding.

In the end both Nicoll and Pawsey received extensions for their 1851 Exhibition Scholarships.

Chapter 7
Research for PhD Thesis at Cambridge, 1931–1934

> [The results of Pawsey and Ratcliffe] contain the important idea that the waves emerging downwards from the ionosphere form an irregular diffraction pattern on the ground. In later work the movements and changes of this pattern were to be studied. (Budden, 1988).

Pawsey's PhD research saw him systematically explore the repertoire of methods (Ankeny, 2019) in ionospheric research. We will see later how this repertoire so strongly shaped his approach to early radio astronomy.

By the close of 1931, Pawsey had begun his thesis project. This was to look for (small) transverse displacements of the signal reflected from the ionosphere. Where Sir Edward Appleton's frequency-change method had been focused on measuring the height of the ionosphere (and he was therefore interested in waves that returned directly downward), Pawsey would instead investigate if some waves did not return as if reflected from a uniform screen, but were instead laterally displaced. Thus lateral irregularities, drifts due to winds and the impact on fading, would be investigated.

Stages 1 and 2: De-Correlated Echoes and Lateral Deviation of Downcoming "Wireless" Waves

Pawsey began by measuring the downcoming wave from broadcast stations at wavelengths of 300 to 480 metres (625 kHz to 1 MHz). He built all his own equipment to do so, an invaluable hands-on experience. He provided his parents with some basic details with a letter to them in Australia on 26 May 1932 after being in Cambridge for little over half a year:

Supplementary Information The online version contains supplementary material available at [https://doi.org/10.1007/978-3-031-07916-0_7].

W. M. Goss et al., *Joe Pawsey and the Founding of Australian Radio Astronomy*, Historical & Cultural Astronomy, https://doi.org/10.1007/978-3-031-07916-0_7

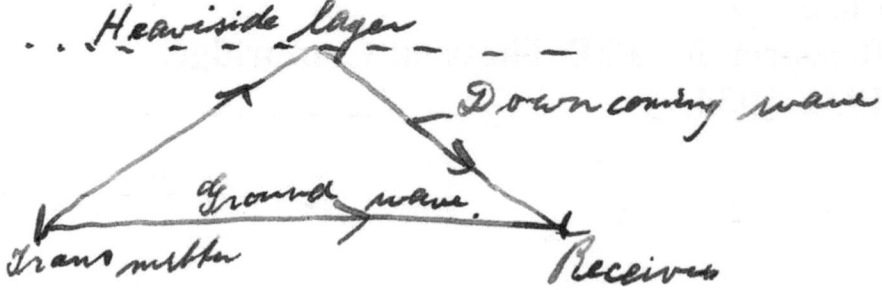

Fig. 7.1 Pawsey's drawing for his parents in May 1932 showing the geometry of his experimental set-up. Credit: Joe and Lenore Pawsey Family Collection

> The wave from a station goes to the receiver by two paths, one direct and one reflected from the Heaviside layer about 60 miles high [presently, the E layer at heights of 90 to 150 km] (Fig. 7.1):

The transmitter and receiver on the ground are shown as well as the ionosphere (Heaviside layer), at an elevation of about 90 to 150 km. The set-up described a Lloyd's mirror with two pathways to the receiver, the ground wave and the sky wave. Interference would be observed due to the sky wave changing due to irregularities and motions in the Heaviside layer- the E layer.

> The ground ray is steady but is attenuated rather rapidly with distance. Thus you get the ground ray from Melbourne stations but not from Sydney.
> The down-coming wave is only present at night. That is why you can hear Sydney then or Brisbane or Melbourne more loudly some times. But this is subject to big fluctuations.
> My record is taken from a station only about 40 miles away but the ground wave is eliminated or should have been by the use of a special aerial system. The ground wave is about 10 times as strong here.

To avoid confusion, we provide a figure of the modern understanding of the ionosphere, to help readers understand the "layers" terminology from the 1930s (Fig. 7.2):

Pawsey carried out a number of experiments to determine the "lateral deviation" of the downcoming waves.

However, for Pawsey, who was interested in very small transverse displacements, a major effort was required to construct aerial systems that excluded the ground wave using a combination of specially polarised antennas that supressed the strong ground wave which could overwhelm the weaker sky wave (see Figs. 7.3, 7.4 and 7.5).

Pawsey provided a simple drawing (Fig. 7.6) of this system to his parents in a letter on 28 July 1932, after being in Cambridge for almost one year:

> I have recently been developing a receiver to receive or suppress as desired a circularly polarised wave coming down from above. This is Dutch to you—it would be difficult to explain. Let me say rather that the current theory predicts the waves reflected from the Heaviside layer to be circularly polarised. My apparatus should receive it for one setting and suppress it for another. So far it is partially successful in that one is much stronger than the

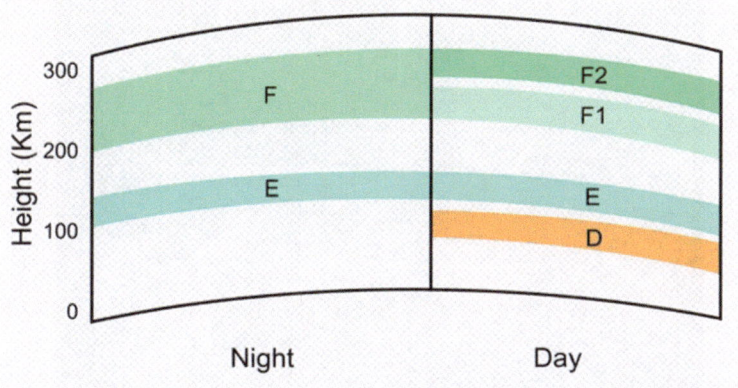

Fig. 7.2 The Layers of the Ionosphere. From https://en.wikipedia.org/wiki/Ionosphere#/media/File:Ionosphere_Layers_en.svg. Credit: IonosphereLayers-NPS.gif: Naval Postgraduate School, derivative work: Phirosiberia, CC BY-SA 3.0 <https://creativecommons.org/licenses/by-sa/3.0>, via Wikimedia Commons

Fig. 7.3 This polarisation analyser is part of the equipment that was used at the Cambridge ionospheric research site (in Cambridge). Credit: Joe and Lenore Pawsey Family Collection

other. I do not know whether or not the weaker one appears because of imperfections in my setting or because the wave is such as to include a component other than the one I intend suppressing. My chief difficulty is that I am now at the same time suppressing the wave which comes along the ground directly and this wave is of the order of 50 times so strong as the one which goes up to the Heaviside layer and is reflected down again (because the station is only 30 miles away so that the ground wave has to travel 30 miles and the Heaviside layer waves about 200 miles up and down). Hence I have to use exceeding care to keep the ground wave suppressed.

Fig. 7.4 Pawsey using the candlestick phone inside the lab at the research site. Credit: Joe and Lenore Pawsey Family Collection

Fig. 7.5 An exterior view of the lab at the research site. Credit: Joe and Lenore Pawsey Family Collection

Pawsey then experimented with using different types of receiving aerials, to see which would work best to identify the very small transverse displacements in which he was interested. They all provided consistent information. Two receivers were placed at variable distances along a line perpendicular to the plane of propagation. He wrote an illustrated letter to his parents on 16 November 1932 [six months later] with a simple description of this experiment:

> I am now trying a new line of work. I wish to discover how much lateral deviation the wireless waves suffer in the upper atmosphere. (Lateral means sideways out of the vertical plane joining transmitter & receiver). The method is as follows. I use two aerials spaced at

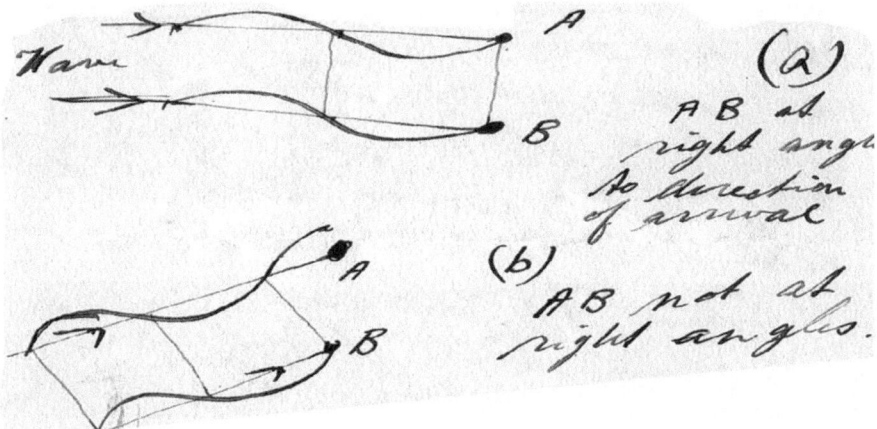

Fig. 7.6 This shows the scheme used to measure small transverse displacements in the E layer of the ionosphere. Credit: Joe and Lenore Pawsey Family Collection

about 20 ft. If these are at right angles to the plane of propagation waves should reach them in phase i.e. corresponding points in the wave arrive simultaneously.

If AB is not at right angles, corresponding points on the wave arrive at different times. A is a little later because the wave has to go farther. This difference in time of arrival is to be looked for. The actual time diff. For 30° is in my case several nano-seconds (billionth of a second).

It can obviously not be got directly but the method I am using is I think quite able to detect it.

I can so arrange a wireless set that the out puts from the two aerials are opposed, & if they are equal & opposite give no signal. Say I do this for the case shown in (a). Suppose then the wave comes in at an angle as shown at (b). Now the two signals are not in phase i.e. the one at B is zero & that at A quite large in my diag. (Previously they were zero together) Hence the two signals when opposed will not annul one another but should give a signal in my set.

I hope you understand this for it is about the simplest thing I am likely to do—the principle I mean.

In the publication by Ratcliffe and Pawsey (1933, p. 301), the authors provided more technical details:

The first problem to be investigated with these receivers was that of finding what is the greatest region over which the atmospheric wave, at the ground, behaves like a plane wave. For this purpose the atmospheric wave was received independently at two points situated on a line perpendicular to the plane of propagation [the two aerials A and B in Fig. 7.6]. We shall refer to this experiment as the **"spaced receiver"**[1] experiment. Similar experiments

---

[1] After WWII Sir Martin Ryle (who will be much discussed later in the book) introduced the term Michelson interferometer for such spaced receiver systems which were the radio analogue of the famous Michelson optical interferometer developed by Michelson in the 1880s. For historical consistency we will also use the term "spaced receivers" before 1946. The lack of correlation between sky waves detected in receivers at larger separation was due to structure in the ionosphere and this is exactly the application to measuring radio sources used by the radio astronomers after WWII.

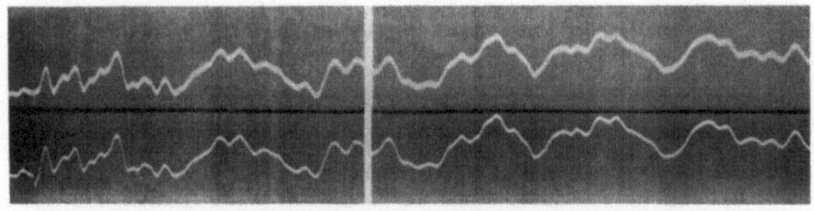

(*a*) Separation = 5 ms. 14/6/32. 2245 ɢ.ᴍ.ᴛ.

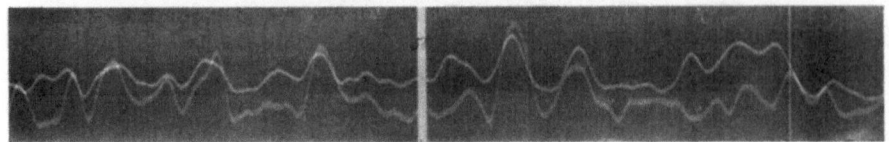

(*b*) Separation = 250 ms. 15/6/32. 2300 ɢ.ᴍ.ᴛ.

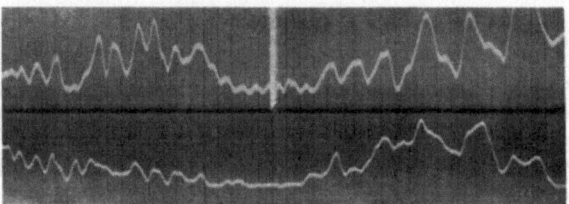

(*c*) Separation = 550 ms. 13/6/32. 2311 ɢ.ᴍ.ᴛ.

Spaced Receiver Records taken on London Regional 356 ms., 56 kms. distant.

Fig. 7.7 Three panels of time series of simultaneous recordings at different separations for the spaced receivers, 5, 250 and 550 metres. The wavelength was 356 metres. Credit: Fig. 7.2, "A Study of the Intensity Variations of Downcoming Waves", Ratcliffe, J.A. and Pawsey, J.L. (1933). Mathematical Proceedings of the Cambridge Philosophical Society, vol. 29, Cambridge University Press. All rights reserved

have been described by previous workers, but they have not always worked with the ground wave suppressed, so that their results may be influenced by interference between ground and atmospheric waves. In practice, it is found that when the two receivers are separated by more than a certain distance there is a lack of correlation between the variations of the two intensities. This gives an indication of the greatest distance over which the wave may be considered to be plane.

The impressive results were shown in Fig. 7.7 as it appeared in the 1933 publication:

Simultaneous records were obtained of the signal intensity received by two similar suppressed ground ray aerial systems. The direct current outputs from the two receivers were conveyed to the same place by wire and the intensities were recorded on the same moving film. [Our Fig. 7.7] shows a typical example of the results obtained. The London Regional transmitter [0.84 MHz or 356 metres] was used. It will be noticed that the intensity variations are quite similar when the receivers are only 5 metres apart, but that important dissimilarities appear when the receiver separation exceeds about 300 metres. The other stations observed have yielded similar results. We therefore conclude that when the receiver separation is of the order of one wavelength the observed signal variations are largely uncorrelated.

The third panel, separation 550 metres, shows that the signals at the two spaced receivers are not correlated. Ratcliffe and Pawsey suggested that a major cause of "fading"[2] was interference of waves, which had been scattered from a series of diffracting centres distributed over an area of radius of at least 20 km on the ground.

Pawsey then undertook an experiment using an Adcock direction finder,[3] which he had to modify. Smith-Rose and Barfield at the National Physics Laboratory (Chap. 5) had improved the Adcock system such that it could measure the direction of signals to a few degrees accuracy for any direction around the horizon. The idea was that it would test the existence of lateral deviation by determining whether or not signals appeared on the aerial when it was arranged to suppress all rays in the plane of propagation. Pawsey used the Adcock direction finder in a non-standard mode (two equal vertical masts) to accurately cancel the strong signal from the direction of the transmitter, enabling the detection of the weaker transverse displaced signal from the ionosphere. This was an excellent example of precision antenna design and experimental technique.[4]

The system consisted of vertical wires 1.7 m high, spaced by 1.5 m. The aerials and the coupling units attached to them were mounted on a rigid frame capable of rotation about the vertical axis of symmetry. To make adjustments, the observer [Pawsey] retired beneath an earthed symmetrical wire-netting screen [unfortunately we do not have a photo of the 189 cm tall Pawsey underneath this apparatus!] which was mounted on this frame at the lower ends of the aerials. Typical lateral deviations of the downcoming waves did not always lie in the propagation plane; on occasion the deviations could reach 15 degrees, in agreement with other determinations. From simple geometry, Pawsey calculated the distance of the point of deviation of the laterally deviated wave from the point where the regularly reflected wave would be deviated to about 30 km for an assumed height of the ionosphere of 100 km (e.g. the lower E layer, present during the daytime).

As always, when explaining a new experiment to his parents (13 February 1932), Pawsey provided a simple description of the scheme:

---

[2]Defined as attenuation of a radio signal with variables such as time, geographical position and frequency.

[3]The Adcock direction finder had been invented at the end of WWI, consisting of four vertical equidistant masts used to transmit or receive directional information.

[4]We note the parallel between the experimental techniques that Pawsey used for this research and those being used in 2019 to develop radio systems that can detect the signal from the Epoch of Reionisation (EoR) in the early very distant universe. In both cases, a strong signal must be removed in order to detect a much weaker signal. For Pawsey, it was the displaced signal from the ionosphere and for the EoR, it is the weak spectral signature of reionisation. The scientific models for both are also related, from a big-picture perspective: the epoch of reionisation is a shell of ionisation at the distant edge of the universe and completely surrounding us, just as the newly discovered ionosphere is a shell of ionisation completely surrounding the earth.

The transmitter & receiver are near together so that you expect the signal to go straight up and down again. I try a scheme which suppresses a signal which comes down vertically but not one which comes at an angle. The result is that I suppress most of the down-coming pulse.

However sometimes I get a bit coming through. Now this might be due to bad suppression or to a signal coming in sideways. I have another bit of evidence though. If the signal comes sideways it has to go a different distance than the vertical.

Stage 3: Use of the Appleton Frequency-Change Method

As discussed in Chap. 5, the Appleton and Barnett "swept frequency" or "frequency change" method determined the height of the reflecting layer from the interference between the direct and reflected signal (ie, "sky" and "ground" waves). Pawsey performed additional experiments in 1932–1934 using this method to determine the number of downcoming waves (and thus the height). In addition the relation between the amplitudes and phases at two spaced receiving stations were determined from the simultaneous observations of the fringes at the two stations. Thus additional information about the lateral structure and the motions in the ionosphere could be obtained. Appleton and Smith-Rose themselves provided transmissions from the National Physical Laboratory (Teddington, Middlesex) at a frequency of 1 MHz. Two receivers, each able to detect the ground wave and the downcoming wave were placed 600 m apart in an E-W line. Observations were obtained at sunrise; many of the records showed the presence of multiple reflections (fringes) and a few showed a dominant wave from the E regions.

Figure 7.8 illustrates an impressive set of fringes from both stations, spaced by 600 m. The two receivers were equipped with simple vertical aerials so the ground and downcoming waves were detected. The outputs were rectified (ie, the intensity was determined) and sent by telephone line to the home station. The data of the independent fringe patterns were then recorded on movie film. The plot shows time from left to right. Note the change in phase of the fringes occurs at point A; previous to A, the fringes from the two stations are almost in phase and after A the fringes are almost in anti-phase. The echo is from the E layer.

The observations were consistent with those found for even smaller receiver separations (from the projects in Step 1 and 2) with the variations likely due to irregular variations of a number of rays from different directions. The least change of angle of a single ray capable of producing the observed phase changes would be 15 degrees; the E layer would be the responsible medium.

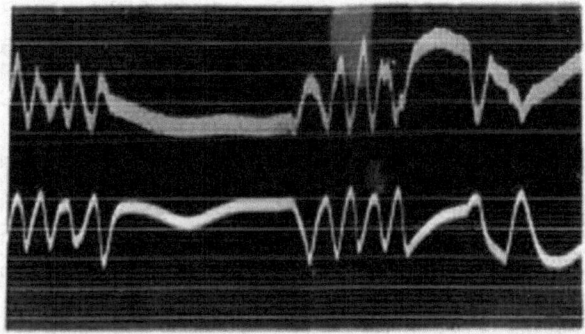

Fig. 7.8 Frequency Change "fringes" obtained at 1 MHz at Cambridge with transmitter 89 km distant. The x axis is time over an 8 sec interval. The y axis is the intensity of the Lloyd's mirror due to interference of the ground wave and the wave reflected from the E region of the ionosphere measured at the two Cambridge receiver stations, separated by 600 m. The frequency of the transmitter was varied during the 8 sec interval. Recorded at sunrise 07 h 29 m GMT on 24 November 1933. Note the change of phase of one station with respect to the other of about 180 deg at point A. Credit: Fig. 7.1, Pawsey, J.L. (1935). "Further Investigations of the Amplitude Variations of Down-coming Wireless Waves". *Proceedings of the Cambridge Philosophical Society*, vol. 31, January. All rights reserved

Stage 4: Tuve-Breit Pulse-Echo Method, Pawsey 1932–1934

At this period, Pawsey used a spaced receiver with elements perpendicular to the transmitter direction to measure the transverse displacements of the downcoming wave. Thus he could determine the cause of the phenomenon of lateral displacement that he had observed previously. Completing his familiarity with the repertoire of ionospheric research, the Tuve-Breit method of pulsed transmission and reception to detect echoes from the ionosphere was utilised so "that the echoes could be studied independently and also aerial systems were used which admitted only one magneto-ionic component." He noted, "the spaced receiver method for the measurement of lateral deviation was chosen as the one least open to error. The intensity variations of one echo were compared with those of the same echo received on a similar receiver at a given distance from the first". The pulse transmission frequency was 1.8 MHz from a transmitter at a distance of about one km from the two receivers. Both the E layer and F layer could be separately detected with the pulse system; the E layer disappeared at about midnight.

Two series of observations were carried out with receiver separations of 140 m and 600 m. Echoes from both the E (90 to 150 km) and the F regions (>150 km) were obtained. The advantage of the Tuve-Breit method was that the identity of the region of the ionosphere could be designated simply based on the time delay. Figure 7.9 shows the remarkable case of simultaneous fading of pulses in the F

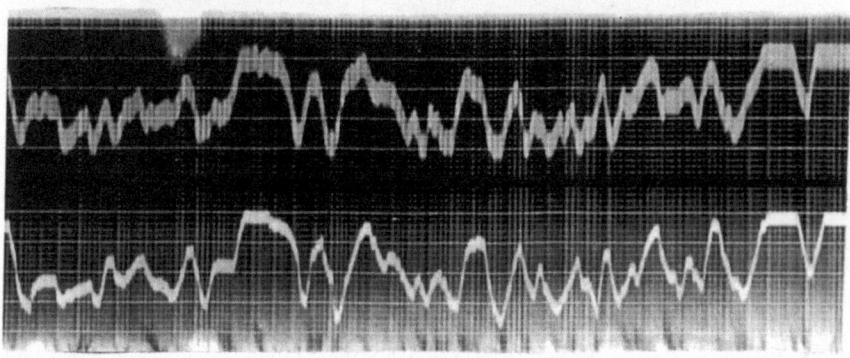

Fig. 7.9 Simultaneous fading pulses based on the Tuve-Breit pulse method. 16 March 1934, F layer at 350 km height. 1.76 MHz or 170 m. Receiver separation 140 m. The lower receiver (eastern receiver) shows a consistent lag of about 1 sec with similar shape (the minor grid spacings are 1 sec). Transmitter displaced by 1 km. The time range on 16 March 1934 was from GMT 23 h 28 min for a duration of only 3.2 min. Credit: Fig. 7.4, Pawsey, J.L. (1935). "Further Investigations of the Amplitude Variations of Down-coming Wireless Waves". *Proceedings of the Cambridge Philosophical Society*, vol. 31, January.

region at a height of 350 km. The lower record from the east receiver (140 m displaced) shows a consistent delay of one sec at 1.8 MHz.[5]

During 1933, Pawsey wrote his family (3 August) with a simple diagram of the Tuve-Breit method:

> I am running automatic records of signals reflected from the upper atmosphere. You see the idea is to send out a short signal and a receiver about a mile away picks up first a signal which has travelled along the ground and then a little later, because it has had to travel 100 or 200 miles, a signal which has travelled up to the Heaviside layer & back. The time difference is short, about 1/1000 of a second, but it is sufficient for the one to be all over before the next arrives.
>
> The transmitter keeps sending out the short signal at regular intervals and the receiver records the time lag between the direct wave and the "echo", [the wave which has been up and down] and from this we can tell how high it went since we knew how fast the signals travel.
>
> Thus if the delay was 1/1000 second, the difference in distance travelled would be 186 miles (the velocity of wireless waves is 186,000 miles sec—the same as light).
>
> The direct wave has to go 1 mile only so the length up and down is 185 miles & the height about 92 miles.

---

[5] White referred to Pawsey's research in his discussion of an adaptation of the Breit-Tuve method in order to record and partially automate the recording of interference phenomena (White, 1934).

The transmitter is left on and the receiver records heights on a moving photographic film—the process is automatic. There are times however when things do not go as they should—today first the transmitter went on the blink and then the film ran out.[6]

Summary of 4 Stages of Research

In summary there were a number of independent methods of estimating the amount of lateral deviations of downcoming waves; all agreed. The foundations for future research on the elucidation of the properties of the ionosphere—height, density, structure and velocity—were laid in the early 1930s. Pawsey was only one of many Australian, New Zealand, UK, Canadian and US scientists who were to play a major role in this new field of ionospheric physics.

During the years 1931 to 1934, Pawsey spent much of his time working at the Radio Research Group's field stations near Cambridge under the supervision of J.A. Ratcliffe, in addition to undertaking his PhD research. He reported these activities in some detail in the letters to his parents. A common theme is that Pawsey's experience of having grown up on a farm stood him well in the physical labour required during his thesis. The construction of not only the electronics but the mechanical construction of the large wooden aerials was a constant chore. An example of a chronic problem was "mud" in the fields where the large aerials were mounted; in the winter months Pawsey was often relieved when a heavy frost occurred causing the mud to freeze. Based on his experience on the farm in Victoria, he could report to his parents on 30 November 1932:

> I have a taste of mud out around our huts these days. One puts the bike—push bike —down into low gear, pedals hard, spins the back wheel and hopes for the best. One is usually justified by results as I have only once stalled in the middle & one other time shipped to sea in one shoe. However, it is not in the same category as is a dairy farm.

The "mud" at a dairy farm had a characteristic that was a great deal more unpleasant than the mud encountered near that aerial hut in the East Anglia countryside.

In early 1933, the spacings between the two receiving aerials that Pawsey was using for his PhD research were up to 10 km apart. Pawsey wrote on 27 February 1933: "I live a very busy life these days. One of my experiments consists of comparison of fading here & some 6 miles out & I have been going out on the motor bike of evenings." Thus he used the motorcycle to travel from one aerial to the other.[7]

---

[6]Until the development of oscilloscope displays and inscription gadgets (mostly in the USA), film— expensive and time consuming—was the only way to record these rapidly changing outputs. See Yeang, 2012.

[7]Some months later, the required distance was only 2 ½ miles, suitable for a bike trip at night. On 26 October 1933 he wrote: "One of my experiments consisting in recording the fading at two points separated by different distances & find how far one must go before the fading from the same station is dissimilar ... I have previously done it for short wave stations about 300 or 400 metres wave length close (say 50–150 miles) and very distant (1000 miles). I am now doing it for a longer wave

In ESM 7.1, Helium Balloons, we summarise the challenges Pawsey had with helium balloons used in an attempt to lift a small aerial above the ground.

Conclusion of Pawsey's Thesis 1935

At the end of his thesis, Pawsey provided a theoretical discussion of his findings of the properties of the deviations of the downcoming waves from the E and F layer of the ionosphere. Based on the amplitude distribution on the reflected waves (a Rayleigh distribution), he suggested that the observed fading was due "largely to an irregular scattering at the ionosphere". As discussed above, the conclusion was that absorption was not an important cause of the variations; however, the mean level did depend on absorption. "There is doubt that absorption plays an important part in governing the difference between night and day signals, and it is possible that it should be an important factor in the usual fading during the night time. But for the nearer broadcast stations and for those studied which showed fast fading there must exist a further most important mechanism", based on the dissimilar fading from receivers separated by some hundred metres. Thus an interference mechanism was indicated. The major conclusion of the thesis was that the cause "must be due to **waves which have been laterally deviated** ... [our emphasis] [The origin had to be] interference between rays reflected from a horizontally irregular region."

The treatment proposed for the interpretation by Pawsey was a straight forward Fresnel zone diffraction treatment (the term "Fresnel zone" was not used by Pawsey).

> ... [Let us use] the mechanism of this fading using the method usual in diffraction problems of dividing the reflecting region into half-wave zones ... These regions are to be defined by the condition that the optical path from the transmitter to the receiver increases by lambda/2 as the apex of a ray moves from the inside to the outside of a region.

From the perspective of more than 80 years, one of the most significant findings of the thesis was evidence for the horizontal drift system in the ionosphere.[8] Pawsey discussed the causes of these deviations (up to 20 degrees), making a number of arguments that they arose from **horizontal** irregularities. "... These irregularities must be continually moving. The cause had to be ion cloud irregularities of sizes of order 100 to 200 m. The cause remained uncertain, unlikely to be appreciable

---

length 1500 metres. At the moment I have one set in my 'digs' [ie, his rooming house] which is very nice except that one has to go and turn off the other set at the range—at the end of a run—which is a ride of 2 ½ miles each way." Likely his two friends S. Falloon (a fellow PhD student) and R. Witty had assisted in the bike or motorcycle trips to adjust the equipment at the remote stations during these years; they are thanked in the Pawsey, 1935 publication "for a great deal of help with the observational work".

[8]Later work by S.N. Mitra at the Cavendish Lab in 1947 provided a direct determination of the motions in the ionosphere, with velocities in the range 15 to 120 m per sec (median about 45 m per sec). (See Mitra, 1986)

changes in the uv light from the sun." A possible origin was due to fluctuations of charged particles from the sun. The most important conclusion from the Pawsey thesis was that there was a drift of the diffraction pattern over the ground. "This should give rise, at two points displaced along the line of drift, to fading similar except for a time lag at one point." However only one record (Fig. 7.9) showed a clear-cut time lag implying a velocity over the ground of 140 metres/second and a wind velocity in the ionosphere of 70 metres/second.

Pawsey concluded:

> It appears that the horizontal movement of these ion clouds of about 100 metres/second which exist in the neighbourhood of the E region is an important cause of fading ... It is suggested that the irregularities are caused by non-uniform showers of charged particles which do not penetrate to the lower levels of the normal E region.

In November 1934, Pawsey submitted his PhD thesis "An Experimental Study of the Intensity Variations of Downcoming Waves".[9] (By this time he had joined the staff at EMI). On 17 December 1934, Rutherford and Ratcliffe wrote reports on the thesis. Rutherford's summary was:

> The thesis ... contains an excellent account of the study of fading in the reception of radio-signals ... This aspect of the question [the irregular nature of the reflecting layer] has been carefully studied by the candidate using a number of experimental methods including a comparison of simultaneous fading in identical receivers spaced some distance apart ... [M]arked lateral displacement of the downcoming waves [is indicated]. This lateral displacement is particularly marked in reflection from the abnormal E region in the ionosphere where it may amount to a deviation of 20 degrees from the normal ... [Pawsey interprets this as] a diffraction problem arising from irregular scattering of the waves from the neighbouring regions of different ion concentrations. The marked lateral deviation in the E region is ascribed to strong winds in that part of the ionosphere. The thesis is well and clearly written and contains results of much interest and importance ...

Ratcliffe offered a similar report, with emphasis on the clever nature of the two systems designed and constructed by Pawsey to determine lateral deviations. He also mentioned that the fading could likely be associated by winds in the ionosphere (at height of over 100 km) at the level of 100 metres/sec. The Ratcliffe commentary did contain some criticism of the thesis's structure, but not of the research itself:

> The structure of the thesis follows somewhat closely that of the two papers[10] which the author has written on the subject, and as a result it is rather too specialised and assumes a

---

[9]Copy of Phd thesis provided by David Green from the Cavendish Laboratory collection (University of Cambridge) PhD theses, 2012.

[10]Even before his PhD was awarded, Pawsey published two papers on his research. One was submitted on 30 March 1933 and "read" on 1 May 1933 in the *Proceedings of the Cambridge Philosophical Society*, vol. 29, page 301: "A Study of the Intensity Variations of Downcoming Waves", by Ratcliffe and Pawsey. This publication consisted of the first part of his research that had been completed since 1931 October. The second paper was a single author publication with Pawsey alone, "Further Investigations of the Amplitude Variations of Downcoming Wireless Waves", again in the *Proceedings of the Cambridge Philosophical Society*, vol. 31, January 1935, received 19 November 1934 and "read" on 26 November 1934, that included work carried out until March 1934. The thesis and the two publications are nearly identical.

considerable previous knowledge on the part of the reader. It would have been more satisfactory if some of the matter had been rewritten for the purpose of the thesis.

The work has involved the design and construction of several complicated aerial systems for special purposes. A very clear account is given of the elimination of possible errors in the use of these systems, and of the best circuit conditions for obtaining the desired information. In my opinion this difficult practical work, which was conducted by the candidate alone, is of high standard and has led to important results …

After an oral examination on 19 December 1934, both Rutherford and Ratcliffe recommended that Pawsey be awarded a PhD. The Degree Committee met on 25 January 1935; the members included Rutherford, Sir William Pope and a number of other scientists including Ratcliffe as well as the controversial and eminent crystallographer John Desmond Bernal (1901–1971). He was approved for the PhD degree on 13 February 1935.[11]

Ratcliffe's Evaluation of Pawsey's Research of 1931–1934, in 1974

In 1974, Ratcliffe wrote two end-of-career review papers that summarised iono-spheric research in the twentieth century: "The Formation of the Ionosphere: Ideas of the Early Years (1925–1955)" (Ratcliffe, 1974b) and "Experimental Methods of Ionospheric Investigation 1925–1955" (Ratcliffe, 1974a).

Ratcliffe provided a thorough summary of the impact of Pawsey's research in the years 1931 to 1934:

The amplitude of the wave received at night on frequencies near 1 MHz from broadcasting stations distant 50 or 100 km was found to vary within times of the order of minutes. In the early days it was realised that these variations were caused by changes in a wave reflected from the ionosphere. At first the fading was ascribed to interference between a ground and sky wave whose phase changed as the height of reflection altered, but soon the "frequency change" experiments demonstrated that the sky wave alone underwent fading. It was then supposed that this was caused by clouds of denser electrons overhead in the absorbing portion of the ionosphere.

When attempts were made to measure the size of these clouds by observation of the fading at two separate locations it was found that the fading was different at places separated by only one or two wavelengths: it was then realised that because the size of a Fresnel zone in the ionosphere was several times this distance, a simple picture was inadequate, and diffraction theory must be used to explain the result.

It was concluded that ionospheric irregularities with sizes of the order of 100 m must be present. These early experiments were conducted in the belief that the irregularities occurred in the absorption of the wave, but later, when waves of greater frequency were used, it

---

[11] A few years later in 1937, T.H. Eckersley explicitly used Lloyd's mirror interferometry at higher frequencies (9.1 MHz) to study the ionosphere (1937, Nature vol 140 p 846 and 1938, Nature, vol 141, p 369). With this background, the visionary remarks published by McCready, Pawsey and Payne-Scott (1947) concerning the determination of multiple Fourier components of the brightness distribution using the sea-cliff interferometer became clear (see Chap. 36).

became clear that the observed changes of amplitude were caused, not by changes of absorption, but by changes of phase as the transmitted waves passed through clouds of smaller and greater electron concentration.

On some occasions it was found that the fading at two places was similar but that corresponding maxima occurred later at one place than at the other, and it was deduced that the ionospheric irregularities drifted horizontally with the corresponding velocity of [of about 70 metres per second].

Chapter 8
After the PhD: Electric and Musical Industries (EMI) and Marriage to Lenore Nicoll, 1934–1939

You may remember me as the excessively tall Australian, a research student of Ratcliffe's with whom you played cricket for the Cavendish . . .

I have always intended returning to Australia at any rate in time to educate my baby daughter in a country free from class prejudice.

–Letter from Pawsey to P.W. Burbidge in New Zealand on 31 October 1938.

We do not know how Pawsey envisioned the best possible life and career for himself as he neared the end of his PhD studies. The evidence suggests that he was not interested in basic research, and he would not become so until after World War II. His correspondence indicates that he considered himself best suited to applied work and wanted to undertake this in an industry context. He had sought such work before coming to Cambridge. Or was the Depression affecting his optimism about a research career?

Pawsey seems to have taken a jaundiced view of the "Ivory Tower". In June 1936, he wrote to his parents:

You had just received a copy of my Cambridge Philosophical Society paper. You need not bother about being unable to understand it. It would be unintelligible to anyone not working along my lines . . . It is the result of this specialisation of modern life. A more cogent criticism would be the remark "What good is it to anybody?" Actually practically none. The same holds of most work published.

The justification is three fold. Firstly, the doing should supply some training in thinking to the authors. Secondly every now and again someone stumbles on to a revolutionary discovery such as those to which wireless etc. are due and because of the training are able to systematise the discovery and reduce it to a useful form. Thirdly the whole background of human knowledge progresses by this means though mainly incredibly slowly.

The original version of the chapter has been revised. A correction to this chapter can be found at https://doi.org/10.1007/978-3-031-07916-0_43

Supplementary Information The online version contains supplementary material available at [https://doi.org/10.1007/978-3-031-07916-0_8].

W. M. Goss et al., *Joe Pawsey and the Founding of Australian Radio Astronomy*, Historical & Cultural Astronomy, https://doi.org/10.1007/978-3-031-07916-0_8

Much earlier, in March 1933, with at least a year to go to complete his PhD degree Pawsey was already looking for employment; he had written then to the Radio Research Board in Australia, perhaps applying for a position vacated by one of the research staff. As a part of the application, Pawsey asked Rutherford and Ratcliffe to write letters of reference.[1] They responded with remarks about his working style.

> Rutherford: Pawsey has attacked his experimental problems with energy and enthusiasm and has made good progress. He has proved a very competent experimenter with good judgement in the interpretation of his results . . . I should judge Pawsey would prove a very useful investigator on radio problems and I am sure he would work well with his Colleagues. I can recommend his claims for consideration strongly for this post. (9 March 1933).
>
> Ratcliffe: . . . I have been particularly impressed by Mr. Pawsey's caution in interpreting the results of his experiments. He is not satisfied with the results until he has checked them in every possible way. This tendency to suspect every result and to criticise his own work, has been of great value to him in the work on which he is engaged, as it is so easy in this work to take measurements which do not correspond to the quantities one is trying to measure . . . I consider that Mr. Pawsey has a very wide knowledge of Wireless matters. Mr. Pawsey will be found very acceptable in any society, and will fit in well with any team of workers. (8 March 1933).

Pawsey must have gone on looking for work, because in March 1934 he announced to Lord Rutherford that he wanted to join the Gramophone Electric and Musical Industries (EMI), along with a few others from the Cavendish Laboratory including Ted Nicoll. He required special permission because he had not satisfied a residency requirement for a PhD, missing out by one term. Rutherford wrote Priestley, the Secretary of the Board of Research Studies at the University of Cambridge on 16 March 1934:

> J.L. Pawsey is leaving to take a post at the end of this term so he is one term short of the three years required for the PhD. As he did 2½ years work at research in Melbourne before he came to Cambridge, and also published a paper, I recommend he be allowed to proceed to the degree without completing the last term.

Pawsey also wrote Priestley on 2 May 1934 with this information[2] and pointed out that he intended to submit a PhD thesis on his Cambridge research later in 1934. On 16 May, the Board of Research Studies granted Pawsey's request for an exemption.

And so Pawsey left Cambridge to begin his new position at EMI, Hayes, and he travelled from Cambridge to Hayes, London, 115 km, on a bicycle with two suitcases. As he explained to his parents:

> I "came down" [left the university as a postgraduate] from Cambridge today. I am no longer a student . . . Tomorrow I go to the works [EMI] to start work. It all seems queer does it not? I came down here equipped with a push bike, my small leather suitcase full of clothes and my other little one with results of work in Cambridge. I have to write it up yet and I did not wish

---

[1] These two letters are the first entries of Pawsey personnel file (NAA AH8520/PH/PAW/1Part1, 1933 to 1947).

[2] He also submitted a reprint of the publication "Accurate Measurement of the Frequency of the Carrier Waves of Victorian Broadcast Stations" to indicate his Australian research.

to lose it or to take a chance of doing so. The two made quite a load on a bike and I guess I looked a trifle gauche as I came through the centre of London—up Piccadilly and by Hyde Park Corner and so on.

However, the chief thing I want to talk about is this. I have this job. I am writing a thesis in my spare time. I am not sure how long this will take. I hope to finish by June but may find it takes longer than I think and there is a chance of it being carried over till October. During this time my time will be pretty full and I do not want any avoidable distractions.

It therefore looks to me as if the time after I finish my thesis is a time at which it would be a good idea for you two to come over to England. It looks as if I shall be fairly free and also not saving up money for any definite purpose. I would like to live with you again for a while and I do not think there will be another chance. I am not going back to Australia—for two years at any rate—and if we leave it much longer then I shall probably be getting married and preventing it [a visit of his parents to England] that way.

In early April 1934, Pawsey followed other Cavendish colleagues in joining the group of E.C. Cork at EMI in Hounslow, 17 km west of Charing Cross in London. This was at an extremely exciting time for EMI, a company that had emerged only 3 years earlier from a merger between two gramophone recording businesses. At the same time, the British government had set up a Television Committee and put out tenders for a new "high definition" service (defined by them as being a system of 240 lines or more) to be run by the BBC. EMI was one of the two tenders offered for an experimental period and established at Alexandra Palace ("Ally Pally"). Public excitement over the outcome of this rivalry between EMI-Marconi and Baird (the BBC broadcaster until this time) was considerable. And the pressure for EMI's employees must have been not inconsiderable!

Radio astronomer Sir Bernard Lovell (1913–2012) has described Pawsey's EMI experience in a thorough text in his Biographical Memoir for the Royal Society in 1964:

Pawsey's work at EMI was dominated by the preparation for the television tests at Alexandra Palace using the EMI system. He was a member of the group led by E.C. Cork in I. Shoenberg's (later Sir Isaac) research department. Cork's group was dealing with a number of electronic problems associated with the early television development at that time, but Pawsey was concerned with the aerial and feeder design.[3] At the commencement of this work, Pawsey spent much time on field measurements of the polar diagrams of dipoles, reflectors and characteristic impedance of feeders. The number of lines to be used in the tests had not then been settled, and it seems that the late A.D. Blumlein was the first to draw attention to the effects of mismatched feeders on the transmitted picture. Two diffi-culties were recognised at that early stage. First, it was necessary that the impedance of the system should be matched at all frequencies within the television side-bands, so that no part of the signal should be attenuated. Second, since in the proposed installation at Alexandra Palace, and in any foreseeable practical system, a long feeder would be needed between the transmitter and the aerial, the time of transmission would be significant compared with the

---

[3] In a letter to his parents, Pawsey wrote on 12 March 1935 pointing out that he worked for EMI who was responsible for the "special television problems" while Marconi was responsible for the "wireless transmitting end." On 20 March 1935, he emphasised to his parents that he was "back on aerial work. The transmitting aerial, which is the Marconi Co's job by rights appears to not be as efficient as it might be. We did some tests checking this up and now are trying to produce a substitute."

shortest interval resolvable on the television picture. Any mismatch would cause reflexion along the feeder between the transmitter and aerial and eventual radiation with a time delay which would cause a double picture.[4]

Apparently the impedance measurements made on the first aerial-feeder system erected on the roof of the research block at Hayes were most alarming and it was with the measurements and modifications to this system, which eventually gave rise to the Alexandra Palace aerial, that Pawsey was primarily concerned. The vision channel was to be on 45 MHz, with 405 lines, 25 pictures per sec and interlaced scanning at 50 frames per sec giving a side band requirement of 2.5 MHz. The feeder length was 450 feet so that the travel time of 0.5 microsec corresponded to a frequency in the side-band range. Pawsey designed apparatus to measure imped-ances at 45 MHz to a few per cent. This requirement was somewhat similar to the problem Pawsey had encountered in his PhD work at Cambridge, where he had had to supress the strong direct signal in order to measure the ionospheric reflection, which was only a few percent of the direct signal. At an early stage of this work it was necessary to solve the difficult problem of making a terminating resistance which would be constant and nonreactive over the frequency range 43 to 47 MHz. The solution of these various problems eventually led to the double ring of full-wave dipoles with the mast in the centre, erected at Alexandra Palace in the summer of 1936.

A most detailed and clear account of this work is given in the paper read before the Wireless Section of the I.E.E., by Cork and Pawsey (Cork & Pawsey, 1939) on 7 December 1938: "Long Feeders for Transmitting Wide Side-Bands, with Refer-ence to the Alexandra Palace Aerial-Feeder System." The big challenge for Pawsey would have been going from MHz frequencies to 45 MHz for TV. The antennas used at Cambridge for ionospheric research were simple aerials and all the structures were much greater than the wavelengths used. But many new issues would have arisen at EMI working at 45 MHz because lengths of feed lines were now on wavelength scales. While some skills such as building radio finding devices to precise standards would still apply, many new problems would be encountered.[5]

As Lovell identified, a major achievement of this era was the **Pawsey stub**, the eponymous invention that Pawsey and Cork made in the 1930s. The relevant patent by Cork and Pawsey was Patent Number 462911, "Matching Aerial and Feeder". The device was developed for the 45 MHz Alexandra Palace television transmitter erected in August 1936, so was critical to EMI's challenge to the Baird company for the BBC tender. The main purpose of the Pawsey stub is to prevent the feeder from acting as an additional antenna. At EMI, Pawsey was working with much shorter

---

[4] See Chap. 6 for remarks concerning how this scientific problem is closely analogous with current (2020) experiments to detect the EoR (Epoch of Recombination) signal. The problems are still exactly the same, but at quite a few orders of magnitude lower levels. The EMI work was even in similar frequency ranges.

[5] The authors thank CSIRO antenna engineer Alex Dunning for these insights.

wavelengths than had been the case for his PhD research at Cambridge.[6] At these shorter wavelengths, the distances between the connectors matter, since a wave could be reflected from any imperfections in the wiring. These kinds of reflections need to be cancelled out. Here Pawsey could immediately utilise the same principle he had drawn on in his PhD research. Then, he had supressed the ground wave, which was the unwanted signal. In this case, Pawsey stub is a clever structure which is exactly 1/4 wavelength long and when unwanted currents (created by connecting an asymmetric structure such as a coaxial cable to a dipole antenna) are reflected by the stub, they have the opposite direction and are cancelled. The patent application reads: "This reflected current diverts power away from the antenna elements turning the cable itself into an antenna as well; this works by making the aerial connections look 'balanced', as the ± voltage phases appear to radiate equally from both halves of the balanced ¼ wave lines and hence cancel." It acknowledges that "[T]hanks are due to Mr. A.D. Blumlein,[7] who foresaw the effects of mismatched feeder on the transmitted picture".

A simple Pawsey stub, a quarter wave balun (balanced to unbalance transformer), as would be used by radio amateurs in the modern era is shown in Fig. 8.1.

The ham radio literature contains numerous references to the elegant and simple Pawsey stub with no reference to J.L. Pawsey and no reference to the history of the device. We note that the same Pawsey stub principle is used in microwave oven door seals, since the metallic connection between the door and oven cannot reliably stop microwave leakage. The Pawsey stub turns a deliberate physical open circuit into a good short circuit path for all currents circulating inside the oven, so none leak to the outside surfaces (Meredith, 1998, p. 126).

In 1936 and 1937, there were fleeting references to EMI work-related experiences in Pawsey's letters to his parents. On 22 and 29 January 1936, he reported on his participation in the testing of a new 200-foot (61 metre) experimental mast at EMI, Hayes. "It is now ready to hurl to the top and erect." The aerial was finally erected on 25 January 1936 with success. The output power of the TV transmitter was increased by a factor of two due to the fact that the previous tower was only 37 m high. A major concern was the distortion of the TV images; much work remained. By 15 July 1936, much progress had been achieved. There had been a big rush to get the new BBC TV station up and running:

---

[6] Dunning has pointed out that at MHz frequencies and below concepts like Pawsey stubs would have been irrelevant but become essential for high quality TV research.

[7] Alan Dower Blumlein (29 June 1903-7 June 1942) was an English electronics engineer, notable for his many inventions in telecommunications, sound recording, stereophonic sound, television and radar. He received 128 patents and was considered one of the most significant engineers and inventors of his time. He died during WWII, 7 June 1942, aged 38, when the Halifax bomber on which he was testing an H2S airborne radar system (used to provide air crews an electronic map image of the ground below) crashed in Herefordshire. See https://www.newscientist.com/article/mg12617215-100-forum-mystery-of-the-missing-biography-a-look-at-the-life-of-alan-blumlein/#ixzz6DaRdYoES.

THE QUARTER WAVE 1 : 1 BALUN. (PAWSEY STUB)

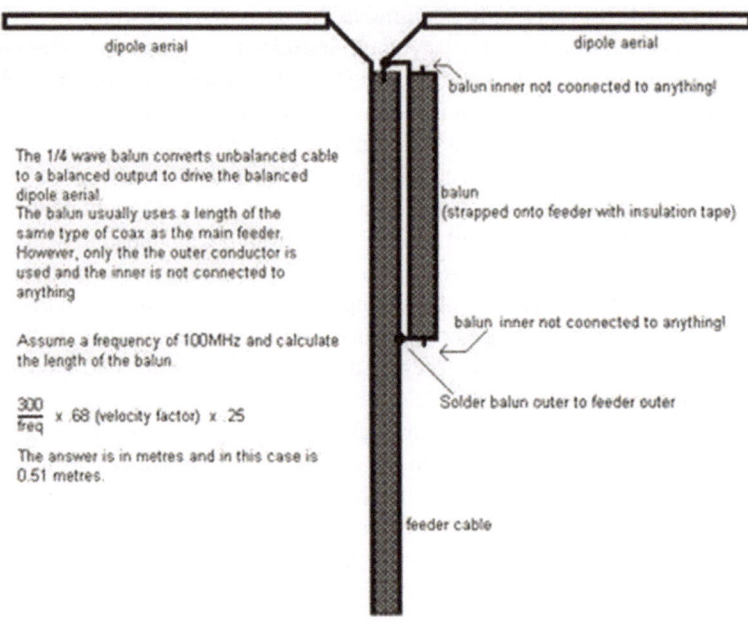

dipole aerial   dipole aerial

balun inner not coonected to anything!

The 1/4 wave balun converts unbalanced cable
to a balanced output to drive the balanced
dipole aerial.
The balun usually uses a length of the
same type of coax as the main feeder.
However, only the the outer conductor is
used and the inner is not connected to
anything

balun
(strapped onto feeder with insulation tape)

balun inner not coonected to anything!

Assume a frequency of 100MHz and calculate
the length of the balun.

Solder balun outer to feeder outer

$$\frac{300}{freq} \times .68 \text{ (velocity factor)} \times .25$$

The answer is in metres and in this case is
0.51 metres.

feeder cable

Fig. 8.1 A Pawsey stub from http://www.gareth.net.nz/nrgworkshop/half_wave_dipole_aerial.htm

Cork and I are responsible for the aerial measurements and adjustments at the new BBC
television station at Alexandra Palace. The aerial is in course of erection now. We expect to
begin tests about the beginning of next week (20 July 1936). There will then be a wild rush to

get the thing finished by about Aug. 10th. This date being fixed by a wireless show at
Olympia on about Aug. 20th at which it is desired to show television sets.

Not surprising in the end, the EMI-Marconi company were the standout winners of
the rivalry at Alexandra Palace. Exclusive broadcasting with the 405-line electric
Marconi-EMI system became the standard for all British TV broadcasts until the
1960s.

In 1937 (12 January), Joe and Ted Nicoll's sister Lenore watched an evening's
series of broadcasts. Their assessment was hardly favourable: "The quality of programs is pretty poor. I guess it is hard to make them good."

Pawsey was involved with 29 patents made at EMI from 1934 to 1939; colleagues
involved in addition to Pawsey were Alan Blumlein, Cork, Bowman-Manifold and
E.L.C. (Eric) White. Pawsey also authored or co-authored 12 reports at EMI during
the time from 23 May 1936 to 14 December 1938, again with Cork and Bowman-
Manifold. The earliest report from May 1936 was "The Centre-Fed Dipole
Antenna", and the last in 1938 was "The Cylindrical Current Sheet Antenna".[8] His
expertise and insight in antenna design was undoubtedly consolidated at EMI. His
work there showed the same integration of practical and theoretical perspicacity. We
note that here, too, in a paper by Cork and Pawsey (1939), the use of Fourier
methods of analysis was evident: "An alternative method of considering the phenomenon is in terms of the Fourier components of the signal." Although he was the
author, or co-author, of 12 EMI reports on aerial and feeder designs, the I.E.E. paper
is the sole published account of Pawsey's work during the EMI period. Under other
conditions the work described in these documents would form the subject of several
published papers, but various factors of EMI policy, and the competition of the Baird
system, led to the restriction on publication.

J.L. Pawsey: Courtship and Marriage

After arriving in the UK at the end of September 1931 and "coming up" to
Cambridge on 5 October 1931, Pawsey started an active work schedule. In addition,
he had a busy social life, initially governed by events organised by the Dominion
Services and Students' Hospitality Scheme, organised by Lady Frances Ryder and
Ms. Celia Macdonald of the Isles (see Chap. 6). Pawsey had a number of casual
relations with several young women in 1931–1933, all with an Australian
connection.[9]

---

[8] NAA C3830 D26/4. A number of the reports were concerned with details of the Alexandra Palace
television transmitter, e.g. filters for separation of sound and vision frequencies and the question of
interference of Overseas Broadcast radio signals at Alexandra Palace by the local transmitters.

[9] Alison Bedggood from London was a distant relation of Betty Bedggood, a friend from Melbourne. Of Alison, Pawsey wrote to his mother (22 December 1931): "Alison is the attraction—by
the way—she is one of three daughters—the middle one—about my age. They are fairly homely
people—the girls work—Alison and Madge are in offices in the city."

Pawsey had an interesting friendship with Ysonde Guilbert of Tasmania for a short period in early 1932: "[She is] interested in the same things as I am, clever, she got one of the top scholarships in Tasmania on leaving school to go to the University and full of pluck and initiative e.g. her trip over here. She expects to get married when she goes back which brings your hope of a spot of gossip to the ground and opens the question of whether or not a married woman may have men friends."[10] Pawsey and Ysonde clearly had engaging conversations: "I have been amazed at the similarity of our views, the more so since we have not spent any appreciable time together nor have we even moved in similar circles of friends. It is definitely not due to passive acquiescence on either part for as you know I am argumentative and she is as independent a thinker as I am." (24 February 1932).

Joe had remarkably frank discussions with his parents, starting on 8 December 1932, about his goals for companionship:

> I am "on the bust" at the moment. Lady Frances's parties form a nucleus and an excuse for a visit to London . . . Tea and a dance on Monday—a visit to the Ford works . . . I had decided to do my best on the previous evening to pick up a nice girl at the dance and take her out the next day. However, I reached the end of the last dance without having made the attempt. I suppose it was a case of being too particular. You see I wanted: (a) Dominion [e.g. Australian, New Zealand, Canada or South Africa] girl, (b) more or less hard up (financial reasons), (c) good looking (aesthetic reasons) and (d) intelligent. I was handicapped by only meeting two satisfying (a). One was a former friend Mary Martin who satisfies a,b,c and d but was booked up. Incidentally she is Catholic and we differ on most subjects but are content to differ and be friends. The other failed to qualify on condition (c).

He elaborated on the theme of "Dominion girls" two months later (letter, 15 February 1933):

> You commented on my preference for Dominion girls. The reasons are various. Firstly I get on best with Dominion people in that our manners are more nearly the same—an apt saying [attributed to a Dr. Murray] expresses it, "You do not know whether or not an Englishman would prefer your absence to your presence". Anyway, you know better [what matters] to me . . .

In an undated letter from the period 1932–1934, Pawsey elaborated:

> [As regards people from the Dominions], I have a first class introduction to overseas people both as common exiles and in that we are more interested in the familiar things of common place to the English. Thirdly when we are in a strange country we miss various conveniences etc. we are used to and do not give credit for the others which replace them. Consequently, with two people of different country of origin in the home country of one, the other always has a grouch against life which the other cannot understand. Fourthly Dominion people in England [have been chosen in their home countries to have special abilities].

Finally on 26 July 1933,[11] Pawsey continued his discussion of aspirations for marriage with his mother:

---

[10]From Pawsey's letter to his mother on 10 February 1932. Ysonde left London for Tasmania (where she was to be married) on 26 February 1932.

[11]At this time, he was involved with Helen Borland; within a few months Lenore appeared on the scene.

I was also interested in Mother's philosophy of happiness. I agree pretty well—you work to get success because you think educational success and eminence will make you happy. But it won't alone. I have felt that there is a tremendous enhancement of pleasures in friendship and I think I, at present, do not look much further ahead than to a super friendship with a wife someday. I wonder shall I have the ability to pick a winner. If I don't I shall be in the soup rather—my hobby is not beer.

Pawsey was clearly looking for something more permanent. In 1933, he began a more serious relation with an acquaintance from Melbourne, Helen Borland, now teaching in Edinburgh.[12] But this did not last either (ESM 8.1, Helen Borland), as he informed his parents in letters that directly addressed his views of marriage and sex (complete extract in ESM 8.1):

It is quite hard enough without the usual restrained wording. I have, tentatively at any rate, decided not to give you my own views on the subject in general for fear of rubbing you up the wrong way, even though I am an idealist in the matter . . . I believe that the most effective way to discuss sex is to try to avoid possible innuendo by making all statements as precise as possible. Loose wording suggests more than a bare statement in the same way that a naked person is less sexually exciting than one in semi- transparent drapery.

Courtship (1933–1934) and Marriage (1935) to Lenore Nicoll

In October 1933, Ted Nicoll's older sister Greta Lenore Nicoll (1903–1974) was on a tour of Europe[13] and the UK.[14] Up to this time, Joe had already met Ted's mother, Mabel Edwards Nicoll (1876–1971), and Ted's brother Hastings (1916–1943). Lenore wrote in her diary:

Joe wrote his mother on 26 October 1933 with a first mention of Lenore to his parents:

Last Friday week Ted Nicoll roamed over and announced that he was off to London as his sister was coming over from Paris. So I gathered up the odd excuse and went down the next day on his motor bike (we only registered one motor bike this quarter and I use his). We, that is Ted and Lenore (his sister) and another Canadian girl [Kate Neatby] and I went to a show. The next day we went to a service in Westminster Abbey and I returned in the afternoon.[15] Such services are singularly cold lifeless things. Lenore came back [to Cambridge] with Ted and is staying at his digs. She intends to stay for a month or so. She is or was a teacher, her health was not too good so she chucked it and came over here for divertissement. She is now whiling away the idle hour by learning typing and shorthand. She is older than Ted, 27 [in fact 30] I guess, very nice but I do not think quite up to Ted's standard.

---

[12] Her father, Dr. William Borland, was the pastor of the prominent Presbyterian Scots' Church, on Collins Street in Melbourne.

[13] She had traveled from Southampton by train to Paris, then toured around France before returning to Paris in mid- October.

[14] Lenore Nicoll diary.

[15] Pawsey implies the meeting was on Saturday 16 October 1933, clearly incorrect by two days.

The next mention of Lenore was made by Joe on 1 November 1933, when he announced to his parents that Lenore would stay longer than "a month or two": "Lenore Nicoll is staying here indefinitely. Supplies quite good company about the place—quite an addition to Cambridge life in fact."

We owe a coherent narrative of the developing relationship to an account Pawsey sent in a long letter written to his parents on 29 August 1935. In this letter he summarised Lenore's history. Lenore had been the first girl from Battleford, Saskatchewan, to go to university and for many years was known as the youngest female graduate of the University of Saskatchewan. ("Took degree youngest,", his letter stated).

> Taught school at various places till 1932 Christmas. Owing to some obscure illness, probably run down, took a year off. Stayed with friends in Eastern Canada [Ontario] and in September 1933 [arrived Southampton 24 September, having departed from Halifax, Canada, on the SS *Westernland*] drifted over to Cambridge to live with Ted a while.

He added: "During the rest of the year Ted and I and a few others implanted in her the general outlook which she should have picked up in university if older. I saw an awful lot of her but reiterated to myself that I was not interested emotionally in such effect that I believed it."

As the busy Christmas season approached with several parties organised by Lady Frances Ryder, Joe and Lenore took part in several events (letter 5 December 1933):

> I went down to London last weekend for some Lady Frances Ryder parties. They culminated in a gorgeous dance at the Goldsmiths Hall. There were three of us from Cambridge from the Cavendish. Bob Chipman,[16] Bill Henderson and I and Lenore Nicoll. We all stayed at the old place in Kensington Garden Square so we had a merry time. After my experience two years ago when Ted and I at a similar dance were almost the only ones in dinner jackets we decided to hire tails for the occasion and all did so and appeared in full grandeur.[17] Went to bed about ¼ to 5 and came back here the next day feeling somewhat forlorn. I am intending to stay up here over the Xmas vacation. The idea is to do some work.

In contrast to the Christmas sessions of 1931 and 1932, Pawsey did not go to a home of one of Lady Frances's friends. He, Ted and Lenore stayed in Cambridge, cheered on by a Christmas shipment from Australia, arriving just in time (letter to his parents on 28 December 1933):

---

[16]Born 1912, Winnipeg, Manitoba, died 2008, Chester, Pennsylvania. Degrees from University of Manitoba, McGill and PhD, Cambridge, 1939, where he worked under the supervision of Ratcliffe. During his career he worked at Acadia University, Queen's University, McGill and Toledo University.

[17]Ted Nicoll wrote his family on 15 December 1933 (Nicoll Family Archive) with a humorous description of the mix-up, after being told initially that a dinner jacket was the dress code. ". . . [W]hen Pawsey and I arrived in dinner jackets we and three others were the only ones in them, everyone else of the 150 men had full evening dress (tails) . . . [W]e were sort of 'black sheep'—as we all had black ties instead of white . . . I don't know what effect it had on Lady Frances but at any rate tried to forget about it and had a very good time. We danced from 09:30 to 02:30, the hall can hardly be described it was so beautiful . . ." Joe and Ted accompanied the Ainsworth sisters from Sydney.

I received your "box of happenings" on Xmas eve. I was coming home with Ted and Lenore from the Mears so we three came along here and unpacked it. There were a most intriguing lot of things inside—I expect you know most of them, crystallised pineapple, tomato juice, tinned meats etc. Ted and Lenore were batching at the time so we used some of the stuff for Xmas dinner, tomato juice, turtle soup etc. I am going to keep some of the stuff for camping days, give some to Peggy Mears and some to my landlady Mrs. Long. It was a jolly good collection.

On 9 May 1934, Joe explicitly mentioned Lenore for the first time in 1934, after his mother had asked about her in an earlier letter:

PS Answer to a question. Lenore Nicoll is a **Canadienne**. Ted Nicoll's sister. I went to my first motor bike trip [in 1932] with Ted and share digs with him now.

On 23 May 1934, Joe wrote his parents a long letter describing a long Whitsunday (19 May) weekend trip to Stratford and Oxford. Joe was trying to persuade his parents that Lenore was another passing fancy:

The next day [20 May1934] I went on to Stratford-on-Avon to join Ted and his fiancée and Lenore (Fig. 8.2). [Lenore met the group there; she had already begun her long 4 to 5-week bike tour of North Wales and Ireland, see below.] ... I suppose you have been wondering who and what etc. is Lenore Nicoll. She is Ted's elder sister—a school teacher at home— over here because she liked the idea—living on her savings—cheaply like we all do. She is quite good company—**but I had best kill the romance you are weaving into it all now—I do not anticipate you having her for a daughter-in-law.** [our emphasis].[18]

On 30 May 1934, he wrote to his parents: "Lenore has gone on a cycle trip through Nth Wales and has gone to Dublin. Rather game don't you think?" On 20 June 1934, Joe reported again to his parents about a recent trip to Cambridge:

Last weekend I went up to Cambridge. Ted was taking out a PhD [about 15 June]. Lenore was returning from her long cycle trip in Ireland and England. She started off from Cambridge, went to Stratford where I told you Ted and I met her one weekend [20 May] and then went on through Wales to Holyhead, took boat to Dublin, went down to Killarney by train, returned by bike and then went up to Gretna Green on the way home. It was an amazing trip to do alone.

On 4 July 1934, Joe had a confession for his parents: "I had a jolly fine weekend with Lenore walking round the Isle of Wight last weekend. [1 July]. I suppose you are dutifully shocked."[19]

On 11 July 1934, there was more: "We get our yearly holiday in about 3 weeks' time—one miserable fortnight. It seems as if I shall be going over to the Continent with Ted and Kate Neatby and Lenore. We are thinking of taking a car over."

---

[18] This sentiment was similar to a letter written at the end of March (31 March 1934) as he "came down" (ie, left Cambridge as a student) moving to the new job at EMI. Lenore was not mentioned explicitly as Joe tried in a half-hearted manner to convince them that he was still a "free man": "I am not thinking of getting married—I mean I have no one possible in view which is a condition of affairs which may not last indefinitely ...".

[19] In this letter Pawsey also detailed his discomfort with Helen Borland's negative impact on his self-awareness, see ESM 8.1, Helen Borland.

Fig. 8.2 Joe Pawsey, Lenore Nicoll and Kate Nicoll at Stratford on 20 May 1934. Photo taken by Ted Nicholl. Credit: Joe and Lenore Pawsey Family Collection

Joe Pawsey, Lenore and Kate on Stratford hike

Events took a new turn as Ted and Kate made a major decision in the next days of late July 1934. They announced that they would marry on 2 August in the registry office in Cambridge with witnesses Lenore and Kate's brother. The new couple would not participate in the European trip. Joe wrote to his parents on 1 August 1934: "Ted and Kate are getting married tomorrow and going to Cornwall so Lenore and I are all that is left of the Rhine party."

On 12 August 1934, Pawsey wrote a postcard to his parents: "Have come via Oberammergau and the Austrian Tyrol. All going well. The car giving no trouble. Have climbed about 4000 ft. this morning. We are now going on by Zurich to the Black Forest and then to Havre where Lenore sails for Canada and I to Southampton." On 29 August 1934, more details were sent. The highlight of the two-week trip to Europe was the Passion Play at Oberammergau in Germany, "A wonderful moulding of an old play in a modern, ultra-simple stage setting. In particular the blending of bright colours was very fine."

Lenore returned to Canada at the end of the trip. Pawsey explained:"Previously to this trip she had accepted a position in a little high school in Lashburn, Saskatchewan [110 km west of Battleford as an English and French teacher[20]]." He concluded: "Both our boats sailed the next night about midnight [20–21 August]—mine a packet to Southampton—hers the *SS Ascania* to Montreal and as I said before I did not enjoy the parting."

In the omnibus letter of 29 August 1935 to which we have already referred above,[21] Joe also revealed the story of the turbulent development of their relationship in 1935. Both Lenore and Joe had major episodes of doubt as they decided their future. This letter (a typed letter from Joe to his parents), provided a series of "telegraphic style" brief phrases that provided a coherent summary of the difficult exchanges between Joe and Lenore starting in August 1934. Here is the sequence of events:

As Lenore left from Le Havre on 20–21 August 1934 to return to Canada after almost a year in Cambridge with her brother, Joe sent a telegram to the ship signed "Love, Joe". On 23 August he continued: ". . . Feel as you do. Love, Joe." On the same day she replied from the ship: "Thanks and love."

The next telegram exchange began the following May (12 May 1935) as he sent a telegram to her in Saskatchewan: "Ted recovering well. Letter following answering letter in favour your visiting England." She replied 31 May 1935: "Sailing *Athenia* June 29 if possible otherwise July Love Lenore."

Then something went wrong. On 2 June there was no mention of "love". Lenore succinctly wrote: "Visit cancelled. Lenore." We do not have more detailed knowledge of her doubts, or his, during this time.

However, after almost two months, on 25 July, Joe issued Lenore an ultimatum: "Missing you. If ever coming come now. Trip or stay [at home in Canada?]. I pay fare. *Empress of Britain* sails Montreal August 3. Telephone me daytime Southal 2468 'personal' if doubtful otherwise telegram reply paid."

And Lenore decided to come to the UK: "25 July To Pawsey. Hounslow. Sailing 3 August. If you think one week in England worth price of fare. School starts Sept 1 anxious to see you cable immediately Lenore." [That is—she would have to return immediately to Canada to begin the new school year.] On the same day Joe replied: "Come want you stay. Meeting *Empress Britain*. Cabling money Joe."[22] On 29 July 1935, Lenore confirmed: "Sailing *Empress of Britain*." On 1 August, Joe replied to

---

[20] She remained in Lashburn until May 1935.

[21] The letter was sent via airmail, a recent innovation for post from the UK to Australia. On the same day, Joe wrote his parents a letter via the conventional "sea post": "I have completely exhausted tonight on a long air mail letter announcing my approaching marriage to Lenore. Result 11.40 pm. Hence may I retire from the letter business?"

[22] On 26 July 1935, Pawsey wrote the National Bank of Australasia in London with a request to send a sum of money to "a lady in Canada with utmost efficiency." He wanted the transfer to occur by the next day. The funds were to be used for "rail and steamer fares" from Canada to the UK. Pawsey hoped that the transaction could be organised by letter but was willing to come in person to the bank.

Lenore on the *Empress of Britain*: "Wanted to see you so badly seemed futile act otherwise. Meet Southampton."

During this period, Joe had kept his parents in the dark about the turbulent events in his relationship with Lenore. Longer extracts from the correspondence discussed here can be found in ESM 8.1, Lenore Nicoll. The first mention of Lenore's plan to return to England in 1935 appeared in a letter that he sent on 31 July, after Lenore had set sail. His information was written with a level of ambiguity:

> The chief news this week is that Lenore is coming over for a summer trip. You may spend a lot of time guessing what the ramifications of this may be. That is what I am doing at present. However we shall see . . . My summer holidays begin the day after tomorrow. I do not know what I shall do except that I am going with Lenore and by car. Probably I shall remain in England. It is quite a time since I had an English holiday.[23]

Joe then confessed to his parents that some of the fault lay with him: "You will gather from the above a tale of rather unfortunate vacillation on my part."

We now return to the omnibus letter of 28 August 1935 for what Pawsey called "the next act", which:

> began with the commencement of my holiday August 3 [1935] when I went to Oxford to stay with [his cousin Frances Lade Ward] for a week [at her home] before meeting the *Empress* at Southampton. From Southampton, Lenore and I came up to London to dump our luggage with Ted. Incidentally, neither Ted nor Kate knew of Lenore's visit before our appearance. We then went on to Oxford and stayed the night with Frances before setting off the following night in the Fiat[24] for Scotland. We got as far as Gretna Green [in Scotland, famous for runaway marriages] the first night though without any serious result [ie, no marriage!]. The rest of the time was taken up by a rather fevered trip up through the Highlands to the extreme corners, John o' Groats and Durness and a three day run down to the great North road. The net mileage was that I burned my boats and Lenore and I are to be married next Saturday week [7 September 1935].[25] I have been back about ten days now [at EMI]. Lenore is staying with Ted and Kate. We are busy looking for a place to live in the neighbourhood so that Kate and Lenore may do the feminine of fraternise during the day if they so desire. We have a small place in mind which calls itself a flat . . . We wish to live pretty cheaply and blow our money on other things when we wish to blow it. Our present choice is at 25/ a week, two-year contract.

Joe now provided additional information about the Nicoll family: Lenore was the oldest and Ted the next child. There were a younger sister Bessie Nicoll[26] and a younger brother Hastings Nicoll. Pawsey wrote:

---

[23] On 7 August 1935, Pawsey told his parents he would meet Lenore the following day. "I shall give you more information on this when details are available. At present they are not."

[24] Pawsey's car called "Lucresia", bought for £35 in July 1935.

[25] In a letter from Pawsey to his family after their marriage on 27 September 1935, he admitted to his parents: "[In the evening of 19 August after our return from Scotland], Lenore and I had a most nerve wracking discussion on this problem of getting married . . . The occasion was one of about the highest nervous strain[s] in the course of those hectic weeks."

[26] Pawsey had met Bessie the previous month while she toured Europe (letter from Joe to his parents 24 July 1935). "Last Sunday I went up to Cambridge with Ted . . . and Bessie Nicholl, his other sister. She is an exceedingly nice girl. While up there we said goodbye to Bill Henderson who sails for Canada tomorrow. He is the last of my real friends to leave Cambridge."

We are getting married right away because Lenore has come over here from her work and is quite at a loose end till she is married. So we decided to get a house and have it fixed in a rudimentary fashion sufficient to just live in before getting married. I do not believe in its reality. It is not a case of getting excited about it, surprising little so in fact. We intend to get married in a local registry office. If possible we shall have Ted and Frances [Lade] as witnesses and to give us the necessary push off. I have arranged to have Saturday, Monday and Tuesday off. I am not sure yet where we shall go for a brief honeymoon. PS. Difficulty with nomenclature. Lenore knows me by the name of Joe.

On the day before the wedding, Joe wrote a short letter to his parents:

Tonight completes Book I of a life in the twentieth century. The individual chapters so far have been fairly good. It is hard to predict what will come. Marriage—War threatening—a century of social revolution. Work in London—Australia—or where. I have not time to really write.

We got your telegram and appreciated it very much—not so much your sending a telegram which I expected but I liked your wording very much. Lenore did too. I think she felt that somehow it conveyed that she was being really welcomed—that you were trying to look on it from the point of view of acquiring a daughter and not losing a son.

Finis.

Then next day Pawsey sent a short letter describing the next step, their marriage.

Book II 7 September 1935.

Married at Uxbridge Registry Office 12.15 pm [Ted Nicoll and Francis Lade—soon to marry Eric Ward—were the witnesses]. Now in train to Lyme Regis. Lenore looked fine in new dress for occasion—extent of celebration.

Love

Lade

Just a note to tell you both how much I liked being included in the telegram you sent. It was a very lovely welcome into the family. Joe is a dear and I know we will be very happy together. I will write to you at greater length before long.

Love

Lenore Pawsey

The wedding had been held at the Registry Office on 7 September 1935; the new couple went to a coastal resort on the English Channel in West Dorset, Lyme Regis. They stayed three nights with a total bill of £3 18 6.[27] Two images of this day have been preserved in the Pawsey Family Archive (Figs. 8.3 and 8.4).

After the wedding Joe and Lenore moved to a new house, 14 Tudor Way in Hillingdon, Middlesex. They were setting up house with a bare minimum of household furnishings, including wedding gifts.[28] On 27 September, Joe continued the discussion with his parents about the "surprise" wedding. He appeared to be feeling somewhat guilty that the parents had not been informed earlier:

I was surprised at your great surprise at my getting married. You know Lenore and I were pretty thick last year. Also having got her to come over and leave her job and having got

---

[27] Pawsey Family Archive.

[28] Letters from Joe to his family 11, 25 and 27 September. After the wedding, Joe wrote five letters in the rest of 1935. The rate of letter writing decreased in 1936 (eight letters) and in 1937 only three letters, all before Margaret's birth in April 1937.

Fig. 8.3 From their wedding day 7 Sept 1935. Credit: Joe and Lenore Pawsey Family Collection

Fig. 8.4 Off on the train on their honeymoon, 7 Sept 1935. Credit: Joe and Lenore Pawsey Family Collection

engaged it seemed fairly obvious that we should get married right away. Anyway we did it and I see no reason to question its wisdom.

Also there was to be another wedding in the Lade family; Frances Lade was to marry Eric Ward, an English farmer who had been on a visit to Australia and was now back in England. Likely he had met Pawsey's cousin Frances Lade during this visit. Ward had run his father's farm and was now taking over a farm near Oxford.

Family orientation and concerns were growing. In April 1935, Ted Nicoll had become seriously ill. He and Kate had just moved from "the doleful district of Hounslow to a place called Ickenham" to a much nicer district (Joe's letter to his parents on 27 March 1935). Joe reported to his parents on 24 April that Ted was quite ill, taken to the hospital on 23 April. "This means that Kate is left alone in their house which is not too good as she is about to have a baby. I had a try to get Frances Lade [Joe's cousin, a nurse] to come and stay with her but she is booked up." A few days later (30 April 1935), Joe reported that "[Ted's] ... malady is now diagnosed to be pleurisy and he seems to be in for a fairly bad time but should pull through quite all right." A week later (8 May 1935), Joe reported to his family again: "I have been living a rather hectic life recently in that Ted is sick in one part of London and now Kate has just had a baby [Brian Frederick Nicoll, born May 1935, died September 1936] in another part and is in a nursing home with it. Kate is OK but Ted is still pretty poorly."

Fortunately, Ted was recovering two months later (2 July 1935) and Joe reported that: "The Nicolls are now all well. Ted is back at work. The nipper rejoices in the initials B.F. Nicoll. However Ted assures me that Canadians don't know what this means."[possibly he suggested that BF signified "Bloody Fool"?].

On 9 October 1935, Pawsey mentioned in a letter to his parents that he and Lenore had been babysitting for Ted and Kate Nicoll, looking after their nephew Brian. "[He] is a magnificent spectacle—specimen I meant to say ...".

In January 1936, both Joe and Lenore wrote to the Pawsey family in Victoria with descriptions of their reactions to the death of King George V on 20 January 1936. Lenore went to London on the day of the funeral (28 January 1936) but saw only massive crowds of spectators. Joe went to the railway station at Hayes (on the Great Western Railway) and watched the passing funeral train. In late August 1936, Lenore and Joe went on a cycling holiday in Czechoslovakia, visiting Prague and Bohemia; this was to be their last holiday without children.

On 10 November 1936, Joe discussed Lenore's pregnancy in a letter to his parents (apparently they already knew that she was four months pregnant): "Lenore is now feeling very well again but there are odd symptoms. Skirts are requiring to be let out a bit. It is a queer business is it not? However, she is still quite strong." Lenore's younger sister, Bessie,[29] had arrived in July and had been traveling around England on bicycle. At Christmas 1936, she stayed with Lenore and Joe; all were to go to her brother and sister-in-law (Ted and Kate) for Christmas dinner. In 1937, Bessie started working as a librarian, apparently giving up her teaching career.

---

[29] As a typical Canadian, Bessie had joined a women's ice hockey team, the "Wembley Lambs".

By 25 March 1937 (letter from Joe to home), the impending arrival of the new baby dictated many chores. "My major job is the construction of a stand for a baby basket. I have made it of the dinner wagon construction with little wheels so it can be moved about with great ease. Lenore meanwhile does the frills." The doctor reassured them that the baby was ready to be born "in good order and at any time from now on." Margaret Lenore Pawsey was born on 16 April 1937.[30] Almost exactly two years later (20 April 1939), a son, Stuart Frederick Pawsey, was born while the family was living in Ickenham, Middlesex, London borough of Hillingdon.

Parenthood changes most of us; perhaps becoming a parent brought the worrisome international situation more urgently into Joe and Lenore's thoughts. In 1938 Pawsey was actively seeking work back in Australia. His correspondence indicates that even with the fear of impending war, he wanted to raise his family outside Britain.

Seeking Employment outside the UK

Pawsey might well have been confident in finding a position in Australia, as there had been occasional earlier interest in employing such an able researcher. For example, in 1935, Geoffrey Builder, who had worked with Pawsey at the Radio Research Board in Melbourne in 1930, wrote to him about a possible job offer In Australia. By 1935, Builder had moved to AWA (Amalgamated Wireless Australasia) as head of their Standards Laboratory and wrote to Pawsey on 12 February 1935, who was by then at EMI[31]:

> [To Pawsey] I heard a rumour that you are working for the Gramophone Company but that you might like to come back to Australia. If so, would you let me know by return air mail what sort of salary you want to come out here, and other details such as what you are doing now and have been doing recently. In fact, make it a formal application for a job as research scientist. There should be ample opportunities in this Firm and if you do want to come out here and take up the commercial side of radio, you could scarcely do better.

A few weeks later (28 February 1935) Pawsey wrote to Builder with a guarded response. He explained that he was involved in television research. He provided a short summary of his work environment:

> [To Builder]: I like the work and the men with whom I am working who are an extremely capable crowd. Also I am gaining valuable experience both in classical wireless and the new art. So I feel that I should not wish to leave here unless I were offered a considerable increase in salary, say to about £900, together with reasonable holidays and a contract for a year or so. If this should appear high you will realise my position and in any case I thank you sincerely for letting me know of the position. If at any future date I should desire to return to Australia may I write you to enquire as to whether you have any position vacant at the time?

---

[30] Margaret was to die at age 40 in London, 20 December 1977.

[31] Builder correspondence, located in the Pawsey Family Archive.

I have been down from Cambridge a year now working on various problems connected with directional aerials, and feeders, and measurements on interference and field strengths at wavelengths of from 3 to 7 metres (100 MHz to 43 MHz). I had a bit of a grind at first in writing up the last of my Cambridge work on medium wavelength fading and lateral deviation which I published in the Cambridge Philosophical Society and on my PhD thesis, in the evenings. However that is satisfactorily concluded.

I am surprised to hear that you had left the Radio Research Board ... Anyway it sounds like an interesting job and I wish you the best of luck in your new venture ...

Circumstances had changed greatly by 1938, and Pawsey was very keen to move. On 21 September 1938, Pawsey wrote[32] to David Rivett enquiring about a possible position with the CSIR in Australia, in particular with the Radio Research Board. Pawsey reminded Rivett that he had worked with Laby of the University of Melbourne and Munro and Huxley of the RRB on Atmospherics. He reported on his PhD project with Ratcliffe and then the position at EMI:

You may know this company as the builders of the BBC Television Station, London ... I have always intended to return to Australia and though I have a very interesting job over here with a salary of £ 500 I feel I would like to make the break now and take my wife and my baby daughter to Australia to make a home. If ... you have any vacancies on the RRB or know of any other positions which you think might suit me I should be very grateful if you would let me know.

On 6 October 1938, Rivett responded to Pawsey. He had sent a copy of the letter to Prof J.P. Madsen, Chairman of the RRB, asking whether there were any openings:

With regard to other opportunities in physics, I am afraid that the CSIR cannot offer as much as we had hoped to be in a position to do at this stage. Everyone seemed satisfactorily for an excursion by us into chemical and physical problems associated with secondary industries; but unfortunately the necessity for very heavily increased expenditures on defence preparations has led to a reversal of policy and an instruction to us to delay our programme indefinitely. We are proceeding with the erection of an Aeronautical Research Laboratory [in Melbourne] and also with a Standards Laboratory [in Sydney], but this will, for the time being, be the full extent of our venture into applied physics and engineering. I doubt whether in either of these two laboratories there will be any post of interest to you. However, we advertise all openings and I shall ask Mr. Cook [Assistant Secretary of CSIR] to send [any future relevant] notices ...

Rivett wrote Madsen in Sydney (8 October 1938) with a cautious message about Pawsey: "It is just possible that he might be a useful addition at some time to our team." But nothing was likely in October 1938; a year later the prospects were to change rapidly as war approached.

Pawsey also had written earlier (22 September 1938)[33] to his previous University of Melbourne advisor Prof T.H. Laby a revealing letter about his future plans. He was asking Laby's advice about a possible position in Australia. He was pleased that the EMI group was composed of a number of his colleagues from Cambridge: "The laboratory [working on the new television system] is almost an old Cavendish club; there are at least 8 men here who were research students at the Cavendish in my

[32] NAA AH 8520 PH/PAW/1 Part 1.

[33] Pawsey Family Archive.

time." Pawsey was proud of the group's achievements at the high frequencies of 180 MHz and 400 MHz; it was possible to determine the impedance of the system with a precision of a few per cent. At these high frequencies the ionosphere played no role in the propagation characteristics ("no evidence of skip distance phenomenon; I imagine that the signals are due to some sort of scattering in the atmosphere. But we know of no observation evidence to elucidate the point.") There were also major questions of the choice of polarisation; the evidence was that the standard use of vertical polarisation for television transmission was less efficient than horizontal. Major questions remained between the behaviour over short distances and long distances (greater than 40 miles).

Pawsey wrote to Laby concerning his satisfaction with an industrial environment:

On the whole I have found the industrial work here quite interesting. There are advantages which are the way in which one tends to half do a job and then be rushed off to another and also in that we are not allowed to publish anything except on very rare occasions. [They were allowed to publish a paper on the Alexandra Palace for the purpose of "advertisement".] . . . I now feel that I want to take my small family out to Australia to make a home so I am thinking of resigning here and looking for a job in Australia. Do you know of any positions going which might suit me? The sort of thing I was thinking of was a radio research job. I do not think I should take on academic life; I have been away too long. If there is any prospect of television developments at home I should very much like to get in on that. If I were to apply out there would you please be so kind to allow me to give your name as reference. When I do manage to get back to Australia, whenever that may be, I am looking forward to seeing you again.

There is no evidence that Laby responded. On 17 October 1938, Rivett wrote mentioning that the New Zealand Radio Research Board might have a position for Pawsey, who replied gratefully, but said he was reluctant to pursue the uncertain position in NZ while he had a "good job" in the UK.

Nonetheless on 31 October 1938, Pawsey wrote[34] Prof P.W. Burbidge of Auckland University College Physics Department (of the University of New Zealand) about a possible position "on short wave work" with the NZ Radio Research Board. Pawsey had met Burbidge in the summer of 1933 at Cambridge: "You may remember me as the excessively tall Australian, a research of Ratcliffe's with whom you played cricket for the Cavendish." He was explicit to Burbidge:

I have always intended returning to Australia at any rate in time to educate my baby daughter in a country free from class prejudice. Recent events have made us wish to accelerate the move in the hope of removing my wife and baby from the battle zone. I presume you can recommend New Zealand as a place in which to make a home . . . If there is this job I should be very much obliged if you could give me an idea of conditions of work, prospects etc.

Burbidge replied on 19 December 1938 with discouraging news: "[I] regret that at the immediate present, there is no opening under the NZ RRB, since we have entered into negotiations for the services of a man in the short-wave branch." Burbidge listed the areas of research in NZ: (1) survey of field strength of broadcast stations, (2) ionospheric work in Christchurch and Wellington under the direction of

---

[34] *Ibid.*

Professor F.W.G. White [Fred White, formerly a student of both Ratcliffe and Appleton] of Christchurch and (3) short-wave work at Auckland. He did, however, hint at the likelihood of positions in the future due to "strategic reasons" [presumably the approaching WWII]: "[t]here are possibilities on the horizon."

Pawsey wrote back on 18 January 1939. "I [think] I should be most useful in connection with television, ultra-short wave measurements of directional aerial systems since this is the most recent [work I have done]." He also pointed out that he was still interested in additional research on the "degree of horizontal irregularity" in the ionosphere, a continuation of his PhD thesis research.

On 13 April 1939,[35] when the outbreak of war was more imminent and Pawsey must have been feeling increasingly anxious, a fascinating correspondence began with C.W. Hansell of Radio Corporation of America (RCA) Communications, Inc. of Rocky Point, New York (engineering department).

On this date (13 April 1939), Pawsey wrote Hansell with a request to be considered for a RCA staff position (at RCA at Rocky Point, New York). He referred to their 1938 meeting at EMI. Hansell had sent photos of the RCA installation on the Empire State Building for the new television system in New York. Pawsey was impressed with the severe weather conditions the aerial had to withstand (elevation about 450 metres) as well as "the almost complete absence of reflected waves on the feeder over a very great range of frequencies ..." Pawsey mentioned that Ted Nicoll had now joined the RCA research staff in the receiving cathode ray tube section in New Jersey.

> His going to the States has been encouraging my own ideas about leaving England. I am an Australian ... I have always intended getting out to the Dominions, or the States, before it was time for my small daughter to go to school, I have never adjusted myself to the English class system ... However, international affairs today suggest that it would be good thing to get my small family out of England before ill befalls. What would you say, are the possibilities of my getting a job on the RCA staff? ...

On 16 May 1939, Hansell replied. He gave Pawsey the names of five RCA sections and one NBC (National Broadcasting Company) division which might need a man of Pawsey's talents. Hansell thought the best opportunity would be the antenna division of RCA manufacturing. His own division seemed unlikely since he had not had any new colleagues in the last 10 years. The main obstacle was the requirement for US citizenship: "the reasons for this are probably that RCA does considerable confidential work for the Army and Navy and that some of its activities, such as broadcasting, are particularly sensitive to political pressure." Hansell also provided Pawsey with the names of five additional companies that he might contact (e.g. General Electric). He ended the letter with a description of the status of television in mid-1939. Regular transmissions had started on 30 April 1939. Receivers were being sold; "our main difficulty technically is multipath transmission, which is very bad in the metropolitan area of New York. It is not so serious in

---

[35] *Ibid.*

outlying districts so that it may be capable of giving good service to several million potential viewers."

Several months later (6 August 1939), Pawsey replied to Hansell. Pawsey had "got wind of a likely opening elsewhere [likely in Australia] and was hoping to hear something more definite before replying to your letter. However, I am still waiting which leaves the situation much as it was when I wrote you . . ." Pawsey described in detail the second television station in the UK, at Birmingham with links to London at 170 MHz by cable or radio links. There was much discussion of methods to improve the picture quality.

Pawsey expressed his misgivings about the signs of war in Europe:

International politics are still as crazy as ever. There is a [cheap] book on the subject which appeals to me. (*Why War?* By C.E.M. Joad—English philosopher and broadcast personality). The author states his faith in human rationality very aptly.

> "I maintain that man is rational in the sense that, if a proposition is true, and if it is presented to him often enough and persuasively enough, then, though he will reject it again and again, he will in the end accept it, and when he has accepted it long enough, he will begin to act upon his acceptance."

On 15 December 1939, Hansell continued his correspondence with Pawsey. He never succeeded in getting anyone at RCA to be interested in hiring Pawsey: ". . . [I]t is somewhat difficult to get them interested enough to offer a job to a man in another country when they are receiving application [sic] constantly from good men closer to home."[36] Hanson sent Pawsey a copy of a letter that one of Hansell's colleagues (Beverage, a Chief Research Engineer at RCA had written a colleague at another RCA lab in New Jersey) on 7 December 1939 trying to sell Pawsey's services to another branch of RCA[37]:

> Mr. Hansell had met Pawsey [in 1938 at EMI in the UK] and was very much impressed by him and we would have liked to have employed Pawsey very much but since we are an international public service business, one of our requirements is that all of our employees must be American citizens . . . Pawsey is an exceptional and outstanding engineer in the field of antennas and transmission lines and is just about the type of man you have been looking for.

Nothing came of this effort.

In the letter of 15 December 1939, Hansell continued as he expressed concern about the early course of the war (this was before the Battle of Britain 10 July 1940 to 31 October 1940) and the Blitz (early September 1940 to mid-May 1941):

---

[36] Pawsey Family Archive. Hansell also discussed the sudden cessation of television developments due to the war in 1939. "We understand that British television has been discontinued, due to the war. Ours is operating a few hours a week but the public hasn't taken to it in a big way. That may prove to be fortunate in the long run for I believe television and all other services on frequency above 30 MHz will eventually use frequency modulation [FM]. Therefore, before the television standards are frozen, the system should be changed to use [FM]."

[37] From Pawsey Family Archive: letter sent by H.H. Beverage, RCA Communications, Inc. to a colleague (Paul Godley of RCA in New Jersey, US), 7 December 1939.

Naturally the progress of the war is a subject of considerable interest here. Sentiment in the US is overwhelmingly in favour of the democracies against the dictatorships. Feeling against Russia invading Finland is very strong, no doubt especially so because as we understand Finland was the only country which had been paying its debts to the US contracted [in WWI]. Over here some of us have been predicting that, in the end, if Russia is at all successful as an aggressor, England, France and Germany will forget the quarrel between them to deal with Mr. Stalin. Our own "reds" and "pinks" have found themselves badly confused since Hitler and Stalin became pals . . . No doubt the war has increased the need for men with your qualifications so that keeping employment is no problem for you. I hope it has not at the same time made living too uncertain and difficult, or separated you [from] your family.

Pawsey's reply to this letter was not sent until February 1940. As we will see in the next chapter, this was because, at last, it was sent from Sydney where he had a job working on WWII radar at the Division of Radiophysics at CSIR.

Part III
WWII 1939–1945

Chapter 9
Pawsey's Role in Australian Radar Research in World War II, 1939–1945

> There has been through the ages, always existed [sic] a vital interplay between war and contemporary scientific discovery. The more highly organised the world becomes, the more drastic the adjustment necessary to absorb the impact of new techniques.... the nature of radar embodied a battle of wits, not only between fighting men, but also between contending scientists, at an intensity not previously experienced. It involved a new level of sophisticated skills, and created its own new industry. All at once, physicists with electronic training found themselves involved in warfare in a new way, with an enhanced standing, in the nervous strain of conflict. Evans (1970, p. 232).

Joe Pawsey played a critically important role in the development of radar in Australia. His leadership contributed to the success of the Council for Scientific and Industrial Research, Division of Radiophysics—RPL—in 1939–1945. More than anyone else at RPL, he exemplified, and cultivated in the scientific staff, the combination of practical, engineering expertise and know-how, with a thorough understanding of the physical principles of the radar equipment that underpinned the Australian achievements across the war years. His ability to navigate personalities and social systems constructively was equally critical to RPL's successes.

There are several published accounts of Australia's wartime radar research program. These histories identify that RPL's success was founded on the decade of work undertaken by the CSIR Radio Research Board (RRB) (Chap. 4), under the leadership of Sir John Madsen, where Pawsey had begun his career. The successful research endeavours of the young RRB physicists such as David F. Martyn, F.W.G. White, A.L. Green, George Munro, Leonard Huxley, J.H. Piddington provided the scientific expertise needed for the radar research program. The instruments used in ionospheric research provided the basis for military radar after 1939. The most successful Australian defensive weapon of WWII was the Air Warning/Light Weight 200 MHz radar, planned and placed in operation in 1942. This system made

Supplementary Information The online version contains supplementary material available at [https://doi.org/10.1007/978-3-031-07916-0_9].

W. M. Goss et al., *Joe Pawsey and the Founding of Australian Radio Astronomy*, Historical & Cultural Astronomy, https://doi.org/10.1007/978-3-031-07916-0_9

major contributions to the Australian-US victory in the war against Japan in Papua New Guinea. Pawsey was a major informal contributor to this system.

In this chapter we provide a brief overview of wartime radar research in Australia and of Pawsey's role at RPL. Radar research involved most of the scientists who would enter radio astronomy when the war ended, and these figures are introduced here. We discuss the emergence of Pawsey's leadership, as someone who could respond constructively to difficult personalities—David Martyn's among them—and effectively manage the challenging liaison between scientists, the military and other government agencies. We also outline the technical challenges that the radar scientists faced. Many of these were due to Australian isolation, the absence of expected British support, and difficulties in accessing equipment and supplies. These social, geographical and technical features of wartime radar research experience would influence post-war radio astronomy.

Interested readers can access 8 chapters that provide an in-depth extended analysis of these topics, including some technical details that indicate how different ionospheric research instruments and questions influenced early radar research, in **the electronic supplementary material: ESM 9.1. Radar History; ESM 9.2, Radiophysics Laboratory 1940; ESM 9.3, Difficulties; ESM 9.4, Applied Science; ESM 9.5, Light-Weight; ESM 9.6, Microwave Radar; ESM 9.7, Golden Year; and ESM 9.8, Radar and Victory.**

Radar History: An Australian Perspective, 1930s

The story of radar during World War II has been told by many people and at length.[1] It's a gripping story: how such a swiftly developed technological innovation provided the slender margin by which the UK, and therefore the Allies, survived the Battle of Britain (Brown, 1999) and by which the Americans successfully waged the war in the Pacific. Radar is even sometimes exaggeratedly referred to as "the weapon that won the war".

The dominant story is centred on the creation of successive impressive weapons such as the Chain Home Link system, which Britain had ready in 1939 in barely enough time to make the margin of difference needed for the Battle of Britain; John Randall and Harry Boot's discovery of the cavity magnetron in 1940 in (Sir) Mark Oliphant's laboratory at Birmingham; and E.G. "Eddie"[2] Bowen's carrying it in secret to America to get it manufactured and to then help found the "Rad Lab" (ie,

---

[1]References used in this chapter include Bowen (1998), Brown (1999), MacLeod (1999), Jones (1978), Watson-Watt (1957), Hartcup (1970), and Watson, Jr. (2009).

[2]Bowen's nickname as a youngster in Wales was "Eddie". He later became known as "Taffy" as a nod to his Welsh roots.

Radiation Laboratory) at MIT.[3] This story is centred on the UK and the US (it is possible, but harder, to find accounts of radar in Japan, Russia, and other Allied and Axis nations).[4] Arguably an equally important version of 'the story of radar' is this: radar was more successfully developed in settings where open communication fostered innovation and enabled rapid innovation.[5] Australia was one such setting.

The story of radar in Australia, and the central role played by Joe Pawsey in its development, is different, and is one of adaptation and collaboration. This Australian perspective provides a form of "history from below".[6]

Radar: British Secrecy and Australian Developments, 1930s

Who discovered or invented radar? In his comprehensive, well-narrated account, Louis Brown (1999) points out that this is the wrong question. Whether the question is "who first thought of using echoes of radio waves to find such things as ships and airplanes", or "who first thought of how to create a usable device that could do this", the answer is, dozens of people (Brown, 1999, p. 456). Martyn, Green and others of the Radio Research Board in Sydney (Chap. 4), who "became aware of increases in signal strengths at their Liverpool receivers, when planes flew overhead", were among the many people around the world who observed radio signals being reflected from boats and aircraft in the 1920s and 1930s (Evans, 1973, p. 282). Naturally many of these observers realized that such echoes could be used for what would become "Radio Direction Finding" (RDF) or "RAdio Detection And Ranging" (RADAR—the acronym developed by the Americans in 1940).[7]

David Martyn and Jack Piddington were aware of early British radar research in the mid 1930s. But the British kept this work an official secret. Piddington, then working with Appleton in the UK, wrote Martyn on 22 July 1937: "There is another job looming, which is hard to write about. It concerns Watson-Watt's present

---

[3] Bowen established close connections with key scientists in the US: Vannevar Bush, Lee DuBridge, I.I. Rabi and Alfred Loomis.

[4] We found Louis Brown's *A Radar History of WWII: Technical and Military Imperatives*, 1999, Institute of Physics Publishing, Bristol and Philadelphia, a superb account, including important detail about radar in Axis nations, but still very centred on the US. Brown's analysis is incisive and convincing.

[5] As Brown (1999) details so exhaustively, secrecy and hierarchical institutional structures hampered both technical development and effective deployment, in Russia, Germany and Japan.

[6] Those interested in "history from below" will find some excellent precis of this as methodology by using google, and also by consulting works from E.P. Thompson's famous *The Making of the English Working Class* and Marcus Rediker's many prize winning books. The term has inspired many works and several disputes within academic history.

[7] See Brown (1999) page 83. Word invented by S.M. Tucker of the US Navy. The British used this term after 1 July 1943.

transmitter and our use of it as a blind for other users. This is strictly hush-hush, naturally."

The Australians tried to get involved early, with Piddington explicitly pursuing some early trials once he had returned to Sydney later in 1937. But British secrecy, and Australian habits of dependence on British lead, retarded their efforts. In reporting on a trip to the Radio Research Board in 1936 (written in early 1937), Martyn suggested that H. Wimperis (1876–1960), be consulted about the probable defence applications of low-region ionospheric research during his (Wimperis's) 1938 visit to Australia. Wimperis, Director of Scientific Research of the Air Ministry, had set up a committee of scientists under the chair of H.T. Tizard (1885–1959) to undertake new defence research, leading to early British efforts in radar. But Wimperis claimed to be surprised when Madsen raised Martyn and Piddington's guess at the focus of British defence research during his visit, and he made no disclosures. Thus the Australians lost precious time due to British secrecy.

It was not until February 1939 that the well-known invitation was issued for Dominion governments (Australia, New Zealand, Canada and South Africa) to come to the UK to receive details "of new developments in defence applicable particularly to air." David Martyn, the pre-eminent ionospheric physicist with extensive scientific networks in Britain, was chosen as the logical leader of the Australian radar program. He made a flying boat trip to the UK in March 1939, visiting defence establishments.

Martyn arrived back in Sydney on 7 August 1939. Within a short period the CSIR Division of Radiophysics (the name was intended to be non-descriptive, a camouflage term invented by Martyn) was founded on the University of Sydney campus. The building was to be completed in April 1940 as an extension of the CSIR National Standards Laboratory. Martyn was appointed as chief of the new division in September. The work of the proposed laboratory would fall into three categories: (1) Research, Development and Instructional (CSIR), (2) Construction, Installation and Maintenance of equipment and buildings (mainly the Postmaster General (PMG)) and (3) Operational, the three Australian Armed Forces: Army, Navy and Air Force (Civil Aviation and Meteorology would also be involved).

Pawsey's Recruitment to RPL

Pawsey was recruited to RPL in mid-April, 1939. His expertise on antennas at 40 MHz (from the EMI television research) was invaluable. Martyn had met Pawsey during his trip to the UK in March, as he reported to Madsen back in Sydney in a letter from 16 April 1939:

> I saw Pawsey yesterday. His work is on ultra-short waves with EMI and especially on aerials and feeders. He would be a key man on our scheme. He is keen to come to Australia—a salary of about £530 would get him if prospects for good research were seen—as they are. He has a contract [with EMI] for about another 18 months, but thinks it only applies to taking up a job with another company—not the Government. I suggest he be approached

immediately, and have suggested he compile a list of such special testing apparatus, etc., as would be required if he came to Australia on ultra-short wave work. It would be wise not to stress Defence work in approaching him, as he is a Methodist with pacifist leanings. He would make a good man in charge of a group—he is thoroughly sound and now experienced in production, etc.

It is interesting to speculate on what the moderate, pragmatic, and not especially political Pawsey might have said to Martyn, to produce this assessment of his politics! Pawsey was hardly a "Methodist with pacifist leanings." Pawsey later made his willingness to take an active part in defence activities explicit, in correspondence with Martyn and Madsen.

Pawsey was more anxious to find work away from the UK once war had been declared. The Pawseys' home was only about 25 km from the centre of London, and the likely target of German bombs. Lenore, Margaret (called "Mar" by her parents in 1939) and the baby Stuart went to the farm of Eric and Frances Ward, "Elm Tree Farm", Settrington, Malton, Yorkshire (30 km to the NE of York); Frances Ward, nee Lade, was Pawsey's cousin. Eric and Frances had been living there since the early 1930s.

So it was a relief when formal offer was made to Pawsey, by Martyn, on 22 September 1939, three weeks after the start of WWII. Pawsey resigned from EMI on 31 October. He was to spend a few months visiting Watson-Watt's group in the UK and also to purchase some electronics. Pawsey departed with his wife and two young children via ship on 22 December 1939 for the dangerous trip through the Suez Canal to the Indian Ocean on the way to Australia.

The day before departure, Pawsey provided Martyn[8] with a complete report of his activities prior to departure, including a carefully worded report on the problems of finding all the items on the "special apparatus" list.[9] "I have been traveling all over Great Britain as the work in which we are interested has been considerably decentralised since you were here [due to the war] but I have been able to see a little of most of it."[10] He had ordered a signal generator that worked from 150 to 300 MHz.

With the balance of the measuring gear money—about £200—I have bought various components with the idea partly to build definite bits of measuring gear in Sydney and partly to get a small stock of short-wave components for general experiment which may be hard to get in Sydney. Unfortunately, I am ignorant of facilities for purchasing such things so that I may be bringing "a few coals to Newcastle".

A major reason for the letter to Martyn was then apparent:

---

[8] Joe and Lenore Pawsey Family Collection. Joe made a common mistake addressing his new Chief as "Martin", instead of "Martyn".

[9] 21 December 1939.

[10] Joe and Lenore Pawsey Family Collection. In late November 1939, Pawsey submitted a travel expense report to Australia House with trips to Kettlewell (Yorkshire), Watchet (near Bristol), Christchurch (near Portsmouth), Woolwich (near Greenwich) and Southend (east of London) for trips on 11, 15, 20, 21 and 24 November. The total expense was £10.

Now to the principal point, to me, of this letter. My parents and various relations live in the vicinity of Melbourne and have not seen me for about nine years. Also I am bringing my wife and two children whom they have never seen. Consequently, I am very keen on having a week or two at home soon after my arrival. There are two possible ways of doing this. Firstly, that I could stop at Melbourne on my way to Sydney with the boat and then after seeing you and discussing things return to Melbourne to see my people . . . I look forward to seeing you . . . It seems a future full of interest.

Martyn replied that regretfully, the exigencies of wartime precluded a stay in Melbourne. "What I suggest you do therefore if you feel so inclined and do not mind the extra expense which you may incur is to leave the ship at Melbourne, stay a day or two there and arrive down at Sydney approximately at the same time as the ship does. I think we might turn a blind eye if you happen to be day or so late."

We do not know which option Pawsey chose. Likely the family did get off the boat in Melbourne and stayed a few days with the family before moving on to Sydney. In any case the family was back in Sydney in late January 1940[11]; Pawsey started work on Thursday 2 February 1940, a career with CSIR and CSIRO that was to last 22 years. He joined a research staff of 10, mostly former colleagues from the RRB.

An extended discussion of the development of radar in Britain and elsewhere, of the difficult relations between Britain and Australia that resulted from unwarranted British secrecy, and of Pawsey's reflections on leaving Britain, can be found in ESM 9.1, Radar History.

Radiophysics Laboratory, 1940–1941: Shore Defence, the T/R Switch and the Buggery Bar

Pawsey started work in early 1940 on a Shore Defence radar, in collaboration with the Australian army. At this time, the expectation was that the UK would develop and supply parts and designs, to be adopted and adapted by the Australians at the Radiophysics Laboratory. In this early stage of WWII, before Japan attacked in the Pacific in December 1941, the expectation by the Australian military planners was that any attacks would be conventional naval attacks by ships. The idea was to defend the main ports and population centres. A carrier borne attack was considered unlikely. Of course, the fallacy of this mind-set would be exposed in the Pearl

---

[11]Martyn was worried about Pawsey's knowledge of the security of the new Division of Radiophysics activities. He wrote Cook, Secretary of the Radiophysics Advisory Board, on 24 January 1940. "Pawsey should be in Melbourne quite soon now on the *Strathmore* and may be coming through by train. It is possible he is uninformed on the subject, who, in Australia, are initiated in Radiophysics matters and who are not. I think it would be a good idea if you could contact him before he visits Melbourne when VERBUM SAPIENTI SAT EST (a word to the wise is enough)." However, Cook was out of town and only arrived back in Melbourne on 5 February 1940.

Harbor attacks of 7 December 1941 and the Darwin attacks (by almost the same Japanese aircraft carriers) on 19 February 1942.

The first RPL shore defence radar was completed in March and installed in July 1940; tests were carried out at Dover Heights in Sydney in May 1940 at ranges of 3 to 8 miles. A major factor contributing to the success of Shore Defence radar (a "scanning" radar in contrast to the "floodlighting" system used in the Chain Home in the UK) was the collaborations established with Col. (later Major General) J.S. Whitelaw, the commander of coastal defence in the Army's Eastern Command. Whitelaw would remain a RPL supporter throughout the war.

This support was important because the biggest challenges at RPL were not scientific. Rather, they arose from the difficulties of liaison with the various branches of the military and with other government agencies. For example, manufacture of any parts or prototypes was designated the responsibility of the Postmaster-General's (PMG) department. An endless series of misunderstandings between Radiophysics and PMG resulted in lengthy delays in equipment production, hampering scientific work. One of the complicating factors was secrecy, preventing better collaboration across two good working groups. (Minnett in MacLeod, 1999, p. 424): "There were severe delays as the PMG and the RPL struggled within their individual areas ... to resolve the problems of production under conditions of rigorous secrecy." Pawsey made a frank remark to Marjorie Barnard in 1945: "[With the PMG contract], there were delays and mutual recriminations" (Barnard, 1946).

Pawsey's main engineering success of 1940 was the planning and execution of a Transmit/Receive Switch. Thus a single antenna (in place of two) could be used for a radar system; the T/R switch turned off the high power transmitter during the small time interval when the radar echo returned from the target. Pawsey and Harry Minnett had a successful version working on a single antenna at Dover Heights. In 1999, Minnett wrote: "[Pawsey] had a brilliant intuitive feeling for physics and a profound grasp of radiation and transmission techniques at ultra-short wavelengths. For the younger members of RPL, his knowledge more than made up for the lack of textbooks on the subject."

However, Pawsey had to learn to confront the gap between scientific development and end-user needs in relation to the equipment he helped invent and produce. The fate of one particular invention provides insight into the tensions and difficulties between Radiophysics and the Military.

In 1941, in order to carry out the experimental work on the Shore Defence system and, at the end of the year, the Air Warning (AW) radar then in hasty development, Pawsey developed an accurate impedance measuring device. This coaxial unit was devised and later manufactured for Army use by the PMG. The 200 MHz battery oscillator which energized the unit could be used to determine the aerial directional patterns and the approximate power gain.[12]

---

[12] In 1943, Pawsey and Kerr published an update to the problem of determining impedances, especially at higher radio frequencies. RP 163 was published 6 January 1943 "Connecting Networks Between Impedance Measuring Gear and Unknown Impedance".

The RPL Impedance Measuring Set served an important function in 1940, 1941 and 1942 as it was used by RPL personnel to match and optimise the Shore Defence and Air Warning radio direction finding systems. But radar mechanics in operational areas would need to match the AW and Shore Defence aerials. This required positioning the 36 dipoles and cutting the connecting cables to the correct length to match the system for maximum gain. Radar mechanics were provided with instructions in the form of a document prepared by Pawsey on 10 June 1941. "Concentric Feeder Measuring Equipment (200 Mc)"[13] was complex: 18 pages of text filled with equations, with a liberal use of hyperbolic sins and cosines, followed by the two figures above. The intended audience was clearly physicists and electrical engineers, rather than hastily trained radar operators. Pawsey was perhaps naïve to think they could master it.

As a result, many of the RAAF had major problems using the IMS, and using the instrument often baffled the radar mechanics. According to one memoir, "[t]he Impedance Measuring Set was colloquially and universally called the 'buggery bar', allegedly because Wing Commander Pither (ESM 9.4, Applied science), the irascible RAAF Officer then in charge of radar development as Head of the Directorate of Signals, exploded at a meeting with RPL scientists and said 'It is useless, you can't get within buggery of the required result.'" (MacKinnon et al., 2009). At some stations, a buggery bar was not even supplied.

In many cases, problems with using the "Buggery bar" were resolved by radar mechanics who had been radio "Hams", who resorted to their experience in maximising performance in short wave transmission (MacKinnon et al., 2009, p. 7). Pawsey was eventually able to confront the problem as the LW/AW system (ESM 9.5, Light-Weight) became prevalent after late 1942, with a new design: aerials with open wire transmission lines that required no adjustment of cable lengths for matching of the aerials, leading to an optimisation of the power transfer from the transmitter to the aerial.

The rapid need for Air Warning instead of Shore Defence that occurred in 1942 used the expertise developed in 1940, to produce the aircraft detection radars. Schedvin (1987, p. 251): "[T]he building of the ShD system yielded many of the skills necessary for the development of light-weight air warning (LW/AW) sets which played such a vital part in the later stages of the war."

Australian Isolation: Other Developments in Radar

Unfortunately, the Australian group was destined to invest heavily in technology that was already outdated before they could adapt it for the local conditions. Radio direction finding equipment was needed in aircraft, usable as pilots engaged the

---

[13]RP 96–1.

enemy directly. A Shore Defence transmitter antenna was tall and heavy; how could such a device be adapted to the dimensions needed for aircraft?

As is well known, John Randall (1905–1984) and Harry Boot (1917–1983), physicists in Australian expatriate Mark Oliphant's (1901–2000) research group at Birmingham University,[14] provided the answer in February 1940 with the invention of the cavity magnetron. This was a high-powered vacuum tube that generates microwaves from the interactions of streams of electrons with a magnetic field while moving past open metal cavities that produce a resonate frequency.

Although the cavity magnetron opened the possibility of building equipment in the dimensions needed for installation in aircraft along with higher resolution detection of smaller objects, this information was useless as the British did not have the capacity to manufacture a new weapon at scale. These circumstances brought into being the famous "Tizard Mission" in September 1940 (during the Battle of Britain), a trip to the then-neutral USA to offer a whole series of British military secrets in return for the US undertaking their manufacture and production. The delegation included UK military leadership along with Tizard and Edward (at that time, "Eddie") Bowen (see ESM 9.2, Radiophysics Laboratory 1940; ESM 9.3, Difficulties; ESM 9.6, Microwave Radar). Bowen travelled with the cavity magnetron—later termed by one US historian as "the most valuable cargo to reach our shores" (Baxter, 1946)—in a lead-filled box, designed to sink should their boat be torpedoed in the crossing. The Tizard Mission also brought with them designs for rockets, superchargers, Frank Whittle's jet engine, and the VT proximity fuse. They also carried the famous Frisch-Peierls memorandum (written by two German Jewish physicists likewise working in Mark Oliphant's laboratory), describing the feasibility of building an atomic bomb, which was given to Enrico Fermi in the US. As is known, the USA accepted the agreement. Further: within a month, millionaire physicist Alfred Loomis had brought the "Rad Lab" into existence at MIT, with the collaboration of Bowen.

Evans (1970, pp. 52–56) comments:

> All this rapid new development [in radar overseas] was by no means an unmixed blessing to the isolated Australian [radar] team. Although the potentialities of microwaves were immensely exciting, almost over-night Australia found itself way behind in the design situation ... The hunt had to be started all over again with the magnetron.

In 1941, this situation stimulated RPL to send Pawsey to the US to gain information about the magnetron (below). In addition, the agreements to obtain the latest radar equipment from Britain, made by Martyn in the UK in 1939 and Madsen in 1940, were now irrelevant—under siege itself, Britain had neither the materials nor the

---

[14] Randall would later go on to lead the King's College, London, team that worked on the structure of DNA, employing famous crystallographer Rosalind Franklin (1920–1958) who first observed the double helix; his deputy Maurice Wilkins (1916–2004) would share the Nobel prize for discovering this structure with Watson and Crick. Randall and Boot are often credited with 'discovering' the magnetron (and the story of how and why they did so is interesting in itself). But in reality, as had been the case with radar itself, many versions of magnetrons had been developed, by many different people, since 1910. Readers can find a summary in ESM 9.6, Microwave Radar.

human resources to supply the Australians. As Evans (1970, p. 52) summarised: "The original channels of communication arranged by Madsen on his previous visit to Europe were now largely outmoded." Scientific liaison in both the US and the UK became critical to the success of the Australian research program, but required stationing physicists with a strong background in electronics overseas, to learn from and transmit back major new aspects of radar research, such as the use of the magnetron, and cm wavelength radars.

Extended discussion of the development of the Shore Defence system, the excitement of the first local radar successes and the challenges of liaison and of the Buggery Bar (illustrating Harry Minnett's engineering skills), can be found in ESM 9.2, Radiophysics Laboratory 1940.

Difficulties at Radiophysics, 1941

At RPL, Martyn was the leader of a research team that consisted of a number of radio-engineers and physicists from the Radio Research Board. This included Prof Leslie Martin (1900–1983), Pybus and G. Brown in Melbourne. By March, 1940, the research staff included J. Piddington, J. L. Pawsey, H.J. Brown (also from EMI UK), O.O. Pulley and L.G. Dobbie from Australasian Wireless Amalgamated (AWA), George Munro (from the Radio Research Board), and electrical engineering graduates Victor Burgmann, G. Tangie, J. Warner, Ron Bracewell (1921–2007), L. Hibbard and Harry Minnett (1917–2003). Martyn and other members of the Radio Advisory Board kept an eye out for talented and qualified students. Frank Kerr, born 1918, completed his MSc in Physics at the University of Melbourne in 1940 and was immediately recruited to join the Radiophysics Laboratory, where he found a mentor in Pawsey. Among the new recruits were two talented female physicists, Joan Freeman (1918–1988) and Ruby Payne-Scott (1912–1981), then working for AWA (Goss, 2013). Both women would go on to have stellar careers in physics, Payne-Scott in radio astronomy and Freeman in nuclear physics. For both of them, wartime research provided an unprecedented opportunity at a time when their careers were severely limited by sexism (see ESM 9.1, Radar History). Pawsey's support for both was extensive.

By December 1940, the staff consisted of 65 individuals, including 27 research staff (all levels), 18 workshop personnel, 5 clerical staff and 4 "Commonwealth Peace Officers" (security staff). By June 1941, the staff had expanded to 41, and now included several new engineering and a few physics graduates, including Brian Cooper (1941 graduate). RPL lacked a hierarchical culture and physicists and engineers worked collaboratively. In Britain, physicists dominated; Watson-Watt did not want electrical engineers involved too early because he considered them more likely to be conventional. Given that the American and German radar devices were largely engineer-led, the evidence of history does not seem to support this concern. Indeed, later the British radio astronomer Hanbury Brown remembered being impressed by the better techniques of the EMI engineers (Pawsey's former colleagues) whom he encountered in radar work In the UK in 1939 (Brown, 1999, p. 461).

But despite the excellence of the personnel and the consuming, urgent work, RPL management was plagued by chaos and uncertainty. Martyn, who had generated often severe conflicts in many of his collaborations through the 1930s, had no interpersonal skills for management and exacerbated and inflamed many of the daily conflicts with other agencies such as PMG. Perhaps partly to smooth these troubled waters, Madsen decided to send Martyn to the UK in early 1941. Instead, tragic events ensued.

In late 1940, David Martyn began an unfortunate liaison with Mrs. Ella Horne, a German divorcee.[15] The Commonwealth Investigation Branch and Military Intelligence became aware of these events and concerned about potential security implications. Although their (markedly discriminatory) investigation cleared Horne of being a Nazi spy, military intelligence did become aware that Martyn had boasted about his position at CSIR and was clearly indiscrete.

When the indiscretion was made known at RPL, it was feared that the Defence Forces would no longer wish to participate in radar research if they considered the research conditions to be insecure. The Radiophysics Board discussed the matter at their meeting on 17 April 1941, having been informed by Military Intelligence one or two weeks previously and it was decided to send Madsen overseas for scientific liaison in place of Martyn. As Schedvin (1987, p. 258) put it: "Madsen must have judged that there was no one else with sufficient seniority to lead the team."[16] Madsen departed for New Zealand immediately on 25 April 1941. He stayed for 2 months in North America, later five months in the UK.

Martyn was humiliated, furious and bitter, and became implacably resistant to attempts to patch up relations. His reactions served only to convince others of his instability. As a result, throughout 1941 work at RPL had to proceed without clear leadership, working around Martyn amid increased distrust from the Military. We presume that Pawsey's quiet leadership—he maintained constructive relations with Martyn throughout the war years—was increasingly developed, and felt, at this time.

It was F.W. "Fred" White who found the path forward. Born in New Zealand, White had moved to the UK in the same year as Pawsey and began work as a Demonstrator in Physics with Sir Edward Appleton. There he became acquainted with Edward Bowen, before commencing a PhD with Ratcliffe at the Cavendish in 1934, the year Pawsey was completing his PhD[17]; the two men met there. White returned to New Zealand in 1937 as Professor of Physics at Canterbury College, University of New Zealand, Christchurch. Interestingly, he became friends there with another new Professor who arrived in the same year: the celebrated philosopher of science Karl Popper, whose Jewish ancestry and connections had led him to flee

---

[15] These events and their consequences are described in detail in the NRAO ONLINE.7 text.

[16] Martyn had also been criticised for his troubled relations with both the PMG (Postmaster General, the manufacturing arm for radar equipment) and the military services.

[17] See Fox, K. (2018).

his native Austria three years after publishing his famous treatise *The Logic of Scientific Discovery* (1934), which set out the radical theory of scientific progress through falsification (Gattei, 2008 and Jarvie et al., 2006). In Christchurch, while Popper published the passionate and influential work *The Open Society and Its Enemies*, White developed gunnery radar for the New Zealand navy with the outbreak of war.

In 1940 White had been invited to come to Australia to fill in for Martyn as Chief of Division during the planned trip to the UK. He arrived in March—now filling in for Madsen—and was "thrown blind into a maelstrom within a few weeks." But, as Schedvin put it (197, p. 239): "The unmaking of one man is often the making of another." White turned out to have all the social and emotional abilities for leadership that Martyn so lacked. He had an imperturbable temperament, a capacity for considerable attention to day-to-day operational details, and excellent communication skills. He engineered a major reorganisation of Radiophysics, in which Martyn's role, Chief of the Division, would be abolished and Martyn placed in a research-only role.

Conflicts between Martyn and colleagues at CSIR (later CSIRO) would continue for three decades—placing a significant constraint on the early radio astronomers, who badly needed a brilliant theoretician such as Martyn. Despite working at some physical distance from the rest of the scientific community, Martyn continued to carry out cutting edge and highly cited research in ionospheric physics through the rest of his often admirable career,[18] which also featured extensive efforts in support of scientific internationalism. But the traumatic events of WWII cast a long shadow. He would experience psychosis in the 1950s (Chap. 26) and tragically died from suicide in 1970. We provide the first complete analysis of these events in ESM 9.3, 1941: Difficulties.[19]

Scientific Liaison Overseas

Madsen's 1941 trip (end April to early December) to North America and the UK was successful as he organised the Australian Scientific Liaison Groups and met a number of key collaborators. The advent of the magnetron had led to increased complexity in overseas liaison. In Washington, he organised to bring George Munro from London to be the Australian representative in the US and Canada. Madsen met with colleagues at the Naval Research Laboratory and at the Radiation Laboratory at MIT in Boston. He cleared the way for Pawsey a few months later to visit research groups in the US and Canada for centimetre radar discussions.

---

[18] See Graham et al. (2020).

[19] The first exhaustive account of the Horne affair and its later tragic impact on Martyn—who became the subject of the first electronic surveillance conducted in Australia—is provided in NRAO ONLINE.7.

In the UK, Madsen met Oliphant as plans were initiated for a visit by the Australian-British scientist to Australia in 1942 (see ESM 9.6, Microwave Radar). Madsen wrote: "I found him quite interesting and full of enthusiasm and his lab [Birmingham physics] is turning out some excellent work, restricting itself to fundamental issues and passing over applications completely to other bodies." This visit let Madsen know that "[h]e now has well in sight the production of a megawatt magnetron . . . [The magnetron] has brought about a completely new form of technique as compared with radiation at lower frequencies."

In the meeting, Oliphant expressed considerable interest in Pawsey, whom he had known earlier at Cambridge.[20] Madsen wrote:

> I cabled immediately to Munro [in Washington] to make arrangements for Pawsey to meet him upon [Oliphant's] arrival in America. Oliphant has promised not only to give Pawsey all the information he possibly can upon micro-waves, but is anxious to keep in touch with him during his visits [in the US] to some of the important laboratories. This helps to make good . . . the disadvantage of Pawsey not being able to come across to England.[21]

In December, Madsen had a remarkable trip from Hawaii back to Sydney by flying boat, departing only a few hours before the Japanese attack on Pearl Harbour on 7 December 1941.

In July 1941, Pawsey was sent to the USA for scientific liaison, returning in the first days of October. He spent six-seven weeks in Boston and three-four in Ottawa. He also visited Bell Labs in New Jersey, meeting microwave engineering pioneers Harald Friis and George Southworth. In addition, he met Karl Jansky, who discovered radio waves from the Milky Way a few years earlier. Finally, on the way to the west coast, Pawsey visited his wife's family in Battleford, Saskatchewan (Canada) on 27 and 28 September 1941.

The leadership crisis in 1941 delayed the development of a successful air warning system. War with Japan was only a month in the future by the time Martyn's situation was resolved. Attacks on Darwin would begin in mid-February 1942.

Air Warning, 1941–1942: Applied Science and Wartime Bureaucracy

A major problem at the end of 1941 was the lack of interest in the Australian military to initiate action of radar for warning of sustained air raids. White had pushed for installation of three radar sets for air warning earlier in 1941. But even Madsen was

---

[20] Both Pawsey and Oliphant (who was from Adelaide) were 1851 Exhibition Fellows. Oliphant had received his PhD at Cambridge in 1929, became assistant Director of Research at the Cavendish in 1935 and departed for Birmingham in October 1937, thus overlapping at the Cavendish Laboratory in Cambridge.

[21] We do not know if this meeting between Pawsey and Oliphant took place in the US as intended.

doubtful of "concerning ourselves … with mass aircraft attack." RPL persisted that this neglect be addressed, playing a role in bringing about a change in outlook.

RPL was able to rapidly respond in the week after Pearl Harbour (8 December 1941 in Australia), providing makeshift air warning protection for the Sydney area in only five days. "There is nothing in radar history to compare with this feat for speed linking development to full production and then into action," Brown commented (1999, p. 221). Jack Piddington and colleagues (including Brian Cooper and Len Dobie) were able to modify the existing Shore Defence radar at Dover Heights, an Australian Army site. They sacrificed the range resolution and accuracy needed for gun laying for defence against ships to gain enough range and accuracy for air warning. They lengthened the pulse by a factor of 13. By Saturday night (13 December) they could detect aircraft out to about 65 miles. If they had used the original ShD system, this range would have required a power output of 100 kw, compared to the available 10 kw. This experimental radar was operated by army personnel and maintained by CSIR RPL staff, providing around the clock protection against air attack for many months. White and Madsen had been at the CSIR administrative offices in Melbourne and only were informed of the events on their return to Sydney.

After the December success at Dover Heights, efforts were quickly begun to manufacture 3 sets of the new system called AW Mark I, air defence mark I, for delivery in early 1942. See Fig. 9.1 One of these was to be shipped to Darwin. This shipping was badly mishandled -a major problem was that the AW aerial of 6 tons was not intended to be transported by air. An example of the confusion was that the dipole elements were loaded on the final flight. The result was the disaster of the air raids on Darwin on 19 February 1942, launched by aircraft carriers (four of the six Japanese carriers that had been at Pearl Harbour in early December). It was not until March that Brian Cooper and Jack Piddington arrived at the site and had the AW radar working within four days. On 22 March, a large raid was detected with 31RS at a distance of miles, the first Australian radar to detect the enemy. Soon the Kittyhawk aircraft of the US Army were successively intercepting the Japanese attackers.

Readers interested in the challenges of technological innovation in wartime can read the details at ESM 9.4, Applied science. We particularly explore the role of Wing Commander A. George Pither of the RAAF, in 1942 in charge of the Radar Section of the Directorate of Signals of the RAAF. Pither was an obstacle rather than a supporter for RPL—he considered that giving the scientists freedom to pursue their projects as they thought fit, had produced an indulgence of curiosity and no actual reliable outcomes—but was also committed to accelerating military use of radar. Minnett et al. (1998a) later wrote: "the relationship between [Pither and RPL] would prove a troubled one for years to come."[22]

---

[22]NRAO ONLINE.10 has a discussion of Pither's historical account. Additional material about the Darwin events of 1942 is included in NRAO ONLINE.9 ("Darwin Radar Station 1942"), NRAO

Fig. 9.1 The Air Warning (AW) radar of 1941–1942. The modified 6 by 6 broadside array at the NSWGR circa 1941–1942. Located at the New South Wales Government Railways (NSWGR) annex in Wilson Street, Redfern near the Eveleigh Railway Workshops. Negative is no 6D from RP 201 by White 1943. Credit: CSIRO Radio Astronomy Image Archive JP09–1

RPL stepped up with rapid innovation when it turned out that military strategy was devastatingly incorrect. This rapid innovation was a collective achievement— the whole team at RPL played crucial roles. This likely influenced Pawsey's approach to scientific development in the future.

These events also mark a significant shift away from Australia's orientation to, and dependency on, Britain. RPL was already as much interested in American scientific developments as those in Britain. Australian science was in any case now focused on the very different Pacific war needs, and developing its own

ONLINE.11 ("Moran, Threat from the Air 1941–1942") and NRAO ONLINE.12 ("Epilogue, Darwin Radar Failure, 1942").

solutions to these challenges. We concur with Minnett's analysis of the significance of the development of Air Warning radar (in MacLeod, 1999, p. 425):

> Within two weeks of Pearl Harbour, RPL had an experimental air warning system of unique design operating successfully at Dover Heights ... Such a swift response was only possible *because of the availability of appropriate parts of the ShD technology and the experience gained in developing them.* [Our emphasis, identifying Pawsey's contribution.]

> ... One final innovation was crucial to the success of the air warning venture. A radiated power of ten times that of the ShD transmitter had seemed the only way to achieve the needed detection range of 100 miles ... An inspired adaptation of the ShD technology by Piddington, born of a basic understanding of system design, enabled the desired range to be achieved without increasing transmitter power. The new AW Mk I equipment was to be the first of a series of uniquely Australian long-range, air warning radars.

Light-Weight/Air Warning (LW/AW) Radar, 1942

The most successful defensive weapon in Australia during WWII was planned, prototyped and placed into service within a few months after mid-1942. The main participants were Worledge, Bert Israel, Pawsey and Bullock. The success of the LW/AW radar arose from its developmental sequence: Shore Defence, AW and then AW/LW. The AW system was a 6-ton structure, very awkward to transport and move in the tropics, components hardly fitting in a DC-3 or a Catalina flying boat. Pither consulted J.G.Q. Worledge,[23] leader of the NSW Government Railways (NSWGR) radar structures group, about a new light weight structure (Minnett et al., 1998a, p. 458). Pither wrote:

> The problem confronting us at the moment and in the future is to make available an aerial system which, when used with the AW equipment, can be packed into an aircraft, flown to a new aerodrome and erected in the minimum of time, in order to give warning of enemy attack. The deciding factor in this problem is the aerial system. In order to get an adequate range, a large aerial system is necessary, but this is naturally very heavy, and we must balance the problems of range against weight.

But Pither did not consult with RPL, considering the 'boffins' both impractical and intractable. This might have cost the Australians an important defensive weapon, since Pither likely did not understand the consequences of his suggestion to truncate the vital aerial array (MacLeod, 1999, p. 458). The range on aircraft would have been reduced by about 20%. "The members of Worledge's group were not radio engineers and depended upon RPL [Pawsey's group] to carry out electromagnetic design of a new aerial." The day was saved however by Flying Officer Bert Israel of the RAAF, a "Mr Fix-it" man who acted as a very successful interface between Pither, Worledge and Pawsey. Israel had been associated with the radio industry. "He established a

---

[23] Graduate of Glasgow University. "This Scotsman ... had a genius for simplicity and a keen appreciation of service needs. On many occasions he laid down [the] specifications for the equipment he had to design." Porter (1988, p. 172)

good rapport with RPL, and in particular, with J.L. Pawsey, who was widely known for his mastery of the theory and practice of aerials and transmission lines . . . Pither committed the RAAF to a risky technical venture without the benefit of expert advice." (Minnett et al., 1998a, p. 458).

Israel later told Minnett that he would ". . . not have dared to proceed without his [Pawsey's] advice." The new design (4 × 8 dipoles as the CHL) had only a range loss of 6% of the British CHL device he had seen in Singapore. Thus in the end, Pither's key role in initiating the project, Israel's persistent liaison, Worledge and Bullock's design (below) and Pawsey's technical advice, produced a lighter weight, simpler antenna that was this new aerial combined with the existing AW radar.

Worledge and E.M. ("Ernie") Bullock, a 1942 engineering graduate of Sydney University, discussed the design. Bullock started work on 20 July 1942; the proto-type was erected on time on 13 September 1942. The final weight of the aerial was about a ton.[24] The new aerial was later named the **Worledge aerial** system by the Air Board as an appreciation of the designer's work. See Figs. 9.2 and 9.3.

By mid-October the first equipments were being shipped to Papua New Guinea where they were an immediate success. It was possible to move them within a day or so to new locations by air, as the military situation changed. In 1942–1943, the LW/AW was used effectively as a defensive weapon in Papua New Guinea. Hal Porter summed up the situation (1988, p. 113) at a later era: "During the latter half of 1942 and the whole of 1943, an immense chain of stations was constructed in the Australasian area . . . By the end of 1943 radar had passed from the defensive stage to the offensive, both on the ground and in the air." Interested readers will find a fuller account of the technical aspects of these developments in ESM 9.5, Light-Weight (including Additional Note 2, with an account of the exciting successes of these early endeavours in the Pacific war in Additional Note 3).

Emerging Leadership and Microwave Radar in Australia

Aside from providing informal advice to Israel—and to nearly all projects within the laboratory—Pawsey's main role in 1942 was to lead the microwave research at RPL. After his trip to Canada and the US in 1941, he had a valuable collection of information on the new magnetron and microwave radar.

The crisis surrounding David Martyn during 1941, led to new leadership at RPL, from 1942. White's major overhaul of the organisation created three divisions. The first division provided a role for Martyn; he was then seconded to the Army to investigate problems associated with radar (and made many practical contributions

---

[24]The first power supply (to generate electricity) was a small engine used on farms, a two-cylinder air-cooled engine. This turned out to be unsatisfactory. The replacement engine was a Ford 10 petrol engine which was much more reliable. However, the weight was substantial at 1500 pounds and two were required. This weight represented a substantial increase that had to be manhandled through surf and steep bluffs.

C.S.I.R.
RADIOPHYSICS   LABORATORY

Fig. 9.2 Dover Heights, September 1942. The third stage of assembly for the first prototype LW/AW Radar. The mast and aerial had been raised into position and the fourth leg of the "A" frame had been placed into position. The roof and frame were to be secured at this stage. An additional half of the flange was to be added later as it would surround the mast. RP 201 19 March 1943. F.W.G. White. Credit: CSIRO Radio Astronomy Image Archive JP09–2

working in "operation research"). The second division was Liaison with PMG and Services. The third division was for the research and development of S band equipment at 10 cm, and, after Melbourne Prof Leslie Martin declined to move to Sydney, J.L. Pawsey was appointed its director.

The major project was the first S band (10 cm) Army Shore Defence set (CD No.1 series). The set consisted of two 1.2 m aerials with an imported magnetron. The tests at Dover Heights were successful as a 6000-ton ship was detected at a distance of 70 km. The bearing accuracy was 2 degrees and the accuracy in range 450 metres. The set was to be used at coastal stations and minesweepers.

Fig. 9.3 The initial aerial (LW/AW) in late 1942.The aerial was a 32-element array of half-wave end-fed dipoles. The feeder system used 330 ohm twin transmission lines (developed by Pawsey). The modest canvas tent contained the transmitter and receiver, with cramped quarters for the operators. The tent was made at Chullora Railways Workshops, near Lidcombe, a Sydney suburb. Credit: CSIRO Radio Astronomy Image Archive JP09–3

Freeman's autobiography (1991), *A Passion for Physics, the Story of a Woman Physicist,* provides many delightful anecdotes that illustrate Pawsey's talents in his leadership role. As noted, she had joined RPL in June 1941, working initially with Frank Kerr. Of that period, she wrote:

> Another event... particularly important to me, was the appearance of Dr. J.L. Pawsey (whom everyone affectionately called Joe) ... [H]e greeted me with a warm, ingenuous smile. I sensed a quickening spirit throughout the Lab. Although he had a quiet, gentle presence, Pawsey's personality and influence seemed to reach out to everyone; his natural enthusiasm and drive were unbounded and infectious. I soon fell under his spell and found myself learning steadily from his example and thriving on his encouragement ... Pawsey was very helpful, stimulating me in his inimitable style to think for myself, and encouraging me to build up my self-confidence.

These features became the core of Pawsey's leadership style.

Freeman later moved to Pawsey's microwave group. She later recalled how staff were selected to work on the various components of the system: the magnetron transmitter, the modulator, the aerial, the klystron local oscillator and frequency mixer, the intermediate frequency amplifier and the display system. "Pawsey gave us a superb introductory course of lectures: on aerial, electromagnetic wave transmission, and the properties of wave guides and cavity resonators, providing us in clear and simple terms with all the background knowledge we needed to pursue our developmental work," Freeman wrote (Freeman, 1991, p. 79). She described her role with the microwave radar 10 cm system thus (p. 83):

> Meantime [in 1942], our microwave group was pressing ahead as fast as ·possible with the components for a 10 cm radar set. My klystron oscillator was ·performing. Satisfactorily, but it became evident that great stability was necessary in its high voltage supply. A special variable, voltage-stabilized power supply was required. Pawsey said that I should design and build this. "But I knew practically nothing about circuitry of this sort", I protested. "This will give you a good opportunity to learn", he replied with a smile.
>
> I think Pawsey could have done the whole job himself in the time he gave to guiding me, but carried on with characteristic patience, feeding me with suggestions at the appropriate moments, and then leaving me to develop them. He insisted on my doing the job logically and thoroughly, working out all the necessary theory. I did indeed learn a great deal from that exercise, and gained much satisfaction when the completed power supply worked exactly as it should. Then, at Pawsey's request, I wrote a full report on it. This was typical of the way in which Joe Pawsey operated. His own publication list is not very long; but there must be many papers written by people who have had the benefit of ideas, bearing the Pawsey stamp.

Many from the radio astronomy years would echo these sentiments.

In 1942, Pawsey had his own reflections about what kind of leadership, and what kind of institutional structure, would best suit radar development. They were not centred on himself. Some were prompted by the five month (30 May to 26 October) visit to Australia by ex-pat scientist Mark Oliphant, in whose laboratory the magnetron had been developed. Oliphant's visit had been organised by Madsen during his trip in 1941 (ESM 9.3, 1941: Difficulties). Oliphant had offered his services to the defence work in Australia.

Perhaps unfortunately, a number of factors limited Oliphant's impact. Firstly, his voyage in both directions was unexpectedly long. He departed from Birmingham on a very tedious sea trip from the UK (from Glasgow on 20 March, arriving in Western Australia on 27 May). Later in 1943, the sea trip back to the UK (with his family) was even longer, 26 October 1942 to 1 March 1943. For Oliphant, the loss of time due to six months at sea was frustrating given that he enjoyed only five months in Australia.

Secondly, despite their invitation, Madsen and the RPL management had not defined a clear role for Oliphant. In the end he took no active role in the microwave work underway at RPL, though his visit was a morale boaster for the radar researchers in Sydney.

Oliphant's visit also provoked fresh responses to the perennially vexing question of how best to manage radar research and production. As is evident in Pither's account, relations between the military and RPL could be mistrustful and strained. Madsen had lost some of the support of the military, and in fact over the next year the

Radiophysics Advisory Board (RAB) would become less active, and Madsen would resign (extensive details are provided in ESM 9.6, Microwave Radar).[25] As the nature of the problems of the complex management of radar design, prototyping, manufacturing, testing, full scale operations and then improvements became more acute, the coordination of the various players—CSIR RPL, PMG, Department of Munitions, external firms such as AWA and HMV, Navy, Army and Air Force—became more severe, with frequent conflict and inefficiency.

Oliphant raised some controversy by suggesting that a "dictator of radar" be created—and that he himself could take on this role. Pawsey strongly supported this—so much so that he wrote a letter outside of official channels (from home, located in the Joe and Lenore Pawsey Family Collection) to Rivett, 5 September 1942:

> I am writing you on the subject of endeavouring to retain the services of Prof Oliphant in the hope that [my own] opinion, from one of the research staff of RPL added to those you have already heard from—those in control—may help you in reaching a best decision. I understand that it is now accepted that the Lab should act as a research establishment as opposed to functioning primarily as a prototype production centre ... I believe that the maintenance of a strong research section in Australia is most desirable. To mention only one aspect, a proper research section can [create major] reductions in the work of production by simplification of design ... [At present], the RPL, because of the diversion of effort to production, is not a fully efficient research organisation. [However, it is one] which could be transformed into a [efficient research lab].
>
> In order to make it fully efficient it requires two things: (1) effective coordination with the whole of the RDF effort in Australia and (2) inspiring leadership. The former seems to me to be best realisable through the appointment of a sort of dictator of RDF in all aspects for Australia, a project which scarcely appears realisable because of the difficulty of arranging such an appointment. But the latter could be achieved if Oliphant could be induced to still further extend his stay in Australia.[26]
>
> Oliphant is one of the leading physicists in the world engaged in this type of work. His power lies in his well-balanced appreciation of the operational problems involved in the application of scientific equipment to war, his very brilliant qualities as an experimental physicist, and his ability to inspire his subordinates. Further, I think he would have the courage to persist with ideas he considered valuable for simplification or improvement despite strenuous opposition from official quarters. In all these respects, I think he is outstanding among the men available in Australia today.

No answer from Rivett has been located. Although White told Rivett on 23 July 1942 that he would ask Oliphant to collaborate with RPL in organising the centimetre research work, no action was taken and Oliphant spent the last 3 months of his visit working on the plan to move the group of Leslie Martin and E.H. Burhop (the valve production laboratory) from Sydney University to the University of

---

[25] See NRAO ONLINE.15 for details of Madsen's resignation, which was not an easy transition. The structure of the RAB under the leadership of a civilian scientist was no longer relevant for the task of large-scale production, required as the war reached a new phase in 1942 with full scale mobilisation. In January 1944, the RAB ceased to exist.

[26] These suggestions for a "radar dictator" were similar to the proposal made by General Whitelaw and Commander Buchanan reported by Evans (1970, p. 120) in a letter of 9 July 1942 from White to Rivett. But we have no reply from Rivett, and no such "dictator" was created.

Melbourne (Laby's laboratory). The move was a success with the creation of prototype klystron and especially magnetrons that were given to industry for manufacture.

So to Pawsey and Oliphant and doubtless others, Oliphant's the visit was a disappointment.[27] Interested readers will find a greatly extended account in ESM 9.6, Microwave Radar. But it did stimulate microwave radar development. As MacLeod has analysed: "Oliphant argued that the Allies would need radars for the coming counteroffensive, and particularly in amphibious landings. Such radars would have to be easily transported, quick to put into operation, self-contained, and built so as to survive humid tropical conditions." (1999, p. 413): And his departure left leadership more to Pawsey himself.

1943- a "Golden Year" in Australian Radar: Changes in Outlook

"The year 1943 was a decisive year for the Australians in the Pacific war. During the latter half of 1942 and the whole of 1943, an immense chain of radar stations was constructed in the Australasian area ... By the end of 1943, radar had passed from the defensive stage to the offensive, both on the ground and in the air." (Porter, 1988, p. 113). Simmonds and Smith (1995, p. 64) have described this period as the "Golden Year", with the number of personnel trained the highest at any time in WWII. The early operations in New Guinea served as a proving ground for the lightweight "air transportable" radar, with experience against the Japanese in 1942 leading to refinements in design and packaging that increased the flexibility of the LW/AW Mk I set. See Fig. 9.4, a high frequency system working at 10 cm (S band) used by the Army for coastal defence.

This radar system had robbed the Japanese of the advantage of air surprise as air interceptions by fighter aircraft became more certain. Radar enabled the Allies to choose the time and place for major engagements such as Coral and Bismarck Seas and Midway. As airborne radar located targets, the air-sea battles took place by remote control.

Pither[28] (1946, p. 51) was proud of the RAAF achievements of this period:

---

[27] See NRAO ONLINE.13 for details of Oliphant's visit in 1942; Part I is a description of the visit, while Part II summarises a document about the future of physical science in Australia, written on the long 4 month return sea voyage to the UK in 1942–1943: "The Physical Sciences in War and Peace".

NRAO ONLINE.14 details the transfer of the valve laboratory to Melbourne.

[28] On 23 August 1943, a major conference was held of the stake holders of RAAF radar activities in Australia, the first such meeting in Australia. The one week meeting was organised by Group Captain G.P. Chamberlain of the Royal Air Force (from the UK) who had just arrived in Australia to replace Pither who was to go to the UK on a UK-Australian air force exchange for one year. Simmonds and Smith (1991, p. 20) reported that the "timing was excellent because the drive

Fig. 9.4 Coastal Defence Number 1 (A 272 MkI), a coast watching radar set (Army). Two antennas used in order to improve bearing determination. This radar could function as a coast artillery directing set providing precise azimuth and range determinations of ships up to 45 miles. Location likely Northern Territory. Pawsey is the tall man fourth from the right. Credit: Joe and Lenore Pawsey Family Collection

The end of 1943 had marked the completion of an era of development in radar. The advent of the LW/AW and the successful program of the ASV programme brought to a conclusion a period of frantic development, the outcome of which was a system of radar which was adequate to cope with the threat of the Japanese at the time. In point of fact, this system was really adequate for the rest of the war. With the defeat of the Japanese in the Solomon Islands and at Milne Bay (Papua New Guinea), the war looked like taking a turn for the better, and it began to appear we were entering a new stage.

In this new stage, the increasing shift in outlook from Australian scientists was more evident. In late 1943 (July–December) Fred White and Lt Col S.O. Jones (Director-ate of Radio and Signal Supplies, Ministry of Munitions) paid a long visit to the USA and UK, to seek advice from the two main Allies on production problems. Clearly White was also looking for assistance as he considered a reorientation of RPL research and policy, with victory in both Europe and the Pacific now forecast in the not too distant future.

A major goal was to ask the US and British radar colleagues for assistance as the Australians were focussed on "special problems of the South West Pacific Area—SWPA". The Australians needed advice on radar warfare in the tropics. The US

towards Tokyo was about to commence." The meeting was said to have been "extremely beneficial to all concerned and the service in particular". Relations with RPL and the RAAF improved during Chamberlain's long visit to Australia. Pither returned to Australia December 1944 after playing a major role at the D Day invasion in France in June 1944.

response was favourable. White likely contacted Vannevar Bush, the head of the US Office of Scientific Research and Development, who put him in contact with Karl Compton, the head of the Office of Field Service of OSRD and President of MIT.

Compton was quite receptive to providing assistance. After all, the US was also heavily involved in the war in the Pacific, as both the Australian and US military branches were fighting the common foe Japan. Out of this initiative, the American Group Radiation Laboratory at RPL would develop in mid 1944 (ESM 9.8, Radar and Victory). Compton himself would visit Australia in early 1944 to organise the visit of the US group later in 1944.

But the major achievement of this visit was that on his own initiative, White approached Bowen at the Radiation Laboratory (Rad Lab), suggesting that Bowen join RPL in January 1944. Bowen, who became known as "Taffy" in Australia, would be appointed Chief of Division in 1946. ESM 9.7, Golden Year, contains Hanbury Brown et al's account of Bowen's recruitment.

In the UK White received no more help than a suggestion that a prominent radar scientist from TRE (Telecommunications Research Establishment) might be sent to Australia (for the details see ESM 9.7, Golden Year). This suggestion was soon after withdrawn, and an invitation to the Australians to send a scientist to TRE instead. This was typical of British relations over all the war years. The expectation that Australian scientists would travel to meet their overseas counterparts, but not the reverse, would colour the first decades of radio astronomy also.

However, in 1944, Henry G Booker (1910–1988), a theoretician at TRE who had completed a PhD with Ratcliffe in 1936, and hence known Pawsey at Cambridge, visited Australia. He provided input into research that Pawsey was now working on. Earlier in 1944, CSIR had taken over the field of super-refraction or anomalous propagation (Evans, 1970, p. 169).[29] Under the leadership of Pawsey and John Jaeger, this group aimed to measure radio transmission conditions over known paths, (2) make precise determinations of the meteorological conditions over these paths and (3) correlate and interpret the results (Evans, 1970, p. 169). Further details are provided in ESM 9.8, Radar and Victory. Booker would remain an important professional connection throughout the remainder of Pawsey's life.

Entirely separately to White's endeavours, Henry Tizard, the Chair of the scientific committee that first instigated radar research in the 1930s, was in Sydney from 28 August to about 1 October 1943. He did not meet White who was in the UK at this time.[30] Chiefly he observed the frustrating environment hampering CSIR's relations with the services. Tizard was frank: "I am very much afraid that the good work of the RPL will fail to have its full effect on the Australian Services unless the human problems are solved."

---

[29] See the Additional Note at the end of ESM 9.8, Radar and Victory, D.F. Martyn in 1944, Seconded to the Commonwealth Solar Observatory, for Martyn's activities in 1944, as he moved to Mt. Stromlo.

[30] See NRAO ONLINE.17 for details of the Tizard visit; White was in the US or the UK from 15 July to early December 1943.

The bottom line was presented by Rivett to White on 8 December just as White had returned from the US and the UK:

> Of the correctness of your view that the CSIR Laboratories can contribute a very great deal indeed to the success of the Pacific War, there can be no doubt at all. When Tizard was here, we had several discussions about the best way in which we could develop our usefulness. Much ... depends on our power to convince the Services that we really can contribute something; they seem just a trifle slow in appreciating this possibility.

But amid these negative assessments there were in fact several initiatives to improve communication across agencies, which were felt to have some effect. This, and further technical details about radar development at this stage of the war, are provided in ESM 9.7, Golden Year.

Radar and Victory in the Pacific, 1945

At the Radiophysics Laboratory, the new level of sophisticated skills developed through 1942 and 1943 had created its own new industry by the end of the war. The pathway to the rapid growth of radio astronomy in 1945 had been laid.

Pawsey's radar program was mature and flourishing by 1944. At this time he expanded his research program to include the development of a 25 cm advanced air warning height radar (location Georges Heights—Middle Harbour—near Mossman). This was built around the Australian magnetron (developed in Melbourne) and was "essentially Australian in design and engineering" (Evans, 1970, p. 228). Pawsey's colleagues B. Mills and R. Payne-Scott—soon to be early radio astronomers—had worked on the issues of calibration and signal visibility of Plan Position Indicator detection with radars. In 1999, Mills wrote[31]: "Finally, looking back, I see the rapid and successful development to have depended on the foresight of Joe Pawsey in setting up a program to study the basics of signal visibility."

Orders for this "outstanding technical wartime achievement" (Evans 1970, p. 228) were cancelled, with the advent of peace in August, 1945. See Fig. 9.5.

As peace approached, White and others made further trips overseas to discuss new approaches in both science and policy. Details are provided in summary in ESM 9.8, Radar and Victory. Within the first month of 1945, White left Sydney and RPL to become a member of the CSIR Executive. His task was to assist David Rivett as the Assistant CEO of CSIR, with responsibility for the physical sciences.

Pither (1946, p. 94) provided his assessment of the evolution of radar since 1939:

> [R]adar, which started from zero in 1939, became the greatest scientific development of the war. In conjunction with fighters, it stopped the Japanese raids on Darwin, and the tremendous Japanese losses at bases without radar cover in the islands are an indication of what would have happened to Allied bases in Northern Australia and New Guinea in the absence of radar warning.

---

[31] MacLeod (1999, p. 65), also Goss and McGee (2009, p. 60).

Fig. 9.5 The prototype LW/AWH Mk II. Goss and McGee,2009, page 60. Both the figure in the Goss and McGee volume and the CRAIA image are reversed left to right. Here the orientation has been corrected. Credit: CSIRO Radio Astronomy Image Archive B1362

What were the ingredients that provided the foundation for radio astronomy's remarkable growth in 1945–1950? (1) A thorough knowledge base in metre-wave and microwave physics, (2) the existence of networks between Australian, US and UK radio scientists, (3) the distinguished careers in academia, government science and industry that occurred after 1945,[32] (4) Numerous personnel trained in electronics, (5) the pioneering research by the Radio Research Board from 1927 that continued at the Division of Radiophysics in 1939–1945 and (6) individual scientific careers developed as "programmes of research came to be moulded on individuals, rather than the reverse."[33] We provide an extended analysis of each of these points in ESM 9.8, Radar and Victory.

---

[32] A prominent example is Edward M. Purcell, who had been at the MIT Radiation Laboratory during the war and began an illustrious career in the Physics Department at Harvard. He and his PhD student H.I. "Doc" Ewen discovered the HI line at Harvard on 25 March 1951 (Chap. 20). In 1952, he shared the Nobel Prize in Physics for the discovery of nuclear magnetic resonance using techniques learned during the war.

[33] Evans (1970, p. 224) elaborated further: "Given a broad measure of personal inclination in the choice of projects, the emphasis of investigational patterns often tended to take unexpected directions, which assumed an individuality of their own. It was unlikely that any of those planning the post-war research of the Radiophysics Laboratory could have had any notion of what was to happen within a few years to the character of peace time research."

Pawsey's rapid post-1945 success is a special case. What did Pawsey's wartime experience bring to this unexpected new line of research? We cannot doubt that the experience of war profoundly shaped J.L. Pawsey and his colleagues at Radiophysics. Intellectually speaking, wartime research had raised several interesting issues and lines of inquiry in radio research, much of which could not be pursued until the arrival of peace. One of these was investigating extraterrestrial sources of radio emission.

The social impacts of war were also substantial and shaped the working culture, attitudes and views of those at RPL into the future. Pawsey's son Hastings has remarked that his father developed a lifelong dislike of secrecy after his wartime experiences. It is well recognised that many people, scientists certainly included, were deeply affected by the war and gave considerable thought to how society could change to avoid such wars again. Sociologist Robert Merton wrote a famous article in 1942, "A Note on Science and Technology in a Democratic Order" that argued that science was structured by a system of moral values organised around impersonal, unbiased and impartial commitments to factual knowledge. This scientific "ethos" stood in contrast to the partisan and prejudiced beliefs that had led nations into war. While we do not know if Pawsey read this article, we will see in subsequent chapters that such values resonated at RPL.

Pawsey had honed his capacities as an applied scientist by drawing on his extensive understanding of physical theory and concepts to solve the various technical challenges that arose in the rapid development of new radio direction finding technology. And even more, he had honed his capacity to develop the skills of his team.

Chapter 10
Transition to Peace, 1945–1946

... [T]he RPL became a major actor in the changing relationship between government science, the universities, and the military in Australia. Before the war, the CSIR had worked closely with the Australian universities in the sciences of primary production, in veterinary sciences and in agriculture. [After 1945], the RPL became a microcosm of a new military-academic-scientific complex in physical and engineering sciences, the like of which Australian had not seen. (MacLeod, 1999, p. 413).

Wartime can create significant changes in people and institutions. New opportunities are envisaged. Australian scientists imagined many new scientific projects, and they also dreamed of improvements to society and culture to which science might contribute. Many people emerged from wartime with significant, passionate commitments to building a better future. As Schedvin (1987, p. 334) wrote of the immediate post-war period:

For those ... who were attempting to anticipate the shape of the post-war world, the application of scientific rationality seemed the only way to curb the madness of political rivalry and the social division created by economic instability. Science and education would join forces in overcoming the self-destructive impulses within civilization.

These general remarks were certainly true of most of those who worked on radar during the war and became part of radio astronomy in the post-war decades. The importance of building capacity in Australian science—independent of the UK, and able to thrive nationally—was a priority for many. At the same time, the importance of developing scientific internationalism, which often seemed like the obverse of the populist authoritarianism that had underpinned the dreadful brutality of WWII, was also keenly felt.

Source material includes NAA C3823 E12/2 (Radar- Future Post-War Activities 1943–1945) and C3830, D1/1 (Future Program of RP 1945).

Supplementary Information The online version contains supplementary material available at [https://doi.org/10.1007/978-3-031-07916-0_10].

W. M. Goss et al., *Joe Pawsey and the Founding of Australian Radio Astronomy*, Historical & Cultural Astronomy, https://doi.org/10.1007/978-3-031-07916-0_10

Home (1988a, p. xvi) has summarised the new era that was beginning:

By war's end, it was clear that Australian science had undergone an irreversible change in
line with the general industrialisation of the nation's economy. The number of scientists
working in all fields had greatly increased, and the demand for their services did not decline
with the coming of peace … [M]any looked forward to Australia making its own, independent contribution to the new, scientific age … No longer, however, did [CSIR and later
CSIRO] confine itself to the applied research envisaged by its creators.

Post-War Planning in 1943

Even amid the exigencies of war, people look ahead. In CSIR, planning for the future
began as early as 1943.[1]

Fred White, already well ensconced, predicted the expected post-war conditions[2]:

1. Australia has relied almost entirely upon scientific engineering and radio research
   in other countries, with only very feeble attempts to build up research in these
   fields within the Commonwealth.
2. In spite of the wealth of natural resources within the Commonwealth, a policy of
   financial expediency has produced an engineering and radio industry almost
   entirely dependent upon imported materials and parts. Under emergency conditions this state of affairs has had to be changed and in the post-war world it is to be
   expected that a new outlook will obtain.

White predicted that the major technical developments of the war would continue, as
well as the availability of technical expertise provided by the vast numbers of men
and women that would be released from the Services. To take advantage of these
factors, local research in scientific areas should continue in order to make Australia
more self-sufficient. In essence, White proposed that the successes of WWII should
lead the way in contributing to post-war reconstruction.

In early March 1943, the proposed transition plans were based on four major
themes: (1) two-way collaborations with industry, (2) university collaborations such
as training of young scientists and engineers at CSIR research centres with lectures
by host scientists, (3) applied science projects that were closely related to wartime
research (e.g. aviation navigation and long distance propagation studies) and (4) first
gradual steps to the recognition of the importance of fundamental research. It was
expected that there would be continued relations of industry to the RPL, but in
addition, there would be enhanced contact between CSIR and the universities; the

---

[1]As Home (1988, pp. 147–165) has pointed out there had even been a short-lived premature
exercise in February 1941 to begin laying plans for reconstruction; this had come to an abrupt
end as the Pacific War began in December 1941. See also (Home, 1988b)

[2]NAA C3823 E12.2 "Report to the Executive Committee on Post-War Reconstruction". The
secretary of the CSIR, G. Lightfoot had requested this report on 2 March 1943 for the meeting
in May.

expectation was that universities would, for the first time, play a major role in Australia-wide scientific research.

White presented a report on 22 March 1943 to the CSIR Executive Committee on the topic of the future. One of the documents that contributed to this report was written by Pawsey, identified with his initials.[3] Pawsey's document covered similar points as other contributors with elaborations:

> The most vital post-war problems of reconstruction will be of a sociological nature. As far as this affects CSIR and RP in particular the relation between research in private industry and by a government body is probably the most important question ... It is suggested that an attempt be made to set up machinery to deal with this problem to allow CSIR to examine physical problems associated with production.
>
> Specific fields of research which suggest themselves are: aerial navigation and blind landings, ship navigation ... special communication problems, problems of physics applied to medical science, biology, agriculture etc., and various special peacetime applications of RDF [radar].

White presumably compiled this contribution with those of others to produce his report, which began:

> No scientific institution can flourish unless it is encouraged to participate in pure scientific research and maintain a close contact with, and an interchange of personnel with, the universities. It must be recognised, therefore, in providing staff and finance, that something like **50%** of the activities of CSIR Laboratories should be of this type. It is difficult to assess the value of such work in terms of direct financial return to the community, but it may be judged by other scientists, by the standing and reputation of the work of the institution in relation to other purely scientific work throughout the world.

In addition to university collaboration with existing CSIR laboratories, it was also desirable that CSIR would directly assist universities to set up their own facilities, without in any way trying to dictate policy to the academic colleagues. White saw the possibility of building up an Australian cohort of scientists with common interests, in order to combat the constant threat of isolation from colleagues in Europe and North America.

What were the spheres of endeavour that would in the end benefit Australia? Communications via HF (3–30 MHz) radio and civil aviation were clearly areas relevant to a large sparsely populated country such as Australia; both activities were related to the wartime success of RPL. The safety and efficiency aspects of expanded civilian air travel also were essential projects.

Specific projects that RPL might undertake were (a) study of the ionosphere with applications to the forecasting of radio communications, (b) applications of radio to both civil aviation and shipping at sea, (c) radio development leading to television,

---

[3] The National Archives collection (NAA C3823 E12/2) provides a thorough summary of the planning for post-war reconstruction in 1943 and 1944. The first two documents are single pages consisting of hand-written notes. The author of page one has not been identified; on the page there are several additions, apparently in White's handwriting. Based on the context, a possible author could be L.G. Dobbie, the chairman of the Association of Research and Technical Officers of Radiophysics. This group provided a two-page summary of their assessment of future plans to White; his final report of 22 March 1943 has incorporated their suggestions.

FM radio broadcasting, and microwave communication over land and (d) continued investigations for the armed Services of Australia. There was a special category for "physical studies" consisting of five categories: (i) cyclotron for production of artificial radioactive substances of importance in medicine, (ii) electron microscope, (iii) millimetre and short radiowave studies, and (iv) application of radio methods in biophysical work, such as high frequency electrical methods for extermination of weevils in grain and radio methods of humidity measurements. Only category (iii) evolved into one of the main activities of the post-war RPL.[4]

White discussed the expertise of the staff. This consisted of 34 scientists "of whom about eight are men [in fact there were only two women in this cohort, Ruby Payne-Scott and Joan Freeman] who have had either experience abroad or considerable experience in industry and in the Universities of Australia". The problem was that the ratio of "seniors to juniors is too small. If effective research is to be undertaken . . . some of the junior staff must be given experience abroad in research institutions . . . [Finally], the wartime development can be maintained in peace time only by a progressive policy in which research laboratories are essential."

White's proposals influenced several concrete steps which began after the May 1943 meeting of the Council. The CSIR Executive Committee started to address "first practical problems" (Home, 1988, p. 244) such as the establishment of new divisions (e.g. Division of Building Research) and aligning the tasks of some divisions for post-war activities. But, the end of the war was still two years distant. The detailed planning for the post-war RPL was still in its infancy as 1944 began.

Post-War Planning in 1944

During the course of 1944, the end of the war in Europe and the Pacific was expected with increasing confidence.[5]

In July 1944, White and colleagues prepared three additional proposals concerning post-war reconstruction. The first was a report of RPL activities on 10 July 1944, written to be a component of a comprehensive document for the Minister for the CSIR, John Dedman.[6] The report, "CSIR—Ten Years of Progress", contained sections on each of the CSIR divisions. A second comprehensive report was submitted to the CSIR Secretary on 14 July 1944, intended to be discussed at the

---

[4]Readers may enjoy reflecting on these suggestions from the perspective of 70 years later. Ionospheric research and forecasting continued; application of radio research to aviation resulted in an aircraft landing system; millimetre and short radiowave studies were strong; and a cyclotron eventually happened.

[5]"[T]he sense of planning for a peaceful future [in Australia] strengthened." (Home, 1988, p. 245).

[6]NAA C3623 E12/1. Letter from Lightfoot to White. Dedman had requested this report in early June 1944 after a visit to the Fisherman's Bend laboratories of CSIR (Industrial Chemistry and the Aeronautical Laboratories). "[Dedman] said he would like to have a report setting out the case for CSIR after the war." The report was due on 3 July 1944.

CSIR Council meeting on 31 October 1944 in Sydney: "Present and Future Activities of the Commonwealth Physical Laboratories" by Frederick W.G. White, Norman A. Esserman (Officer in Charge (OIC) of the National Standards Laboratory), George H. Briggs (OIC Physics) and David M. Myers (OIC Electrotechnology). It proposed that four entities, RPL and the three components of the National Standards Laboratory, would merge into a single large institute, housed at the University of Sydney campus. Ironically, this proposed merger was never completed; however, the institutes shared the same building until 1968 (now known as the Madsen Building at Sydney University).

The planned post-war RPL activities described in the 14 July 1944 proposal were mainly concerned with specific interests in Australian civil aviation (navigation and landing aides) and in propagation studies (ionosphere at HF, MF and LF frequencies—30 kHz to 30 MHz—anomalous propagation and atmospheric noise due to thunderstorms).

The third document was written by White and Frank Nicholls (Australian Scientific Research Liaison Office). "A Discussion of CSIR" (19 July 1944) provided an analysis of the organisational structure of CSIR and the way this structure could be reformed to effect post-war reconstruction. Two aspects are especially striking: the balance of applied and fundamental and applied research, and the degree of autonomy that the individual divisions would maintain.

In a rubric "Fundamental-Applied Research Relationship", White and Nicholls wrote:

> It is quite obvious that in some fields of work, a desire to solve an applied problem will result in the solution of a fundamental problem. In some cases fundamental work will be undertaken so that a particular scientific laboratory in CSIR will have at its command a full appreciation of the science which it professes to follow ...
>
> If effective assistance is to be given in any technological field, suitable staff and facilities for handling applied science must be provided. Nevertheless it is obvious that applied science itself can prosper only if the fundamental aspects of the science are sufficiently appreciated and investigated. Any laboratory within CSIR must therefore devote part of its effort to fundamental investigations and part to applied work. The relationship between the fundamental work and applied work may be referred to as the "fundamental-applied relationship", and the solution of this relationship will vary considerably in different laboratories concerned with different departments of science.
>
> In technology, Australia has relied too long upon fundamental scientific work carried out overseas. It is important that those not associated with science should be made to appreciate that unless fundamental work is actually going on in Australia it is very difficult for technologists to realise the significance of scientific discoveries or to get the full data they need. This is particularly true, of course, in any field where local conditions may affect the results of scientific investigations.

White and Nicholls stressed that the balance between applied and fundamental research required close liaison with universities in Australia. Schedvin commented that White was to continue the Rivett tradition of support for basic science that was so central a part of the CSIR ethos (Schedvin, 1987, p. 314). He added that while basic science was also supported by the Chiefs of Divisions in the biological science, the strongest commitment came from RPL and physical chemistry. Biologists were less committed to the classical theory of the growth of knowledge; amid the complexity of the living world, theoretical gains were often instrumental. "It was the philosophy of physics, with its clear distinction between pure and applied research and confident prediction about the best strategy for promoting the growth of knowledge, which carried the day." G.A. Julius retired and David Rivett gladly relinquished his burdensome role as Chief Executive Officer (CEO) to become Chairman in place of Julius. The new CEO (Richardson) was in poor health, and given the increased size of the organisations, two new assistants were appointed to support him. One of these two was Fred White. John Briton became Chief of RPL, in White's stead as of 31 January 1945; "Taffy" Bowen remained as Deputy Chief of the Division.

Bowen was given the task to develop a formal plan for the post-war future of RPL. This proposal was to be discussed by the Executive Committee of CSIR (mid-June) and presented to the CSIR Council in its meeting of 11 July 1945.

On 28 February 1945, Bowen prepared a draft of the transition plan for White. The first text, "The Transition of RPL Activities from War-Time to Peace-Time Applications", outlined the manner and timescale of a transition plan (five pages). An appendix of one page outlined research planning, "Summary of Post-War Programme of RPL", containing the same list of research problems as in 1944. The process of change was envisaged as gradual, with effective military support being supplied to the RAAF until the end of the war while "allowing the introduction of as much post-war work as possible".

On 27 April 1945,[7] Bowen wrote White a long letter. Its main purpose was the eventual status of his short-term appointment at RPL. His period of secondment from the Ministry of Aircraft Production (UK) was to end on 1 July 1945. Bowen was carefully considering his own future, which might well include rejoining premiere research groups in the USA and UK. But he preferred the challenge and perhaps the autonomy that RPL offered, writing, "[M]y experiences at the RPL during the last 18 months are such that I would be anxious to stay for a further period of perhaps three to four years." He was to remain Chief of the Division until 1971![8]

Bowen added, "The time during which development work can still make a useful contribution to the war is rapidly drawing to a close, and may have ceased already. It is therefore necessary to draw up afresh the research aims of the Laboratory, and the continued expenditure of funds at the same rate as in the past must be justified in relation to a very specific programme which must be prepared and approved by the Executive. This I regard as my most important single duty during the month of May." Bowen, White and Rivett were likely also motivated by the desire to find a project that could keep their extraordinary team intact.[9]

By 2 July 1945, a thirty-page document appeared: "Future Plan of the Division of Radiophysics," to be presented to the 35th meeting of the CSIR Council on

---

[7] NAA C3830 Z1/7/B/1.

[8] He became RPL Chief from 3 June 1946.

[9] Sullivan (2009, p. 122), reports Frank Kerr's retrospective comment that "it was possible to sell the idea to the authorities that the group should be kept in existence as a 'national asset'."

11 July.[10] In it, Bowen also addressed the question of fundamental versus applied research, commenting:

> The danger of concentrating entirely on applied problems is well known and need not be enlarged upon. It is not generally realised that a similar danger exists from concentrating on pure research, and for this reason it is important to maintain a correct balance between them. Due to a policy which has been dictated by the war, the Division of Radiophysics has in the past been forced to give too much attention to applied problems, and it is most important that this be remedied in post-war years by strengthening the fundamental side.

The report asked:

1. Are the techniques which have been developed primarily for military purposes capable of fulfilling useful peacetime needs?
2. What is the magnitude of the laboratory programme which is needed to meet these needs?
3. What redistribution of scientific efforts is required, and in particular what change of emphasis should take place between fundamental and applied research?

Bowen wrote:

> In considering the contents of a future programme for the Division of Radiophysics, it becomes obvious, however, that the stress of wartime conditions has seen the development of remarkable new techniques in the fields of radio and radar, which are perhaps as far reaching as the development of aircraft during the last war or the introduction of gunpowder in a previous era. Radar techniques have at least as many applications to the peacetime activities of mankind as they had in war, and having underwritten extensive development work in wartime, it is to the greatest economic advantage of Australia to make the maximum use of it in the immediate post-war years. In making a choice between the possible applications, it is patent that first choice should be given to those which provide greatest benefit to the people of Australia. Finally, the specialised techniques themselves can make a practical contribution in many fields of fundamental investigation. Examples which have already been explored in a preliminary way are the application of radar techniques to atmospheric physics research and the use of both pulse and radio frequency techniques for the acceleration of elementary particles.

It would turn out that the most significant part of the Future Programme document was a minor portion of the report (one page out of thirty), a small section titled, "Radio Noise":

> ... [A] knowledge of [the level of the radio background] and its variation with place, time and wavelength is of as much practical importance as a knowledge of the propagation of radio waves. At the same time, the **origins of radio noise** are of great scientific interest. On wavelengths from the longest used down to a few tens of metres, lightning flashes are the

---

[10] The draft of the report had been sent to White and Rivett on 5 June for discussion at an Executive Committee meeting during the week of 11 June. Also on 5 June the draft plan had been discussed at RPL by senior staff, Pawsey, Pulley, Piddington, Alexander and Burgmann. Bowen apologised to White on 5 June: "[The report] is perhaps too popular in content and might reasonably be objected to by the Executive on these grounds." White replied on 15 June 1945; he and Rivett were generally satisfied with the draft. He stressed that the Council would be interested in hearing why the particular programme had been selected, the cost of the programme, the level of urgency and the follow-on monetary value to Australian industry.

dominant source of noise. For a short range of wavelengths a type of noise appears which is thought to originate in the **stars or in interstellar space**. This is called "cosmic noise". Then in the microwave region the principal source of noise can be identified as radiant energy in the far infra-red radiated according to the classical laws of heat transfer. [our emphasis].

Bowen was of course referring to the observations of extraterrestrial (mostly solar) radio emission that occurred during the war—including those of Pawsey and Payne-Scott in 1944, and of Elizabeth Alexander from New Zealand in 1945 (Chap. 11). The confusing title of this line of work was "Study of Extra-thunderstorm Sources of Noise (Thermal and Cosmic)", that is, the study of noise outside of (or, not caused by) thunderstorms. Bowen wrote:

> Little is known of this noise and a comparatively simple series of observations on radar and short wavelengths might lead to the discovery of new phenomena or to the introduction of new techniques. For example, it is practicable to measure the sensitivity of a radar receiver by the change in output observed when the aerial is pointed in turn at the sky and at a body at ambient temperature. [This was a reference to the wartime trials in March and April 1944 conducted by Pawsey and Payne-Scott, Chap. 11).[11]
> The aerial receives correspondingly different amounts of radiant energy (very far infra-red) in the two cases. Similarly, the absorption of transmitted energy in a cloud can be estimated in terms of the energy radiated to the receiver by the cloud. None of these techniques is at present in use.

There was no mention of the, by then reasonably well-known, results of Karl Jansky and Grote Reber's investigations of "cosmic noise" from the 1930s and 1940s.[12] We note that, as was natural at that time, the presence of cosmic background was lumped together with atmospheric emission.

As Sullivan (2009, p. 124) has written: "It was this enigmatic paragraph ... that would develop into RP's radio astronomy program! It surprisingly did not explicitly mention solar noise, but instead proposed an exploratory program of 'very far infra-red' radiometry wherein antennas would be pointed to different parts of the sky."

It is interesting to compare Bowen's document with Joe Pawsey's ideas for post-war research. In 1945 (unknown month and day), Pawsey drew up a new list of "research possibilities"; there was some overlap with his 1943 list, with a number of new items. He began:

> Collection of ideas some of which might be worth follow up.
>   Classification of (1) Broad files and (2) items of 1.

There were 15 categories:

---

[11] We agree with Sullivan (2009) that this text refers to the March and April 1944 10 cm observations of Pawsey and Payne-Scott of the behaviour of an S band receiver (10 cm).

[12] We have assumed that readers will be familiar with Janksy's and Reber's discoveries in the period before 1945, since they are mentioned in virtually every history of radio astronomy, are on many websites, etc. We refer interested readers to Sullivan, *Cosmic Noise*, 2009, for a complete and thorough discussion of this early history.

1. Radar and atmosphere including meteorological echoes, horizontal irregularities in propagation, abnormal microwave propagation, and the Watson-Watt, Bowen and Wilkins effect.[13]
2. Effect of centimetre waves on biological molecules and organisms.
3. Discharge in gases, including T/R (transmit/receiver switch for radar sets) circuit theory and break down for DC and HF pulses.
4. Circuit theory including clear statement of T/R theory, triode oscillator theory and elements of theory of transmission lines stop filters.
5. Ultrasound (high audio frequencies), including examination of materials for structural flaws using radar techniques, effects of ultrasound on biological materials and organisms.
6. Prediction and control of climate including study of heat exchanges with object of utilisation of solar energy and refrigeration through radiation.
7. Electrical production of music.
8. Pulse techniques applied to acceleration of ions and electrons.
9. Electrostatic generation of power including the theory of variable condenser electrostatic dynamo.
10. Far infra-red energy levels including technique of measuring noise levels more accurately and extension to clouds at microwaves, extension to other wavelengths and energy from sun.
11. Lightning observations.
12. Aerial theory.
13. Radar, including the search for numerous problems: of increased range, increased resolution (e.g. shorter wavelength and shorter pulses) and improved display by adding a CW transmitter in order to determine the speed of the target by Doppler shifts. Also solutions to problems of power supplies and energy generation for the entire radar station.
14. Electron optics.
15. Aerials.

We note how significantly this list was oriented to the "applied" problems and their potential extension that had arisen in Pawsey's wartime work. Item 10, "Far Infra-Red", was destined to become radio astronomy. There were no references in this document of the earlier detection of the galactic radio background by Jansky and Reber either.

In ESM 10.1, Paul Wild, we describe a lecture given by Paul Wild in 1965 with his vision of the origin of radio astronomy in Australia in 1945, with special emphasis on Bowen's July 1945 document.

---

[13] See ESM 9.1, Radar History, and NRAO ONLINE.5.

Radiophysics and the Military Part Company

A major consideration remained for the RPL in 1945: what level of support would RPL provide to the armed forces of Australia in the post-war era? Bowen was pointedly critical of a continuation of active collaboration with the RAAF. His opinion was based on experience of 10 years starting with the group of Watson-Watt in the mid-1930s and continuing with the Tizard Mission of 1940, the Radiation Laboratory in the US (1941 to 1944) and RPL in Sydney (since January 1944).

In regard to possible future support of the Australian Armed Forces, Bowen wrote:

> It is considered that in the post-war period it will be unprofitable, both for CSIR, and for the nation to undertake the day-to-day development problems of the Armed Forces. The reasons for this need not be stressed, but such proceedings in the past have stifled research and seldom produced effective assistance to the Armed Forces. It is considered that there are only two profitable ways in which useful activities can be undertaken on behalf of the Armed Forces:
>
> 1. By senior officers of CSIR and its Divisions having regular and close contact with the Chiefs of staff and their military problems in order that they will in the future continue to be fully informed of the course of military science … In order that this might be effectively carried out, it is necessary for senior members of CSIR to be formally accredited as scientific consultants to the various Boards of the Armed Forces.
> 2. By considering the trend of military developments and the form of warfare likely to be encountered in 20 or 30 years' time, [it is necessary to institute] a research programme based on this trend. A good example in the present day is the very definite tendency in the Air Forces toward the use of unmanned reaction-propelled missiles. Missiles of this type, which combine the potency of rockets with the accuracy of radio and radar control, are likely to form a great bulk of the offensive weapons in the next war. At present, there is practically no defence against them and only by intensive investigation and development of the principles involved will an effective defence be found. Work of this description would undoubtedly make a great contribution to the future effectiveness of the defence services, and is the only type of activity which should be contemplated by a research organisation like CSIR.[14]

His prescience concerning the future challenge of long-range ballistic missiles was striking.

On 26 July 1945, the CSIR informed Bowen, who was Acting Chief while Briton was overseas, with excellent news: "That in the opinion of this Council a programme on the lines indicated by Dr. Bowen is of importance and should be taken as the basis of a post-war programme for the Radiophysics Division."

Within the next month the war in the Pacific was over: 15 August 1945 was "Victory in the Pacific" Day. Two days later, a meeting was held at RPL to discuss

---

[14] Already in December 1945, Bowen (1945, p. 33) had emphasised the importance of civilian control of wartime research and the necessity of combining applied and pure research. "[During a war] … there is as much need for basic work as at any other time … [It is also essential] to observe a basic principle, the provision of complete freedom to the scientist and the avoidance of military control."

"Activities of RPL Following Conclusion of Hostilities". The attendees were Bowen, Pulley, Piddington, Pawsey, Higgs, Eagles, Kerr and McCready.[15]

Main projects for the RAAF would continue such as the AWH Mk II (L band air warning height) with the same staff and timescale. The idea was to only complete the prototype. In addition, the preparatory work on the GCI Mk III would continue at a slowed pace; in the end, only a design was completed.

The work at RPL would continue with respect to propagation work, valve research and civil aviation development. The Laboratory also planned for the disposition of all technical publications which would be soon de-classified. The Australians would await the publication of US and UK papers on radar; RPL did publish *A Textbook of Radar* in 1947, edited by Bowen.

On 11 August 1946, Bowen submitted his second future plans documents to the CSIR Executive.[16] Although similar to its predecessors, by mid-1946, some of the proposed projects from mid-1945 were already being de-emphasised, such as work for the RAAF.

The items in mid-1946 for the future were: (1) radio propagation and atmospheric physics research, (2) measurements and standards at radio frequencies, (3) vacuum research (valves), (4) radar aids to civil aviation, (5) radar aids to surveying and (6) research and development for the Armed Forces. Topic 1.1 was "super-refraction and the temperature-humidity structures of the lower atmosphere" while a new topic appeared, 1.2, "solar and cosmic noise at radio frequencies". The latter topic was the result of the successful observations at Collaroy (see Chap. 11) followed by Dover Heights in early 1946.

From CSIR to CSIRO: Organisational Change

Even in 1944, White and Nicholls had written of the necessity of reform of the governance and managerial aspects of CSIR.[17] The rapid growth since 1940 had strained the CSIR management structure, inherited from an organisation with a staff five-times smaller compared to 1939.[18] Their major concern was to balance the level of autonomy of the "fundamental unit of the organisation", that is, the divisions:

---

[15]NAA C3830 D1/1.

[16]NAA C3830 D1/2. The submission was about a month before his departure overseas to North American and the UK, his absence from Australia September to December 1946. An earlier version of the future plan document on 24 May 1946: "As would be anticipated, plans for 1946/7 differ in many respects from those laid out in 1945, [just before the end of the war]."

[17]"A Discussion of CSIR" (19 July 1944). C3823 E12/2.

[18]One issue was the fact that the large Council (including state members, meeting only a few times per year) was the governing body and not the Executive Committee of 3 to 5 members. During the War, the EC had become the *de facto* governing body; this problem was rectified when the CSIRO was formed in 1949.

> In all such organisations as CSIR the decision must be taken at some time as to whether the
> organisation is to be controlled from a central headquarters or whether a policy of **decen-**
> **tralisation** [our emphasis] is to be followed in which the larger units of the organisation are
> given greater responsibility and opportunity for independent initiative. In scientific work, the
> latter course appears to be very desirable. The Council [of CSIR] should endeavour to place
> in the divisions responsible officers who are thoroughly versed in particular scientific
> subjects ...
>   Any action to bring the administration of general affairs of CSIR more into the hands of
> the chiefs of divisions will lead generally to a better understanding of the organisation as a
> whole and should result in more efficient running of the divisions themselves.

During 1948 and 1949 the structure of CSIR would be reshaped and the organisation would become the Commonwealth Scientific and Industrial Research Organisation, CSIRO on 19 May 1949. Many factors were at play, including, as we explain below, concerns that the organisation was associated with Communism.

During this time, "centralism" was a major issue, with different actors making contradictory claims as to its benefits or problems. As Schedvin (1987, p. 348, 357) explains:

> Administration and research were kept separate. Administration was centralised and highly
> constrained [at the CSIRO Head Office, Melbourne]; research [at the division level] was
> unconstrained and decentralised ... The CSIRO Executive retained tight control of policy
> issues such as staffing, financial allocation and publication. As before, research strategy was
> the responsibility of chiefs of divisions ... Within their areas of responsibility, chiefs ...
> were given more freedom. Expansion was encouraged and initiative rewarded; the early
> history of CSIRO was the age of the entrepreneurial chief.

E.G. ("Taffy") Bowen was to become a prime example.

Aversion to Secrecy at CSIR

Following the war there was a marked aversion to secrecy in relation to research, a sentiment widely held by the scientists. This aversion was emphatically and explicitly discussed in correspondence between Bowen, White and Rivett at this time. Their aversion was one of the reasons RPL was uninterested in substantial, continued association with the military.

Rivett was passionate about the importance of complete freedom for scientists to pursue their research in an entirely open environment. For example, when concerns were raised about the need for secrecy in relation to nuclear research, he said:

> If national sovereignty demands the right to prepare secretly for the destruction of other
> sovereignties, let those who take the responsibility of making a decision to that effect ...
> keep their projects clear of [those] national scientific institutions in which the traditional
> freedom of science must be maintained. (Schedvin, 1987, p. 329)

Unfortunately his public statements to this effect resulted in concerns in the UK and USA that CSIR, by refusing to admit that secrecy was ever acceptable, would be a security risk. (Schedvin, 1987, p. 338) Reorganisation of CSIR was taking place amid turbulent post-war politics during 1946–1949, and attacks on Rivett and on CSIR appeared in the tabloid press.[19] There were insinuations that Rivett was "soft" on Communists as a result of his committed internationalism (Schedvin, 1987, p. 339). Amid this fracas, for some time in 1948, the government intended to move both RPL and the Division of Aeronautics to the Public Service Board. Thus, these groups would be government departments located within the Department of Supply, entailing, among other challenges, security requirements due to the onset of the Cold War. Had this transfer occurred, the development of radio astronomy in the post-war era would likely have been substantially limited. During this difficult time, opposition to secrecy became even more evident among the scientists.[20]

We can gain more insight into the scientists' perspective at CSIRO and the culture dominant at RPL, from correspondence of this period. For example, Bowen wrote to Rivett about these issues on 29 September 1948, amid the stress of negative public scrutiny:

> The politicians also seem to forget that it isn't enough to have a bag of money and a laboratory full of scientists. Just as important is the existence of a problem, the appropriate stimuli and the right kind of atmosphere. Where the real problems exist there is no difficulty in getting scientists to tackle them. Just at the moment there don't seem to be any real problems in the defence field. I have kept myself generally in touch with the defence matters here and overseas and nearly all the problems which exist could be handled by competent radio engineers . . .
>
> Stimuli too are important. They are coming from many directions at the moment but not from the defence people. Rather the reverse. Since the end of the war the Division of RP has done some applied work on navigation aids for civil aviation. The results of this work have definite application to Air Force problems and the Air Force has taken an interest in the work. They have quite failed to make use of the developments, however, because of their lack of trained personnel. What is the good of making gadgets for them if they can't make use of those which already exist?[21]

---

[19] Fred White has written a personal, first-hand account "CSIR to CSIRO—The Events of 1948–1949". Additional information can be found in Schedvin (1987), "The Culture of CSIRO" and Sullivan (2009, pp. 122–123).

[20] A discussion of Pawsey's own views on these issues is provided in NRAO ONLINE.36.

[21] Bowen had written to Rivett on 1 October 1948, with a note of support concerning harsh criticism of Rivett in the Parliament. ". . . [H]ow sorry I am that your name and that of CSIR have been used so badly by a minority in the House [of parliament]. [We] are aghast at the statements being made. Need I say that we are wholeheartedly behind you and the point of view you have taken [in defence of the criticism that CSIR was 'soft on communism']". Rivett replied on 4 October 1948: "That defence work involving secrecy and military security precautions should be placed under Defence control can, I think, hardly be disputed, and recent events add to one's convictions that that is the proper course. Tragedy will arise if other parts of our work are similarly taken from us on any wrong assumption that they, too, belong to defence. The situation is threatening. I cannot see us carrying on CSIR in the atmosphere that is proper and inevitable for studies in war technology. There could be only one justification for it, namely, that international relations justified the declaration of a state of emergency. It is not for us to say that any such stage has been reached."

Towards the Sun

The additional topic of "solar and cosmic noise" that appeared in Bowen's planning document for August 1946 contained a precis of the enormously exciting developments that had occurred within a few months of Victory in the Pacific Day. This precis summarised Pawsey and Payne-Scott's contribution to the RPL Annual Report from June 1946, which commented, presciently:

> The subject is clearly of fundamental importance to astrophysics since it provides a method of investigating extra-terrestrial phenomena by means other than light. The work is still in its infancy but the success already achieved gives strong grounds for hoping that important astrophysical discoveries relating to the outer atmosphere of the sun and stars will derive from further study.[22]

---

[22]NAA C3830 D1/2.

Part IV
Hot Corona

Chapter 11
Beginnings of Solar Radio Astronomy, 1944–1945

My impression from Woolley was that the sun took an impish delight in setting traps for young players.[1] J. L. Pawsey, 24 August 1946

The Forerunners

The very early history of radio astronomy is well known, and here we merely briefly recapitulate the events for readers.[2]

In 1932, Karl Jansky (1905–1950), an engineer at Bell Laboratories, was investigating atmospheric sources of static using an antenna that could be rotated to identify the direction of sources of radio interference at 20 MHz. He observed a "hiss", whose location of maximum intensity rose and fell on a cycle of 23 hours and 56 minutes, the period of the Earth's rotation relative to the stars (rather than to the sun). From correlating his observations with astronomical maps, Jansky suggested this "star noise" came from the Milky Way and was strongest from the galactic centre, specifically, from the constellation Sagittarius. In 1933 he published two papers: "Radio Waves from Outside the Solar System" and "Electrical disturbances apparently of extraterrestrial origin". However, the papers aroused only limited notice and Jansky did not get support to pursue this line of investigation at Bell. Astronomers did not have the familiarity with radio engineering and ionospheric research to engage with his finding.

Jansky's papers inspired radio engineer Grote Reber (1911–2002), who built his own antenna, a 9 m parabolic dish, in Wheaten, Illinois, confirming Janksy's

---

[1] Sullivan archive, letter of 24 August 1946 from Pawsey to Sydney E. Williams of the University of Western Australia. Pawsey introduced the "impish" quote with the following text: "[M]ay I pass on a caution to you which was aimed at me by Woolley in my own first burst of enthusiasm six months ago. What is the probability of a chance coincidence?"

[2] Exhaustive coverage and analysis are provided in Sullivan (2009).

© The Author(s) 2023
W. M. Goss et al., *Joe Pawsey and the Founding of Australian Radio Astronomy*,
Historical & Cultural Astronomy, https://doi.org/10.1007/978-3-031-07916-0_11

observations using a 160 MHz receiver in 1938. In 1940 he turned to making a radiofrequency sky map at 160 MHz, and in 1943–1944, he made and published contour maps showing the brightness of the radio sky, finding hints of strong sources in Cassiopeia A and Cygnus A for the first time. Reber interpreted his observations as arising from thermal emission from hot electrons. As Sullivan (2009, p. 63) has commented: "[O]ver the next decade thermal radiation become one of the leading explanations, albeit a troublesome one, for the galactic background."

Prior to 1942, no one observed radio emission from the sun. There were some near-misses in the 1930s, where radio emission from the sun might have been detected by professional radio physicists.[3] In astronomy, there was in fact, a long history of study of the relations between solar activity and terrestrial effects, such as impacts on the earth's magnetic field. But, as Sullivan (2009, p. 89) comments:

> The solar maximum of 1937 was the first with any significant amount of traffic in shortwaves and therefore a decent chance for detection and identification of solar radio waves. Yet this did not happen. Even though radio specialists often had the sun on their minds, they too did not have the directional antennas needed to pinpoint the sun, and they tended to think in terms of indirect or particle effects. And although solar astronomers in turn also had radio on *their* minds, they knew insufficient radio physics to pay attention to the hiss as anything more than an ancillary phenomenon.

It was perhaps inevitable that people associated with radar during WWII would make serendipitous observations of extraterrestrial radio emission. Two of these are of particular importance. The first was the observation of sunspot emission by James Stanley Hey (1909–2000) in February 1942. Hey was investigating radar jamming following the escape of three German warships up the English channel, not prevented by radar detection—though this should be attributed mostly to human factors in the form of problems in military bureaucracy and communications which delayed the relay and interpretation of radar detection of the vessels (Sullivan, 2009, p. 90). Hey correlated reports of severe noise jamming with sunspot activity after checking with the Royal Observatory. He concluded that a sunspot region, which was believed to emit streams of energetic ions and electrons in magnetic fields of around 100 G (gauss), could emit metre-wave radiation. These findings could not be shared or published due to wartime secrecy, and controversially were delayed by Sir Edward Appleton in order to claim priority in 1945 (Sullivan, 2009, p. 92).

The second significant observation also occurred in 1942, by Bell Laboratories radio engineer George Clark Southworth (1890–1972). Southworth is remembered for his development of wave guides in the 1930s, but he had been interested in the sun and had been designing occasional experiments to understand solar radiation since the 1920s (Sullivan, 2009, p. 92). In June 1942, at the advice of Karl Jansky's former boss at Bell (Harald Friis) Southworth trialled some directional experiments at 3.2 cm that detected small excess noise. He quickly determined that he and his assistant[4] were measuring *direct* solar radiation, because their 2° beamwidth allowed

---

[3] Including a study in Tokyo by Minoru Nakagami and Kenichi Miya.
[4] Archie P. King.

them to locate the origin of the excess noise. This was the first detection of "normal" or thermal radiation from the sun.

Southworth's results were written up in 1942 in two memoranda-like files labelled "Confidential" due to wartime concerns (that they would give the enemy insight into the state of radio research in the USA) and circulated to his radio colleagues at Bell. In 1943 and 1944, Southworth made many contacts with the astronomical community, but despite considerable interest among them, this line of work was not pursued. It took considerable effort from Southworth to obtain the security clearance that would allow publication of the reports in April, 1945 (Southworth, 1945). Although Southworth attempted to get Jansky transferred to his team to continue these studies, Bell Laboratories continued to show a lack of support for radio observation of the sky. With Southworth engaged in writing a book on wave guides, solar research at Bell came to an end in late 1946.

Southworth calculated the intensity to be expected from thermal emission from the sun, found almost exactly the value he had measured in the summer of 1942 and confirmed this with repeated observations. Sullivan describes in detail the irony that, in fact, his results were a matter of chance. They contained an error not discovered until it was identified following publication by Payne-Scott in Sydney[5] and by a Bell Labs colleague, Charles Townes.

Southworth concluded from his observations and calculations that radio emission from the sun (measured at microwave wavelengths) did fit Planck's (thermody-namic) radiation law.[6] In 1943 he made more observations at 1.25 cm and 9.8 cm. The 9.8 cm intensity was lower than that at 3.2 cm, as predicted by the Planck relation. But the 1.25 cm observations were lower than expected. Southworth speculated that the diameter of the 1.25 cm sun was larger than the optical sun or that atmospheric absorption or effects had influenced the data. It is noteworthy that Southworth, who interested nearby optical astronomers in his results, saw himself as providing an extension to solar infrared observations (hence "far" infrared) and along with them, confirmation for Planckian theory at microwave frequencies (Sullivan, 2009, p. 98).

The RPL scientists likely did not know of the wartime discoveries of Hey and Southworth prior to their publication in 1945. But challenges in improving radar receivers led them to their own interest in radio observation of the sky.

---

[5]Letter from Pawsey to Southworth on 7 December 1945.

[6]Sullivan (2009, p. 93) records that the irony was that this agreement *did* turn out to be a matter of chance. The emergence of synchrotron theory will be discussed in Chap. 34.

From Applied Science to a New Field

Solving the practical problem of building better receivers during the war led Pawsey and Ruby Payne-Scott to radio astronomy within two months of the end of the war in August 1945.

Harry Minnett, the young electrical engineer who worked so carefully with Joan Freeman on microwave radar, recalled, in an interview, the challenges of 34 years earlier[7]:

> Receivers were getting more and more sensitive and we were concerned with the whole thermodynamic theory of their noise level and its relationship, through the antenna, to space—if the antenna were in an enclosure at three-hundred degrees, what would be the noise level? This was different from the purely circuit approach that had been worked up by Nyquist and others.

As radar research had progressed, Pawsey and RPL physicist Ruby Payne-Scott recognised that the capacity to improve receivers would rely on developing a complex physical understanding of the effect of the many sources of radiation that existed outside the receiver. Since antenna theory had developed from understanding and building circuits from the 1880s, research had concentrated on developing physical theory to understand the internal components of these systems, such as how they generated noise themselves as their noise temperature (effectively their sensitivity) changed, in association with their power. The paradigm for this research was thermodynamics, a paradigm well established in the nineteenth century to understand the relationships between heat, radiation, energy and matter. The mathematical basis for thermodynamics was sophisticated and well established by the late nineteenth century.

Radar research had shown that developing more sensitive receivers would require a better understanding of how the range of external factors—specific sources, the earth, the sun's radiation, the complexity of atmosphere and ionosphere, that is, the "antenna in space"—influenced an antenna. Hey discovered this in September 1944. By then Hey was investigating why the installation of more sensitive receivers on radars that warned of the approach of devastating pilotless V-1 missiles failed to improve their performance, when his colleague J.M. Scott suggested that it might be external ("cosmic", called "Jansky noise") noise rather than electronics that were limiting system performance (Sullivan, 2009, p. 100). Hey and colleagues would discover the first radio source (called "radio star") in May 1946, the source Cygnus-A (Sullivan, 2009, pp. 101–105).

The problem of improving receiver sensitivity, which would require a strong theoretical, that is, mathematical understanding of the antenna in space, had attracted attention from Pawsey and Payne-Scott in 1943 and 1944. How to quantify the radiation output of the "antenna in space" was a theoretical challenge for this period. Payne-Scott worked out the mathematics of converting the output of the receiver into

---

[7] Sullivan (2009, p. 126, interview 1978).

absolute units, becoming an expert on the new concept that captured this quantification, "antenna and brightness temperature". A prominent example of her understanding can be found in the three-page memo "Derivation of Incident Radiation from Resistance Noise Concept" (unpublished memo from December 1945 by Payne-Scott, Radiophysics Archive).[8] Payne-Scott also took up the challenges of how to calibrate receivers and transmitters during her wartime radar research. For example, when she was involved in the calibration of the prototype microwave radar systems at RPL,[9] she built a noise tube that could provide an absolute scale leading to a determination of the system temperature and hence sensitivity of the new radar receivers.

Of the turn in attention to the "antenna in space", Minnett recalled (Sullivan, 2009, p. 126): ". . . it obviously occurred to Ruby Payne-Scott and Joe Pawsey that radiation from objects might possibly be seen. I remember that Ruby had a small paraboloid [of four feet diameter] poking out a window at certain objects in the sky to see how the noise level varied."

These measurements were the first radio astronomy undertaken in Australasia. The purpose of these S-band tests in March 1944 was not only curiosity about radio observations of the sky, but also to characterise the performance of the new S-band receivers that were to be a part of the new radar systems. At this time, the microwave frequency of 3 GHz (10 cm) was much higher than the conventional LW-AW P-band (200 MHz or 1.5 m) radars that had been perfected at the frequency used by the famous LW/AW (Light Weight Air Warning) radar in WWII (see Chap. 9 and ESM 9.5, Light-Weight).

Pawsey and Payne-Scott conducted a number of observations that were summarised in a report, RP 209, "Measurements of the Noise Level Picked up by an S-Band [10 cm] Aerial", 11 April 1944, which of course was classified until after the war. Based on the report we can infer that Pawsey and Payne-Scott were trying to answer a number of simple questions: (1) What did the sky look like at 10 cm? (2) What was the source of microwave radiation at 10 cm? (3) How were the RPL staff to calibrate the absolute level of the S-band signals? (4) Was the level of S-band "noise" so extreme that it would influence the sensitivity of radar systems? (5) What was the relation of the "cosmic" static detected earlier by Jansky (20 MHz) and Reber (160 MHz) to the S-band sky? (6) Could the RPL group detect the Milky Way at S-band?

In the report, Pawsey and Payne-Scott mentioned a major dilemma:

Since air is relatively transparent [at 10 cm], we expect the principle objects [emitting at S-band] to include ground, clouds, possibly ionosphere and matter in space. If the only

---

[8] Sent to Southworth on 7 December 1945 by Pawsey. The memo contained a direct simple derivation of the radiometer equation (the sensitivity of a radiometer based on the system noise, bandwidth and integration time). Payne-Scott noted that Burgess (1941 and 1946) had derived the equation "in a less direct fashion" (Burgess, 1941 p. 293, and 1946, p. 313). The memo also located in NAA C3830 A1/1/5 Part 1, November 1945 to December 1946.

[9] Goss and McGee (2009), have a detailed description of these efforts.

source of noise power is thermal energy,[10] we might expect the noise temperature of an aerial looking into the sky to be very low.

However, there may be other noise sources. The noise from electrical disturbances in the earth's atmosphere appeared to be of very low intensity at centimetre wavelengths, except in the vicinity of lightning. At about 20 MHz a further source of noise, called "cosmic static" and apparently originating in the Milky Way, has been investigated by Jansky (1932, p. 1920) and Reber (1940, p. 68). If Reber's suggestion as to the nature of this noise is correct, it will vary inversely with frequency [weaker at 10 cm] and will be negligible at cm wavelengths, while if Jansky's suggestion is correct,[11] it is of the nature of thermal noise from objects at very high temperature and hence will result in values of [intensity] very much greater than ambient temperature. Thus, the equivalent noise temperature of free space at centimetre wave-lengths may have any value from almost zero to some hundreds of times ambient temperature, depending on the nature of the noise sources.

Despite the theoretical and practical interest attached to measurements of ultimate noise levels, the authors are not aware of any reported measurement of received noise powers in the centimetre range of wavelengths.[12] These described here are of a preliminary nature, and the authors hope to extend them further.

Sullivan (2009) has provided a vivid description of the process followed by Pawsey and Payne-Scott:

They used a 20 by 30 cm horn connected to a receiver with a system temperature of 3500 K, one person pointing the horn around the room or out the window in various weathers, the other taking a reading on a meter. Changes of 20 to 300 K [in antenna temperature] were noted, and they were particularly struck by the apparently low absolute temperature of the sky, less than 140 K. Moreover, they noted a "most unusual" consequence of this: inserting attenuation between the horn and receiver actually *increased* the measured signal![13]

Pawsey and Payne-Scott also attempted to detect 10 cm radiation from the southern Milky Way by pointing a 4-foot dish out the window towards the direction of the constellation Centaurus (near the Southern Cross) and then towards another displaced position. Nothing was detected with limits of less than 10 K antenna temperature, "very much less than that observed by Jansky and Reber and so small as to have no observable effect on noise in present observations." The latter statement implied that military radar 10 cm receivers would not be degraded by background "cosmic" radiation.

The 1944 RP 209 report also stated: "No attempt was made to observe the sun." Sullivan (2009, p. 128 and private communication 2008) has wondered why not—

---

[10]Thermal emission is an ideal body or surface that completely absorbs all radiant energy falling upon it with no reflection and that radiates at all frequencies with a spectral energy distribution dependent on its absolute temperature. It is related to blackbody radiation.

[11]Pawsey and Payne-Scott seem to have interpreted Jansky's results as implying the signal scaled as a black body, proportional as frequency squared. Reber also made this assumption in his first vain attempts to detect the Milky Way in 1938 at 3300 MHz; only when he observed at 160 MHz did he succeed in detecting the galactic background.

[12]Pawsey and Payne Scott had not received Southworth's classified document from the US.

[13]An explanation was provided by Pawsey and Payne-Scott: "This last measurement independently confirms the low equivalent temperature of sky radiation [at 10 cm]."

given that they mentioned it. The idea might have been mentioned at RPL, since Piddington and Kerr recalled making transient trials to detect Jansky's "star noise" also.[14] They were of course not aware of Hey's report of sunspot radio emission from 1942, nor of Southworth's observations.[15] Additionally, the window from which they pointed their 4-foot antenna had a southern exposure, precluding an observation of the sun in March from Sydney.[16] Sullivan has calculated that they *would* have detected the sun if they had tried to do so: "Pawsey and Payne-Scott would have detected an antenna temperature ... of about 150–200 K, well above their sensitivity to changes of about 20 to 30 K. This type of dish was in fact very similar to that employed by Southworth in 1942–1943."

The summary paragraph of the April 1944 memo by Pawsey and Payne-Scott pointed out that the availability of a 10 cm "cold sky" would enable engineers to calibrate the sensitivity of the receiver, using the on-off system with a known noise source:

> A possible practical application of this work is in the measurement of noise factors of receivers without elaborate equipment. If it is established that the equivalent temperature of an aerial pointed at a clear sky is consistent, then it should be possible to devise simple procedures, either pointing the aerial alternatively at clear sky and constant temperature enclosure or inserting attenuation in the feeder, which yield an accurate result without the use of the exceedingly cumbersome signal generator equipment at present required.[17]

From these actions directed at the applied problems of calibration and receiver sensitivity, the first recorded observations of a new branch of basic science began.

Radio Astronomy in New Zealand and Australia

It was not developments in the US and the UK, but independent wartime radar observations from New Zealand that gave impetus to investigating the "very far infrared" in the post-war period (Chap. 10). Sullivan comments (2009, p. 128):

> What galvanised Bowen and Pawsey into jumping onto solar noise was the "Norfolk Island effect"—solar radio bursts observed by New Zealand military radar stations from as early as March 1945 and analysed by Elizabeth Alexander, an English radar researcher at a RPL-type lab in New Zealand ... When Bowen learned of these observations in July 1945, he was entranced.

---

[14] Goss and McGee (2009), "Neophyte radio astronomer: Ruby Payne-Scott", page 66. There are two possible earlier attempts with no published evidence, Piddington and Martyn in 1939) and later by Kerr (Sullivan, 2009, p. 86 and 128 for descriptions of the two unpublished attempts).

[15] "The Microwave Sun", see Appendix G, Goss and McGee (2009).

[16] Sullivan (2009, p. 128), from Minnett to Sullivan October 1986.

[17] Payne-Scott described this system in a series of memos in 1943 and 1944 (Goss & McGee, 2009, p. 53); the method is now known as the "Y factor" method of determining the "noise figure" of a receiver using a high-power noise source.

The "Norfolk Island effect", a large increase in noise lasting about 30 minutes at sunrise and sunset, was detected on 27 March 1945 by Flying Officer Hepburn of the Royal NZ Air Force and by additional colleagues including Roy Stewart and Keith McPhail[18] operating the radar station on Norfolk Island.[19] The observations were reduced in Wellington by Dr. Elizabeth Alexander (1908–1958) of the Radio Development Laboratory of New Zealand's Department of Scientific and Industrial Research. Alexander was a British geologist who married a New Zealand physicist, Norman Alexander, whom she had met in Cambridge; there, the Alexanders also met and became friends with Pawsey. The Alexanders moved to Singapore in the late 1930s. Alexander undertook some work in radio direction finding there from 1939, before being evacuated with her three children, to New Zealand in January 1942. (Norman Alexander remained, was captured, and survived Changi prison.) Elizabeth Alexander was immediately appointed Senior Physicist and Head of the Operational Research Section of the Radio Development Laboratory in Wellington.

The data from Norfolk Island was crude (only one station used a meter) and variable. All the NZ radar units were equipped with aerials of fixed elevation (towards the horizon) and could only be moved in azimuth. The opportunity at some of the stations, especially Norfolk Island, at sunrise and then at sunset was advantageous. Alexander was thorough in her analysis and confident that the effect she saw was real. Alexander realised that radio frequency radiation at 200 MHz had been detected during both sunrise and sunset; the location of the antennas was the peak of Mt. Bates (321 metres, with a view of the sea over 360 degrees in azimuth) on the north side of Norfolk Island. During the next month, Alexander coordinated follow-up observations at four sites on the North Island: Piha, North Cape, Maunganui Bluff and Whangaroa.

Alexander sent "Report on the investigation of the 'Norfolk Island Effect' from July 1945"[20] to her friend Pawsey in August 1945. But the news had reached Bowen and White in July. As it happened, a young airman, Tim Marsden, was the Commanding Officer at Whangaroa, and he sent the data from his station to his father, Ernest Marsden, Director of Scientific Developments Department of Science and Industrial Research, New Zealand, who sent it on to his former colleague Fred White.[21]

On 27 July 1945, Bowen wrote to Marsden with an enthusiastic response[22]:

---

[18]Goss interviewed [the late] Roy Stewart in 2011 and Keith McPhail in 2013, both in New Zealand.

[19]Although a part of Australia, during WWII the NZ military were responsible for the defence of Norfolk Island.

[20]RD 1/518.

[21]Marsden apparently did not realise that White had left RPL, transferred to CSIR Head Office as Assistant Executive Officer in Melbourne January 1945; the letter was intercepted by Bowen and then sent on to White (1 August 1945).

[22]All correspondence with RPL with NZ colleagues, NAA C3830 A1/1/11945–1946 Part 1.

I took the liberty of reading your letter ... to White before sending it along to him in Melbourne. We were very interested indeed to hear about the radar [sic] observations made by your son at Norfolk Island [Bowen had misunderstood, the confirming observations from Tim Marsden were from Whangaroa, not Norfolk Island], and we will attempt to repeat them here in Sydney. We are quite mystified by the results because it appears that, while thermal noise from the sun is expected at radio frequencies and is actually received on 10 and 3 cm equipment, one would not expect to be able to detect it on C.O.L. ["chain overseas low-flying" radar] equipment at 200 MHz ... If we are able to duplicate your son's results in Australia, you may be quite certain that we will keep you fully informed of the work.[23]

Then, in a letter to White on 1 August 1945, Bowen wrote:

I have taken the liberty of sending [Marsden] a reply ... [He repeated to White the surprise experienced at RPL when they heard that the sun was detected at 200 MHz] ... I have heard rumours of the same thing happening in England [presumably the Hey detections from 1942], but as far as I am aware, the subject has never been followed up.

As Sullivan has pointed out, Bowen's letters indicate that by July 1945, Bowen and Pawsey were aware of the thermal microwave radiation detected by Southworth at Bell Laboratories in New Jersey during the war.[24] Commenting at the end of that year, Payne-Scott wrote that the Australians "were inspired by the almost simultaneous arrival of three reports in the laboratory"[25]: those of Reber, Hey and Alexander. The RPL group already knew of Jansky's pre-war publications.

Alexander's letter to Pawsey provided a thoughtful and thorough analysis on 1 August 1945:

As far as the 200 MHz radiation from the sun—Report RD 1/58 [from the Department of Scientific and Industrial Research, Wellington, dated 1 August 1945, "Report on the Investigation of the 'Norfolk Island Effect'"] attached, described our present and proposed investigations of the problem. I think the main differences between Southworth's latest results and ours are first his work in the centimetre band fits more or less with black body theory and ours shows definitely too much energy on 200 MHz for theory. Sir Edward Appleton has also taken measurements on 200 MHz and confirms our thinking.[26] He suggests that at times of increasing sun spot activity there is an increase in energy at both ends [this statement is quite unclear regarding the high frequency portion of the spectrum] of the sun's spectrum, and has encouraged us in our efforts ...

The official report of 1 August 1945 (RD 1/518) provided numerous details with descriptions of the observations made at four sites on the North Island of NZ,

---

[23] Bowen asked Marsden to ask his son if he could "confirm that the noise is received from the sun at any angle of elevation, not only at sunrise or sunset, by observing at a fixed azimuth at sunrise or sunset". Apparently, Tim Marsden did not continue his radio astronomy work. Pawsey and Bowen may have heard rumours of the Hey detection of low frequency solar bursts from the 27 and 28 February 1942 wartime detection on the south coast of the UK at wavelengths of 4–6 m (55–76 MHz) (Hey, 1973, p. 14, and in *Nature*, 1946).

[24] Goss and McGee (2009) discuss the timing of the recognition of these detections carried out by CSIRO in 1945–1946.

[25] The December 1945 summary report, discussed in Chap. 12.

[26] In fact, as discussed by Sullivan (2009), Appleton had no new data at this period.

including Piha on the west coast, across the isthmus from Auckland.[27] The conclusions were uncertain:

> The results so far obtained are too few and insufficiently accurate for foundations for any kind of theory. There is a strong suggestion, however, that there was an increase in solar radiation on 200 MHz observable in the New Zealand area at the end of March and during April of 1945.

During the following year, a more mature account was written by Alexander for a popular NZ electronics magazine *Radio and Electronics*, vol 1, no 1 April 1946. Alexander provided more details about her interpretations of the origins and consequences of the NZ research since March 1945:

> Early October [1945] the "Norfolk Island Effect" became very strong and F/O [Flying Officer] Brook in charge of the Piha unit had a Yagi aerial constructed and so mounted that it could be rotated in both azimuth and elevation. Though the gain of the Yagi was much less than that of the standard broadside array,[28] the [solar] noise could still be observed, its intensity remaining the same throughout the day whenever the aerial was directed towards the sun, showing that the earlier observations were limited by the limited directivity of the aerials.[29] During the periods of observations, March to December 1945, there were two periods of sunspot activity and it was noted that these coincided with periods of intense solar "noise".
>
> At an earlier stage in the programme, reports were sent to the UK, USA and Australia. Enquires were made whether it had been observed on longer wavelengths outside the NZ area. The reply from Australia stated that such radiation had not been observed ... but later (October 1945) the RPL in Sydney took accurate measurements at 200 MHz, and found close correlation with sunspot activity.

After speculating about the mystery that although "it is well-known that in the visible and infrared regions of the spectrum the sun radiates very nearly, though not exactly, like a black body", the Norfolk Island effect at 200 MHz was clearly much in excess to the expected for a thermal body at the temperature of the sun, 6000 K. Alexander concluded:

> ... anyone [including radio amateurs] who cares to build an ultra-shortwave receiver can be fairly sure of collecting useful information [about solar radio emission]. The time is appropriate since the sun is just entering a new phase of activity and sunspots may be expected with increasing frequency over the next years.

Alexander compared the unexpected results at 200 MHz with the Southworth and Reber data at cm wavelengths and at 160 MHz (1.9 metres). "Both these observations detected radiation from the sun, in amounts agreeing with black body theory." But actually, this statement was not quite accurate. Sullivan (1982, p. 44) has pointed out:

---

[27] The site of one of the Bolton-Stanley locations during the Cosmic Noise Expedition of 1948.

[28] Ie, the COL 200 MHz system used in NZ radar.

[29] The aerials were fixed at zero elevation, towards the horizon, movable only in azimuth.

Modestly tucked away at the end of the paper[30] is the first published radio detection of the sun, although Hey and Southworth had both earlier detected the sun, but had been prevented by wartime secrecy from being openly published. Reber notes that the measured solar intensity is "rather surprising", but does not indicate that his measurements imply an astonishing brightness temperature of about a million K for a disk the size of the optical sun.

Reber had not detected solar bursts but the hot corona!

Bowen's response to Marsden (6 November 1945) provides additional proof of the decisive nature of the New Zealand news of the "Norfolk Island" effect on the Australian plans for the post-war era:

> I am writing you to inform you further of our plans for research [on radio frequency noise from the sun and stars] and to let you know of the interesting results which we have been getting.
>
> We have always been interested in the broad question of extra-terrestrial radio noise, but it was not until your report on noise from the sun at sunrise and sunset came to hand that we carried out a series of practical measurements . . .
>
> Our present idea . . . is that we should work in collaboration with the Observatory at Mount Stromlo and set up facilities for taking measurements of the intensity of the noise received on 200 MHz on the sun and stars. We then thought of extending the wavelength range . . .
>
> I shall be interested to hear from you as to whether you consider that New Zealand should carry **out similar work. I suggest that we should make arrangements so that we do not overlap.** (our emphasis).

However, circumstances in NZ were not favourable in 1945 for sustained fundamental scientific research. The scientific leadership in this small country (of 1.7 million) were not willing to commit to a long term programme. On 24 November 1945, Marsden replied (to Bowen) negatively about the New Zealand plans for the future: "In the circumstances we shall not continue this line of work."

Also on 2 November 1945, Pawsey finally replied to Alexander's 1 August letter:

> We have become very interested in radio noise from the sun, or "cosmic static" as Jansky christened it, when it was attributed to the stars and not the sun. We have been carrying out other observations at 200 MHz equipment and obtained very good correlation with solar activity as evidenced by sunspots. We are in a very good position for such investigations as we can work in collaboration with [Stromlo] and consequently we are planning to proceed with an active programme of investigation. [He enclosed a copy of the paper submitted to *Nature* on 23 October 1945, to be published in February 1946. See Chap. 12).

On 23 November 1945, Alexander responded. She was not optimistic about the New Zealand continuation of the "Norfolk Island effect":

> I am glad you are working on radio-frequency radiation from the sun. I do not know whether any work will be done here on the subject after the end of the year when our COL [Chain Home Overseas Low—200 MHz aircraft warning radar antennas] close down. We've taken some measurements with Yagis [at Piha] and so had a look at the sun throughout the day. The aerial gain is too small to take measurements except when the radiation is very strong (as in the middle of October). Our results are not worth much so far but though the elevation of the sun makes little difference at times [i.e. the Yagi antennas could observe at any

---

[30] Reber (1944a).

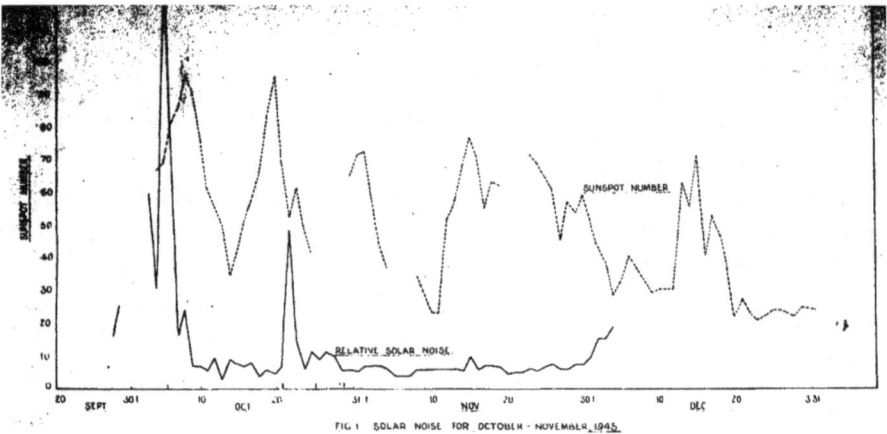

Fig 11.1 Correlation of sunspot number and the relative solar intensity in September to December 1945 from Norfolk Island and Piha (west of Auckland on the North Island). The sunspot data was obtained from the Carter Observatory in Wellington, NZ. Courtesy of the National Archives of Australia, New Zealand Millar report 20 Feb 1946, NAA A1/1/1 Part 1 1945–1946

> elevation in contrast to the COL's, which were fixed to elevation zero for sunrise and sunset.] At other times there does appear to be concentration at low angles which may be either an atmospheric or ionospheric effect or both. I doubt that New Zealand will be able to put sufficient effort into building aerials adequate to investigate the phenomenon. It is a large-scale job, if it is done properly . . .

Final unpublished accounts of the New Zealand data were prepared by Alexander on 17 December 1945 and J.G. Millar on 15 January 1946 and 20 February 1946.[31] The Millar report of 20 February 1946 provided the last available description of the New Zealand solar work of 1945. Millar offered a few details of the observations from Norfolk Island and Piha:

> Norfolk Island reported no measurable increase in noise level from 24 July until 28 September. During October and November, regular reports were received from two units [Norfolk Island and Piha] . . . The agreement between the two sets of readings was good, and there was no significant difference between sunrise and sunset figures at the unit reporting both. [Norfolk Island had a full view of the entire diurnal path of the sun, but could only observe at sunrise and sunset, close to the horizon.]

In Fig. 11.1 (from the New Zealand Millar report of 20 February 1946) we show the correlation between sunspot number and relative 200 MHz flux density from the end of September to the end of 1945. In Chap. 12, we show a comparison of this figure with the Australian data. Millar continued:

> It would appear that in general the solar noise remains at a fairly constant level, relieved by "bursts" lasting several days. [Norfolk Island RNZAF personnel] reported "the evening of

---

[31] Sullivan archive and Grahame Fraser (Christchurch NZ) archive. The Millar February report is RD 1/536 from the Radio Development Laboratory.

October 5[32] was the only occasion since readings began that a return to the abnormal conditions of last March–April was seen". During the quiet periods, the needle of the noise meter would remain steady, but during bursts the needle generally fluctuated violently, at times striking the end of the scale . . . The bulk of evidence does not support the suggestion that there is a "focusing" effect at sunset and sunrise . . . [From a comparison of the NZ solar noise at 200 MHz and the occurrence of radio fade-outs], it is seen that the solar noise bursts occurred on most occasions when a fade-out was observed, and that during the July–September periods when no solar noise was observed, there was only one weak fade-out . . . It appears very likely that the bursts of metre radiation are closely associated with the shorter radiation [ultra-violet] surges causing ionospheric fade-outs.

Reunited with her husband, Alexander and her family departed New Zealand in early 1946. Before then, Alexander continued her reports of the "Norfolk Island effect", starting from late July 1945 (RD 1/518):

The Norfolk Island Station has kept watch daily for solar radiation using the meter [calibrated signal generators and noise meters] since July 24th, 1945. Owing to various difficulties the other stations were unable to take readings until September 1945. [The observations ceased at the end of the year when the stations were closed.]

During the period March to October . . . two periods only of intense solar radiation on 200 MHz were observed. These periods were centred on 27 March and 5 October which were also periods of great sunspot activity. On 26 March there was a complete radio fade out [presumably at short wave or HF, 3 to 30 MHz]. Radio data for October are not to hand . . .

During the period of intense activity early in October a Yagi aerial was mounted [at Piha, NE of Auckland], and observations were taken throughout the day. Noise signals from the direction of the sun were observed. The signals fluctuated rapidly but did not completely disappear until sunset.[33]

Such evidence as we have so far in New Zealand points to a direct correlation between sunspot number and solar noise. During the period of intense noise observed about 5 October [in Sydney, the most intense signals from the sun during October 1945 were detected at Collaroy on 5 and 6 October], violent surges of noise were observed at irregular intervals. These surges were of momentary duration and sent the noise meter needle hard over. Although we have no absolute measure of the power received, there is strong evidence that these were during periods of intense activity.

She concluded: "long wave solar radiation is far removed from black body radiation."

---

[32] The RPL group at Collaroy had only been observing daily since sunrise 3 October.

[33] It is not clear whether the New Zealand scientists realised the importance of this observation, possibly the first-time solar signals had been observed for an entire day. With their low gain, the Yagi antennas could only detect the sun during periods of intense solar activity.

Chapter 12
Serendipity: Sunspots at Collaroy, 1945–1946

... By the "serendipity pattern" in research ... empirical facts aid in the initiation of theory: The serendipity pattern refers to the fairly common experience of observing an unanticipated, anomalous and strategic datum which becomes the occasion for developing a new theory or for extending an existing theory ...

Chance discoveries, then, depend on an impressive list of estimable qualities in a scientist: enterprise, courage, curiosity, imagination, determination, assiduity, and alertness. There is "nothing fortuitous" in so-called discoveries. (Merton and Barber, *The Travels and Adventures of Serendipity*, 2004, p. 196 and 178).

Serendipity

The beginnings of radio astronomy in Australia were marked by serendipity. Within days of beginning investigations, inspired by the reports from New Zealand, Pawsey and his colleagues experienced extraordinary results.

The Radiophysics group were amazingly fortunate to have begun their initial observations on 3 October 1945. Not only was the most prominent solar cycle of the modern era just beginning, but in addition, a local maximum in the solar noise at 200 MHz occurred during the first days of October. For example, the sunset observation of 4 October had a brightness temperature of about 13.5 million degrees.[1] The observations they made then established their place, their confidence and their fascination with this surprising research.

Robert Merton, a prominent sociologist of science, explored the close relationship between serendipity and discovery in science. Merton thought that serendipity in

---

Supplementary Information The online version contains supplementary material available at [https://doi.org/10.1007/978-3-031-07916-0_12].

---

[1] Based on the temperature of a blackbody of the same size as the sun (0.5 deg. diameter), that would produce thermal radiation corresponding to the observed signal.

© The Author(s) 2023   169
W. M. Goss et al., *Joe Pawsey and the Founding of Australian Radio Astronomy*,
Historical & Cultural Astronomy, https://doi.org/10.1007/978-3-031-07916-0_12

science—the finding (or realisation) of things that one was not (quite) in quest of—
had a pattern in which accident combined with sagacity. Pawsey's early solar
research exhibits those dimensions of "sagacity" —experience, skill, tacit
knowledge (including a fine engineering knowledge of his team's instruments),
and theory-informed assumptions—that enabled the "serendipity pattern" of unexpected, anomalous and sometimes strategic discoveries.

From our vantage point, one striking feature about Merton (1910–2003, Professor
at Columbia University) is that he was only two years younger than Pawsey. They
shared the experiences and outlook of their generation. Their similar strong commitments to what Merton articulated as "the ethos of science" arose from similar
reactions to the experience of WWII. Merton's sociology of science perhaps fits the
early years of radio astronomy in part because it described the science of its time—
and Pawsey was the archetypical Mertonian scientist.

Action in Sydney at the End of WWII

After learning of the Norfolk Island Effect just days before Victory in the Pacific
Day, Pawsey "swung into action" (Sullivan, 2009, p. 129). He had several advantages to exploit. The first was the group's location in the Southern Skies, which
would enable them to obtain unique data regardless of developments in Britain and
the USA—the advantage inherent to Australian astronomy. The second was his
extremely fine knowledge of his aerials, having been directly involved in their
design, use and construction, and of ionospheric research, which enabled confidence
in distinguishing signals of interest from instrument or atmospheric effects. This
"tacit knowledge" was a key component of Pawsey's sagacity; he also involved RPL
radar engineer Lindsay McCready as a collaborator. The third was his expert
understanding of antenna physics, which enabled him to begin considering possible
explanations for the observations of Hey, Alexander, Jansky, Reber and Southworth,
and to consider what results might be obtainable with the bigger antenna of the radar
equipment available for use in Australia at 200 MHz. The Collaroy and Dover
Heights aerials had just the suitable sensitivity (a few times 10^4 Jy in one sec) to
easily detect the quiet sun (10^5 Jy at 200 MHz). The fourth was Ruby Payne-Scott's
expertise in the quantification of antenna and brightness temperature and in calibration of the signals, which might allow more than qualitative observations to be made.
Without these factors, the accident of observing at a time of major sunspot activity
would not have launched Pawsey's research program with the impact and success it
enjoyed in late 1945.

The stage was set for a new era at CSIR Division of Radiophysics in early
October 1945. Pawsey and Bowen saw an opportunity. In addition to filling RPL
trucks and cars with surplus military radar equipment, often in unopened boxes, they
used their military relationships to good effect. Royal Australian Air Force—RAAF-
Radar Station No 54, Collaroy Plateau (54RS) a distance of about 30 km north of the

RPL at the University of Sydney[2]—was particularly important. During the latter years of WWII, RPL personnel had become familiar with the conditions at 54RS, and a warm spirit of cooperation had been established between its personnel and the staff at RPL. Ted Dellit has provided the story of 54RS.[3] The aerial was a COL 200 MHz radar set, as had been used by the New Zealand Air Force (Chap. 9 and ESM 9.6, Microwave Radar). Dellit wrote:

> The presence of competent and experienced Radar Mechanics and Operators would also assist the scientists and engineers from RPL when conducting their experiments. Relationships between RPL and RAAF personnel at Collaroy Plateau were most amicable, although it is quite probable that many of the RAAF men were not aware of the significance of the work they were called on to do to assist RPL staff . . . [T]he first visit from RPL personnel for experiment purposes was on 23 February 1943. These experiments continued through the life of the Station and, even when it was officially disbanded in February 1946, two Radar Mechanics, who were left on site, made daily recordings of the random [!] radiofrequency radiation emitted by the sun at sunrise [no mention of sunset] to supplement measurements at Dover Heights.[4] **This experiment [at Collaroy] was the beginning of Radio Astronomy in Australia.** [our emphasis].

Collaroy 54 Radar Station observations would occur with few breaks from 3 October 1945 to 15 March 1946. An informal agreement (possibly brokered by Pawsey) apparently allowed the observations to begin, for formal agreement was only requested from the RAAF by Bowen on 9 October 1945.[5] During the month, Flight Lieutenant Blumenthal from Number 54 sent the data from each day's observations to Payne-Scott. The only significant break in the data collection occurred from Christmas Eve (24 December 1945) to 2 January 1946.

As with all the RAAF AW radars, 54RS operated at 200 MHz; the system was the British Chain Overseas Low (COL) system (See ESM 12.1, Collaroy, Fig. 12.1). The antenna was an array of 40 half-wave dipoles restricted to looking at the horizon. Observations could be taken every 2° of azimuth sweeping along the horizon. The Minutes of the PC, the Propagation Committee of RPL,[6] from 14 September 1945, in a meeting chaired by Pawsey[7] reported: "Miss Payne-Scott is going to look for 200 MHz signals from the sun at sunrise and sunset. Such signals, at a level greater than suggested by blackbody theory, have been reported on COL sets in

---

[2] A brief history of this station and the close connection with RPL is described in ESM 12.4, Collaroy.

[3] Warringah Library, LA940.544 DEL, 2000. "Who are they?: The Royal Australian Air Force on Collaroy Plateau in the Second World War" by Ted Dellit.

[4] The Dover Heights solar noise work began well after the Collaroy RS 54 observations had begun on 3 October 1945. The Dover Heights observations of the sun started in mid to late January 1946. The Australia Day observations of 26 January 1946 (McCready et al. 1947), the first interferometry in radio astronomy of a celestial object (see the text later in this chapter), were carried out at Dover Heights by Ruby Payne-Scott.

[5] NAA C3830 A1/1/11945–1946 Part 1.

[6] After April 1949, the Radio Astronomy Committee, see Goss and McGee (2009).

[7] NAA C3830 B1/1 Part 1. Payne-Scott was not in attendance. Pulley, Piddington, Kerr, Iliffe, Yabsley and three RAAF officers were also present.

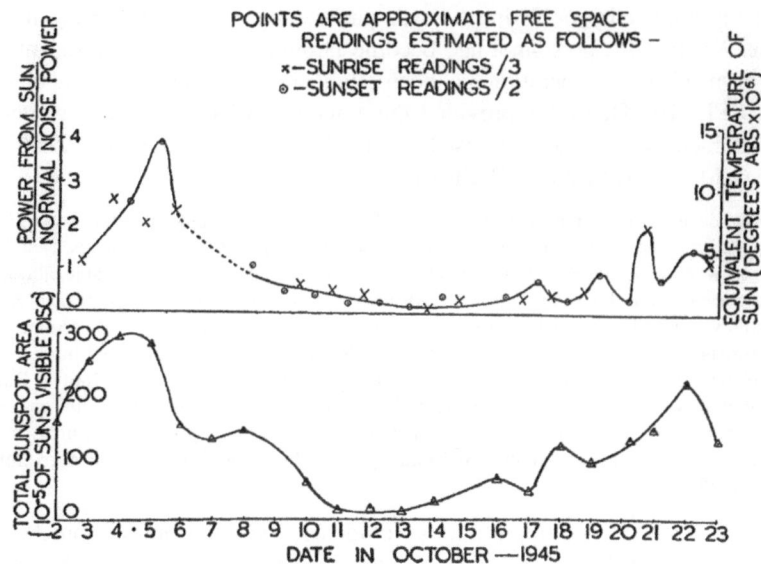

Fig 12.1 Collaroy data from the CSIR -54RS observations in October 1945 (top panel). Bottom panel shows the Mt Stromlo data provided by C.W. Allen of the Mt Stromlo Solar Physics (sic) Observatory showing the total sunspot area. Credit: "Radio-Frequency Energy from the Sun" Pawsey, Payne-Scott and McCready, *Nature*, 9 Feb 1946 vol 157. All rights reserved

New Zealand." This is likely the first mention of solar radio astronomy at a Propagation Committee meeting.

As noted, a local maximum in the solar noise at 200 MHz occurred during the first days of October 1945. For example, at Collaroy the sunset observation of 4 October 1945 had a brightness temperature of about 13.5 million degrees.[8] The next largest peak in the first weeks of October occurred on 21 October 1945 with a peak of seven million degrees. During the first three weeks of the month the average brightness was 3.2 million degrees.[9]

Within a week, Pawsey noted that the sunrise and sunset intensities were quite variable. But there was a definite minimum in intensity; the level did not go to lower values. This base level was significantly larger than the uncertainty in the signal level. As the short-term variable emission (soon to be correlated with sunspot area) decreased, the base level was maintained. This might well express the radio emission

---

[8] Based on the temperature of a blackbody of the same size as the sun (0.5 deg. diameter) that would produce thermal radiation corresponding to the observed signal. The basic intensity unit was (in modern terms) a million Jansky; 10^{-21} watts/m²/Hz). Typical values for the daily intensities were 5.4 units at sunrise on 3 October and on 4 October 8.2 at sunrise and 10.8 at sunset. The maximum values were obtained on 6 October at sunrise, 11.2 units. Minimum values were 0.6 units at sunrise on 14 October 1945.

[9] In ESM 12.2, Marjorie Barnard, we describe the impressions of Lindsay McCready, as presented to Marjorie Barnard in October 1945.

from the sun, *without* the additional emission that it seemed likely was correlated with sunspots.

Sea-cliff interferometry—interferometry was a method familiar to Pawsey from his PhD research—was tried at Collaroy starting in October 1945.[10] However, the group observed no fringes in 1945. McCready, Pawsey and Payne-Scott (1947) wrote: "In initial observations early in October [1945] no interference pattern [between the direct radiation from the sun and reflected radiation from the sea surface] was observed ... there was a wide distribution of spots over the sun's surface.[11]".

After five successful days of data gathering,[12] John N. Briton (chief of RPL from 31 Jan 1945 to 17 May 1946) summarised the status to the Air Force command in Sydney on 9 October 1945:

> RPL and RAAF personnel have, over the last few days, observed on the COL at Collaroy a substantial increase of noise level, similar in effects to noise jamming, when the aerial points towards the sun as it may [fixed elevation of the aerial at 0 degrees] at sunset or sunrise. We do not know the cause of this but suspect it to be an irregularly occurring phenomenon originating in the sun.

Pawsey's initial assumption that the variable radio signals could be correlated with sunspots was based on Hey's report (50–75 MHz) from the 1942 radio outburst (Hey, 1946) which had been associated with a rare, prominent sunspot during solar minimum. To obtain information about sunspots, Pawsey contacted Clabon W. "Cla" Allen (1904–1987) of the Mt. Stromlo Solar Observatory, who sent him measurements of sunspot activity for the same time period.[13] In Chap. 14, we discuss the close connection between Mt. Stromlo and RPL.

Allen's data included variations in sunspot size. Pawsey could plot his observed levels of variable solar noise against the size of the sunspots. The correlations were striking. Pawsey informed Cla Allen of the results straight away on 15 October 1945: "It looks as if there is every chance of a cause and effect relation between solar

---

[10] Sea-cliff interferometry, discussed in Chap. 13, is the use of an antenna pointing over the sea (east or western horizon, east in this case) as a Lloyd's mirror interferometer. The direct ray from the celestial object interferes with the reflected ray from the sea. Twice the cliff height is the baseline of the interferometer.

[11] The paper continued: "Towards the end of January [1946], a compact sunspots group dominated the sun and for this reason an attempt was made to detect a lobe pattern on the morning of 26 January at Dover Heights. [85 metres]. A regular series of maxima and minima was observed, with the expected period and with very deep minima [implying a small angular size]." Fringes were only detected from the higher cliff site—120metres—at Collaroy on 8 February 1946.

[12] NAA C3830 A1/1/11945–1946 Part 1. During the three-week period of 21 days, the RAAF missed seven sunrises and four sunsets; perhaps the airmen were more reluctant to be on station at about 05:30 am for sunrise than for sunset, at a civilised time of 06:00 pm. (Five of the seven missed days were on a Saturday, Sunday or Monday, days related to weekends; the pubs in Collaroy would have been an attractive destination for the peace-time Airmen during the weekend, at least Saturday).

[13] NAA C3830 A1/1/11945–1946, 11 October 1945.

activity as evidenced by sunspots and this noise effect [at 200 MHz]. I enclose a graph of variation of 200 MHz noise over this last week."[14]

Bowen, still Deputy Chief of RPL, was anxious to highlight the success within CSIR. On 17 October 1945 (only two weeks after the first observations at Collaroy) he wrote to Fred White, now the Assistant Executive Office of CSIR:

> I thought you would be glad to hear that some of Pawsey and Miss Payne-Scott's noise measurements are bearing considerable fruits. You will remember the interest which was aroused some weeks ago by the observations in New Zealand of noise from the sun at sunrise and sunset. Random[15] observations of a similar nature have been made in the UK and elsewhere for a number of years, but no one has been able to spare sufficient time to make a consistent set of observations, and the only comment which has been made is that noise shows considerable variation from time to time and sometimes does not appear at all ... Pawsey and Miss Payne-Scott have now organised a series of measurements at sunrise and sunset every day for some weeks and have established a curve of variation of noise level with time, which correlates almost exactly with the Wolf [International] Sunspot Number activity obtained from Stromlo ... [We plan] to report the present observations in *Nature* [and to] discuss with Mt. Stromlo the question of their taking on routine noise observations which would be associated with their normal [optical] solar measurements.

The resulting publication in *Nature* (McCready, Pawsey and Payne-Scott, 1946, p. 158) stated:

> It is apparent that the peaks of 1.5 metre [200 MHz] coincide with peaks of the sunspot area curve and with the passage of large sunspot groups across the meridian. This strongly suggests a physical relation between the two phenomena, as suggested by the British Army observers [Stanley Hey and colleagues].

The main results are shown in Fig. 12.1, with the radio data from 54RS in the top panel and the sunspot area in the bottom panel. The correspondence with the unpublished New Zealand data (Fig. 11.1, Chap. 11) is striking.

The Original Data from Collaroy: First Post-War Radio Astronomy Records

In Figs. 12.2, 12.3 and 12.4, we show the original data from 1945–1946; Pawsey's handwriting is apparent on each document. Figure 12.2 is his annotated worksheet for the Collaroy data of October 1946. The hand-written data refers to the sunspot area on each day (provided by Cla Allen at Stromlo) and the intensity of the 200 MHz radiation at sunrise and sunset in db (relative to the fiducial value of 10^{-21} watts m^{-2} Hz^{-1}). Fig. 12.3 is the worksheet called "Probability", the results of the "ogive" analysis that is displayed in Fig. 12.4 or the data from both sunrise

---

[14] *Ibid.*

[15] The significance of the word "random" is not clear; perhaps this refers to the serendipitous nature of the observations made in 1942 by Hey and colleagues in the UK. Systematic observations were begun in the UK starting in the era 1946–1947.

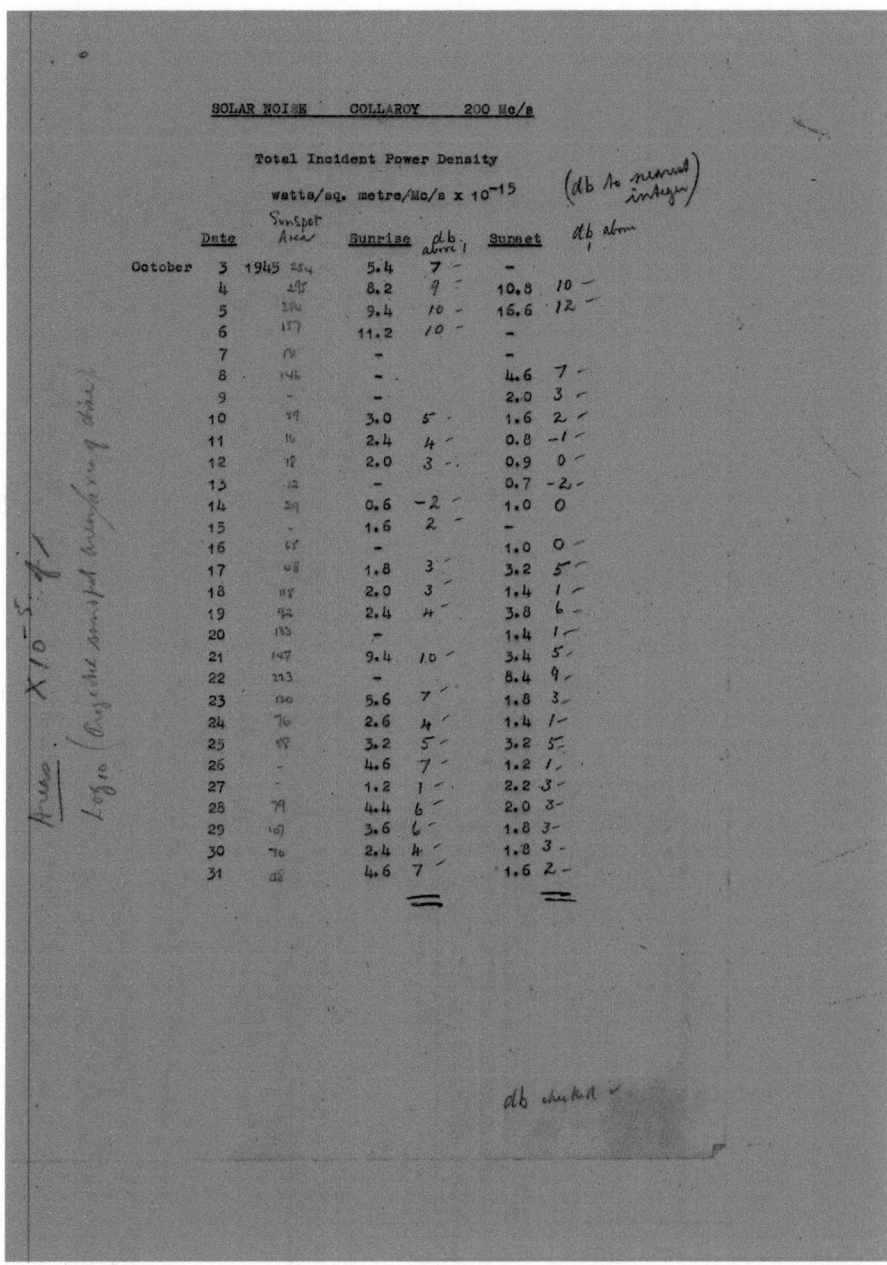

Fig. 12.2 Annotated worksheet with Pawsey's handwriting, October 1945 for Collaroy. The recording technique consisted of pencil and work sheet. An average solar intensity was determined by looking at the display and estimating an average value. (From 6 to 9 February 1946 and onwards from 27 Feb through March a recording milliammeter was used.) Both sunrise and sunset data were recorded (about 30 min each); the sunspot area is also listed for each day as provided by Cla Allen at Mt Stromlo. Credit: CSIRO Radio Astronomy Image Archive JP12-2

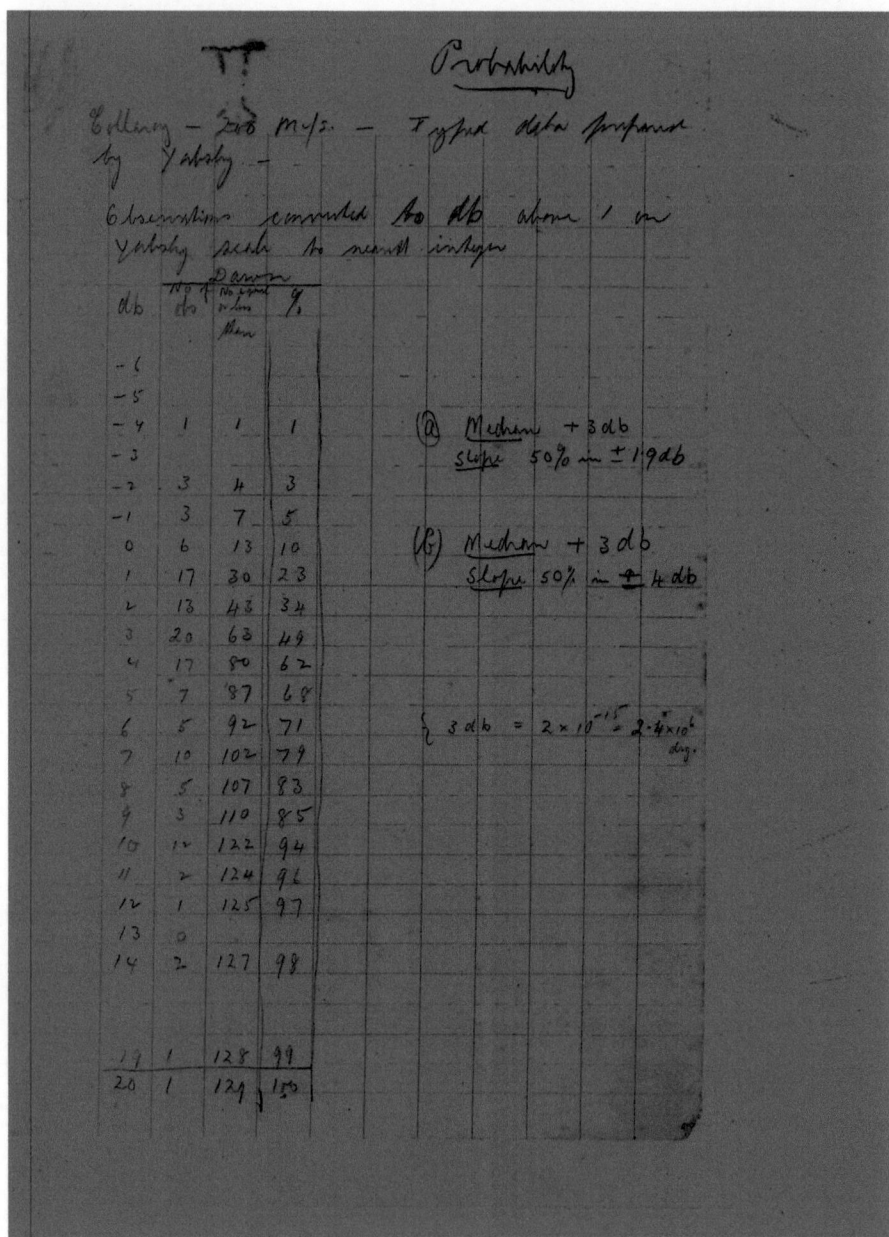

Fig. 12.3 "Probability" calculation leading to the plot of the ogive, the probability of the occurrence of each observation. For each level, the number of points with a certain signal (or less) than the db value. A histogram was determined as well as the percentage. The calculations on the right side (e.g. Median value and slope) were done by Don Yabsley. Credit: CSIRO Radio Astronomy Image Archive JP12-3

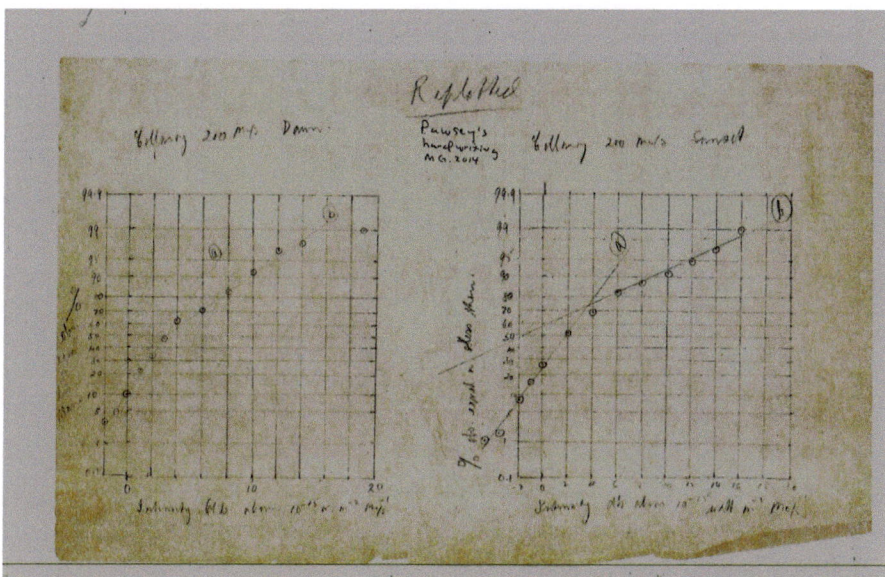

Fig. 12.4 Ogive for sunrise (left) and sunset (right); the two straight lines represent two distinct distributions. The slope of the line provides the dispersion of each distribution. The fainter distribution (the quiet sun, **a**) has a roughly constant value with a smaller dispersion. The more intense distribution, (**b**) represents the Type I activity correlated with sunspots. Credit: CSIRO Radio Astronomy Image Archive JP12-4

(dawn) and sunset from Collaroy, October 1945 to March 1946. The ogive plot (Fig. 12.4) is a statistical scheme used to present the cumulative distribution of the observations. It determines how many data values lie above or below a set reference level.

In Pawsey's plot (Fig. 12.5) the x-axis is intensity (db, thus a logarithmic scale 10 times the log of the ratio with respect to the fiducial value, 10^{-21} Watts/m^2/Hz or 10^5Jy) and the y-axis is the percentage of the total observations included up to the given intensity ("per cent observations equal or less than the value"), also on a logarithmic scale. Thus, the top of the graph refers to 100 per cent of the data and the bottom to one percent. With a log-log scale, a normal probability distribution appears as a straight line. The distribution appears to consist of two intersecting lines, each representing a distinct distribution. The slope of each represents the dispersion of the distribution. The more intense distribution (up to 16 db, a factor of 40 above the fiducial value) has an approximate normal distribution with a wide dispersion, and the fainter one has a roughly constant value with a somewhat smaller dispersion. The former represented the intense "sunspot" noise and the latter the weak steady level arising from the quiet sun. Thus, Pawsey fitted a model to all the data based on two components of the 200 MHz solar emission, the ogive representation combining all the data leading to a characterisation of the properties of each component. Since the sunspot noise was present about two-thirds of the time, the

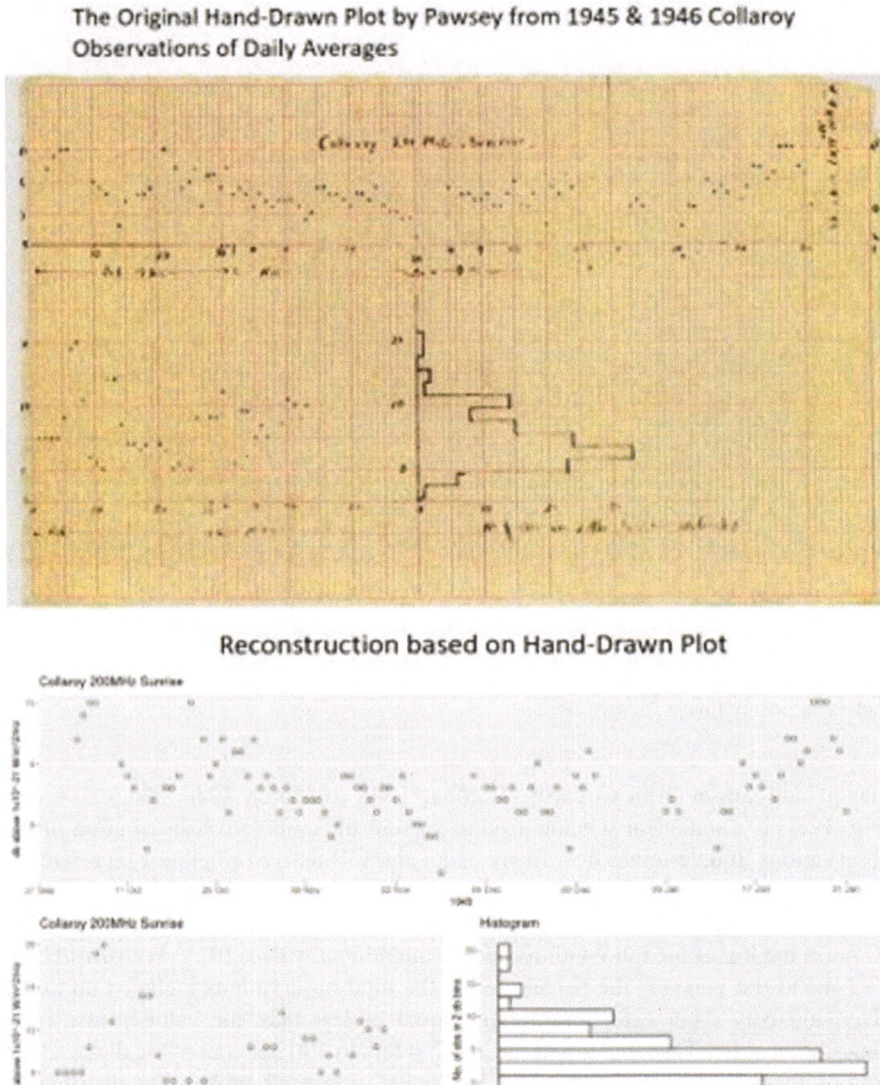

Fig. 12.5 The histogram of Collaroy data at 200 MHz from October 1945-March 1946 as plotted by Pawsey. The time series at the top (in this faint plot done by Pawsey in 1946) covers October, November, December 1945 and January 1946. The middle and bottom plots are a reconstruction of the Pawsey plots done by Harry Wendt in mid-2020, The middle plots covers the time range October to December 1945 with the bottom left plot is February and March 1946 when observations ceased at Collaroy. The time series covers daily observations over the six-month period. The bottom right histogram shows the number of observations in 2 db steps starting at 0 db (the unit is 10^{-21} watts m^{-2} Hz^{-1} or 10^5 Jy, the total flux density of the quiet sun at 200 MHz.) Credit: CSIRO Radio Astronomy Image Archive JP12-5 and Harry Wendt

ogive method was an efficient method to separate the two components, especially since a single detection of the base-level was difficult due to poor signal-to-noise for a single record. In the end, all data was used to fit the two-component model.

In Fig. 12.5, the time series and the resultant histogram of intensities are shown for the Collaroy campaign (October 1945–March 1946), as plotted by Pawsey. The histogram has been plotted using a base level 1×10^{-21} watts m^{-2} Hz^{-1}, or 10^5 Jy or 10 solar flux units. The top of the plot shows the time series (plotted by Pawsey) with time running from left to right (October, December in the middle and March at the right extreme). The histogram at the bottom has a log plot (in db) and the intensity is also on a db scale (See the published example in Fig. 10.2, Chap. 10).

Pawsey, Payne-Scott and McCready rushed off a letter to *Nature* on October 23 (including data from that very day!) by telegram. (It must have been a let-down when the paper was not published until February 1946). It contained four references only: Jansky, Reber, Hey (the restricted UK Army observations of 1942), and Alexander (the restricted report from New Zealand of the Norfolk Island effect of 1945). The instrumentation description was terse, including the important fact: "Since the aerial rotated about a vertical axis, we could only direct it to the sun near sunrise and sunset."

The short paper contained new facts about the sun:

> We observed, from the direction of the sun, a considerable amount of radiation having the apparent characteristics of fluctuation "noise" when observed on a cathode-ray oscillograph or head-phones [they did **listen** to the sun, listening to an audio representation of the detected radio frequency signal]. However, the output meter reading fluctuated considerably, a characteristic which is not typical of normal thermal agitation "noise". The variation of apparent azimuth of arrival and of intensity with horizontal rotation of the aerial and sun's elevation was qualitatively consistent with the assumption of radiation from the body of the sun modified by the known directional characteristics of the aerial.

Pawsey et al. also mention a marginal 25 cm detection made from Middle Harbour on 4 and 5 October using a L band radar antenna. "On 25 cm, a small effect was observing corresponding to about 6000 K, the actual temperature of the sun". This was a detection of the photosphere of the sun at 6000 K; later results by the RPL group would find a temperature of about 10,000 K.

Pawsey and his colleagues suggested that "cosmic noise" was likely to be the sum of emissions from all the stars in the galaxy, all imagined as being sun-like, if varying in intensity. Sullivan calls this conceptual model of galactic emission "a collection of stars".

The intensity of the radiation correlated with sunspots could not be explained by thermal radiation, a fact also suggested by Alexander.

Pawsey et al. suggested that because the emission they detected showed rapid fluctuations, they could not arise from the whole sun, but must come from something smaller and hotter. Given Pawsey's background and the group's radar experience in WWII, they speculated that it was perhaps something analogous to a lightning strike:

> In view of observations of such intense bursts of radio from the sun at the wave-lengths at which cosmic static is known [from Jansky's work at 20 MHz and Reber's at 160 MHz], it appears desirable to question the suggestion that the latter originates in the interstellar space.

It seems more reasonable to attribute it to similar bursts of radiation from stars which, because of their large number, could yield an approximately constant value for any one area in the sky. Furthermore, because of the very high levels relative to expected thermal radiation (a maximum equivalent temperature of more than 10^7 degrees in our case) and the observed short-period meter fluctuations, it seems improbable that the radiation could originate in atomic or molecular processes, but suggests an origin in gross electrical disturbances analogous to our thunderstorms.

The problem with using a solar analogy, the "collection of stars", to explain the background of the Milky Way as observed by Jansky and Reber, was pointed out by (optical) astronomers within a few months. In *Nature* of 15 June 1946, Greenstein, Henyey and Keenan published a short note, "Interstellar Origin of Cosmic Radiation at Radio-Frequencies" (submitted 16 April 1946):

> We wish to point out some considerations which seem to vitiate this interesting new suggestion [by Pawsey, Payne-Scott and McCready]. In the first place, it is necessary to consider the fraction of the area of the sky covered by stellar disks (the "dilution" of stellar radiation) . . . Stars similar to the sun . . . all have similar dilutions of 10^{-14} . . . If hotter stars are considered, the larger dilution results in a discrepancy of 10^{-12} without considering the time-average effects. In other words, a hot star would have to suffer disturbances 10^{12} times as intense as those observed in the sun . . . if such bursts are to account for the cosmic radiation at radio frequencies. It would seem probable that the solar observations represent a new and different type of phenomenon from that observed by Reber.[16]

We explore why this critique had little impact in Chaps. 33 and 34.

In ESM 12.3, Distortions, we provide a description of the distorted international news report from the Associated Press from January 1946, just before the *Nature* paper of Pawsey, Payne-Scott and McCready was published in February 1946. The press account asserted that "radar contact" had been made by the Australian group. The CSIR management tried with little success to dampen the excitement.

Planning the Next Phase of Research, December 1945

In early November 1945, a few weeks after the Collaroy article had been submitted to *Nature*, Pawsey distributed a memo about the future of radio astronomy at RPL, "Solar and cosmic noise investigations".[17] In addition to a continuation of the solar

---

[16]Reber and Greenstein published the first review paper in radio astronomy "Radio-Frequency Investigations of Astronomical Interest" in the February 1947 issue of *Observatory,* including a discussion of the possibility of observing the HI line at 21 cm. They carried out another calculation of the stellar dilution factor suggesting an even less favourable case of a factor of 10^{14} requirement for the mean enhancement of stellar radiation. In an addendum, Greenstein discussed the million-degree corona results of Pawsey from November 1946. See Chap. 14.

[17]NAA C3830 A1/1/11945–1946 Part 1. The document is not dated; the estimated date is based on both context and references to this meeting in the Propagation Committee minutes of the following week on 12 November 1945, the regularly scheduled monthly meeting of the Propagation Committee. Pawsey was chair; Frank Kerr, Don Yabsley, Pulley, Price, Iliffe, Parker and Ruby Payne-Scott (apparently attending only during the "Solar Noise" discussion) were in attendance. Pawsey

noise work, Pawsey suggested "[we should] map the intensity levels over the southern sky—complementing and overlapping Reber's work in northern hemisphere. Observe if intensity fluctuates or not." Additional major projects were to investigate the frequency spectra of both cosmic and solar noise as well as possible polarisation.

In the document "Solar and Cosmic Noise Investigations" Pawsey was especially keen to attack a major problem, the localisation of the solar noise, "[Does solar noise] originate in a single small area?" For solar research there were three main items: (1) continue Collaroy and Stromlo observations, (2) "explore possible observations on ShD [Shore Defence radar at Dover Heights] with view to simultaneous observations at two points and use of precision DF [direction method, i.e. sea-cliff interferometry] method", and (3) carry out additional observations on existing radar sets at 50, 25, 10, 3 and 1.2 cm. Pawsey proposed that the new observations be carried out at 200 MHz with a polar mounting that could be moved manually or even automatically, which allowed them to follow the sun and observe for many hours beyond sunrise and sunset for the first time, as well as using a recording milliammeter (chart recorder). The personnel were to consist of Don Yabsley (in charge) with the continued collaboration of Payne-Scott.[18]

A summary statement was prescient: "For subsequent work it is desirable to obtain a large, say 30-foot diameter, paraboloid suitably mounted".

At a meeting of the Propagation Committee[19] the following week (12 November 1945), Pawsey set out the next two steps: (1) "Miss Payne-Scott will write a report[20] on work to date in TI form [technical report]. (2) Steps will be taken to start observations on the sun and Milky Way from the Dover ShD station."[21] Item (1) was something of a joint endeavour with Pawsey. They were prompt; by the following meeting of the Propagation Committee on 10 December 1945, it was reported "Miss Payne-Scott has completed a survey of the subject to be issued as an internal report."

Payne-Scott began the report with a summary of the "theory of thermal radiation and the concept of equivalent temperature", influenced by the two papers by R.E. Burgess from 1941 and 1946 "on receiver noise, which have done much to clarify our ideas". The report included derivations of basic relationships about blackbody radiation, antenna temperature and noise, from which Payne-Scott could predict that the sun would be more easily detected at shorter, and the Milky Way at longer, wavelengths. These ideas fitted known data.

---

was planning for the future as he proposed three themes: (1) some possible lines of investigation, (2) a general plan, (3) specification of equipment required and (4) the required personnel. He divided each topic into the desired plan of attack for both "cosmic" and "solar" research.

[18]"Yabsley to take over in a few weeks ... In the meantime continue exploratory work [at Collaroy and soon at Dover Heights] on catch-as-catch-can basis."

[19]NAA C3830 B2/2 Part 1.

[20]To be SRP 501 December 1945, "Solar and Cosmic Radio Frequency Radiation, Summary of Knowledge Available and Measurements Taken at Radiophysics Lab. to December 1, 1945".

[21]This is the first news that the group would move to Dover Heights.

Payne-Scott then discussed the October Collaroy observations, also including a five-hour observation of the sun on 31 October 1945 with a polar mounted aerial. The solar intensity was constant, adding confirmation to the view that fluctuations correlated with sunspots. She had also attempted a whole sky map of the 200 MHz radiation, as recommended by Pawsey, with a movable Army SLC 4 (searchlight control) Yagi antenna. This also enabled the group to check that their observations were indeed of solar radiation and not due to atmospheric effects that might particularly influence measurements taken at the horizon. She described the contour map, Fig. 12.4 (unfortunately missing in the copy of the report found at RPL; only Figs. 12.1, 12.2 and 12.3 were located).

> It will be apparent that, in addition to the radiation from the sun, there appears to be radiation from a more diffuse area centred approximately on the centre of the Galaxy. As no sensitive equipment[22] capable of being tilted at all angles is available, the sky cannot be thoroughly explored. However, over the last few weeks [late November 1945], the centre of the Milky Way has been rising at [about] the same time and about 15 degrees south of the sun, and the RAAF Collaroy [54RS] have extended their sunrise observations on the COL set to cover these bearings. Radiation has been obtained over a fairly diffuse area, covering about 20 degrees in bearing and roughly centred on the centre of the Galaxy ...There are not yet sufficient results to produce a clear picture, and a number of puzzling variations have been observed; it is possible that some of these are due to absorption in the clouds of matter that cause the dark patches observable in the Milky Way.[23]

The conclusions reached by Payne-Scott in the December 1945 document are striking:

> (1) Radiation, at present indistinguishable from thermal radiation, has been detected from the sun over a frequency range of 44 MHz to 30,000 MHz. (2) From 30,000 MHz to 1200 MHz this radiation is such as would be emitted by a blackbody of the same size ... as the sun with a temperature of [6000 K]—possibly up to three times greater. (3) At 200 and 160 MHz the amount of radiation received, if produced as a blackbody radiation by the sun, would require a temperature of at least 5×10^5 K. (4) This radiation is not constant but has long period fluctuation that appears to be correlated with sunspot activity and that results in effective temperatures up to 10^7 K.

These insights clearly influenced Pawsey's plans for 1946. The suggestions for "future work" were foretelling:

> It is hoped to soon begin here a programme of more exact work [in solar noise research], in conjunction with the Stromlo Observatory. Among questions to be investigated are the frequency dependence of the radiation, its polarisation, further study of the long-term variations and an investigation of the short-period [one to several seconds] fluctuations. It is also hoped to make a survey of the Southern sky; Sydney is almost at the antipodes of Reber's stations, so that we can survey areas inaccessible to him; in particular it will be interesting to see whether any radiation can be detected from the Magellan Clouds.[24]

---

[22] Implying that the low gain searchlight system could only be moved in azimuth.

[23] Payne-Scott did not recognise that interstellar dust is completely transparent at 200 MHz.

[24] Fred White was clearly very impressed with Payne-Scott and Pawsey's work. On 13 December 1945, he wrote to Pawsey ("Dear Joe") after reading the Payne-Scott December report: "... I do not agree that this paper is really a compilation of available knowledge. You have put into it a good deal

The minutes of the Propagation Committee show the formalising of this new research direction, stating, by 15 January 1946[25]:

> A cosmic and solar noise section has now been specifically established under Dr. Pawsey and Mr. McCready. The section will include Miss Payne-Scott, Mr. Yabsley and a technical officer and a technical assistant ... The new section [cosmic and solar noise] is getting into swing ... [A]n attempt will be made to obtain simultaneous fading observations at two sites, to determine whether the fading is due to solar or atmospheric effects.[26] Also the variation of intensity as the sun rises at daybreak will be closely studied [Dover Heights could only be used for sunrise observations]. Observations at Collaroy are continuing, though these are difficult to interpret at present as the sun is now in the region of the sky occupied by the centre of the Milky Way.[27]

On 23 January 1946, Bowen sent an enthusiastic report to Sir Edward Appleton in London, since he was a kind of "clearing house" of research reports in ionospheric and radio physics (Sullivan, 2009, p. 90)[28]:

> This Laboratory [has] obtained what I think is the first direct experimental verification [of the correlation of solar noise with sunspot activity] during October 1945 when, over a period of three weeks, we measured solar noise on 200 MHz ... A letter to *Nature* has been concocted ... Miss Payne-Scott, who with Pawsey and McCready has been largely responsible for the work here in RP, has written an internal report summarising our latest ideas on the subject of solar and cosmic noise.[29] I am enclosing a copy as I am sure you would like to read it. After adding some further experimental results, we propose publishing it in one of the journals.

Unfortunately, the December 1945 document was never published. Observations at Collaroy would continue until 15 March 1946, and the RPL group were even successful in detecting fringes using sea-cliff interferometry at the 120 m high cliffs at Collaroy on 8 February 1946. But by then, even more exciting results were being recorded at the group's new field station at Dover Heights.

---

of material connected with your own experiments, in much more detail, of course, than this was given in [*Nature* article submitted in October]. In reading it through one gets the impression that the references to others' works are also an essential part of the introduction to your own work and also essential to the general discussion of the whole subject which the papers contain ... I thoroughly agree with the suggestion that is it should be published. I am not sure whether the *Astro Physical* [*sic*] *Journal* is the correct medium; maybe it is since it might more readily accept this type of material than the *Physics Society* or the *Proceedings of the Royal Society* ... I must say I was very pleased with this work—it is very interesting indeed ... I would like to congratulate you, Miss Payne-Scott and McCready for an excellent piece of work." NAA C3830 A1/1/11945–1946 Part 1.

[25] NAA C3830 B2/2 Part 1.

[26] If the fading were due to atmospheric or ionospheric effects, the received signals would be different at two well- separated sites. On 19 March 1946 the PC minutes reported: "Simultaneous fading records have shown the fading to be similar at Dover and Collaroy." Later comparisons (6 May 1946) between Dover Heights and Stromlo showed similar agreement (separation about 260 km.)

[27] While the mid-summer sun in December–January was in the constellation Sagittarius.

[28] NAA C3830 A1/1/11945–1946, Part 1.

[29] Goss and McGee (2009) and Sullivan (2009) discussed the Payne-Scott "summary paper" of December 1945.

In ESM 12.4, Bowen, Collaroy, we provide a description of a surprising account of the first radio astronomy in Australia, written by Bowen in a popular Australian science journal in December 1945. He suggested that atomic bombs in the vicinity of sunspots might explain the excess radio energy observed by the RPL group at Collaroy.

Chapter 13
Sea-Cliff Interferometry: Dover Heights, 1946

I am sorry that Appleton is making a song and dance about our letter to *Nature*, but I suppose he is just expressing his well-known "ownership" of all radio and ionospheric work.— Bowen to White, 26 April 1946[1]

By mid-November 1945, Pawsey had begun planning for an observational programme in the post-54RS era. Although Pawsey was clearly appreciative of the excellent cooperation of the RAAF at Collaroy, the radio group at RPL now needed a site closer to the Laboratory at Sydney University, with accessible and convenient public transport.[2] In addition, RPL needed a field station site under their own control, allowing modifications on the equipment to be made by their staff. The obvious solution was to start work at the Dover Heights ShD (Shore Defence) or C.D. (Coastal Defence) station of the Australian Military Forces. This station had been used by RPL during WWII for radar development and was only 10 km from the lab and reachable by public buses. The station was called CA No. 1 (Coastal Artillery).[3] As we have seen, Pawsey had begun planning for the first observations (January 1946) the previous November. An image from the WWII era of the Shore Defence aerial at Dover Heights is shown in Fig. 13.1.

Letters and reports from NAA C3830 A1/1/11945–1946 Part 1 and A1/1/5 Part 2, Propagation Committee minutes C3830 B2/2 Part 1.

Supplementary Information The online version contains supplementary material available at [https://doi.org/10.1007/978-3-031-07916-0_13].

---

[1] NAA C3830 A1/1/11945–1945 Part 1.

[2] For example, Ruby Payne-Scott could not drive a vehicle, likely due to poor eyesight. She always took the bus to Dover Heights from Central Station or Bondi Junction; the journey to Collaroy was much longer.

[3] As discussed in Chap. 9 and ESM 9.2, Radiophysics Laboratory 1940, Dover Heights was one of a series of sites for Sydney's coastal defence. Dover Heights was the main command post.

© The Author(s) 2023, corrected publication 2024
W. M. Goss et al., *Joe Pawsey and the Founding of Australian Radio Astronomy*,
Historical & Cultural Astronomy, https://doi.org/10.1007/978-3-031-07916-0_13

Fig. 13.1 The Shore Defence 200 MHz aerial during WWII. Elevation above the sea was 85 m. Used in January 1946 for the first radio astronomy interferometry of the sun on Australia Day (26 January); the small size of the radio emission (Type I burst) indicated a brightness temperature in the range 0.5 to 100 million K. Credit: CSIRO Radio Astronomy Image Archive JP13-1

By mid-January 1946, the solar observations had begun at Dover Heights. For example, on 24 January 1946, Briton wrote the Commanding Officer (Australian Military Forces, the army), Headquarters Coast Artillery in Watson's Bay, Sydney:

During the ensuing weeks we intend to conduct simultaneous fading tests on extra-terrestrial radio interference (from the sun during sunspot activity) at No. 54 RAAF AW Station and your Dover Heights CD [Coastal Defence] station which you have already kindly placed at our disposal. [RPL required a dependable telephone communication between Dover Heights and Collaroy; thus, the use of the "emergency range-finding lines" were required.] We have discussed the matter with Lieutenant Clark-Duff, who can arrange the emergency link to the CD station … if you concur … [Q]uite apart from the fading tests [between Collaroy and Dover Heights] it would be very convenient for us to have access to an outside line from the CD station.

A few weeks later (7 February 1946), a Staff Captain of the NSW Fixed Defences replied that the request was granted. Clark-Duff would arrange the necessary link. Actually, the first successful joint fading tests (between Collaroy and Dover Heights) occurred on this date as shown in the publication by McCready et al. (1947): clearly the use of the special telephone connection had already been organised previously, based on informal discussions.

Later in the year, Briton wrote Colonel P.L. Moore, Commanding Officer, Fixed Defences, at South Head in Sydney with a major concern (15 May 1946):

… [W]e have been using the aerial … at Dover Heights in our investigations on radio frequency energy radiated by the sun. These investigations are yielding results of considerable scientific interest and we wish to continue them for a much longer period … The facilities which you have provided [200 MHz aerial and hut] located on the cliff edge and at a place of easy access [close to RPL] are ideal for our purpose. However, I understand that plans are in hand to scrap the equipment and dismantle the aerial. I should be very sorry to lose the use of the aerial at this stage, and I wonder if some arrangement could be made to allow us the use of it for another year. It may be possible to proceed with the conversion of other stations prior to Dover and so allow us the necessary time, or, alternatively of CSIR's taking over the aerial in situ from the Army, if this conformed with Army requirements.

A month later, Lt Colonel Moore replied (13 June 1946) with good news: the station would be made available until the end of the year. Apparently, in the course of 1946, the condition of the aerial had deteriorated considerably. On 13 December 1986, John Bolton wrote W.T. Sullivan[4]:

By the time I became interested in Dover Heights—about November 1946—the antenna had been almost destroyed by vandals and only the basic steel work was left. As this was largely rusty by this time, Stanley and I cut it up with an oxy torch and dropped the bits over the cliff [!] around February 1947.[5]

Clearly the Australian Military Forces had lost interest in the Dover Heights site; this location was transferred to CSIR and used by RPL until December 1954 (Slee, 1994). The last solar data recorded by Payne-Scott was during the major sunspot of July 1946. The data were published in an internal report: "A Study of Solar Radio

---

[4] Sullivan archive.

[5] From an environmental point of view in 2020, this activity is hard to imagine!

Frequency Radiation on Several Frequencies During the Sunspot of July-August 1946."[6]

Breakthroughs

After the move to Dover Heights and the closure of Collaroy 54RS in mid-March 1946, the behaviour of the radio sun during early 1946 favoured the novice Sydney radio astronomers. One of the largest sunspots appeared in early February 1946 (Newton, 1955) during the most prominent solar sunspot cycle of modern times (solar cycle 18, 1944–1954). The largest sunspot of the modern era, with a maximum area of 6150 millionths of the solar area, occurred a year later, 7 April 1947; the maximum area of the large sunspot observed by RPL in early 1946 (central meridian transit February, day 5.7) was 5250 millionths. In subsequent sunspot cycles, the maximum sunspot sizes have been much smaller.[7] For example, the large sunspot of 24 October 2014 (Active Region 2192 solar cycle 24, the weakest solar maximum in a century) had a maximum area of 2740 millionths, only roughly the 30th largest sunspot in the modern era.

The publication that presented the results of the Dover Heights research of January–March 1946 was authored by McCready, Pawsey and Payne-Scott.[8] It contained numerous breakthroughs: (1) first successful interferometry in radio astronomy; (2) the elucidation of the principle of aperture synthesis; (3) continued determination of the correlation of solar noise with sunspot area over a six-month period; (4) detection of sudden increases (bursts) extending from a second to some minutes with similar characteristics at observing sites spaced up to 250 km; (4) typical rise times of a factor of a hundred within a second with an occasional increase of 10^8 Jy per second; (5) a limit of about 6.5 arcmin established for the radio source size; and (6) location of radio emitting region coincident with the prominent sunspot of 6 February 1946, using the sea-cliff interferometer technique. The paper is included in the Sullivan publication of *Classics in Radio Astronomy*, 1982, "Solar

---

[6]RPL 9, the date given only as August 1947. Goss and McGee (2009) have discussed these results in detail, page 108. Based on this data, Payne-Scott detected seconds of time frequency delays of Type III bursts, later confirmed by Payne-Scott (1949) at the Hornsby RPL field station. Also, NRAO ONLINE.20, and NRAO ONLINE.23.

[7]The fourth largest sunspot (27 July 1946) was 4720 millionths in size, observed later in 1946 by Payne-Scott at Dover Heights. The sunspot of March 1947 was number five in this ranking, observed by Payne-Scott, Yabsley and Bolton as they discovered a giant Type II burst (10^{11} Jy at 60 MHz, one of the largest extragalactic signals yet detected), accompanied by aurorae in Sydney a few days later. See NRAO ONLINE.20.

[8]In ESM 13.1, Historical Introduction, we present portions of the original text from this paper. The referee (likely Appleton) required modification to the text leading to loss of valuable historical information about the sequence of events leading to the research. The controversial use of the two WW II reports in the original version of the 1947 paper and especially in the initial RPL solar noise paper in *Nature* on 9 February 1946 is described in ESM 13.2, Fracas.

Radiation at Radio Frequencies and its Relation to Sunspots". Here we consider these achievements in more detail.

The First Fringes: Australia Day, 26 January 1946

The McCready, Pawsey, and Payne-Scott paper began by setting the scene:

> The discovery of radio-frequency radiation, with the characteristics of fluctuation noise, arriving at the earth from the direction of the Milky Way, was announced by Jansky (1933a, b). This discovery is potentially of fundamental importance to astrophysics, since it provides a source of information concerning extraterrestrial phenomena other than that obtained through the use of light. Up to the present, however, the interpretation of such observations has contributed little to astrophysics, and it appears that more complete observational data are necessary. Jansky's original work on cosmic noise was confirmed and extended by himself and others to cover the frequency range 15 to 160 MHz,[9] but, at first, no measurable radiation was observed from the sun. It was therefore suggested that the radiation originates not in the stars but in collision processes in the residual ionised matter in interstellar space. The development of microwave radar suggested the possibility of detecting at these wave-lengths the blackbody radiation from the sun to be expected on Planck's law, assuming the optical temperature of 6000°K. The intensity of this radiation per unit frequency range is proportional to the square of the frequency at radio frequencies. It is too small to be detected in the ordinary short-wave region but should be detectable at centimetre wave-lengths. In 1942 Southworth detected centimetre radiation from the sun (Southworth, 1945) and showed that it was of the order to be expected from the Planck formula.[10]

Due to the poor resolution (primary beams 10°), a major problem with these early solar radio observations was the accurate determinations of the location and size of the emitting region, assumed to be located over a small region of the solar disk (optical diameter of about 30 arcmin).[11] The RPL group used sea-cliff interferometry, a technique they had trialled without success at Collaroy. Sea-cliff interferometry had been perfected in WWII with radar aerials located on a sea–cliff; the system was a Lloyd's mirror based on interference between the direct reflection from an aircraft and the reflected radiation from the sea.[12] The technique is illustrated in Fig. 13.2. Many groups in Australia, the US and the UK had used this technique to

---

[9]The authors made no explicit reference to the 160 MHz data of Reber from 1940 and 1944.

[10]To consider radio observation of the hot corona in historical perspective: Southworth was observing at a few cm and could **only** detect the photosphere. Hey's antennas at several metres were not large enough to detect the corona. Reber had detected the corona, but could not calibrate his signal; as we have seen, he did not identify what he had detected.

[11]At 200 MHz, the solar size is somewhat larger, about 40 arcmin; the range during a solar cycle is 35 to 45 arcmin.

[12]As pointed out in Chap. 7, Pawsey had carried out Lloyd's mirror interferometry during his ionospheric research at Cambridge. The low frequency system showed interference between the direct wave from the transmitter and the reflected wave from the reflection from the ionospheric layer.

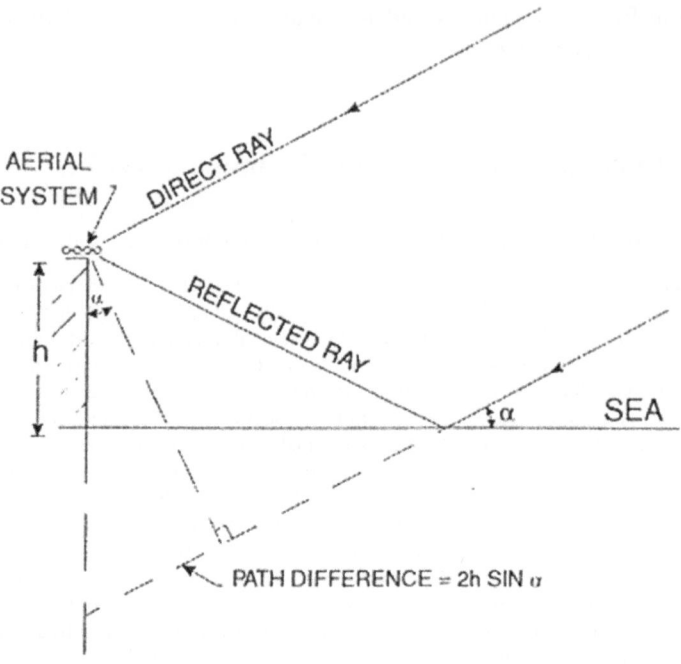

Fig. 13.2 A schematic diagram of a sea-cliff interferometer (Fig. 7.5 from ANZAAS conference). The effective baseline of the virtual interferometer is twice the cliff height. The direct ray from the radio source and the reflected ray interfere to form an interferometer. Credit: CSIRO Radio Astronomy Image Archive B1639-4

determine the height (in addition to range and azimuth) of incoming aircraft for low-frequency radars (frequencies less than a few hundred MHz).

In the publication of the results, McCready et al. (1947) wrote:

> An attempt was next made to elucidate the connection between sunspots and the radiation by means of accurate directional measurements. Because an aerial of about a mile in aperture would be required to produce a beam narrow compared with the half-degree angular diameter of the sun, the direct-scanning method is not feasible. An alternative is a method involving the use of a steerable minimum. In practice, such a method may be realised rather simply by recording the intensity variations as the sun rises over the sea. Interference occurs between the direct and reflected rays, leading to a series of maxima and minima familiar in radar as "lobes", or in optics as "Lloyd's mirror" interference fringes. Since the angular separation of the lobes on our equipment is about equal to the sun's diameter [30 arcmin at Dover Heights with a height above the sea 85 m, and 20 arcmin for the higher elevation at Collaroy, 112 m above the sea], clearly defined maxima and minima will not be expected unless the radiating source is considerably smaller than the sun itself. In initial observations early in October [1945] (Collaroy), no interference pattern was observed [due to the distributed nature of sunspots at that time]. Towards the end of January, a compact sunspot group dominated the sun, and for this reason an attempt was made to detect a lobe pattern on the morning of 26 January [26 January 1946, Australia Day]. A regular series of maxima and minima was observed, with the expected period and very deep minima which were less than the limit of detection [thus implying a small angular size].

We have no record of the circumstances of the observations at sunrise (about 5:20 a.m. Eastern Australian Standard Time, Saturday 26 January). It seems likely that Cla Allen at the Mt. Stromlo Solar Observatory would have telephoned Pawsey the previous days with news of a prominent sunspot (area 1050 millionths). In an interview with Goss,[13] the late Peter G. Hall (1951–2016, son of Ruby Payne-Scott) remembered his mother's excitement as she detected fringes from solar radio emission for the first time: "[She was excited] by the realisation that the [compact] radio emission was associated with sunspots; quite late in her life, the excitement was still with her."[14]

Based on her experience in the absolute calibration of radio telescopes, Payne-Scott would have realised immediately that the size upper limit of 6.5 arcmin implied that the size of the radio source was much less than the solar diameter 30; i.e. less than 4% of the area, implying a brightness temperature of the order of 10^9 K. From McCready, Pawsey, and Payne-Scott: "Consequently, though thermal radiation will be present, it is overshadowed at 200 MHz by radiation due to some other mechanism, probably gross electrical disturbances as suggested [earlier]."[15]

The Giant Sunspot of Early February 1946

Australia Day 1946 was a rehearsal for the exciting events of a fortnight later as the giant sunspot of 7 February 1946 appeared; the flux density at 200 MHz increased by a factor of 10 (from 10^6 Jy on 26 January to 10^7 Jy on 7 February). Figure 13.3 shows some of the data starting on 7 February 1946 at Dover Heights and at Collaroy. Sea-cliff interferometry was now used at Collaroy to good effect: a recording milliammeter was taken there and used from 6 to 9 February and then from 27 February until the end of data collection at Collaroy on 15 March 1946. Note the faster fringe rate observed at Collaroy due to the increased height above the sea, about 120 m compared to 85 m at Dover Heights. The radio fringes appeared on the chart 6 min before optical sunrise; radio refraction is about 1° at the horizon compared to optical refraction, 0.5°; thus the radio fringes were observed some minutes before [optical] sunrise. We can only imagine the excitement of the 200 MHz observers seeing the radio fringes before the sun arose over the Pacific!

---

[13] 12 February 2007.

[14] Of course we cannot be certain that these memories refer to Australia Day 1946 specifically.

[15] The current understanding is that Type I bursts (observed in early 1946) are thought to arise from the fundamental plasma emission process, "due to the coalescence of Langmuir waves with low-frequency waves (e.g. ion-sound waves or lower-hybrid waves) ... The short duration of individual bursts suggests local acceleration of electrons to a few times the thermal energy ...The long life of a [Type I] storm points to continuing local energy release in the columnar source region, which is probably related to magnetic field recombination after new flux intrudes into existing fields." (*Solar Radiophysics* (McLean & Labrum, 1985), chapters "Metrewave Solar Radio Bursts" by McLean and "Storms" by Kai, Melrose and Suzuki).

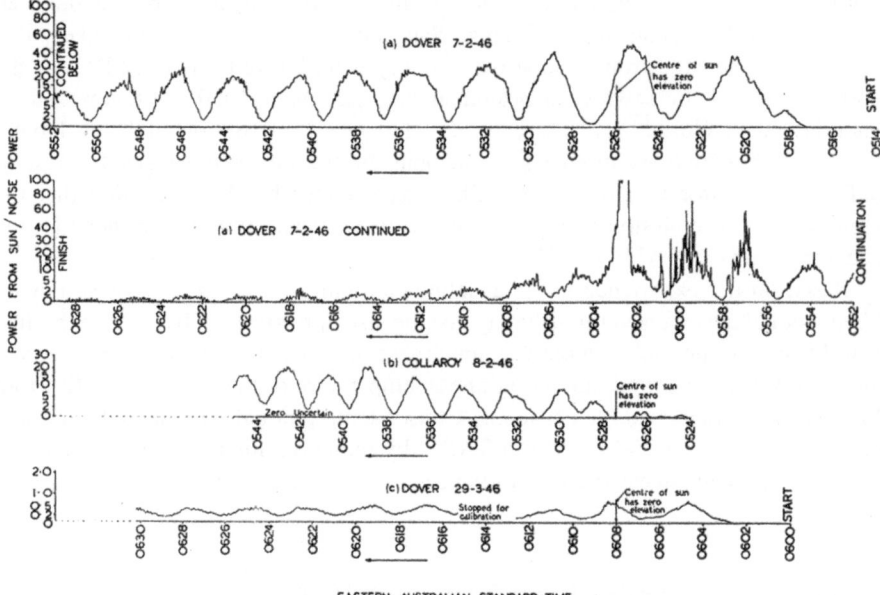

Fig. 13.3 Solar observations taken with the cliff interferometer at Collaroy and Dover Heights showing the interference fringes and strong solar activity on 7 Feb 1946. Credit: Fig. 5 in "Solar radiation at radio frequencies and its relation to sunspots", McCready, L. L., Pawsey, J. L., & Payne-Scott, R. (1947) *Proceedings of the Royal Society of London. Series A. Mathematical and Physical Sciences*, 190(1022)

Not surprisingly, the one-dimensional position of the radio source agreed well with the optical position of the giant sunspot as shown in Fig. 13.4.[16]

---

[16]For a discussion of the complex role of atmospheric refraction in the determination of positions see Goss and McGee (2009), Appendix L, p. 322. Refraction effects impact the positions to first order with a sea-cliff interferometer in contrast to a "spaced interferometer" (two-element interferometer). In the latter case, for a plane parallel atmosphere the two paths to the source are equal, while for the sea-cliff interferometer the path from the direct has a shorter path length than the reflected wave from the sea.

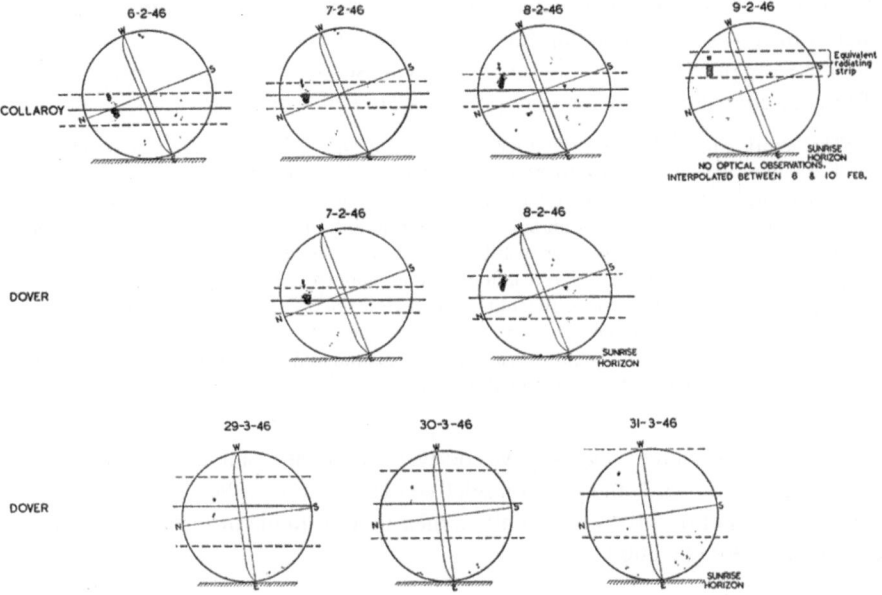

Fig. 13.4 Diagrams illustrating the determination of the location of the radio burst in relation to the position of the sunspots. Note the equator is indicated by a thin parallelogram-EW. Credit: Fig. 7 in "Solar radiation at radio frequencies and its relation to sunspots", McCready, L. L., Pawsey, J. L., & Payne-Scott, R. (1947) *Proceedings of the Royal Society of London. Series A. Mathematical and Physical Sciences*, 190(1022)

Principle of Aperture Synthesis[17]

A major contribution of the McCready, Pawsey, and Payne-Scott publication was the discussion of the principle of aperture synthesis.[18] Based on the discussion of Goss and McGee (2009, p. 101),[19] the proposal arose from Pawsey; he had initially used Fourier techniques in his MSc thesis at the University of Melbourne (Chap. 6). However, the major influence had been his use of Fourier theory to interpret his PhD work (also using a Lloyd's mirror technique, Chap. 7) on the structure of the ionosphere that would have led directly to the ideas expressed in the McCready,

---

[17] See Chap. 36 for a more detailed discussion.

[18] The mathematical details in the publication were derived by Payne-Scott, "the significance of the shape of the interference pattern [of the sea-cliff interferometer]". Examples were derivation of the size of the emitting source based on the observed ratio of minimum to maximum intensity. A key equation provided the relation between the total power of the interferometer signal and a term "in the form of a Fourier cosine transform . . . As [the phase of the pattern] varies, this term varies sinusoidally with an amplitude given by the modulus of the component of the Fourier transform of [the true power distribution] at unit angular frequency."

[19] The discussions with the late Kevin Westfold have clarified the roles of Pawsey and Payne-Scott.

Pawsey and Payne-Scott paper. Pawsey's association of Fourier synthesis and radio interferometry was a major step forward in 1946–1947.

In the text of McCready, Pawsey, and Payne-Scott, after the mathematical details of the sea-cliff interferometer were presented, the authors wrote:

> [This term] is in the form of a Fourier cosine transform ... Since an indefinite number of distributions have identical Fourier components at one frequency, measurement of the phase and amplitude of the variation of intensity at one place at dawn cannot in general be used to determine the distribution over the sun without further information. It is possible in principle to determine the actual form of the distribution in a complex case by Fourier synthesis using information derived from a large number of components. In the interference method suggested here Δ [phase] is a function of h [height] and λ [wavelength], and different Fourier components may be obtained by varying h or λ. Variation of λ is inadvisable, as over the necessary wide range the distribution of radiation may be a function of wave-length. Variation of h would be feasible but clumsy. A different interference method may be more practicable.

The width of the source on the sky was derived by the ratio of maxima and minima of the fringe pattern, while the position relative to the centre of the sun was "calculated from the times of occurrence of minima, measured from the time when the centre of the sun has zero elevation".

Variations Are Intrinsic to the Sun, Typical Bursts Non-thermal

In Fig. 13.5 we show the results of spaced receiver observations at about 16 km separations (Dover Height to Collaroy] obtained at sunrise 7 February 1946; all the sharp dips and peaks agree to within 1 s. (The comparison of Dover Heights and Mt. Stromlo, where observations were also being made, at a distance of 260 km, also agreed.) "It is highly improbable that variations having such a high degree of correlation at widely separated sites should be due to any effect in the atmosphere, and it seems certain that most of them are extra-terrestrial, and presumably solar, in origin."

Based on first-ever size limits determined in early February 1946 with the sea-cliff interferometer, striking lower limits on the brightness temperature were determined:

> This would mean a blackbody temperature of about 3000 million degrees, which is impossibly high compared with any known temperatures on the sun. The known temperatures range from 6000° at the visible surface to about a million degrees in the corona and a few tens of millions at the centre. Consequently, though thermal radiation will be present, it is overshadowed at 200 MHz by radiation due to some other mechanism, probably gross electrical disturbances as suggested in our previous communication (Pawsey et al., 1946). The occurrence of short-duration bursts favours this hypothesis.

The summary paragraphs of McCready, Pawsey, and Payne-Scott provide a snapshot of two major conclusions as viewed from mid-1946 at RPL:

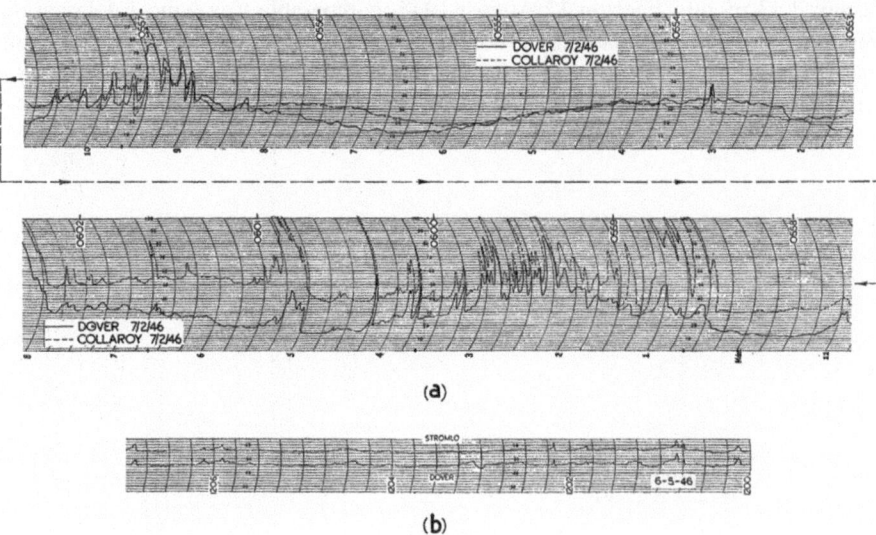

(a)

(b)

Fig. 13.5 Comparison of highly correlated bursts of solar radio emission made at Collaroy and at Dover Heights 16km away. Credit: Fig. 2 in "Solar radiation at radio frequencies and its relation to sunspots", McCready, L. L., Pawsey, J. L., & Payne-Scott, R. (1947) *Proceedings of the Royal Society of London. Series A. Mathematical and Physical Sciences*, 190(1022)

> The connection between this radiation and sunspots is established by two independent lines of evidence, the correlation of intensity with sunspot area and the coincidence of direction of origin with that of sunspot groups. No evidence is yet available as to a particular visible solar phenomenon, associated with sunspots, which gives rise to the radiation.

Apparently, Pawsey and Payne-Scott were not aware of the publication by Greenstein, Henyey and Keenan from *Nature* 15 June 1946, concerning the dilution factor of the proposed stellar emission as we discussed in Chap. 12. Stellar radiation was quite unlikely.

McCready, Pawsey, and Payne-Scott concluded with the following paragraph, an improbable prediction:

> Cosmic noise was originally attributed to radiation from interstellar matter, rather than from stars, at a time when similar radiation from the sun had not been detected. The discovery of solar noise raises the question as to whether the cosmic noise is due to similar processes in stars. The basic difficulty remains that the intensity of cosmic noise is vastly greater than it should be if the stars emitted the same ratio of radio-frequency energy to light as does the sun. Nevertheless, the great variability of solar noise suggests the possibility of vastly greater output from stars differing from the sun and it seems that data at present available leave the question completely open.

Within a short time, radio astronomers realised that a new mechanism for the galactic background non-thermal emission must be found, as we describe in Chap. 34.

The McCready, Pawsey, and Payne-Scott publication was communicated to the Royal Society by Sir David Rivett on 22 July 1946 in person. Publication in print occurred 13 months later on 12 August 1947, comparable to a typical delay at this period for this journal of about 10 months.[20] The successes of Pawsey's RPL group in 1945–1946 formed a solid foundation for radio astronomy as a new discipline in Australia, within only 12 months after the end of WWII.[21]

---

[20] Goss (2013) page 127–128: "Likely in this post-war era, there was a substantial backlog of research output that had been delayed by the fact that many scientists had been involved in wartime research." The complex history of the publication of McCready, Pawsey and Payne-Scott was also discussed by Goss and McGee (2009), Chap. 8. Goss and the late John Baldwin (2010) have independently evaluated the publication delays in the *Proceedings of the Royal Society* in the post-war era.

[21] See also the detailed discussion in Goss and McGee (2009, p. 127) and Goss (2013, Chap. 7).

Chapter 14
The Million Degree Solar Corona, 1945–1946

I think Martyn might get a mention in the section on the discovery of thermal radiation, I am all for a quiet life and the theory was a vital part of the discovery and I think I could suggest something like "the observations of JLP and the theoretical studies of Martyn are in agreement in showing that event the thermal radiation from the sun is greatly in excess of the 6000 K black-body value over a certain range of [cm] frequencies."—Pawsey to Bowen from London, 16 August 1948[1]

Introduction

Pawsey realised that there was more to extract from the first 3–4 months of solar data; and the results were close to revolutionary. As we have noted earlier, yet again a remarkable piece of good fortune occurred: the Collaroy and Dover Heights aerials had just the suitable sensitivity (a few times 10^4 Jy in 1 s) to easily detect the quiet sun (10^5 Jy at 200 MHz).[2]

In 1945, Pawsey must have recognised an extraordinary aspect of the times' series (intensity versus time, see figures in Chap. 12); from 11 to about 14 October 1945 there was a marked **minimum**. But the intensity did not drop below this value.[3]

Supplementary Information The online version contains supplementary material available at [https://doi.org/10.1007/978-3-031-07916-0_14].

---

[1] C3830 F1/4/PAW1 Part 2.

[2] We have additional insight into this achievement as a result of the surprising discovery of his original records in May 2013, in the house of the late Sally Atkinson (Bowen's personal assistant for many years and later RPL archivist). These are described in detail in ESM 1.1, Additional Details.

[3] Several other groups observing the sun in 1945–1946 did not have the sensitivity to detect the quiet sun. Ryle and Vonberg and Lehany (1.75 m or 80 MHz) and Lehany and Yabsley (200 MHz) were exceptions. See Pawsey and Yabsley (1949), who carried out a re-analysis of the Ryle and Vonberg data, with confirmation of the presence of the quiet sun contribution.

© The Author(s) 2023
W. M. Goss et al., *Joe Pawsey and the Founding of Australian Radio Astronomy*,
Historical & Cultural Astronomy, https://doi.org/10.1007/978-3-031-07916-0_14

By mid-1946, Pawsey became convinced that this "base level" represented "another jewel from his [and Payne-Scott's wealth] of data." As Sullivan (2009) summarises: "He noticed that his large set of daily values of the 200 MHz solar flux density had a peculiar distribution … With a sharp lower limit corresponding to an equivalent brightness temperature for the solar disk of about 1×10^6 K."

In this chapter we revisit this exciting development, Pawsey's realisation that he had observed radio emission from the hot corona. The story is set out in detail by Sullivan, and retold here with an appreciation of Pawsey's handling of what turned out to be difficult circumstances.

Understanding the Sun

Since 1939–1941, Grotrian and Edlén had recognised that the previously unidentified coronal lines observed during total eclipse could be identified with highly ionised lines of iron, nickel and calcium, the forbidden lines. The lines arose from atoms stripped of 10–15 electrons; if the corona was in thermal equilibrium, a surprising temperature of 10^6 K was indicated. In addition, the line widths indicated a large temperature. A major theoretical problem remained: to provide a mechanism to transition from a photosphere temperature of 6000 K to a temperature of one to two million degrees K over a short distance from the photosphere through the hotter chromosphere to the outer corona. Even with a number of proposed solutions, the problem of the source of heating of the hot corona remains an active region of research today.

This was the state of knowledge in 1946. It was known at Mt. Stromlo Solar Observatory, whose Director, (later, Sir) Richard Woolley (1906–1986), was an enterprising Englishman who arrived in Australia in 1939 and remained in touch with developments in astronomy overseas.[4] The Observatory also had David Martyn on staff from December 1944, placed there and allowed to return to his ionospheric research, given that the end of the war was anticipated.[5]

There was an open communication between Mt. Stromlo and RPL, particularly fostered by Pawsey; we have seen that he was working closely with Cla Allen from Mt. Stromlo in October 1945, and that was excited by the work. Pawsey maintained a friendly correspondence with Martyn, who also was kept informed of the RPL results and, like Allen, was excited and interested in them. Reflecting back, on 29 January 1947 Bowen wrote to White:

> As you know, Pawsey began the solar noise work almost entirely on his own in 1945. It became obvious that we would gain immensely by discussions with Stromlo, and we took the question up with Woolley and Allen. Martyn became interested, and he revealed an

---

[4] Woolley would return to the UK in 1957 to become the 11th Astronomer Royal and Director of the Royal Greenwich Observatory. (Gascoigne, 2012).

[5] See NRAO ONLINE.7 and Home, 2000.

unexpected talent for interpreting solar phenomena. He saw the importance of the work and rapidly became our main contact at Stromlo. Since that time, we have adopted a policy of maximum collaboration, having full discussions on our experimental results as they appeared, making equipment for Stromlo ... This has undoubtedly been very profitable and allowed both of us to proceed much more quickly than would otherwise have been possible.[6]

In response to the enthusiasm of Allen and Martyn, Pawsey arranged for a 200 MHz Yagi system to be supplied to Mt. Stromlo so that observations could be made there—a friendly gesture that Woolley received in somewhat bureaucratic terms.[7]

In 1945, both Martyn and Woolley began to realise the implications of what was known of the hot corona for the new radio observations. On 22 November 1945[8] Martyn wrote to Pawsey ("Dear Joe"):

I am beginning to get some definite ideas on a mechanism which may be responsible for the results you have been getting in solar radiation at 200 MHz. I think it quite possible that the radiation [of the enhanced levels of solar noise] may be polarised, and the purpose of this note is to suggest that, if at all possible, you might care to make observations on this point.

In fact, Martyn himself detected the circular polarised radiation at Stromlo using the RPL supplied equipment.

Sullivan summarises:

Why the corona was so hot was not all understood, but the evidence was there. Martyn realised he could apply standard techniques in ionospheric theory to calculate the expected radio emission from the sun. Once he had adopted likely values for the electron densities in the corona, he found that the corona was opaque at Pawsey's frequencies [200 MHz]. The observed radio waves were therefore not at all from the 6000 K optical surface (photosphere) of the sun, but from well above, out in the million-degree corona. When the sun was quiet [in the radio], this coronal emission constituted the entire solar signal; when active, the coronal emission was dwarfed. (Sullivan, 2009, p. 135).[9]

---

[6] Bowen to White, 29 January 1947, NAA C3830 Z1/7/B Part 1.

[7] Woolley wrote to White on 16 April 1946: "The set Pawsey mentioned has actually arrived, though the basis on which it has come does not appear to be quite clear. May I suggest that it ... be placed on loan to the Observatory [Commonwealth Observatory] by Radiophysics. The Observatory certainly desires to borrow an instrument for observing the sun in radio wave lengths." Bowen, having read this letter, replied to White on 18 April 1946: "We must admit that most of the arrangements with Mt. Stromlo have been done informally between members of staff without any communication between the Chiefs. This is an arrangement of which I am very much in favour, as long as it works." It seemed to Bowen that Woolley preferred "the more formal method" of communication between Chiefs. Details of the complete letters are presented in NRAO ONLINE.24 "Martyn Pawsey Bowen-Controversy over Million Degree Corona 1946".

[8] All correspondence from NAA C3830 A1/1/1 Part 1 and C3830 A1/1/5 Part 1 and the Sullivan archive.

[9] Sullivan (2009, p.135) has pointed out that Vitaly L. Ginzburg had also made similar calculations about the hot corona while considering the possibility of solar radar echoes. But this work was completed in Russia and RPL only became aware of it in 1948, long after the October 1946 paper on the hot corona had been published. In a letter from Bowen to Pawsey in London (NAA C4659 Part 8) on 26 July 1948, Bowen wrote: "I have just seen a remarkable paper by Ginsburg [from March 1946] ... The amusing thing is that it describes the possibility of a million degree radiation being

At this time, the separation of the different sources of emission (corona and "sun-spot") had not become clear to Pawsey and Payne-Scott; nor had they much knowledge about the sun. In the summary report of December 1945 (Chap. 13), Payne-Scott recorded:

> The radiation [of the sun] may come from the corona, which has recently been shown to have a very much higher temperature than the photosphere, and which, although transparent to visible light, may well be opaque to long radio waves. Dr. D.F. Martyn, of the Stromlo Observatory Canberra, has suggested this origin. This theory does not seem to account for the greatly increased radiation when at the time the sunspots actually cross the meridian [the enhanced emission].

Informal discussions continued in 1946, and Pawsey visited Mt. Stromlo in early 1946.

Preparation to Publish Radio Observation of the Hot Corona

By late July 1946, Pawsey prepared to publish the RPL results in *Nature*.[10] The plan was to publish a joint paper with Martyn, combining his observations from Collaroy and Dover Heights with Martyn's theory.

Pawsey wrote **three** internal reports summarising the status of his determination of the base-level and its interpretation. The first was dated July 1946—after the Collaroy observations were completed in March 1946 and some months after the observations at Dover Heights had started in January 1946. This document, sent to Martyn, was titled "Notes for Preparation of Letter Concerning Radio-Frequency Solar Radiation and Corona Temperature."[11] In it, Pawsey summarised the core content:

> (3) Your [Martyn's] concept: What should be expected from existing astrophysical data. Concept leads to shells with two properties. (a) Nothing inside gets out. (b) Radiation should not go below thermal limit for shell.
>
> (4) Pose question—Does the level [of solar noise] come down to this thermal level—If so, have powerful tool for investigating structure of solar atmosphere. [In Pawsey's hand writing in pencil] "In particular can define approximate upper bound to T [local temperature] for any electron density level."
>
> (5) Present evidence for 200 MHz. (a) Anticipated thermal level. (b) Radio-frequency results show a fairly definite low limit. (c) Correspondence between values shows that, if data in (a) correct, the low limit is in this case a fair measure of thermal radiation. (Use 6 months RP data +any available Stromlo.)[12]

---

received from the sun and makes a calculation of the depth of penetration of radio waves." The translation of the abstract (from Russian to English) had occurred only a short time earlier.

[10] Submitted to *Nature* on 18 September 1946.

[11] NAA C3830 A1/1/5 Part 1.

[12] Likely anticipating that Collaroy RS54 would cease operation in March 1946.

... (7) Strong suggestion that similar low limits on other wavelengths which at any rate define upper limit of temperatures, which may yield data on temperature and electron density in sun.

(8) Requires large number of measurements to go further—Suggest this is fit subject for applications of results by many laboratories.

These plans were evidently known to Bowen, White and Woolley. But the path to publication would quickly prove to be complex.

On 29 July 1946 as Bowen wrote Woolley[13]:

> Dr. Martyn recently discussed with Dr. Pawsey and myself the desirability of writing a letter to *Nature* on the concept of a shell surrounding the sun . . . [W]e now have six [actually close to eight] months observations of intensity levels on 200 MHz. These show a pronounced tendency to occur at a level corresponding to a temperature of about 2×10^6 K ... with a fairly sharp cut off below 10^6 K and a wide dispersion on the upper-side up to 10^8 . . . The concept originated at Stromlo, but Martyn thinks that the publication would be improved by including Pawsey's data. As regards authorship, we suggest ... the inclusion of Pawsey, an arrangement which is agreeable to both Martyn and Pawsey . . .

Woolley replied on 5 August 1946 suggesting instead that he [Woolley] and Martyn publish a mathematical paper dealing with the transfer of radiation . . . in the corona . . . It appears to me that this subject can be most profitably written without reference to any observational data about the intensity of the "noise" radiation actually received . . . but with an eye to the application of the mathematical results to observations of noise at a later date when the noise observations have been studied further than at present.

Before Pawsey received this surprising letter from Woolley, he had written (8 August 1946) to Martyn with a detailed plan of the proposed paper. He emphasised that the 200 MHz data was solid, showing a sharp cut-off in the distribution of intensities at the low end. The data at 60 and 75 MHz were consistent with a cut-off at about a million degrees but was of poor quality due to confusion with "cosmic" noise and also ionospheric effects at low elevation at sunrise at Dover Heights. A new fact was also emphasised by Pawsey: "On 200 MHz the interference pattern [with the sea-cliff interferometer at Dover Heights] has been observed to show shallow minima when the intensity is low . . . [consistent] with the idea of a distributed thermal source."

He added:

> I see no reason we should not proceed to write a letter to *Nature* as originally agreed. If it is not done soon it will almost certainly miss the bus as other workers will start to investigate causes for high radiation levels. Further, no immediate significant progress appears feasible on the observational side, and on the theoretical side a detailed study is required. Publication in this manner would have advantages, other than the claim to priority, in that it would provide an opportunity to collect low intensity observations from other workers, and could also be used to give notice that a theoretical paper was in . . . preparation.

Pawsey then wrote a second report, as the data from both Collaroy and Dover Heights were available (October 1945 to March 1946, Collaroy, and March to

---

[13] In NRAO ONLINE.24 the text of the entire letter is presented.

May 1946, Dover Heights). This is presented in ESM 14.1, Text of Pawsey. A final summary of the data in 1946 was written on 19 August 1946 with a short description of the limited data at 60 MHz, "Solar Noise at 60 MHz: Estimated Values of Steady Level [his emphasis] Solar Noise for the Period 8 to 12 August 1946".[14] The data was difficult to interpret due to the frequent presence of Type I storms. The equivalent temperature was in the range from 0.6 to 1.2×10^6 K, with an average of 0.8×10^6 K.

Pawsey and Martyn continued to discuss the merits of joint publication in correspondence during August 1946 (letter from Pawsey to Martyn on 8 August and a reply on 15 August).[15] Their detailed planning meant that it was a shock when, on 4 September 1946, Martyn announced that he had prepared a paper on his own— and had submitted it on the same day to *Nature,* with no input from Pawsey! Martyn gave his reasons:

> I have made considerable progress with the theory of quiet sun temperature [thermal] radiation since I wrote to you last. I have got out the effective solar temperature and variations of limb brightening over the radio spectrum ... It was my original intention to present this in full detail only in a paper, but I am concerned about the consequences of Bowen's discussion of my basic ideas in England.[16] In the circumstances, it seems wise to publish a summary of the main theoretical conclusions at once in *Nature*, and I have prepared the enclosed note for this purpose ...
>
> The way is now open for anyone (including yourself and team) to publish observational material confirming (or controverting) my theoretical finding.

Pawsey rang Martyn as soon as the surprising letter as well as the text of the paper arrived on 9 September 1946. He immediately saw a way to salvage the situation: send in a second paper, a proposal accepted by Martyn. Pawsey wrote on 10 September 1946:

> As agreed in our telephone conversation of [9 September, Monday], I am enclosing a draft of a proposed letter to *Nature* for publication immediately following the one you are drafting. This change of plan from the original proposal of a joint letter of less content that you now envisage is acceptable to me, subject to your making certain small alterations in the [final section of your paper] dealing with the observational verification ... I think the material is first-rate and should be a decided stimulus to observation and interpretation.[17]

---

[14]NAA C3830 A1/1/6 Part 2.

[15]See NRAO ONLINE.24 for details.

[16]NAA C3830 Z1/9/1946. Bowen was in the UK, US and Canada from September to December 1946 giving talks on radio astronomy—e.g. at the Cavendish Laboratory. There was little interest in Cambridge concerning the detection of the quiet sun.

[17]Later a number of CSIR colleagues, and especially Bowen, criticised Martyn for having failed to include any reference to the base-level in the radio data (in the paper of 2 November1946), caused by the hot corona. Ironically the first version of the paper (NAA C3830 A1/1/5 Part 1) did in fact include a paragraph that Pawsey had initially suggested. The key sentence that was dropped in the second version was: "... [T]he results of Pawsey et al. *Nature* paper with Collaroy data of 9 Feb. 1946] at 1.5 metres gives a temperature of slightly less than 10^6 K at periods of negligible sunspot activity." On 10 September 1946, Pawsey asked Martyn to drop this sentence since all the observational evidence was to be presented in the second (Pawsey) paper. Presenting the material

Although Pawsey had salvaged the situation, Bowen was now viewing relations with Martyn very sourly. Bowen had left on a trip overseas (September–December 1946) just before Martyn's letter announcing unilateral publication arrived (4 September 1946), leaving Pawsey as Acting Chief of RPL. On 26 September 1946, Bowen, presumably informed by White of the circumstances, wrote a handwritten letter ["Dear Joe"][18]:

> I am terribly sorry you have had so much trouble with Martyn. He has played exactly up to form and I have written a strong letter to White on the subject. Under the circumstances I am sure it is better to hold up publication of both his and your letters rather than let Martyn get away with it ... I have asked the Executive to give serious thought to stopping Martyn's letter ... I take the strongest possible view therefore that the letter in *Nature* should contain [his theoretical suggestion and your experimental verification], written by Martyn and yourself with the right acknowledgements. Martyn can then, to use his own words, be free to go ahead and publish any amount of further speculation on the subject under his own name. Let me say again how sorry I am that this has happened and how much I hope it can be rectified.

Pawsey replied to Bowen while reporting on RPL activities[19]:

> I am very sorry that you had to become involved in this rather unsatisfactory business. When the dispute originally arose I decided to divide the material into two letters and be rather punctilious about not including any ideas in my letter which could possibly be attributed to Martyn ... White[20] discussed the matter with me and I could see no alternative but to consistently back my own decision. I feel this was the best thing to do and since Lewis [a Liaison officer] had already passed my letter into *Nature* I think the matter can now be left.

These events had a significant impact into the future, as Bowen was unprepared to tolerate Martyn further and clashes occurred between the two men. In 1947–1948 he wrote to White on more than one occasion to complain about or to curtail Martyn's activities at RPL, while Martyn felt that as a CSIRO Officer, he ought to have full access to RPL's resources and data.[21] As a result, after the early 1950s, relations with Martyn precluded collaboration. There were few RPL publications on ionospheric research. Pawsey's last such publication was the 16-page paper "Ionospheric Thermal Radiation at Radio Frequencies" by Pawsey, McCready and Gardner in 1951 (NRAO ONLINE.22). And after mid-June 1947, Martyn wrote only one more paper directly related to solar noise research, "Solar Radiation in the Radio Spectrum-I. Radiation from the Quiet Sun" in 1948, an extensive elaboration of his *Nature* paper of November 1946. These troubled relations effectively left the RPL group

---

in two papers rather than one was not inherently unreasonable, but Martyn's handling of the situation was upsetting.

[18] Joe and Lenore Pawsey Family Collection.

[19] NAA C3830 Z1/9/1946.

[20] Who had received a letter from Bowen, letter not located in the archive.

[21] Details are provided in NRAO ONLINE.24, correspondence (from NAA C 3830 Z1/7/B, the Bowen series "Collaboration with Mount Stromlo on Solar Noise the Radio Research Board") with pencil messages on the draft: (1) **"NOT SENT"** in Sally Atkinson's handwriting in pencil and (2) **"Do not destroy but bury deep"** in Bowen's handwriting (in pencil) at the top of the page.

without a good theoretician, with many unfortunate consequences over the subsequent decade.

ANZAAS 1946 and URSI, Paris, 1946

The million-degree corona was presented at two conferences in 1946. The 25th yearly meeting of the Australian New Zealand Association for the Advancement of Science (ANZAAS) was held in Adelaide from 21 to 28 August 1946, with 1400 delegates, Prof P. Marshall as President.[22] Payne-Scott and Pawsey attended, giving two 15-minute presentations on 19 August 1946, "Discovery of Cosmic and Static Noise, Observations as described in RPR. 24" [the *Publications of the Royal Society* paper which was then in press] by Payne-Scott and "Interpretation of Observations"[23] by Pawsey. A summary of the "sunspot related enhanced emission" was discussed with special emphasis on:

> the evidence given of coincidence in direction and simultaneous occurrence, that sunspots can be associated with the production of high levels of radio frequency radiation. This does not imply that the spots themselves generate the radiation since sunspots are associated with a number of other forms of solar activity. So far, we have heard rumours, but have not seen evidence for an association with separately recognisable phenomenon, but [have seen no] real evidence.

The "thermal radiation"[24] was also summarised, describing the evidence that was to be presented in the paper submitted the following month to *Nature*:

> A simple hypothesis explaining the skewness (see Fig. 14.1) is that the observed intensity is the sum of two causes, with differing distributions, viz:
>
> (1) One with a wide range of variation but symmetrical distribution—correlated with sunspots, and
>
> (2) A constant source of intensity equal to that at cut off at about 10^5 Jy, corresponding to a blackbody temperature of about 10^6 K.
>
> Now although on the sun the photosphere and regions just above it are at temperatures about 6000 K, it has been recently shown that the corona had the incredible temperature of about 10^6 K. Among other evidence, the most direct is the identification by Edlén of certain coronal lines as being those of ionised atoms with hundreds of volts ionisation potential. This potential corresponds to thermal energies at about 10^6 K.

---

[22] Pawsey was in Adelaide from Wednesday 21 August to the following Wednesday 28 August 1946 during the ANZAAS conference. He travelled via Melbourne where he spent a full day (20 August) and a partial day (28 August) visiting his mother. The travel time was about 3 h by air from Sydney to Melbourne, a comparable time from Melbourne to Adelaide. (Joe and Lenore Pawsey Family Collection.)

[23] NAA C3830 A1/1/11945–1946 Part 1, a summary of the two talks.

[24] Jim Roberts has frequently mentioned to Goss in the last four-plus decades his objection to the confusing use of the term "thermal radiation" by Pawsey to describe "free-free" or "thermal bremsstrahlung" from the sun. "Thermal radiation" has a connotation of blackbody radiation from a solid body such as the moon.

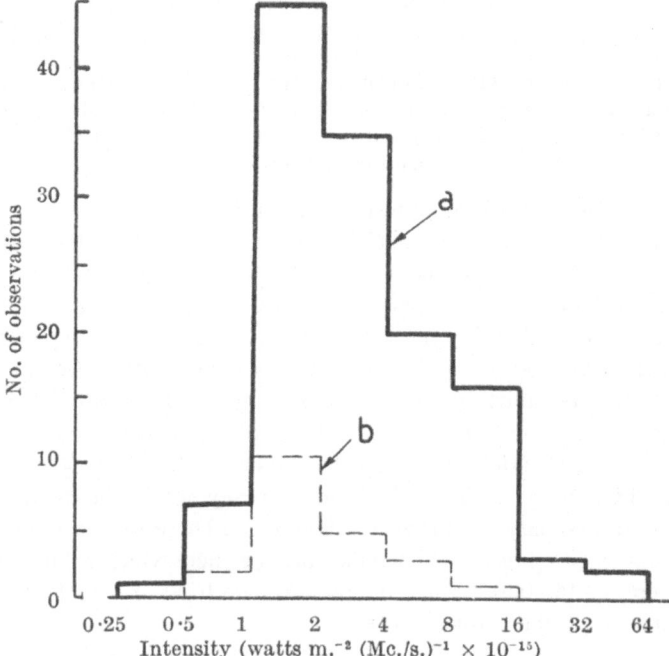

HISTOGRAMS SHOWING DISTRIBUTION OF DAILY VALUES OF SOLAR
RADIATION INTENSITY AT 1·5 METRES WAVE-LENGTH (INCREASES
OF A FEW SECONDS ARE NEGLECTED). (*a*) DAILY : OCT. 5, 1945–
DEC. 12, 1945 ; JAN. 1, 1946–MARCH 15, 1946 (R.A.A.F.
OBSERVERS). (*b*) SUNDRY DAYS, MARCH TO MAY 1946 (LABORATORY
OBSERVERS)

Fig. 14.1 Histogram showing the daily values of solar intensity at 200 MHz. The minimum value corresponded to an effective temperature of 0.6 to 1.2 million K. In his initial analysis, Pawsey had used a 3 day average; since about two-thirds of all days exhibited enhanced levels, this averaging tended to mask the lower limit of intensity. Pawsey's method may be described as a "matched-filter", a linear filter for maximising the signal-to-noise ratio (SNR) in the presence of additive stochastic noise. This is a common technique in radar. Credit: "Observation of million degree thermal radiation from the Sun at a wavelength of 1.5 metres," Pawsey, J. L. (1946). *Nature* 158, no. 4018: 633–634. All rights reserved

Despite the high temperature, the visible radiation is small owing to the exceedingly low density and consequent transparency of the corona. But as was pointed out to me [Pawsey] by D.F. Martyn, at radio frequencies the electron densities in the corona are sufficient to make the corona opaque in a manner analogous to the terrestrial ionosphere for adequately low frequencies ...

Thermal radiation on 200 MHz should therefore originate in a region at a temperature of 10^6 K. It is striking that this is equivalent temperature of our assumed constant source and is strong evidence that the received radiation is actually thermal in origin.

Pawsey completed his presentation at ANZAAS with a philosophical discussion of RPL's participation in the formation of a new field of science:

This work is a new branch of astronomy. This science, and astrophysics in particular, has made amazing progress in the face of what appeared to be a crippling difficulty in obtaining data. Because of this difficulty, new observational tools have an unusual importance ... **Consequently, it is reasonable to expect that the discovery of this radiation [cosmic and solar noise] will come to be recognised as one of the fundamental advances in astrophysics**. [our emphasis] The first stages after such a discovery are those of general exploration of the phenomenon. To these our work belongs.[25]

Bowen's trip in September 1946 allowed him to represent RPL at the first post-war International Union of Radio Science (URSI) conference in Paris during that month. The ionospheric researcher had long attended URSI meetings, so URSI was familiar and important to RPL scientists, none of whom were members of the International Astronomical Union (IAU) as yet.[26] Unfortunately, he had to depart before the session on solar noise occurred. However, on 20 January 1947, Bowen submitted a paper to URSI (presumably for the proceedings of the 1946 URSI), "Recent Australian Researches on Solar Noise".[27] Most of this paper is concerned with a summary of sunspot noise and "bursts" (time scale of 1–60 s increases), summarising the publication from McCready, Pawsey and Payne-Scott (Chap. 13). There is a short summary of "thermal radiation ... These results suggest that the 10 cm radiation is being received from the chromosphere which is known to be at a higher temperature [56,000 K] than the photosphere [6000 K] while the 200 MHz radiation comes from the corona [10^6 K]."

"Hot Corona" Published in *Nature*, 2 November 1946

Pawsey's paper "Observation of Million Degree Thermal Radiation from the Sun at a Wavelength of 1.5 Metres" following Martyn's "Temperature Radiation from the Quiet Sun in the Radio Spectrum" were published in the 2 November 1946 *Nature*. Pawsey's succinct paper had only 393 words and one figure (Martyn's paper, 1410 words, two figures). Figure 14.2 here shows Martyn's distribution of radio brightness at frequencies from 20 cm to 30 m. In the single published figure in the Pawsey article, the majority of the data points (about 130) are from Collaroy 54RS and about 20 points from Dover Heights (Fig. 14.2).

A few months after the publication, Reber and Greenstein wrote the first "review paper" on radio astronomy: "Radio-Frequency Investigations of Astronomical Interest" in the February 1947 Observatory. An addendum, written by Greenstein on

[25] This text is comparable to a talk Pawsey gave at the (Australian) URSI meeting in Sydney (January 1950) —ANCORS—Australian National Committee of Radio Science—"Proper Fields for Radio Astronomy": "... The outstanding deficiency in such solar-noise studies to date is that no one has yet 'seen' the phenomena producing solar noise, so that the conflict between such theories cannot yet be resolved by appeal to observation ...".

[26] See NRAO ONLINE.51 where the importance of URSI is discussed more fully.

[27] Sullivan archive.

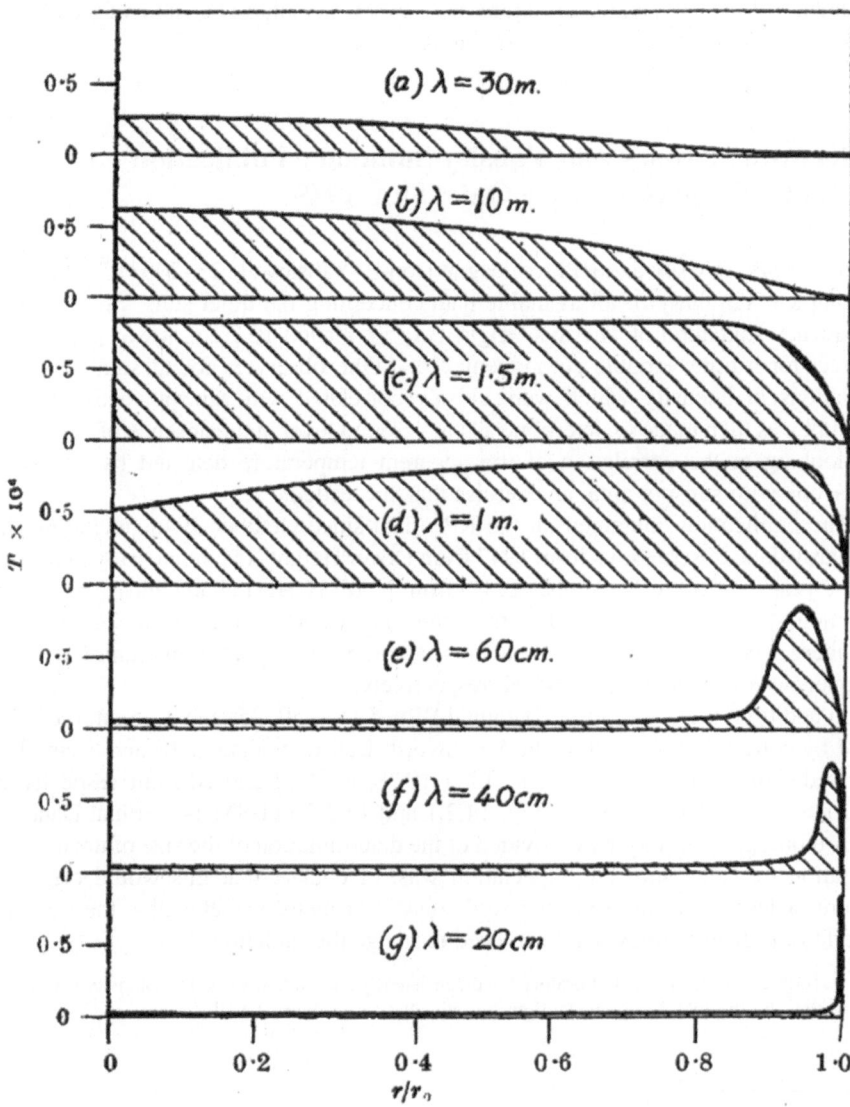

Fig. 14.2 The variation of radio brightness as a function of solar radius at 7 wavelengths from 15 m to 20 cm. Prominent limb brightening was predicted at 20, 40 and 60 cm, later confirmed at CSIR/ CSIRO, NRAO ONLINE.20. Credit: Fig. 2, "Temperature Radiation from the Quiet Sun in the Radio Spectrum", Martyn, D.F. *Nature,* 158, no. 4018: 632–633.

10 January 1947, was added to summarise Pawsey's results. The value of the *Nature* paper was being recognised by a leading astronomer.[28]

Afterword: Pawsey and Yabsley Summary Publication on Radio Properties of the Quiet Sun, 1949

In June 1949, Pawsey and RPL colleague Don E. Yabsley (1923–2003)[29] published a complete summary of all available data concerning the quiet sun: "Solar Radio-Frequency Radiation of Thermal Origin". At 50, 20 and 10.7 cm, it was possible to correct for a time variable component: "... [T]he observers found small but real variations in intensity which were closely correlated with sunspot activity." An attempt was then made to determine a "background level in the absence of sunspots" by looking at the correlation of the apparent temperature detected by the radio telescope with sunspot area, an exercise that succeeded.

The publication contained all the RPL data since October 1945, including the Collaroy data, observations from 1947 from Dover Heights (March to May) and later in the year (August to November 1947) from North Head (Lehany and Yabsley). In addition, data at 80 and 175 MHz from the Cambridge research group of Ryle and Vonberg (see Chap. 16) were re-analysed, inferring a quiet temperature at these frequencies of 1.3 and 0.6×10^6 K, respectively.

In addition, the publication contained RPL data at 50, 25, and 1.25 cm (contributed by colleagues) as well as the Southworth Bell Labs data at 10 and 3 cm. The derived temperatures were 5, 1 and 0.7×10^5 K at 50, 25 and 10.7 cm, respectively and 1×10^4 K at 1.25 cm. See Fig. 14.2.1 and 14.2.2 in ESM 14.2, Final Data.

A thorough summary was provided of the determination of the size of the thermal radiation from the sun; the expectation was, of course, that this would originate "more or less uniformly over the sun's disk". At metre wavelengths, the sea-cliff interferometer in Sydney had been used to image the radiation:

> ... [O]bservations at this Laboratory have consistently shown that when the intensity is low, so that the thermal component might be expected to be dominant, the source is found to

---

[28]The note read: "Pawsey has tried to eliminate the variable solar noise from the emission from the 'quiet' sun. At 200 MHz he observed solar radiation over a period of 140 days; for half this time, when outbursts were small, the mean solar radiation showed an enhancement of 400 over that expected for a black body at 6000 K. On no occasion was the intensity less than 50 times the blackbody value. He analyses the frequency distribution of the observed intensities into two components and ascribes one to a quiet coronal source at a temperature near 10^6 K and the other to the highly variable enhanced radiation."

[29]Don Yabsley trained in engineering at Sydney University and joined RPL during the war. He spent the remainder of his career undertaking engineering tasks with the radio astronomy group at CSIRO, including resurfacing the Parkes telescope (GRT), enabling its continued use at a fraction of the cost of building a new instrument. He retired in 1987. Information taken from his funeral brochure, supplied by Harry Wendt.

approximate the whole solar disk ... [A]t shorter wavelengths at nearly all times, solar radiation is mainly thermal in origin ... The outstanding conclusion concerning the nature of the solar atmosphere which may be drawn from these observations is that they offer direct confirmation of the existence of **kinetic temperatures** of the order of a million degrees in the corona.

Priority Disputes and Scientific Discovery

In Chap. 12, we applied sociologist of science Robert Merton's concept of "sagacity" as key to serendipitous discovery to analyse Pawsey's role in the beginnings of radio astronomy. In the difficult events of 1946, we can see another commonplace feature of the social organisation of science about which Merton would famously write only a decade later: priority disputes (Merton, 1957). In 1942—horrified by war—Merton had formulated his key thesis that science (unlike the appalling political sphere he was witnessing) produced knowledge successfully because it had a foundational "ethos" expressed through five "norms" or rules of conduct, including the open sharing of original results (Merton, 1942). He explained the common occurrence of priority disputes as an outcome of the fact that the main reward in science is recognition. Due to competition over priority, the norms of sharing data, techniques, and knowledge may break down, being replaced by acrimonious accusations of breaches of correct scientific behaviour (Pinch, 2015). Merton's thesis provides a useful explanation of the fragility of both Martyn and Bowen; and we can see in Pawsey, the "ethos" of science that Merton saw as the hallmark of the scientists of his day.

Science, often represented as the scientific method, consists of theoretical predictions that are then confirmed or refuted with experimental or, in this case, observational results. Martyn's unilateral decision to publish his theoretical paper separately was in keeping with this model. Indeed he wrote later: "In point of fact it was developed ... before the facts were known. It is a theory of prediction rather than explanation, and perhaps has correspondingly greater weight because of this."[30]

But this is not accurate either, because in reality, there is often *not* a neat division between prediction and empirical confirmation. Rather, as we have seen in this case, ideas are developed in tandem with the emergence of empirical observation, and revised and reviewed over time, with inferences running in both directions, through interaction and conversation. The ethos of science is practiced as much in the

---

[30]Martyn to Appleton 27 October 1948, Sullivan archive, original in the Appleton archive. We note that this offers a charitable interpretation of another issue, the fact that Martyn's *Nature* paper did not acknowledge Pawsey or the RPL; he thanked Woolley and Allen "for much advice on solar data". Pawsey's paper acknowledged that "I am indebted to Dr. D.F. Martyn for pointing out to me the probable existence of high-level thermal radiation, and to members of the Royal Australian Air Force and of the RPL who took the observations." To acknowledge Pawsey would diminish the value of the paper as a theoretical prediction. But to not do so distorted the imbricated development of the outcome.

qualities and nuances of such interactions as it is in the public record of the outcomes.

We agree with Sullivan that Pawsey's own words most accurately tell the story. He wrote Appleton on 8 September 1948 from the UK[31]:

> The actual sequence of events … was as follows: (a) observation of considerably high and very variable effective temperatures, 10^6–10^8 degrees on 200 Mc/s—J.L.P. and colleagues. (b) Suggestion of high-temperature coronal thermal emission [before December 1945[32]]— D.F.M. and colleagues. (c) Successful search for 10^6 degree base level on 200 Mc/s—J.L.P. (d) Detailed theory—D.F.M.

---

[31]NAA C3830 F1/4/PAW/1 Part 2.

[32]Based on Payne-Scott's "Summary Paper of December 1945", see Chap. 13.

Part V
Connections

Chapter 15
Horizons, 1944–1947

In [his] case I think there is every reason for CSIR to offer him [Pawsey] the greatest inducement to stay in their employ.—Taffy Bowen to Fred White, 28 June 1945

Introduction

The serendipitous successes and enticing possibilities of the first year of radio astronomy at Sydney had set Joe Pawsey on a quite unexpected scientific path. He was suddenly established at the lead of a new research field in basic science, with a clear sense of scientific purpose. But just a few months before observations at Collaroy began, Pawsey had not been at all clear about what he would like to pursue in the post-war world. This chapter concerns his personal contemplation of that future, and provides context for the long trip overseas that he undertook in 1947 and 1948.

Before Solar Radio Astronomy at Collaroy

In late 1944, when it was clear that the Allies would emerge as victors against the Axis Powers, Allied scientists began to plan their post-war activities. In many cases, these plans were based on experience and innovations simulated by wartime research. In contrast to the UK and the US, an Australian scientist's future at a local university was uncertain. There was no established tradition of research at an international level in astronomy, physics or electrical engineering, notwithstanding the successes of individuals. In any case, Pawsey did not especially envision himself in a University role; his strength and experience were closer to the interface between physics and engineering.

© The Author(s) 2023, corrected publication 2024
W. M. Goss et al., *Joe Pawsey and the Founding of Australian Radio Astronomy*,
Historical & Cultural Astronomy, https://doi.org/10.1007/978-3-031-07916-0_15

In 1944, Marcus "Mark" L.E. Oliphant, who, as we know, connected well with Pawsey during his wartime visit to Australia (Chap. 9 and ESM 9.6, Microwave Radar) offered him (Pawsey) a position at the University of Birmingham. He was impressed by Pawsey's desire to "work ... in a laboratory ... mixing modern physics and radio techniques." [1] Oliphant told Pawsey that he had been promised by Henry Tizard funding for a number of senior fellowships at the University of Birmingham. "I wonder whether you would consider accepting one of these in order to help us with our problems?" [2] The expectation was that it would be possible to use pulse techniques (as used in radar transmitters) to achieve particle energies in accelerators up to one billion electron volts (GeV).

Within the next month, A.C.D. "David" Rivett (1885–1961, then CEO of CSIR), Fred White (1905–1994, then Assistant CEO), John Briton (Chief of RPL) and E.G. "Taffy" Bowen (Deputy Chief of RPL) were all involved in assessing the practicalities and desirability (or not) of Pawsey taking up the Birmingham offer. Rivett wrote to Pawsey on 19 February 1945: "... I let Oliphant know that we were always glad to seize opportunities for giving our men the chance to gain experience abroad ... [Y]ou can rest assured that there will be no difficulties put in your way by CSIR." [3]

Over the next few weeks, however, CSIR began to have serious doubts about the offer from Oliphant. On 19 March 1945, Rivett wrote to Pawsey: "... I ought to let you know that a good deal of disappointment was expressed about this plan for, after all, you are one of the people that we have to rely upon to develop sound fundamental work in radiophysics when once we get free of the handicaps and shackles imposed by war ... [W]e shall not stand in your way if you really feel you ought to go; but I want you to be quite certain in your mind that we shall be far from happy about your departure." [4] Apparently, White and Rivett could see that if Pawsey were to go to the UK to work for some years, [5] it would be unlikely that he would return to CSIR. The CSIR Executive envisioned Pawsey playing a major role in the post-war revitalisation of Australian science. Pawsey replied to Rivett on 28 March 1945 [6] that he was surprised by the negative reaction:

> I consider my particular field of work to be the application of high frequency radio techniques. This is consistent with my experience over the last 10 years, first in television and then in radar. I have not yet a definite idea of the particular form of work I should wish to

---

[1] Joe and Lenore Pawsey Family Collection.

[2] Ibid.

[3] Pawsey Personnel File CSIR and CSIRO. A8520, PH/PAW/1B. Part 1. Pawsey had written Rivett on 17 February 1945 that he had just received a letter from Oliphant suggesting an extended visit to Birmingham. "I am very keenly interested in the possibility and should like to discuss it with you [during an upcoming visit to Melbourne the following week]."

[4] Ibid.

[5] Oliphant offered a three-year fellowship. From the Joe and Lenore Pawsey Family Collection. 30 April 1945 Oliphant to Pawsey.

[6] Pawsey Personnel File.

undertake after the war and one of the chief reasons for my wishing to go overseas is to assist me in forming my ideas on this subject.

His goal was to gain a new perspective by visiting colleagues in the UK and the US.

> I could profit most through an extended trip of a year or so, most of my time being spent working in a suitable laboratory and the remainder in visiting others ... If I went I should wish my family to go also, at any rate as far as Canada to my wife's relations, as my wife's health has been poor in Sydney and she would be most unhappy if left here alone ... I appreciate your compliment in saying I am one of the people you have to rely on to develop sound fundamental work in radiophysics after the war, but I suggest I could do this a lot better after a period spent overseas. [7]

Rivett was explicit as he wrote Pawsey on 6 April 1945 [8]: "... I am much impressed by White's view that you will probably go further if you follow certain lines of work which he has in mind than if you join Oliphant's team and possibly become chiefly interested in developing apparatus ... I can assure you that I am not going to put any difficulties in your way if you feel you ought to go to Birmingham when the offer comes." On 30 April 1945, the offer came. Pawsey replied to Oliphant by cable on 16 June 1945 that he would likely accept, with an expected arrival in the UK in early 1946. The final decision was to await the outcome of discussions with David Rivett. [9]

Not surprisingly, CSIR continued to be concerned with the prospect of losing Pawsey. Bowen (at that time Acting Chief) wrote White on 28 June 1945, urging that they offer some "inducements" to keep him, beginning hopefully:

> ... Dr. Pawsey has become rather less interested in the Birmingham proposals or perhaps more interested in a future in CSIR and the Division of Radiophysics. The main attraction to Pawsey is the opportunity it provides of undertaking interesting work in an overseas laboratory in a field which may be able to make use of techniques in which he has become the master. Pawsey's keen desire for continued contact with overseas work is most under-standable ... With the departure of D.F. Martyn he has spent the major part of his time on all aspects of work connected with the propagation of radio waves. He has been instrumental in reorganising this programme in a manner which is leading to very effective research being done, and he has made a remarkable difference to the morale of the officers engaged on [this] work.

Bowen recommended that Pawsey spend about a year in the UK at the Cavendish Laboratory with Ratcliffe and Booker and a few months in the US. [10]

After a discussion with Rivett that same day, Pawsey sent a telegram turning down the offer made by Oliphant to go to the UK. [11] In a letter from 1 July 1945, Pawsey [12] explained the reasons for accepting the CSIR offer to remain in Sydney: "... [T]here would be a very good chance of being sent [to Europe] by CSIR shortly.

---

[7] Ibid.

[8] Ibid.

[9] Joe and Lenore Pawsey Family Collection.

[10] Pawsey Personnel file.

[11] Joe and Lenore Pawsey Family Collection.

[12] Ibid.

If I had accepted your [Oliphant's] offer the 3 year period would have put me right out of the picture in the development of this laboratory ..." Oliphant had no illusion about the process that had been used to force a CSIR counter-offer to Pawsey (3 July 1945): "I am glad, however, if the result of my action in offering you a fellowship has been to improve your position with CSIR."

Planning for Travel

Although CSIR was committed to providing Pawsey an overseas trip, White told Bowen on 10 July 1945 that the trip could not occur in 1946. [13] Pawsey had agreed to this delay in consultation with Rivett. Both were aware that, if Pawsey had left Sydney in 1946, the progress in what was then termed "solar and cosmic noise" at RPL would have suffered. The trip to North America and Europe was only to occur in late September 1947, 2 years in the future.

Nonetheless, on 17 July 1945, Pawsey began discussions with his PhD supervisor Jack Ratcliffe [14] about his anticipated visit to the UK. Ratcliffe was in the process of moving back to his university post at Cambridge; the move from his wartime position as one of the leaders at TRE (Telecommunications Research Establishment) at Great Malvern occurred in August 1945. Pawsey wrote:

> Fred White ... second in command [of CSIR] argued [sic] me into staying here with an unofficial promise of about a year overseas at CSIR expense in the near future and with an agreement on my part to carry on with the propagation work. [Radio astronomy was not yet underway at RPL; this was to occur starting in October 1945.] It looks as if I am getting conservative in my old age [37 years!] but White's alternative looks thoroughly interesting. In consequence I am interested to plan a sort of sabbatical year ...

Pawsey was interested in whether Ratcliffe would be his host during the sabbatical. Ratcliffe wrote back within a month (14 August 1945) with an airmail letter, [15] which took 3 months to reach Sydney! Ratcliffe told Pawsey about his plans to start up the "team on Radio" at the Cavendish with ex-TRE staff including Booker, Finlay and Weeks plus Ryle (pre-war from Oxford). "We plan to do work of the old kind i.e. to use radio as a tool in physical investigations, rather than to make bigger and better radios [i.e. aerials] ... We should also welcome you very much, as our team will be somewhat 'in-bred' from TRE and a senior man from elsewhere will do us a lot of good."

---

[13] Pawsey Personnel file. Also, Bowen recommended on 28 June 1945 that the travel expenses for the entire family would be covered by CSIR. However, on 2 July 1945, Rivett pointed out it was impossible to cover the family's expenses as this would lead to a reduction in the number of scientific visits sponsored by CSIR.

[14] Joe and Lenore Pawsey Family Collection.

[15] Joe and Lenore Pawsey Family Collection.

After the delayed arrival of the letter, Pawsey responded on 5 November 1945. [16] There was no news on his trip to the UK but he did have exciting news of the Collaroy 200 MHz solar data (Chap. 12). He explained that there was a semi-steady signal from the sun of

> ... [an] astonishingly high level, corresponding to blackbody radiation at a temperature of millions of degrees ... [The] level [also] changed from day to day in a manner which correlated exceedingly well with total sunspot area ... [T]he fluctuating character, variations in fractions of a second often, intrigues me. It is not our [earth's] atmosphere then [sic] I think we have something pretty important in interpretation. I presume the solar phenomenon is the same as "Jansky's cosmic static" originating in the stars.

Family Life

As Pawsey had mentioned, there were significant personal as well as professional motivations for this trip—Lenore missed her childhood home and family, and her parents, equally, missed her. Lenore's mother Mabel corresponded frequently with her children, Lenore, Lenore's brother Ted (who, with his wife and children was now established in Princeton, a considerable distance from Battleford, Saskatchewan) and Lenore's sister Bessie, now living in England. Mabel's letters to Lenore remain extant and are warm, engaged with the many details of domestic life, and replete with solicitude for Lenore's health and wellbeing, as well as admiration and affection for the Pawsey children, Margaret, born 1937, and Stuart, born 1939, and at the end of the war, Hastings, born 1945.

From these letters, a consistent and considered portrait of Pawsey as a family man emerges—as for example in this vignette, written amid the intense wartime pressures of 1943, Mabel to Lenore:

> We would like to have been close to Stuart to catch his joy at seeing the four white horses in the "Cinderella" pantomime. The stage must have been roomy ... You and Joe must have very clever children or you are very good teachers, perhaps the two make a strong combination. We just can't get over Mgt [Margaret] installing the electric light in her doll's house. Joe has the right principle, that of demonstrating on paper and then leaving it to her to work out. What a thrill she would get when the light would turn on. It would mean much more to her than if Joe had installed it. It required some patience on Joe's part. I do admire parents who take time to teach in that way for busy parents usually find it quicker to do the thing than to demonstrate. [17]

Lenore, who had moved to an entirely new country with a baby and a toddler during wartime, not infrequently found the war years a struggle. In 1941 (ESM 9.3, Difficulties), Pawsey had made a lightning side visit to Lenore's parents in Canada during his necessary trip to the USA and the UK; while this connection was warmly remembered by Mabel Nicoll, his absence was difficult for Lenore. In 1941 and 1942

---

[16] Joe and Lenore Pawsey Family Collection.

[17] Letter from Mabel Nicoll to Lenore Pawsey, 6 March 1943, Joe and Lenore Pawsey Family Collection.

Margaret and Stuart suffered measles and chickenpox, with a resulting period of deafness following for Margaret.

1943 became a truly distressing year in the Nicoll and Pawsey families. Lenore experienced a miscarriage and then a prolonged period (months) of illness, during which care for her two young children had to be distributed between her mother-in-law (who came from Melbourne for a brief visit) and friends who took them in. Pawsey's father, Joseph Andrews, who was in fragile health, died on 30 June, at the family farm, "Glencoe", in Victoria. Not long after that, while Lenore was still unwell, Lenore's younger brother Hastings was hospitalised in Regina, Saskatchewan, due to a congenital condition, then termed Bright's disease, which affected his kidneys. He died very shortly afterwards, on 7 August 1943. Lenore wrote to her mother more than once during the following months of her grief and regret at not being closer to him. [18] Lenore developed asthma and spent further time away from her family to recuperate, her health remaining fragile during the rest of the year. The following year she acknowledged an ongoing depression. [19] Lenore Pawsey's experiences of physical and mental ill health can be attributed to her circumstances of stress, isolation from siblings and parents and also close personal networks, the constraints of women's roles and expectations, and the physical impacts of the limited diet, exercise and healthcare of the era, and were shared by many women of her generation. [20] It was fortunate that the family were able to celebrate a promotion at work for Joe, the one happy occurrence of 1943. [21]

It must have been particularly sad for Lenore when her father, Jack, died the following year, on 11 December 1944, since she had not seen him since her marriage. "I hope you got my letter saying that you mustn't be grieved if Daddy or I slipped away quietly some time. Well Dad has just done that," her mother wrote. "Though I could see that Dad was failing I still thot [thought] he might be with me for some long time ... I got up while it was still dark and when he didn't speak first as he usually did I just came down and got breakfast and when he wasn't down by nine I went up to him but he was gone." [22] Lenore was pregnant with a child she hoped, and eventually did, name Hastings (born 1945).

Throughout these tribulations, family life remained harmonious. Joe supported Lenore when she took up work teaching French during the later war years and provided nursing and domestic care in periods of illness. "How pleased we are to have you write so many compliments of Joe," Mabel Nicoll wrote to Lenore on 1 April 1944. "You are not given to making extravagant statements so when you say

---

[18] Mabel replied, "Now, Lenore you must not regret not having written to Hastings (Nicoll). You were exceptionally kind and thoughtful of him. It's difficult to keep up a correspondence when it's mostly one-sided and his was with you, Ted and Bessie." 26 October 1943, Joe and Lenore Pawsey Family Collection.

[19] Mabel Nicoll to Lenore Pawsey, 1 April 1944, Joe and Lenore Pawsey Family Collection.

[20] Allon (2014), Davis (1989), Grimshaw (1980), Holmes (2016).

[21] Letter from Mabel Nicoll to Joe and Lenore Pawsey, 29 October 1943, Joe and Lenore Pawsey Family Collection.

[22] Mabel Nicoll to Lenore Pawsey, 13 December 1944, Joe and Lenore Pawsey Family Collection.

'Joe was simply marvellous' we know it is true. That everyday constant kindness is worth a hundred times more than the spasmodic attention some husbands give."

Throughout the period 1944–1947, the Pawsey family was indecisive about the issue of whether their children, including baby Hastings, would accompany Joe and Lenore Pawsey during their planned visits to North America and Europe. Pawsey suggested [23] that the three children would accompany the parents as far as Battleford, Saskatchewan, Canada, Lenore's home, where they would be looked after by Lenore's mother. On 17 July 1945, Pawsey wrote Ratcliffe that he would bring the family to the US and UK with "ports of call" in Princeton, New Jersey, US, to visit Ted and Kate Nicoll, and in London, where Lenore's sister Bessie Whittard lived. But the Pawseys also worried about the disruption to the children's schooling. [24]

In the end the children were left in Sydney during the parents' absence from September 1947 to October 1948 in the charge of the two grandmothers, Margaret Pawsey and Mabel Nicoll. Mabel Nicoll had arrived in Sydney by ship from Canada in August 1946 (when Hastings was only 16 months old). She stayed in Sydney until 3 October 1949, when she moved to the UK by sea to be with her other daughter Bessie Whittard.

Towards Departure

On 25 February 1946, Bowen sent a memo to CSIR Head Office with a request for funds for Pawsey's trip overseas [25]: "It is some time since he was abroad and he recently turned down a very attractive offer from Professor Oliphant." A sum of A£ 2500 was requested for the trip.

On 9 October 1946), Fred White [26] wrote Ratcliffe ("Dear Jack"): ". . . [Pawsey] has done excellent work during the war and is still doing so, and in spite of competition from your group, he seems to be holding his own in this work on solar noise." White, the CEO at this time, was worried about the guidance of the solar noise group during Pawsey's absence ("Most of [the solar group] however, need some guidance . . ."). White asked Ratcliffe if someone from the Cambridge group could come to Sydney in an exchange. [27] Ratcliffe replied that the senior staff had only recently begun their post-war duties at the Cavendish and an exchange was

---

[23] 28 March 1945, Pawsey Personnel archive.

[24] 3 March 1947. Pawsey Personnel archive. Bowen writes: "Due to interference with their schooling he [Pawsey] is reluctant to take the two elder children and may be able to arrange for them to stay here in Sydney." However, 2 months later (28 April 1947), Bowen said that the children would be taken to the US and placed in the care of the Nicoll family in Princeton.

[25] Pawsey Personnel archive.

[26] Pawsey Personnel archive.

[27] Pawsey Personnel archive.

impossible in the near future. [28] Personnel from the Cavendish Laboratory would not be able to visit RPL for long term visits until Peter Scheuer came to Sydney for a few years some 13 years later.

Three months after White wrote Ratcliffe, Bowen started the ball rolling for Pawsey's trip to North America and Europe. The CSIR Executive approved the trip on 17 March 1947. By 28 April 1947, Bowen had proposed a detailed itinerary to the CSIR Executive with a departure in late September to San Francisco by sea. After 5 months in the US and Canada, Pawsey would depart by sea for the UK in March 1948. He was to attend two international conferences in Scandinavia in mid-1948, URSI in Stockholm and IUGG (International Union of Geodesy and Geophysics) in Oslo. The return trip to Sydney was to be by sea from the UK starting September 1948. In the UK, the "home base" would be the Cavendish Laboratory. The final itinerary was approved by the CSIR on 13 June 1947. The approvals by the Acting Minister of the CSIR, W.J.R. Riordan, and the Prime Minister of Australia, J. B. Chifley, were subsequently obtained, and bookings were made on the SS *Marine Phoenix*, with sailing set for 25 September 1947 from Sydney to San Francisco.

---

[28] From the Joe and Lenore Pawsey Family Collection. A month later (12 November 1946), a letter to Pawsey from Bracewell arrived. Ron Bracewell (from the RPL) had recently arrived at the Cavendish Laboratory to work on a 16 kHz ionospheric investigation with Ratcliffe. Bracewell was aware of the possible exchange suggestion and doubted if it were possible. However, Bracewell was certain that Pawsey would be welcome during the proposed visit to Cambridge. Bracewell's handwritten letter began with the salutation: "Dear Doctor Pawsey".

Chapter 16
A New Field of Science, Postwar

The time was ripe in Australia for good radio research to go ahead. It was ripe because about 10 or so years previously, about the end of the 1920s, the CSIR had set up a radio research board and had got people interested in radio research. It was then research on the ionosphere mainly...

The thing that is exceptional is to have a group of people who don't have research commitments. They usually have the equipment or something like this which having got they are morally bound to use. We were in the position where we didn't have anything to tie us down.

--The interview with Joseph Pawsey and Ron Giovanelli (Moderator George Baker)
Australian Broadcasting Corporation, television programme HORIZONS, 1960, Nov 11

By the time Joe and Lenore Pawsey were ready to embark on their voyage to North America and Europe in September 1947, the scientific importance of the trip had increased substantially. In the 2 years since the conclusion of the war, radio observations of the sun had produced dramatic results. The first discrete radio source had been discovered by Hey in the UK,[1] and early data from radio observations of the cosmos promised to lead to startling new discoveries.

The story of the first 10 years of the field that would so soon become "Radio Astronomy" is very well known, with many fascinating details provided by Sullivan

---

Supplementary Information The online version contains supplementary material available at [https://doi.org/10.1007/978-3-031-07916-0_16].

---

[1] Discovered at 64 MHz by Stanley Hey, S. John Parsons and James W. Phillips in the UK, published in *Nature* on 17 August 1946. Stanley Hey, associated with the Army Operational Research Group of the British army had a major impact on early radio astronomy. Hey, however, was less well known than the other UK groups; he did not have the prestige of the university groups at Cambridge and Manchester and, in the military, he did not have the same ability to pursue a self-directed research program. However, he made three major discoveries: (1) Cygnus A, (2) the radio bright sun associated with sunspot activity (February 1942 and published 17 October 1945) and (3) radar echoes from meteor trails. (See his book *The Evolution of Radio Astronomy*, 1973.)

© The Author(s) 2023
W. M. Goss et al., *Joe Pawsey and the Founding of Australian Radio Astronomy*,
Historical & Cultural Astronomy, https://doi.org/10.1007/978-3-031-07916-0_16

in his comprehensive history of this period. This chapter provides a summary overview. It is intended to provide readers who have *not* read of these events elsewhere with a guide to the people and places of significance to the new field. It provides context for Pawsey's decisions and choices as leader of the solar and cosmic noise group at RPL in the years up to 1955. In some cases, we provide a few hints of events even later in the twentieth century, so that interested readers can trace the continuities back to these early beginnings.

Other Developments at RPL: Cloud Physics

As Pawsey commented:

> [A]t the end of the war the decision was made by CSIR, as the organisation then was, to carry on in peacetime work, and we had the proposition to find worthwhile objects of research. We were not bound by anything and we experimented in quite a lot of different avenues. Two of those in the radiophysics division were the physics of rain and possible rainmaking, and radio astronomy. There were half a dozen others which were not very successful, and we gradually built up into the two successful ones which were radio astronomy and rain physics and rain making.[2]

While Pawsey began exploring solar and (soon) cosmic noise, Bowen led a team to explore cloud seeding to make rain—a new technique whose aim was to somewhat control and improve rainfall. In 1946 USA researchers I. Langmuir and V. Schaefer reported that rain could be induced by seeding clouds with dry ice. Bowen immediately instigated a trial in eastern New South Wales using RAAF aircraft, which bore early success in February 1947.[3]

The rainmaking project provides a telling contrast to radio astronomy. Unlike solar and cosmic noise, it fitted the CSIR's objectives to undertake applied science in areas of significant need for Australian agriculture (and industry). The foundational concept of cloud seeding was plausible and pathways for investigation straightforward. But it led neither to the hoped-for benefits for agriculture, nor to visible developments in science in the manner of radio astronomy.[4] We also draw attention to the stark contrast between the leadership styles of Pawsey and Bowen. The difference between Bowen, the decisive and entrepreneurial leader and Pawsey, the highly respected and thoughtful scientist and communicator, becomes increasingly evident throughout the rest of this book.

This cloud physics research program continued for 24 years. While the program failed to provide rain to the very dry continent of Australia, the considerable

---

[2] Australian Broadcasting Corporation, television programme HORIZONS, 1960, the Nov 11 interview with Joseph Pawsey and Ron Giovanelli by Moderator George Baker. Details of these programs are provided in Sullivan (2009), pp. 123–124.

[3] https://csiropedia.csiro.au/cloud-seeding/.

[4] For Pawsey's impressions of CLOUD PHYSICS in 1947–1948 see NRAO ONLINE.25.

members of the group involved at RPL became among the international leaders in the field.[5]

RPL Sydney: Pawsey Builds a "Learning Organisation" for a New Field of Science

By contrast with rainmaking, radio observations of cosmic and solar noise were a form of speculative, basic science, with no clear lines of investigation, no apparent applied benefit and no conceptual anticipation of where discovery might have impacts! And yet the radio group was spectacularly successful—and not only in terms of discovery, but also in unpredicted and unpredictable applied outcomes. An outstanding example was the development of Wi-Fi 40 years later involving many researchers whose mentors were project leaders from Pawsey's groups. The links from Wi-Fi back to radio astronomy and the value of the intangible networks which had evolved out of this basic research are described in "Four Pillars of Radio Astronomy", Chap. 7 (Frater et al., 2017).

To pursue research in this new field, Pawsey developed what was essentially an early form of what was termed, in the management literature of the 1990s, a "learning organisation", as discussed below. The development of early radio observations of the sun and cosmos proceeded flexibly, with promising researchers supported to develop independent lines of research. This model arose from, and suited, his leadership style; it also reflected the "British" model of investing resources in individual scientific research leaders, preferred by Ratcliffe and Rivett. As their work progressed and each member began to design new instruments with which to pursue it—drawing on the stores of RAAF, USAF and Navy surplus radar related gear that Bowen was able to salvage prior to disposal in 1945—the group spread out over a number of different field stations spaced across greater Sydney.

---

[5] An excellent study of the RPL rainmakers has been provided by R.W. Home, "Rainmaking in CSIRO: The Science and Politics of Climate Modification," pp. 66, in *A Change in the Weather: Climate and Culture in Australia*, Sherratt et al. (2005). An additional description of RPL cloud-seeding efforts has been provided by Helen Sim in an unpublished MSc thesis at the University of New South Wales (Sim, 1995): *The Rise and Fall of the Rainmakers: A History of the CSIRO Cloud Seeding Experiments, 1947–1981*, available at NRAO ONLINE.50. See also "Guidelines for the utilisation of cloud seeding as a tool for water management in Australia", http://www.cmar.csiro.au/e-print/open/cloud.htm.

The Sun and the Radio Stars

Radio observation of the sun remained a large and significant component of research at RPL. But from 1948 onwards, the dominant story became the exciting possibilities of a newly discovered phenomenon—the "radio star", that is, a strong source of radio emission coming from far away in the galaxy.

In these 7 years a program of surveys and positions of discrete radio sources using the sea-cliff interferometer at Dover Heights was led by a new post-war recruit at RPL, John Gatenby Bolton (Robertson, 2017). Bolton (1922–1993) was to become one of the most eminent of the first generation of radio astronomers, famous among the Australia public as the first scientific director of the Giant Radio Telescope (GRT), commonly termed "the Dish", at Parkes, NSW. Bolton was a British naval radio engineer who remained in Australia after being stationed in Sydney in 1945. A self-reliant, bright young man whose Cambridge studies had to be completed in a short two-year period due to the beginning of the war, Bolton successfully applied for the first Research Officer position advertised by RPL after the war and commenced in September 1946. His first assigned task from Pawsey was to investigate the polarisation of the newly identified sunspot radiation, first designing and building an antenna for the purpose. The antenna was installed at Dover Heights in November 1946 with the help of another new recruit, Bruce Slee (1924–2016). Slee was an RAAF Radar Operator during WWII and while posted at an RAAF Radar station near Darwin he had observed the solar radio emission using the same COL MkV radar system used by Pawsey, Payne-Scott and McCready for their first solar observations at Collaroy. After reading a news article about these radio solar observations, Slee wrote to Radiophysics to report his wartime observations and to inquire if positions were available for such work. He was invited to meet Pawsey who was so impressed that Slee was hired on the spot and started work in November 1946. Slee continued his very active research career in radio astronomy at CSIRO until his death at age 92 in 2016.

Bolton and Slee were fascinated by Jansky's reports of "cosmic noise" and, using an astronomy textbook and a star atlas borrowed from the local municipal library, unsuccessfully attempted to detect radiation from astronomical objects as they rose above the horizon at Dover Heights. Slee commented that "Pawsey was not amused by our deviation from the solar work". However, in a few months, Pawsey (who stated in a recorded interview on ABC TV[6] that there was value in using "another job number" for unapproved risky projects) gave Bolton, Slee and Gordon Stanley (an engineering graduate who had been recruited to RPL in 1943 for radar research) a free hand, along with equipment that had been prepared for a proposed (then cancelled) expedition to Brazil to observe a solar eclipse.[7]

---

[6] Australian Broadcasting Corporation (1960) TV interview at 19:42.

[7] The cancellation was due to logistic challenges, see NRAO ONLINE.21, "Eclipse Expedition Failure 1947."

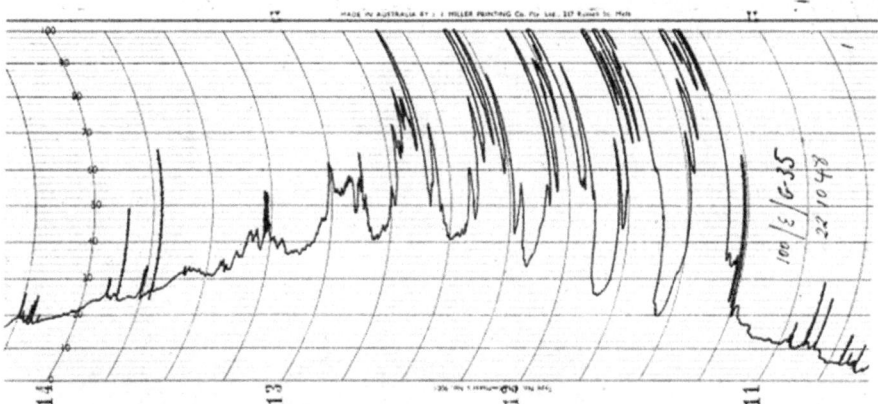

Fig. 16.1 A sea-interference pattern from Cygnus A similar to the first Cygnus A fringes observed in June 1947. Time is increasing from right to left. Occasional interference spikes can be seen before Cygnus A rises. Credit: CSIRO Radio Astronomy Image Archive B1639-2

In 1946, J. Stanley Hey in the UK reported observations of fluctuating radio emission from the Cygnus constellation (Hey et al., 1946). This was interpreted as the discovery of the first "radio star", now known as Cygnus A. Pawsey attempted but could not repeat these observations.[8] In February 1947 John Bolton and Gordon Stanley started using the new 100 MHz receiver as a cliff interferometer at Dover Heights to look for the interference fringes—not scintillations—expected from any discrete sources such as found by Hey in the Cygnus constellation (Slee, 1994). In June 1947 they detected clear fringes from the direction of Cygnus thus confirming that Hey's variability was the result of a discrete radio source (Bolton & Stanley, 1948b). Figure 16.1 is a Cygnus A recording, and the rapid fluctuations in intensity first noticed by Hey are also clearly visible at the fringe amplitude maxima. Figure 16.2 shows equipment similar to that used at Dover Heights. The investigation of the cause of the rapid fluctuations in intensity (scintillation) is discussed in Chap. 18.

By November 1947 the group of Bolton, Stanley and Slee started searching for weaker sources using a receiver with improved stability. They quickly discovered 3 new discrete sources and Bolton introduced a new naming convention: the constellation name followed by a letter A, B, C ... in order of the brightness of the source in the constellation. This naming convention has been used for the brighter radio sources ever since, and the first source found by Hey became Cygnus A. The new sources found were Taurus A, followed by Virgo A (originally called Coma Berenices A because the position had a large error, placing it in an adjacent constellation) and Centaurus A. These three next strongest discrete radio sources

---

[8]He wrote to Woolley at Stromlo on 11 Sep 1946 reporting fluctuations similar to solar bursts. However the description of being like solar bursts makes it more likely that they were radio frequency interference and not the type of variability seen by Hey; Sullivan (2009, p. 138) concludes that the early attempts to confirm the new "radio star" were unsuccessful.

Fig. 16.2 From left, John
Bolton, Gordon Stanley and
Joe Pawsey in early 1954
with some equipment
typical of the era. Image
made at the RPL lab
building at the University of
Sydney campus. Credit:
CSIRO Radio Astronomy
Image Archive B11833-6

were to have a major impact when their positions were determined well enough to
find the optical counterparts as we discuss in Chap. 18. By the end of January 1948,
they had found a total of 6 sources and Bolton wrote a paper for *Nature* (Bolton,
1948) announcing a new class of astronomical objects. Cygnus A was not alone but
was the brightest member of this new population of "radio stars". Like Cygnus A,
none of the new discrete sources had outstanding stellar identification but the
positional uncertainties were many degrees.[9]

Thus began Bolton's research program focused on finding similar discrete
sources of radio emission and measuring their positions with sufficient accuracy to
make identifications with known, optically identifiable, astronomical objects (Sullivan, 2009, pp. 335–351). This program would be the focus of much of the excitement—and controversy—that defined the field of radio astronomy as it developed in
the 1950s, as discussed in many of the subsequent chapters of this book.

---

[9]This story is beautifully told by Peter Robertson in his biography of Bolton. (Robertson, 2017).

Groups, Stations and Projects

In the first 7 years of research, and in addition to Bolton's group at Dover Heights, the main programmes were:

1. Led by Pawsey: exploration of solar bursts. Field stations were in operation at Dover Heights, Hornsby Valley, Georges Heights and later Potts Hill. As we have shown in Chaps. 12–14, radio observation of the sun proved to be an immediately exciting and fruitful area of research. Within 18 months, Pawsey, Payne-Scott and McCready had outlined the principles of aperture synthesis and made the first suggestion that galactic emission might be a related process. The researchers involved were Lindsay McCready, Ruby Payne-Scott and Don Yabsley. Lindsay McCready (1910–1976) was an engineer specialising in receiver design who had joined RPL during the war years. He was a loyal supporter of Pawsey and was placed in temporary command of the solar and cosmic noise group in 1947–1948 during Pawsey's absence, but struggled to manage the strong personalities of Payne-Scott and Bolton (Goss & McGee, 2009). McCready would later work with Paul Wild on the development of solar dynamic spectrograph at Penrith.[10] In Fig. 16.3, we show the Fritz Goro image of March 1951[11] with Pawsey at Potts Hill, a total power 98 MHz crossed dipole Yagi antenna for total power observations of solar bursts.

2. Led by Paul Wild: solar spectroscopy, initially using a field station at Penrith, then replaced by the new Dapto solar spectrometer which became operational in August 1952. Like John Bolton, Paul Wild (1923–2008) was born in Sheffield (UK), studied physics and mathematics at Cambridge, and then served as a naval radar officer during the war. During his many wartime visits to Sydney, Paul met and became engaged to Elaine Hull, who insisted that their marriage depended on his permanent relocation to Australia after the war. Wild applied unsuccessfully for the Research Officer position won by Bolton in 1946, but won the next advertised position. He joined Pawsey's solar group and started building instrumentation with Lindsay McCready. Between February and June 1949 they built one of the first radio spectrographs to study the dynamic spectra of solar bursts.[12] They observed the spectrum of bursts of radiation from the sun over a wide spectral range of frequencies (70–130 MHz) with each scan in frequency extending 1/3 s. Wild started observing at the Penrith field station at the foot of the Blue Mountains outside Sydney, a very sparsely populated area at the time, so there was a reasonably low radio noise level. There, observations of frequency and time resolved solar bursts led to the classification of the different types of

---

[10] In NRAO ONLINE.20, we present an extensive review (Precis) of metre-wave solar work at RPL from 1945 to 1960, from the immediate post-war era to the construction of the Culgoora Radioheliograph.

[11] See Additional Note 1 in NRAO ONLINE.23.

[12] The time-frequency dependence of the solar emission.

Fig. 16.3 Fritz Goro photo of Pawsey at Potts Hill in March 1951, 98 MHz yagi antenna at the north end of the reservoir. It was used for total power observations of solar bursts. Note the grating array aerials (10 foot diameter, used at 20 cm) to the south. Credit: Getty Images, Fritz Goro, The *Life* Picture Collection, licensed by Getty November 2021/ Licence organised by Shutterstock, Inc., New York, NY (5 Nov 2021) Original, Premium Editorial All Media

solar bursts into Type I, Type II and Type III solar bursts (Wild and McCready, 1950).[13]

Through the decade of the 1950s, Wild's solar group became the international leaders of solar radio astronomy and Wild went on to become the world's foremost solar radio astronomer, designing and building the famous radio heliograph at Culgoora near the town of Narrabri in NSW (1967–1974).[14] He became Chief of the CSIRO Division of Radiophysics in 1971 and Chairman of CSIRO from 1978 to 1985.

3. Led by Bernard "Bernie" Mills: radio source surveys. Mills started at the Potts Hill field station using Ruby Payne-Scott's interferometer in the evening when

---

[13] See Chaps. 33, 34 and NRAO ONLINE.20.

[14] See Chap. 37.

they were not doing solar work. He then moved to Badgery's Creek and later Fleurs.

Both Bernie Mills and Ron Bracewell (1921–2007) graduated with second class Honours degrees in Engineering in 1943. (Mills, who, like Payne-Scott, Christiansen and others, had generated a considerable education in the progressive left wing politics prevalent at the time, was later pleased to point out how a "second class" Honours result could presage a world-leading career in science.) The entire graduating class of 6 were immediately recruited to radar research at RPL. There, Mills was amongst those most transformed by working under Pawsey. "In the short lecture courses that [Pawsey] gave on transmission lines and antennas, he promoted a physical understanding, rather than the highly mathematical approach to which Bernie had been exposed during his studies" (Frater et al., 2013). When the war concluded, Bowen directed Mills towards other research until an early diagnosis of tuberculosis enforced a 6-month rest. He seized the opportunity of transferring to Pawsey's group when he returned to work in 1948. Mills's choice to develop an independent research program by studying the discrete radio sources that John Bolton had first identified led him to specialise in undertaking surveys of radio emission across the southern sky in the 1950s using antennas designed for this purpose. We discuss Mills's work extensively in Chaps. 35, 36.

4. Led by Frank Kerr (1918–2000) and involving field stations at Hornsby Valley and Potts Hill: moon echoes and the hydrogen spectral line (Chap. 20). Kerr obtained a Bachelor of Science degree from the University of Melbourne and joined RPL in 1940, working on the magnetron in the radar group under Pawsey (Chap. 9). After WWII, Kerr was interested in radar possibilities for astronomy, working together with Alex Shain (1922–1960). They first studied radar echoes from meteor trails (see comments on research at Manchester, later in this chapter), then worked on moon echoes from the Hornsby site going to very low frequency (20 MHz) to see the ionospheric effects. Kerr and Shain explored the possibilities for detecting echoes from the sun, but the equipment that would be required proved to be too expensive. Alex Shain continued an independent program of investigating low frequency radio emissions later, building a cross array at Fleurs. He tragically died of cancer at age 38 (1922–1960).

Kerr led an HI group (see Chap. 20), which built a 36-ft dish at Potts Hill taking measurements of galactic structure. He moved to the US and later became director of astronomy at the University of Maryland (1966–1985). One of Kerr's close collaborators was engineer John Murray (1924–2019), who joined Radiophysics in December 1947. Murray was involved in the 1948 and 1949 eclipse expeditions and worked on the radio spectrograph at Dapto. In 1953 he was assigned by Pawsey to work on a new multichannel HI receiver in a separate group based at a new Murraybank field station in West Pennant Hills. Pawsey was very worried about the design issues with the 4-channel Potts Hill HI receiver that Kerr was using and therefore asked Murray to come up with a better design. The 48-channel Murraybank HI receiver was very successful and was later installed as the first HI receiver for the Parkes radio telescope. After a few years in Leiden in the early 1960s, he contributed extensively to HI research using the Parkes telescope (see Dramatis Personae).

5. Led by Wilbur Norman "Chris" Christiansen (1913–2007): developed instruments for solar imaging research at field stations Potts Hill, and later Fleurs. Christiansen graduated with a Bachelor of Science from the University of Melbourne in 1934. In 1937 he joined Amalgamated Wireless Australasia (AWA), working with Geoffrey Builder and A.L. Green, ionosphere pioneers. Christiansen worked on antenna designs used for overseas shortwave communications. In 1948, he was offered a position in Pawsey's group at RPL and began by making observations at 50 cm of a partial solar eclipse in 1948 in Sydney. These observations showed that the regions on the sun associated with sunspots were about 3 arc min in size (compared to the photosphere diameter of 30 arc min).

   This discovery, and the frustrating inefficiency of depending on eclipses for high angular resolution observations, led to the development of the "grating array" concept: an array with many equally spaced small dishes. The final design was a 32-element east-west array of 6-foot dishes built on the side of a reservoir at Potts Hill, Sydney. Later a north-south array of 16 dishes was added. The first 21 cm array (EW) was operational by 1952 as the distribution of radio brightness across the sun could be studied with a resolution of only 3 arc min. This was used to make the first earth rotation aperture synthesis image (Chap. 37). The image of the quiet sun obtained after eliminating the "slowly varying component" regions associated with sunspots, showed prominent equatorial limb brightening.[15] A staged photo of the Potts Hill array was made in March 1951 by Fritz Goro,[16] with Pawsey adjusting the cables of the grating array at the edge of the reservoir, Fig. 16.4. Amid these solar observations, in 1951, Christiansen (Fig. 16.5) and J. V. Hindman carried out the important crash programme to confirm the detection of the 21 cm line of hydrogen made by Ewen and Purcell at Harvard (Chap. 20).

   Following his visit to Meudon in 1954 Christiansen went on to design and build the "Chris Cross" at Fleurs. This was developed into Australia's first major aperture synthesis telescope, the Fleurs Synthesis Telescope (FST) (see Chap. 37) which continued operations until 1988.

6. Led by Jack Piddington (1910–1997) and Harry Minnett (1917–2003): microwave research using field stations Sydney University grounds and Potts Hill. During WWII, Piddington and Minnett were senior staff working on radar (see Sullivan, 2009, p. 277): "In 1948 they gained exclusive rights (at RPL) to extraterrestrial noise research at the highest technically feasible radio frequencies—their mandate was to explore the sky at the shortest wavelengths (less than 25 cm and even at 24GHz or 1.25cm) and see what turned up." Bowen commented (5 May 1948[17]) that "[m]ost of the work and ideas come from

---

[15]NRAO ONLINE.23, the review of cm solar work (Precis) from 1945 to 1960.

[16]Additional Note 1 in NRAO ONLINE.23.

[17]NAA F1/4/PAW/1.

Fig. 16.4 Fritz Goro photo of Pawsey "adjusting" the 21 cm Potts Hill grating array at the edge of the Potts Hill reservoir in March 1951. Credit: Getty Images, Fritz Goro, The *Life* Picture Collection, licensed by Getty November 2021/ Licence organised by Shutterstock, Inc., New York, NY (5 Nov 2021) Original, Premium Editorial All Media .This image is reversed left to right with respect to the image provided by Getty based on the orientation of the dishes with respect to the reservoir

Harry Minnett ... but [they are] getting quite interesting results on the moon [at 1.25 cm]." Piddington and Minnett (1949) discovered a lag between the optical and radio lunar phase which was explained as a thin layer of surface dust with very different thermal conductivity.[18] Without many sources strong enough to detect with small dishes at these very high frequencies, Piddington and Minnett moved down in frequency leading to their discovery of the relatively flat spectrum galactic centre source at 1210 MHz as discussed in Chap. 23. By the mid-1950s high frequency radio astronomy was completely dominated by the US groups. However, the 600 MHz continuum survey continued at Potts Hill and was used in the redefinition of co-ordinates of the Galactic Plane (see ESM 26.5, New Galactic Coordinates).

7. Theory group including Stefan Freidrich "Steve" Smerd (1916–1978),[19] with Kevin Westfold based at the Sydney University grounds and David Martyn who had been moved to Mt. Stromlo in 1944. Martyn continued to work on iono-spheric and solar theory with the CSIR group. Steve Smerd (1916–1978) had

---

[18] Sullivan (2009, pp. 277–280) has a detailed discussion of these observations and their impact on the NASA moon landing mission.

[19] http://adb.anu.edu.au/biography/smerd-stefan-friedrich-11716.

Fig. 16.5 Goro's photo of
Chris Christiansen in March
1951 at Potts Hill.
Christiansen was immersed
in data from the Potts Hill
grating array of 32 dishes
used to image the sun. As
shown in Fig. 16.4, the
dishes were located along an
EW (and later NS)
orientation at the southern
edge of the Potts Hill
reservoir. The first earth
rotation aperture synthesis
was made with this array.
Credit: Getty Images, Fritz
Goro, The *Life* Picture
Collection, licensed by
Getty November 2021./
Licence organised by
Shutterstock, Inc.,
New York, NY (5 Nov
2021) Original, Premium
Editorial All Media

arrived in Australia in May 1946 from the UK.[20] His first few years at RPL were
troubled, and both Pawsey and Bowen were disappointed with his performance.
Kevin C. Westfold (1921–2001) joined the Radiophysics group in 1948 to work
with Martyn and Smerd on solar radio emission theory. Westfold was critical of
some of Martyn's views and Pawsey decided to move Westfold to John Bolton's
Dover Heights group. He then worked with Bolton and Slee on the survey at
100 MHz. He and Bolton published papers on a radio star-based model for
galactic radio emission (Jarrell, 2014, p. 2311). Westfold moved to the University
of Sydney in 1951 but kept in contact with Bolton and eventually joined Bolton at
Caltech in the late 1950s. While at Caltech he published the definitive paper on
the polarisation of synchrotron emission at radio wavelengths (Westfold, 1959).[21]
He later returned to the new Monash University in Australia, becoming Deputy
Vice-Chancellor (1982–1986).

---

[20] Sullivan (2009, p. 288) has provided an account of the arrival of Smerd, an Austrian refugee in the
UK in 1938. During the war, he worked on radar research at the University of Birmingham and the
Admiralty. He was recruited by Pawsey after the war to work at RPL, starting in 1946.

[21] See also Chap. 34.

Managing an Emerging Learning Organisation

This organisational structure, with many independent field stations built up by Pawsey in the late 1940s, bears a remarkable resemblance to the "Organisational Learning" concept promoted by Peter Senge of the MIT Sloan School of Management in the 1990s (Senge, 1990). Management research in the 1990s identified that small independent groups are best equipped to adapt to new information in a rapidly changing environment. Broad strategic goals are useful, but must be matched with adaptability and flexibility. But because small, independent groups can create silos without knowledge exchange, a horizontal transfer of knowledge is required in order to develop collective as well as individual knowledge. In the Radiophysics group this was provided by Pawsey, as he moved around all the groups and shared information. Because he was a good listener and developed a network of trust, this transfer of knowledge proceeded smoothly. Groups were encouraged to experiment, and diffuse knowledge was therefore based not just on successes, but also failures. An acknowledged success of Pawsey's approach was the emergence of a cadre of future leaders: Wild, Christiansen, Mills, Bracewell (all "Four Pillars"—see Frater et al., 2017), Kerr and Bolton (Robertson 2017).

Developments in the UK

As is well known, the initial three research groups in what would become "radio astronomy" were Sydney (RPL), Cambridge (the Cavendish Laboratory) and Manchester (Jodrell Bank).

The Cavendish Laboratory, Cambridge: Martin Ryle's Group

RPL and the Cavendish were the dominant radio astronomy groups throughout the 1950s. This success was owing to the research leader, Martin Ryle (1918–1984), but also to Jack Ratcliffe, Pawsey's PhD supervisor, whose role and talents were similar to Pawsey's. Ratcliffe gave Ryle the space he needed to work, relieving him of nearly all administrative and teaching duties and finding administrative support. Like Pawsey, he edited all the group's papers, made extensive comments and suggestions and was available as "a sort of grey eminence . . . a source of sane and quiet thoughts, who would ask awkward questions" (Sullivan, 2009, p. 171). Along with Ryle, he recruited Francis Graham Smith (1923-) and Derek Vonberg (1921–2015).[22] By

---

[22] Sullivan (2009) provides a summary of the formation of this university research group starting in 1945, which was staffed initially mainly by ex-TRE (Telecommunications Research Establishment) personnel returning to academia from wartime service.

1954 the radio group had expanded to include Bruce Elsmore, John Baldwin (1931–2010), John Shakeshaft (1929–2015) and Peter Scheuer (1930–2001) who was a young student doing his PhD with Ryle in 1954.

From an upper class, academic family, Ryle was idealistic, intense, informal, acutely intelligent, scientifically uncompromising, and a marvellous experimenter. His complete commitment to his work won not only respect, but strong loyalty, from his research group. Ryle combined extremely flexible hands-on engineering capability with swift and inventive problem solving and the capacity to creatively use the resources at his disposal—all features of the radar-jamming and engineering work that had characterised wartime service; and he attributed much of his success to these skills forged in the crucible of war (Sullivan, 2009, p. 170).

Ryle and his group started doing solar work, but with the invention of the phase switch (which enabled them to work on much weaker sources) they shifted their main focus to the extra-solar discrete sources. For this they developed the aperture synthesis technique using horizontal interferometers, accumulating spacings over time and computing Fourier transforms. This was a cheaper approach than the multi-element arrays being built in Australia. While effective, for the non-variable extra-solar radio sources, it did not work well for the solar observations. Thus, radio source surveys became the main research focus. Following the aperture synthesis success, source structure became the main game and led to one half of the Nobel prize in 1974 (for Ryle). Following up on interplanetary scintillation (a successor to ionospheric scintillation first seen in Cygnus A) led serendipitously to the discovery of pulsars, and to the other half of the Nobel prize (for Hewish).

The Cavendish was more poorly resourced than was RPL, despite its access to discarded wartime radar equipment; this reflected the comparative economic circumstances of Britain and Australia in the post-war era. Despite this, the elite status of the Cavendish did not alter. It retained its high scientific status, and due to this, to the students who studied at Cambridge, and to its existing networks in many areas of physics (and other disciplines), better access to research networks.

As we shall see, the relations between RPL and the Cavendish group were often not easy, but not for lack of Pawsey trying to establish more collaborative connections. As Sullivan and others have described, Ryle's group were often regarded as secretive, protective or defensive. They were not infrequently perceived as mistrustful of others' research findings, and had a tendency to cite their own work more than that of others. Some have commented that they were also upper class and socially elite, making it culturally difficult for the Australians to connect with them. See ESM 16.1, Proposed Coordination, for details of the correspondence with Cambridge related to solar work in the late 1940s and the attempts to generate collaboration.

At the University of Manchester: Bernard Lovell's Group

The University of Manchester appointed Bernard Lovell (1913–2012) to start a research program looking for radio echoes from the ionisation trails produced by cosmic rays (Davies et al., 2016). He started using surplus radar equipment at the

University's botany research station at Jodrell Bank, 20 miles south of Manchester. Searching for the cosmic ray trail radio echoes was not successful, but Lovell was able to start a series of observations of echoes from meteor trails. Lovell went on to build a 218-ft transit dish still hoping to detect cosmic ray trails but the real success was the use of this dish for radio astronomy by Robert Hanbury Brown (1916–2002) and Cyril Hazard. Lovell's group enjoyed an extensive and highly collegial relationship with Pawsey's.

The relative impact of these first three radio astronomy groups has been analysed by Edge and Mulkay (1976, p. 43) Tables 2.1, 2.2, and 2.3. By 1949 the Manchester group had published 21 papers, mostly on meteor astronomy, while the smaller Cambridge group had published three. However, the Sydney group, with 24 papers and more research scientists, was dominating the new field of radio astronomy. Another measure used by Edge and Mulkay was the number of citations in reviews, with 36 references to the Sydney group, 15 to the Manchester group and 19 to Cambridge by 1950.

Developments in the USA

The development of radio sstronomy in the USA lagged behind that of Australia and the UK until the early to mid-1950s. Edge and Mulkay (1976, p. 52) illustrate this dramatic difference in their Fig. 2.1, and they estimate a 4 year delay before the US matched the fraction of radio astronomy publications in the UK.

Ken Kellermann's history of the National Radio Astronomy Observatory, *Open Skies* (Kellermann et al., 2020), describes the development of radio astronomy in the US and discusses why commencement was slower there. One reason was simply that nuclear (particle) physics was a more exciting option in the aftermath of war. A second was that the US effort was distributed over multiple competing university-based groups. Indeed, science research in the US was dominated by university groups, in physics often funded through military research programmes. This had three consequences. The first was that radar researchers had university jobs—and jobs in industry—to entice them at the end of war. By contrast, the research group at RPL did not have such attractive career alternatives. The second was that there was no concentration of exploratory effort such as occurred at Sydney, Cambridge and Manchester. A third was that optical astronomy was very strong in the USA—and this had the dampening effect that initially those interested in radio observation had little influence and limited funding.

An additional serendipitous factor was that US wartime radar development had been at high frequencies—but the initial, easy radio observations that were so striking elsewhere, were at low frequencies. The celebrated "Rad Lab" was the centre of US microwave and radar research during its brief existence between 1940 and 1945. And indeed, a number of its cadre of superb researchers knew of, and were interested in, radio observation of the extra-terrestrial. However, unlike

RPL, the Lab was disbanded at the conclusion of the war and its functions dispersed, particularly to industry.

In late 1947, there were only a handful of radio observation groups in North America: the National Bureau of Standards (Reber) in Washington, D.C., the group of Robert Dicke at the Radiation Laboratory in Boston and later Princeton,[23] the Naval Research Laboratory, in Washington, Cornell in Ithaca New York, and the National Research Council in Ottawa, Canada. At this time there were a number of optical astronomers who recognised the value of developing links to the new radio astronomers (Mt Wilson and Palomar Observatory and the California Institute of Technology, Yerkes Observatory of the University of Chicago, the David Dunlop Observatory of the University of Toronto and Harvard College Observatory).

In the later years of the 1950s and early 1960, a number of additional institutes began active radio astronomy programs: the Department of Terrestrial Magnetism at the Carnegie Institution of Washington (1952, Merle Tuve), Caltech (Bolton, 1955), Stanford (Bracewell, 1955), University of California, Berkeley (Weaver, 1958) and University of Illinois (Swenson, 1957). Pawsey met these groups in his visit to the US in 1957–1958 (Chap. 28). Merle Tuve (1901–1982) (Director of DTM 1946–1966) was a major sponsor of Pawsey's long visit; he became a multi-year colleague and advisor for Pawsey (Chaps. 28, 38 and 40).

Below we provide a short synopsis of the four main groups in the US.

1. *Grote Reber, National Bureau of Standards (NBS) Central Radio Propagation Laboratory, Field Station Sterling, Virginia, June 1947 to Spring 1951.*

   Pawsey was, of course, familiar with Grote Reber's wartime radio maps of the sky (Chap. 11). Sullivan's comprehensive history of this period, *Cosmic Noise*, devotes a chapter to Grote Reber's pioneering work (Sullivan, 2009, p. 54, "Science in your back yard"). The section "Reber Beyond Wheaton" describes his work in the later 1940s. In 1947 Reber moved to the National Bureau of Standards (NBS) at a site near Washington which is now Dulles Airport. As Sullivan has reported, Reber's sojourn at NBS was not a happy one; he would later confide to Sullivan (2009, p. 72): "NBS was a peculiarly reactionary and backward agency. Only sure (not pure) science was possible. Anything speculative was stamped on because a failure would bring discredit to the Bureau!"[24]

---

[23] Dicke moved as a new Assistant Professor of Physics to Princeton in mid-1946. The post-war radio astronomy (using the famous Dicke radiometer) at the high frequency of 1.25 cm occurred at a partial solar eclipse on 9 July 1945. The moon was detected in October. Pawsey met Dicke in late 1947 and early 1948 in Princeton (see Chap. 24).

[24] Sullivan reported that when NBS announced the CRPL group would move to Colorado in early 1951, Reber left for Hawaii where he remained until his permanent move to Tasmania 3 years later in 1954. Starting in January 1950, Reber, Stanley, Bolton and Pawsey (NRAO Reber archive) corresponded about the severe difficulties of sea-cliff interferometry from high mountains such as Haleakala (3100 m) on Maui, where Reber carried out a series of observations in the era after 1951. A major problem was the effect of the earth's curvature (non-flat sea as observed from 3100 m above sea-level). Two years later (4 July 1952), Pawsey wrote Reber again in Hawaii with additional warnings about the "troublesome" earth curvature effect. "It is simply due to the

2. *Robert Dicke, Princeton (formerly of the Radiation Laboratory, MIT)*
     R.H. Dicke (1916–1997) had moved to Princeton as a young faculty member
when the war ended. Like Pawsey, towards the end of the war he had become
interested in potential astronomical applications using the technologies evolved
from WWII radar research. In 1946, he published two "classic" papers, one on his
eponymous "Dicke radiometer" and one applying this technique to a number of
difficult short wavelength (1.25 cm) astronomical observations (Sullivan, 1982,
papers 11 and 22).
     The first publication has had a lasting, decisive influence on radio astronomy
in the last 75 years. Sullivan (1982, p. 105) summarised:

> [Dicke described] both basic thermo-dynamic principles and practical hardware concerned
> with microwave measurements. In addition, accurate and detailed derivations are given for
> the influences on overall system sensitivity of various receiver parameters such as band-
> widths and integration times.
>   It turns out that the noise power from an astronomical source at these relatively high
> frequencies is usually considerably less than that arising in the receiver itself ... [Dicke's
> switching] scheme thus produced greatly-improved accuracy and effective sensitivity.

We note that the same load switch concept had also been developed in Cambridge in
1946 (see ESM 18.1, Exchange of Letters) as a variation on their phase switch. We
do not know if these developments were independent.
     The second paper, on microwave radiation from the sun and moon, provided
observations at the extreme frequency range of K band (22 GHz or 1.25 cm) in 1945.
Sullivan wrote (1982, p. 217):

> Not only [was] their experimental method accurate enough that their measured values agree
> well with [modern data], but they also established the following "radio firsts": (1) first solar
> eclipse observation, first **absolute** [his emphasis] intensity calibration [in radio astronomy],
> first measurement of a size for the sun and first measurement of the lunar brightness.

Ironically, Pawsey would not meet Dicke until late 1947 in Princeton, after Dicke's
first phase as a radio astronomer had ended. At Princeton, Dicke turned his attention
to atomic physics, radiation processes, gravity and later on, cosmology. The latter
led to a renewed contact with radio astronomy in the 1964 era, with Dicke taking a
role in the discovery of the 3 K background radiation by Penzias and Wilson in 1964
(Penzias & Wilson, 1965).

3. *Hagen, Haddock, and Mayer, Naval Research Laboratory (NRL)* (Sullivan,
     2009, p. 206)

---

reflection ray being weaker owing to reflection at a curved surface." In summary, Sullivan noted
(p 73): "Reber spent 3 years working on Haleakala and ended up with little to show for it." Reber's
publications for this era are Reber (1959), "Radio Interferometry at Three Kilometers Altitude
above the Pacific Ocean. Part I. Installation and Ionosphere. Part II. Celestial Sources".

In 1947 Pawsey met John P. Hagen[25] (1908–1990) and F.T. "Fred" Haddock[26] (1919–2009) at the Naval Research Laboratory. Haddock would leave NRL to initiate radio astronomy at the University of Michigan at Ann Arbor in 1956. The first solar radio observations were made with a 10-foot antenna at 3.2 cm; a strong burst (the first detected burst at this high frequency) was detected in July 1948 (Schulkin et al., 1948). Later a burst was detected at a higher frequency at a wavelength of 8.5 mm by Nannielou Hepburn (1926–2014) (Hagen & Hepburn, 1952), who worked at NRL from 1951 to 1955.[27]

Solar eclipses were important in the early years of solar radio astronomy as a method of achieving high angular resolution, determining the nature of the corona and the association of radio emission with sunspots. The NRL group was to go to the ends of the earth following solar eclipses in May 1947 (Brazil), September 1950 (Alaska) and finally 1952 (Sudan Africa). See ESM 16.2, NRL solar eclipse, for details. Later the group had great success with a new 50-foot dish, with its 25 arc min beam at 9 cm. The first observations ushered in a new era of galactic radio astronomy as the thermal radio emission from gaseous nebulae in the galaxy (HII regions) became prominent at the higher frequency (see Sullivan, 2009, p. 208 for a summary of this exciting period at NRL).[28]

4. *Cornell, Ithaca New York Department of Electrical Engineering 1946.*[29]

The first university department in the new field ("Microwave Astronomy Project") was formed in the Electrical Engineering Department at Cornell University in 1946, headed by Charles Burrows (1902–1970), with Charles Seeger (1912–2002) and two instructors in the astronomy department, Ralph E. Williamson (1917–1982) (see NRAO ONLINE.26) and Donald MacRae

---

[25] Hagen had joined NRL in 1935, becoming the head of the CM Wave Research in WWII. He started radio astronomy at NRL in the post-war era as a cm solar radio astronomer. From 1955 to 1962 he was in charge of the ill-fated Vanguard Earth Satellite Project.

[26] Haddock joined NRL in 1941.

[27] Nannielou Hepburn was the first woman in the US involved with radio astronomy. She studied astronomy as an undergraduate at Goucher College (1948) and obtained a PhD at Harvard (1958) in B.J. Bok's group, with a thesis based on HI studies of the galaxy M33. Her publications are listed under the name she preferred, Nan Dieter Conklin. For an autobiography see *Two Paths to Heaven's Gate* by Nan Dieter Conklin (2006) and *Nan Dieter Conklin: A Life in Science*: National Radio Astronomy Observatory/Associated Universities, Inc. Archives, http://www.nrao.edu/archives/Conklin/conklin.shtml. Dieter-Conklin was one of Goss' s advisors during his PhD research at Berkeley in 1967.

[28] In 1958, Mayer's collaboration with Charles Townes proved especially fruitful. In his biographical memoir for Cornell H. "Connie" Mayer (1921–2005), V. Radhakrishnan wrote in December 2006: "In 1958, Connie collaborated with Charles Townes and his students at Columbia in the first application of the maser to astronomy. When Townes received the 1964 Nobel Prize for the invention of the maser, he asserted that Connie's desire to improve receiver sensitivity was influential in his work and shared a portion of his prize money with him." (Alsop et al., 1958, p 301).

[29] A thorough review of the early history (1946–1962) of Cornell radio astronomy is presented by D.B. Campbell in *Journal of Astron History and Heritage* (2019).

(1916–2006). Although both Williamson[30] and MacRae departed within a short period, they continued their association with the Cornell group. In January 1947, Martha Stahr Carpenter joined the astronomy department with associations with the radio astronomy project. Also in 1950, due to conflict with Burrows, Seeger left Cornell for a year in Sweden at Chalmers Institute of Technology. Then he spent a decade working with the radio group in Leiden under the leadership of Oort (below).

Under the leadership of Bill Gordon, the 305 m Arecibo telescope was conceived and designed during the 1950s, opening in 1963.

Developments in Canada

National Research Council in Ottawa

The National Research Council (NRC) in Ottawa was involved in the WWII radar developments from 1939 based on the 10 cm wavelength magnetron. After the war the Radiation Laboratory was dissolved but one of the physicists, Arthur E Covington (1913–2001) was able to start a research project on cosmic static at 10.7 cm wavelength.[31] In 1947 Covington started a successful solar monitoring program and also measured the size of the radio emission from a sunspot during a solar eclipse.[32]

In Pawsey's report to CSIR in April 1948, "Solar and Cosmic Noise Research in the US and Canada", he summarised his opinion of Covington: "At Ottawa, Covington is a young and inexperienced man working in relative isolation. He has got some thoroughly good results by good honest work and perseverance."

David Dunlop Observatory in Toronto

The remarkable theoretician Ralph Williamson had moved to Toronto from Cornell in 1946. Williamson can be credited in playing a major role in initiating radio astronomy in Canada.[33] After his visit to Toronto in 1947, Pawsey tried for over a year to convince Williamson to come to Sydney as a staff member. The failure to

---

[30] Details of Williamson's career and his association with Pawsey in 1947–1948 are presented in NRAO ONLINE.26.

[31] Covington, 1983 "Early Radar Research and a beginning of Radio Astronomy" in Kellermann and Sheets (1984).

[32] See also Sullivan (2009), pp. 211–213 and NRAO ONLINE.26.

[33] Williamson had written a credible review article in the *Journal Royal Astronomy Society of Canada* that had impressed Pawsey. Many details about Williamson are included in NRAO ONLINE.26.

recruit Williamson was a disappointment as he would have brought a level of astrophysical understanding that could have been a great asset to the Sydney group.

The Netherlands: The Bridge Between Radio and Optical Observation

There was no radio astronomy group in the Netherlands in this period but celebrated astronomer Jan H. Oort (1900–1992), from the University of Leiden, had very broad astronomical interests and saw the potential of radio observations.[34] He had written to Reber for advice on how to establish an experimental radio astronomy facility in the Netherlands (Sullivan, 2009, p. 404). Oort, along with Walter Baade (1893–1960) and Rudolf Minkowski (1895–1976) from the US (see Chaps. 22, 23), had been in communication with John Bolton following his suggestion that the radio source Taurus A might be identified with the Crab nebula. They knew that this nebula associated with the supernovae of AD 1054 was one of the most remarkable objects in the entire sky. Its association with one of the strongest radio sources in the sky was highly significant and indeed would eventually be the key to our understanding of the radio emission mechanism (Chap. 34).

It is likely that this interaction between Oort and the Australians led to the invitation of Bolton and Westfold to visit the Netherlands in July 1950 and give a series of lectures on radio astronomy. The resulting close connection between Hendrick C. "Henk" van de Hulst and the Australian radio astronomers is exemplified by a long review paper written in the journal for amateur astronomers in the Netherlands, *Hemel en Dampkring ("The Sky and Atmosphere")* of 1 December 1951. Van de Hulst gave a thorough account of the pioneering radio astronomical work (solar and radio source research) at RPL, including an innovative figure illustrating the sea-cliff interferometer.

Leading the World from "Down Under"?

Thus we can understand Pawsey's desire was to embark on a lengthy journey to ensure a global scientific network to support, and make visible, radio astronomy research at RPL. John Bolton was just beginning to look for discrete sources at Dover Heights when Pawsey departed.

The Sydney group was, perhaps arguably, the leading group in this field around the world, but the long travel time to other groups in the UK and interested colleagues in the US threatened to obscure or marginalise their results. It was difficult to publish results in an area of such novelty, yet publishing locally might

---

[34] See *Jan Oort: Master of the Galactic System*, van der Kruit (2019).

doom major discoveries to virtual obscurity. (Bolton soon developed a preference for publishing a short letter in *Nature* for priority, and then a longer paper in the new *Australian Journal of Scientific Research*, to confront this dilemma, see Robertson, 2017). The contacts that Bowen, White, Pawsey and others had nurtured during the war, now would be expanded and developed.

Chapter 17
Pursuing "Radio Astronomy": Pawsey's Travels to North America, the UK and Europe, 1947–1948

> I like the term 'radio astronomy' much better than Burrows's efforts ['microwave astronomy'] and we might very well consider adopting it generally.—E. G. Bowen to J. L. Pawsey, 20 February 1948

J.L. Pawsey's long visit to North America and the UK was a masterpiece of international scientific networking. Pawsey was careful to ensure that he visited all the leading researchers and research groups across the breadth of specialisations relevant for his budding research groups: radio and communications engineers, microwave physicists, military and naval research laboratories. The result was an extremely tight and relentless schedule. N.A. Whiffen of the Australian Embassy in Washington (Australian Scientific Research Liaison Office, ASRLO[1]) did much of the organisation for the visit in the US, including resolving the issue of military clearance at a number of US institutes, not at all an easy task. Here, we provide an overview of the places and people visited, to give a sense of both the importance of the network Pawsey cultivated, and the effort that doing so required.[2]

It was during this trip that Pawsey came up with a new name for the disciplines of "cosmic noise" and "solar noise": **radio astronomy**. Probably inspired by a proposal for a conference on "micro-wave astronomy", he used the new term in correspondence with Bowen and with a UK colleague in January, 1948. As Sullivan (2009, p. 424) has documented, others, including Martin Ryle from Cambridge, had independently proposed the same name, which became widely taken up over the next 2 years (see ESM 17.1, Co-Invention of Name.)

Supplementary Information The online version contains supplementary material available at [https://doi.org/10.1007/978-3-031-07916-0_17].

[1] After 1949 the ASLO—Australian Scientific Liaison Office.

[2] Additional details in NRAO ONLINE.59 and NRAO ONLINE.60.

© The Author(s) 2023, corrected publication 2024
W. M. Goss et al., *Joe Pawsey and the Founding of Australian Radio Astronomy*,
Historical & Cultural Astronomy, https://doi.org/10.1007/978-3-031-07916-0_17

However fruitful the trip from the point of view of ensuring the visibility and resourcing of Australian radio research, Pawsey's absence had significant costs. Without its leader, the group functioned poorly. One problem was the chronic conflict between John Bolton and Ruby Payne-Scott, which is summarised in Goss and McGee (2009).[3] Pawsey's absence from Sydney meant that his constant support of Payne-Scott was not possible (see ESM 17.2, Payne-Scott Bolton Conflict). A second and deeply significant problem was the absence of leadership to stimulate the group, discuss ideas, encourage and motivate output, and provide direction for the next lines of investigation. Both Kevin Westfold[4] and Lindsay McCready wrote to Pawsey with concerns regarding the solar noise programme in Sydney. This correspondence reveals how dependent the group was on Pawsey's intellectual oversight. It also offers insight into the details of the scientific challenges facing them at this period as they tried to make sense of partial and imperfect data. Pawsey's analysis of these challenges, incorporating reflections on the nature of science itself, are discussed in Chap. 33 and ESM 26.5, New Galactic Coordinates.

On 25 September 1947, Joe and Lenore Pawsey left Sydney on the converted US cargo ship, *SS Marine Phoenix*.[5] They arrived in San Francisco and visited the campuses of the University of California, Berkeley and Stanford, but strikingly, no visits were made to the Berkeley Astronomy Department. He did visit the Griffin Observatory in Los Angeles.

Most importantly, Pawsey visited the California Institute of Technology (Caltech) and the Mt. Wilson Observatory (Carnegie Institution of Washington). Here he was enthusiastically welcomed by Caltech President Lee Dubridge,[6] Prof Robert Millikan (1868–1953, Nobel Prize in Physics 1923) and William Pickering (a New Zealander), Professor of Electrical Engineering. Millikan was so impressed by Pawsey that he suggested that the originally scheduled colloquium on 29 October be postponed in order to accommodate Pawsey's talk, "Solar Noise".

Pawsey's numerous discussions with famous astronomer Rudolf Minkowski[7] and the Mt. Wilson staff[8] were of critical importance, both during his visit and as the basis for ongoing productive and crucial collaboration into the future. An additional important astronomical fact was also passed on to Pawsey: the precession of the equinoxes that caused the positions of stars to move as the earth's axis moves slowly over time (Pawsey was learning elementary astronomy fast).

---

[3] See Goss and McGee (2009), Chap. 9, p. 129 and Goss (2013), p. 150.

[4] 1921–2001, a young theoretical physicist at RPL, see Chap. 16.

[5] C4-class ship, T-AP-195 built by Kaiser Co., Inc., Vancouver, Washington, USA. Completed 9 August 1945. Later used during the Korean War as a troop carrier.

[6] WWII Director of the Radiation Laboratory at MIT, the main centre of civilian radar research under the direction of the OSRD (Office of Scientific Research and Development), under the leadership of Vannevar Bush.

[7] See Dramatis Personae; Minkowski would play a significant role connecting RPL with optical astronomy in the 1950s.

[8] The details of these are described in ESM 17.3, Bolton Pawsey Correspondence. Seth Nicholson was an important contact.

The Mt. Wilson astronomers were interested in a possible optical identification of the radio source Cygnus A based on the early Bolton position. Pawsey wrote back to Sydney: "They immediately searched out the region given by Bolton and Stanley [for a possible Cygnus A optical identification] but found nothing ..." Due to the large offset (1.2° mainly in declination) in Bolton's position in late 1947 and the positional accuracy at that time (see Sullivan, 2009, p. 318, Fig. 14.1), the lack of identification with such a faint galaxy is no surprise. Mills and Thomas and Smith were to publish much improved accurate positions in 1951 that would lead to the identification of Cygnus A in 1954 by Baade and Minkowski (see Chap. 18).

Of equal importance, Pawsey also visited the Yerkes Observatory of the University of Chicago, by arrangement with former Director Otto Struve (1897–1963), and hosted by the new director Gerard Kuiper (1905–1973).[9] Struve was also keen that Pawsey meet Jesse Greenstein, who "is also interested in problems of radio static. [Thus] I hope that you will be able to arrange your schedule in such a way as to include Williams Bay [Yerkes Observatory]."[10] But Pawsey missed Greenstein, who was in the process of moving to Caltech. This was particularly unfortunate, because it meant that Pawsey did not get key technical information about the hydrogen line (HI) prediction, which could be used to investigate a major constituent of the interstellar medium in the Milky Way line. Instead, he was only partially informed about this (see below), with consequences that we discuss in Chap. 20. By good fortune, two prominent European astronomers were also present, long term visitors Jan Oort (1900–1992), Director of the Leiden Sterewacht, who was thinking of starting radio astronomy in the Netherlands in collaboration with the Dutch Phillips Corporation (see Chap. 16), and Bengt Strömgren (1908–1987), Director of the Copenhagen Observatory. The impact of this 1947 trip can be judged by the fact that Minkowski in the USA and Oort in the Netherlands would provide the two major connections between radio astronomy and optical astronomy over Pawsey's career.

In the USA, Pawsey visited, presented to, and dined with an A-list of US physicists and engineers, including: Ernest Lawrence (Nobel Prize in Physics 1939, at Berkeley); Professor William W. Hansen (1909–1949, at Stanford[11]); Walter Orr Roberts (Harvard; he took Pawsey to the Solar Observatory at Climax, Colorado); and Carl Borgmann (Dean at the University of Nebraska, and a friend of

---

[9] By this time, Struve was the Chair of the Astronomy Department of the University of Chicago. Struve moved to Berkeley in 1950 as Chair of the Astronomy Department there. In 1959 he became the first National Radio Astronomy Observatory Director for two-and-a quarter years and was succeeded in this position by Pawsey in 1962.

[10] Otto Struve, letter to Pawsey on 11 October 1947. For Greenstein's interest see "Radio Frequency Investigations of Astronomical Interest" from February 1947 with Grote Reber in *Observatory*, vol. 67, p. 15).

[11] Hansen, a pioneer in the field of modern microwave electronics, became a professor at Stanford at age 36 and along with the Varian brothers (Russel and Sigurd), was one of the co-founders of Varian Associates, a pioneering firm in microwave components. Pawsey described in a letter to Bowen an ingenious proposal that Hansen had to measure the speed of light using microwave circuits. The suggestion was that this could lead to a factor of 300 improvement in the determination of the value of the speed of light.

Pawsey's from Cambridge postgraduate days,[12] who would later play a key role in organising the Ford Foundation's grant for the Culgoora Radio Heliograph in 1962[13] (see Chap. 40). He met with Howard Dellinger (Chief of the Central Radio Propagation Laboratory), Ross Bateman (ionospheric scientist and meteor astronomer) and J.F. Denisse[14] (who was to play a major role in the post-war development of radio astronomy in France), John Hagen (1908–1990), Fred Haddock (1919–2009),[15] and Prof J.A. Stratton of MIT.[16] He spent a considerable amount of time at Harvard, with Shapley, Donald Menzel, W.O. Roberts and Bart Bok. At Cornell, he visited J.R. Burrows, Charles Seeger and Martha Stahr, later Martha Stahr Carpenter, who would later come to Australia. At Princeton, he visited both the Physics Department (R.H. Dicke) and the Astronomy Department (Lyman Spitzer, Martin Schwarzschild and John Stewart).

He also visited important research centres, including the Naval Research Laboratories (NRL) and the Central Radio Propagation Laboratory (CRPL) of the National Bureau of Standards, along with a number of other University research groups (eg McMath Hulbert Solar Observatory, University of Michigan).

Those who attended his talks in Washington, D.C. included A.R. Beach, Alan Shapley[17] (ionospheric physicist), Jack Herbstreit (tropospheric propagation[18]), Thomas J. Carroll (microwave research), Herman Cottony (antenna research)[19] and Morris Schulkin of the Naval Research Laboratory (later an expert on underwater acoustics). He attended the Institution of Radio Engineers Convention in New York City from 22 to 25 March 1948 where 15,000 attendees were at the Hotel Commodore and the Grand Central Palace. Keynote speakers were Wiener (1894–1964), Shannon (1916–2010), von Neumann (1903–1957) and Rabi (1898–1988).[20] Pawsey fitted in one last talk at the Institute of Radio Engineers at

---

[12] Borgmann had received a degree in Chemical Engineering in 1927 from the University of Colorado.

[13] Borgmann became the Program Director of Science and Engineering at the Ford Foundation.

[14] Pawsey wrote "M.J. Denise" in his correspondence.

[15] John Hagen was the head of the Naval Research Laboratory Centimeter Wave Research Branch; Haddock was the number 2 man.

[16] Stratton (Jules, A., 1901 to 1994), President of MIT 1957–1966 and key staff member of the Rad Lab in WWII, where he likely met Pawsey in 1941. Stratton was a physicist and electrical engineer.

[17] Son of the prominent astronomer from Harvard Harlow Shapley.

[18] Herbstreit published a paper in *Nature* 1948 about the spectral index of the galactic background between 25 and 110 MHz (Herbstreit & Johler, 1948). Pawsey pointed out that Herbstreit and Bateman had visited RPL during the war, likely as part of the of US scientists at RPL in 1944.

[19] NRAO Reber archive. In addition, Robert Hayward has provided an annotated version of this Reber material with details about the scientists whom Pawsey met in 1947–1948.

[20] These prominent colleagues were: Wiener, mathematician at MIT, Shannon, mathematician at Bell Labs, and von Neumann, mathematician at Princeton, said to be the most prominent mathematician of the mid-twentieth century. Rabi had been the assistant Director of the MIT Radiation Laboratory in WWII and as President of Associated Universities would hire Pawsey in late 1961 as the NRAO Director (see Chap. 40). Rabi was a Nobel Laureate in Physics (1944); he was the Associated Universities President from 21 April 1961 to 19 October 1962.

the Radio Corporation of America in Princeton on the evening of 25 March 1948, 2 days before his departure for the UK.

The trip also included a number of visits and talks concerned with rain making and cloud seeding, the focus of Bowen's research group; this included discussions with Irving Langmuir (1881–1957, Nobel Prize Chemistry 1932) and Vincent Schaefer (1906–1993) at General Electric. Pawsey sent an informal report to Bowen on 20 December 1947 from Boston and a formal report on cloud seeding in the US, "Informal Notes on Rain Making in the US", on 13 January 1948.[21]

One notable feature of the trip was Pawsey's contact with the two pioneers of radio astronomy. He visited Karl Jansky (1905–1950), along with Harald Friis (1893–1976), W.M. Sharpless and A.B. Crawford (1907–1990), at Bell Labs at Holmdel.[22] Pawsey reported that Jansky had dropped out of cosmic noise research; but Jansky expressed his continued interest in the topic to Pawsey, who sent him frequent preprints until his death in 1950 at age 45. At NBS he met with Grote Reber several times, evidently to the great satisfaction of both (Fig. 17.1). Reber told Pawsey about the 21 cm hydrogen line, in a conversation that left a deep impression on Pawsey and was reported to Bowen in a letter of 23 January 1948 (see Chap. 20 and ESM 20.1, A review of the recollections).

In his report, "Solar and Cosmic Noise Research in the US and Canada" of April (see below), Pawsey provided a detailed description of Central Radio Propagation Laboratory activities. Reber had three 7.5 metre Würzburg aerials, originally provided by the US Army Signal Corps from post-war Germany (ESM 17.6, Additional Images). In 1947–1948 one was ready for use with a polar mount for observations at 480 MHz. The others were to be used at 51 and 160 MHz for solar monitoring. As Sullivan (2009, p. 71) has pointed out, these antennas collected copious data over many years, resulting, however, in minimal output in publications or even internal reports.[23]

The trip was of personal importance to the Pawseys as well. It included a visit to Battleford, Saskatchewan, Canada, Lenore's hometown. This was the first occasion that Lenore had to visit Canada since she left in the mid-1930s to travel to the UK, and she spent a much more extended time in various parts of Canada during the trip. Additionally, the Pawseys stayed more than once with Lenore's brother Ted Nicoll and his family (his wife Kate, and their four (living) children, Pat, Ruth, Roger and

---

[21] See NRAO ONLINE.25.

[22] Friis was a pioneer in radio engineering. In 1944, he invented the term "noise factor" or "noise figure" to characterise the sensitivity of a microwave receiver (Friis, 1944). He had assisted Karl Jansky in the design of the 20 MHz system used to detect the galactic background in the early 1930s. He and colleague Alfred Beck designed a horn reflector used by the military in WWII. This system is related to the Hogg or horn reflector antenna that became the famous antenna used by Arno Penzias and Robert Wilson in 1964 to detect the 2.7 K cosmic background radiation at 7.35 cm. They were awarded the Nobel Prize in Physics in 1978. This meeting is also partially covered by Sullivan (2009).

[23] Reber was to leave NBS in 1951, moving to Hawaii where he remained for a few years before moving semi-permanently to Tasmania in Australia.

Fig. 17.1 5 December 1947, photo of Grote Reber made by Pawsey at the Sterling, Virginia, field station of CRPL. Credit: Joe and Lenore Pawsey Family Collection

Matt). Figure 17.2 shows Pawsey at Harvard College at this time. Given Joe and Lenore's intimacy with Ted and Kate in the UK, it is unsurprising that spending considerable time with the Nicolls was of high importance during this trip.

Pawsey also cultivated professional connections in Canada. He met Arthur Covington (1913–2001) at NRC in Ottawa. Covington had started the first radio astronomy measurements in Canada using WWII radar technology as mentioned in Chap. 16. He spent several days visiting colleagues at the University of Toronto David Dunlop Observatory at Richmond Hill: Dr. Frank Hogg (Director) and Dr. Helen Sawyer Hogg. Pawsey also met a young (31 years old in 1948) US astronomer who had completed a PhD with Chandrasekhar at Chicago (see Chap. 16). Ralph E. Williamson[24] (ESM 17.4, Ralph E. Williamson) impressed Pawsey more than any of the other young astronomers he met in North America, and

---

[24] Pawsey described Williamson to Bowen on 15 March 1948: "He is a likeable, young theoretical astronomer, who is enthusiastically trying to get in to this solar and cosmic noise field ... He would be delighted to visit Australia, but could not leave at short notice as the Toronto University is short staffed."

Fig. 17.2 Pawsey at Harvard College in mid-December 1947, Cambridge, Massachusetts. Credit: Joe and Lenore Pawsey Family Collection

he spent considerable effort over some years to recruit Williamson to spend some time in Australia,[25] in the end to no avail.[26]

The visit to Canada concluded with a visit to Niagara Falls with Lenore. It is fortunate that the North American component of this trip was so joyous for Lenore, for she became very ill with pneumonia during the stormy voyage to the UK in late March 1948 (ESM 17.6, Additional Images). She suffered extensively from various illness throughout the Pawseys' UK visit. This significantly constrained Pawsey's capacity to maintain a hectic professional schedule. During this period the Pawseys were based at Lenore's sister's (Bessie Whitford) house in Iver, 30 km west of London in Buckinghamshire, close to Slough. Pawsey spent much of the trip—over 2.5 months—travelling from there to the Cavendish Laboratory, Cambridge, spending weekends with Lenore. He confessed to Bowen in July, "[t]he whole business

---

[25] For details of Pawsey's connection to Canada, see NRAO ONLINE.26 and Chap. 20.

[26] In 1949 Williamson turned down the offer.

has been worrying and disturbing to me so that I have not appreciated England as I should have done."[27]

Soon after arriving in the UK, two important short conferences were held at which Pawsey spoke. The first was the Institution of Electrical Engineers (IEE) which met on 7–8 April 1948 in London. The second was a special meeting of the Royal Astronomical Society on 23 April 1948, at which the most recognised names in the fields of ionospheric research and "radio astronomy" presented: Sir Edward Appleton, Sydney Chapman (1888–1970, Oxford) and D.F. Martyn among the former, and Pawsey, Hey, Ryle, Hoyle and Stratton (F.J.M., astrophycist at Cambridge University) among the latter.[28]

Despite Lenore's illness, Pawsey visited many colleagues, including the Telecommunications Research Establishment (TRE, the main WWII radar research institution, Air Ministry), Hey's Army Operational Research Group[29] in West Byfleet, Oxford University, and Appleton's DSIR group.

Pawsey visited Jodrell Bank at least twice in the period from April to September 1948. In 1948, Bernard Lovell and colleagues Clifton Ellyet (UK then New Zealand and Australia, 1915–2006), John Clegg (UK, 1913–1987), Nicolai Herlofson[30] (Norway then UK and Sweden, 1916–2004) and Victor Hughes (UK then Canada, 1925–2001) were busy planning and constructing the 218-foot (66 m) fixed "great mirror".[31] In Fig. 17.3, we show Pawsey's sketch of the "Great Mirror", which remained the largest dish in the world for a decade and had been operating for about a week when Pawsey visited. This 218-ft transit dish (66 m) operated at 72 MHz and could observe a region of the sky within plus-and-minus 12° from the zenith. The focal length was 127 feet (39 m), the design based on the distorted catenary principle. The group was planning to use this antenna for three main purposes: (1) meteor radar echoes, (2) possible echoes from cosmic ray showers as had been earlier predicted by Blackett and Lovell in 1941 and (3) cosmic noise.

Pawsey wrote: "Both [1 and 2] seemed doomed to failure so I guess Lovell will simply discover something new. The possibility of drawing a blank I regard as both inartistic and unlikely. Cosmic Noise is the present objective."[32] As Sullivan (2009, p. 191) has commented, Pawsey was uncannily correct in this prediction. The cosmic noise work was to be the primary use of the 218-foot telescope, even though in 1948 Pawsey found that "the cosmic noise work is not being pushed vigorously". The glory days of the 218-foot dish were to occur after the arrival of Hanbury Brown and

---

[27] 6 July 1948, Pawsey to Bowen. All references for the UK component of the trip are either to NAA C3830 F1/Paw/1 Part 2 and/or NAA C4659 8. Many letters are in both locations.

[28] Possibly Pawsey met Hey and Ryle for the first time.

[29] See Pawsey's report to follow.

[30] Norwegian scientist who worked at Oxford and Manchester in the 1940s. Later, Director of the Plasma Physics Laboratory at the Royal Institute of Technology in Sweden. Close colleague of Alvén.

[31] Robert Hanbury Brown was to join the group later in June 1949.

[32] Letter from Pawsey to Bowen on 11 June 1948, "Notes on Radio Astronomy in Europe". C3830 F1/4/PAW/1.

Fig. 17.3 Pawsey's drawing of the "Great Mirror" at Jodrell Bank. Credit: Courtesy of National Archives of Australia. NAA 4659/4 from 1948

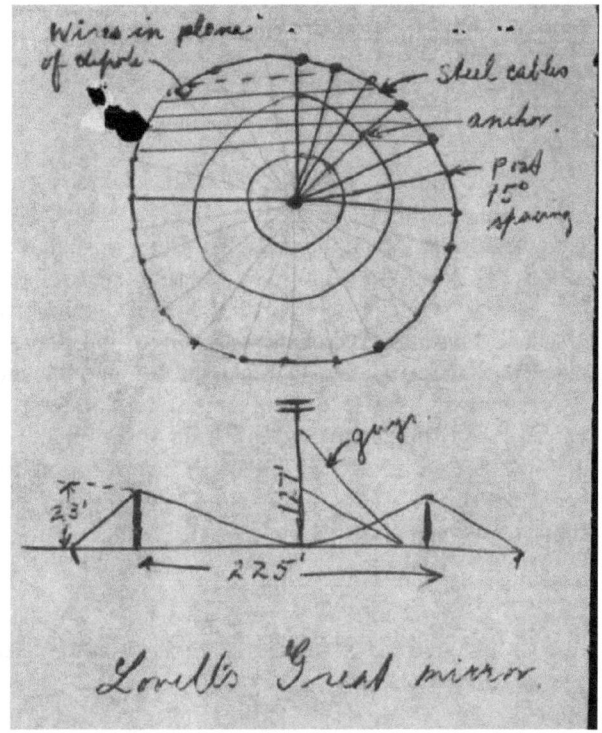

his graduate student Cyril Hazard over the following years. The noteworthy detection of the Andromeda Nebula by Brown and Hazard in 1951 was to be carried out with this instrument.

At Jodrell Bank Pawsey spent some time in discussions with Nicolai Herlofson concerning the theory of scattering off the ionised tails of comets: "Herlofson could not explain it on physical grounds nor relate it to any familiar analogy. Later, he and I discussed the possible analogy of a short-t tuned dipole which scatters nearly as effectively as a half wave one. Perhaps the cylinder shows an equivalent transverse resonance."[33]

Pawsey continued with his negative opinion of the 218-ft dish, based on claimed detection of variable flux density determinations of discrete sources and observations of occasional large disturbances of unknown origin.[34] Pawsey wrote:

> The chief lesson for us to learn from the Manchester experience is that there is great difficulty in interpreting records of sporadic happenings observed on an aerial of fixed

---

[33] NAA C3830 F1/4/PAW/1.

[34] Pawsey noted that these excursions did not show the characteristics of radio frequency interference.

diameter [single dish]. I do not think we should plan to use such an aerial for any exploratory purposes without facilities for some sort of supplementary check.[35]

Likely this impression remained with Pawsey in the early 1950s as the discussions of the Giant Radio Telescope (GRT) began; an auxiliary interferometer might be required.

After late April, Pawsey attended four additional conferences, three in Scandinavia. In Stockholm he attended the International Union of Radio Science (URSI), 12 to 22 July 1948, and the meeting of the Consultative Committee on International Radio (CCIR) from 12 July to the end of the month. The next conference was 4–7 August 1948 in London, the British Commonwealth Conference on Radio Research. Fortunately, Lenore was able to join Pawsey on the last major trip of their UK-Europe visit, the International Union of Geodesy and Geophysics (IUGG) in Oslo, Norway, from 19 to 28 August 1948. Pawsey wrote Bowen on 27 August with a report (from Oslo) of his presentation of Frank Kerr's lunar radar experiments at 20 MHz (Kerr et al., 1949). Pawsey also reported his impressions of sessions on "radar for survey" and "rain making"—hardly any discussion of the rain making trials in Australia occurred.

Contacts with Australian Post-Graduate Students at Cambridge

An important aspect of Pawsey's visits to Cambridge was to meet the "Radiophysics- Cavendish" people. A report on this was sent to Bowen on 18 May 1948—a day filled with copious correspondence, as we will see below!

Both Joan Freeman and Ron Bracewell had come to the Cavendish Laboratory in 1946, leaving Sydney together on the SS Orbita in August 1946.[36] For both Bracewell and Freeman, Pawsey gave glowing reports of progress at Cambridge. Ron Bracewell had already been promised support for a third year from CSIR. He was engaged on low frequency ionospheric studies at the wavelength of 19 km (a frequency of 16 kHz). He had developed into an experimentalist, complementing his theoretical skills accrued earlier at RPL during and after WWII. Freeman had begun on the studies in nuclear physics that would lead her to extraordinary successes in her UK career.

---

[35] NAA C3830 F1/4/PAW/1.
[36] Joan Freeman's A Passion for Physics,1991, for additional details.

Presenting Australian Results

Pawsey was able to present Australian radio astronomy results soon after arriving in late March 1948 in the UK. At IEE, he presented the preliminary position of Cygnus as well as the upper limit of only 8 arc min for the angular size. The Australian position was at this time 1.2° north of the true position while the Cambridge determination was about 1° south. It was to take a few years before Mills and Thomas (1951) and Smith (1951, 1952b) sorted out this discrepancy.

Pawsey mentioned that the source was not moving across the sky (no detectable parallax) over a period of several months. He showed an overlay of the radio position on a photograph of the Milky Way (presumably the overlay shown in the 1948 publication of Bolton & Stanley, 1948a). "There is nothing we can recognise as peculiar in that part of the [sky]. It is an ordinary part of the Milky Way." The intensity of Cygnus was:

> ... surprisingly high. It is the same order at 100 Mc/s as that of the sun at quiet periods despite a presumed vastly greater distance. The facts, together with that of rapid variations, scarcely fit the hypothesis of the origin in thermal radiation from vast clouds of interstellar gas. There must exist localised regions emitting vast amounts of radio frequency energy. Perhaps they may prove to be new types of astronomical bodies.

Pawsey could not have realised in 1948 how prophetic this prediction would become within a few years as Cygnus A was identified with a galaxy at redshift 0.056, an implied distance of 211 Mpc or about 700 million light years.

Pawsey then provided a "teaser" to the audience at the IEE conference. "Finally the Cygnus source is not unique. It is only the most studied of the discrete sources of cosmic noise." Bolton had just submitted his paper to *Nature* on 30 March 1948 "Discrete Sources of Galactic Radio Noise". This was published in late July 1948, including the new sources detected at Dover Heights by Bolton and colleagues, Taurus A, Virgo A and Centaurus A. The optical identifications of these three with the Crab Nebula, M87 and NGC 5128 would be proposed the following year after the successful conclusion of the New Zealand Cosmic Noise Expedition.

At the Royal Astronomical Society meeting 3 weeks later, Pawsey began his contribution: "I'm afraid I shall be accused of collusion with Mr. [Fred] Hoyle when I show my first slide, which gives the observational evidence ... that he asked for concerning the relation between temperature and wave-length of observation."

Pawsey gave a description of his observational evidence at 200 MHz for the million degree corona. He showed that the radiation had a thermal origin: (1) the wave form was consistent with that expected from fluctuation noise, (2) the area of the emission was that of the entire sun and (3) the radio emission was shown to have no circular polarisation.

Ryle and Hoyle, in separate talks, discussed models for the radio bursts originating in a thermal process from gas at a temperature of 10^6 K in the corona and 10^{10} K in sunspots.

Martyn began his presentation:

> The discussion has shown that there is agreement on the quiet sun so in this respect I propose to let sleeping dogs lie! With the disturbed sun, there is a divergence of views on the possible mechanism of production of radiation. Mr. Hoyle and Mr. Ryle incline to the thermal origin [with extreme electron temperatures]. I suggest ... that we should therefore look to plasma oscillations, which are a more efficient source in a limited wave-length range than thermal radiators.

Hoyle invoked an existing theory of magnetic storms in the ionosphere due to Chapman and Ferraro from 1930 in which "... the motion of a slab of ionised material perpendicular to the magnetic field" produced "oscillation of the electrons on the walls of the slab".

It was given to Appleton to provide a summary of the afternoon meeting, and he did so by noting the growing strengths of the field—now drawing in others from established areas of research.

Considering "Radio Stars"

During his 2.5 month visit at the Cavendish, Pawsey was fascinated, among other things, by the short-term variations in Cygnus A; his term for this was "wobbles". In a hand-written addition at the end of a preliminary report on UK radio research (10 June 1948), Pawsey wrote: "'Wobbles' on Cygnus A source vary from day to day and on 'wobbly' day on Cygnus, Cassiopeia is quite steady. Strong evidence against ionospheric origin."

Pawsey had made the inference that both "radio stars" were of a similar nature; the fact that one showed variability and other did not suggested that the time variations were an intrinsic feature of each source. As the investigators were to discover in the next few years, this conclusion was incorrect, as is explained in the next paragraph.

In mid-1948, the nature of Cygnus A and Cassiopeia A were unknown. The identification of the former with a high redshift galaxy and the latter with a galactic supernova remnant would only become known after 1951 (see Chap. 18). The angular sizes would be recognised in late 1952 (Chap. 21) with Cygnus A's size of 1.5 arc min and Cassiopeia A with a size of about 6 arc min. It was then understood why the smaller source (Cygnus A) would show ionospheric scintillation while the large source did not. The analogy was often made that "stars twinkle" and larger angular size planets do not.

Pawsey's visit enabled Ryle to learn more about the new sources being discovered by Bolton, Stanley and Slee after the New Zealand Cosmic Noise Expedition of mid-1948. The two groups were also aware of the problem of the vastly different positions being determined for Cygnus A by the various groups (see Chaps. 18

and 21). Pawsey provided further details of the work at Cambridge in letters and reports to RPL, which we discuss below.[37]

Reporting Back

Pawsey sent back two comprehensive reports of his travels. The first reported on his North American tour, and was posted by Pawsey from the UK on 15 April 1948. It was received in Sydney well before 18 May 1948, when Bowen responded to it. A preliminary report on his UK visit was written on 11 June 1948 (after 2.5 months): "Some Notes on English Radio Astronomy", followed by an extensive report prepared on the ship when he returned to Australia on 10 October 1948 (after 6 months): "Notes on Radio Astronomy in Europe".[38]

The first report, "Solar and Cosmic Noise Research in the United States and Canada" was a 10-page summary which included chapters on "Personalities" (a remarkably frank assessment), "Work in Progress" at various institutes and "Research Problems", followed by a listing of the 18 institutes which had been visited with an accompanying list of personnel.

Pawsey's main conclusion was that the US astronomers and astrophysicists were impressed by the Australian success story in radio astronomy:

> Since my arrival [in the US] I have been struck by an anomaly. Astronomers and physicists have displayed a great interest in our work, but have not undertaken similar work themselves. Stromlo and RPL has [sic] not . . . had any serious competition in the solar field. In the cosmic field our work has supplied a very vigorous stimulant to work which was progressing slowly, chiefly under the impetus given it by an amateur, Reber, working alone in his spare time. The position now is that the astronomers of the US, who form a group who maintain exceedingly close contact with one another, have now become thoroughly interested in the implications but have not yet taken the plunge of tackling a totally new technique. Meanwhile, the physicists, who at the close of the war had the skill and inclination to undertake the radio side but failed to interest the astronomers then, now have other interests. The result is that we have a first-class opportunity to establish the lead which we at present hold.

Pawsey pointed out that the greatest attraction was the Cygnus source, a "new astronomical entity".

---

[37] Later, Graham Smith's newly accurate radio position from Cambridge for Cygnus A would be published on 29 September 1951 (Smith, 1951). This would lead to the optical identification of Cygnus A with a high redshift galaxy in 1952–1953, published by Baade and Minkowski in early 1954.

[38] NAA C4659/4. In fact Pawsey did not visit any of the non-UK radio astronomy institutes in Europe. In the complete report he discussed J.L. Steinberg's group in Paris, having started observations of solar noise using the ex-German giant Würzburg antenna. He also mentioned J.F. Denisse, whom he had met in Washington, a "guest-worker" at NBS. The strong friendship that developed between Denisse and Pawsey continued for the next two decades.

However, I feel myself that this subject is merely a part of the whole and that the solar observations are not only of interest in their own rights but have a good chance of supplying keys to the interpretation of the cosmic noise. The most fundamental problem which is already apparent is that of the structure of the galaxy. Cosmic noise studies may give the answer, but we are still ignorant of basic mechanisms [of the non-thermal emission process-see Chap. 34].

The report from the UK began by contrasting US and UK radio astronomy:

In contrast to the American work, English research in these fields is already well established. It is in the hands of people with initiative, skill and drive ... [For] the three primary discoveries in the whole field, that of cosmic noise itself [Jansky], that of the intense non-thermal solar radiation, which I rank as the least capable of prediction from prior knowledge, Hey was responsible for [the latter] two. Hey's facilities are however [now] limited and he has not been able to follow up his discoveries adequately. This has been done mainly by us in Australia with Ryle working independently a few months behind us.

Naturally, the Cavendish Laboratory at Cambridge figured largely in the report. Pawsey began by providing details of the ionospheric research in Ratcliffe's group. He summarised the 15–50 kHz research of Ron Bracewell (on a CSIR Fellowship) and colleagues. Other groups were working on: (1) 50–500 kHz propagation and (2) short waves (1.6–30 MHz). "Of these investigations the work at 15–500 kHz and at short waves all usefully associate with solar noise work. They give data on radio fadeouts and allied phenomena." He then included detailed discussions of both solar and cosmic noise research at Cambridge. Martin Ryle and Derek Vonberg (1922–2015) had recently published a paper in the *Proceedings of the Royal Society*: "An Investigation of Radio Frequency Radiation from the Sun (Ryle & Vonberg, 1948). Interferometer techniques, as well as solar and cosmic noise research, were discussed in detail during Pawsey's 6 months visit. Ryle had extended the interferometer baseline to 600 feet (183 m) with arrays of four Yagis at the ends. Pawsey reported:

[I]t gets beautiful records of discrete sources at 80 MHz. The prominent ones are that in Cygnus and a new one in Cassiopeia below [the Sydney] horizon ... Ryle uses direct interference by means of coaxial cables of a superlative German type.[39] I think the direct method may be superior to that involving frequency conversion at both aerials and interference at the intermediate frequency. (Ryle & Smith, 1948[40]).

Pawsey included additional details about the "noise adding radiometer" method of observation described by Ryle and Vonberg in April 1948. This consisted of a system to balance the input noise in the aerial against noise from a diode (a type of Dicke switch). This method enabled the sun to be observed over long time intervals in the presence of the variable galactic background. Pawsey pointed out that the two-element Michelson interferometer (with a variable baseline) was straight forward to use, leading to a determination of source angular size and polarisation.

---

[39] These were the so-called "Jerry" [WWII slang for "German"] cables based on a km of captured war material. This was far superior to any contemporary UK produced cable. (Sullivan, 2009, p. 160).

[40] Published shortly after Pawsey's visit.

Within a short period, positions could also be determined based on the absolute phase of the interferometer fringes.

Solar work was being carried out at both 80 and 175 MHz, with plans to move to lower frequencies. The intensities, sizes and the circular polarised state of the solar emission could be determined on a daily basis. Pawsey made a striking distinction as he described the differences between the two solar groups. The Cambridge group "depend on statistical treatment of all observations. This contrasts with the treatment we usually adopted of deriving results from outstanding cases [the prominent example was the huge 8 March 1947 Type II event published in *Nature* by Payne-Scott, Yabsley and Bolton on 23 August 1947]." An example of the Cambridge method was an attempt to look for a periodicity in the solar radio emission close to the 27-day solar rotation period based on the auto-correlation method of the intensity as a function of time delay. If this were to be confirmed, this would indicate "directed emission of the radiation with [a finite] angular distribution which was different at the two frequencies [80 and 175 MHz]." Thus "spots within a day of meridian passage may be identified as [radio] sources in many cases [since absolute positions had not been determined by Ryle and Vonberg]." Pawsey thought the statistical techniques could offer the possibility of:

> obtaining objective results ... and [we should] be prepared to use them. However, the interpretation is very tricky and I feel sure that the desirable procedure is a combination of the direct method utilising outstanding cases and the statistical one. The first method introduces subjective uncertainty in the selection of the data, the second in the physical interpretation of statistical facts.

We return to these issues in Chaps. 35 and 36.

Pawsey's Attempts to Recruit Theoretician Colleagues

Pawsey's trip failed in one objective—that of recruiting a theoretician to RPL. As mentioned, he did not in the end prevail on Ralph Williamson to come to Sydney (ESM 17.4, Ralph E. Williamson).[41] Bowen suggested to Pawsey that he contact Hannes Alfvén (1908–1995, Nobel Prize 1970) who had expressed some interest to Bowen in spending a year in Australia. On 6 April 1948, Pawsey responded to Bowen; he would look for suitable candidates in England and would talk to Alfvén in Stockholm. In the end nothing came of this proposed suggestion.

On 28 July 1948, a few days after returning from URSI in Stockholm, Pawsey reported to Bowen that he had met Olaf Rydbeck[42] of Chalmers University of Technology in Gothenburg, Sweden. Since Rydbeck planned to visit India in

---

[41] See also NRAO ONLINE.26.

[42] See Radhakrishnan (2006). O. Rydbeck (1911–1999). Founder of the Onsala Space Observatory, in the early 1950s.

1949, he had asked Pawsey about combining this with a long visit to Australia. Pawsey wrote:

> ... in view of Rydbeck's abilities and interests in the field of the mathematics of waves being associated with ours, Westfold in particular, I felt that he could be considerable stimulus to the laboratory if he could spend a [long time with us] ... [His visit] would give Westfold and Pearcey someone to talk their own language and I feel could be a strong stimulant. Also Rydbeck is one of the senior Swedish scientists and as such would give us a worthwhile intimate contact with [an excellent] scientific group in Sweden.

Again, nothing came of this suggestion. Pawsey's final suggestion for an eminent visitor, Henry G. Booker, was made to Fred White of the CSIR Executive. Booker had been a student of Ratcliffe (at the Cavendish), about 2 years after Pawsey. He then became a prominent radar scientist in WWII at TRE, and visited RPL in October 1944.[43] In the post-war era, Booker was at Cambridge University where he and Pawsey met in 1948. Booker was not satisfied with his position and had been offered a position at the National Bureau of Standards in Washington, D.C. Pawsey wrote, "there is a certain reluctance to leave the British Commonwealth and he appeared quite interested in the possibility of joining CSIR in Australia." White was encouraged to visit Booker personally during an upcoming trip to the UK.

But again, the Australian group lost the opportunity to recruit one of the more creative ionospheric physicists of the mid-twentieth century. Charles Burrows at Cornell University made an offer in late 1948 that preceded any Australian CSIR offer. Booker was involved in the Arecibo radar project for many years. From 1965 to his death in 1988, he was a leader in the newly founded Department of Applied Electro-Physics at the University of California San Diego.

In the 1950s, Pawsey did succeed in bringing recent PhD astronomers to join the RPL staff: Colin Gum of Mt. Stromlo in 1956[44] and Campbell Wade from Harvard in 1957.[45] Nonetheless the absence of strong theoreticians continued to hamper the Australians. This absence is a sustained theme in this book and was the most prominent aspect of the continued "tyranny of distance" experienced in Australian science.

The Costs of Absence: Bowen's Review of the Laboratory, May 1948

Even though the extensive trip did provide invaluable contacts for the Australians, RPL paid a high price for Pawsey's absence as his indispensable personal leadership was missing for a 13-month period.

---

[43] NAA A8520 PH/PAW/1 Part 1. See NRAO ONLINE.59.

[44] ESM 26.5, New Galactic Coordinates and NRAO ONLINE.58.2.

[45] Chap. 40.

Pawsey and Bowen exchanged numerous letters about the current and future status of RPL. Pawsey had left Sydney barely 2 years after the end of WWII; many new activities had started, leading to new duties for the existing staff and the desirability of recruitment of new scientific personnel. As expected, many problems and challenges arose in this period. Extensive correspondence was no substitute for day-to-day contact. Lindsay McCready, left in charge, could not fill the gap left by Pawsey in 1947–1948.

Bowen kept Pawsey informed of activities at the Laboratory through correspondence, reporting on the activities of the scientists there. Many of his reports were positive—particularly as regards John Bolton. In this period Bowen had decided that John Bolton was to cease solar work and concentrate on "cosmic noise" research (see ESM 17.3, Bolton Pawsey correspondence). But there were negative assessments from Bowen also, particularly in relation to the lacklustre performance of theoretician Steve Smerd,[46] who was "called on the carpet" by Bowen in late February or early March 1948 (see also ESM 17.5, Difficulties at RPL). On 7 April 1948, Pawsey wrote pragmatically to Smerd:

> ... You have now been on the job [theoretical solar research] long enough to form a fair opinion and it is clearly the time for you and Bowen to discuss your program and decide whether or not you are in a job in which you can be successful and happy. If either of these is negative, you should get out and try something else.

There was acceleration; Smerd produced two papers in 1950[47] followed by an impressive review co-authored with Pawsey and published in a book edited by Kuiper in 1953.[48] In addition, Smerd and Westfold (1949) and Smerd (1950b) provided improved solar models.[49]

Within 1–2 years after Pawsey's return to Australia, Pawsey realised the value of Steve Smerd. He became one of the key scientists in the Sydney solar group, succeeding Wild in 1971 as the head of the solar group at Radiophysics. By 1950, Smerd had become the repository of a mass of information about the sun. Pawsey called him a "walking encyclopaedia" (Wild, 1980[50]) on solar matters. Smerd died on 20 December 1978, while undergoing heart surgery.

---

[46] Sullivan (2009, p. 288) has provided an account of the arrival of Smerd, an Austrian refugee in the UK in 1938. During the war, he worked on radar research at the University of Birmingham and the Admiralty. He was recruited by Pawsey after the war to work at RPL, starting in 1946.

[47] The major paper was finally submitted on 30 August 1949 to the *Australian Journal of Scientific Research*, "Radio Frequency Radiation from the Quiet Sun" (Smerd, 1950a).

[48] *The Sun* (Kuiper, ed., 1953), chapter by Pawsey and Smerd, "Solar Radio Emission."

[49] "The Characteristics of Radio-Frequency Radiation in An Ionised Gas, with Applications to the Transfer of Radiation in the Solar Atmosphere" and "A Radio-Frequency Representation of the Solar Atmosphere", respectively. [NRAO ONLINE.20].

[50] Paul Wild (1980, "The Sun of Stefan Smerd") has provided a masterful account of his colleague: "I do not think [he will be remembered] especially for his writings and publications. Although these included some that were definitive and highly significant, they were rather few in number ..." Wild asserted that Smerd was the catalyst for solar radio astronomy in Australia. "To his Sydney-based colleagues by far his most famous writings were universally known as 'the unpublished works of

The correspondence between Pawsey and Bolton during this period is also revealing, showing Bolton's fast developing confidence as a researcher and with it, the beginnings of tension between himself and Pawsey. An overview is provided in ESM 17.3, Bolton Pawsey correspondence.

RPL Awaits Pawsey's Return, 1948

Bowen concluded the 18 May 1948 letter concerning research plans with another topic, Pawsey's return to Sydney:

> In my review of the work of the laboratory last week I didn't say how much we are feeling your absence. For a long time the solar and cosmic noise work went along exceedingly well under its own momentum but there have been signs of slackening off since Christmas. There is no lack of keenness or enthusiasm, it is only that members of the Group are lacking the stimulus of day-to-day contact with someone like yourself who is completely on top of the job. Some of the setback might be due to the fact that, following the burst of activity before Christmas, people have got back to instrumental development. This is particularly true of interferometry, the spectrum analyser and the big aerials for Dover and Georges Heights. In that sense, perhaps, things will fit in very nicely. I expect everyone to be making observations with new equipment around August, that is they will have begun getting results before your return and will, therefore, have a lot of things waiting for you to examine.

The same sentiment of feeling Pawsey's absence had been expressed by Kevin Westfold and by Lindsay McCready in correspondence with Pawsey—for example, in Westfold's April 1948 letters requesting Pawsey's input for research directions, which had prompted a reply from Pawsey in which he had "let off a lot of steam" (Chap. 33). On 10 June 1948, Pawsey indicated to colleagues in Hobart, Tasmania, that he expected to return in September. On 17 June 1948, Bowen wrote Pawsey about the "boys" in the radio astronomy group: "The boys in the radio astronomy group are feeling your absence quite keenly, but I am taking the view that their present gropings [sic] are part of their education. I feel sure they will all be better men after having to fend for themselves for a time."

Even Lindsay McCready was concerned about Pawsey's return. On 25 June 1948 he wrote Pawsey:

> We will all be greatly looking forward to your return and hope radio astronomy will not be tapered off or closed down for a while yet. I think you should enjoy yourself immediately on your return—you should have a lot of entirely new and better engineered tools to play with. Next year [1949] should see less time on equipment design and more on planned observing, etc.

S.F. Smerd'—these, the mighty efforts that never quite came to the public eye, were voluminous indeed." (Also NRAO ONLINE.20).

Publications Concerns at RPL

Among the difficulties that arose in Pawsey's absence were publication problems among the solar and cosmic staff, as these newcomers attempted to present their ground breaking results to the scientific community.

In March 1948, Pawsey became aware of the "publication" tensions at RPL via his "backchannel" in Sydney, frequent hand-written letters from Lindsay McCready. On 24 March 1948, McCready wrote[51]: "... Taffy [Bowen] is demanding a very high standard in report writing—so much in fact that it is ... causing serious delays in publication. Again one's efficiency, morale and interest in his own paper falls off more or less in proportion to the number of times it has to be rewritten or typed up ...".

McCready told Pawsey of papers that went through 10–12 drafts. The situation deteriorated, leading to a major rift in the publication office: "Miss Plunkett resigned ... [due to] too much pressure from Taffy in getting reports out." McCready admitted: "We can all stand criticism in relation to the way we express ourselves in print." McCready asked Pawsey to bring back information on the publication process in other laboratories in the UK. McCready continued: "Was it necessary to have a uniform style—so much so that the author's personality is completely submerged? For example, was the royal 'we' to be 'verboten'?" McCready did concede that most of the "general criticism on papers is fair—but the detailed criticisms are far too severe and, of course, cause too many hold-ups ... I have stimulated Taffy to give us a pep talk on what he really wants."[52]

On 17 June 1948, Bowen responded to the correspondence on research directions between Pawsey and Westfold in April and May (alluded to above). In addition to the "content of [research] work", Bowen was concerned:

> ... [With] writing it up. It is true that those of us who have had a fair amount of experience can give a lot of help in choosing problems for the young people, keeping their sights on the target and in helping them snatch the odd pearl out of the tangled mess, but I am quite sure that what we are suffering from in the laboratory is not that there is too little of this help but too much [our emphasis]. With few exceptions our youngsters have not learnt to stand on their own feet and go for a line of their own. When they have done so, Bolton for example, the results have been exceptionally good.[53]

However, a month later (23 July 1948), Bowen's perception of the counter-productive effect of too much criticism had changed. He was frustrated with the number of papers rejected by the newly formed *Australian Journal of Scientific*

---

[51] The nature of the "backchannel" was clear. McCready wrote: "I would not like you to let [Bowen] know I am writing you." During 1947–1948, McCready sent his (hard to read) letters every month or two to Pawsey. We thank Harry Wendt for assistance in deciphering these airmail letters, due to McCready's indistinct handwriting.

[52] Pawsey responded on 6 April 1948. He was also concerned about the difficulty and slowness of paper preparation. He was already considering various options that would speed up the publication process.

[53] This self-critical message was an unusual admission for Bowen.

Research.[54] Bowen wrote to Pawsey that the failure rate was due to the fact that the manuscripts had not been well written. "So far we have failed lamentably to [submit well written papers] and are getting ourselves a bad name in the process." The criticisms were (a) a lack of clear indication of the expected contents and (b) lack of logic in the treatment and (c) inclusion of extraneous conclusions not justified by the results.

> All [of] this is a sorry story and reflects on the management. I have tried hard to get people to see the light and although they are beginning to see it, there is not much improvement. I am sure I am following the right course in the long run by preferring to write people's papers for them. I intend continuing this policy for another 3 or 4 months at least but the day may come when we will have to take the short view for the sake of getting some good papers out. As I keep on telling the chaps, there is no doubt about the excellence of the work being done in the laboratory but **the writing up is awful** [our emphasis].

Bowen continued a week later on 30 July 1948:

> I have been making myself very unpopular about publishing papers and am being freely criticised for holding them up without justification. My purpose has simply been to ensure that [our] published work ... is of a high standard and to protect [RPL] from outside criticism. The time has come to protect myself. The only way I have of doing this is to send papers out for publication for judgement by the referee. I am doing this to an increasing extent with the result you know—a large proportion are being rejected. I am sorry to be so despondent about this but see no alternative. The solution is in the hands of the authors and I keep on reminding them of it.

Homeward Bound

As we have seen, one of the major rationales for Pawsey's trip of 1947–1948 was to overcome the continued effects of distance on scientific development in Australia. His publicity efforts in the US, Canada and Europe in 1947–1948 were in the end successful. The connections he built with the international nascent radio astronomy community and, of course, with the ionospheric community, were valuable during this period. The trip also helped to familiarise the RPL staff (Pawsey included) with a number of astronomical concepts.

On 23 September 1948, the Pawseys left the UK on the *SS Orontes* for the long trip home via the Suez Canal. On 22 October 1948, they were in Adelaide (after a stop in Perth) visiting Charles Duguid.[55] On 25 October the ship was in Melbourne for a short period: Pawsey met Bolton who was on his way to the Tasmania solar eclipse of 1 November 1948, along with Gordon Stanley. On 29 October 1948, after a trip of 13 months, the Pawseys arrived at Circular Quay in Sydney. They were met by Lindsay McCready and the two older children Margaret (age 11) and Stuart (age 9). The two grandmothers, Mabel Nicoll and

---

[54] NRAO ONLINE.27.

[55] Charles Duguid, a well-known South Australian educator, a prominent advocate for Aboriginal Australians. His wife Phyllis was J.L. Pawsey's first cousin. See NRAO ONLINE.55.

Fig. 17.4 The Pawsey family: front row Stuart, Hastings and Margaret; back row Lenore, Mabel Nicoll, Margaret Pawsey, and J.L. Pawsey. Credit: Joe and Lenore Pawsey Family Collection

Margaret Lade Pawsey remained at home with Hastings (age 3). (Figs. 17.4 and 17.5).

Fig. 17.5 J.L. Pawsey's children Margaret, Hastings and Stuart. Credit: Joe and Lenore Pawsey Family Collection

Chapter 18
Scintillating Relationship with Cambridge, 1948–1951

I think it is true to say that we are leading the world in investigations of radio waves from the sun and stars and in the interpretation of the results. The astronomers in Europe and America are intensely interested and follow the progress of the work practically on a day-to-day basis . . . I understand that a large part of the time of the 100-inch at Mt. Wilson and the 200-inch at Palomar[1] is spent in looking at regions of the sky where radio frequency waves originate.— [Bowen to White, 22 October 1951]

Pawsey returned from his travels clearly determined to maintain the RPL's leading position in "radio astronomy" internationally. His trip had shown him that no other group was quite so advanced as RPL—and had also underscored the reality that maintaining this leadership required the additional effort of keeping the work at RPL prominent in the awareness of groups of scientists at a significant distance. As with most forms of structural disadvantage, the costs of distance were visible to those who bore them (ie, the Australians) but not so to those who benefited (Cambridge).

In the chapter title we use the word "scintillating" metaphorically to describe the flickering between collaboration, competition and controversy that occurred in the relationship between Sydney and Cambridge in the years 1948–1955. It is an apt metaphor. In the years 1948–1953, these groups (along with Hey's at the Army Operations Group and Lovell's at Jodrell Bank) were questioning whether observed scintillations in the newly-discovered "discrete" sources or, as they were then termed, radio stars, were intrinsic to the source, or a consequence of passing through an intervening medium (interstellar, solar system or atmospheric). Similarly, those

---

Associated with this chapter: NRAO ONLINE.22, "Pawsey Ionosphere Research 1947–1954".

Supplementary Information The online version contains supplementary material available at [https://doi.org/10.1007/978-3-031-07916-0_18].

---

[1] Bowen's assertion about the observing time at the 100 and 200-inch telescopes was likely an exaggeration.

© The Author(s) 2023, corrected publication 2024
W. M. Goss et al., *Joe Pawsey and the Founding of Australian Radio Astronomy*, Historical & Cultural Astronomy, https://doi.org/10.1007/978-3-031-07916-0_18

involved were attempting to work out whether intentions of collaboration or competition were intrinsic to scientists in other groups, or whether controversial failures were instead the result of distance and consequent poor communication.

During this period Pawsey worked hard to establish communication systems with Cambridge that would support collaboration. But the insularity of Cambridge largely defeated these attempts, at the cost of some priority and credit for the group at RPL. We begin with a brief discussion of the failed attempt at collaboration in investigations of radio source scintillation.

Scintillation and Cygnus A 1946–1950; Bolton, Stanley and Slee in New Zealand

Numerous questions about the cause and nature of the intensity fluctuation of the radio source in the constellation of Cygnus (Cygnus A) were the focus of investigations in the late 1940s. This ironic saga is recounted in detail by Sullivan (2009, pp. 324–325). Initially most radio astronomers favoured an intrinsic cause for the variations based on the similarity of solar bursts and radio star scintillations; in addition both were stronger at low frequencies and associated with stellar-like objects, as perceived in 1946–1949. In 1949 this began to change with the identification of discrete radio sources with a supernovae remnant (Crab nebula) and external galaxies (NGC 5128 and M87), as discussed later, but the Cambridge group stayed with the radio star model until the optical identification of the radio source with a high redshift galaxy Cygnus A, in 1954 by Baade and Minkowski. This is discussed in more detail in Chaps. 34–36. However, even in 1947 Oort had suggested that the twinkling might be a local effect caused by the ionosphere.

In addition to the search for new radio stars, spaced aerial observations were started. This involved simultaneous observations at separate sites so the time series could be compared. If the variations were intrinsic to the source there would be near perfect correlation of variations observed from the separate sites. Slee (1994, p. 521) recounts that they found the fluctuations well correlated but the baselines available for these initial observations were quite short, only 2 km in projection, so the variations could either be intrinsic or propagation effects due to irregularities larger than 2 km in the ionosphere.

Observed accuracies of the positions of the new sources was a major problem with errors of some degrees, and observing with another position angle on the sky for the determination of the coordinates in 2D was desirable. To obtain better positions they looked for other possible sites on Lord Howe Island, Norfolk Island and in New Zealand in mid-1948. They wanted an east facing and a west facing site within 100–200 km on the North Island of New Zealand. The two sites chosen were Pakiri Hill near Leigh (east), 86 km north of Auckland and the site of the former Royal New Zealand Air Force station at Piha (west--Log Race Road) 40 km west of Auckland. The primary goal of the New Zealand observations was to measure the positions and hence obtain identifications for the newly discovered discrete radio sources as discussed at the end of this chapter, but the same data taken simultaneously from two well separated sites (Sydney and New Zealand) could also be used to test whether the variability was intrinsic or local.

Pawsey had discussed Bolton's six new discrete sources with Ryle during his visit to Cambridge in 1948; the New Zealand expedition occurred as he travelled home. In the spirit of intended cooperation, Bolton wrote directly to Ryle on 12 November 1948 with his findings, likely the first news of this remarkable discovery to reach the UK. He proposed, as it turned out correctly, the first identification of a "radio star" with a galactic supernova. Bolton to Ryle:

> ... However, east and west coast observations in New Zealand have given us some fairly reliable results and an almost independent determination of refraction. The most interesting of these Taurus A—whose position coincides with that of the Crab Nebula—which has a suitable mechanism for million-degree radiation. I hope to publish my results on this fairly shortly, and will send you an advance copy when available.
>
> A further result of the NZ Expedition was that the fluctuations were "local"—there was no correspondence between Dover Heights and NZ records ...
>
> In view of the more or less simultaneous work on the variation in apparent positions from the two sites, Dr. Pawsey has suggested that we could present a very strong case in, say, a joint letter to *Nature* on this subject. We both have experienced much the same extent of variation and both have the "control" in the other sources. If you are agreeable, perhaps we could exchange some of the salient data for further consideration.[2]

After a delay of 2 months, Ryle responded,[3] with the letter sent to Pawsey instead of Bolton, on 12 January 1949. Ryle's response completely ignored Bolton's suggested identification of the radio source Taurus A with a supernovae remnant. This was so removed from his strong preconceived ideas about the nature of the radio stars that it would not have seemed credible. Ryle was also quite cautious in his response to both Bolton's assertion that the scintillations had an extrinsic cause, and to the suggestion of coordinated publication (Ryle to Bolton):

> I have had a very full discussion with Ratcliffe on the question of the early publication of the results on Cygnus. The problem is further complicated by Lovell's recent work and his wish to publish some of his results. While these do not relate directly to the Cygnus source it brings up the question of whether some account of recent work should not be published by all three teams. Your point about publication by some American observer is a very strong argument![4]
>
> (Ryle to Pawsey continued)
>
> We did not feel very happy about a joint letter to *Nature*, both because of the complications of writing a co-operative account by remote control and also because of inevitable differences in belief (e.g. **we are not yet convinced that the fluctuations we observe in**

---

[2]NAA C3830 A1/1/1 Part 3.

[3]NAA C3830 A1/1/1 Part 4.

[4]This point has not been found in the correspondence from Australia to the UK. Ryle mentioned elsewhere in his letter to Pawsey: "In addition to being a convincing story (e.g. for the astronomers) it would make it clear that all three teams [Cambridge, Jodrell Bank and Sydney] were obtaining important results [thus guarding against the hypothetical American]."

Cygnus are due to local effects) [our emphasis].[5] The suggestion that we should like to make is that in the near future all three teams should write short separate notes for publication in *Nature*. The three contributions should then be forwarded together for simultaneous publication in the same issue. It seems that apart from describing work in the two hemispheres such a procedure would have the advantage that three independent observers using quite different techniques could give their own accounts of the phenomena and thus produce a more convincing story of the effects than if combined in a single account. In addition to being a convincing story (e.g. for the astronomers [acknowledging that the "radio astronomers" are not yet considered to be real astronomers!]) it would make it more clear that all three teams were obtaining results (thus guarding against the hypothetical American).

At a later date, say in six-months' time, we should probably all be wanting to publish a fuller account of our work. It might again be desirable to arrange for simultaneous publication, but we can decide that later.

Ryle also proposed some immediate experimental implications arising from Bolton's communication:

> ... I am very anxious to know your views on the New Zealand results. As I said in my last letter to Bolton [arrived in Sydney on 11 November 1948 but not located in the archives], the results, if true, are most significant. The possibility that fluctuations are introduced by the troposphere or ionospheric refraction seems so important that we would like to see some experiments carried out at normal incidence [the source directly overhead at an elevation close to 90 degrees, not 15 degrees as in Sydney or New Zealand]. **We feel so sure that the fluctuations that we see are genuine that we would like to repeat the experiment using normal—overhead—incidence** [our emphasis], and I have suggested to Lovell that we might do some such experiments jointly. Although the distance is only about 200 miles, any local effects [whether troposphere or ionospheric] should be different at that distance ... Perhaps you would let me know what you think of these suggestions and whether you think any other information need be published at the moment besides the Cygnus variations and possibly the position of other small sources. I have not yet discussed this with Lovell, but he has written indicating that he would like to publish a short note.
>
> It was very kind of you to think of sending us a food parcel [during the strict rationing in the post-war UK]; it has not yet arrived, but we are looking forward to it.

Pawsey responded to Ryle on 24 January 1949 before he could talk to Bolton, who was away on summer holiday. He wrote:

> I am sure Bolton would be quite happy to join with you in simultaneous contributions to *Nature* as you propose. I shall ask Bolton to draft a letter when he returns.
>
> .... Bolton has plans to repeat the experiment at other distances some time in the future. [This was to be later in 1949 in order to influence the joint publication with Jodrell Bank and Cambridge.] My [Pawsey's] own attitude is that I am very keen to see important results established by independent means. Hence, I should like to see you try the experiment at normal incidence.[6]

---

[5] Sullivan (2009, p. 325) suggests that Ryle was reluctant to give up on his initial belief that scintillations were intrinsic. Sullivan: "The existence in Cyg A of 'genuine' scintillations was also important to Ryle because it constituted one of his key reasons for favouring the idea that radio stars were stellar-like objects." Ryle was clearly reluctant to credit results contradictory to his own observations and mental model of the phenomenon.

[6] Sullivan (2009, p. 325) points out that Ryle told Lovell on 22 December 1948, that he [Ryle] did not believe the Australian scintillation results from New Zealand.

On 17 February 1949, Ratcliffe wrote Pawsey, perhaps indicating that he was concerned about the future of the collaboration:

> ... After some detailed discussion with Ryle I suggested that he should work out the details direct with you and he has kept me in touch with the decisions that you are making between you[rselves] ... I should like to hear at any stage if you feel that the contacts which you have direct with Ryle **are in any way other than satisfactory** [our emphasis].

On the same day (17 February 1949) Bolton wrote Ryle: "I think your proposal of simultaneous letters to *Nature* is an excellent one." He pointed out that there was no correspondence in amplitude between NZ and Sydney. "I shall be pleased to hear results of your joint effort with Lovell, also if Lovell sees amplitude variations with his parabola."

Seven months later, on 28 September 1949, Ryle wrote to Pawsey about the scintillation projects.[7] The experiments had proved confirmatory:

> The main conclusions so far are that the greater part of the recent fluctuations is different, and **therefore agrees with your New Zealand-Australia results**; we must therefore conclude that some of the fluctuations are caused by some relatively local effect ... **There is no doubt that your original experiment was most important in showing the existence of an uncorrelated component—I think otherwise everyone would have assumed that it must be due to the source.**[8] (our emphasis).

The decisive publication that partially elucidated the nature of scintillations appeared in *Nature* on 18 March 1950. There were two back-to-back papers by the two UK groups, "Origin of the Fluctuations in the Intensity of Radio Waves from Galactic Sources: Cambridge Observations" by F.G. Smith (submitted 9 December 1949) followed by "Jodrell Bank Observations" by C.G. Little and A.C.B. Lovell (submitted 30 November 1949). Each group carried out joint observations from 1 May 1949 to 31 October 1949 at 81 MHz with a separation of 210 km. In addition, each group carried out observations independently. The main conclusion was that the lack of correlation in the fluctuations at the two sites implied that "the origin of the fluctuations must be fairly local, and probably in the earth's atmosphere or iono-sphere."[9] Only the first paper by Smith contained an acknowledgement of Bolton's efforts as reported in a private communication by "Dr J.L. Pawsey" of the joint observations between Australia and New Zealand.

---

[7] NAA C3830 A1/1/1 Part 4.

[8] Sullivan (2009, pp. 325–326) discusses the ironic twists to Ryle's viewpoint as scintillations moved from an intrinsic property to a feature of the local, terrestrial environment.

[9] There were two papers since the Jodrell Bank team did not believe the mysterious observations by Ryle's group of simultaneous, short bursts at 45 MHz at sites displaced up to 160 km. The direction of the bursts was not known; the beams were so large that both Cassiopeia A and Cygnus A were in the antenna response. Many other astronomers were sceptical; Hanbury Brown remarked in a letter to Bowen on 30 April 1950: "Cambridge have bursts which they think come from the sources. We think that this may be true but it would take a hell of a lot of proving! In radio astronomy it is only too easy to ascribe cosmical significance to what is, in effect, activity in the local tramway system." [NAA C3830 A1/1/1 Part 4]. These bursts were seldom observed and never confirmed. They disappeared from the scene within a few months.

Sullivan (2009, p. 326, footnote 37) described the reaction in Sydney: "[T]he Australians were furious ... [by] the lack of any opportunity for them to publish alongside[10] after they had expressly delayed publication." Perhaps the worst impacts fell on John Bolton and created the biggest rift in Sydney-Cambridge relations. Bolton (1982, p. 352) wrote later that his first knowledge of any results [about the scintillation collaboration] was to read a joint Cambridge-Jodrell Bank Letter to *Nature* on his arrival in London in February 1950.

And worse was yet to come. In 1950 it was Bolton's turn to make a trip to the UK and Europe—a journey that acknowledged his swift and extraordinary achievements in the few short years since he had joined RPL, and would serve to establish his own leadership in international scientific circles. (The trip was important personally as well as professionally, since Bolton, a Yorkshireman by origin, was quite literally returning "Home".) Bolton's short period of study at Cambridge at the beginning of the war had revealed the singular richness of physics and research at Cambridge (and we remind the reader of the long impacts of the rich traditions of mathematical physics to be found there (Chap. 5) as well as the opportunities for cross disciplinary conversations (Chap. 37). But instead of the immersive visit of several weeks to which Bolton looked forward, he received a rather abrupt note from Sir Lawrence Bragg who wrote that "because [his work] is at an exciting and formative stage, [Ryle] ought to be free from [visitor] interruptions" (Robertson, 2017, p. 116), and made it clear that Bolton would be welcome to visit for a day or two or to spend a year in Cambridge as a research fellow—but that a 3 week visit would constitute an unacceptable distraction to Ryle. Responding with condolences to Bolton, Pawsey commented from his own experience that Ryle was "egotistical, impetuous and superficially at any rate, extremely confident of his own work." Pawsey considered that Ryle was a good experimenter, but still "immature" with respect to interpretation. These were astute and largely correct insights, and, with typical charity, Pawsey added that Ryle was "Nevertheless quite likeable". His view was that Ratcliffe simply left Ryle to run the radio astronomy research on his own terms—rather as he was himself doing with his budding researchers in Sydney.

That this issue was specific to Ryle and to Cambridge was emphasised by Bolton (1982) who wrote: "At the URSI conference in Zurich that year, Bernard Lovell very graciously apologised for the form of this publication, for he had not been told of our prior work!" Indeed, Lovell had never been involved in the discussions between Cambridge and Sydney.

Bruce Slee (1994) summarised the missed opportunity for RPL in his contribution to the Bolton memorial symposium of 9–10 December 1993:

---

[10]We note that Sullivan (2009, p. 325) suggests that Ryle was suspicious that Bolton's scintillations with the sea-cliff interferometer, necessarily taken at very low elevation angles, were not indicative of intrinsic variability, that Bolton had once been concerned about such possibilities, and that the Australians "therefore" held off publication until corroboration could be sought in the UK; but this was not the case; Bolton was very confident about his results.

We did not get the opportunity to announce this important result because we tried first to enlist the aid of the Cambridge Radio Astronomy group to perform some experiments with baselines of up to a few hundred km in order to define the scale size with some precision. Cambridge and Jodrell Bank performed the experiment and published the result (scale sizes were 5–10 km) without acknowledgment.[11] We published our result belatedly in Stanley and Slee (1950), **but the irony is that we could have done the experiment easily ourselves, at coastal sites within 200 km of Sydney** (our emphasis).

Pawsey expressed his dissatisfaction to Ratcliffe[12]:

> I may also mention that the previous collaborations between Ryle and Bolton concerning the "twinkling" of radio stars left Bolton dissatisfied. You will remember that Bolton told Ryle of his spaced receiver experiments and your people and Lovell followed it up, leading to the companion Cavendish and Manchester papers in *Nature* showing the effect to the be primarily ionospheric ... After the original letters, the next Bolton knew of it was through seeing the articles in *Nature*.

Fortunately, in fact, Stanley and Slee had submitted a manuscript of their own on 4 November 1949, about a month prior to the submission of the UK papers, and it was published in June 1950. However, this paper was in an Australian journal that took a long time to reach scientists overseas, and did not carry the impact and status of a paper in *Nature*. The scintillation data from 1947 to 1951 also was summarised and discussed in a longer paper by Bolton et al. (1953, "Galactic Radio Radiation at Radio Frequencies. VI. Low Altitude Scintillation of the Discrete Sources" [Tau A, Virgo A, Cyg A and Cen A]).

Time variations of radio stars, scintillations, became a tool for studying the ionosphere. In the course of 1950, a major problem in early radio astronomy had finally been sorted out. Sullivan (2009, p. 327) has aptly described the strange turn of events at the beginning of 1950:

> Thus, it was that by 1950–1951 the cause of the scintillations had been solidly ascribed to the ionosphere. Most radio astronomers, whose focus lay increasingly on the radio stars far beyond, therefore lost interest in any further pursuit of the phenomenon, except in understanding its deleterious effects on their data. Moreover, observations on the whole were steadily moving to higher frequencies where the scintillations were weak and of no practical importance. The very scintillations that had given birth to the **radio star phenomenon now ironically left the scene** [our emphasis].

In 1953, Pawsey summarised the end of the scintillation controversy in a very generous reconstruction of history written for Oort in Leiden, thanking him for his influence on the Australian perception of scintillation already in 1947[13]:

> ... I don't suppose you realized your share in uncovering the cause of the twinkling of radio stars. You may remember that in 1947, I met you and Dr. Strömgren at Yerkes [Chap. 17] and we discussed Bolton's recent discovery of the radio star in Cygnus. You asked persistently about the possibility of the twinkling effect being due to the ionosphere ...

---

[11] In fact, the Smith paper of 1950 did contain an acknowledgement, thanking Pawsey.

[12] NAA C3830 Z3/1/III, 30 October 1951, Pawsey to Ratcliffe.

[13] J.H. Oort archive, University of Leiden [J. "Jet" K. Katgert-Merkelijn, earlier Jeannette K. Merkelijn] *The Letters and Papers of Jan Henrick Oort*, 1997).

He [Bolton later on in 1948] took simultaneous observations in Sydney and New Zealand which showed independent twinkling. This he communicated to Ryle, and Ryle and Lovell clinched the matter by the systematic series of observations, which were published in *Nature* [in 1950].

Further Disadvantage

Bowen was as keenly aware as Pawsey of the importance of direct contact with Cambridge, Jodrell Bank and AOG (Hey's group) in order to maintain the status and visibility of RPL. Frequent travel for Pawsey and members of his team were prioritised in the budgets for RPL. It was Bolton's turn to travel in early 1950; later that year, Pawsey flew to Europe (see Chap. 19); in 1951, it was Bowen who went from mid-March to the end of November, leaving Pawsey to play a major role as Acting Chief of the Radiophysics Division. He wrote comprehensive letters to Bowen with reports of all activities of the Laboratory, including the Cloud Physics research.

But on-the-ground relations with Cambridge had their unexpected "scintillations" between collaboration and exclusion, just as much as relations carried out through correspondence. The differences in perspective between the visitors and those within Cambridge again show the social phenomenon in which the systems that create disadvantage are more visible to those who experience the costs. What to Cambridge scientists was an understandable and not very important tendency to insularity, was to the Australians yet another set of missed opportunities that occurred in a long history of scientific marginalisation.

Ryle, mostly unintentionally, contributed to this situation, by his assumptions about how science worked best and by certain features of his personality and working style. As discussed in Chap. 16, Ryle was at the time just emerging as the extraordinary scientific leader he was soon to become; one who commanded great loyalty from his own group, and one whose many successes could be partially attributed to qualities that he shared with John Bolton in Sydney: a powerful work ethic, meticulous attention to detail, a very fine instrumentalist and ruthless dedication to his particular research interests. But he could be emotional and defensive, which led him to sometimes over-react to certain incidents, and to shield, rather than share, Cambridge investigations prior to publication. He strongly preferred and trusted Cambridge data, since it came from instruments whose design and function he was intimately acquainted with and invested in; consequently, he failed to fully see, let alone acknowledge, the contributions of others. He was unwilling to ever acknowledge being wrong, when this occurred (see Chaps. 35 and 36). He also clearly underestimated the cost and difficulty of travel for the Australians as he never travelled outside Europe and North America. In Ryle's biographical memoir, Smith (1986) noted: "Ryle was not, however, one of the modern generation of assiduous travellers and conference attenders. He was President of Commission 40 of the International Astronomical Union (I.A.U.) in the 1967 General Assembly at Prague, but subsequently played little part in international scientific unions." This situation was thus very different from the role played by Pawsey and other Australians from a distant land.

Pawsey encountered Ryle's defensive reactivity during his visit in late 1950, this time in relation to B.Y. Mills's early work investigating positions for radio sources. In ESM 18.1, Exchange of Letters, we summarise a precursor interchange in 1946 (during the lead-up to the failed RPL solar eclipse in Brazil of 20 May 1947) about 5 years earlier.[14] While the issue did not seem to raise concerns at the time, the issue arose again in 1950.

On 1 May 1951, Pawsey wrote to Bowen, who was then in the UK:

I should also like to tell you about the funny business with Ryle ... [also mentioned by Mills] ... When Bernard was about to set up his gear at Badgery's Creek, R.W.E. McNicol [an Australian colleague and ionospheric scientist who had been at the Cavendish] came here and told us that Ryle had a new method of recording radio stars which was much more sensitive than previous methods. It involved some sort of a beam-switching technique and discussions led to the conclusion that it involved switching a half wavelength of feed into the lead to one of the aerials [many times] per second. From this information, Bernard evolved his present scheme which has proved very successful.[15] It has turned out to be very similar in principle to Ryle's actual scheme, and I think it is fair to say that the hint given by McNicol was of definite use.

When I was in Cambridge [in 1950], I told Ryle and Ratcliffe of this, with the intention of trying to keep the game as clean as possible, and saying that we would be very pleased to make acknowledgements to Ryle of first using the method. To my surprise, Ryle attacked me for **pirating** [our emphasis].[16] His view was that they had invented a method which was particularly good and they would have liked the opportunity to employ the method themselves. Ratcliffe took a halfway point of view and I said that it had been my intention not to publish for some time and to make acknowledgements of the method when published[17]; I thought this constituted a fair return for the information received. I also told them that I would get in touch with them at about the time we were considering publication. This time has now come, and I have written a letter to Ryle along these lines.

[14]The events of 1950–1951 were described by Frater, Goss and Wendt (2017, p. 27).

[15]In contrast to the hardware switch in the antenna feedlines (Cambridge), Mills's solution was to use an electronic switch in the preamplifier. A common concept and different implementations produced an equivalent result.

[16]As we see in ESM 18.1, Exchange of Letters, the exchange of 1946 had been a mutual sharing of ideas on the topic of lobe-switching. The contretemps of May 1951 did follow the confusion with Bragg, Ratcliffe and Ryle over the visit of John Bolton to the Cavendish in May 1950 (see Robertson 2017, p. 116). In 1950, after an initial frosty reception, Bolton had finally been invited to stay for 3 weeks by Bragg (letter in C 3830, F1/4/BOL/1 from Bolton to Bowen on 18 May 1950). In the end, Bolton spent four days with Ryle, two with Ryle's rival Fred Hoyle and then a day with Tommy Gold.

[17]When Mills published his survey in 1952 (p. 260), he acknowledged that the "method, the principle of which is due to Ryle, has some advantages" over the previously used total power systems. Mills included a reference to the extensive paper of Ryle (25 pages), 1952, "A New Radio Interferometer and its Application to the Observation of Weak Radio Stars" in *Proceedings of the Royal Society*.

... Ryle's viewpoint seemed to be that the particular method was markedly superior to all others and that the advantage which they might hope to have from inventing the method was in having superior observations for a year or so.[18]

The contretemps was rapidly resolved as Pawsey wrote Ryle on the same date as his letter to Bowen (1 May 1951) with a conciliatory message[19]:

... [As I told you in 1950] that we would not wish to publish hurriedly, so that in our publication we would be able to give credit to you for having first used this method [lobe switching]. The time has now come to consider publication as Mills has nearly completed his work. [As shown above, Mills did give full credit to Ryle.]

Ryle wrote a friendly letter in return on 19 June 1951[20] to Pawsey; the ill will of 1950 had dissipated. He provided two references with descriptions of the "beam-switching" scheme.[21]

Pawsey's primary orientation was to the constructive; he prioritised functional and continuing working relationships and achieved the capacity to continue connections with Ryle, much as he had done earlier with David Martyn.

The Cosmic Noise Expedition, New Zealand, and the Identification of Taurus A, Centaurus A and Virgo A

We have already discussed the impact of the observations of scintillations from well separated sites in New Zealand and Australia, but the primary goal of these observations was to improve the positions of the new sources that had been found at Dover Heights.

To measure more accurate positions, the scientists needed higher cliffs which would provide higher fringe rates and reduce the deleterious effect of total power/ gain variations, preferably cliffs with both easterly and westerly aspects. Two ideal sites in New Zealand, Leigh and Piha, were selected[22] and a portable 100 MHz array mounted on an ex-Army gun-laying radar trailer was shipped to New Zealand in June 1948. Positions obtained from these observations had errors of only about 10 arc min. Based on these positions Taurus A was identified with the Crab Nebula, a

---

[18] In this letter, Pawsey ends his message to Bowen on 1 May 1951 with a description of Stanier's solar work at Cambridge using the Fourier synthesis method with no reference to the 1947 paper of McCready, Pawsey and Payne-Scott. He ended the letter: "However, I don't think anything will come of this dogfight." See Chap. 37.

[19] NAA C3830 Z3/1.

[20] NAA C3830 Z3/1. (Pawsey personal correspondence).

[21] Ryle also asked Pawsey about possible 20 cm line work, observing the recently detected (March 1951) hydrogen line at Harvard by Ewen and Purcell and in Holland. He asked if the RPL group would attempt any HI observations. Ryle concluded: "It is a most important advance."

[22] The Dover Heights cliff in Sydney faced to the east at a height of 79 m, about a factor of 3.4 lower than the site at Leigh in New Zealand which faced east at a height of 279 m and the site at Piha in New Zealand which faced west at a height of 265 m.

remarkable supernova remnant, Virgo A was identified with the galaxy M87 and Centaurus A with the galaxy NGC 5128 (Cen A). Bolton et al. (1949, "Positions of Three Discrete Sources of Galactic Radio-Frequency Radiation") is one of the more important publications of post-war radio astronomy. These were all among the brightest sources in the radio sky but the optical counterparts were fainter than many thousand brighter stars and nebula. This radio sky provided a very different view of our universe. The impact of these identifications on the nature of the radio emission is discussed in Chap. 34. Radio astronomy was now becoming part of the "traditional" astronomy community. But as we discuss later, Cambridge distrusted the reality of Bolton et al. (1949) identification of Taurus A with the Crab Nebula.[23]

The Positions of the Brightest Two Radio Sources, Graham Smith—1951: Cygnus A and Cassiopeia A

The position of Cygnus A obtained from the New Zealand observations was still not good enough to make an identification and disagreed with the position measured at Cambridge. Improving positions using a cliff interferometer has intrinsic limitations due to the required refractive corrections, so Mills embarked on a new program to obtain more precise positions using a horizontal Michelson interferometer (Chaps. 21 and 22). Mills obtained a position for Cygnus A with the error reduced to 5 arc min and he wrote to Minkowski[24] on 16 December 1949 suggesting an identification with a nebulous object near this position. To Mills's surprise (he was quite confident that their source was a Galactic nebulosity because of its location so close to the Galactic plane[25]), Minkowski replied (29 December 1949) that he did not "think it was permissible to identify the source with one of the faint nebulae in the area" and emphasised that more accurate measurements were needed. We now discuss how this developed into another major controversy with Cambridge.

F. Graham Smith submitted a publication to *Nature* on 6 August 1951: "An Accurate Determination of the Positions of Four Radio Stars" (Smith 1951). The paper was published on 29 September 1951 after being submitted on 6 August 1951. The derived positions of Cygnus A had errors of only 15 arc sec in right ascension

---

[23] Edge and Mulkay (1976) have discussed the prevalent distrust that the Cambridge group had in the 1950s for the reality of the Taurus A with the Crab Nebula proposed by Bolton et al. in 1949. For example on p. 103 of Edge and Mulkay, Ryle is quoted from April 1951 (Boyd, 1951): "I think the coincidence of one of the radio stars with the Crab Nebula should not necessarily be taken too seriously ... I think that the present evidence cannot be regarded as suggesting a general origin of this type." Elsewhere in Edge and Mulkay (p. 241), an unidentified Cambridge staff member remarked that "it would not have occurred to us to put Bolton's work on the identification of Taurus A as the Crab Nebula as being one of the outstanding early achievements of radio astronomy, and we should have done." See Smith's remarks in the Ryle Biographical Memoir.

[24] NAA A1/3/1a, RPS, 16 December 1949.

[25] https://arxiv.org/pdf/1306.6371.pdf p. 5.

and 1 arc min in declination; the Cassiopeia A errors were comparable. The Australians knew nothing about the Smith publication, and the identification was the same nebula that Mills had proposed in December 1949. Based on Smith's position, Baade and Minkowski confirmed the optical identification of Cygnus A in 1954[26] (Chap. 22). On 19 November 1951, Bowen was at Caltech, writing a letter to Pawsey in Sydney. Baade and Minkowski had told him: "Exactly under our latest Cygnus position [from Smith] there are two nebulae in collision. These are extremely small and it may be only a coincidence that they happen to be there."

In this case Pawsey protested. On 29 August 1951, he wrote a stern letter of complaint to Jack Ratcliffe ("Dear Jack"). The main complaint originated when the RPL group received (posted in the UK on 14 August 1951) a preprint of the *Nature* letter by Smith (1951). The complaint was that Mills had received a letter from Smith inquiring about the positional determination in Australia a year earlier, 30 October 1950[27]:

> ... This is a subject on which Mills ... and Smith ... have been exchanging information for some time ... I am rather staggered that at this stage a communication ... should have been contemplated—let alone dispatched—in view of the fact that private information was being exchanged in, I thought, a spirit of perfectly friendly collaboration. My criticism of the ethics of the matter is based on the fact that no reference whatever is made in Smith's letter to *Nature* of the published work of members of this Laboratory (and it exists), or to unpublished information which Mills has freely given him, and on the fact that no intimation at all was given to Mills that Smith was contemplating publication of his results so soon or in such a form. There is the further point that *Nature* seems to me to be inappropriate at this stage for such an announcement when other accurate determinations have already been published, and this rather suggests that Smith has chosen this rapid but definitely inadequate form in an effort to beat Mills into print.

Pawsey continued in the 29 August 1951 letter to Ratcliffe. He intensified his criticism pointing out that after the 30 October 1950 letter from Smith to Mills:

> The [letter of 30 October 1950] suggested cooperation, asking for accurate positions of sources [quote from Smith letter in 1950: "We would be very pleased if you would give us your position for Cygnus and for any other sources ... for comparison with our results"; he enclosed the Cambridge positions at that date for the three sources. Mills replied (28 November 1950) welcoming the collaboration and sending his accurate position for Cygnus [this was the final position that was published by Mills and Thomas in June 1951] ...
>
> The advantage of exchange of information is, of course, that a suspicion of inaccuracy is raised by a discrepancy and the measurement repeated under better conditions. Confidence is

---

[26]The initial results were presented in 1952 and only published in 1954.

[27]Sullivan (2009, p. 174) has reported in his chapter "Ryle's Group at the Cavendish" that he found an undated letter from Ryle to Pawsey from August 1951; since the letter has not been found in the A1/1/1 Part 6 at the National Archives of Australia, the likelihood is that it was not posted. In the letter, Ryle referred to mistaken coordinates sent to Mills in the previous year. The appreciable error was due to the use of the solar day instead of the correct sidereal day (ratio 1.00274). Thus, there had been no news from Cambridge to Sydney since the previous year; the Australians were never informed of the error. Pawsey "cried foul" in the 29 August 1951 letter (which was posted). Sullivan wrote: "Ryle [was] even more convinced that informal communication of preliminary results [could] only lead to trouble." Apparently, the Australians were never aware of this mistake.

similarly given by agreement. From a comparison of the Cambridge results in 1950 with corresponding Radiophysics ones, it is clear that Smith was in a position to profit by this.[28]... Smith's new result is **nearer the centre of gravity of the Radiophysics ones than his 1950 one** [our emphasis].

Pawsey referred to Cambridge's earlier distrust of Bolton et al. (1949) identification of Taurus A with the Crab Nebula discussed in the previous section. He complained to Ratcliffe that there were no references at all in the Smith paper to earlier work in Sydney, giving the impression that the Smith paper described the first accurate determination of coordinates of radio stars: "This [omission] is particularly noticeable in the case of the Taurus source which is the one tentatively identified with the Crab Nebula, an identification with which I understand your people disagree but which is widely discussed among astronomers."

Pawsey summarised his conclusions in the 29 August 1951 letter to Ratcliffe:

What I believe should have taken place in connection with this collaboration is that both Mills and Smith should have proceeded in the normal manner to publish full papers on their work. Each should have acknowledged the collaboration which took place and checked with the other that the form of acknowledgement was satisfactory. In the course of this, the other person would have learned of the forthcoming publication.

... I fear that unless you can do something to smooth the troubled waters we shall pretty well lose all collaboration between the radio astronomy sections in our two laboratories. There is already too much mistrust of your people in our group. My own private view is that the *Nature* letters were better suppressed, but that is for you to say on the merits of the case. I am hoping that you will be able to improve the present situation.

Pawsey ended the 29 August 1951 letter on a positive note: "Finally, may I add that I had formed a high opinion of Smith and should be ready to believe that he acted in the way I criticise without taking proper thought."

Much later, in March 1953, Pawsey raised the issue of Mills's priority for discovering the Cygnus A position with Baade and Minkowski: "You will remember that when Mills and his collaborators were beginning their work on Cygnus, Mills wrote you (16 December 1949) asking about the nebula which is **now** [our emphasis] identified as the source. He picked it as the only nebulous object in his field of confusion and, consequently, a possibility since the three tentative identifications to that date (the Crab, M87, NGC 5128) were all nebula. You replied that you did not "think it was permissible to identify the source with one of the faint extra-galactic nebulae in the area" and emphasised very rightly that what was wanted was more accurate measurements."[29] He suggested that Baade and Minkowski acknowledge the early contributions from Bolton and Mills (by correspondence) in their

---

[28] An enclosure showed the comparison of the Smith (1950) results for Cygnus A with the final *Nature* published results of 1951 as well as the Mills position of 1950. The Sydney positions had poorer accuracy (1.6 arc min in right ascension and 3 arc min in declination) than the Smith (1951) coordinates (errors in right ascension 11 arc sec and declination 1 arc min). However, the Mills position was close to the correct position. See Sullivan (2009, p. 318 and 337) for a detailed description of the evolution of the position of Cassiopeia A and Cygnus A in the years 1948–1952.

[29] NAA C3830 F1/4/MIL/1 and A1/1/1 Part 8, 1953. The letter was written by Pawsey to Minkowski, but the final letter went out under Bowen's name on 30 March 1953.

forthcoming paper on optical identifications (the 1954 "bible", see Chap. 22). Baade and Minkowski did accept these suggestions. The main text of the paper (p. 209): "When in November, 1951, J.G. Bolton sent a list of certain sources with appreciable angular diameters, a search for these sources was made by Minkowski on 48-inch Schmidt plates. An extended visual object was found only in one of the positions, that of Puppis A." Mills was given credit in footnote no 17.

Pawsey's interpretation of Graham Smith's actions as thoughtless was accurate. Retrospectively, Smith wrote in his Royal Society Biographical Memoir of Ryle (1986, see also Chaps. 35 and 36).

> ... [W]e took very little notice of any publications, either in journals or textbooks, and relied on Ryle's insight. We were indeed guilty of underestimating, for example, Bolton's work on identification of four radio sources, and Pawsey, McCready and Payne-Scott's work on Fourier analysis of the brightness distribution across the sun. But we were in the full flood of discovery, and we were self-propelled.

As was expected, both Sir Lawrence Bragg (Director of the Cavendish Laboratory) and Jack Ratcliffe responded with fulsome letters on 12 and 13 October 1951.[30] Bragg deplored the "bad blood between the two laboratories" and was certain that the misunderstanding was likely due to the vast distances between the two groups.

> I believe I was primarily responsible for the publication of several short notes by Ryle's group. When Ratcliffe was away in America [at the Carnegie Institution of Washington] this summer, I found that Ryle was worried because he had a large backlog of unpublished work, and was at a loss how to get it all written up with a long break in his experiments. I suggested that he and his men should get the gist of it off their chests in short notes, which could be amplified later if necessary ... Ratcliffe endorsed this policy on his return. Several notes were sent to *Nature*.

Ratcliffe's message was similar: "Let me say at once that there was nothing sinister at all about the publication of Smith's letter. It was sent at the direct personal instigation of both Bragg and myself."

Clearly, a contributor to the problem was the slowness of the postal service via sea-mail. The journals arrived with a delay of some months in Australia. Ratcliffe did fault Ryle for not having kept Pawsey and the RPL group informed of their progress. But the basic hurdle was the tyranny of distance from the UK to Australia: "If we could discuss these matters verbally from time to time I don't think misunderstanding would arise. I hope that I may possibly get out to Australia for URSI [in 1952]. If I do I should like to discuss it all with you. In the meantime I will try to see that we send you more frequent information about what we are doing."

Pawsey was not at all satisfied with the response about mutual obligations for collaborations. He tried once again (letter to Ratcliffe, 30 October 1951) to set out the ground rules of collaboration: (1) exchange technical information, (2) publication

---

[30]NAA C3830 Z3/1 (all Pawsey correspondence with Ratcliffe and Bragg in 1951). Ratcliffe had already prepared a short, handwritten letter on 6 September 1951 informing Pawsey that the Cavendish Laboratory was closed for a summer break. He promised that he would look into the matters raised in the 29 August 1951 letter from Pawsey in the next few weeks.

of relevant material should only occur with mutual consent, and (3) refer to published work (not controversial) as well as to unpublished work insuring that the other party "should feel that he has been treated properly". Pawsey then attempted to apply these three criteria to previous cases of collaboration between the two groups.

But Pawsey, in a letter to Ratcliffe, also accepted:

> The basic difficulty, as you say, is that we are 18,000 miles away and cannot talk things over. I think that for the immediate future we would be well advised to let things go on without any attempt to produce a closer contact in the field of radio astronomy. Let us clarify our own ideas as to mutual obligations and, when the need for closer collaboration arises, make the move then.

To Bragg (30 October 1951 from Pawsey, C3830 Z3/1) the message was similar with a few additional ideas:

> The real difficulty lies in collaboration without clearcut arrangements as to what is expected from each party and no opportunity for personal discussions. I think that, in view of the current disagreement, it is advisable that the two Laboratories should not attempt any further collaboration in the field of radio astronomy until Ratcliffe and I decide on fairly explicit rules for the game. I have written him at some length and with any sort of luck may see him in Sydney next year at URSI.[31]

On 23 November and 7 December 1951, Ratcliffe and Bragg wrote to Pawsey, bringing this lengthy series of letters to an end. They both agreed that a personal meeting was required to sort out the complex rules of engagement. Ratcliffe wrote: ". . . I agree with your suggestion that we should let things go on without any attempt to produce a closer contact in the field of radio astronomy."[32] Bragg extended an olive branch: "[I think our] people slipped up on this end in not sending you a bunch of information when it was first available, but I am sure it was an oversight. They are very good people, and I am sure they wanted to make the collaboration a success."

The conflicts between Sydney and Cambridge were fundamentally located in the *structural disadvantage* created by not only distance alone, but the ongoing legacies of colonialism—the absence of the long standing embedded networks, collegial resources and social status (of various kinds, old school tie among them), that Cambridge could so effortlessly command. And like other forms of structural

---

[31] In August 1952, Ratcliffe did in fact visit New Zealand and Australia, attending URSI from 11 to 21 August 1952 at the University of Sydney. (see Chap. 21). In spite of the fact that Pawsey was ill during the latter part of URSI, he and Ratcliffe seem to have had time for frank discussions just before Ratcliffe left on 28 August. He wrote a letter on the BOAC—British Overseas Aircraft Corporation—from Sydney to Singapore: "I much appreciated our frank talk [on 27 August 1952] I hope you will consider the possibility of the visit [presumably to the Cavendish] I mentioned, it would be most valuable I believe." Unfortunately, no record of the details of the "frank" discussion have been located. Perhaps, the visit to Cambridge would be a further opportunity to improve communications between these two major radio astronomy groups.

[32] Ratcliffe told Pawsey that he was to give a lecture on 24 November 1951 at the Royal Institution, "Friday Night Discourse" on radio astronomy. He would discuss the work of Pawsey, Ryle and Lovell.

disadvantage, this was obviously visible to the Australians, who suffered from the consequences, and invisible to those in Cambridge, who could afford to allow themselves to be swept up in their discoveries and to lack awareness of the work of others. In fact, those in Cambridge definitely saw themselves as being at a disadvantage—the group was smaller and less well resourced than that in Sydney in the first 5 post-war years, and their contributions were more limited. In 1950 Sydney had published 33 papers, Manchester 29 and Cambridge only 11. But they would not have even been aware of just what an advantage being immersed in the rich culture of the Cavendish Laboratory in Cambridge was.

By comparison, the relations between Sydney and Manchester were excellent despite the distance and there was no strong rivalry or competitiveness. So not only distance, but the well-known insularity of Cambridge, constrained cooperation and collaboration between the two major groups.

In summary, structural disadvantage (Kanbur & Venables, 2005; MacLeod, 1980) formed the justification for the procedural approach to Pawsey suggested as a means of generating both fairness and clarity—and the ease with which these attempts could be ignored in Cambridge.

Unsurprisingly, "scintillations" in this relationship were to continue.

Part VI
Quiet Leadership

Chapter 19
Consolidation: Leadership at RPL, 1950–1951

In many cases we observe, by radio, things which are invisible optically so that we obtain
new clues to the nature of the universe . . . [I remind you] that in the detective story, which is
science, clues add up non-linearly: two and two can add up to more than four. Pawsey,
2 May 1957, "Vistas in Astronomy", the first Matthew Flinders Lecture.

Introduction

The 1950s was a decade of wonderful flowering and exploration for the CSIRO
Division of Radiophysics. From 1950, solar noise and cosmic noise scientists
became "radio astronomers". They were present as such in international astronomy
as well as radio science meetings. The key achievements of the first 5 years of
observing and categorising solar bursts and detecting and measuring discrete radio
sources, had produced a set of challenging new research questions. Some were
conceptual and theoretical—including, the nature of the unknown non-thermal
radiation mechanism that underpinned cosmic radiation emission. Some concerned
the complexities and challenges of instrumentation—the early 1950s saw swift
innovation of instruments and methods, in order to resolve new research questions.

Sydney now had two firmly established research programs. The program of
detecting and measuring the positions of discrete sources of radio emission as a
prelude to identifying them with (by preference) optically-known astronomical
entities, led by John Bolton and also, from late 1949, by Bernie Mills, came to
dominate global scientific as well as public imaginations. But in the early 1950s,
Pawsey especially developed the rich set of research programs in solar radio physics
beginning at Collaroy and moving to Dover Heights and then to a series of other field
stations.

As the group's leader, Pawsey was conscious of what was needed to simultaneously grow this new field of radio astronomy around the world, while maintaining
Sydney's leading position within it. Internally, his leadership style continued to

© The Author(s) 2023
W. M. Goss et al., *Joe Pawsey and the Founding of Australian Radio Astronomy*,
Historical & Cultural Astronomy, https://doi.org/10.1007/978-3-031-07916-0_19

emphasise giving his key researchers immense latitude to independently develop their own lines of research. Outside that, Pawsey focused his attention on a set of needs that remained throughout the 1950s: the need for good theorists as part of the Sydney group; the need for good connections with optical astronomers, and especially the importance of attracting to Australia an experienced optical astronomer with interest in collaboration; the need to make Sydney's achievements visible and important in global scientific circles. He was also acutely aware of the need to establish a pipeline of students to provide a steady source of new personnel for new research programmes in radio astronomy as they developed. Pawsey was very conscious that the successes of radio astronomy itself were possible only because the investment in "basic" ionospheric and radio research under the Radio Research Board through the 1930s, had produced the technically experienced and scientifically trained human resources needed for wartime radar research and now radio astronomy to occur.[1] This chapter discusses one attempt to address the pipeline issue.

The difficult relations with Cambridge, discussed in the previous chapter, continued through 1950 and 1951. But in these years, Pawsey and Bowen maintained RPL's wide set of international connections through attending international meetings. On the international stage, the transition to "radio astronomy" meant increased participation in the International Astronomical Union in addition to their initial more natural scientific home based on the technology, URSI (International Union of Radio Science) (See NRAO ONLINE.51). Not surprisingly, in the early 1950s, the Australians gave special attention to participation in the URSI General Assemblies of 1950 (Zurich), 1952 (a high point, since the General Assembly was held in Sydney[2]), 1954 (The Hague) and 1957 (Boulder, see Chap. 28). In 1955 an important meeting of radio astronomers occurred at Jodrell Bank 25–27 August (see Chap. 26), preceding the Dublin IAU GA from 29 August to 5 September 1955.[3] A number of Australians were present at these two conferences in 1955, including Pawsey, Wild and Bolton. Pawsey played an increasing role in both the leadership of the radio astronomy commission in URSI (later Commission J) and Commission 40 (Radio Astronomy) in the IAU; he was the chair of Commission 40 for two consecutive periods from 1952–1955 to 1955–1958.

---

[1] Transcript, NRAO ONLINE.54.

[2] See Chap. 21 and Goss and McGee (2009, *Under the Radar*, Chap. 10) for more details.

[3] An earlier Jodrell Bank symposium on radio astronomy had occurred in 1953 (13 to 15 July), surprisingly with no participants from Australia. (*Observatory*, vol 73, October 1953, p. 185). See ESM_26.3, A Symposium on Radio Astronomy.

To Europe

In 1950, Pawsey travelled overseas to Europe for his second visit as a radio astronomer (see Chap. 24). For the first time since 1941, he travelled by air leaving on 16 August, arriving in London 20 August 1950. He stayed with his sister-in-law Bessie Whittard near London. During the following 10 days, Pawsey visited Stanley Hey at the Army Operational Unit at Byfleet (Surrey) to catch up on the latest radio astronomy activities of this group. He was somewhat disappointed: "[Their] experimental work is very slight [since my last visit in 1948]." At a visit to TRE (Telecommunications Radio Establishment) he showed a film[4] about post-war activities at Radiophysics. On 30 August 1950, Pawsey travelled to Birmingham for a meeting of the British Association for the Advancement of Science. He only summarised one presentation he attended in the conference: the "Evening Discourse" on 4 September during which Lovell praised the work of the Sydney radio astronomy group as he gave an overview of the activities at Jodrell Bank.

A Textbook for Radio Astronomy

Such had been the speed of growth of radio astronomy that within a few years, the idea of an advanced textbook for the field had become attractive. By 1950, Pawsey was searching for a publisher. While he was in the UK, he visited at least two publishers in person, Chapman and Hall, and Oxford University Press (Clarendon Press). Pawsey wrote Bowen on 8 September 1950 from London with a report of the reaction of the two publishers. Chapman and Hall already had a series of books under the general editorship of Lovell with the first book to be on radio astronomy by Lovell himself; this book (*Radio Astronomy* in the Frontiers of Science series) was subsequently published in 1952 by Chapman and Hall with second author J.A. Clegg.[5] Thus Oxford University Press remained a choice. Here there was a complication since Appleton, the editor of the series (*International Monographs on Radio*), had assumed that "Stanley Hey might write a book for an Oxford series on . . . radio astronomy and they could make no decision until they had discussed this with [the editor]."[6]

---

[4]The 1949 film https://www.youtube.com/watch?v=BKxMXPFX5RU.

[5]In 1957 a new book was published by Chapman and Hall by Hanbury Brown and Lovell, *The Exploration of Space by Radio,* subsequently published in the US in 1958 by John Wiley.

[6]Cambridge University Press was also mentioned by Pawsey; it is not clear if they were approached. There was a major dilemma as Pawsey wrote Bowen on 8 September 1950: ". . . it is quite clear that this thing cannot be done without telling people what you are about because the respective publishers immediately refer the matter to technical people such as Appleton, Lovell and Ratcliffe."

Fortunately these discussions could happen straight away—at the 1950 URSI meeting in Zurich.

URSI 1950, Zurich: 9 to 22 September[7]

Pawsey, graduate student Jim Roberts, Bolton and Westfold from RPL in Sydney attended the URSI General Assembly of 9 to 22 September 1950 in Zurich. Roberts deserves a brief introduction, partly due to his role in the events of the 1950s, and also because his unpublished memoir,[8] now available online, informed this book. Roberts had worked in CSIR-RPL during his vacations in 1945 and 1946. He was awarded a CSIR overseas fellowship and had intended to work with Ryle, but by the time Ratcliffe had replied that the Cavendish would accept him, Roberts had already decided to work with Hoyle, with neither CSIR nor Pawsey involved in the decision. (Roberts had the impression that Australians were not welcome in Ryle's group). When Roberts completed his fellowship, he returned to RPL in 1952, only to be placed in the cloud physics group; he then transferred to Wild's solar group at Dapto. In the mid-1950s Pawsey asked Roberts to write a review paper on the entire field of radio astronomy. Also in this period, he and Bracewell wrote one of the key papers on aperture synthesis. Later in the 1950s, Roberts worked with Bolton's group at Caltech and then returned to work at Parkes (Chaps. 30–32 and NRAO ONLINE.4 9).

The main representative of CSIRO at URSI 1950 was David Martyn (CSIRO Mt. Stromlo), who led the discussions on the ionosphere.[9] Martyn's expertise was recognised at this meeting by being made President of the new Commission V, Radio Astronomy.

Roberts wrote,[10] "On the Saturday afternoon the shy young student [Roberts's description of himself] went for a walk with Pawsey, and on the Tuesday there was a conference tour, the Three Pass Tour." Later, Roberts wrote Pawsey (9 October 1950) a letter of thanks: "Once again I must thank you for getting me to URSI and your kindness to me there. I'm sure the contacts of the conference are going to [be] extremely valuable."

---

[7]NAA C3830 Z3/1, NAA C3830 F1/4/Paw/2, AH 8520 PH/Paw/1B/Part1 provide source material for this chapter.

[8]Roberts (2002) unpublished memoir *Have Gen, will Travel*. This text is presented in NRAO ONLINE.49.

[9]Bowen was unable to attend but earlier, on 2 May 1950, he had proposed to the CSIRO Executive that Pawsey represent the Australian radio astronomers. Roberts wrote in his report to CSIRO and Pawsey: "My notes list Pawsey, Martyn, Massey, Herlofson, Twiss, Macfarlane, Rydbeck and Alfvén as present at the conference, and there is a note about Massey's discussion group."

[10]In his 2002 unpublished memoir *Have Gen, will Travel*, Roberts described his experiences at URSI in 1950, p. 59. NRAO ONLINE.49.

As was customary, Pawsey wrote Bowen on 28 September 1950 a report of the URSI conference (with copy to Fred White, CEO of CSIRO)[11]:

> The URSI Conference was very interesting. You may be interested in certain political aspects first. Dr. Martyn has invited URSI to hold its next Conference in Australia in 1952, and the offer has been accepted. We sincerely hope that a reasonable number of people may be able to attend. Martyn has also been elected one of the three Vice Presidents of URSI ...
>
> Martyn gave an account of his theories of magnetic storms which was also of great interest and debated widely. He and Alfvén each have theories and I think the points that are common are probably correct ... Alfvén's theory is that aurorae are manifestations of electric discharges; Martyn's is that they are actually due to high velocity particles generated in the vicinity of the earth from the streams of particles coming from the sun. [Martyn's theory is the currently accepted theory of aurorae.] ...
>
> In the field of radio astronomy, meteors were included within the Commission 5 [of URSI]. Lovell gave a good account of the English work [on meteors], and there was a lot of correspondingly interesting American work also. There were no English representatives of the workers on solar noise [e.g. from Cambridge], but only those from France, and Cornell [Burrows's group] besides ourselves.
>
> As a result, I think that our contributions dominated this part of the Conference, but it is unfortunate that the conference was not fully representative. A very great deal of interest in our work was expressed by a large number of people present at Zurich.

Pawsey wrote afterwards to Appleton that "URSI must be one of the best of the international unions."[12]

Pawsey did not report an unfortunate aspect of the final day of the conference— that due to their imminent departure for Copenhagen, Bolton and Westfold were only able to present their Dover Heights galactic data and not, as had been promised, new results from Wild (swept frequency data of Type I, II and III bursts from Penrith) and Payne-Scott (Hornsby data on time delays for Type III bursts). Thus, the new exciting solar noise observations were not described at the 1950 URSI.[13]

After the URSI GA, Pawsey went to Paris for a 2 day visit with Marius Laffineur at the Institutut d' Astrophysique and Denisse at the l'Echole Normale Superieure. He then returned to the UK, spending 3 to 4 days visiting Jack Ratcliffe and colleagues at the Cavendish Laboratory and giving a brief invited presentation to

---

[11] From the ASLO Office in London—an example of how useful the Office was.

[12] NAA C3830 Z3/2/I. 6 October 1950. Pawsey wrote: "I enjoyed the URSI Conference very much though life was a little too hectic at times. There were some very stimulating discussions. I feel that URSI must be one of the best of the international unions. One thing I regretted was that in Commission V [radio astronomy] we did not have any representative from Ryle's or Hey's groups so that the English work did not get as much mention as I think it deserved. [Only the Jodrell Bank work was presented.] I hope that next time it may be possible to have someone from these groups— please do not misunderstand this as criticism of Lovell [Jodrell Bank] who did an excellent job." Appleton responded to Pawsey on 15 December 1950 (in the letter discussing Pawsey's newly approved book project with Oxford University Press on radio astronomy): "I agree with you about the absence of representatives of Ryle's and Hey's groups, but the Royal Society funds are limited and we try to give everybody a look-in."

[13] Goss and McGee (2009, appendix D "Ryle, Payne-Scott, Bracewell and Bolton: 'Solar Bursts from Aircraft'").

the RAS (Royal Astronomical Society) on Friday 13 October 1950, returning to Sydney on 18 October 1950.[14]

At the Royal Astronomical Society meeting (RAS), Pawsey outlined the research at the RPL, highlighting work in the fields of radar, the physics of rain, "computing machinery" and radio astronomy. Of a scientific staff of about 40, 14 were engaged in radio astronomy: Payne-Scott, Mills, Wild, Bolton, Christiansen, Westfold, Slee, Stanley, Little, Smerd, Piddington, Minnett, Shain and Kerr. Pawsey dwelt on the latest radio astronomy developments in a summary of recent research. What had started as a series of radio discoveries was now coalescing into a coordinated research program in solar and cosmic astronomy:

> The next phase of radio astronomy consists of the utilisation of radio observations in planned attempts to gather information concerning the physical world about us. One example of this is Bolton's study of the shape of the Galaxy from observations of the direction of arrival of cosmic radio waves. A second example is the measurement of electron density and temperature in the solar atmosphere from thermal radio noise.[15] Finally, we have the study of the relation between solar flares, radio "outbursts" [the Type II events] and magnetic-storm particles. Continuous records of noise intensity on differing wave-lengths, recorded as a function of time, show that outbursts have a sharp cut-off on the low frequency side of the spectrum [Paul Wild's ground-breaking research at Penrith] ... The disturbance itself is postulated as the ejection of magnetic-storm particles. From the shape of the drift, the time-height graph can be constructed, and the velocity of the particles estimated at between 300 and 600 km/s. Direct confirmation of this outward movement has been obtained from directional observations employing a new interference technique [Payne-Scott at Potts Hill with the swept lobe Michelson interferometer at 100 MHz]. For example, the outburst associated with the solar flare of 1950 July 17 showed an outward velocity of the same order. **Evidence is accumulating to show that we are "seeing", using radio waves, the genesis of terrestrial magnetic storms** [our emphasis].

The Textbook Is Contracted

Pawsey wrote Appleton on 6 October 1950, shortly before his departure for Australia on 14 October 1950. He referred to the discussions the two had held concerning a potential textbook during the URSI conference in Zurich. Their talk had evidently been constructive. Pawsey said he "welcomed your encouragement very much and I propose to go ahead and try to write one. I have written to Mr Wood of the Oxford University Press asking if the Press would consider publication of such a book."

This letter to Wood was sent the same day, 6 October 1950. Pawsey laid out the plan for the book:

---

[14] The results of Pawsey's short presentation which followed the George Darwin Lecture ("The Electrical Photometry of Stars and Nebulae") presented by Prof Joel Stebbins of the University of Wisconsin, pioneer in photoelectric photometry, are summarised in *The Observatory*, vol. 70, page 203 (December 1950).

[15] See NRAO ONLINE.20.

I am writing to ask if Oxford University Press would consider publication of a book on Radio Astronomy to be written by myself either as sole author or in collaboration with one of my colleagues in Sydney.

As you may know, a wide variety of investigations in this field has been carried out under my general direction by various members of the Radiophysics Laboratory, Sydney, and if it would be relevant to your consideration of this proposal I could send you a list of our published papers on this subject.

[Pawsey sent an outline of the seven proposed chapters; the published book of 1955 consisted of 12 chapters.] ... [T]he proposal is for a book covering the application of radio techniques to astronomical investigations, including meteors, the moon, sun and galaxy. It is intended to be written for those interested in developments in new branches of physics but with astronomers and radio-physicists in mind as more specialised classes of readers. I should aim at a length of about 300 pages of moderate size ... [The final book was to be 356 pages.]

I think this book could be written in about a year.[16]

Shortly after Pawsey's return to Australia in late October 1950, he heard from A.M. Wood of the Clarendon Press at Oxford:

The Delegates of the Press have now considered your proposal for a book on Radio Astronomy. You will, I am sure, be pleased to know that they have agreed to encourage the project. I can, therefore, propose terms ... If you [would] like to submit a chapter in draft we could perhaps help with comments on it at that stage so that systematic imperfections of presentation could be removed. You may be sure we will help in any way we can.

On 15 December 1950, Appleton, the editor of the series and the Vice-Chancellor of the University of Edinburgh, wrote a glowing letter of congratulations. "I am really delighted about the prospect of this, for you write easily for the reader whatever the cost in blood, tears and sweat."

It would, however, take until 1955 for the book to be published, as we discuss in Chap. 24.

Resourcing Astronomy in Australia, 1951–1952

Pawsey returned to the increasingly complex task of leading and managing the growing suite of research projects at RPL. He could see that RPL lacked the particular advantages that Cambridge enjoyed—and that radio astronomy groups in the USA would enjoy as they grew: namely, a pipeline of new students to train in radio astronomy; contact with colleagues from other areas within physics and mathematics and other fields of science, to stimulate new thinking; and deeper connections with astronomy itself. Pawsey was not only conscious of the importance of these resources but also generous and constructive in his outlook. For this reason he differed from Bowen and White, in proposals for more connections with the Commonwealth Observatory at Mt. Stromlo in 1951.

Robertson (1992) has provided some background:

---

[16] Pawsey was guilty of misplaced optimism, a common reaction experienced by numerous authors.

Despite some tension this early collaboration [since 1946] with the Commonwealth Observatory undoubtedly benefited the Radiophysics group. The radio scientists, turned radio astronomers, came into contact with Australia's leading astronomers at the time when the Sydney group was only learning the basics of the science. Clay Allen [sic, Clabon 'Cla' Allen] in particular provided a steady flow of information to RP on solar phenomena and astronomical subjects. The association between Mt. Stromlo and RP was the first major collaboration between optical and radio astronomers anywhere in the world.[17]

Cla Allen had played a major role in the Stromlo/RPL connection since he began advising Pawsey and Payne-Scott in 1945–1946 on solar physics (see Chap. 11). Paul Wild noted, "At this time nobody knew anything about astronomy. We spoke a different language to the astronomers. We owed a lot to C.W. Allen for gently guiding us into the ways of conventional astronomy".[18]

But by 1951, Allen was planning to leave Australia to take over the astronomy department at the University of London. He departed Australia in October 1951.

The year before, Mark Oliphant had returned to Australia to take a position as Director of the School of Physical Sciences at the Australian National University (ANU), located in the nation's capital city, Canberra. ANU, one of Australia's leading research Universities, was an insignia of post-war optimism and Australian pride and independence. Established in the late 1940s with direct involvement from the Prime Minister and world leading researchers such as penicillin discoverer (Sir) Howard Florey (Nobel Prize 1945 Physiology of Medicine), ANU was envisaged as a new University appropriate to a nation that was now equal to or surpassing the former "Mother country" in economy and ideas; a nation in which bright students would no longer need to return "Home" for postgraduate study with the world's leading scientists. Mark Oliphant, who had visited RPL during the war (Chap. 9 and ESM_9.6, Microwave Radar), was an eager participant in this endeavour.

Given Oliphant's deep involvement in UK radar research and development during the war, and his keen eye for exciting new developments in physics, it is perhaps not surprising that he thought of developing radio astronomy at his new national University. It is also a commonplace phenomenon that managers, concerned with their own status and impact and carried away with their own visions, forget to consult with the stakeholders involved—and Oliphant seems to have omitted consulting with the Australian radio astronomers. Thus it came as a surprise at RPL when, apparently under pressure from his Board of Visitors, Richard Woolley, Commonwealth Astronomer, wrote to Fred White that a decision recommended a year earlier was "reaffirmed": to start "radio-astronomy activity" at Stromlo. Woolley reported that Oliphant had "offered to assist the Observatory in setting up a radio telescope with his [ANU] electronics resources". He asked for the collaboration of RPL and hoped to "have the benefit of discussion with Pawsey and others of the RPL staff". The plan was to "investigate the galactic structure of radio noise, since the

---

[17] This episode is recounted by Robertson on (1992, pp. 108–111), and explores issues relevant to the building of the GRT (see subsequent chapters). Here we add Pawsey's role to this history.

[18] 15 October 1965 lecture by Paul Wild on the origin and growth of radio-astronomy in CSIRO (NAA C3830 D5/4/X58 973842).

primary interest of this observatory [is] in the structure of the galaxy". (Ironically, the HI 21 cm line would be discovered at Harvard a month later; this would lead to the most important advances in knowledge of galactic structure in the 1950s.) An open skies policy was anticipated: "If the facilities here are of any use to RPL, we shall of course be only too pleased to offer them access." The Observatory would be required to appoint someone on the same level as Allen to be in charge of the new department of radio astronomy.

As expected, White was not pleased. Eleven days later he responded to Woolley (23 February 1951), beginning in a conciliatory fashion:

> Thank you very much for sending me your letter of 12 February in which you refer to the advice given by the Board of Visitors on radio astronomy and of the interest of the National University [ANU] in the possibility of your Observatory entering this field. It was very good of you to give us this early information about your intentions, although I note that representations have already been made to the Department of the Interior (the parent body of the Commonwealth Observatory).
>
> You can rest assured that there will be no disagreement on our part about the Observatory interesting itself in radio astronomy. I know too that RP will look upon the extension of our work in this field as providing an opportunity for closer cooperation with you. There are clearly problems in radio astronomy which need close association between radio and visual observations, and I can well understand that the information resulting from the work done by the radio physicists is of tremendous importance to astronomers.
>
> I may have misinterpreted the brief statement that you propose to erect a radio telescope of large resolving power because of your interest in investigating the galactic structure of radio noise. This suggests to me that you might be thinking of embarking on a project which RP has been actively pursuing for some considerable time.
>
> In 1948 RP put up to the Executive [of CSIR] proposals for the consideration of a large aerial system for solar and cosmic noise. The Executive agree that this was obviously the next step in the development of radio astronomy at the RP Laboratory ... You will appreciate that, since a very large aerial system of this sort has never previously been designed and constructed, many novel experimental ideas will have to be incorporated and a great deal of originality will be required if the larger system, which we hope ultimately to construct, is to be successful ...
>
> I think RP has established, as far as Australia is concerned at least, a degree of priority for this particular project, and you will agree, I hope, that it would not be in the best interest of collaboration for the Commonwealth Observatory to enter precisely the same field.

White was worried that the Australian government would not look favourably on additional funding for radio astronomy, as he explained in a letter to David Martyn. Martyn had written to assure White that he was not himself the instigator of the proposal.[19] White, reassuring Martyn, explained his main concerns: "[a]s you perhaps know, [a large radio telescope] is ... the next big step to be made and RP

---

[19] Martyn wrote that the Board of Visitors (Vonweller, Madsen, Hartug, O'Connell, Bullen and Kerr Grant and now Oliphant) had "forced" Woolley into this new direction. Martyn had been asked for his opinion and insisted that "... [A]lthough the Observatory might think it could go ahead without affecting RP, when the positions are advertised the only likely candidates would be from RP" due to the lack of suitable expertise in Australia. Martyn had only told Woolley that there should be joint consultation. White replied that he knew about the pressure Woolley faced with the demands from his Board of Visitors.

has planned it for some time now ... [T]he [CSIRO] Executive has, in fact, given approval in principle to this work at ... RPL. I think they have established a certain degree of priority to this phase of the programme and I would hate to see Woolley embark on it in parallel."

Shortly after, Bowen left on a long overseas trip that would occupy most of 1951. Thus, the direct negotiations with the Commonwealth Observatory were handled by White and Pawsey, who was Acting Chief of Radiophysics in Bowen's absence in 1951. Bowen was kept informed with frequent letters from both.

On 22 March 1951, Mark Oliphant weighed in with a letter to Mr. W.A. McLaren, the Secretary of the [Federal] Department of the Interior, the managing agency of the Commonwealth Observatory:

> Professor Woolley tells me that you are troubled by the possible duplication of effort if the Commonwealth Observatory undertakes a program of work in the radioastronomy of the galaxy, since [RPL of the CSIRO] is already engaged upon an ambitious program in this field. I hope that these doubts will be dispelled and that work in radioastronomy will be established at Mt. Stromlo ...
>
> It is right and proper that [RPL] should enjoy the fruits of its pioneering by carrying out its observations further... It is essential that radio-observations be correlated closely with optical observations. They are complementary methods of obtaining information about the heavenly bodies and an observatory is incomplete without either ...
>
> [It is possible] that the interest of the Division in radioastronomy will fade rapidly if some newer and more existing application of radio should be discovered or if the national situation demands that it revert to the development of radar and communications for the military. Astronomical observations, to be of value, must be continued for long periods of time in any one part of the sky, while the complete exploration of the heavens is too great a task for any one set of observers.

Oliphant also made a surprising claim: If RPL was to make any progress in the "advance of astronomy", it was necessary to have available in Sydney powerful optical telescopes, "equivalent to that available on Mount Stromlo, and for a very limited programme" (his emphasis). This was hardly relevant, due to the poor observing conditions in Sydney!

On 29 March 1951, Pawsey played his first active role in the Stromlo-RPL discussions,[20] when he made a visit to Canberra. Pawsey sent an extensive letter to Bowen in London on the following day summarising his activities. He had attended a discussion of the Stromlo plan for radio astronomy with White, Woolley and Arthur R. Hogg of the Observatory. White was despondent; Woolley was insisting that the project move ahead. It was clear to Pawsey that Oliphant was a driving force in the push for radio astronomy since "[he] has a fairly strong finger in the pie ..." but was not available personally since he was in Pakistan.

Pawsey was much more prepared than White to see constructive possibilities in the suggestion. He raised an issue that he had discussed for some years: optical observations of the sun and coordination with the solar programme of Paul Wild at Dapto. Pawsey to Bowen, 30 March 1951:

---

[20]NAA C3830 Z1/9/1951 Part 1.

Woolley's point of view is that he wishes to concentrate on the things which are peculiar to the southern hemisphere,[21] in optical astronomy on the parts of the heavens south of declination −30°, and likewise in the [solar] radio field it is clear that his viewpoint is that it is unlikely that sufficiently good observations could be carried out to make a marked advance. It is very hard to get round this because, if he has no faith, he will have no success. This left the matter rather at a stalemate with White insisting that the setting up of a separate organisation did not seem to be a good thing.[22]

The next suggestion was unexpected. Pawsey suggested cooperation by Stromlo with some of the RPL galactic experiments. "Broadly, we think in terms of sending up one group to carry out experiments at Stromlo, but after your return." He concluded: "My main reaction to the whole scheme is that there are distinct possibilities of our achieving better collaboration with astronomers through such an arrangement and I am prepared to go ahead with it."[23]

The official summary[24] of the meeting was produced with great care; this was to be sent from White to McLaren on 10 April 1951. White wrote a draft on 3 April which was heavily amended by Pawsey and sent back to White on 6 April 1951. Pawsey suggested that there were two complementary lines of research, optical and radio, and "the best possible arrangement would be for the specialists in each line to stick to their own field but co-operate."

White suggested that the planned collaboration would have two aspects: (1) The groups should try to avoid duplication of efforts in radio astronomy since this would lead to "competition between the Observatory and Radiophysics for skilled staff ... I understand that Dr Woolley does not visualise that the Observatory should duplicate and compete with the work of our Division. He is ... interested in a continuous programme of continuous observations of a more Observatory character ..." (2) RPL was keen to have a close collaboration since "... [I]nterpretation of the results obtained by radio methods requires close comparison with those obtained by optical methods."[25] Although White knew Woolley had no enthusiasm for optical solar physics, he did mention again that this was a possible area of collaboration. A more likely area of collaboration was to be in the field of galactic astronomy. In fact, the groups had already started discussions in this area.

---

[21] In optical astronomy, where Australia was only a small part of the global effort, this plan was a sensible strategy. In radio astronomy, where Australia was a dominant force, no niche advantage occurred from the location in the Southern hemisphere.

[22] *Ibid.*

[23] The final sentence of Pawsey's letter indicates that White, Pawsey, Woolley and Hogg spent some time trying to find a role for D.F. Martyn in the new arrangement; there was no conclusion.

[24] CSIRO A9588 KE/12/11.

[25] The final letter of agreement sent to McLaren was titled: "Research in Radio Astronomy". White expressed his disagreement with Oliphant's claim that RPL would need to operate an optical observatory: "... [W]e would regard our entry into the optical side of the investigations as a gross overlap of the work done at Mt. Stromlo." Also, the issue of a national emergency causing RPL to alter their programme of research was irrelevant since "the same national emergency would have a considerable effect on the work being done at Mt. Stromlo, as was the case during the last war."

After Pawsey's letter of 30 March 1951 to Bowen describing the negotiations of 29 March, Bowen responded from London on 17 April to White with a very critical assessment. His arguments were[26]:

> In a sparsely populated country like Australia the only way to make real progress in research is for one compact group to seize on a problem and go for it as hard as they can. This is how we have succeeded in Radio Astronomy and are beginning to succeed in Rain Physics ...
>
> I would be happier about the whole thing if it did not savour so much of jumping on the band waggon [sic]. When small boys do this it does not matter very much, but when grown men indulge in it they ought to haul off and spend some time taking a very critical look at themselves.

We have interpreted the different position taken by Pawsey in comparison to Bowen and White as relating to different priorities as well as to different styles. Pawsey was personally inclined to see more opportunities and less threat in Oliphant's suggestion, and we know that he had had a particularly high regard for Oliphant for many years. But in addition to these views, Pawsey had a particular reason to look for closer connections with the Universities: access to students. The need to build a pool of quality future potential radio astronomers would grow steadily greater.

Addendum: Long Visions

The negotiations between CSIRO and the Commonwealth Observatory came to an abrupt end on 7 May 1952, in a meeting held at RPL in Sydney of the CSIRO Executive and the Board of Visitors of the Commonwealth Observatory,[27] when Woolley reverted to his lack of enthusiasm for radio astronomy (he was focused on preparations for the 74-inch telescope, which opened November 1955). But, as Robertson (1992, p. 112) notes, Oliphant's lobbying for a National Facility was an idea far ahead of its time. It was not until the construction of the Australia Telescope Compact array in 1988 that CSIRO agreed to manage all of their radio telescopes as a National Facility (ATNF) under Australian Government guidelines.[28]

Pawsey's 1951 vision of a collaboration between MSO and RPL was not realised until a decade later. Collaboration between RPL and Mount Stromlo was renewed when Bart Bok was appointed as the new director of the Mt. Stromlo Observatory. Bart and Priscilla Bok arrived in Australia in March 1957 and remained until March 1966, establishing close contacts with many colleagues at RPL, especially the Pawsey family (see Chaps. 26, 27, 29 and 40).

---

[26]NAA C3830 Z1/9/1951 Part 1.

[27]Sullivan archive. In attendance from CSIRO, Bowen, Pawsey, White, George Briggs (Chief of the CSIRO Div of Physics, 1945–1958) and Ron G. Giovanelli (Solar physicist, later the successor to Briggs from 1958 to 1976).

[28]*Guidelines for the Operation of National Research Facilities, A report to the Prime Minister by the Australian Science and Technology Council (ASTEC)*, Australian Government Publishing Service, Canberra 1984.

In 1961, a new agreement between ANU and CSIRO made it possible for ANU PhD students to do radio astronomy projects supervised by CSIRO RPL staff. One of the first students was Marcus "Marc" Price, who came to Australia as a Fulbright scholar in August 1961 to work with Pawsey.[29] He became one of the first ANU students to start a PhD in radio astronomy. One of the co-authors (R.D. Ekers) was the third graduate student at ANU in radio astronomy, 1963–1967, with John Bolton his RPL supervisor.

The impact of Pawsey's vision is well summarised by Hyland and Faulkner (1989):

> Tragically, the personal interaction between Bok and Pawsey was brought to a halt by Pawsey's untimely death in 1962, but it was largely as a result of their shared vision that the idea of radio and optical astronomers working in close collaboration became well entrenched in the ethos of Australian astronomy.

---

[29] Price (2011) in "Science with Parkes @ 50 years". After Pawsey's illness in 1962, Price was supervised by John Bolton.

Chapter 20
Finite Resources: Pawsey and the HI-Line, 1948–1960

The utility of such a line, should it be found, is obvious. It opens up the possibility of determination of constitution of matter and of Doppler velocities in a manner analogous to optical spectroscopy.—J.L. Pawsey, 15 April 1948

Nevertheless, given its resources and technical expertise, the fact remains that RP surely would have soon succeeded in detecting the interstellar 21 cm line if it had ever made a serious effort. As it turned out, RP *did* make first-rate contributions to 21 cm hydrogen observations in the early 1950s but only after others had taken the initiative.—Sullivan (2009, p. 126)

Introduction

It is well known that the group at RPL had the equipment and the talent to have achieved one of the ground-breaking discoveries of early radio astronomy—the detection of the HI line in 1951. Yet they did not. In this chapter we revisit the question of why not, adding additional insights into the chain of events, including some revision of earlier versions of these events.[1] As the group's leader, this missed opportunity must be, and was, attributable to Pawsey; and yet not for the reasons that some retrospective accounts have suggested (see ESM 20.1, A Review of Recollections, for additional details of the imperfect memories of Bowen and Bolton.).

Supplementary Information The online version contains supplementary material available at [https://doi.org/10.1007/978-3-031-07916-0_20].

[1] See also the detailed timeline of these events compiled by RD Ekers and WM Goss and available NRAO ONLINE.54.

1948 and Pawsey's First Realisation of the Importance of the HI Line

Numerous authors have published detailed descriptions of the early work on the HI (hydrogen) 21 cm line. Van de Hulst, a Dutch astronomer, had predicted the existence of the 21 cm hyperfine line of neutral interstellar hydrogen in 1944 (van de Hulst, 1945). This paper was published immediately after the end of the war, but it was published in a somewhat obscure journal (to non-Dutch scientists) and was in Dutch. This significantly delayed realisation of the possibility outside Holland.

While Jan Oort's astronomy group in Leiden had the knowledge and motivation to search for the HI line, the Dutch technology was in bad shape after the war, and the Netherlands had not been involved in wartime radar developments (during WWII, van de Hulst had been a student of M. Minnaert Utrecht; later he was advised by Oort). Despite heroic efforts, well described by Sullivan (2009, pp. 404–409) they did not make the first detection.

Pawsey's radio astronomy group in Sydney had the best instruments for this purpose, but with less astronomical motivation and no well-focussed effort they also did not make the detection. Sullivan (2009, p. 125, 396–403) has provided a comprehensive account of the line's discovery in 1951 by the atomic physicists Ewen and Purcell at Harvard. Swiftly following confirmatory observations were then made in the Netherlands and Australia. Details of the Australian activities are presented by Wendt et al. (2008) and Wendt (2011); here we add some additional insights.

As we have seen, the idea that it might be possible to observe the 21 cm hydrogen line was first introduced to Pawsey by Grote Reber. Pawsey first met Reber in December 1947 during his trip to North America (Chap. 17), and they discussed the 21 cm hydrogen line in January 1948. The Australians had not noticed the brief mention of the 21 cm hydrogen line prediction by van de Hulst in the review article "Radio-Frequency Investigations of Astronomical Interest" by Reber and Greenstein (1947)[2] which was published in *The Observatory* in February 1947. Receipt of a copy of the Reber and Greenstein paper was formally acknowledged by the Radiophysics Office in July 1947 but there is still no evidence that either Bowen or Pawsey read it since neither make any reference to it in their correspondence. In 1947, cosmic and solar noise researchers were not sufficiently sensitised to the astronomical importance of a spectral line to have taken note of this brief suggestion.[3]

---

[2] Reber and Greenstein quoted van de Hulst's "The Origin of Radio Waves from Space" published in *Nerlandsch Tijdschrift voor Natuurkunde*, Dec 1945.

[3] As Sullivan discussed, researchers and later papers tend to reference Shklovskii's (1949, p. 10) paper, in which he re-derived the theory from first principals (due to Shklovskii's inability to access van der Hulst's paper).

But following Pawsey's meeting with Reber in January 1948, it is clear that Pawsey was very interested in this suggestion and introduced the idea to RPL via a letter to Bowen written from the US on 23 January 1948[4]:

> ... Mr. Reber [then at National Bureau of Standards] gave me some valuable information. He tells me that there is an absorption line[5] of hydrogen atoms on a frequency of 1420.4 MHz; and one for deuterium at 327 MHz. This is derived from theory and laboratory work which he thinks is published; but which I have not yet seen. It may be in the *Physical Review* and is probably by the Columbia University people. If this is correct, there may be very considerable interest in searching for either cosmic or solar noise absorption or emission bands at this frequency.

The wording in this letter gives us more insight and what information about the 21 cm line was conveyed to Pawsey by Reber in their meeting in Jan 1948. Van de Hulst had met Reber in the summer (Northern Hemisphere) of 1946 and had tried to persuade Reber to search for a line at 21 cm (Sullivan, 2009, p. 396). As Sullivan wrote (p. 397) Reber did make an attempt to build a suitable 21-cm receiver; but in a later interview with Sullivan, Reber indicated that he was not very impressed at the time. However, Greenstein also had discussions with van de Hulst, as acknowledged in his addendum to the Observatory review, and this was almost certainly the origin of the discussion and reference to van de Hulst in the Reber and Greenstein review in *The Observatory*. The words used by Pawsey, in his letter to Bowen, suggest that Reber did not tell him specifically about the Van de Hulst prediction. In an earlier letter to Greenstein (10 Nov 1946),[6] Reber had enquired about details in the van de Hulst paper, so it is clear that Reber did not have a copy of it. Pawsey was therefore misled into thinking that these ideas came from the Columbia University group, who had been conducting laboratory experiments to determine the frequency of this transition. When Pawsey consulted with the Columbia group, he received a more pessimistic view of the detectability. Unfortunately, as discussed in Chap. 17, Pawsey's attempt to meet with Greenstein during his US visit had failed; and it seems that he was not made aware of the published paper by van de Hulst until a year later. In February 1949 Wild does mention the Reber and Greenstein review in his internal report RPL 33.

After leaving the USA, about 15 April 1948, Pawsey wrote a summary report, "Solar and Cosmic Noise Research in the United States and Canada", in which the importance of the HI Line was emphasised, but also the challenges of a search for it, "The Search for Atomic Spectral Lines in Noise":

> The utility of such a line, should it be found, is obvious. It opens up the possibility of determination of constitution of matter and of Doppler velocities in a manner analogous to optical spectroscopy. I mentioned the possibility of detection of atomic hydrogen lines at 1420.47 MHz and deuterium at 327.38 MHz in my letter of January 23rd.

---

[4]NAA C3830, A1/1/1 Part 1,2,3,4, F1/4/PAW/1, C4659/4 and C4659/8, A1/3/17 Part 1, A1/1/17-Box 3.

[5]Van de Hulst had not discussed whether an emission or absorption line was favoured.

[6]10 Nov 1946, Sullivan (2009, p. 71, footnote 43).

> Since then I have tried to elucidate the matter in the hope of being able to send you a complete statement of the hydrogen radio spectrum including all lines, intensities, Zeeman and Stark shifts etc. I have learned that it is a complex spectrum and I have not progressed far. A lot of people know scraps of it but it is not coordinated. This represents the Columbia [University in New York] position. My most helpful contact was at Toronto with [Ralph] Williamson and a research student Reeson. In fact, I left Williamson with the promise that he would attempt a survey of the subject and let me know. (See NRAO ONLINE.26).
>
> ... The position is therefore quite uncertain. Lamb of Columbia [Willis E Lamb, Nobel Prize in Physics 1955], for example, did not expect we should be able to find lines owing to low probabilities of emission and absorption and "smearing" due to changes due to magnetic fields and so on.[7]

While Pawsey's opening paragraph is enthusiastic, his following "elucidation" is not so clear cut; so it is, perhaps, unsurprising that a month later, when Bowen responded to Pawsey (to ASLO in London on 18 May 1948) he was rather negative. As noted by Bowen in ESM 20.1, A Review of Recollections, summarising the recollections of Bolton and Bowen, he had previously been made aware of the possibility of spectral lines at radio frequencies by his colleagues at Columbia.[8] Bowen discussed a number of topics (e.g. cosmic point sources and the value of doing an all sky survey) with particular emphasis on spectral lines. And he expressed his doubts about the value of searching for them:

> This possibility is certainly an interesting one, but in view of the present state of knowledge I doubt very much whether we should yet devote a special effort to it. A search for the atomic hydrogen and deuterium lines could be made with the Georges Heights equipment but this would involve dislocation of other work which is scarcely justified at present. At the moment Harry Minnett is chasing up the references you supplied and we are hoping that Williamson will live up to the promises he made you to let us have a survey of the whole subject. [No further news was reported in the next months from Minnett's effort.][9]

We note here that it was Pawsey who introduced the idea of searching for the HI line to Bowen, and it was Bowen who was the more reluctant to pursue it.

It was on the same day (18 May), that Pawsey also wrote to the whole group at RPL in response to Kevin Westfold's letter of concern of 20 April 1948 (Chap. 17).

---

[7]RPL were not the only group who learned of the predicted HI Line from Pawsey. The Canadians learned it likewise: Covington ("Beginnings of Solar Radio Astronomy in Canada") wrote: "When Pawsey was being shown, sometime in 1949 [sic, actually 1947–1948] the 10–30 cm broadband radiometer with its horn antenna under construction for the absolute [solar] flux determination, he told me about the 21 cm hydrogen line prediction and wondered whether or not I could make, or would plan to make, any observations for its confirmation. As it stood, the instrumentation was hardly suitable. This was the first time that I had heard of the prediction and is one occasion when I realised the magnitude of the difficulties of switching from one promising area to another. I readily gave a negative reply and realised that I would be continuing solar noise work." Cited in Sullivan (1984, p. 317).

[8]E.g., Nafe et al. (1947) measured the hyperfine line of atomic hydrogen in the laboratory.

[9]We note that in a later account of these events, Bowen (1984) pointed out that he had not been aware of the theoretical prediction made by van de Hulst until after the discovery of the H-line in 1951. We now realise that even though Pawsey did not know this in 1947, he still recognised the significance of searching. Bowen does not mention his own opposition to launching a speculative search in this period.

The focus of this letter concerned the balance of research planning based on "theoretical lines" versus "exploratory lines" and on the necessary conditions for success.[10] Pawsey then provided an analysis of the conflicts with the solar group in Sydney, in particular, the emphasis on observations over theory. He suggested the group move forward on three key projects. The third project was added as an afterthought. Written in Pawsey's distinctive handwriting: "<u>Can we observe the atomic hydrogen spectral lines or other lines</u>?"

Mills and HI Line in 1949

During Pawsey's absence and without Bowen's interest or support, there was no attempt to look for the HI line in Sydney. As Bowen had said, each of the group were thoroughly preoccupied with their own projects. In 1949, having returned from overseas, Pawsey seized the next available opportunity—he tried to convince the young Bernard Mills to look for the HI line as he began his career at RPL in the late 1940s.[11]

But he did not succeed. In a 1976 interview with Sullivan, Mills described his choice between two exciting projects to launch his career[12]:

> One was a search for the hydrogen line. Pawsey was very interested in it at the time. And the other was trying to locate very precisely the positions of radio sources. And it was a difficult decision to make. I eventually chose the precise positioning because I was more familiar with some of the techniques, and it looked as if it was something that would lead to an immediate result, whereas the other was extremely speculative ... [The technical difficulties] appeared rather forbidding. One knew one had to get right down to the absolute maximum theoretical sensitivity, because the thing was probably going to be faint ... And I'm pretty sure we knew that the Dutch were doing it, too.

In 2006, Mills provided a perspective on his decisive choice of 1949:

> If I had been a trained astronomer and therefore aware of the possible great importance of the H line, no doubt this would have been my choice. But I looked on it as merely a technical challenge, whereas I was intrigued by the mystery of the discrete sources and had no hesitation in choosing this option. This did ensure some friction within the group as John

---

[10] At the end of April, Lindsay McCready (acting as a stand-in for the absent Pawsey) had also written a chatty letter to Pawsey with some gossip. "Westfold has written to you letting off some steam on the subject of our carrying out a lot of experiments without adequate theoretical basis ... It may be in some people's psychological make up to seize on a point of criticism and exaggerate grossly ... [But] a good theoretician in the lab would relieve you of a lot of detail and assist in devising crucial experiments. Any luck in enticing anyone outside of RP to fulfil the gap? ... Smerd and Westfold produce lots of math and talk a lot without publishing much. They seem too concerned in proving Martyn wrong on points of details. Fred Lehany likens them to terriers attempting to trap a lion. Fred did his best and I think he improved Westfold ..."

[11] See Frater et al. (2013), Mills's autobiographical (2006) and Sullivan (2009, Chap. 7, p. 125 and Chap. 16, p. 398).

[12] Text from 1976 Sullivan interview.

Bolton had made discrete sources his own, following his use of the cliff-top interferometer to discover the first such source [in 1948] and to establish the existence of this class of object by finding several others. However, Pawsey knew that the future lay with the use of horizontal baselines and Bolton was still making effective use of the interferometer that had proved so successful for him previously.

Sullivan (2009, p. 126) has provided a concise summary of the impact of Mills's choice:

Mills's decision not to pursue the hydrogen line can hardly be called a managerial mistake, for he went on to do leading research on discrete sources. Nevertheless, given its resources and technical expertise, the fact remains that RP surely would have soon succeeded in detecting the interstellar 21 cm line if it had ever made a serious effort. As it turned out, RP *did* make first-rate contributions to 21 cm hydrogen observations in the early 1950s but only after others had taken the initiative.

Lack of understanding of the astronomical significance in Sydney was one of two crucial factors that led to the Sydney group's missing this important discovery. The other was simply personnel: like Mills, none of the group of new researchers, who were already involved in a plethora of new discoveries wished to put aside their current projects in a speculative attempt to detect the HI line.

Paul Wild, Ruby Payne-Scott, John Bolton and the HI Line

Pawsey had more success in getting Wild to study the possible radio frequency lines of atomic hydrogen. This resulted in two internal reports by Wild: The Radio Frequency line-spectrum of Atomic Hydrogen: (1) "The Calculation of Frequencies of Possible Transitions" (RPL 33, February 1949) and (2) "The Calculation of Transition Probabilities" (RPL 34, May 1949). Wild commented on the possibility of radio frequency emission from atomic processes (Reber & Greenstein, 1947), but his real interest was in solar emission (Saha, 1946). He made no reference to the van de Hulst paper and in his interview with Sullivan (3 March 1978) he noted that they never managed to get a copy. Following the detection of the HI line in 1951, Wild "dusted off his internal reports" and published his analysis in the *Astrophysical Journal* (Wild, 1952) and this became one of the classics in the field.[13]

An additional participant in the discussions of possible observations of the HI line in the era 1948–1950 was Ruby Payne-Scott (Goss, 2013, p. 243). John Murray (interview 26 January 2004) reported to Goss that Payne-Scott was a frequent

---

[13] As Purcell pointed out in the *Astrophysical Journal* of November 1952 (p. 457), Wild (1952, p. 206) had incorrectly predicted the intensity of the HI fine structure line at about 10,000 MHz between the $2^2S_{1/2}$ to $2^2P_{3/2}$ in atomic hydrogen. The line intensity in absorption was expected to be about 100 times less than the prediction of Wild. Purcell was not too critical of this mishap: "The [fine structure line of HI] is only one of several topics treated in Wild's paper, which is concerned with the general problem of line emission in radio astronomy." [We thank Shri Kulkarni for pointing out the Purcell 1952 publication.]

proponent of a search for the HI line when Pawsey asked about new projects in 1948–1949 at meetings of the Propagation Committee (1944–1949 and later Radio Astronomy Committee, 1949–1954). Murray reported that Payne-Scott was not necessarily interested herself, and the idea was also not received with enthusiasm by Christiansen and Mills on the grounds that "the group was extended fully".

In the 1980s, John Bolton, as well as Taffy Bowen, published recollections of this period, in which both criticised Pawsey for the missed opportunity of the HI line detection. Bolton has suggested that his group requested to be allowed to search for the HI line in 1949, but were turned down.[14] As we detail in ESM 20.1, A Review of Recollections, these recollections are not substantiated by the archival record. Given that both Payne-Scott and Bolton were quite preoccupied with their own projects at this time, it seems likely that there was no RPL scientist willing to put aside their current work in order to search for the line.

HI Line Detected and Confirmation in Sydney 6 July 1951

Sullivan's (2009) account of Purcell and Ewen's detection of the HI line in March 1951 is recommended reading and captures the excitement and drama of the discovery—complete with Ewen's unrolling a long section of strip chart along the hallway for Purcell in a morning demonstration of success. (In fact the HI detection project was by way of a "holidays and weekends" pursuit by Ewen, using equipment borrowed from his day job at the Nuclear Laboratory, resulting in the line being first detected on Easter morning (Sullivan, 2009, p. 403). A number of photos of Ewen and his equipment used for the HI detection of the HI line on 25 March 1951 were taken by photographer Fritz Goro. (See NRAO ONLINE.23 Additional Note 1 for details about Goro.) After the discovery, Purcell decided to hold off publication until he could receive confirmation from either the Dutch or the Australians. Sullivan records that this was appropriate as Purcell recognised how close the Dutch were to detection—and because Bowen was a "wartime crony".

Wendt et al. (2008) and Sullivan (2009, p. 409) have described the successful confirmation of the HI line by the RPL group of Christiansen and Hindman following the 25 March 1951 detection at Harvard. This event was followed by the Muller and Oort detection at Kootwijk in the Netherlands on 11 May 1951.[15]

In a sign of how the centre of gravity in scientific research was shifting towards the USA, Frank Kerr's presence at Harvard during the events of 1951 provided considerable aid to the involvement and ultimate success of the RPL group at this point. Kerr had spent a year at Harvard completing a Masters' degree in

---

[14] John Bolton to Don Morton (then Director of the Anglo-Australian Telescope in Australia) 3 June 1985; letter provided by Morton to Goss. However, Wild doubted the accuracy of this statement.
[15] Letter from Kerr to Pawsey of 1 June 1951—NAA C3830 A1/3/17 Part 1. As noted by Sullivan, Kerr incorrectly dated this discovery as 17 May, perhaps confusing the Dutch script for 7 and 1?

Fig. 20.1 Left to right: Ed Purcell, Taffy Bowen and Doc Ewen. Purcell won the Nobel Prize in Physics in early November 1952 for his work on nuclear magnetic resonance. The striped flask is a crude model showing the earth's orientation with respect to the plane of the Milky Way. Credit: Credit: Getty Images, Fritz Goro, The *Life* Picture Collection, licensed by Getty November 2021./ Licence organised by Shutterstock, Inc., New York, NY (5 Nov 2021) Original, Premium Editorial All Media

1950–1951.[16] He wrote to Pawsey with the exciting news of the detection on 30 March 1951, with a sketch of the chart recording of the discovery obtained by Ewen on Easter Sunday, 25 March 1951.[17] Bowen visited Purcell at Harvard in Oct 1951 (Fig. 20.1). Purcell wrote Pawsey[18]: "Incidentally, Bowen came over last week to see the line and Doc [Ewen] succeeded to bring it in on schedule."

A key meeting was held in Sydney at RPL on 12 April 1951, with attendees Pawsey, Arthur Higgs, Piddington, Christiansen, Wild and Bolton. The purpose was a discussion of the RPL "verification of the Harvard result." Pawsey wrote in the minutes:

> It was agreed that parallel investigations to check detectability of lines were desirable in order to obtain independent checks but that in order to avoid cut-throat competition, the groups who were experimenting in the same field, e.g. Piddington, Christiansen and Wild, should consider themselves, at least for the 1420 MHz line, as a single group and possible publication should be joint.

---

[16] In addition, H.C. van de Hulst of the Leiden Observatory was at Harvard in this period, providing useful astronomical advice to Ewen and Purcell; he also provided a communications channel with Muller and Oort in Leiden, the Netherlands.

[17] Kerr took leadership of the new HI group when he returned to Australia.

[18] Purcell letter to Pawsey 22 Oct 1951 A1/3/17 Part 1

> ... Christiansen and Bolton outlined schemes for attempting to detect the 1420 MHz line
> with which they were proceeding ... They hope to have equipment for tests to start in a week
> or two. Piddington outlined a different scheme with which he was proceeding.[19]

On 20 April 1951, Pawsey wrote Ed Purcell with news of the Australian efforts:
"...[B]ecause of the great potentialities of Ewen's result in this field, two separate
groups here are attempting to check the result." By time of the radio astronomy
[group meeting] of 8 May 1951 (attending Pawsey, Shain, Payne-Scott, Mills, Wild,
Murray, Gardner, Piddington and Hindman), the groups of Christiansen and
Piddington described their progress. There was no mention of Bolton, likely due
to the fact that he was seriously ill in Prince Henry Hospital with both pneumonia
and an undiagnosed illness of the kidney and bladder.[20] In the same letter from
Pawsey to Bowen in London, Pawsey told Bowen that on 18 May 1951 he would go
to Potts Hill to see the first tests carried out by Christiansen. Pawsey was about to go
on leave for a fortnight and thus Christiansen might be in touch with both Purcell and
Bowen if there were any results of the HI observations in the intervening period.

By 7 June 1951, Pawsey wrote Bowen[21] that Bolton had recovered his health. He
had been seriously ill for 1 week and then rested at home for a week. Then he had
gone on holiday "in the country". Thus Bolton had missed the exciting period during
which the line was detected. Bolton was present at future radio astronomy meetings
on 18 July and 4 September 1951.

Sullivan (2009, p. 409) has described the two groups' activities at RPL during this
period June to early July 1951:

> Although the Australians had earlier decided against a campaign to search for the line, within
> 2 weeks of receiving Kerr's letter, two separate groups were lashing together 21 cm receivers
> to check the Harvard discovery. One was initiated by Jack Piddington and carried out by
> James V. Hindman, who had previous experience with solar microwave observations and
> quickly was able to assemble a sensitive 21 cm receiver. But Hindman was having trouble
> determining his precise operating frequency, an absolute necessity. What he needed was a
> top-quality signal generator and he learned that the man who had one was his colleague
> "Chris" Christiansen. When Hindman approached Christiansen about a loan, he discovered
> that Christiansen too was frantically working on a 21 cm receiver. He had temporarily
> dropped his own solar work after receiving a rush assignment from Pawsey. Christiansen
> was having problems with the front-end of his receiver, however, and so it was natural that
> they should [later] join forces.

In the same letter of 7 June 1951 to Bowen, Pawsey expressed embarrassment that
the line had not been detected. He described the two groups of Christiansen and
Hindman, working together but using different equipment. He reported that the
Leiden group had now detected the line on 11 May 1951.

---

[19]The composition of the groups was complex; apparently Pawsey decided to have at least two
groups and perhaps even three working on the project; this scheme would enhance chances of
success.

[20]NAA C3830 Z1/9/1951 Part 1. 18 May 1951.

[21]NAA C3830 Z1/9/1951, Part 2.

Frantic efforts were made by the combined group at Potts Hill using the 16 by 18 foot paraboloid antenna; Hindman was injured on the aerial at this period and Chris Christiansen carried out the final activities on his own. The line was detected in Sydney at Potts Hill on 6 July 1951 and discussed at the radio astronomy meeting on 18 July. There was some confusion in early July about the reality of the detection; Pawsey wrote Minnaert in Utrecht on 9 July that the line had not yet been detected! However, 4 days later Pawsey amended his message: "Since writing you a few days ago we have obtained confirmation of the existence of the hydrogen line at 1420 MHz in the galactic spectrum."[22]

Pawsey wrote an enthusiastic letter to Bowen in London on 13 July 1951 announcing the discovery[23]:

> We have at last succeeded, after much trial and tribulation in identifying the hydrogen spectrum line at 1420 MHz in galactic noise from the region near the centre of the galaxy. Purcell has already sent a communication to *Nature* telling of the Harvard discovery and saying in his letter to the Editor that we might be sending a confirmatory contribution. Purcell's letter was written on June 14 and, in view of the delay, I don't think it is proper to try to get an ordinary letter across and I have in consequence sent a telegram telling the Editor of our confirmatory results and asking him to put in an editorial note to this effect if this can be done. I enclose a copy of the telegram.
>
> I am also enclosing a photo of the first record showing evidence of the effect [Fig. 20.2]. It is a little hard to interpret but the bumps on the record occur systematically when the frequency sweeps through that of the hydrogen line. The evidence for existence I consider quite conclusive but we do not know anything quantitative about the line. We propose to continue with a crude exploratory survey using the existing equipment and then will probably stop and make first-class equipment in light of knowledge concerning the phenomenon. Christiansen has worked like a [the "N word" was used—likely signifying "slave" in 1951] for the last 2 months trying to get this gear working and it is a very credible performance on his part. This line is really exceeding weak and it is necessary to make the right compromises all along the way in order to make the spectrum line evident.

Purcell and Oort also received letters at this time with reports of the detection at Potts Hill. Oort replied with a letter of congratulations on 20 July 1951, written by hand from his summer holiday on the Frisian Lakes where there was no typewriter. "It will be extremely important to have northern as well as southern observations." Oort recognised the astronomical value of getting observations of the part of the galaxy only visible from the South.

Pawsey wrote Ewen on 31 July 1951 after receiving a microfilm copy of Ewen's recent PhD thesis *Radiation from Galactic Hydrogen at 1420 MHz*: "I should like to express my hearty congratulations to you on your success in this work. I believe[24] you have initiated a revolution in radio astronomy. I know enough about the difficulties to appreciate the skill you put into this."

The publication of the two papers by the Harvard group and the Leiden group occurred on 1 September 1951 in back-to-back papers: "Observation of a Line in the

---

[22]NAA C3830 A1/3/17 Part 1

[23]NAA C3830 Z1/9/1951- Part II.

[24]NAA C3830 A1/3/17 Part 1.

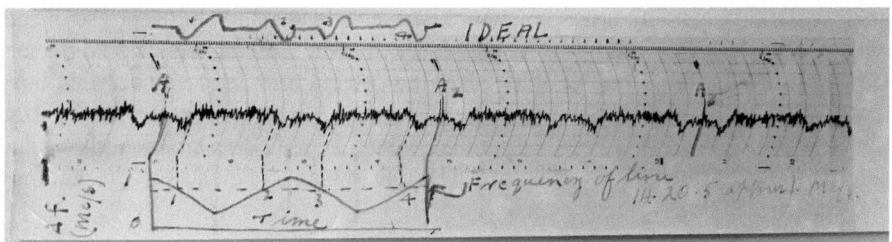

Fig. 20.2 Photo of the first record showing evidence for the Australian detection of the HI line, produced by Christiansen and Hindman at Potts Hill and sent by Pawsey to Bowen, 13 July 1951. The photo of the record was paper clipped to the letter in the archive. The explanatory annotation in blue has been added to the photo, in Pawsey's handwriting. However, the "IDEAL" signal it illustrates corresponds to the later published observations and not to this record. In the published observations the signal is seen changing from positive to negative as the frequency sweeps the HI signal across both frequency switched bands (Christiansen & Hindman, 1952). Credit: Courtesy National Archives of Australia. NAA C3830 Z1/9/1951- Part II

Galactic Radio Spectrum: Radiation from Galactic Hydrogen at 1,420 MHz", by Ewen and Purcell and then "Observation of a Line in the Galactic Radio Spectrum: The Interstellar Hydrogen Line at 1420 MHz, and an Estimate of Galactic Rotation" by Muller and Oort. The papers were followed by a cable of 12 July 1951 from Sydney, sent by Pawsey. Christiansen and Hindman were mentioned as having carried out the observations.

> The following cable dated July 12 has been received from Sydney, N.S.W.
>
> Referring Prof. Purcell's letter of June 14 announcing the discovery of hyperfine structure of the hydrogen line in galactic radio spectrum, confirmation of this has been obtained by Christiansen and Hindman, of the Radio Physics Laboratory, Commonwealth Scientific and Industrial Research Organization, using narrow-beam aerial. Intensity and line-width are of same order as reported, and observations near declination 20° S. show similar extent about galactic equator.
>
> J. L. PAWSEY

The success of the Australians[25] in the following months was noteworthy:

> Although the Australians came last to the 21 cm problem, they stuck with it most doggedly once they had a detection. For three solid months Christiansen and Hindman surveyed the southern sky with their square dish, milking all they could out of their makeshift receiver. Their profiles were more reliable because of their larger scanning range of 1000 kHz and larger frequency-switching interval of 160 kHz (although still inadequate, as they were aware). They slowly tuned in frequency while the sky drifted through their 2.3° beam, obtaining an independent frequency profile every two to four beamwidths. (Sullivan, 2009, p. 411)

Pawsey's quiet and careful leadership style was oriented primarily around support for young scientists to develop their own lines of investigation, often taking risks.

---

[25] Christiansen and Hindman (1952). Note that the scanned copy in ADS has the wrong antenna illustration (plate 1). The bound volume and the scanned version on the publisher's www site are correct.

Mills was a case in point. It was not part of Pawsey's leadership philosophy to dictate what project Mills, or Wild or Christiansen or the other researchers, should pursue. Having said that, the scramble to confirm the detection in 1951 showed that busy though they were, it was in fact possible to redirect RPL efforts when needed. Had this been done even 1 or 2 years earlier, the Australian radio astronomy group would surely have made the first detection of the HI line, owing to their superior technology and expertise.

Chapter 21
No More Radio Stars! 1952

> The important conclusion is that the sizes [of the radio sources] are minutes of arc, definitely not those of single stars. Radio astronomy has thereby lost a graphic term; "radio star" must be replaced by "radio nebula".—Pawsey (1953, p. 137)

The year 1952 was an important turning point for radio astronomers; during the year it became clear that none of the discrete radio sources were "radio stars" and that the first identifications of radio sources with galaxies and supernovae remnants were typical for the population. The year also produced one of the few fully collaborative efforts: a Christmas special appeared in *Nature* on 22 December 1952, "Apparent Angular Sizes of Discrete Radio Sources" (p. 1061) as papers by the Jodrell Bank team of Hanbury Brown, Jennison and Das Gupta, Mills (1952a) in Sydney and Smith (1952a, 1952c) in Cambridge were published.

This collaboration may have only succeeded since the three groups met together in Sydney in August 11–21, 1952 for the URSI General Assembly, one of the greatest coups for Australian science of the early 1950s.

The End of the "Radio Star" Model: Measuring Radio Source Sizes

In late 1951 and early 1952, Mills was the first to obtain results from new era of radio source surveys (see Chap. 35 for the context and descriptions of the instruments). Mills had submitted his paper on the Badgerys Creek 101 MHz survey on 16 October 1951 (Mills, 1952b). The identifications of Bolton et al. (1949) for Taurus A (Crab Nebula), Centaurus A (NGC 5128) and Virgo A (NGC 4486) were confirmed and positions were measured for Cygnus A, Hydra A and Fornax A. The survey interferometer (see Fig. 21.1) had two-baselines (60 and 270 m) and this allowed Mills to also determine, a rough estimate of the angular size of the sources.

© The Author(s) 2023
W. M. Goss et al., *Joe Pawsey and the Founding of Australian Radio Astronomy*,
Historical & Cultural Astronomy, https://doi.org/10.1007/978-3-031-07916-0_21

Fig. 21.1 Two of the three elements used for the 101 MHz all sky survey made by Mills in 1950–1951. The three elements were spaced by 60 m and 270 m, giving lobe spacings of 3° and 40 arc min. This is the 60 m spacing. With two spacings a rough estimate of the angular size of the radio sources could be detected. The same antennas were used for the sky survey as discussed in more detail in Chap. 33. Credit: CSIRO Radio Astronomy Image Archive B2594-3

Surprisingly, Mills found that a number of sources showed large angular sizes. There were two galactic sources: Puppis A with an equivalent diameter of 33 arc min and the galactic centre with a size of 35 arc min. Two sources identified with galaxies, Centaurus A and Fornax A, had sizes of 25 and 20 arc min. Bracewell (1952) published a short paper in *Observatory* "Radio Stars or Radio Nebulae?" drawing attention to Mills's angular size results.[1] Thus, by mid-1952, finite sizes for a number of discrete radio sources had been obtained, all clearly many orders of magnitude larger than stellar diameters.[2] The move away from the "radio star" model had major implications for the understanding of the radio emission mechanisms as we will discuss in Chap. 34. The unexpected discovery of double structure when measuring the angular size measurement for the northern source Cygnus A will be described in detail later in this chapter.

---

[1] The Mills paper was published in a less visible Australian journal and the rough source sizes tabulated in the paper were not emphasised.

[2] For Centaurus A, the outer lobes with total extent of 5–10° had not yet been detected; the angular extent of 25 arc min refers to the inner lobes with separation about 10 arc min. An additional paper was published by Mills in 1953 "The Radio Brightness Distributions over Four Discrete Sources of Cosmic Noise". With baselines up to 10 km (lobe spacing about 1 arc min), angular sizes in the range 2–6 arc min were derived for Centaurus A, Taurus A and Virgo A.

Fig. 21.2 URSI 1952 Sydney. Welcome on the *SS Strathmore* on 8 August 1952 in Sydney. Left to right D.F. Martyn, Col E. Herbays (URSI Secretary, Belgium), Sir Edward Appleton (Vice-Chancellor of the University of Edinburgh Nobel Prize 1947), E.G. Bowen and J.L. Pawsey. Credit: CSIRO Radio Astronomy Image Archive 2842-1

URSI Comes to Sydney

The 10th General Assembly of URSI, 11–21 August 1952, was a momentous occasion. It was not only the first time that URSI was held in Australia, but one of the first times an international scientific congress had been held outside Europe or North America at all. From this perspective it was a profoundly important symbol of the maturing of Australian science. That it occurred at all was testimony to Australia's leadership in radio astronomy specifically—It was still difficult for Australia to attract international meetings in other fields of science. And despite this leadership, Robertson notes that until URSI, not *one* radio astronomer from the groups overseas, came to visit Sydney (Robertson, 2017, p.111). It took some of the delegates to URSI 1952, a month to travel to Sydney (Goss & McGee, 2009, p. 185). It was momentous from a domestic perspective also, for it was the first international congress to be held in Australia since the Second Pan-Pacific Science Congress almost 30 years earlier, in August 1920, in Melbourne and Sydney (Goss & McGee, 2009, p. 184). In Figs. 21.2, 21.3, 21.4, 21.5, 21.6, and 21.7, we show a selection of images from the General Assembly.

Fig. 21.3 URSI 1952 Radio Astronomers at URSI, Sydney, 13 August 1952. Front row left to right: Chris Christiansen, F. Graham Smith (UK), B.Y. Mills, S.F. Smerd, C.A. Shain, R. Hanbury Brown (UK), R. Payne-Scott, A. G. Little, M. Laffineur (France) and J. G. Bolton. Second row: J.P. Wild, J.L. Steinberg, J.V. Hindman, F.J. Kerr, C.A. Muller (Netherlands) and O.B. Slee. Third row: C.S. Higgins, J.P. Hagen (USA) and H.I. Ewen (USA). Back row: J. H. Piddington, E.R. Hill and L. W. Davies. Individuals with no country designation are Australian. Pawsey not in photo due to illness. Credit: CSIRO Radio Astronomy Image Archive B2842-43

RPL owed the decision to hold URSI 1952 in Sydney to David Martyn, whose reputation resulted in his election, in 1950 at Zurich, as President of the newly formed Radio Astronomy Commission (Commission V) of URSI. He promptly, and successfully, invited URSI to hold its next General Assembly in Australia, although the necessary finance had not yet been obtained (Piddington & Oliphant, 1971).

The details of URSI 1952 have been provided by several authors, among them Ron Bracewell, who was secretary of the Sydney organising committee, and recently by Goss and McGee in biographies of Ruby Payne-Scott (Goss & McGee, 2009, p. 185 and Goss, 2013, p. 201). We refer interested readers to these comprehensive accounts.

Sixty-three overseas delegates from 13 countries attended. URSI began by a welcome on board the ship on which URSI's President, Sir Edward Appleton, had travelled (Fig. 21.2). Eminent delegates included many of Pawsey's and Bowen's connections, among them Ratcliffe, Balthazar van der Pol from Geneva, Burrows

Fig. 21.4 The "21 cm HI Club" at URSI Sydney August 1952 from left Frank Kerr, Paul Wild, Jim Hindman, "Doc" Ewen (Harvard), Lex Muller (Leiden) and Chris Christiansen. All but Ewen and Muller from CSIRO RPL. Credit: CSIRO Radio Astronomy Image Archive B2842-45

from Cornell and Dellinger from RCA in the USA. For the Australians it was the first opportunity for the majority of the scientists to meet overseas colleagues.[3]

Appleton's response to the formal welcome offered by Sir John Madsen paid tribute to Australia's many successes in its wartime radar research program. He then summarised recent key developments in radio astronomy, summarising the recent discovery by Bolton and Stanley of the small angular size of Cygnus A as determined at Dover Heights, David Martyn's work on tides in the ionosphere, and the discovery of the HI line.

Commission V (Radio Astronomy) held 6 sessions, 4 of them scientific, at which there were 19 talks with 9 given by Australians. One notable feature was Payne-Scott's absence from among the speakers. Payne-Scott had retired from RPL

---

[3]The IAU General Assembly was held in Rome from 4 to 13 September 1952. The Australian representation at the Rome meeting was meagre since the meeting was only 12 days after the URSI Assembly in Sydney. J.A. Roberts was present, sending a report to Pawsey; he was on his way back to Australia from his three-year period completing a PhD in Cambridge with Fred Hoyle. M. Laffineur (Chief of Radio Astronomy Laboratory at the Institut d'Astrophysique de Paris) attended both conferences.

Fig. 21.5   Tour of Potts Hill, the 32-element 21 cm Grating Array. From right: Appleton, Bathalsar van der Pol (Phillips Laboratory), Fred White, unknown, Chris Christiansen (tour guide). Credit: CSIRO Radio Astronomy Image Archive B2842-R66

13 months earlier due to the approaching birth of her first child[4] (and following miscarriage of an earlier pregnancy), at age 40. Her marriage had been hidden from CSIR /CSIRO for some years (though it was known to her colleagues), due to the rule applied to all public servants, that women were required to resign from employment upon marriage. This "marriage bar" was not lifted until 1966. Nonetheless Pawsey went to some lengths to encourage Payne-Scott to give a presentation on her research, writing to her personally more than once and visiting her at her home (as was common in the early 1950s, she had no car). But in the end Payne-Scott did not present. With the childminding help of RPL Chief Secretary Sally Atkinson, she did attend at least one of the sessions for Commission V, and is present in more photographs of the event than Pawsey (Goss & McGee, 2009). This brief attendance at URSI constituted her last professional activity as a radio astronomer.

Ironically, Pawsey played very little part in the proceedings of URSI 1952: both he and Bracewell were absent from much of the congress because of a severe illness,

---

[4]This child became famous theoretical statistician, Peter Hall 1951–2016; Payne-Scott's second and last child, Fiona Hall, born 1953, is an internationally celebrated artist.

Fig. 21.6 At URSI Sydney August 1952. From left Taffy Bowen, and Robert Hanbury Brown (Jodrell Bank) Credit: CSIRO Radio Astronomy Image Archive B2842-13

probably influenza.[5] By the last week of August, Pawsey was able to provide tours of the Dapto field station for overseas visitors to the URSI congress.[6] Despite his illness, Pawsey made an enormous effort to extend hospitality to his visitors and to foster collaboration and connection with overseas groups, as we show below. But during the conference, David Martyn organised an informal meeting of all those present who were Fellows of the Royal Society (FRS), for the purpose of nominating Pawsey to be elected as a Fellow himself. As Martyn wrote to Pawsey: "At a recent informal meeting of a group of Fellows of the Royal Society it was suggested that you be invited to let your name go up for election, and I was asked to act as proposer."[7]

Pawsey responded immediately with two letters, a handwritten letter of acceptance on 1 September 1952 and a more formal typed letter from 3 September. He wrote: "I am a little diffident as to my qualifications but if, as you mentioned in your letter, a group of Fellows favoured this, I should be foolish not to accept your offer. I am very grateful indeed to you for taking this interest in my career."

We discuss the eventual success of this proposal in Chap. 25.

---

[5] Goss (2013). This was fortunately 5 years before the global influenza pandemic of 1957, of which we are conscious as we write during COVID-19 in 2020!

[6] John Murray, interview 2007.

[7] NAA A9874/60, "Royal Society, London. Dr. Pawsey's Qualifications" and Joe and Lenore Pawsey Family Collection.

Fig. 21.7 At URSI Sydney August 1952. From right: Father Pierre Lejay (Director Bureau Ionospherique Francais, Paris, President URSI 1952–1957), Letty Bolton, van der Pol, Nicolai Herlofson (Royal Institute of Technology Stockholm) and John Bolton (RPL). Credit: CSIRO Radio Astronomy Image Archive B2842-12

Excitement at URSI: The Angular Size of Sources

Perhaps the most significant outcome of URSI 1952 for radio astronomy was the acceptance that the discrete radio sources could not be identified with stars in our galaxy, with many even outside the galaxy.

A major contributor to this shift in thinking was the eventual measurement of the angular size and complex structure of the first discrete radio source discovered, Cygnus A. The central role was played by the Jodrell Bank group of Hanbury Brown (1916–2002), Roger Jennison (1922–2006) and Mrinal Kumar Das Gupta (1923–2005), and this quest was aided by, and in turn cemented, collaboration with RPL.

Soon after joining Jodrell Bank in 1949, Hanbury Brown began his multi-year quest to determine if "radio stars" really had an angular diameter as small as optical stars, as was generally assumed at this time. The problem was that the upper limits for most radio sources was some 5–10 arc min in size, tens of thousands of times the angular sizes of optical stars (0.01 to 0.005 arcsec). This resolution at radio wavelengths would require huge baselines of some thousands of kilometres.

In February 1950, Hanbury Brown had the idea of correlating the intensity fluctuations, a comparison of the source noise which is inherent to the radio emission process rather than the amplitude of the radio waves (see also discussion in Chap. 37. The intensity interferometer was invented, with a key role played by Richard Twiss (1920–2005), a mathematical physicist. The idea was that phase could be ignored; as with a Michelson interferometer, the correlation observed would decrease for a source of a given size as the baseline increased. The main disadvantage was that only very strong sources could be observed since the source noise had to be larger than the receiver noise. In 1950, the only two sources that were feasible were Cassiopeia A and Cygnus A in the northern sky. A most important aspect was that the method would work during the presence of strong scintillations and electronic phase instabilities; this meant very long baselines would be possible.[8] The initial planning of the system to determine the angular diameter of Cygnus A and Cassiopeia A had envisioned that a baseline of at least 100 km would be required. The system design and construction included a radio link system[9] to bring back the signals to the main station at Jodrell Bank where the correlator was located. In the autumn of 1950, two postgraduate students at the University of Manchester began a research project with Hanbury Brown to develop a prototype intensity interferometer.

The full-scale instrument was built in 1951 and observations of Cassiopeia A and Cygnus A began in the summer of 1952, just before URSI.[10] The initial observations occurred after Hanbury Brown was in Australia for URSI. A site 4 km from Jodrell Bank (close to Lovell's house) was used for the initial observations. At this baseline, Jennison and Das Gupta found (from Jennison, 1994) "Cassiopeia was completely resolved whilst the correlation for Cygnus fell to 79 percent," implying sizes greater than 2 arc min for the former and a N-S size for Cygnus of about 0.5 arc min. The results were sent by cable to Sydney to be presented by Hanbury Brown.

Hanbury Brown (Sullivan, 1984, p. 228) later explained in colourful language: "There was really no need to have developed the intensity interferometer; we could have done the same job with a conventional interferometer in half the time and with half the effort. We had built a steam-roller to crack a nut."

As can be imagined, these results generated considerable excitement at URSI. John Bolton (1953, p. 23) provided a graphic account of the exciting results presented during the sessions of Commission V:

---

[8] Detailed accounts of the adventuresome preparations for the intensity interferometer are presented by Sullivan (2009, p. 351–360) and Hanbury Brown (in Sullivan, 1984, pp. 226–232 *The Early Years of Radio Astronomy*).

[9] Note Mills also used a "radio link" system for his interferometer to obtain long baselines in perhaps the first use of this type of arrangement.

[10] But by the time of URSI, the realisation that there would be no need for long baselines began to emerge. At the first spacing used for testing (300 m), there were indications that there was slight de-correlation of the signals from Cassiopeia A. Also Baade and Minkowski had written that the optical identifications of Cygnus A and Cassiopeia A suggested that possible optical sizes might be in the range 0.5 to 5 arc min, not sub-arc seconds.

Measurements of the angular diameters of several of the stronger sources have been made by
Mills, Smith and Hanbury Brown, using different techniques. The Cygnus source, which
was observed by all three observers, apparently has an angular size of the order of 1 min of
arc, and the Cassiopeia source, which was observed by Smith and Hanbury Brown, a size of
several minutes of arc … Detailed agreement between observers was not good, but in
discussions following the formal sessions it was realised that the results could be reconciled
if complex distributions of surface brightness were assumed. The writer [Bolton] gave
evidence for the existence of a number of objects with angular widths of more than a degree
and with sharp central concentrations. Supporting evidence was given by Mr. Mills's work
in the case of two sources—one provisionally identified with the galaxy NGC 5128. **It seems
that the term radio "star" may be a misnomer.**[11] [our emphasis].

Collaborations at and After URSI 1952

During the URSI 1952 meeting, the three groups investigating radio sources tried to
sort out the discrepancies of their different determinations of the angular size of
Cygnus A by invoking various scenarios such as complex brightness distributions
and spectrum variations over the source (the frequencies varied from 100 to 210 MHz).
During URSI and the weeks afterwards, numerous informal discussions were held in
Pawsey's office as well as the field stations, and were invaluable. Hanbury Brown
expressed his hope for increased cooperation between Jodrell Bank and RPL in a letter
to Pawsey written after his departure from Sydney in early September 1952. Both
Smith and Hanbury Brown returned to the UK via the US, the latter departing from
Sydney by air on 10 September 1952. Hanbury Brown travelled to Pasadena, visiting
Caltech and the Mt. Wilson and Palomar Observatories.[12] Pawsey wrote with similar
warmth of Hanbury Brown to Lovell on 10 September 1951: "I think his visit has been
really worthwhile in establishing contact and friendship between members of our two
groups." Hanbury Brown was to move to Australia in 1962 as leader of the Narrabri
Stellar Intensity Interferometer at Sydney University.

The idea was floated that adjoining papers would be submitted to *Nature* after a
number of checks were carried out. Ryle and Smith were reluctant since the Smith
paper in *Proceedings of the Physical Society* (Smith, 1952c) had already been
submitted. But after the URSI meeting there was a flurry of letters as the joint
publication was organised. Pawsey suggested[13] that if there were to be a three-way
publication in *Nature*, then Mills should be included. He wanted to avoid a repeat of
the 1950 imbroglio that only included Cambridge and Jodrell Bank with an omission
of Bolton (Chap. 18). For Pawsey the main point was: "this result, finite size of
previously unresolved sources, is of sufficient general interest to warrant [such] a

---

[11] Pawsey made the same point in his review paper of 1953, "Radio Astronomy in Australia":
"Radio has thereby lost a graphic term; 'radio star' must be replaced by 'radio nebula'."

[12] NAA C3830 Z3/1 Part 4 Hanbury Brown wrote Pawsey on 3 October just as he returned to
Jodrell Bank: "I had a very interesting session [in Pasadena.] The main outcome was that my
education in matters of astronomy was very greatly increased."

[13] Pawsey to Lovell, 10 September 1952, NAA C3830 Z3/1 Part 4.

preliminary (our emphasis) communication to *Nature*." He agreed, however, that "the discrepancies between observers are sufficient to warrant delaying further publication until one or two checks are applied." He requested that Hanbury Brown could coordinate the joint publication, waiting until the additional tests were carried out.

Two days later, Pawsey wrote a similar letter to Ryle[14]: "The tumult and shouting of URSI have at last subsided and we can settle down again ... It was a great pleasure having Smith here and I hope you will tell him how much we enjoyed his visit." Pawsey again asked that Mills be invited to participate and that the publication be delayed for 1–2 months to allow checks to be made in Sydney and Manchester, adding "I have just heard from Mills that his check confirms his result."

Pawsey's concerns were soon allayed, first, on 18 September 1952, by Lovell, who wrote: "It was clear to me that if any publication in *Nature* was intended, then it would be only fair for three notes to come from Sydney, Cambridge and Jodrell Bank simultaneously ."[15] And on 27 September 1952,[16] Ryle replied to Pawsey— the first such correspondence from Ryle. He was happy to contribute a note for the joint publication: "I think your suggestion for the arrangements would be ideal ... The URSI meeting seems to have been a very great success, and Ratcliffe has come back full of admiration for all the very fine work that you are doing. I am very much looking forward to hearing all the details from Smith."

In the course of October 1952,[17] Hanbury Brown (we assume) duly organised a joint submission of the three articles to *Nature*: (1) "Apparent Angular Sizes of Discrete Radio Sources" by Hanbury-Brown et al. (1952), (2) "Observations at Sydney" by Mills (1952a) and (3) "Observations at Cambridge" by Graham Smith (Smith, 1952a). To some extent, the papers by Mills and Smith represented the results presented at URSI the previous August. But as Sullivan has commented, the hurry to produce results for URSI had its costs for Mills. He lacked the critical spacing in his observations between 1 and 5 km that would have allowed him to detect the double structure of Cygnus A (Sullivan, 1982, p. 274).

The Jodrell Bank publication mainly gave new data that was obtained after August with three new baselines at four new orientations and three baseline lengths.[18] The data for these initial observations of Cygnus A showed a very elongated source, see the sketch at the bottom left hand corner of Fig. 21.8. The minor axis had been determined in August 1952 while the cross-cut 4 showed the

---

[14] NAA C3830 A1/1/1 Part 7.

[15] *Ibid.*

[16] NAA C3830 A1/1/1 Part 7.

[17] NAA C3830 Z3/1 Part 4. On 2 October 1952 just after his return to Jodrell Bank, Hanbury Brown wrote to Pawsey, reporting that the Cygnus A and Cassiopeia A observations were progressing well. The expectation was that the paper would be ready in a few weeks; he had written to the Cambridge radio astronomers inquiring if they were willing to participate in the joint enterprise.

[18] The original data presented at URSI was the 3.99 km baseline, almost N-S. Ironically, the fringes were almost perfectly aligned along the minor axis for these initial observations of August 1952. The new observations had baselines of 0.30, 2.16 (2 orientations), and 3.99 km.

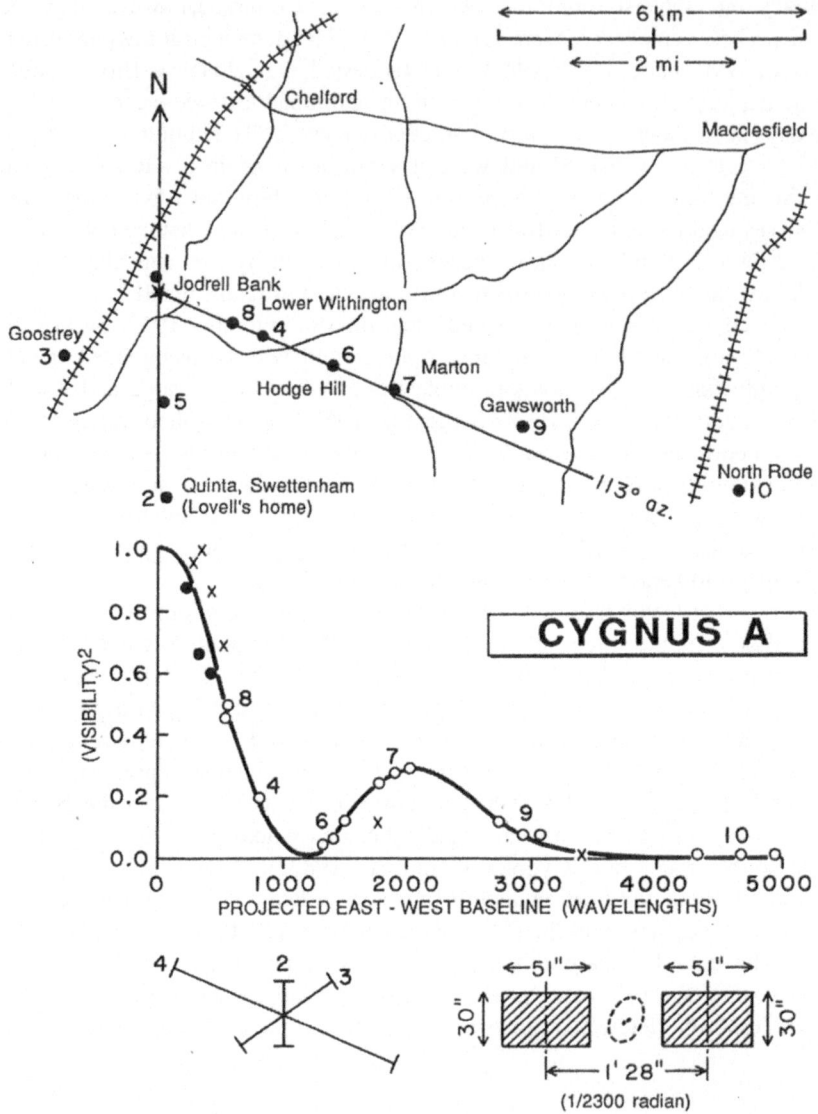

Fig. 21.8 It appears in *Cosmic Noise* as Fig. 14.20 with this caption: "The graph (adapted from Jennison and Das Gupta) shows for Cyg A the square of fringe visibility (or in the case of the intensity interferometer the equivalent correlation) versus projected east–west baseline, measured in wavelengths. The 125 MHz Jodrell Bank data (open circles) were taken with baselines between Jodrell Bank and the remote sites indicated on the map, numbered in the chronological order in which they were used over the period July 1952 to early 1954. Data at 210 MHz by Smith (1952b, c) (filled circles) and 101 MHz data by Mills (1952c, 1953) (crosses) are also shown. At the bottom (left) are the initial angular sizes derived from each of three crosscuts (site numbers indicated) measured by Hanbury Brown et al. (1952), and (right) a model of radio brightness distribution based on more complete data (Jennison & Das Gupta, 1953); also sketched is an outline of the faintest optical emission reported by Baade and Minkowski (1954a) from their colliding galaxies (with two dots representing nuclei as seen in Fig. 14.13). The Fourier transform of this model radio distribution is the solid line in the plot." Credit: © Woodrow Sullivan III, Fig. 14.20, *Cosmic Noise: A*

almost E-W major axis of about 2 arc min with a correlation coefficient of only 0.3.[19] For Cassiopeia A (later identified as a galactic supernova remnant), the source was remarkably symmetrical with an equivalent angular size of 4 arc min. With a baseline of 4 km (fringe spacing about 2 arc min), the source was hardly detectable, "resolved out".

Hanbury Brown et al. (1952, p. 1061) concluded:

> These preliminary measurements establish two major points. First, the apparent angular size of the two most intense radio sources is thousands of times greater than that of the visible stars and is of the order of a few minutes of arc. Secondly, the source in Cygnus exhibits a pronounced asymmetry in angular size [not a circularly symmetric], whereas the source in Cassiopeia appears to be roughly symmetrical. The measurements are not yet adequate to define satisfactorily the shape of the sources or the distributions of intensity across their disks [presumably implying that the extended source was a "disk".] Further observations are now being made with the present apparatus using different baselines.

Roger Jennison noted that "the stage was now set for an all-out attack on the structure of Cygnus A" and provided an account of his thinking in a 1976 interview with Woody Sullivan (Sullivan, 2009, p. 356):

> Cygnus didn't make sense—there was something peculiar about the first results. The three measurements did not fit together, but we had confidence in them—we were sure we'd done them right, so the fault wasn't ours. I know this sounds like Archimedes, but it's perfectly true: 1 day I took a long bath and I was laying back in the tub thinking about this distribution. And then all of a sudden it clicked that if Cygnus were *two* blobs instead of one, then I could get this peculiar difference between the projections in different directions.

Thus after October 1952, Jennison and Das Gupta continued their work at 125 MHz with additional observations of Cygnus A; the purpose was to check Jennison's hypothesis that the source could be double. Observations at seven additional spacings at position angle 113 degree (the direction of the elongation) were carried out at slightly longer baselines (up to 5.4 km, see Fig. 8 LOWER right-hand side). Closely spaced data was obtained at baselines near 1500 wavelengths. Sullivan (2009): "They wished to check ... whether the [visibility] curve actually had passed a minimum and was rising, as one would expect if Cyg A had at least a strong second component." They did this by varying the observing frequency; thus, they were able to determine the gradient of the visibility curve. A secondary maximum was obtained at 2000 wavelengths. They also obtained data at a baseline of 12 km (5000 wavelengths) where no third maximum was detected (Fig. 1 middle panel).

←―――

Fig. 21.8 (continued) *History of Early Radio Astronomy*, Sullivan, W. T., III. (2009). Cambridge University Press, Cambridge, UK, p. 357.

---

[19]Hanbury Brown et al. (1952): "A preliminary analysis indicates that the results are incompatible with a source of simple elliptical shape and constant surface intensity, and that a more complicated model must be used."

Thus the source was a symmetrical double source; the source was compact along the minor axis (<35 arcsec).[20]

Jennison and Das Gupta submitted their paper to *Nature* ("Fine Structure of the Extra-terrestrial Radio Source Cygnus I") on 4 November 1953 (published 28 November 1953) providing a succinct summary and pointing out the dilemma that the shape of the optical galaxy had little direct resemblance to the radio source:

> The simplest distribution which will yield the transform ... consists of two components of equal intensity, each of length 51 s, separated by 1 min 28 s ... The additional information supplied by the results obtained on bearings of 179° and 58° indicates that the source has a very small minor axis, less than 35 s in a position angle of approximately 180°, and it is apparent that the components forming the source must be distributed in a narrow strip as shown.
>
> The accurate position of the radio source in Cygnus has been determined by Smith. It coincides with an extra-galactic object photographed at the 200-in Hale telescope by Baade and Minkowski [published later, 1954a, b]. This object has a maximum diameter of 30 s in position angle 150° and shows a high excitation emission spectrum with marked signs of tidal distortion. It has been interpreted by Baade and Minkowski as two late-type spiral galaxies in collision. Although it represents the most compact distribution of the radio source which can be derived from the measurements shown ... it is nevertheless much larger than the visual object described above. The two components of the radio source straddle the visual object with little overlap between the regions of optical and radio emission. If this identification of the radio source is correct, there would **appear to be no direct correlation between the radio emission and the visible light from the colliding galaxies.** (our emphasis).

This basic insight into how optical and radio astronomy might differ, beyond just "seeing further" than optical astronomy opened up new questions and ways of thinking from the late 1950s. The complex, often double lobed, structure of the powerful radio sources associated with distant galaxies like Cygnus A triggered the development of greatly improved radio source imaging techniques. The Cygnus A experience made it clear that far more extensive coverage of baselines would be required in future.[21] In Chap. 37 (Figs. 37.19 and 37.20) we illustrate how the technology and image processing techniques have developed since 1953 using the best Cygnus A images over the following 65 years as an example.

---

[20] In July 1953, Jennison presented a paper at a "Symposium on Radio Astronomy at Jodrell Bank" (reported by Hanbury Brown in *Observatory*, 1953, vol 73, p. 185). He presented the double model of Cygnus A.

[21] One of these studies involved an innovative use of the Parkes dish and a small (60 foot) antenna moving on a rail track to provide a continuously changing baseline as the radio source was tracked across the sky. This innovative technique was due to John Bolton and was the PhD thesis project (1963–1967) for one of the authors (rde).

Chapter 22
"Radio" is Part of Astronomy, 1947–1961

Radio observations are completely complementary to optical observations. There's only one astronomy and you observe in any way you can, and what's significant about radio astronomy is simply that it's coming over a short period of time and it's a time when you can rapidly go ahead because other people haven't done the ground work already.—J.L. Pawsey[1]

Introduction

In many ways, the year 1953 represented a year of consolidation for the radio astronomers at RPL under the leadership of Pawsey and Bowen. The solar research program was flourishing (see Chap. 25). Bolton, Stanley and Slee completed their survey of discrete sources at Dover Heights, by now finding 104 such sources. Mills was now developing new instruments to investigate these sources further.

In many ways, 1953 was the year in which radio astronomy ceased to be a strange specialisation of radio engineers and became embraced as a leading part of astronomy as a discipline. This coincided with the growth of radio astronomy groups in various parts of Europe, and of course in the US, to which the engine of research development would now slowly shift.

In this chapter we explore Mills's growing achievements, stories reflected in the correspondence that Pawsey, Mills and Bolton enjoyed with Walter Baade (1893–1960) and Rudolf Minkowski (1895–1976) from the Mt. Wilson-Palomar Observatory in California during this year. It is this correspondence that most clearly signals that radio astronomy was now being embraced as part of astronomy proper.

---

[1] Australian Broadcasting Corporation, television programme HORIZONS, 1960, Nov 11 interview with Joseph Pawsey and Ron Giovanelli by Moderator George Baker.

© The Author(s) 2023, corrected publication 2024
W. M. Goss et al., *Joe Pawsey and the Founding of Australian Radio Astronomy*,
Historical & Cultural Astronomy, https://doi.org/10.1007/978-3-031-07916-0_22

Radio Astronomy in 1953

As Pawsey wrote to Bart Bok: "[o]ur main ventures at present include: (1) High resolution cosmic surveys ... at metre wavelengths, (2) Cosmic source brightness distribution, (3) HI at 1420 MHz, (4) High resolution solar data at 21 cm and (5) Metre-wavelength solar disturbances–dynamic spectra from 40 to 240 MHz and simultaneous directional work."

Many projects at RPL moved forward in 1953.

Cosmic radio observations:

1. At Dover Heights, John Bolton, Gordon Stanley and Bruce Slee completed the 100 MHz survey with the sea-cliff interferometer, detecting 104 discrete sources (see Chap. 34).
2. The 400 MHz survey with the Dover Heights hole-in-the ground transit survey began (see Chap. 23).
3. The Mills Cross prototype (see below) at Potts Hill was tested by Bernie Mills and Alex Little, later leading to the construction of the full Mills Cross at Fleurs (See NRAO ONLINE.37).

HI line observations:

4. Frank Kerr took leadership of a new 21 cm HI group including Jim Hindman and Brian Robinson. The construction of a 36-foot transit telescope at Potts Hill was completed (started June 1952). A program to survey the southern galactic plane was started as part of a long-term collaboration with the Dutch who had commenced a survey of the Northern galactic plane. The joint effort produced the first HI image of the entire galaxy, showing its spiral structure. Preliminary HI observations were also made of the two closest external galaxies: the Large and Small Magellanic Clouds.

Solar radio astronomy observations:

5. Paul Wild, John Murray and Bill Rowe published a paper in *Nature* with evidence of harmonics in the spectra of Type II and Type III bursts.[2] The extended frequency coverage of the Dapto swept frequency radio telescope made this significant discovery possible.
6. the N-S arm of the Potts Hill Grating array was added to the existing E-W arm (Chris Christiansen was on leave for some of this period at Meudon in France,[3] see Chap. 25).

---

[2] NRAO ONLINE.20.
[3] NRAO ONLINE.23.

B.Y. Mills

By now, Mills's research was flourishing. Mills had initially used a swept-lobe interferometer developed by Payne-Scott and Little, at Potts Hill, to examine Cygnus A, with results that are discussed below. In 1950, increasing levels of radio interference at Potts Hill and the need for a longer baseline had prompted him to move to Badgery's Creek, which is where he invented a phase switch similar to that created by Cambridge as a result of the 'hint' provided by R McNicol (Chap. 18). There he produced a survey of discrete sources whose results were at odds with those of Cambridge, as discussed in Chap. 35, and a closer study of Cygnus A, Taurus A, Virgo A and Centaurus A, with the results discussed at URSI 1952. As discussed in Chap. 21, the Cygnus A results were later published in the amicable collaboration with Graham Smith from Cambridge and Hanbury Brown from Jodrell Bank.[4]

Mills was now clearly established as a research leader, and it is unsurprising that in 1953 he was invited to spend 6 months in the United States visiting the California Institute of Technology (Caltech) by Jesse Greenstein and the Department of Terrestrial Magnetism Carnegie Institute of Washington by Merle Tuve (Mills, 2006). Looking back, Mills commented, "The invitation came at an awkward time, but to decline was unthinkable".

Correspondence Between Mount Wilson/Palomar with RPL: 1953[5]

The Mt. Wilson and Palomar Observatories were both the creation of astronomer George Ellery Hale, who built the world's largest telescope four times in succession in the first decades of the twentieth century, emblematic of the turn towards "big science" in the US in the decades *before* WWII,[6] and who also founded the California Institute of Technology, Caltech. The Mt. Wilson Observatory was funded by the Carnegie Institution of Washington in 1904 and soon comprised a number of telescopes. The staff at the Mt. Wilson Observatory included celebrated astronomer Edwin Hubble (1889–1953), also the first to use the newly-largest-in-the-world telescope at the new Palomar Observatory when construction was completed in 1949.

From 1931, the staff at Mt. Wilson included German astronomer Walter Baade, previously of Hamburg University. In 1933, observing the terrible constraints being imposed in Germany by the National Socialist party, Baade held a staff position open

---

[4] https://arxiv.org/pdf/1306.6371.pdf.

[5] NAA C3830 Z3/2/I.

[6] See Peter Galison, in Galison et al. (1992).

for his young Jewish protégé at Hamburg, Rudolf Minkowski, who was induced to accept in 1935.[7] Despite the economic, social and political turmoil of the time, the 1930s was an exciting decade in astronomy and Mt. Wilson was often at the centre of the excitement. Hubble, of course, was famous for identifying, in 1929, that the Universe is expanding. New ideas from subatomic physics were prompting new discoveries as they were applied to astronomical phenomena. In 1930, Subrahmanyan Chandrasekhar predicted the violent collapse of white dwarf stars with >1.44 solar masses, and within 3 years, Baade and his colleague Fritz Zwicky had identified supernovae as a new class of astronomical object and ascribed their formation to the neutron star that results from white dwarf star collapse. Neutron stars were to become a big part of radio astronomy after the discovery of pulsed radio emission.[8]

At Mt. Wilson, Baade and Minkowski enjoyed a fruitful collaboration in the late 1930s. Minkowski, whose German research focused on subatomic physics, began systematic studies of supernovae in California with Baade. During World War II, Baade took advantage of the blackout conditions to resolve stars in Andromeda for the first time, and then to propose two distinct "populations" for stars, "Population I" and "Population II", whose characteristics would come to bear on the identification of the Galactic centre, discussed in the next chapter. Baade and Minkowski were thus well established as international leaders in astronomy by the end of WWII.

And in fact, in 1952, at the Eighth General Assembly of International Astronomical Union in Rome (4–13 September and thus competing with URSI 1952 in Sydney for attendance), Baade used his wartime research to announce his recalculation of the size of the known Universe, doubling that of Hubble in 1929, to the stunned audience. Ryle was also present at this meeting and as Sullivan has pointed out (2009, p. 350) the presentations on optical identifications by Baade and Minkowski had a profound impact: "After discussions with Baade at the 1952 IAU meeting in Rome, Ryle returned home to Cambridge shorn of doubt he had had about the validity of some of the claimed identifications. In a colloquium he described these developments as 'dramatic—a turning point in radio astronomy— the completion of the first stage of radio star observations.'" Baade presented tentative data on the identifications made by himself and Minkowski from 200-inch Palomar data. He described Cassiopeia A (galactic source "resembling the 1604 Kepler Nova") and the extragalactic source Cygnus A ("colliding galaxies").

Along with Jan Oort in the Netherlands (Chap. 16), Baade and Minkowski were relatively early in the astronomy community to become interested in the new radio

---

[7] See Osterbrock (2002), http://www.plicht.de/chris/12minkow.htm and https://web.archive.org/web/20090908004424/http://www.mwoa.org/hale.html.

[8] A popular book has been published describing the fascinating history of this topic "Neutron Stars—The quest to Understand the Zombies of the Cosmos" by Katia Moskvitch, Harvard University Press 2020.

observation methods. Pawsey tried hard, unsuccessfully, to convince Oort to visit Sydney. In 1953 Pawsey wrote:

> [We at CSIRO RPL have heard] that there is considerable doubt about your [Oort] being able to come out to Australia. I am writing in the hope that a further plea may help you to find a way to come. We in the Radiophysics Laboratory would particularly welcome a visit from you. It is quite clear that we should gain a lot from discussions with someone who has thought deeply about the implications of radio astronomy, as you have ... The real point is that we have here the most extensive series of radio astronomical observations in the world and we should very much like the opportunity of discussing their implications with you.[9]

Unfortunately, Oort travelled little, although he was happy for visitors to come to him (Van der Kruit, 2019). The contrast between colleagues overseas and in Australia was stark. Australia had few optical colleagues working in extragalactic astronomy who were able to collaborate with the radio astronomers.

For several years, Baade and Minkowski had collected information from Cambridge, Jodrell Bank and Sydney. And naturally they were among the experts to whom these groups turned for help in finding optical identifications for their mysterious discrete radio sources.

On 16 September and later on 3 December 1952, Pawsey wrote Baade. From the former letter Pawsey summarised his understanding of the identification of the discrete radio sources:

> The sources on which there are reasonable clues at present ... are (1) Cygnus, (2) Perseus (NGC 1275), (3) Virgo ("probable colliding galaxy" [sic]), (4) Taurus (Crab), (5) Tycho's Supernova (old supernova), (6) Cassiopeia, (7) Puppis (peculiar nebulous object), (8) Centaurus (peculiar galaxy). The unifying thought is the probable existence of violent motion.

Then in late 1952, Pawsey contacted Baade in regard to obtaining radio source optical identification information for the textbook he was contracted to write, *Radio Astronomy* (see Chaps. 19 and 24). As he said, "since so much of the optical work comes from Mt Palomar it is clear that your help could be of very great value."[10] (He was clearly not expecting to find the help he needed from astronomers in Australia who had no access to the highly advanced optical instruments available in the US.)

This letter from Pawsey sparked a considerable correspondence between Mt. Wilson/Palomar and Sydney before Mills left for his visit to the US in August 1953.[11] This correspondence helped produce a key paper for the decade on optical

---

[9] Pawsey to Oort 17 July 1953. Oort archive Leiden, University Library. Oort has written (in Dutch) at the bottom of the Pawsey letter that he responded later to Pawsey on 12 August 1953. Oort noted that he told Pawsey (in Dutch "in de selfde trend als mijn antwoorden aan Woolley"; "in the same manner as my [earlier] answers to Woolley".) Clearly, Oort turned down both invitations to Australia, to RPL and to Mt. Stromlo. The letter to Pawsey has not been located. Oort was to visit Australia for the first time in 1963 for the IAU Symposium, after Pawsey's death the previous year.

[10] 16 September 1952, Pawsey to Baade. Baade and Minkowski had just presented their results on Cygnus and Cassiopeia A at the IAU in Rome.

[11] Except where noted, from NAA C3830 F1/4/MIL/1 and A1/1/1 Part 8, 1953.

identifications. On 3 March 1953, Minkowski sent Bowen a draft of the two papers on optical identifications that would be published in January 1954 in the *Astrophysical Journal,* "Identifications of the Radio Sources Cassiopeia A, Cygnus A and Puppis A", (Baade and Minkowski, 1954a, p. 206), and "On the Identifications of Radio Sources", (Baade & Minkowski, 1954b, p. 215). In NRAO ONLINE.37 we summarise some aspects of Mills's activities in 1953: (1) correspondence with Baade, (2) construction of the prototype Mills Cross, and (3) conference in Boston end of December 1953.

Baade and Minkowski commented: "We are afraid that the generous way in which we have been supplied with unpublished information may have led to misquotations and misinterpretations on our part. We would like very much to be corrected before these papers are sent to the *Astrophysical Journal* in about three or four weeks." (The papers were submitted 3 months later on 19 June 1953.)

After some weeks, Mills, Bolton and Pawsey had all provided comments on the two publications from the colleagues in Pasadena, California. Pawsey wrote to Minkowski (the final letter went out under Bowen's name on 30 March 1953)[12]: "[We] are most impressed with the tremendous advances that you [Minkowski] and Baade have been able to make in this subject." He gently requested that the paper acknowledge the positions that Bolton and Mills had, at different times, proposed in correspondence in the late 1940s, particularly Mills's early identification of Cygnus A, which had not been accepted until Smith's later improved position (Chap. 18). Pawsey:

> I felt that it definitely added to the interest of the paper that you should describe the way in which you were led to the study of the Cygnus and Cassiopeia regions with the 200-in. telescope by a letter from [Graham] Smith. I wondered if it would not add to this interest if you mentioned something of the earlier discussions with Mills concerning the same nebula . . . I am not sure how many of these historical snatches should go in the paper.

Pawsey concluded: "I should like to add in conclusion my appreciation of the collaboration which has gone on between yourselves and the various members of [our] Laboratory. Your interest and advice has been a tremendously important factor in helping us in our research in radio astronomy." Baade replied that he "hope [d] very much that the cooperation with the different radio groups [Sydney, Cambridge and Jodrell Bank] which has been so fruitful will continue in the future."[13]

The two papers (Baade and Minkowski) of January 1954 became the touchstones of optical identification of radio sources. Sullivan wrote that they became "the bible" on optical identifications: "They were authoritatively written and filled with photographs, spectra, historical details, and copious notes gained from circulating drafts to the three major radio groups [Sydney, Cambridge and Jodrell Bank] in March 1953."

---

[12] The initials at the top of the letter read: "JLP:DJB", dictated by JL Pawsey to the secretary "DJB".

[13] Baade to Pawsey on 30 March 1953: "I was very glad to learn that our draft paper on the identification of radio sources found your approval. Minkowski and I had necessarily to trust the astronomical aspects [of your input] since neither of us is a specialist in the radio field. But it makes us very happy that our views and those of the radio [experts] seem to be in essential agreement . . ."

The Mills Cross, March 1953

While Pawsey and Bowen were writing to Minkowski, Mills was writing to Baade, on 20 March 1953.[14] He wrote with some details of his own upcoming publication, "The Radio Brightness Distribution over Four Discrete Sources of Cosmic Noise" which discussed the four sources: Cygnus A, Taurus A, Virgo A and Centaurus A. At the time, Mills was preoccupied with various issues of detail in the publication—such as an error of orientation.[15] Reflecting on such issues from the vantage point of 50 years later (in his 2006, in an autobiographical text in *Annual Reviews of Astronomy and Astrophysics*[16]), Mills commented on the problems of the Centaurus A observations of 1953: "[The simple two-dimensional models] gave very misleading information for Centaurus A because of the complexity of the source, the limited number of observations, and the absence of phase information."

Mills also sought Baade's advice concerning his next work. Should he continue with observations to determine more precisely the detailed structures of the four bright sources or should he consider a survey "with larger aerials . . . with the object of obtaining rough angular sizes of considerably more sources . . . ?" Mills told Baade that he favoured the second option. "I would be very interested to hear your opinion which is likely to be more rewarding from the optical point of view."[17]

Mills was seeking Baade's views on the new instrument he was planning—what would become the Mills Cross. The need for this instrument had been under consideration since the previous year. The understanding cemented at URSI in 1952, ie, that many radio sources were extended and likely to be nebulae and other objects rather than stars, was connected with Mills's musing about challenges in using spaced interferometers for survey work. After the survey carried out by Mills in 1951, "The Distribution of the Discrete Sources of Cosmic Radio Radiation" (1952, p. 266) that yielded 77 sources with the Badgerys Creek instrument at 101 MHz (see Chap. 35), a new survey was indicated. Mills hoped for an increase of an order of magnitude in the number of detected sources. High resolution rather than sensitivity was (in his view) key to source survey research at metre wavelengths. "By then, I knew that collecting area was relatively unimportant; the important thing was a large overall size to give high resolution."[18] However the higher angular resolution

---

[14] Minkowski archive University of California, Berkeley, courtesy W.T. Sullivan, III. The content of this letter was independent of the correspondence occurring in March 1953 concerning RPL comments on the draft publication from Baade and Minkowski regarding the identifications of Cas A, Cygnus A and Puppis A.

[15] See discussion by Robertson et al. (2010), "Early Australian Optical and Radio Observations of Centaurus A", p. 402.

[16] Mills wrote an outstanding autobiographical text in 2006 "From Engineer to Astronomer"; this was the first autobiographical text written by a radio astronomer in *Annual Reviews of Astronomy and Astrophysics*. (Vol 44, page 1).

[17] We include details of this hitherto unpublished correspondence between Baade and Mills in NRAO ONLINE.37.

[18] Frater et al. (2013).

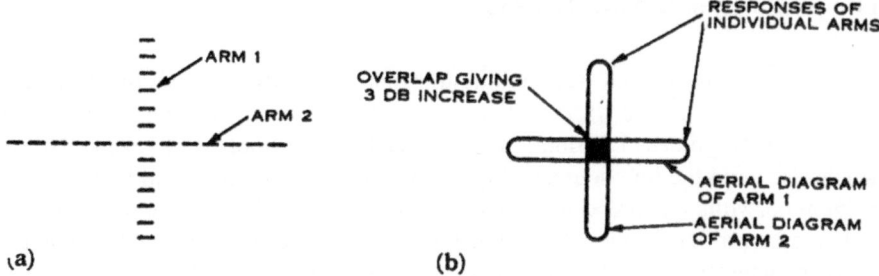

Fig. 22.1 Mills cross conceptual design; (**a**) the crossed arrays of dipole elements, (**b**) the resulting beam patterns from the two orthogonal arms and the overlapping area (filled) resulting from the product of the two fan beams. Credit: Fig. 1 from "A high-resolution aerial system of a new type", Mills, B. Y., & Little, A. G. (1953), *Australian Journal of Physics*, 6(3), 272–278

was to have a downside as we discuss in detail in Chap. 36. The impact of the resolved extended sources on the counts of radio sources contributed to the great survey controversy between Sydney and Cambridge.

As a filled array seemed "wasteful",[19] the solution was constructing a partially filled antenna, as Mills later explained; we include the full quotation here because it elegantly sums up the thinking leading up to the Mills Cross (Mills, 2006, p. 6):

> By now it was clear that the spectrum of nonthermal emission was such that at low frequencies the performance of a very large antenna would be dominated by resolution rather than sensitivity. Accordingly, I began thinking of constructing partially filled antennas such as rings, squares, and crosses, but all suffered from severe problems with unwanted [sidelobe] responses. A solution occurred to me after discussing the imaging problem with Christiansen who was using two grating arrays along the sides of a reservoir to produce maps of the Sun by the first application of earth rotation synthesis. However, fast imaging was really needed because of the variable solar emission, quite apart from the inconvenience of carrying out Fourier transforms when no computer was available. With my thoughts concentrated on linear arrays I soon realized that a solution to both our needs was an antenna in the form of a symmetrical cross, with the outputs of the arms combined through a phase reversing switch as then used in my interferometer systems. Only the signals received in the overlapping area of the fan beams would produce a modulated signal that could be picked out with a phase-sensitive detector to produce a simple pencil beam response or, in the case of grating arrays, an array of pencil beams. This process effectively multiplied the two antenna responses (see Figs. 22.1 and 22.2).

In hindsight, although Mills's deductions are logical, we can see that his hidden assumptions and biases led him to make at least two erroneous assumptions: (1) about the nonthermal spectrum and (2) that computers would not improve.

All the exotic radio sources being discovered before the mid-1960s had spectra which made them stronger at low frequencies, but low frequencies meant either low angular resolution or high spurious responses (sidelobes), so Mills adopted the concept of large diameter arrays which were sufficiently filled to suppress the

---

[19] See footnote 16.

Fig. 22.2 Mills Cross Fleurs 1954, 85.5 MHz, 25 October 1994. Looking south along the north-south arm. The area to the south and slightly to the east (left) was the site of the future Chris Cross (late 1957) and Fleurs Synthesis Telescope (1975). Credit: CSIRO Radio Astronomy Image Archive B3476-3

spurious responses but didn't "waste" collecting area. In retrospect we see how the situation changed in a number of ways. A rare but very exotic new class of radio source was found, the quasars (see Chap. 32), and these were often stronger at high frequency. Unanticipated improvements to receivers made it possible to achieve higher sensitivity at higher frequency where the background noise is lower. Perhaps the biggest unanticipated surprise was the need for flexibility, eg to change frequency for new spectral lines (eg the OH transition at 18 cm). This flexibility is not possible for an array with many elements.

At this time Ryle's group at the Cavendish Laboratory in Cambridge were also experimenting with unfilled aperture concepts (see Chap. 37). Their paths were starting to diverge in ways which would make a dramatic difference some 5 years later. Ryle's group were measuring individual Fourier components one spacing at a time and were becoming very dependent on computed images. Mills however did not consider "carrying out Fourier transforms when no computers were available" credible so his unfilled apertures had enough elements to form a 2D beam in real time. These were formally equivalent procedures but the Mills approach was entirely analogue. In retrospect it was the enormous expansion in computational capacity and need for flexibility that eventually made the Mills Cross a dead-end.[20] We are in 2021 at this same crossroad in the current era when SKA and other large arrays have

---

[20]Ryle and Hewish (1960, p. 220) noted: "Apart from a considerable economy of structure this method [aperture synthesis] avoids some of the difficulties associated with the physical achievement of a graded excitation of amplitude and phase which is required in the case of large extended arrays such as the Mills Cross. Besides allowing greater collecting areas to be realised, the shape of the reception pattern can be adjusted, by computation alone, to suit different types of observation.

to resort to Mills style beam forming to observe fast transient events (such as pulsars) which far exceed the capacity of our fastest supercomputers.

The proposed new Cross did not meet with immediate approval. Doubts and criticisms were expressed, notably from Bowen, but these were related to costs and risk management, and not the design concept. But as Mills wrote retrospectively, "Pawsey supported me and gave approval for the construction of a small experimental model to explore the technique. He also assigned the laboratory's brightest young Technical Officer, Alec Little, to help, and this was the beginning of a long and fruitful association." A quickly constructed prototype (at Potts Hill) confirmed all expectations, even detecting continuum radiation from the Large Magellanic Cloud for the first time (Mills & Little, 1953). The full Mills Cross was constructed on a disused WWII airstrip at Fleurs (near Mills's previous instrument at Badgery's Creek).[21]

As Mills's departure date for the US in early August 1953 approached, he prepared a detailed proposal for the full cross and presented it to Pawsey for approval. Mills's initial cost estimate was £1500 without including internal labour; the final estimates made a few weeks later, including internal labour, would be increased to £2500 or A$85,000 in 2018 dollars. The costs did not include cost of the site, fencing or painting. The minimum time to cover the observable sky was 3 months. The beam swinging would be ±1.5° with rapid scanning and ±40° in discrete steps; the adjustment of each major step in declination required half a day's efforts.[22]

On 1 April 1953, Pawsey approved the proposal. The key players in the subsequent construction were Alec Little, responsible for the aerial, and Kevin Sheridan, responsible for the receiver. Other participants were Alex Shain, Alan Carter, Keith McAlister and Arthur Watkinson. Shortly before Mills's departure to the US, detailed responsibilities were discussed by Pawsey: "Organisation of work during Mills's absence in USA". Mills commented that he had "no qualms about leaving the supervision of construction [of the Fleurs Mills Cross] in the capable hands of Alec Little." (Mills, 2006, p. 10) The completed Mills Cross at Fleurs in 1954 is shown in Fig. 22.2.

The method necessarily involves considerable computation, but this does not present a serious problem with the large electronic computers now available."

[21] The new solar site at Dapto was not flat enough and was "too far away" from Sydney. The new site would require 1500 feet (460 m) for each of the orthogonal arms. The Fleurs site will be memorialised in the proposed Western Sydney Airport museum, now under construction.

[22] Slightly later in June 1953 a tentative plan was made to extend the size of the Cross. A sketch was made on 9 June 1953 with extensions of 750 feet to the E, W, N and South (230 m). However, this doubling of the resolution was never completed. NAA C3830 A1/1/1 Part 9.

Mills's Visit to the US August 1953–February 1954

Mills left for the US in early August, 1953, with the transition from California to Washington, D.C., starting 20 November 1953. He returned to Sydney on 1 March 1954, via Hawaii with a visit to Grote Reber in Maui.

Mills described the major impact of this visit on his career:

> This visit was well worthwhile as the few months spent at Cal Tech marked a turning point in my grasp of astronomy and astrophysics. Discussions with some of the leading astronomers and astrophysicists of the day (particularly the iconoclastic Fritz Zwicky), attendance at colloquia, and even a postgraduate course on stellar structure all helped to fill in some of the numerous gaps in the knowledge I had managed to acquire. I returned home in early 1954 with my mind full of plans for observational programs. (Mills, 2006, p. 10).

Of course, as all the Australians did while overseas, Mills attended several conferences during his visit. The first was an American Association for the Advancement of Science meeting, "Symposium on Radio Astronomy" Section D-Astronomy, 26–27 December 1953, Boston. Bart Bok was the chair of Section D. (See NRAO ONLINE.37 for details) We note that Mills wrote to Pawsey on 30 September 1953, anticipating further difficulties with Cambridge (see Chaps. 18, 35, and 36):

> I have agreed to speak and will be sharing the platform with Smith to talk on "radio sources". Unfortunately, they wish to bring out a book of the symposium papers.[23] The question is whether it is OK to contribute to a publication of this sort. Naturally I am not too keen to waste valuable time preparing a formal paper which will be necessary, particularly as it will have to be fitted in with what Smith is to say, which will undoubtedly cause endless trouble. However, if a book is to be brought out I suppose the best plan is to contribute a paper. What are your reactions to this? I have been in touch with Smith and we have agreed to cut it short and omit controversial points if it should come to publication.

From 4–6 January 1954, the National Science Foundation, the Carnegie Institution of Washington, and the California Institute of Technology organised "The Washington Conference on Radio Astronomy-1954".[24] Bowen and Mills were both present.

Mills discussed "CSIRO Results on Shapes, Sizes and Spectra of Radio Sources" and "The Galactic Noise Background" (including the 18 MHz data obtained by Alex Shain), describing his results, but with no mention of the ground breaking research done on optical identifications in 1949 by Bolton, Stanley and Slee. He also gave presentations describing the prototype Mills Cross at Potts Hill and the plans for the complete 85 MHz Mills Cross completed later at Fleurs in 1954.

Bowen presented a series of papers reporting the latest on the solar research at RPL: "Some Recent Results in the Study of Radio Emission from the Sun", "Solar Research at CSIRO", and "A New Phenomenon in Solar Radio Noise". He also

---

[23] The planned publication was to have been by the Harvard University Press; however, the book was never published.

[24] Burke (1954, p. 149) and *Science,* 30 Apr 1954: Vol. 119, Issue 3096, pp. 576–588.

signalled something of the future at RPL with a presentation on "A New Radio Telescope Design: the Big Antenna versus the Interferometer Array".[25]

The title of this last talk of Bowen's gestured to a fracture that had already taken place at RPL—John Bolton's frustration at the failure at RPL to build the instruments in which he was interested (and by default, with the apparent priority given to arrays with a large number of small elements such as the new Mills Cross). The "learning organisation" model was reaching its limits.

---

[25] Described in NRAO ONLINE.39.

Chapter 23
The Galactic Centre, 1951–1954

Now to the object in the centre of the Galaxy, the contour diagram of which you kindly included in your letter. Frankly, I jumped out of my chair the moment I saw what it meant. I have not the slightest doubt that you finally got the nucleus of our Galaxy!! 16 Feb 1954, Baade to Pawsey—a handwritten letter.

Introduction

The discovery of radio source Sagittarius A (Sgr A) and its association with the centre of the Milky Way is a fascinating story, involving RPL personnel and prominent US and Dutch astronomers. When we say "discovery", however, we do not mean a single event. Contra the conventions in science that award prizes, professional respect and that very nebulous (!) phenomenon of "historical recognition" to individuals, discovery is a lengthy process involving many actors, many different kinds of contributions, and many events. This was understood by one of the actors involved in this story: the famous astronomer Hendrick "Henk" van de Hulst, who had predicted the existence of the HI line at radio frequencies. Reflecting back on the history of radio astronomy, van de Hulst suggested the concept of "nanohertz astronomy"– that is, history of astronomy on a longer timescale.[1]

Of "nanohertz astronomy", van de Hulst wrote:

Nanohertz astronomy [is] the art of registering the coming and going of astronomical convictions in periods of the order of 10^9 s = 30 years. This approach is complementary to the common one, where the history of science is described by focusing on the sudden

Supplementary Information The online version contains supplementary material available at [https://doi.org/10.1007/978-3-031-07916-0_23].

[1] See ESM 23.5: Van de Hulst's shared interests in the history of astronomy with W.M. Goss and experience during WWII, Project Window Aluminium Foil.

discoveries and the rapid breakthroughs of insight. This complement is as necessary as are the added measurements at very short spacings in the Fourier synthesis of extended sources. Otherwise a broad underlying valley or elevation might be misjudged and the basic structure misinterpreted. And—to continue this metaphor—the historical development of science is indeed such an extended source with a highly complex structure.[2]

The narrative of the discovery of the Galactic centre in the terms of nanohertz history includes many events. It began with Piddington and Minnett at RPL, was quickly moved forward by John Bolton, Bruce Slee and Kevin Westfold, with major contributions from Dick McGee and Joe Pawsey under the excited influence of comment from optical astronomers Otto Struve (1897–1963), Walter Baade, Jan Oort and Henk van de Hulst himself. This laid the groundwork for successive events over the ensuing decades, culminating in the award of the Nobel Prize in 2020 for the confirmation of the black hole at the Galactic Centre.

As with many discoveries, almost all the conditions laid out by sociologist of science Robert Merton and Elinor Barber (2004) came into play in the early stages of this chain of events, those with which we are concerned here. Serendipity (the focus of Merton and Barber's book) played a big role in Bolton, Slee and Westfold's celebrated discovery: (1) they had built an inexpensive transit telescope by digging a parabolic hole in the ground and lining it with conducting wire mesh. This could best observe the region of sky directly overhead; by chance the centre of the galaxy passed almost overhead at the southern latitude ($-34°$) of Sydney; (2) the centre of the galaxy has a rather flat spectral index, so more prominent at higher frequencies. Piddington and Minnett and Bolton were observing at higher frequencies. Bolton had pushed to these higher frequencies to obtain sufficient angular resolution with his 72-foot dish. By comparison, at Fleurs with the new Cross, Mills was using a long baseline array and beam forming with many small elements to obtain high angular resolution at lower frequencies (80 MHz). However, the radio source at the centre of the galaxy disappears at lower frequencies due to thermal absorption in the plane of the galaxy.

Soon after Bolton's team started the sky survey with the "hole in the ground" antenna, the prominent source in the Galactic Centre was apparent, confirming the earlier suggested identification by Piddington and Minnett (1951a, b) of strong continuum radio emission which might be associated with the centre of our galaxy. At this point, Merton and Barber's observation that knowing too much can hinder new discoveries came into play: the optical astronomers had used observations of the stars in the Milky Way to determine that the Milky Way was a flattened disk. Thus they had determined the location of the centre of this disk using stars. Unfortunately, the strong absorption of starlight by dust in the centre of the galaxy implied that the initial position was 30 degrees away from the strong radio source (with no extinction due to dust), existing at the Galactic Centre. In the 1950s, astronomers aware of the then-presumed optical position would likely have questioned the association of this radio source with the centre of the Milky Way. Bolton and McGee jumped to the

---

[2]H. C. van de Hulst, "Nanohertz Astronomy", in Sullivan (1984) p. 385.

conclusion that their observation was of the Galactic Centre; but the more conservative members of Pawsey's group were cautious.

Fortunately, strong support for the Galactic Centre interpretation came from both Baade at Mt. Wilson/Palomar Observatory and Oort and van de Hulst in Leiden, but for very different reasons. Baade was looking for support for his idea that the galactic bulge region was the real centre of the galaxy and, based on the (outdated by 1954) "radio star" model of the galactic emission, it should be a strong radio source. The identification of the Sgr A radio source with the Galactic Centre was just what he had been looking for. Very different evidence came from Oort and van de Hulst in the Netherlands. New observations of 21 cm hydrogen line doppler shift in velocity had indicated the position of the anticentre and this was exactly 180 degrees away from the position of the radio source Sgr A, not consistent with the old optical position of the Galactic Centre.

In this chapter we discuss the roles of Pawsey, McGee, Bolton and Piddington and Minnett in the discovery of the Galactic Centre, drawing on new material outlined in ESM 23.1, Discovery of Galactic Centre.[3] This renders the contributions of Piddington and Minnett as well as Pawsey more visible, and affords them the recognition that featured so strongly in Merton's analysis of reward systems in science.

The Piddington and Minnett (1951) Observations and Interpretation

Piddington and Minnett (1951a, b), had observed at the relatively high frequency of 1210 MHz, using a 10-foot and later an 18 by 16-foot prime focus antenna at Potts Hill Reservoir (see Fig. 23.1a, b, see also NRAO ONLINE.23 Additional Note 1. They detected a prominent discrete radio source in the Sagittarius constellation. The larger aerial had been used earlier by Lehany and Yabsley to extend observations of the solar disk to higher radio frequencies. At 1210 MHz, this aerial had a beamwidth of 2.8°, comparable to the 2° resolution of the later-developed 80-foot hole-in-the-ground aerial used by McGee.

Piddington and Minnett describe a "... new, and remarkably powerful, discrete source" at 1210 MHz, the "Sagittarius-Scorpius Source"[4] (the position was close to

---

[3] We use primary material that supplements the sources available to Bland-Hawthorn and Robertson (2014, p. 194–199).

[4] Palmer and Goss (1996) in "Nomenclature of the Galactic Center Radio Sources" have pointed out that most of the early papers used the terminology "the Galactic Centre Source". The earliest reference to "Sgr A" is likely the paper by Kraus, Ko and Matt in 1954, a report on the 250 MHz all-sky image made at Ohio State and discussed at the June 1954 American Astronomical Society Meeting at Ann Arbor, Michigan. In this paper we use the name "Sgr A", although this was not the contemporaneous name in Australia. In the 1950s in Australia, the name Sgr A was not used,

Fig. 23.1 (**a**) The 18 by
16-foot aerial at Potts Hill.
Credit: CSIRO Radio
Astronomy Image Archive
B2649–2. (**b**) Ken Nash,
RPL photographer, photo of
a long exposure star trail on
8 August 1952 with the
18 by 16-foot aerial at Potts
Hill. Credit: CSIRO Radio
Astronomy Image Archive
B2839

the border of the two constellations). These authors suggested a likely identification
with the centre of the Galaxy, a point source with size less than 1.5 degree.

---

presumably because this early position was on the boundary of two constellations. Piddington and
Minnett (1951a, b) referred to the source as the "Sagittarius-Scorpius Source".

Piddington and Minnett (1951a, b, pp. 468 and 469):

> The measured flux density of 26,000 Jy indicates a particularly powerful source (at 1210 MHz), its presence has not previously been reported at lower frequencies.
>
> An interesting feature of the Sagittarius-Scorpius source is that it lies close to the galactic plane and very close (within experimental uncertainty) to the plane defined by the maximum level of radio emission. It also lies very close to the centre of the Galaxy and to the maximum of galactic radiation ... The significance of the position may be considerable. If the source were relatively close to the Sun it could lie in any direction with equal probability. If, on the other hand, it was at a distance from the sun which was a considerable fraction of the galactic diameter, it would be more likely to lie in a direction close to the plane of the galaxy.

Piddington and Minnett also looked at a possible identification with the galactic nebula NGC 6451; they did not give much credence to the association:

> Although, as seen above, the accuracy of location of the source is not high [0.5° in right ascension and 1° in declination], it may be significant that the position found almost coincides with that of the galactic nebula NGC 6451, a loose cluster of about 70 stars extending over 15 min of arc. A much more accurate determination of position is required, however, before the coincidence is given serious consideration.

Thus the suggestion was that the new source was likely at the galactic centre. The possible association with NGC 6451 was not considered again.[5]

Piddington and Minnett then discussed the possible luminosity of the new source, based on an assumed distance of 10 kpc (in 1951 assumed to be the distance to the centre of the Milky Way; modern value is 8.3 kpc). They also noted that the Class I sources (the radio sources close to the galactic plane, see Mills 1952a, b) show a concentration towards the galactic plane: "The implication is that some sources lie at distances much greater than the thickness of the Galaxy in the vicinity of the sun, distances of the order of 10 kpc." The spectrum of Sgr A was also observed by Piddington and Minnett to be flat, between 100 and 1210 MHz; they made the analogy with another flat spectrum source, the supernova remnant Taurus A. They suggested that the spectrum "resembles that of an optically thin, thermally emitting gas", but were aware of the problem with this interpretation in the "pre-synchrotron emission" era. [See Chap. 34].

Since, in Merton's analysis, recognition of priority is a significant component of the reward system in science, we suggest that some recognition might accrue to Piddington and Minnett—a view shared by their colleague Dick McGee.[6] Of all the

---

[5] Our interpretation differs slightly from that of Bland-Hawthorn and Robertson, who suggested that "Piddington and Minnett hesitated in claiming Sagittarius A to be the Galactic centre simply because astronomers only had an approximate idea of its location" and considered that "Sgr A might coincide with the Galactic nebula NGC 6451, in fact an open cluster of about 70 stars." Our view is that Piddington and Minnett's choice of words indicate confidence about the identification rather than hesitation, especially given that even in 1951, it was realised that open clusters were not likely to be prominent radio sources.

[6] In 1996, Goss and Dick McGee were preparing a paper for an IAU Galactic Centre Symposium in La Serena Chile, "The Discovery of Sgr A" (Gredel, 1996). At this time, McGee was convinced that Piddington and Minnett should be given the initial credit for the association of this radio source with the galactic centre. His new observations at Dover Heights at 400 MHz were of the same source

possible detections of Sgr A done before 1954, only the Piddington and Minnett detection had suitable combination of frequency and angular resolution to separate Sgr A from the confusing background due to the intense galactic plane in the region of the galactic centre. Since the spectrum of Sgr A is essentially flat at 1 GHz, the contrast of Sgr A, with respect to the steeper spectrum diffuse galactic emission at 1210 MHz, was considerably enhanced compared to the existing low frequency radio data in existence in 1951.[7]

The importance of Piddington and Minnett's discovery was indicated by the interest shown by Baade and Minkowski, already in November 1951. Bowen was visiting Caltech when he wrote to Pawsey on 9 November 1951[8] inquiring on behalf of the famous Pasadena astronomers. They were interested in obtaining accurate positions of the new galactic centre radio source as well as the radio spectra. Even though the Piddington and Minnett paper was published in the *Australian Journal of Scientific Research*, not widely read by the International community, the paper is well cited for this period. The 24 citations include many influential astronomers: Minkowski, Greenstein, de Vaucouleurs, Haddock and Geoff Burbidge. Thus we recognise that Piddington and Minnett had played an important role in the discovery of the galactic centre radio source.

The New "Hole-in-the-Ground" Telescope at Dover Heights

The story of John Bolton's observation of this same source is an enthralling one. In 1951, Bolton had become frustrated with the effects of confusion as he and colleagues used low frequency interferometers to carry out radio source surveys. He recognised the advantage of using higher frequencies and larger dishes.[9] Famously, late in 1951 he began constructing a 72-foot, 160 MHz hole-in-the-ground antenna at Dover Heights (Fig. 23.2).

As is now well known, the construction of this antenna was at first kept secret from Pawsey because Bolton did not have approval for the project. Until this time,

---

earlier detected at the higher frequency of 1210 MHz. McGee's notebooks in the National Archives of Australia (C4633/3 from 1953 to 1954) are consistent with this assertion.

[7] At low frequencies, the non-thermal supernova remnant Sgr A East is a major component of the radio emission near the galactic centre. At frequencies below roughly 300 MHz, ionised HII gas absorbs Sgr A East (Pedlar et al., 1989, p. 769). At 1210 MHz (Piddington and Minnett, 1951a, b), the contribution of the extended galactic background with a resolution of 2.8° (Piddington and Minnett, 1951a, b) is much less pronounced than at low frequencies. For example, at 1210 MHz the galactic background in a beam of a few degrees is about 20 times less intense than at 400 MHz, a frequency at which the galactic background and Sgr A have comparable intensities.

[8] NAA C3830 Z1/9.

[9] The concept of using larger dishes as the elements in an interferometer finally emerged when Bolton moved to Caltech and built the Owen's Valley Radio Observatory (letter from Bolton to Don Morton 3 June 1985).

Fig. 23.2 In the foreground is the 72-feet hole-in-the-ground built in 1951 for a survey of the region near the galactic centre at 160 MHz. In the background is the 16-feet reflector built in 1950 mainly for instrument development in the decimetre wavelength range. Credit: CSIRO Radio Astronomy Image Archive B2763-1

Pawsey had strongly encouraged many separate small groups to pursue their own ideas. But the previously almost unlimited funding and supply of WWII equipment was drying up; Pawsey had to start prioritising and limiting the number of experiments and sites that could be supported. Tensions were building up in the radio astronomy group as a result.

In this environment Bolton was pushing to keep the Dover Heights site with a small number of variable baseline interferometers on the cliffs, and using larger size elements and the sea-cliff interferometer to beat the confusion which was plaguing the study of the weaker discrete sources. Pawsey did not approve Bolton's proposal for a 72-foot (later 80-foot) "hole in the ground" transit dish. So Bolton, Slee and Westfold built the telescope themselves, with their own labour, at minimum cost (Bolton, 1982). Gordon Stanley hauled ash from the Bunnerong Power Station at Matraville, a distance of about 14 km to the south of Dover Heights, to stabilize the sand at the cliff top at Dover Heights.

Short reminiscences by Slee (Goddard and Haynes, 1994, p. 517) provide a flavour of the excitement of this new endeavour:

> Most of the dish was excavated manually from the sand, with the spoil being used to build up the outer rim. John Bolton and I did most of the work as a lunchtime project over several

months, keeping it largely secret from the rest of the laboratory until we could obtain some new experimental results. It was first used as a 72-foot dish at 160 MHz [6° beam] and had a rather crude reflecting surface of parallel steel strips obtained from packing cases.

Kevin Westfold (1994) recalled that:

> [s]ince the project had not gained a high enough ranking in competition with projects from the other groups, John felt that he had to proceed independently. After laying pegs we stretched out strips of steel packing strip and erected a central tiltable mast to carry the dipole.

The instrument was first put to use in mid-1952. Bolton recalled: "Our first observations were at 160 MHz and the increase in detail shown by these was sufficient to persuade Joe to let us improve the surface accuracy and resolution by going to 400 MHz."[10]

Slee remembered: "After a quick survey of the central strip of the Milky Way (the beam position was changed by tilting the feed mast), we found the results encouraging enough to release the news." Dick McGee's memory concurs (McGee joined the group later); in letters to Goss, he wrote[11] "[t]he original hole-in-the-ground, 72 feet and 160 MHz, was kept secret from Joe Pawsey, but Taffy was shown the operation. If I remember correctly the 160 results were given out at the 1952 URSI in Sydney. Obviously, Joe was made aware of it then."

The only image published from this telescope was of the centre of the Galaxy, Fig. 23.3 taken from Bolton et al. (1954a, b), p. 96.

These encouraging results seem to have been nothing less than what Bolton, Slee and Westfold were sure was a (very crude) map of the Galactic Centre. Westfold: "Gordon and Bruce had already constructed a receiver, so it was not long before we had a map of the Galactic centre for John to wave in front of E.G. (Taffy) Bowen, chief of the Radiophysics Laboratory."[12]

Pawsey's Interest and Bolton's Departure, 1952–1953

Pawsey once commented that:

> the ideal way of starting a new research is to do it on another job number and then you don't have to tell anybody what you're doing and if it fails it doesn't worry anybody; but if it's successful then you can go to your immediate superior and say, "look how good this is" and he sort of scratches his head and says, "perhaps you shouldn't have been doing this," but if

---

[10]Letter from Bolton to Don Morton 3 June 1985 (see ESM 23.2, Roles of Bolton, Pawsey, according to Don Morton, 1985).

[11]Letters to Goss on 2 and 21 January 1996, the former with the "notes" referred to in ESM 23.1, Discovery of Galactic Centre."

[12]McGee's interpretation was: "Such activity seems to be extremely questionable of group loyalty, but it appealed to John enormously and Taffy, being jealous of the various honours heaped on Joe, would be too delighted to take part in the deception." Letters to Goss on 2 and 21 January 1996 (see ESM 23.1, Discovery of Galactic Centre).

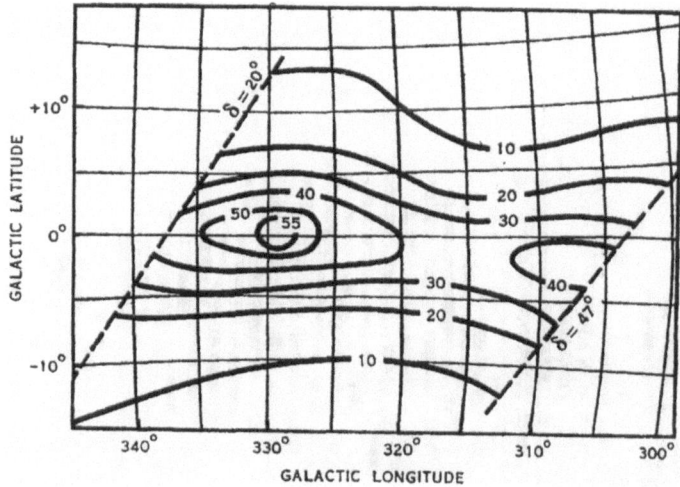

Fig. 23.3 Beam width is 6° for this image of the galactic centre at 160 MHz, made with the 72-foot "hole in the ground aerial" at Dover Heights. Credit: Fig. 1, "Galactic Radiation at Radio Frequencies. VIII. Discrete Sources at 100 Mc/s Between Declinations +50° and −50°", Bolton, J. G., Stanley, G. J., & Slee, O. B. (1954), *Australian Journal of Physics*, 7(1), 110–129

he's a good man he would never mention that … (Australian Broadcasting Corporation Television Interview, 1960).

As he had done in response to Bolton's first observations of the first radio source, Cygnus A, Pawsey did indeed provide resources on this demonstration of success. The group was given approval to upgrade the telescope, work that commenced in late 1952. Westfold: "The next thing I knew was that the paraboloid was being constructed in concrete, incorporating a reflecting mesh, so that a clean map, which showed a strong source in the Galactic nucleus, could be made."

Slee remembered:

[we got] Pawsey's approval to upgrade the construction. This involved extending the dish to a diameter of 80 feet with an accurate concrete surface in which was embedded a chicken-mesh reflector. Gordon Stanley designed a low-noise, 400 MHz preamplifier [2° beam] and Dicke switch to go at the focus … At this stage we were joined by Dick McGee (replacing Kevin Westfold) who, with my assistance, was largely responsible for the 400 MHz survey of the central strip of the Galactic plane. (Figs. 23.4 and 23.5 show the antenna with Dick McGee in the dish).

McGee recalled:

The early part of 1953[13] was taken up with completing the 80-foot telescope and much effort in making Gordon's (Stanley) [radio frequency] cavity switch work (it was never a success) together with some "engineering" to make the famous Stanley-Steamer (a very wideband oscillator invented by Gordon) easily tuneable. (The way Gordon had it tuning was by pure

---

[13] When McGee started working at Dover Heights.

Fig. 23.4 The newer more accurate 80-foot dish at Dover Heights. Dick McGee adjusts the mast. View to the north towards North Head and the Sydney Harbour entrance. Credit: CSIRO Radio Astronomy Image Archive B3150-1

Fig. 23.5 Another striking photo of the 80-foot aerial, possibly with either Dick McGee or Gordon Stanley measuring the position of the feed, view to the east over the Tasman Sea. Credit: CSIRO Radio Astronomy Image Archive B3150-2

chance.) In April the Stanley front end system was replaced [by] Bruce Slee's preamp and electronic switch, and the sky survey began on the 1st of May [1953].

But at this point, shortly after the survey began, the tensions between Pawsey and Bolton came to a head. One cost of an informal "wrong job number" approach is that relationships built on overlooking activities undertaken without permission until they become successful, can become difficult and strained—and this was certainly the case for Pawsey's relationship with Bolton. Bolton was also enflaming things by going behind Pawsey's back and directly interacting with Pawsey's boss, Bowen. As we have discussed, Bolton had long perceived Pawsey as not only overly cautious, but as likely to stifle the most exciting opportunities in cosmic noise investigations, partly as a result of his (Bolton's) being out of favour with Pawsey. In mid-1953 following the success with the hole-in-the-ground antenna, Bolton proposed a new much larger facility at Dover Heights using a large parabolic cylinder as a cliff interferometer at 400 MHz. While the sky survey observations, which indeed included the Galactic Centre, were being made by McGee and colleagues at Dover Heights, Bolton was given the news that his proposed telescope, which was in competition with the Mills cross proposal, would not be funded.[14] It seems likely that Pawsey would have recognised the disadvantage of only observing a single Fourier component with the fixed cliff height. While this may have been a sensitive survey telescope it would not be able to measure source structure. In any case, Bolton had a meeting with Pawsey that evidently became acrimonious, followed immediately by a meeting between himself, Pawsey and Bowen, from which he emerged declaring that he was "out of radio astronomy."[15]

Bolton had developed a closer relationship with Bowen, with whom he evidently shared an outlook prizing "big", high impact or dazzling achievements. He left Pawsey's group to join Bowen's in cloud physics, perhaps temporarily, as we discuss in Chap. 27. Bowen was at the time negotiating a possible move to the USA to build a very large new radio astronomy facility, with Bolton as the leader of its research program. It is also possible that Bowen was motivated to see if Bolton could bring the same kind of success that he had delivered for Radio Astronomy to Bowen's struggling Cloud Physics program.

---

[14] Mills (2006) has given some of the background in the era 1953: "... [W]hereas I was intrigued by the mystery of the discrete sources and had no hesitation in choosing this option. This did ensure some friction within the group as John Bolton had made discrete sources his own, following his use of the cliff-top interferometer to discover the first such source [in 1948] and to establish the existence of this class of object by finding several others. However, Pawsey knew that the future lay with the use of horizontal baselines and Bolton was still making effective use of the interferometer that had proved so successful for him previously."

[15] Gordon Stanley (1994) describing the interaction with Pawsey and Bowen in "Recollections of John G Bolton at Dover Heights and Caltech".

Regardless, from mid-1953 Bolton was no longer actively involved in performing the sky survey being then undertaken by the reconstructed hole-in-the-ground dish telescope, as McGee later wrote to Goss.[16]

Surveying the Sky

McGee remembers that "It took until November [1953] to survey the 32 degrees of sky available to the 80-footer [declination −49 to −17°]. I attempted to measure the lowest sky temperature between 3 and 5 November 1953."[17]

The contour map on which McGee was now working revealed an exciting image, which McGee showed to Bolton. John Bolton later recalled: "Dick McGee and I were the first to see our own map [of the galactic centre] and there was no question in our minds that we were looking at the nucleus of the Galaxy."[18]

Pawsey was immediately interested, too, as McGee recalled:

.... I found the original contour map [in the archives file] in equatorial coordinates (Fig. 23.6). This is the map I was working on when Pawsey called in at Dover[19] and was excited by the appearance of Sgr A. He took the map to get it copied and, if I recollect correctly, said he wanted to send it to Walter Baade [after conversion from 1900 coordinates to galactic coordinates, Fig. 23.7]. It would have been probably mid- to late-January 1954.

We also know from an interview with John Murray,[20] who attended a colloquium given by McGee about the new galactic centre source, that Pawsey was initially reserved about the new result. He was not pleased with McGee's galactic centre interpretation, warning: "you'll lose your reputation if you are wrong." Soon, however, after exchange of letters with Baade (February 1954, below), Pawsey's conservatism evolved to enthusiastic support.

Pawsey wrote Baade (12 February 1954) a long letter with two topics. The first was to report on a visit that Otto Struve had spent in Australia (early 1954, 6 to 12 January) discussing a possible radio astronomy symposium at either the IAU General Assembly of 1955 in Dublin or preceding the IAU at Jodrell Bank; Pawsey

---

[16]McGee wrote, "He [Bolton] took no part in the actual survey" (letters to Goss on 2 and 21 January 1996, the former with the "notes" referred to in ESM 23.1, Discovery of Galactic Centre). The major part of the work was carried out by McGee with assistance by Bruce Slee. In an interview with Peter Robertson (30 November 2006 at his home in Eastwood), McGee acknowledged Bolton's important contribution in the survey design; the authors are very grateful to Robertson for access to this material. The additional materials provided by McGee extends those that informed Bland-Hawthorn and Robertson (2014) and Robertson (2017, p. 131), who drew from an interview with McGee by Robertson in representing the survey as the work of "Bolton and McGee".

[17]McGee was uncertain when John Bolton went to the Cloud Physics group; it was mid-1953.

[18]Letter from Bolton to Don Morton 21 June 1985 (see ESM 23.2, Roles of Bolton, Pawsey).

[19]In late 1953 or early 1954.

[20]Murray was a research engineer at RPL. The colloquium was likely in early 1954. Goss recorded an interview with John Murray in Sydney on 28 March 2007.

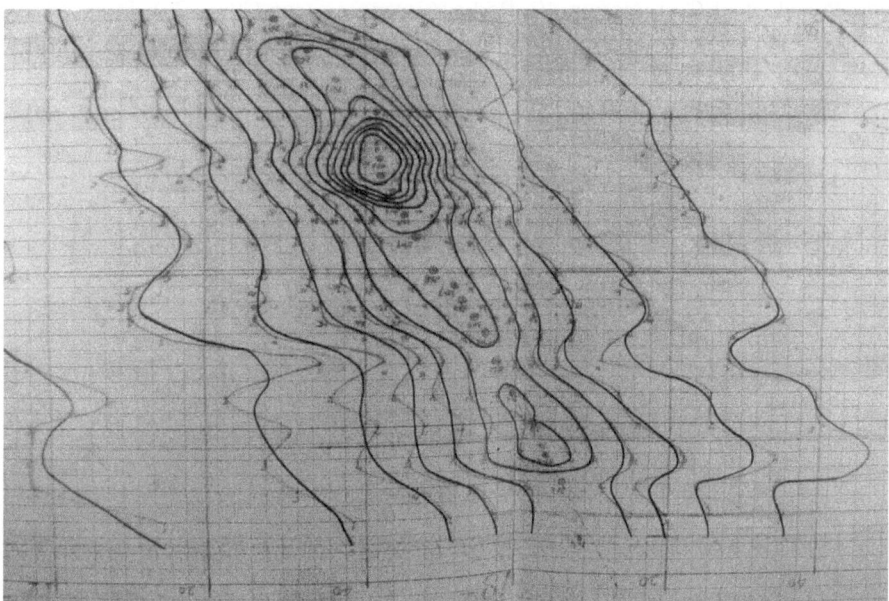

Fig. 23.6 Sgr A region at 400 MHz- beam width 2°. The declination grid is at intervals of about 4°. Dover Heights 80 foot Hole in the Ground antenna. His initial contour image drawn in equatorial coordinates. Constructed late 1953 or early 1954. Credit: Courtesy of the McGee archive in the National Archives of Australia. NAA: C4633/3

Fig. 23.7 400 MHz image of Sgr A in galactic coordinates before sending to Baade in Pasadena at Mt. Wilson Observatory. Beam 2° The grid lines are at intervals of 5–6° galactic longitude (left-right) versus galactic latitude. Courtesy of the National Archives of Australia. NAA C4633/3

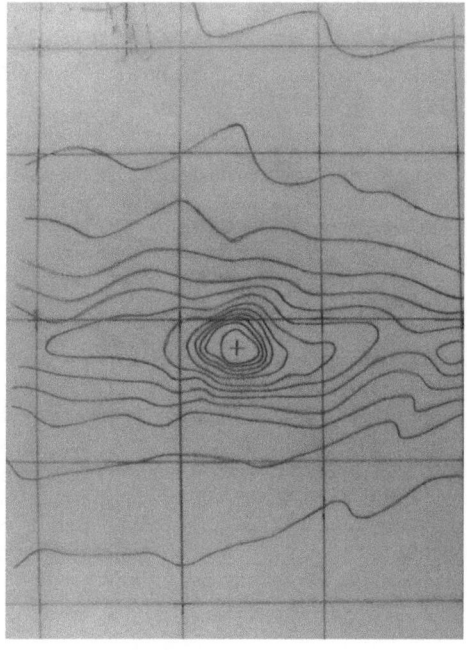

was President of Commission 40, Radio Astronomy (see Chap. 26). Since Struve was president of the IAU from 1952 to 1955, his support for a possible symposium sponsored by the IAU was critical.

At the end of this two-page letter to Baade, Pawsey included the galactic centre result:

> When Struve was looking over some of our radio results, he remarked on a discrete radio source which lies very close to the galactic centre. He told me you had some optical evidence of a peculiar object in that position, and that you might be interested. I enclose a tracing of a contour map which McGee has prepared. It gives measured radio brightness at 400 Mc/s and covers the region of the galactic centre (galactic coordinates).

In sharp contrast to Pawsey's measured transmission of this result, Baade responded enthusiastically via the oft-quoted letter of 16 February 1954[21]:

> Now to the object in the centre of the Galaxy, the contour diagram of which you kindly included in your letter. Frankly, I jumped out of my chair the moment I saw what it meant. I have not the slightest doubt that you finally got the nucleus of our Galaxy!! ... The strongest argument at present is its position which coincides with the expected place of the nucleus. From the Leiden profiles of the 2 lcm radiation of hydrogen, it is possible to fit quite accurately the zero point in the anticentre direction, and according to van de Hulst, this zeropoint falls into longitude 147°.5. 147°.5 + 180° = 327°.5 which checks closely enough with I = 328° as the longitude of your object. On the other hand, one would expect that the latitude of the nucleus on the presently used galactic co-ordinate system (Lund tables) would come out as b = −1° since it is well known that the present coordinate system needs a correction which is in the neighbourhood of I = 328° amounts to Δb = + 1°. **It is very improbable that the coincidence between inferred and observed position of the nucleus is accidental.** (Baade's emphasis) ... Since Dr. E.G. Bowen[22] is here at present [in Pasadena] and Mills is coming tomorrow, I hope to have a chance to discuss these questions with them. [after the sign-off of the letter a postscript followed] ... I showed your tracings to van de Hulst [also visiting Pasadena from early January to late March 1954]. He is also convinced that your object is the galactic nucleus.

Henk van de Hulst wrote to Pawsey on 19 February 1954 (NAA C3830, A1/1/1, Part 91954-55):

> Baade got really excited about your fine observations of the galactic nucleus and shows your plot to anybody who comes near his office. The position agrees well with the best we can do on the basis of the 21 cm [HI] observations ... [based on large scale symmetry of the distribution of HI in Galaxy over a region of 20 degrees in galactic longitude]. The HI

---

[21] More than half of the letter was about the new radio source near the galactic centre. He also suggested that a small radio astronomy meeting of half day's duration be held at the IAU at Dublin.

[22] In 1954, McGee and the "boys" at RPL were given a copy of the Baade letter. On 2 January 1954, Dick McGee wrote Goss: "A triviality—towards the end of his letter Baade says that Taffy Bowen was in Pasadena and that Mills was coming tomorrow. Bernie brought back the story that in what must have been a symposium, Taffy was asked to comment [perhaps on the new galactic centre source] and he said: "Yes, I have had the boys chasing this problem for some time now." At the time, Taffy displayed zero interest in the current radio astronomy activities, and the story caused much sniggering in the Lab.

determined values were longitude 327.5° and latitude −1.5°, compared to the Dover Heights position of 327.9 and −1.0.[23]

Baade also sent the contour image to the Director of the Observatory (Sterewacht), J.H. Oort in Leiden, the Netherlands. Oort responded to Baade (22 February 1954) with a copy to Pawsey. He described in some detail to Baade how the HI symmetry in the Milky Way was used to determine the coordinates of the centre of rotation of the Galaxy; then the comparison was made with the galactic coordinates of the Dover Heights 400 MHz radio source. In a pencilled note (in the very characteristic handwriting of Oort) to Pawsey, Oort wrote: "This [the new 400 MHz image he had received from Baade] is indeed of very great interest. As you will see [from the letter to Baade] I share Baade's opinion that what you have observed may actually be the nucleus of the galactic system."[24]

Confidence and Caution: Publishing the Galactic Centre Discovery, 1954

By the time Oort was replying to Baade and Pawsey, Pawsey had already become sufficiently confident to push for publication. He wrote a memo on 22 February 1954 inviting McGee and Bolton to consider a letter to *Nature* or *Observatory*. He added in the memo: "The subject [galactic nucleus] is of wide interest."[25] This was clearly a response to the interest from his international connections in astronomy.[26]

The first draft of the paper was written by McGee (with no input from Bolton). The first draft was available in mid-March 1954, a handwritten draft that was given to Pawsey. The first page is partly shown in Fig. 23.8, with the title given by McGee "The Galactic Nucleus", which is struck out. In Pawsey's unmistakeable handwriting, and encircled with his red pencil, Pawsey scrawled the confident title: "Radio

---

[23] As a postscript, van de Hulst mentioned an up-to-date 21 cm HI determination (done by Gart Westerhout) of 328.0° for the longitude. "So there is really nothing to worry about." Van de Hulst and Goss (along with Andrew Goss, age 15) discussed this experience during a two-day trip in 1987 to Chaco Culture National Historical Park, where van de Hulst spent at least 30 min lying on the ground looking up at the Supernova Pictograph (possibly a cave painting of SN 1052, the Crab Supernova).

[24] Note that in 1954 there was no concept of special activity with radio emission related to black holes in the centres of galaxies—this was to come a decade later.

[25] NAA C3830 A1/3/1 also in A1/1/1 Part 9.

[26] McGee to Goss (2 January 1996): "I believe a vital document was Pawsey's memo of 22 February 1954 inviting John and me to consider a letter to *Nature* or *Observatory*. This set the ball rolling AFTER [McGee's emphasis] Baade, van de Hulst and Oort had written to him in reply to seeing the Sgr A map."

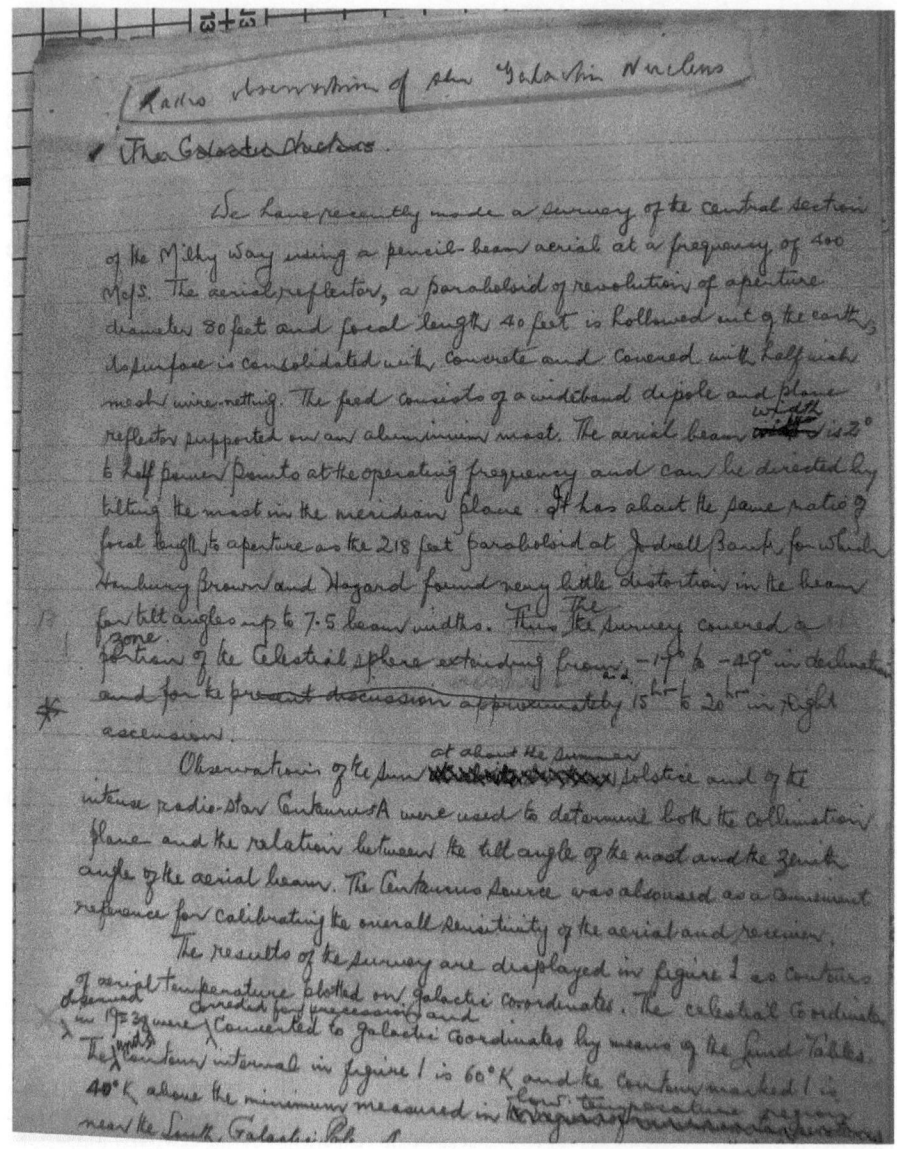

Fig. 23.8 The first page of Dick McGee's first draft with the substitute title written by Pawsey-original with a red pencil. Courtesy of the National Archives of Australia. NAA C4633/3

Observations of the Galactic Nucleus".[27] As McGee recollected: "Then it was ready to go into the RPL review system of which Frank Kerr was chairman."[28]

The review process lasted 18 days with the paper reviewed by Kerr, Mills, Shain, and again by Kerr on 9 April 1954.[29] At this date the title was changed, and the order of authors reversed. The file shows no input from Bolton, whom McGee met in the corridor after the first set of internal referees' comments were available. McGee recalled that although he had put the authors as "Bolton and McGee", "John would not hear of that and wanted me to be the sole author"[30] to which McGee recalled replying, "Cut it out John, I knew nothing about this until you told me what to do,"[31] and a compromise that reversed the author order was agreed.

Bolton remained very confident of the discovery. McGee recalled: "My only recollection of seeing John in the whole production of the paper was that time in the Lab when he looked through the criticisms and blew up. John was "furious with the criticisms" and fumed about "people sitting in the Lab criticising people working out in the field."[32]

One of the critics was Mills. And Mills's criticism is particularly interesting because it was wrong, and it was wrong because of another phenomenon described by Merton: Mills knew too much! Because of this, Mills was certain that the basis of Baade's identification of Sgr A with the galactic centre was incorrect.

To explain this, we remind the reader that Baade had used wartime blackouts for observations that resulted in his classifying stars into two populations, "Pop I" and "Pop II", based partly on stellar age, with Pop II stars being considerably older. Although the centre of our Galaxy is optically obscured by dust, Baade proposed a bulge of Pop II stars around the Galactic centre, analogous to the central bulge in the Andromeda galaxy. In the "radio star" model of discrete radio sources that had been the prevailing assumption until just a year or so earlier, these Pop II stars would have included radio stars. Baade therefore assumed that if a bulge of Pop II stars existed at the Galactic centre, the combined emission from all these stars would be visible in the radio as a strong source of emission. When such a source was identified, Baade

---

[27] In the NAA C4633/3 file, a one-page text written by Pawsey was found (on the top of the page in McGee's handwriting: "Initial comments from Pawsey after first draft"). There are 4 1/2 closely written foolscap pages of which Pawsey had crossed out 1 ¾ pages together with detailed comments on the text. Pawsey: "Present results restricted to region of galactic centre. Theme—First unambiguous evidence of an outstanding radio source in the position of the galactic nucleus, and inference that source is the galactic nucleus." This is followed by a series of abbreviated, cryptic comments, numerous corrections to McGee's text made by Pawsey.

[28] McGee to Goss, 2 and 21 January 1996.

[29] The torturous internal review system is described by Goss and McGee (1996, p. 373) with quotes of some of the critical remarks by CSIRO colleagues.

[30] McGee to Goss, 2 and 21 January 1996. McGee wrote: "There are no comments on any of the drafts by John Bolton and you know John was always very helpful with numerous comments on manuscripts."

[31] Interview with McGee by Peter Robertson, 30 November 2006 at McGee's home in Eastwood.

[32] McGee to Goss, 2 and 21 January 1996.

assumed that this galactic radio source was the emission from the population II stars in a bulge at the galactic centre.

But, as previously discussed, already by 1952, Mills had made a strong case that the radio star model was wrong! Mills's case would be undermined if the radio emission observed was from the Pop II stars that Baade proposed existed at the Galactic centre. So he strongly critiqued Baade's identification. Mills's review stated: "The results should obviously be published quickly as they are important. However, the general tone of the paper seems far too dogmatic—even the title! And positive identification with the Galactic nucleus is impossible at present." Doubtless this did not endear him to his colleague John Bolton. Ironically, we now know that most of the radio emission is a result of the Population I stars (thermal emission from ionised gas and supernovae remnants produced by young stars: thus, while the radio source that Bolton's group found was indeed the galactic nucleus, it would turn out *not* to be evidence for Baade's Population II star model—a resolution of the issue that of course Mills did not anticipate. In ESM 23.3, B.Y. Mills text, April 1996, about the 1954 Controversy, we present an excerpt of a letter from Mills to Goss in April 1996 concerning Mills's experiences at Caltech in late 1953-early 1954 as he interacted with Baade.

Others of the RPL internal reviewers remained cautious. Shain wrote: "The title and the statement towards the end of page 2 indicate that you are making a definite claim that the 'hump' on your contours is the galactic nucleus and, if only by inference, that the position of this 'hump' is the best available determination of the hump." Shain suggested the new title of "Observations of the Region of the Galactic Nucleus at 400 Mc/s" and McGee accepted this new title.

But Pawsey's confidence was not shaken by the internal review. On 14 April 1954, McGee took the paper to Pawsey for final approval. Pawsey immediately changed the title provided by Shain back to "Observation of the Galactic Nucleus at 400 Mc/s", almost his original suggestion. McGee suggested that the word "Possible" before observation would better reflect the internal referee consensus. "But Pawsey, buoyed by the support of Struve, Baade, Oort and van de Hulst, compromised with 'Probable', since he found 'Possible' too weak. Thus the final title was 'Probable Observation of the Galactic Nucleus at 400 Mc./s.'" (Goss & McGee, 1996, p. 375) In 1996, Dick McGee was embarrassed by his reluctance of 1954: "I regard myself as a new boy wimp for introducing the possible-probable reduction of the title."[33]

The paper was sent to *Nature* on 15 March and appeared in print on 22 May, McGee & Bolton (1954). The acknowledgements included Kevin Westfold (University of Sydney), Gordon Stanley, and Bruce Slee of RPL. The last sentence

---

[33] McGee to Goss, 2 January 1996.

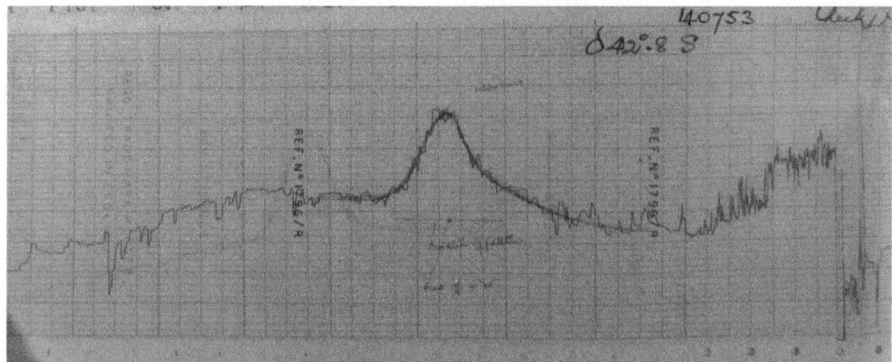

Fig. 23.9 The faint line below the peak gives the size of the beam, 2°; the original scan at declination −42.8°, is close to galactic longitude (lii 342 deg). Credit: Courtesy of the McGee archive in the National Archives of Australia. NAA C4633/3

of the publication was: "The project was originally suggested to one of us by Dr W. Baade, [in 1951]."[34] As was normal for RPL publications, Pawsey's editing was not acknowledged.[35]

An extensive paper was later published in the *Australian Journal of Physics* in September 1955, "Galactic Survey at 400 Mc/s Between Declinations -17 and -49 Deg" by McGee, Slee and Stanley, submitted 16 February 1955. The entire survey, including the galactic centre, was presented. A number of other sources were detected: Centaurus A, Fornax A, Pictor A, the Puppis SNR, and possibly the HII region M17. The weakest source listed in the source list had a flux density of 60 Jy. In Fig. 23.9 we show an original record of a transit at a declination of −42.8°. This is one of the few original data recordings from the Dover Heights hole-in-the-ground antenna.

Exploratory observations of the galactic centre at 760 MHz were carried out with the 80-foot aerial with a beam size of 1.2°; Sgr A remained unresolved at this resolution. No images were shown at this frequency. More details of the galactic centre source were presented with valiant attempts to discuss the continuum spectra from 18 to 3200 MHz; clearly the major handicap was source confusion. The authors were prescient:

> Because of these discrepancies in size and intensity, it appears possible that the nucleus source may have a complex structure with different parts of it displaying differing spectra. High resolution surveys at frequencies near 100 Mc/s and 20 Mc/s can make important contributions towards clearing up the present confusion.

---

[34] Bolton wrote to Morton on 21 June 1985 about the 1951 request from Baade to Bolton: "It was end 1951 in the form 'could we find a counterpart at radio frequencies in our own galaxy for the semi-stellar nucleus he had found in M31?'"

[35] McGee reported to Goss in 1996 that Pawsey had declined McGee's offer to be a co-author of the *Nature* paper in 1954.

This concern turned out to be confirmed; however, a real understanding of the region near Sgr A required high resolution of the galactic centre at cm wavelengths (to avoid free-free absorption at lower frequencies) with resolutions of some arcseconds; these observations would be made with the Very Large Array (US) in the 1980s as we describe at the end of this chapter.

Aftermath, 1955

As Goss and McGee (1996) have indicated, a discussion of the new galactic coordinate system occurred at the Dublin IAU meeting in 1955; Bolton proposed that, in view of the new Sgr A results, the galactic coordinates should be revised. Certainly the new coordinates from CSIRO played a major role. However, the major contributor to the new galactic system was the whole Galaxy HI and continuum surveys, summarised in the five papers in *Monthly Notices of the Royal Astronomical Society* (1960) by Blaauw, Gum, Pawsey and Westerhout, "The New IAU System of Galactic Coordinates".[36] Blaauw et al. wrote: "Adopting the principle that the new pole is to be based primarily on the HI observations, with radio continuum and optical results used for check purposes, a very satisfactory [galactic] pole [can be determined]."

In the final publication in the series of five, Oort and Rougoor, "The Position of the Galactic Centre":

> The position of Sagittarius A has been discussed . . . Tarchis position agrees so precisely with the direction of the galactic centre [determined by HI observations] . . . that this by itself makes it almost certain that Sgr A is situated *at* [their emphasis] the centre of the Galaxy. For Sgr A is not only one of five brightest sources, but it is also unique among known sources, consisting as it does, of a small, apparently thermal core surrounded by a more extensive non-thermal envelope (Westerhout, 1958). It would be an extremely improbable coincidence if this unique source should accidentally lie within 0.1° of the centre without being connected with it.

In ESM 23.4, Dover Heights images on 2 November 1989 at the conference "40 Years of Radio Galaxies, Evolution of Ideas and Techniques", we show 4 images (Figs. 23.4.1, 23.4.2, 23.4.3, and 23.4.4). This event was a portion of a symposium held at the CSIRO Division of Radiophysics in Epping. In the morning, a tour to Dover Heights occurred with the unveiling of a plaque. The pioneers of 1949–1954 were present: Bolton, Slee, Yabsley, McGee, Roberts and Robinson. Two of the authors were also present: Ekers and Goss.

---

[36] Additional authors were Kerr, Oort and Rougoor. Colin Gum had died in a skiing accident in Switzerland on 28 April 1960. The new coordinate system had been recommended by the IAU held in August 1958 in Moscow.

Fig. 23.10 A modern image of the radio intensity at 888 MHz from the Galactic Centre region observed with the ASKAP radio telescope at the Murchison Radio Observatory in Western Australia. This is a single 6 hour observation covering 10 square degrees by mosaicking the 36 simultaneous ASKAP beams using the Pawsey supercomputer in Perth. Credit: CSIRO 2019–2020 Annual Report p15. Credit: Wasim Raja, CSIRO

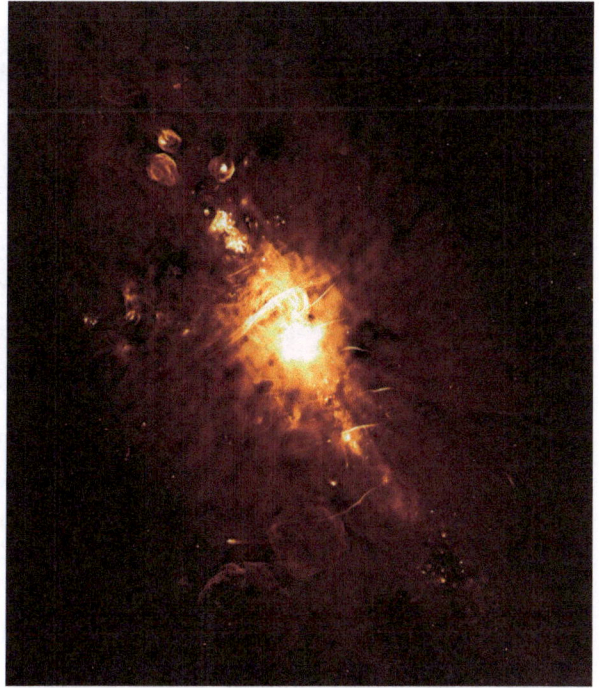

The Big Picture

A modern image of the galactic centre (Fig. 23.10) was made using the Australian SKA pathfinder (ASKAP). This telescope is a direct descendent of the telescopes developed by Pawsey's group in the 1950s, combining the aperture synthesis concept pioneered by Christiansen with a beam forming focal plane array using the principles first exploited by Mills. What we now see is an incredibly complex assembly of astrophysics processes in the centre of the galaxy, most of which were unknown in 1953. There is thermal radio emission from gas heated by young stars and this was understood in 1953 and correctly explained why the Galactic Centre has a flat spectrum and stands out at higher frequencies. The shell structures are all supernovae remnants, like the Crab Nebula which had been identified by Bolton back in 1948, but the synchrotron radiation mechanism was not known in 1953 (see Chap. 34) and even the supernovae association had not been accepted by all astronomers. The smooth, diffuse, non-thermal radio emission, which is the dominant component of Sgr A, is synchrotron radiation from the high density of cosmic rays in the central region of the galaxy but this connection was completely unknown in 1953. We also see striking filamentary streaks and arcs of radio emission throughout the whole region—these we do not even understand today, 70 years after the Galactic Centre was first discovered.

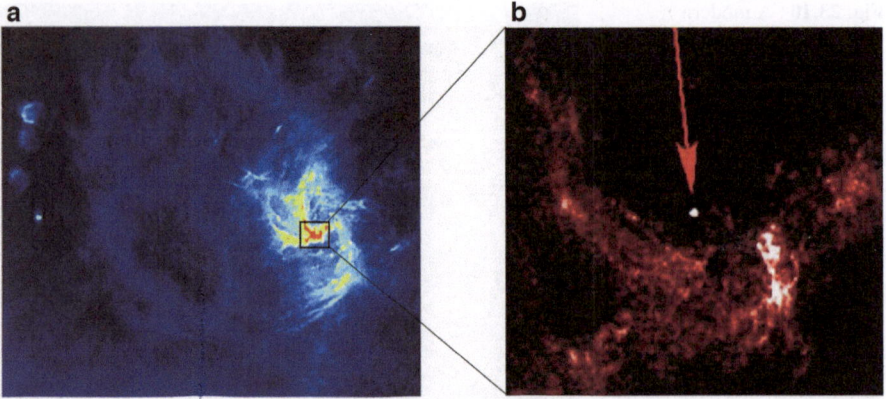

Fig. 23.11 (**a**) (left) VLA image of Sgr A at 5GHz (Killeen-unpublished) showing the supernovae shell(s) of Sgr A East (blue) and the Sgr A West hot spiraling gas (yellow and red). The circular structures on the far left are compact HII regions. Credit: NRAO/AUI/NSF. (**b**) (Right) A higher resolution 1.3 cm VLA image of the central region in Fig. 11a showing the inner spiral gas and the very small diameter Sgr A* source (red arrow) which is the location of the black hole. Credit: NRAO/AUI/NSF

Right in the brightest central region we also find what we now know is the most significant component of the Sgr A radio source. The high resolution VLA image (Fig. 23.11a) shows a supernova remnant (Sgr A east), and hot gas (Sgr A west) (Ekers et al., 1983) swirling around a very small radio source[37] (Fig. 23.11b) which we now call Sgr A* (Sagittarius A star).[38] This is the same as the active galactic nuclei (AGN) found in external galaxies, but it is such a small fraction of the total Sgr A radio emission that it is insignificant in the 1953 images. There was no concept of active nuclei in galaxies at that time.

The Nobel Prize

The Nobel Prize in Physics 2020 included Reinhard Genzel and Andrea Ghez "for the discovery of a supermassive compact object at the centre of our galaxy."

Reinhard Genzel and Andrea Ghez each lead a group of astronomers that, since the early 1990s, has focused on the region called Sgr A* just discussed at the centre of our galaxy. The orbits of the brightest stars closest to the middle of the Milky Way have been imaged in the infra-red with increasing precision. The measurements of these two groups agree, with both finding an extremely heavy, invisible object that pulls the stars in tight orbits around an object four million times more massive than

---

[37] Balick and Brown (1974).

[38] Goss, Brown and Lo (2003) published a discussion of the discovery of A* and its association with the Galactic Center black hole in 2003, "The Discovery of Sgr A *".

the sun in a region no larger than our solar system. The third Nobel prize winner, Roger Penrose, is one of a number of scientists who have demonstrated that such a dense object must be a black hole.

The sequence of discoveries that have led to this understanding about the nucleus of our galaxy provide an exceptional example of a continual discovery process which has spanned the last 70 years. According to Van de Hulst's (1984) conception of low frequency components in the history of science, this would best be understood as a ½ nanohertz (2×10^9 s) astronomy event.

The discovery process started with the first indication from Jansky and Reber in the 1940s that the continuum radio emission was strongest towards the centre of our galaxy. But as Baade noted in his letter to Pawsey in 1954 "I have concluded . . . that there was positively no chance whatsoever to detect the nucleus of our galaxy in the [heavily obscured] optical [wavelength] range." But, Baade, Oort and others realised that the strong radio source observed first in the Piddington and Minnett (1951a, b) image and much more clearly in the McGee and Bolton (1954) image and now called Sgr A, was the first unobscured view of the centre of the galaxy. By 1982 even higher resolution images from the VLA, showed that the Sgr A radio source was itself a composite of hot spiralling gas, supernova remnants, diffuse synchrotron emission from high energy particles and near the centre of this complex an even more compact object. Much higher resolution radio observations had already separated out this much smaller component and named it Sgr A*. This was on the size scale of the energy sources called AGN detected in the nuclei of other galaxies and assumed to be black holes. Sgr A* was also found to be coincident with infrared emission (Becklin & Neugebauer, 1975), and in the 1990s that infrared emission became the target for the observations by Genzel and Ghetz leading to direct evidence for the black hole in the nucleus of our galaxy and the 2020 Nobel prize.

Chapter 24
The Royal Society: Europe and North America, 1954

> We think Pawsey is the soundest candidate in every way, and even he runs considerable risk in view of the short length of his publication list. The overseas prestige of the Radiophysics Division rests mainly on its radio-astronomical work, and the Division is organised . . . along the lines that Pawsey has been deputed by Bowen as wholly responsible for this work in its scientific aspects.—16 September 1952, Martyn to Rivett [Pawsey was elected a Fellow of the Royal Society in March 1954].

In the year 1954 Pawsey was elected as a Fellow of the Royal Society and we provide some details on the nomination process that led to his election. Later that year Pawsey travelled to Europe and North America, leading the Australian delegation to the URSI General assembly in The Hague and re-establishing his International contacts.

Election as Fellow of the Royal Society of London

Pawsey was elected as a Fellow of the Royal Society in March 1954.[1] He was the second radio astronomer to become a Fellow of the Royal Society, after Martin Ryle in 1952. Bernard Lovell followed as the third radio astronomer in 1955. This was not only recognition of Pawsey's own achievements in establishing the radio astronomy group in Sydney but gave formal legitimation to the field. The nomination process is

---

NRAO ONLINE.22

Supplementary Information The online version contains supplementary material available at [https://doi.org/10.1007/978-3-031-07916-0_24].

---

[1] Source material: NAA A9874/60, "Royal Society, London. Dr Pawsey's Qualifications" and Joe and Lenore Pawsey Family Collection.

© The Author(s) 2023
W. M. Goss et al., *Joe Pawsey and the Founding of Australian Radio Astronomy*, Historical & Cultural Astronomy, https://doi.org/10.1007/978-3-031-07916-0_24

an intriguing story and not without controversy. The sensitive issues were (1) the suggestions by Appleton and Rivett to sponsor both Bowen and Pawsey simultaneously for election and (2) Martyn's sharp criticism of the Royal Society's process related to the election of Martin Ryle the previous year.

In August 1952 the URSI General Assembly (Chap. 21) had brought a number of eminent radio astronomers and ionospheric scientists to meet in Sydney. Martyn had taken advantage of this situation and convened a meeting of those who were Fellows of the Royal Society (FRS) to discuss possible new members. This resulted in the suggestion that Pawsey be nominated. On 27 August 1952, Martyn wrote Pawsey: "At a recent informal meeting of a group of Fellows of the Royal Society it was suggested that you be invited to let your name go up for election, and I was asked to act as proposer."

Pawsey responded immediately with two letters, a handwritten letter of acceptance on 1 September 1952 and a more formal typed letter on 3 September. He included an obligatory modest paragraph followed by his own assessment of his achievements (see Chap. 25) which he identified as leading the radio astronomy group at RPL, and in particular applying interferometry to study the sun, ionospheric observation, and contributions to radar techniques (the epigraph for Chap. 25 contains these statements). He added:

> I am a little diffident as to my qualifications but if, as you mentioned in your letter, a group of Fellows favoured this, I should be foolish not to accept your offer. I am very grateful indeed to you for taking this interest in my career . . .

> With regard to the names of Fellows who might support my nomination, I think the following are reasonably acquainted with my work or me: yourself [Martyn], Ratcliffe, Massey, Oliphant, Appleton, Chapman, Rivett, Bullen, Marston.

> . . . I wish my book were finished and published. [publication of *Radio Astronomy* was to be in mid-1955]

A week later, Sir David Rivett FRS raised another issue. On 12 September 1952, he asked: "About Bowen, should he not be put up also? Both Pawsey and Bowen have handicapped themselves somewhat by their generosity in refraining from attaching their names to papers or work which was proposed and guided by them."[2]

On 16 September 1952, Martyn responded to Rivett with a strong letter disagreeing with the proposal to nominate Bowen; and in addition, raising an issue of process related to Ryle's election in 1952. Martyn's letter to Rivett, clearly marked as CONFIDENTIAL, included the following:

> Bowen was also considered. Oliphant, Massey, Chapman, Ratcliffe and I think he would stand little chance of serious considerations in view of the paucity of his "original" contributions to knowledge as shown by published work. Now that Australians may perhaps

---

[2] In a letter from Martyn to Oliphant on 23 September 1952, Martyn suggested that Rivett had been urged by Fred White and Ian Clunies Ross to assist in putting Bowen's nomination forward. Rivett, Clunies Ross and White had discussed the Bowen nomination with Appleton during his visit to Australia in August 1952 (for URSI), asking Appleton to assist. Appleton had passed this news on to Martyn.

begin to take a more effective collective hand in getting people elected some of us think it is wiser to begin with really sound candidates, so that our views may come to have some weight.

We think Pawsey is the soundest candidate in every way, and even he runs considerable risk in view of the short length of his publication list. The overseas prestige of the Radiophysics Division rests mainly on its radio-astronomical work, and the Division is organised . . . along the lines that Pawsey has been deputed by Bowen as wholly responsible for this work in its scientific aspects. Bowen's knowledge of the subject would be adequate for the purpose of giving a popular lecture, say, and there it stops. On the other hand Pawsey could stand up to a searching examination of any aspect of the Division's radio-astronomical work.

However, Bowen's rain work is his own responsibility, and here he has made his original contributions. Unless I am greatly mistaken about it this work is still far short of FRS standard . . .

The only other person I know who would support Bowen is Appleton. But he too thinks that it would be useless to put him up before he has been invited to give an account of the work of the Division to the [Royal] Society. . .

We are all taking a much closer interest now in the elections since last year [this dated 16 Sept 1952], when a serious blunder was made in the election (first time up) of M. Ryle of Cambridge. This one mistake has done more to lower the prestige of the FRS than any other single act I know of. Ryle is a very nice fellow, and has published a great many papers, many of which are now known, and were known last year, to be unsound. He had the advantage however of high social connections and was pushed hard by our friend Appleton and others, for politic reasons, we feel sure. I have found out that Ryle's name was not on the recommended lists put up by the Sectional Committees, but it was adopted by the Council (as is within its right of course) at its first meeting for consideration of candidates. Surely, there's something wrong when a candidate passed over by the Committees gets in on his first time up!

Some of us feel that to prevent this happening again we must pay close attention to candidates in our own fields. The best way to prevent it seems to be to support the strongest candidate to the fullest extent. I've gone into the question you raise at some length because it is important, and I would like you to know that we have been considering it carefully from several angles.

This letter to Rivett seemed to have had an effect. He responded to Martyn on 22 September 1952, beginning in a somewhat whimsical tone:

If in the next world you take up Law and if I happen to get into trouble in the same geographical section, I shall have no hesitation in calling you in to [function as] my referee. Your letter convinces me that a very great addition to the barrister class was lost when you were led into physics!

Many thanks for so clearly putting the cases of Pawsey and Bowen in the matter of prior claims for nomination for the RS. I have signed the relevant paper [today] . . .

It is not easy to assess relative merits and claims in cases like these two and I daresay both of these will be in the RS sooner or later[3] but competition in physics [Rivett was a chemist] must be pretty fierce.

---

[3] In fact, Bowen became a FRS 20 years later, March 1975, well after Martyn's death in 1970.

On 2 October 1952, Martyn wrote Jack Ratcliffe, enclosing Pawsey's Certificate signed by five Australian FRS members (Martyn, Rivett, Bullen, Oliphant and Marston). Martyn had added in the last paragraph: "There is a certain amount of pressure for Bowen to be put up. Oliphant and I think he will need to publish more to stand a reasonable chance. What do you think?" Unfortunately, if Ratcliffe responded, the letter was not located in the archives.

Pawsey was sent a telegram on 18 or 19 March 1954 with a notification of being elected to Fellowship of the Royal Society; a formal letter followed in the post. Given the time interval between his nomination in October 1952 with election in March 1954, Pawsey was likely selected the first time he was on the ballot. Further details of the formal election process and the results are provided in ESM 24.1, Pawsey's Fellowship.

There was limited publicity in Sydney; Pawsey's friend, Professor Ronald L. Aston (Prof of Geodesy and Surveying at the University of Sydney) sent out a circular letter on 5 May 1954 with congratulations:

> The many friends of Dr Joseph Lade Pawsey will have read with pleasure and satisfaction the recent announcement of his election as a Fellow of the Royal Society. The award of this coveted honour is a fitting tribute to his outstanding achievement as a physicist and as an inspiring leader of a research team which has earned for Australia a prominent place in the rapidly developing new field of Science—radio astronomy.[4]

The Significance of Radio as a Field of Astronomy

In the previous chapter, we showed that radio astronomy was becoming part of Astronomy and accepted as such by astronomers globally, led by those in observatories such as Mt Wilson/Palomar, Yerkes, Paris and Leiden. But in Australia there was one standout, Richard Woolley (1906–1986) Director at Mt Stromlo, who continued to show a lack of awareness of how radio astronomy would have a transformational impact on the field, as we show in the small vignette, as told to Lindsay McCready (who wrote it to Pawsey on 27 August 1954[5]) by Stuart Henry Bastow (CSIRO Executive 1949 up to his death in 1964; CSIRO CEO 1957–1959). Bastow had attended a popular lecture on astronomy given by Richard Woolley, "The Future of Astronomy in Australia", which included no mention of radio astronomy. In response to a question (from Ron Bracewell!) about the future of radio astronomy, Woolley was said to have replied "in a gathering of astronomers it was not considered decent to mention radio astronomy."

---

[4] A year later, 12 May 1955, Lovell wrote Pawsey a letter of thanks after Pawsey had congratulated him on being elected as a FRS, "...[A]s you say, I think we have been very fortunate that three of us from this new subject have gotten in so quickly. I am naturally very pleased, and it comes at a time of anxiety and notable depression over the new telescope."

[5] NAA C3830 F1/4/PAW/3. Thirty years later Paul Wild (1987) recalled a somewhat different version "In ten years' time radio astronomy will be forgotten."

Ironically, Woolley was the first president of IAU Commission 40, Radio Astronomy; apparently little was done in Commission 40 until Pawsey became the President in 1952 after the IAU in Rome.

Pawsey's Overseas Trip July–October 1954

Pawsey had not visited North America since 1947; he had not been in Europe since 1950. Since that year, Pawsey had worked continuously to create and then nurture the radio group at RPL. The contacts made in the US in 1954 would lead to his long visit to North America in 1957–1958.

Bowen initiated the planning for Pawsey's overseas trip on 17 December 1953 in a letter to Guy B. Gresford, Secretary of Physical and Industrial Sciences of CSIRO in Melbourne.[6] Bowen wrote:

> In Dr Pawsey's case it is desirable that he should attend the URSI General Assembly [23 August–2 September 1954 in The Hague, the Netherlands] and should spend several months in visits to observatories and other centres of radio astronomy in Europe and North America at about the same time … [Since his last visits overseas, radio astronomy] has advanced immensely and a number of new establishments have come into being. If this laboratory is to maintain its present outstanding position in this field, it is most important that Dr Pawsey should have every opportunity to become familiar with the latest developments.

Also, given that Pawsey was transiting from the Secretary of URSI Commission V (Radio Astronomy) to the President of Commission 40 (Radio Astronomy) of the IAU, coordination at the 1954 URSI was essential as the radio astronomers were becoming more involved in the astronomical community.[7] As we discuss (NRAO ONLINE.51), Pawsey was a key player in the introduction of "radio astronomy" to the IAU.

The CSIRO Executive approved the visit in early 1954. On 6 April 1954, Bowen wrote Gresford that Pawsey was to depart in July for Europe. During some of the visits to various astronomy institutes in Europe (Norway, Sweden and the Netherlands), Christiansen (then on an extended visit to Paris) would accompany Pawsey. Details of Christiansen's visits in 1954 are summarised in ESM 24.2, Christiansen's visit. Pawsey was to be away from Australia from 7 July to 21 October 1954, arriving in the US on 30 September to spend 3 hectic weeks in the US and Canada.

Pawsey described his trip in 1954 as a "Cooks tour", consisting of numerous brief visits of a few days at most with colleagues. The 3-week sojourn in the US and Canada was especially crowded, in some cases a stay of only 1 day [as in the case of the University of Michigan solar observatory with Leo Goldberg and Helen Dodson, and Boulder, Colorado, with the eminent solar physicist, W.O. Roberts

---

[6]Sources for information about the overseas trips in 1954: NAA AH 8520 PH/Paw/1 B/part 1, correspondence, C3830 Z3/1/V 1954, C3830 F1/4/paw/3, C3830 Z3/3/A Travel 1954.

[7]URSI, International Union of Radio Science. IAU, International Astronomical Union.

(1915–1990)]. The longer visits were for conferences: 15–17 July 1954, Liege, Belgium, for the IAU Colloquium on Solid Particles in Astronomical Objects, 20–22 July at Jodrell Bank to attend a Colloquium on Meteors, 16–18 August in Brussels for the Joint Commission on the Ionosphere and Joint Commission on Radio Meteorology. The URSI General Assembly followed a week later, from 23 August to 2 September 1954 in The Hague, the Netherlands. He presented a paper at a small conference organised by Ratcliffe and ionospheric colleagues in Cambridge from 6 to 9 September, "Physics of the Ionosphere" (Chap. 25). After deciding that the schedule was too crowded, Pawsey cancelled his planned participation in the UGGI (International Union of Geodesy and Geophysics) from 14 to 29 September 1954. During this period he visited colleagues in the UK in London and Cambridge.

A number of visits to Institutes in the UK and Europe were also planned. He visited Oxford followed by the Telecommunications Research Establishment (TRE) at Malvern. After a 2 day visit to London, Pawsey began a major tour with Christiansen. The two colleagues began in Paris on 30 July to 5 August 1954, then on to Scandinavia where they visited Svein Rosseland (1894–1985) at the University of Oslo, Olaf Rydbeck (1911–1999) at Chalmers University of Technology in Gothenburg, Sweden, and finally Nicolai Herlofson (1916–2001) at the Technical University in Stockholm. Herlofson played a critical role in the recognition of the synchrotron process for radio emission as discussed in Chap. 34. At this time, Pawsey went to the Joint Commissions in Brussels, while Christiansen continued visiting fellow astronomers in Kiel, Germany. The two colleagues met up again in Leiden, visiting Oort, Westerhout, van de Hulst, and Eric Hill, a graduate student from CSIRO working with Oort and Mueller. Then Pawsey attended the 2 week URSI Conference in The Hague from 23 August to 4 September.

URSI General Assembly in the Hague from 23 August to 4 September 1954

The URSI conference provided a "show-case" for rapid developments in radio astronomy at RPL since the previous conference in Sydney in 1952. Pawsey, Christiansen, David Martyn, Eric Hill and Brian Robinson[8] were the RPL participants at the meeting, presenting a number of new results such as initial data from the new 80 MHz Mills Cross and initial HI data from the 36-ft Potts Hill transit parabola. There was substantial correspondence between the staff back in Sydney and Pawsey,

---

[8] Brian Robinson was an Australian who had been sent to Cambridge in 1954 by Pawsey to work on a PhD at Cambridge in the ionospheric group of Ratcliffe. His thesis was completed in 1957, advisor K. Weekes with close collaboration of Ratcliffe. Before going to Cambridge, Robinson had worked on HI observations of the Magellanic Clouds with Frank Kerr at Potts Hill.

to put together last minute results to be presented at the conference. Christiansen wrote Arthur Higgs on 9 September 1954:[9]

> All our contributions arrived from Australia in good time and were "put over". I think that the Australian contributions were at least as good as any other. Actually, apart from Oort and Hagen, there was no-one who had much to say that we didn't know from the literature. The Cambridge boys had some rather nice occultation results, and some shaky solar work and a claim of about 1700 discrete sources [a precursor of the 2C Mills Cross controversies of the following years, see Chaps. 35 and 36]. The Manchester group [was too busy with their new large aerial and there were no new results at this time]. Hagen's [Naval Research Laboratory] interesting cm galactic observations [e.g. HII regions]. Some preliminary results with the Department of Terrestrial Magnetism Cross (Bernie [Mills] doubtless would call it the **Double Cross**[10])[our emphasis] and quite a bit of radio-optical work on the sun.

Solar Work: Potts Hill

New results were presented at URSI by Christiansen on the distribution of radio brightness over the disk of the quiet sun at both 20 and 60 cm as well as the properties of the slowly varying component at 20 cm. This was the first aperture synthesis image, and it is discussed in the context of the development of aperture synthesis in Chap. 37. Here we put it into the chronological sequence and add some more details from the Australian perspective. The latter topic was summarised in a long letter from Joe Warburton (1924–2005) to Christiansen on 15 June 1954, posted to Christiansen in care of the Institut d'Astrophysique in Paris.[11]

Christiansen also presented plans for a future instrument. The idea was to construct a 21 cm instrument which was a combination of the Christiansen and Mills concepts (to be called the "Chris-Cross"). This would permit the scanning of the sun as in television, based on successive scans of the sun at a series of adjacent declination strips. These raster scans were carried out using the rotation of the earth.[12] This was to be constructed adjacent to the Mills cross at Fleurs and

---

[9]NAA C3830 F1/4/CHR.

[10]Christiansen joked that Tuve's group at the Department of Terrestrial Magnetism had gotten the idea for their newly constructed cross from Mills, during his visit there in 1953–54!

[11]The details of the 1D and later 2D determinations of the 21 cm brightness distribution of the quiet sun and the instrumental and processing details are described in the solar precis in NRAO ONLINE.23.

[12]Christiansen et al. (1961, p. 48) have provided details of the scanning method begun in July 1957. "As the earth rotates, a succession of [scans across the sun are recorded as] ... the sun passes through the pencil-beam ... from east to west. The geometry of the system is such that successive beams cross the solar disk at progressively changing declinations. It is therefore possible to obtain the data for complete two-dimensional 'pictures' of the sun simply by recording the receiver output whilst a succession of pencil beams drift across the solar disk ...[T]he beams are shifted in declination, so as to maintain a space equal to the beam width between successive scans. This adjustment is made by means of a phase-shifting mechanism in the north-south interferometer."

completed in 1957. This telescope later evolved into the Fleurs Synthesis Telescope (FST) in 1973 providing the first arc sec scale imaging in Australia.

Mills Cross: Fleurs

In addition to the new aperture synthesis imaging of the structure of the solar disk using the Potts Hill array, another prominent showcase at URSI in 1954 was the commissioning of the new 85.5 MHz Mills Cross telescope at Fleurs. On 3 August 1954, Bernie Mills wrote from Sydney to Pawsey in the Netherlands with exciting news:

> Needless to say, the news [of the new 20 MHz Mills Cross at Carnegie Institution of Washington Department of Terrestrial Magnetism, Merle Tuve and colleagues] acted as a stimulus here and the remaining dipoles in the Fleurs aerial were connected up last weekend, so that our aerial too is in full operation—except for the rapid scan which will be added in a week or two. The beamwidth now appears less than 1 degree.

Mills was working on a draft of a paper on new observations of nearby galaxies in the continuum as well as an update of progress with the new instrument. On 10 August 1954, Mills wrote enthusiastically as he sent a slide illustrating the observations of the previous week; these data were to be shown at URSI: "You will observe the large amount of detail in the vicinity of the galactic plane." Lindsay McCready wrote a week later (19 August) to Pawsey: "Fleurs is working beautifully—almost complete."

On 13 September 1954, Mills wrote a letter to Pawsey (who was still in London about to leave for the US): "We are progressing well with the present aerial and this week hope to have the rapid scan system in operation. All the phase-changers are in, and working more or less satisfactorily." By this point, Mills was hard at work on his first publications from the Cross. A few weeks later, he submitted a short paper to *Observatory*, "Abnormal Galaxies as Radio Sources" (Mills, 1954, p. 248). Mills announced "a new radio telescope recently put into operation at Sydney. This radio telescope is of a novel type and consists of a cruciform arrangement of two 1500-foot arrays of dipoles." Within the next months, Mills prepared his first major publication with the new instrument, "The Observations and Interpretations of Radio Emission from Some Bright Galaxies" which would be submitted the following year on 12 April 1955 to the *Australian Journal of Physics*. Thirteen of the bright southern galaxies were observed with ten detections. Famous galaxies such as the Large and Small Magellanic Clouds, NGC 55, NGC 253, NGC 300, M83, NGC 4945 were detected at 3.5 m.[13]

---

[13] Hanbury Brown and Hazard had previously detected the external galaxies M31 and M81 in the radio continuum with the 218-ft transit dish (Hanbury Brown & Hazard, 1951, 1953).

HI in the Magellanic Clouds: Potts Hill

The final highlight for URSI in 1954 was a summary of the exciting results from the 36-ft HI studies, based on the new HI data from Potts Hill obtained by Frank Kerr and colleagues in the HI group. The telescope at Potts Hill had been in operation since early 1953; with a beam size of 1.5° and a velocity resolution of 8.5 km/s, it was well suited to a number of crucial experiments. Already in mid-1954, a major discovery had been made with this antenna. Frank Kerr, Jim Hindman and Brian Robinson had detected HI in the Large and Small Magellanic Clouds. These results were presented at the August 1953 American Astronomical Society meeting in Boulder, Colorado; the main paper was submitted to the *Australian Journal of Physics* on 22 February 1954, published in June 1954. The surprising result was that the HI sizes of these irregular galaxies exceeded the optical sizes, arising from the stars. In both cases, the galaxies had a mass of HI gas that was comparable to the mass of all the stars. Based on the observed HI radial velocities, both Clouds were observed to rotate.

Galactic HI: Potts Hill

Pawsey also decided to highlight the new observations of the galactic HI at the URSI conference. Major efforts had already started in 1954 to image the galactic HI visible from Sydney. Close collaboration with the Dutch astronomers at Leiden had begun with the northern Milky Way observed from the Netherlands and the southern galaxy observed by Kerr, Hindman and Martha "Patty" Stahr Carpenter.[14] Patty Carpenter was an author or co-author of three publications: (1) "Observations of the Southern Milky Way at 21 centimeters" from a paper she gave at the 9–12 November 1955 meeting of the American Astronomical Society at Troy, New York (Kerr et al., 1956); (2) "The Large Scale Structure of the Galaxy" by Kerr et al. (1957, p. 677), and (3) the proceedings of the Jodrell Bank August 1955 conference "21-cm Observations in Sydney" by Martha Stahr Carpenter (1957, p. 14). A footnote in

---

[14]Martha Stahr (pronounced "STAIR"; "Patty" to her friends) Carpenter (1920–2013) arrived in July or August 1954 as a visitor to RPL from Cornell University. On 6 August 1954, Kerr told Pawsey, "Patty Carpenter is now working with us, and she has been computing the curves of radial velocity [of HI] versus distance for each longitude, using the models [from Leiden]." Her farewell was on 3 June 1955. In early May 1999, Carpenter showed Goss and Ellen Bouton her farewell gift from RPL: a photo album of Sydney by the famous Australian photographer (and colleague of Shackleton) Frank Hurley. The book was signed by many of her colleagues. Carpenter and her husband left Sydney to return to Ithaca; they travelled via the UK where she gave a paper at the Jodrell Bank Symposium on radio astronomy in August 1955. See the American Astronomical Society obituary https://aas.org/obituaries/martha-stahr-carpenter-1920-2013 and an extensive memoir in https://www.aavso.org/media/jaavso/2838.pdf, based on her long association with the American Association of Variable Star Observers (AAVSO); she had been president in the 1950s for three terms. Carpenter clearly had a major impact on the AAVSO organisation.

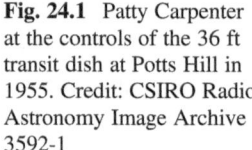

Fig. 24.1 Patty Carpenter at the controls of the 36 ft transit dish at Potts Hill in 1955. Credit: CSIRO Radio Astronomy Image Archive 3592-1

this publication reported that a preliminary result was given at URSI in August 1954. Three images of Patty Carpenter at the 36-ft telescope are shown in Figs. 24.1, 24.2 and 24.3.

In early August 1954, Kerr sent Pawsey a detailed report of the HI results with figures that would be the basis for the presentation by Pawsey at URSI. The emphasis of the HI galactic data was on galactic structure and HI absorption lines. Kerr wrote:

> The galactic profiles, however, are as good as two complete sets [thus observed twice] can make them, they will of course be better when we have four sets. [The spiral arm diagram agreed reasonably well with the Leiden data] and it is possible now to see the overall shape of the galaxy quite nicely.

The Sagittarius and Orion spiral arms were evident in HI. The famous first image of the spiral structure in our galaxy obtained from the combination of the Australian and Dutch observations are shown in Fig. 24.4.

The determination of the HI plane of the galaxy was planned: the latitude range would be extended from a few degrees to 8°. There were many new details such as low level ("long tails") extensions on some profiles; this unexpected result was to be investigated later. Carpenter was working on computing the curves of radial velocity versus longitude for each longitude using the rotation curve of the galaxy derived from the Dutch observations. Kerr described in detail the results of the HI absorption data obtained in the direction of strong galactic plane continuum sources. (Surprisingly, these data were never published.) The Australian group realised that HI absorption data could assist in sorting out the fundamental uncertainty (the

Fig. 24.2 Patty Carpenter waves from the declination axis of the 36 ft at Potts Hill in 1955. Credit: CSIRO Radio Astronomy Image Archive B3592-6

Fig. 24.3 Carpenter with her hand resting on the declination hand-crank wheel of the 36-ft transit dish at Potts Hill in 1955. Credit: CSIRO Radio Astronomy Image Archive B3592-7

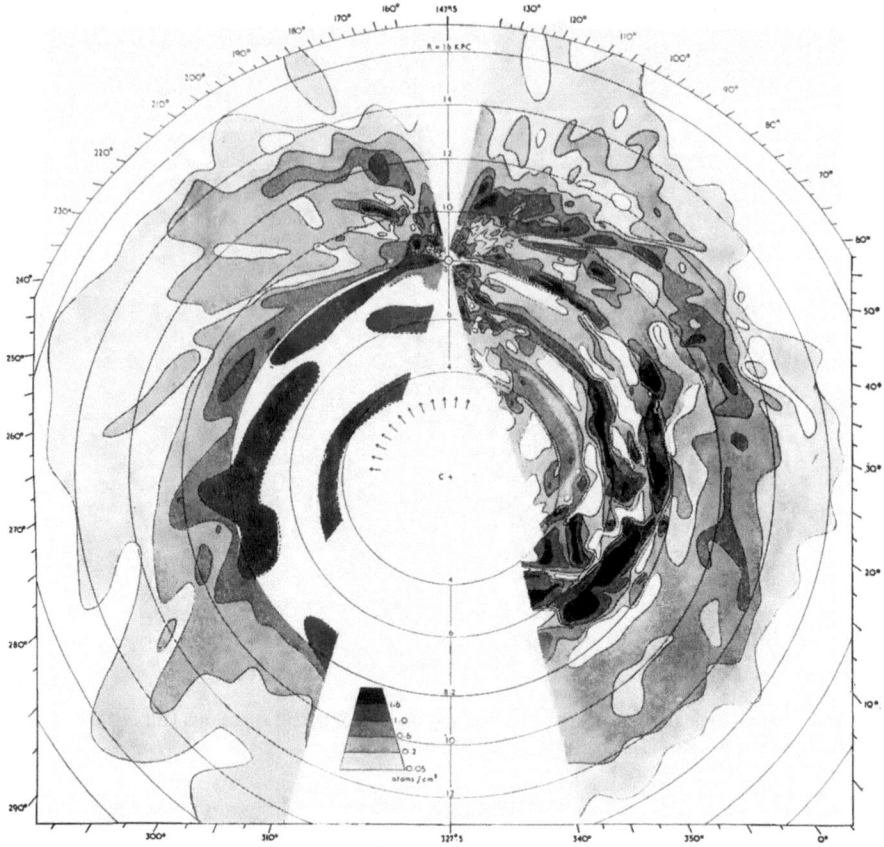

Fig. 24.4 Figure 4 from "The galactic system as a spiral nebula", Oort et al. (1958) MNRAS, vol 118

assumption of optical thin HI emission) as they converted observed emission intensities to HI column density. (HI absorption could assist in assessing the optical depth.)[15]

Finally in the 6 August 1954 letter to Pawsey, Kerr presented a two page memo, "Notes on Calibration of 1420 MHz Receivers". A number of prescient suggestions were described: the use of standard regions for HI observers through-out the world (these would be visible in both hemispheres); with a simple HI profile, observers could check calibration and velocity scales. Kerr wrote, "Except in special cases, the

---

[15] Kerr mentioned a fascinating proposed use of the moon to calibrate the efficiency of the HI system at Potts Hill: "We've not been able to do anything else yet about the proposed method of measuring aerial efficiency by observations of the moon eclipsing part of the galaxy." (It is not clear if the Kerr group ever tried this.). Years later Peter Kalberla and colleagues used this lunar technique in a clever fashion to study the stray radiation of the 100 m telescope at the HI line, "Time Variable 21 cm Lines and the Stray Radiation Problem" by Kalberla et al. (1980).

regions should be chosen where there is a fairly flat distribution in both space and frequency . . . The intensities should be at least moderately high . . ." Kerr ended with a plea for a **clear channel** [our emphasis—a protected band at the HI line, a suggestion that was later adopted internationally]: "Unless the matter has already been settled, URSI should enquire from CCIR (Comité Consultatif International pour la Radio—Consultative Committee on International Radio) about progress in setting aside a clear channel around 1420 MHz."

David Martyn: URSI 1954

David Martyn was a member of the URSI Executive in 1954 and at the URSI conference was elected to be the next President of the Ionosphere Commission.[16] Tragically, URSI 1954 confronted Pawsey with another challenge of an unexpected kind: providing care for David Martyn, who experienced a severe episode of psychosis during the conference, a difficult period in what was clearly a long experience of mental illness.[17] During the conference he became subject to severe paranoid delusions of being under surveillance, which included the notion that his life was threatened. This was perhaps likely a trauma effect resulting from the humiliation and persecution during his WWII period in Sydney in 1941 when Ella Horne, a German woman he had indiscreetly befriended, was accused of spying. Pawsey provided support for him in Le Hague until he returned early to the UK where, fortunately, he could be cared for by the several members of his family who were doctors, and within some days he had recovered sufficiently as to lecture at Cambridge and return early to Australia.[18]

London and Freiberg September 1954

After the URSI conference, Pawsey continued his hectic schedule in the UK and Europe. He returned to London where he spent some time helping ASLO (Australian Scientific Liaison Office) cope with the breakdown of Martyn. On 4–6 September 1954, he met with his former colleague from Mt Stromlo Observatory, Cla Allen of the University of London. Along with the keynote speaker David Martyn, he

---

[16]NAA, A8520, PH/MAR/12 Part 6. Personnel file David F. Martyn.

[17]The WWII events leading up to the tragic breakdown of David Martyn at URSI is described in NRAO ONLINE.7

[18]We have discovered a coincidental connection with David Martyn's brother Dr Allan Martyn, and the radio astronomy community. Later in his career, Allan Martyn moved to a medical practice close to Jodrell Bank Observatory. Prof Rodney D. Davies (FRS, 1930–2015, past Director at Jodrell Bank, colleague of Pawsey at CSIRO in the early 1950s) of the University of Manchester told Goss in 2006 that Allan Martyn was his family physician for many years.

attended a small conference in Cambridge at Corpus Christi College, "The Physics of the Ionosphere".[19] As we discuss in Chap. 25, Pawsey had maintained more than an interest in ionospheric research. Martyn—the pre-eminent ionospheric physicist—was able to travel to Cambridge and the Cavendish Laboratory for a series of lectures during the next week. Pawsey reported to ASLO that Martyn's was the outstanding contribution of the conference.[20] The following weekend Pawsey was again in London.

On 12 September 1954, Pawsey travelled by air to Freiberg, Germany, to visit the Fraunhofer Institute and the Director Karl O. Kiepenheuer,[21] and then returned to Manchester and Cambridge with a final week in London before his departure for New York on 30 September.

New York, Visit to North American October 1954

Pawsey arrived in New York on 1 October and left on 19 October. The short nature of the visit was dictated by financial considerations as explained to Bok in a letter from 5 May 1954: "I should like to make it a much more leisurely visit but dollar shortages require that I spend only two weeks in America [five days of the visit were to be in Canada]. It looks as if, if [sic] I am to have any time in the States, I shall have to find a job there."

Pawsey had put the North American trip together earlier in 1954; several of the hosts, Bok[22] (Harvard) and Goldberg (Michigan) had complained to Pawsey about the brevity of the planned visits. Before the visit to Harvard, Pawsey spent a

---

[19] Pawsey's presentation and publication are described in NRAO ONLINE.22, section D (September 1954). Also two related publications of Wild and Roberts (scintillation observations of Cygnus A from Dapto) from 1956 are described.

[20] Walter Ives, Chief Scientific Liaison Officer, managed Martyn's care from when he departed Le Hague until his return to Australia. His extensive correspondence with Sir Ian Clunies Ross concerning Martyn's illness is held in the CSIR archives. This quote from a handwritten letter of 14 September 1954 from Walter Ives to Clunies Ross. NAA, A8520, PH/MAR/12 Part 6. Personnel file David F. Martyn.

[21] Kiepenheuer (1950) had already proposed the synchrotron model for the galactic radio emission (see Chap. 34) but there is no mention of this remarkably important interpretation in Pawsey's letters.

[22] In mid-June 1954, Bok sent a series of extensive abstracts of four HI papers to be given at the June 1954 AAS meeting in Ann Arbor (papers by Bok and Ewen, Lilley and Heeschen). Pawsey was impressed (in his letter to Bok of 28 June 1954): "I was interested in the changing pattern of papers in radio astronomy. So many of our earlier ones described a new instrument or technique and the bulk of the information to be found with it. Your papers [from Harvard] exemplify the more mature approach; a restricted problem inspired by a known astronomical question. With the vast increase in the detailed information obtained in observations the science must go that way." Bok had suggested that Pawsey stay in Cambridge for some weeks to give a series of lectures. Pawsey pointed out (10 June 1954) that he hoped to return in the future for an extended visit. This was in fact to occur in late 1957 (see Chap. 28).

weekend with his brother and sister-in-law at Princeton (Ted and Kate Nicoll) and had a half-day trip to Washington to visit Merle Tuve as well as the ASLO. Bowen had written Pawsey in London on 22 September 1954 suggesting a last-minute change in plans in the US:

> I would like to urge you to go to Washington to see Merle Tuve. As the prime mover in our quarter of a million dollars [for the GRT proposal] he deserves talking to. Furthermore, his own radio astronomy programme is on a much sounder footing than it was previously and he is also Chairman of the committee which is considering the Associated Universities' project [to form the National Radio Astronomy Observatory (NRAO)] . . .

In the end, Pawsey spent only 2 days at Harvard, 1 short day in Michigan (McMath-Hulbert Observatory at Lake Angelus during the day 12 October) as he visited Helen Dodson, Leo Goldberg and E. Ruth Hedeman.[23] In Boulder, the host was W.O. Roberts with a brief visit of only 21 h.

Canada: October 1954

A major objective in the trip to North America in 1954 was a longer visit to the National Research Council in Ottawa to discuss a long-term visit to Canada in the following year or two. Pawsey began the organisation of this visit earlier in 1954 with letters to Frank Davies (head of the Radio Physics Laboratory of the Defence Research Board in Ottawa) and to his friend Don McKinley (Assistant Director of the Radio and Electrical Engineering Division) of the National Research Council. (Pawsey and McKinley had met in Sydney towards the end of WWII for discussions about cm radar.) Pawsey planned detailed conversations with Don McKinley concerning the proposed visit to Canada for himself and his family. McKinley responded on 12 May 1954 with an enthusiastic welcome. But he was not optimistic about major expansion in either ionospheric research or radio astronomy in Canada. McKinley: "We are planning a program on upper air research touching [only] on those parts of ionospheric and aurora work that are particular to Canadian conditions . . . [We do not contemplate] that we would expand the effort in the extra-terrestrial field."

There was no plan to expand to the same level of support that the Australians had done in radio astronomy or ionospheric research. He expected that the total staff in Canada would not exceed five or six scientists. The emphasis in Canada would remain in engineering fields. McKinley added: "However, the two groups [science and engineering] are sufficiently closely allied that, if both exist, there will be plenty of mutual interplay and inspiration."

On 10 June 1954 Pawsey wrote to Don McKinley again: "I am very much interested in the possibility of an extended visit to Ottawa in the year following if

---

[23] From NAA C3830 Z3/3/A 1954.

we could work out a suitable arrangement. My short visit this year would give me a chance to discuss possibilities."

Pawsey was quite interested in working on "the elucidation of ionospheric phenomena of the auroral zone". He was anxious to try out the partial reflection technique.[24] "I should like to see whether one can, in the auroral regions, observe the thermal background of radiation from the D-region at the appropriate frequency." Pawsey was unclear about radio astronomy prospects for him in Canada; this question would await discussions in Ottawa.

He finished the letter to McKinley with an admission of one of the motivations for a Canadian visit, a family visit to Canada:

> I should also like to spend some time at one or two American Observatories and I am hoping to explore the possibilities of combining this with a spell at Ottawa in a visit to North America which would be long enough to bring the family. Lenore pines for some snow and central heating.

The visit to Ottawa began in the afternoon of 7 October 1954, extending to the afternoon of 11 October, when he departed for Windsor, Ontario, and the nearby University of Michigan. On arrival in Ottawa, Pawsey gave a lecture at the "Science Association" of the NRC, "Some Research Activities of the Radiophysics Laboratory, Sydney". The main emphasis in his lecture was the CSIRO radio astronomy research.

The meeting with McKinley was disappointing concerning the long-term visit. We know the details based on a handwritten letter (on United Airlines letter paper) from Pawsey to his wife Lenore, found in the Joe and Lenore Pawsey Family Collection in 2014.

The letter to his wife contains a number of short statements that illustrate both the feelings of achievement and frustrations of his career in 1954: "(1) At long last I feel I am headed for home ... It will be great to be home. (2) I certainly am in demand [as he described his over-crowded schedule in the US during October 1954] ... Life is one hectic round."

The crux of the letter described the surprising misunderstanding with McKinley:

> This Canadian trip [in the future] is still up in the air. Don McKinley who suggested it started a bit of hedging. The proposition under discussion was a transfer from CSIRO to NRC for about a year—bonds of Empire and all that. This would be ok for both parties. Don wants me to apply for a job with NRC. That I was sure was a back-door method [i.e. Pawsey was to be forced to apply for an open position at NRC] so I went for a private talk with Herzberg (Chief of NRC Division of Physics). He and I saw eye-to-eye and with Herzberg's approval I put my foot down. I wrote Don saying the next move was [in the hands of] NRC—to write Fred White [CEO of CSIRO] asking for me for a year (or else no me). So we shall see what we shall see.

> One good thing which came up was that I got to know Herzberg. I and Herzberg are likely to walk hand in hand. This Canadian possibility has a lot of queries to it, the scientific politics here are not good—but I should venture for a short while. Similarly on the US side there are

---

[24]NRAO ONLINE.22.

rivalries which complicate what I thought was a nice simple situation. However, I can wait and see.[25]

Completing the Textbook *Radio Astronomy*: Caltech and Berkeley, Mid October 1954

After the visit to Canada, the final extended visit in North America was to Caltech, where he remained for 5 days. His host in Pasadena was Jesse Greenstein. The final stop was on Tuesday 19 October with a short 1 day visit to San Francisco; Pawsey made a quick stop at the University of California, Berkeley, to visit Ron Bracewell, who was on a 1-year visit to Berkeley.

Pawsey had chosen Bracewell as his co-author of the textbook he had contracted to write. But as is often the way of such projects, after signing the contract, he progressed very little over the next year, writing to Wood of the Clarendon Press on 26 March 1952: "The writing has progressed much slower than had hoped, largely owing to my being kept very busy with other work during the greater part of last year, but it is now going ahead."[26] By October 1952 he sent the good news that a first draft was now complete.[27] In late 1952 and 1953, Pawsey sought feedback from many colleagues, including Baade and Minkowski from Mt Wilson/Palomar Observatory, Cla Allen from University College London (on the solar chapter), and Len Huxley from the University of Adelaide (on meteor radar). On 2 October 1953, the complete manuscript and figures were posted from Sydney to Oxford. A lengthy letter accompanied this shipment with detailed questions about many issues, such as section numbering, equation numbering, abbreviations, figure captions, section and subsection headings and permissions for figure usage.

In Berkeley, Pawsey and Bracewell discussed completing the book, the proofs of which were being sent to both Berkeley and London at the time (correspondence was complex) and which were due back to Oxford University Press on 13 December 1953. In the end Bracewell did all the proofreading and reported on 13 December: "You will be pleased to hear that the last of the revised proofs have reached me and been duly returned to the Press. They have been coming in at intervals ever since I arrived here and it is a pleasure to have dealt with the last." However there would still be delays before publication.[28]

---

[25] This letter to his wife also exposes the fragile nature of Pawsey's health. He had been ill at both ends of the trip. He had been hospitalised in late May in Sydney for varicose veins before his departure in July 1954. On his return to Sydney, he was again ill with the flu. Christiansen wrote him in Sydney from London on 16 November 1954: "I have heard that you are back in Australia and had celebrated your return by getting gastric flu. I hope you are now fit and well."

[26] References are NAA C3830 Z3/2/ I, II and III.

[27] Pawsey also reported to Wood: "[At URSI in Sydney] I told Sir Edward Appleton of this [completed draft] . . . and he urged us to get on with the job as quickly as possible. We shall do this."

[28] The textbook *Radio Astronomy* by Pawsey and Bracewell finally appeared in June, 1955. Details of the story of this textbook can be found in NRAO ONLINE.53.

Pawsey departed from San Francisco on 19 October, arriving in Sydney on 21 October.[29]

Back at CSIRO in Sydney: Late October 1954

After Pawsey's return to Sydney, there was an exchange of information between the NRC and CSIRO about the plans for Pawsey's Canadian visit. B.G. Ballard, Vice-President for Science at the NRC, wrote to White. Pawsey's insistent conversations with Herzberg had paid off. A reasonable offer was now being made. The NRC group had been impressed by Pawsey, "we would like to have him come again for a longer stay. We would like to suggest a minimum period of one year, longer if you can possibly spare him … [T]here would be a considerable gain from your viewpoint in the broadened experience and additional contacts that would result." They proposed that the NRC would pay Pawsey's salary and the CSIRO would cover travel expenses. White responded on 5 November 1954 in a somewhat non-committal manner: "I know that Dr Pawsey would very much appreciate an opportunity of spending more time in Canada."[30] No details were discussed. On the same day, White wrote Bowen about the Pawsey- Canadian plan, in a somewhat negative tone and with a puzzling final sentence.

---

[29] Pawsey had originally planned to spend about 8 h visiting Grote Reber in Maui on 18 October 1954. On 10 October, Reber wrote Pawsey (in Pasadena) that he was in the process of moving to Tasmania! He and his boxes of equipment were to leave Honolulu by ship on 17 October, the day before. However, he was to arrive in Sydney on 1 November. Thus, Reber and Pawsey would meet later in 1954 in Australia.

[30] McKinley wrote on the margin of the copy of the Ballard letter that he sent along to Pawsey (Pawsey was not sent an official copy): "This is naturally a bit irregular—sending you this copy [of Ballard's letter of 21 October 1954 to White] but I thought you'd like to know the general tenor of the [mutual?] approach. I trust you approve of it. Don". On 22 October 1954, McKinley had written to Pawsey responding to the latter's letter written just as he left Canada for Pasadena in early October. Clearly Pawsey had been angry with McKinley, who tried to pour oil on troubled waters over the misunderstanding. The Canadians were much more bureaucratic in their organisation of short term, temporary visitors than the Australians. Pawsey had strongly objected to the Canadian requirement that he must obtain a Canadian position that required an official application, the Canadian Selection Board's approval and finally the approval of the Cabinet Minister. McKinley had presented the procedure to Pawsey in a heavy-handed manner. He stressed in the 22 October letter that all this process was just a "formality" and that it would not appear to the CSIRO that Pawsey was about to receive a permanent position with the National Research Council of Canada. Pawsey replied to McKinley on 4 November 1954 with a rather neutral tone. Apparently, he had lost his enthusiasm for the NRC visit. Pawsey wrote: "Please do not misunderstand me on my stand re discussions between NRC and CSIRO prior to application forms etc. It was only a question of which came first. I quite recognise that any appointment such as this must be approved by your powers-that-be and that they need to be satisfied as to desirability of an appointee." But in the end, the damage was done and the visit fell through. Instead, Pawsey organised his long visit in 1957–1958 to the US (see Chap. 28).

> I know that Joe is quite keen on the idea of spending some time in Canada in the future, and while I would like to help him do so, I think both you and he will agree that much depends on the way things turn out in connection with the large radio telescope [GRT]. It may be important for Joe not to be away at certain stages of this development if it is to go ahead effectively. **There are probably other considerations too which we can discuss when we meet.** [our emphasis]

This last cryptic remark raises some questions. Does this provide a glimpse of rising tensions in the relations between White, Bowen and Pawsey? As we will see (Chaps. 30, 31 and 38) the relations would steadily deteriorate leading to the break in 1960 as Pawsey was to leave CSIRO.

Chapter 25
The Sun and the Ionosphere, 1946–1955

On looking over my dossier it is fairly clear that my first claim to admission [to the Royal Society] must be based on my share in building up the radio astronomy group here [at RPL]. As you know I started the line of work after the war and I was personally responsible for the initiation of two most fruitful lines of investigation, interferometry applied to the disturbed sun and the observational study of the quiet sun … Once the group was well established I have found it more stimulating to the group to not compete directly in the detailed investigations but to attempt to integrate the work. [i.e. write review papers] …

A second claim concerns the development of original methods for the study of the ionosphere. My 1935 paper described the method of measuring winds in the ionosphere by means of the movement of a diffraction pattern over the ground. This was exploited extensively about 15 years later. My 1951 paper described an entirely new method for determining ionospheric temperatures, by means of thermal emission.

The third claim might be made for contributions to radio or radar techniques.[1]

—J.L. Pawsey

When David Martyn wrote to Pawsey in August 1952 to propose his nomination for the Royal Society, Pawsey replied with a formal typed letter that included an obligatory modest paragraph, followed by his own assessment of his achievements. In his own assessment, his main scientific contributions were in solar radio astronomy (which in 1952, he considered the "most fruitful" investigations at RPL, rather than the new found interest in the discrete sources and extra-Galactic observations) and the development of new methods of ionospheric research. We therefore devote this chapter to a brief summary of research and achievements in these areas up to 1955. The first part of the chapter contains a brief overview of the main achievements in solar radio astronomy, led by Paul Wild. Often it is difficult to identify Pawsey's particular contributions, because, as noted, his leadership style was to visit

---

[1] Source material: NAA A9874/60, "Royal Society, London. Dr Pawsey's Qualifications" and Joe and Lenore Pawsey Family Collection.

© The Author(s) 2023, corrected publication 2024
W. M. Goss et al., *Joe Pawsey and the Founding of Australian Radio Astronomy*, Historical & Cultural Astronomy, https://doi.org/10.1007/978-3-031-07916-0_25

and discuss observations and analyses, to make comments and propose ways of resolving immediate difficulties, but not to lead the research nor to take authorship for such contributions, However many papers do acknowledge Pawsey's advice and criticisms in their preparation. The second part contains an overview of Pawsey's well-crafted but not extensive ionospheric research.

The Status of Ionospheric and Solar Physics in the World of Astronomy

It is fair to say that the excitement of the new possibilities that radio offered for observations at both galactic and extra-galactic scales, dominated 1950s astronomy. Histories of radio astronomy have given less interest to the solar research of the 1950s and 1960s. A good explanation of why this happened was offered, good humouredly, by the pre-eminent solar radio astronomer of the 1950s, Paul Wild:

> I have the feeling that, to most astronomers, the sun is rather a nuisance. The reasons are quite complex. In the first place the sun at once halves the astronomer's observing time from 24 to 12 h, and then during most of the rest of the time it continues its perversity by illuminating the moon. Furthermore I have met numerous astronomers who regard solar astronomy to be now, as always before, in a permanent state of decline—rather like Viennese music or English cricket. Nevertheless, those who study the sun and its planetary system occasionally make significant contributions. There were, for instance, Galileo and Newton who gave us mechanics and gravitation, Fraunhofer who gave us atomic spectra, Eddington and Bethe who pointed the way to nuclear energy, and Alfvén who gave us magneto-hydrodynamics. **Perhaps the point to be recognised is that the sun has more immediately to offer to physics than to astronomy**. [our emphasis][2]

Wild's statement was even more relevant for solar radio astronomy than for traditional solar astrophysics, for the sun is unique at radio wavelengths. The emission at shorter wavelengths, from X-rays to optical and infrared, comes from regions containing dense matter associated with the visible sun as we know it—the photosphere and chromosphere. But the radio emission is generated in the tenuous plasma known as the solar corona (Wild, 1985), and this has no relevance for the stellar astronomers. Even though solar observations and interpretation did provide key information for the studies of other stars, the case still had to be made for its relevance to our understanding of stellar physics (Pecker, 1975).

Ionospheric research, on the other hand, continued post-war for many of the same practical reasons that had generated it in the interwar period—because of the ongoing impact the ionosphere had on various human communication systems. David Martyn, as dominant an ionospheric physicist in the 1950s as he had been in the 1930s, was studying solar tidal effects in the ionosphere, as well as lunar effects in the upper (F2) region. He undertook detailed statistical analyses of a huge body of ionospheric data and accurately determined the magnitudes and phases of

---

[2] Quote from Paul Wild's address to the IAU GA in Sydney, 1973.

tidal effects in the various layers, and from this, developed theories of tides caused by solar heating and lunar gravitation. He also explored a variety of electromagnetic effects and made a study of the morphology of storm geomagnetic variations in the ionosphere (Piddington & Oliphant, 1971).

Researching the impact of the ionosphere on human communication systems was to continue as a major research activity around the world. Solar bursts which affected the ionosphere could disrupt communications and this led to the establishment of an international network of ionospheric prediction services (IPS). The development of the Australian IPS observatory was closely linked to earlier developments in Australia and was co-located with the solar radio astronomers at Narrabri (previously called Culgoora) (Wilkinson et al., 2018).

Ionospheric research merged into space plasma physics as in-situ observations by rockets and spacecraft drastically changed the field and the radio propagation effects have become secondary and marginalised. This is now a huge research field extending from the ionosphere to the outer corona of the sun. It now has little overlap with astronomy.

In the modern era, ionospheric studies have taken on new significance. The same ionospheric irregularities that made the radio sources scintillate also disturb GPS signals and now are the dominant source of error for precision location measurements.

Solar Radio Astronomy in Australia (1947–1955)[3]

Radio Astronomy began with observations of radio emission from the sun, and of course the exciting first 18 months of observations at RPL were of what, by happy coincidence, was a very active sun. Pawsey's most active involvement occurred in the 1940s. He was particularly closely connected to Ruby Payne-Scott's group as she developed observations of the disturbed sun at Dover Heights, Hornsby and then Potts Hill. The details of this work have been laid out in exhaustive detail else-where.[4] From 1950, Paul Wild and "Chris" Christiansen (Chap. 24) led the two RPL solar research groups. Wild's had become, and would remain, the dominant group globally.

Pawsey appointed Paul Wild in 1947; by 1950, he had taken over leadership of the solar group (Penrith and then Dapto) that was exploring the properties of the active sun and the dynamic spectrum of solar bursts. A second solar group lead by Christiansen was exploring the spatial properties of the quiet sun by making 2D

---

[3] This section is based on Paul Wild's Introduction in: "Solar Radiophysics: Studies of the Emission from the Sun at Metre Wavelengths" (McLean & Labrum, 1985). See also Paul Wild in Frater et al. (2017).

[4] See Goss and McGee (2009) and Goss (2013).

images. Christiansen's emphasis was on the spatial imaging technology. He had no spectral information and the two groups continued independently.

As Wild later described it (1985), he was interested in metre observations which revealed "a spectacular range of phenomena undreamt of before their discovery". To explore these new phenomena, a new instrument was built: a radio-spectrograph which could record multiple frequencies at the same time. It was the basic tool used for solar observations at Radiophysics for 7 years (1949–1956), the only one of its kind in the world. Wild later noted that these pioneering studies were made by physicists and engineers with a background in ionospheric research, early TV and radar. They had no background in astronomy or solar physics.

The First Radio Observations of the Sun

The solar emissions had extraordinarily high intensity. At 3 m wavelength the maximum brightness temperature was up to 10^{12} K for some solar bursts. Clearly such radiation was of nonthermal origin; gyro radiation and radiation from plasma oscillations were both suggested as possible radiation processes. At the other end of the intensity scale, Pawsey (1946) found that the minimum brightness temperature of the sun at quiet periods was a little over 10^6 K. This detection of the million degree corona and the implications are discussed in detail in Chap. 14.

The location of the radio bursts was investigated by Ruby Payne-Scott using the 100 MHz swept-lobe interferometer at Potts Hill (1948–1951), see NRAO ONLINE.20 for details.

Paul Wild (1968, p. 117) praised the design concept:

Another Pawsey-inspired experiment was put into operation and brilliantly performed by Payne-Scott and [Alec] Little. The idea was to locate ... the instantaneous position of the dominant source on the sun at any one time.

The instrument would be able to sweep across the sun 25 times a second with a resolution of about 40 arc min and a positional accuracy of about 2 arc min. With this resolution the quiet sun, with a size of 35 to 40 arc min, would be resolved out. This instrument was the first rotating lobe interferometer.

Christiansen continued the observations of the spatial structure of the solar emission at Potts Hill and later at Fleurs. He found that at 21 cm, the emission from the sun arises from the transitional region between the corona and the outer chromosphere. In this region the change-over between the steady optical sun and the spectacularly variable metre-wave sun occurs. His imaging techniques also identified the predicated equatorial limb-brightening at these wavelengths (Christiansen & Warburton, 1955).

Wild (1985) summarised the situation leading up to the 50s:

The situation around 1948, when I joined the Sydney group of investigators led by Joseph Pawsey, was one characterised by mystery, incredulity and intense interest. A whole new field of research lay ahead with obvious objectives: to disentangle the confused conglomeration of phenomena; to interpret and understand them; and to put the results to use in the mainstream of research for solar physics, astronomy and physics.

Before the "origin of species" could be identified there had to be an exercise in taxonomy. Already Pawsey and his colleagues at Sydney had found that, in addition to the polarised storm radiation, there were different kinds of unpolarised bursts; there were large *outbursts* lasting 10 or 20 min which accompanied large flares, and there were short, sharp, "isolated" bursts lasting a few seconds. A new clue was discovered when Payne-Scott et al. (1947) noted systematic time delays in the starting time of bursts, high frequencies preceding low.

For Wild, "the obvious next step" was to develop a *radiospectrograph* to record the intensity of the solar emission as a continuous function of frequency and time. The first spectrograph was built at Penrith using a hand guided rhombic aerial to track the sun, with observations made in the frequency range 40–70 MHz. By June 1949, Wild and McCready (1950) had observed the dynamic spectra of many solar bursts which differed widely and showed great complexity. Based on these observations they began classifying spectral types.

Wild and McCready identified and named three types of bursts, Types I, II and III, distinguished by the way the frequency drifted with time. This work was published in a series of papers in 1950 that became the foundation for all future work on solar bursts. They deduced that the Type II bursts were associated with shock waves coming out through the solar atmosphere at 1000 km/s and were associated, 30 h later, with aurorae in the earth's night sky. They associated Type III bursts with streams of electrons being ejected at a third the speed of light and taking only an hour to reach the earth. The mechanisms proved to be correct and their nomenclature for the phenomena became the international standard. In the late 1950s, discoveries of type IV (1957) and type V (1959) bursts were also made by researchers overseas.[5]

A new and much improved radiospectrograph in a field station located on a dairy farm at Dapto, N.S.W. was completed in 1951 and made the unexpected discovery that Type II and Type III bursts often showed major spectral features repeating with a 2:1 frequency ratio (Wild et al., 1954). This result showed that the bursts were generated at both the fundamental and second-harmonic frequencies. In turn, this led to the conclusion that the emission was from plasma oscillations.

At this point in time it became increasingly clear that further progress would also require spatial information, and the Dapto radiospectrograph was modified to act as an interferometer for this purpose. The first plans were being made for a future major instrument that could make a movie of changing 2-D images of the sun. This would eventuate as the famous circular Culgoora radioheliograph which became operational in 1967 (see Chap. 37). When the radio heliograph was finally shut down in 1984, the CSIRO solar radio astronomy project was considered so successful that it had resolved all outstanding research problems in the field.

---

[5] See Frater et al. (2017), pp. 94–96, and NRAO ONLINE.20.

Ionospheric Research 1947–1954[6]

It is interesting that throughout the first half of the 1950s, Pawsey was involved in a few projects in ionospheric research; likely these were a minor activity in comparison to his leadership of the radio astronomy activities of RPL. In these years he made some modest contributions from typically well-designed experiments and continued an enjoyable correspondence with Appleton.

Thermal Radiation from the Ionosphere, 1947–1953

In 1947, Pawsey began working on a problem concerning the ionosphere that may well have originated during his years at the Cavendish: what is the temperature of the various layers of the ionosphere? At the frequencies used, this probes the ionosphere at a height of 70–80 km. In both 1947 and 1949, he initiated experiments at RPL to answer this question. He "made some false starts being tricked by not identifying man-made noise [which masked the weaker steady thermal emission from the ionosphere]". The main problem was (as Pawsey explained to Appleton): "Natural noise carries no label to distinguish it. I believe that a bit more experience may be helpful."

From August 1949 through June 1950, Pawsey et al. (1951) observed at two sites in a mountain gorge (450 m, 65 km SW of Sydney) at Burrogorang Valley, now part of the Warragamba dam, and at Rankins Springs (450 km W of Sydney). Both sites were well protected from local electrical interference such as motor cars, home generators and industrial plants, which was to be a key factor. The point was to observe during the daytime when the low ionosphere insured that atmospherics from distant thunderstorms were not observed (Pawsey et al., 1951).

A sketch showing the idealised noise record based on observations taken over a whole day is shown in Fig. 25.1. Additional details of the 1951 observations are provided in NRAO ONLINE.22 (section A).

Pawsey et al. (1951) explained their rationale:

> ... the fluctuating base-level at night may be evidence of ionospheric fading effects on noise (atmospherics and other) transmitted from a distance. Similar fluctuations by day appear to be evidence of reception of noise propagated from a distance, and the absence of fluctuations, when observed, suggests that on these occasions interference propagated via the ionosphere is not a main factor.

> An alternative argument for the rejection of atmospherics as the origin of the base-level may be derived from the well-known inaudibility by day of distant broadcasting stations which are readily audible at night. This may be interpreted as showing that the reduction of strength of stations is greater than that of noise. This is consistent with an origin in thermal radiation in the ionosphere which is not subject to such attenuation, but it is most improbable that a

---

[6] See NRAO ONLINE.22 (Section B).

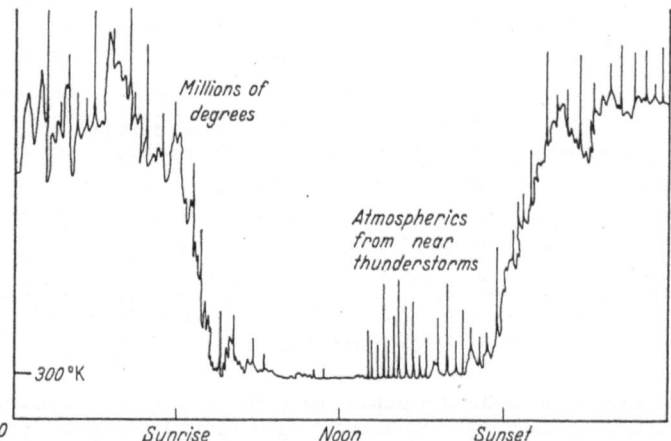

Fig. 25.1 The idealised sketch noise record covering a whole day. Credit: Fig. 4, "Ionospheric thermal radiation at radio frequencies", Pawsey, J. L., McCready, L. L., & Gardner, F. F. (1951). *Journal of Atmospheric and Terrestrial Physics*, *1*(5–6), 261–277

> steady background should result from atmospherics from within a small radius of say, 500 or 1000 km.

Thus, it was clear that the base-level was not due to man-made or atmospherics. There was a parallel with the base level of the steady emission of the solar corona as explained in Chap. 14.

> The observations have shown that the *natural* noise level observed on an aerial at frequencies in the vicinity of 2 Mc/sec during the hours around noon frequently falls to an intensity corresponding to an equivalent aerial temperature between 200° and 300°K. It is not observed to fall below this, and at the times when this low level is observed the characteristics appear similar to those of thermal noise.

> Further, there are reasons for believing that this level cannot be accounted for in terms of the integrated effects of great numbers of atmospherics. These facts are strong evidence for the hypothesis that there is a background source of random noise of this intensity. This background source is identified with thermal radiation from the ionosphere because, as will be shown, the measured intensity agrees, within the limits of the data, with that derived from other sources. This radiation, from a microscopic viewpoint, arises from the acceleration of [electrons due to] collisions [in the plasma].

The measured temperatures of 240–290 K in the D layer at heights of 70–80 km agreed with other observations.

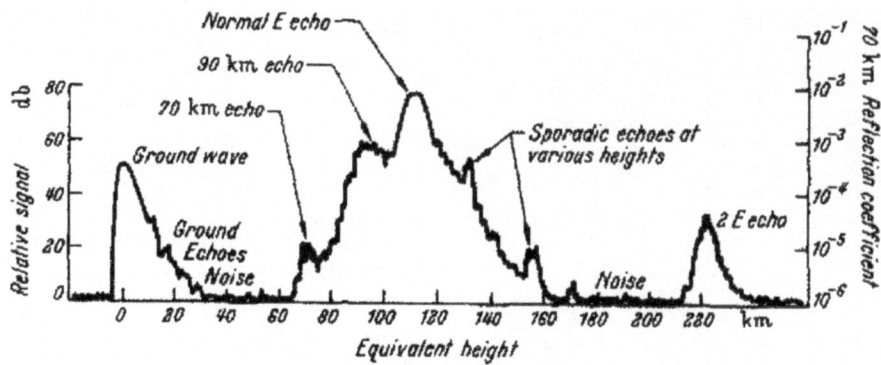

Fig. 25.2   Sketch of typical midday echo pattern. Signal intensity versus range of the ionosphere (0 to 250 km). Credit: Fig. 3, "Study of the ionospheric D-region using partial reflections", Gardner, F. F., and Pawsey, J. L. (1953). *Journal of Atmospheric and Terrestrial Physics* 3, no. 6: 321–344

Study of the Ionospheric D Layer, 1951–1953[7]

In 1951–1953 Pawsey led a study of the D layers of the ionosphere, still observing in the Burrogorang Valley. For the first time the 2.28 MHz pulse echo system was used to investigate the structure of the D region. A normal pulse echo method was employed (ionosonde, now also CHIRP sounder) with transmitter and receiver separated by 20 or 30 km from the transmitter (provided by George Munro of the Radio Research Board); the transmitter had a peak power of 1 kW with a pulse length of 30 μs (9 km). The result of this series of observations was the geometry of the D region, overlaid by the major E layer of higher density. The results are in a highly cited paper by Gardner and Pawsey (1953). Two distinct regions were apparent: a region at about 70 km where distinct strata form by day and a second more prominent region of higher electron density at about 90 km which extended up to the E layer at 110 km. See Fig. 25.2.

Within a few days (7 March 1951),[8] Pawsey had made a hand drawn sketch (Fig. 25.3) and wrote an enthusiastic letter to Fred White, CEO of CSIRO in Melbourne and a well-known "ionomer"[9] describing the first data with the ionosonde at Burrogorang from Sunday, 4 March 1951. The fact that the noise levels at this site were so low (the key to the success of the research of 1949–1950) was decisive. "We have often wondered what could be seen of ionospheric echoes using the high sensitivity which such low noise-levels permit. The answer is a glorious conglomeration extending down to 70 km or a little lower [in the D layer of the ionosphere.]" Pawsey ended his first impressions with the statement: "... and

---

[7] Summary of material in NRAO ONLINE.22.

[8] NAA C3830 Z3/1 and also CSIRO KE 20/2.

[9] A term to describe ionospheric researchers used at Cambridge and, after his return from Cambridge, by Ron Bracewell in Sydney. This new terminology disappeared in later years.

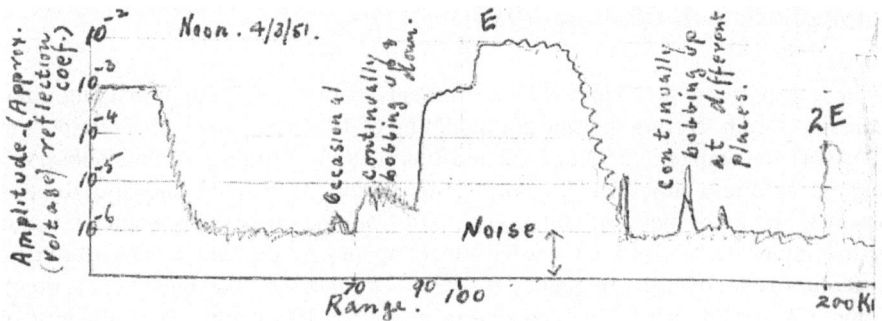

Fig. 25.3 Signal intensity versus range—Pawsey's hand-drawn sketch in letter to Fred White. Courtesy National Archives of Australia: NAA C3830 Z3/1

then a lot of new observational material has a habit of stimulating the associated theory."

But despite this potential, the research was not continued, presumably being of less urgency than the many other demands at RPL. Nor was ionospheric research rated very highly elsewhere, as Pawsey informed Appleton in a letter about his D layer research on 23 July 1953. He told Appleton about Geoffrey Builder (1906–1960), who had completed his PhD with Appleton in 1933,[10] and worked for the Radio Research Board in Sydney in the 1930s and at AWA (Amalgamated Wireless Australia) during WWII. Builder had continued his ionospheric research after the war, joining the School of Physics at Sydney University in 1947, becoming a Senior Lecturer in 1950. Pawsey had been hoping to continue the D layer research with Builder. But, Pawsey wrote, "a storm, in the form of Professor Harry Messel, has hit Sydney University. Messel is a most extraordinary dynamic personality. He does the impossible at times: he recently got promises of £30,000 from Sydney businessmen." But he did not support ionospheric research and Builder had ceased to work in the area. Pawsey wrote: "I am rather sorry, because I think he is a most competent person in this field." In NRAO ONLINE.22 (Section C, "Geoffrey Builder"), we provide additional information about Pawsey's concerns about ionospheric research at the University of Sydney.

Perhaps the absence of collaborators added an extra deterrence to Pawsey's research in the area. Even this limited "ionomy" of his, did not continue.

---

[10]Builder had designed the apparatus for the British ionospheric expedition to Norway during the International Polar Year in 1932–1933, led by Appleton.

Ionospheric Scintillation (1954)

Quite unexpectedly the Dapto solar instruments made some ground-breaking ionospheric research. During the sunspot minimum of 1954 (e.g. number of sunspots in
1954 was 46 compared to 93 in 1952 and 208 in 1955), Wild's group only observed
the sun at Dapto infrequently. During this period, Paul Wild and Jim Roberts
observed the ionospheric scintillation pattern from the far northern source Cygnus
A (maximum elevation 15°). It was visible for about 2 h a day and they measured the
dynamic spectra over the frequency range 40–70 MHz (i.e. frequency versus time).
Some 200 records were obtained over a period of 18 months. The influence of
Pawsey's background in ionospheric physics can be seen in the planning and
interpretation of these observations. In September 1954, Pawsey presented a summary of these data at the Cambridge conference on the physics of the Ionosphere
which was discussed in the previous chapter.[11] Three different types of observations
were obtained: swept-frequency spectroscope to obtain dynamic spectra, a swept-
frequency interferometer to study spatial deviations at various frequencies and a
triangle spaced antenna system to study lateral sizes and motions of the scintillation
pattern on the ground.

The purpose of these investigations is succinctly summarised by the authors in the
second 1956 paper:

> Most of our knowledge of the terrestrial ionosphere has been obtained with the use of radio
> waves transmitted from the earth, reflected from the various layers of the ionosphere and
> received again at the earth. With this method, investigation is restricted to regions below the
> layer of maximum electron density. The discovery of extra-terrestrial radio sources now
> permits the study of the ionosphere by means of radio waves transmitted from outside. This
> method is of special interest, because it may allow us to study regions above the layer of
> maximum ionisation.

The new data of the mid-1950s revealed ionospheric gradients that could act like
giant refracting prisms, leading to focusing by single lens-like structures. These
structures were on the order of 10 km.

Pawsey's summary from 1955:

> These relatively systematic high intensity "ridges" would not be observed if scattering were
> from a number of randomly distributed scattering centres. They are most simply explained if
> each "ridge", single or multiple, is due to focusing caused by a single lens-like irregularity in
> the ionosphere. The duration of the typical "ridge" is from 10 to 40 seconds, the interval
> between, from 30 to 250 seconds. If the time variation is accepted as due to drift of the
> pattern over the ground, the ridge on the ground has a width from 1 to 4 kilometres and the
> spacing between ridges is from 5 to 20 kilometres. The drift speed is commonly about

---

[11] Published by Pawsey (1955b), "Radio Star Scintillation due to Ionospheric Focusing". In 1956,
Wild and Roberts published two papers with details of these results: "Regions of Ionosphere
Responsible for Radio Star Scintillation" (Wild & Roberts, 1956) and "The Spectrum of Radio-
star Scintillations and the Nature of Irregularities in the Ionosphere" (Wild & Roberts, 1956b). The
latter publication provides a more detailed account.

100 metres per second (Pawsey, 1955, *The Physics of the Ionosphere*, "Radio Star Scintillation due to Ionospheric Focusing", p. 172).

This interpretation was counter to the prevailing theory, based on the diffractive scattering caused by small-scale irregularities, which had been developed in the Cambridge group. Frater and Ekers (2012) pointed out that as a consequence, the Wild-Roberts results received little recognition in the next 20 years. It was not until 1975 that the full theory of scattering by a power-law spectrum of irregularities was worked out by Soviet scientists and this made it clear that the diffractive scintillations described by the Cambridge model were modulated by refractive scintillations that caused the focussing effect that had already been seen so clearly in the dynamic spectra of Wild and Roberts taken 20 years earlier. Had the importance of refractive effects been accepted at the time, scintillation theory would have advanced much more rapidly.

Chapter 26
Overseas Again: Jodrell Bank and IAU, August 1955

[During the Jodrell Bank Symposium session in which Pawsey and Ryle spoke], there [was] ample room for controversy of the kind that occurred aplenty on the second day . . . All these theoretical considerations are still highly preliminary, but we came away from Jodrell Bank with the distinct impression that the ice of pure theory had been broken and that we are headed toward an understanding of the very high radio energies of some of the discrete sources.—Bok (1955, p. 21)

Introduction

The Pawseys had been hopeful of spending much of 1955 in the USA and Canada. But late in 1954 Lenore's ill health threatened that plan. Greenstein wrote on 12 January 1955: "I hope your projected visit to the [US in 1955] will come off; and that we will see you."

On 9 March 1955, Pawsey wrote a letter of explanation to Lloyd Berkner, President of AUI in New York, who had invited him to participate in a conference on solar eclipses:

When we met in Europe last autumn [URSI, in the Netherlands] we discussed the possibility of my visiting the US next autumn [in the US, 1955]. This will not be feasible this year as my wife has been ill and [I] should not get involved in such a trip until she is quite recovered. However I hope this is a postponement and not a cancellation and shall get in touch with you when I can see sufficiently far ahead.

It is possible I may go to Dublin for the IAU meeting but if so I shall make it a very brief visit and I should not arrive until the Jodrell Bank Symposium [end August].

There is one, J.P. Wild [who would be a good candidate for your conference on solar eclipses in 1955]. He has an intimate knowledge of solar radio astronomy . . . [His] main work has

Supplementary Information The online version contains supplementary material available at [https://doi.org/10.1007/978-3-031-07916-0_26].

W. M. Goss et al., *Joe Pawsey and the Founding of Australian Radio Astronomy*, Historical & Cultural Astronomy, https://doi.org/10.1007/978-3-031-07916-0_26

been on dynamic spectra of bursts. [I suggest you invite him to the conference on solar
eclipses in London before Jodrell Bank.] He is an excellent speaker.

Berkner wrote with greetings and best wishes for Lenore and also the invitation for
Wild to attend the special 3-day solar meeting in London.

On 14 March 1955, Pawsey, concerned about the qualifications of his staff in the
light of anticipated new projects at RPL, wrote to Oort in Leiden,[1] "As you know we
lack trained astronomers. Eric Hill will be our only example. Kerr, Mills, Bracewell,
Christiansen and the rest are good physicists but their astronomy is scrappy." He
suggested that some of the outstanding young radio astronomers from Leiden might
spend 3–9 months in Sydney visiting RPL, specifically Maarten Schmidt and Gart
Westerhout: "As soon as one of these new projects [80 MHz cross and the 36-foot
paraboloid] gets underway it turns out a wealth of astronomical detail which should
be a wonderful hunting-ground for a good astronomer . . . If, for example, a man like
Schmidt could come out to join the group I feel he would be a tremendous stimulus."
Oort wrote back on 26 March 1955 with the news that Westerhout could not travel to
Australia since he had been given a military deferment as an indispensable staff
member of the Leiden Observatory. Schmidt was already slated to move as a
Carnegie Fellow to Mt Wilson-Palomar in Pasadena. Oort was optimistic that Eric
Hill, a graduate student from Sydney (RPL) working on a PhD with Oort, could fill
this niche; "I think he has learned about various general astronomical problems
during his study here, and will be able to work independently." As we will see
below, this aspiration was not fulfilled.

On 6 May 1955 Pawsey wrote Ryle describing new developments at RPL:[2]

We have at various times discussed difficulties that can arise through possible duplication of
work and we were, I think agreed that it would be helpful if we could inform each other of
the beginnings of projects involving a considerable capital outlay. I am therefore writing to
tell you that we have agreed that Christiansen will proceed with a multiple element cross for
observations of the sun on 21 centimetres.

You will remember that Christiansen outlined the scheme at URSI [in 1954]. It involves a
combination of the Mills and Christiansen ideas, (the "Chris-cross" we sometimes call it)
and permits the scanning of the sun as in television. We chose 21 cm largely because of prior
observations . . .

I expect to visit England for the IAU and shall look forward to seeing you then.

The Textbook Was Published at Last

It was 21 February 1955 when Clarendon Press editor Wood wrote to Pawsey to tell
him that his long-awaited textbook, *Radio Astronomy* by J.L. Pawsey and
R. Bracewell, "is passed for press." Bracewell had managed to correct the proofs

---

[1] Letter from Pawsey to Oort 14 March 1955, Oort archive Leiden and NAA C3830 Z3/1/V.

[2] All letters in 1955 were addressed to "Dear Martin" and "Dear Joe". Correspondence that we have
seen from 1949 to 1952 was the more formal "Dear Ryle" and "Dear Pawsey".

via a complicated correspondence late in 1954. On 11 May 1955, Bracewell wrote Pawsey from Berkeley that the book would be published on 16 June 1955, to be sold at a price of 55/- sterling (£2/15^3). Pawsey replied on 19 May 1955 from Sydney. He was frustrated with the publisher: "It is depressing that the thing has been delayed so long, but we have to accept it." The table of contents of the book is provided in ESM 26.1, Table of Contents.

Prior to his trip, Pawsey ensured that a number of colleagues had been sent complimentary copies. Hanbury Brown (29 June 1955) at Jodrell Bank replied that he looked forward to using the book in his lecture courses.[4] Oort (30 June 1955) from Leiden was effusive in his gratitude: "I must confess that I feel a little embarrassed by this beautiful gift, which I have so little deserved. But I want to thank you most heartily for it. It is certain that I shall use it frequently and to great advantage." C.W. "Cla" Allen (16 November 1955) reported on his successful use of the book in a 12-lecture course at University of London Observatory. He did point out some omissions in the book which he had added in his lectures, such as Stokes parameters (polarisation), the Nyquist theorem, special relativity and the Kramers theory of free-free emission.

At least six reviews of the book appeared from December 1955 to June 1956, most by well-known radio astronomers. The opinions expressed were all generally favourable.[5] However, the field was changing so swiftly and the book had been so delayed between the effective completion of the text in 1953 and its publication in mid-1955, its usefulness was limited by being considerably out of date by the time of publication. Apparently, Pawsey only replied to one reviewer, Graham Smith. On 12 December 1956, he thanked Smith for his (generally positive) review and added that, "I entirely agree with your comments [concerning the book's datedness]." He recapped the "break-down of the elapsed time". The closing date for the manuscript was mid-1952; the complete manuscript was posted to Oxford in October 1953. Then the publication of the book was in June 1955, a delay of 3 years. Many readers did not see the book until mid-1956, a delay of 4 years. Pawsey remarked to Smith: "[The delay] gave me a surprise."

Because it had dated so quickly, Pawsey was already suggesting to Wood that a second edition be published, but despite much planning, this never happened.[6]

---

[3] The cost of the book in the US was $8.80. In the modern era the cost would be about US$100.00.

[4] By a remarkable coincidence, W.M. Goss may have this copy. In 2007 Stuart Pawsey of Berkeley, California, (the son of J.L. Pawsey) gave Goss one of his copies of his father's book. Stuart had bought this on eBay. The book has the signature of Robert Hanbury Brown.

[5] Details are provided in the NRAO ONLINE.53, Pawsey Bracewell textbook 1955.

[6] See NRAO ONLINE.53.

ANZAAS Presidential Address

Just before his departure for Europe on 19 August 1955, Pawsey had a brief visit to the ANZAAS (Australia New Zealand Association for the Advancement of Science) conference in Melbourne. The meeting started on 17 August and ran to 24 August. Pawsey was President of Section A (Astronomy, Mathematics and Physics) and gave the Presidential Address on either the first or second day of the conference, before a return to Sydney and his overseas flight. The address of about 2500 words was later published in the *Australian Journal of Science,* January 1956 (vol 18, No 3A, p. 27).

Pawsey's text provides a valuable summary of the first decade of his career as a radio astronomer. He evaluated the (1) history of radio astronomy in Australia since 1944, (2) the status of RPL in 1955 and (3) the challenges facing his new field. He identified the most important contribution that radio had brought to astronomy as the discovery of "radio stars". He adopted synchrotron emission as the major non-thermal emission process with an association with cosmic rays. And he provided a foreshadowing of the up-coming Cavendish-Sydney controversy over sources counts from the nearly completed 2C and upcoming Mills Cross surveys (see Chaps. 35 and 36). The full version of Pawsey's ANZAAS address is in ESM 26.2, ANZAAS.

Overseas Again

Pawsey began his 7-week trip to the UK, Ireland and the Netherlands on 19 August 1955. He arrived on 21 August 1955,[7] staying a few days with his sister-in-law Bessie Whittard in Iver, South Bucks, a small town 24 miles to the west of central London. On Tuesday, 23 August 1955, he visited Walter Ives, the Chief Scientific Officer of ASLO, in London. On Wednesday 24 August, he went by train to Crewe near Manchester; the Jodrell Bank Symposium was from Thursday to Saturday. He then went to Dublin for the IAU General Assembly which ran from 29 August to 5 September. The following day he went to Glasgow to attend a meeting of the Royal Astronomical Society (as did many other participants from the IAU). His activities in the following weeks are described in NRAO ONLINE.41 and also Chap. 27. He travelled at the end of September 1955 to the Netherlands to meet Oort and Hooghoudt as well as to visit the new 25 m aerial at Dwingeloo. He left Amsterdam on 1 October 1955, arriving home in Sydney on 5 October 1955.

---

[7]In 1955, the air trip in this pre-jet era took about 2.5 days from Sydney to London, with up to seven intermediary stops.

Jodrell Bank Symposium

The Jodrell Bank event, sponsored by the University of Manchester, was held on 25–27 August 1955, the second "international" symposium organised at Jodrell Bank. Two years earlier (13–15 July 1953); Lovell had organised a meeting, "A Symposium on Radio Astronomy" with 48 participants, all from Europe except Bok and Minkowski from the US and Hill from Australia (see ESM 26.3, A Symposium on Radio Astronomy).

Planning for the first IAU sponsored conference[8] on radio astronomy (IAU Symposium No 4) began in October 1954 by Pawsey and Hendrik ("Henk") van de Hulst (secretary of the organising committee) of the Leiden Sterewacht.[9] The proposal was to have a 3-day conference that would precede the General Assembly of the IAU in Dublin, Ireland. The expectation was that the new 250-ft radio telescope at Jodrell Bank would be close to completion. The proposed programme was to begin with discussions of the hydrogen line on the morning of Thursday 25 August 1955, with lectures by van de Hulst and Oort. The afternoon programme would consist of a tour of the new telescope (under construction) followed by a lecture by Lovell. The Friday morning programme would be talks about radio point sources and a presentation on optical identifications by Jesse Greenstein. During the afternoon, galactic and extragalactic structure would be discussed with the main lecture by Hanbury Brown. The Saturday morning conclusion section was to consist of the quiet and active sun with the main lecture by Cla Allen. The conference was to conclude in the afternoon following discussions of meteors, the moon and Jupiter. On Sunday 28 August 1955, many of the attendees were to travel to Dublin for the IAU; a special chartered flight had been organised from Manchester.

Some of the colleagues travelled in a more economical manner; the Cavendish radio astronomers had driven a simple WWII surplus van from Cambridge to Manchester, followed by a ferry trip to Ireland. Figure 26.1 shows the leader of the group, Martin Ryle, changing the tyre. Graham Smith (extreme right) and Bruce Elsmore (extreme left) appear to be assisting while John Baldwin (second from left) and John Shakeshaft are onlookers.

A major problem at this conference was to limit the number of participants; the conference hall at Jodrell Bank could accommodate only 100 people. The committee agonised for some months about the choice of attendees. A quota system was used. For example, the Australians were allocated a total of six, four from RPL[10] and two from Stromlo (G de Vaucouleurs and Woolley[11]). The numbers from Cambridge and

---

[8] NAA C3830 C25/5.

[9] Other members of the organising committee were John Hagen of the US and Laffineur of France. Pawsey clearly had decided that a "policy of specialised colloquia [was desirable]. His notation of a selective symbiosis between radio technique and astronomical wisdom, self-effacing as it may seem, was clearly to the taste of many radio astronomers" (Edge & Mulkay, 1975, p. 64).

[10] Pawsey, Wild, Bolton and Martha Stahr Carpenter (see preceding chapter).

[11] Woolley did not attend.

Fig. 26.1 The trip from Manchester to the IAU in Dublin in 1955. Left to right Bruce Elsmore, John Baldwin, John Shakeshaft, Martin Ryle and Graham Smith. Credit: John Baldwin. Courtesy of and copyright: Cavendish Laboratory, University of Cambridge. All rights reserved

Jodrell Bank were to be 6 and 12, respectively. From France, the quota was five, from Germany four and from the Netherlands eight. A number from the USSR were invited with four scientists on the final list in early August: S.B. Pikel'ner, V.V. Vitkevitch, P.V. Shcheglov and B.M. Chikhachev.[12] Photos from the meeting show Vitkevitch and Chikhachev; Pikel'ner is uncertain (see below). However, the actual Soviet participation at Jodrell Bank remains uncertain. With the help of Leonid Gurvits we assume that Pikel'ner, Vitkevitch and Chikhachev were present; with some uncertainty about Pikel'ner.[13]

Ten papers from the Soviet participants were given, including two by Vitkevitch and one by Tchikhatchev (the German version of Chikhachev). There were three key papers by Shklovskii and one by Ginzburg. The papers by Shklovskii were included in the published volume; Ginzburg's paper "The Nature of Cosmic Radio-Emission and the Origin of Cosmic Rays" was not included in the conference proceedings publication of 1957. We discuss this major oversight and the implications for the

---

[12] We are indebted to Leonid Gurvits for the correct spelling. Vitkevitch was a senior officer in the Soviet Army and a member of the Communist Party. Attendance at an overseas conference was only possible for scientists who were "trusted" by the Communist Party and the KGB. Shcheglov was still a graduate student working on a PhD with Shklovskii at Moscow University. Boris Chikhachev was a radio engineer who was hired for the new Radio Astronomy Institute of the Sternberg Institute in 1953 by Shklovskii.

[13] Pikel'ner was described in Bok's *Sky and Telescope* article as a participant in the conference; possibly he was present. A week later, he was listed as a member of the Soviet delegation at the Dublin IAU.

development of the synchrotron radiation theory in the theme chapter on radiation theory (Chap. 34).[14]

A comprehensive correspondence has been located in the Australian archives, concerning the organisers' attempts to finalise the final list of attendees. In the end, more than 100 participants from 18 countries attended. Each attendee was asked to provide an extensive abstract before the meeting. Van de Hulst and Pawsey spent many hours approving these and getting copies prepared to hand distribute at Jodrell Bank via mimeographs, in the pre-Xerox days. Bart Bok, in *Sky and Telescope*, November 1955, complained about the time constraints: "It was the most concentrated three-day symposium I have ever attended, with some 95 papers that we could have spent a full week discussing and analysing." The discussions were quite limited due to the restricted length of the conference.

Pawsey was completely occupied in the month before he departed for Europe as he prepared the eight abstracts from RPL, of which he would present three. Paul Wild would present three papers, two by himself and one on behalf of Alex Shain. John Bolton would present a paper by himself and Bruce Slee,[15] and Patty Carpenter would present the paper on the HI galactic data from Potts Hill.[16]

Henk van de Hulst had initially been selected to be the editor of the proceedings. He had objected to this suggestion earlier in 1955 (29 April). In a circular letter to the organising committee he wrote "... having only just finished [editing an earlier IAU Cambridge conference, IAU Symposium No 2 *Gas Dynamics of Cosmic Clouds*], I would not be tempted by this prospect." Then on 15 July 1955, Pawsey wrote Ryle trying to convince his Cambridge colleague to take on this onerous task: "... van de Hulst ... rather plaintively said he had had enough" after the 1953 conference.[17] Pawsey realised that the editor would then have to be either himself or Ryle. "You have the advantage of propinquity to the Cambridge Press and also I think you would do it exceptionally well. Against my doing it I can plead geographic isolation and the fact that I shall ... have to deal with Dublin Proceedings (Commission 40 and the Solar Flares Joint Discussion [at Dublin])." Apparently, both Ryle and Pawsey

---

[14] The Paris Symposium presentation of 1958 (Chap. 34) was titled "Radio Astronomy and the Origin of Cosmic Rays", by Ginzburg, paper number 105, published in Bracewell, 1959 Paris Symposium on Radio Astronomy, p. 589. The first sentence of this publication: "A paper ... sent to the Manchester Symposium on Radio Astronomy in 1955 described the views developed earlier ... concerning magnetobremstrahlung (synchrotron) origin of nonthermal cosmic radio emission. Unfortunately, for unknown reasons, it was not included in the Symposium volume." Leonid Gurvits (Private communication in 2019) has suggested that it is unlikely that the famous astrophysicist S.B. Pikel'ner would have been allowed to attend the conference in the UK in 1955. In 1955, this paper's title appeared in a list prepared by Pawsey in early August, "Titles of Contributions submitted by the Russian Delegation".

[15] "Apparent Intensity Variations of the Radio Source Hydra A". Pawsey doubted the reality of this result, which was never confirmed. Slee has presented a frank discussion of this controversy in 1994, "Some Memories of the Dover Heights Field Station 1946–1954."

[16] Carpenter presented her work on galactic HI from Potts Hill. Bok (in the *Sky and Telescope* report, 1955) praised her "steady progress" on the southern HI survey.

[17] NAA C3830 C25/5.

declined. In the end, van de Hulst relented, becoming the editor by default of IAU Symposium No. 4 *Radio Astronomy*. Perhaps this experience contributed to the delay; the book only went to the publisher in April 1957, a delay of almost 2 years. However, van de Hulst had already predicted a long delay in 1955 due to pressures with deadlines with the IAU and Cambridge University Press; the previous Cambridge symposium of 1953 was only published in 1955.

The editing of the book was carried out in a thorough manner, clearly with many changes suggested by van de Hulst. He wrote in a succinct preface:

> A symposium on Radio Astronomy, organised by the International Astronomical Union, was held on 25-27 August 1955 at the Jodrell Bank Experimental Station of the University of Manchester. It coincided with the tenth anniversary of this station; the sessions took place in the control building of the 250-ft. telescope which is under construction.
>
> The symposium brought together 108 participants from: Australia, Belgium, Canada, Czechoslovakia, Finland, France, Germany, Great Britain, Italy, Japan, the Netherlands, Norway, Spain, Sweden, Switzerland, U.S.A., U.S.S.R., and Yugoslavia. The symposium committee consisted of A.C.B. Lovell, chairman; H.C. van de Hulst, secretary; J.P. Hagen, M. Laffineur, and J.L. Pawsey. About sixty mimeographed abstracts distributed beforehand or at the meeting helped to ensure an effective exchange of information.
>
> This volume contains all but two of the papers presented. One contribution, paper 16 [by Townes on possible new radio spectral lines], has been added at the editor's request. Many papers have been improved as a result of discussion at the symposium or by the inclusion of data not available in August 1955. The essential parts of the discussions have been reported.
>
> The six parts of this volume approximately correspond to the respective morning and afternoon sessions, each starting with an introductory lecture. Purely instrumental papers or parts of papers have been omitted from this volume as they fell outside the range of topics outlined for this symposium, as also did scintillation and purely geophysical problems. It may be fairly said that this volume gives an almost complete report on all other research in radio astronomy at the time of the meeting.[18]
>
> The editor wishes to express his thanks to all authors for their cooperation. Dated April 1957, Leiden, by H.C. van de Hulst

For Bart Bok (*Sky and Telescope*) the highlights of the conference were the HI data from the Netherlands (Leiden, on day one of the conference), the solar data from Sydney and the discussion of the galactic and extragalactic continuum radio emission. Bok awarded unofficial blue ribbons to the group from Leiden and Sydney. Bok was enthusiastic as his optimistic report on the status of radio astronomy in 1955 showed:

HI: 25 August 1955

> At Jodrell Bank, the first half of the 21-cm session dealt with galactic structure, and here the Leiden group deserves the blue ribbon. Under the inspiring leadership of J.H. Oort and van de Hulst, significant advances have been made in two important areas of investigation. G. Westerhout has obtained 21-cm profiles at positions 2 degrees apart along the Milky Way

---

[18]Our emphasis. There is no mention of the missing paper by Ginzburg (Footnote 14). Eighty papers were published.

from Cygnus to beyond Sirius and from 10 degrees south to 10 degrees north of the equator of the galaxy. He demonstrated an impressive plastic three-dimensional model of this arc of the Milky Way, which shows nicely the separate spiral arms and how they branch off in spots, very much as do the spiral arms in galaxies outside our own.

For the solar results, Bok was impressed by Christiansen's Potts Hill observations of the radio brightness distribution across the solar disc and also with the Dapto dynamic spectrograph data from Paul Wild and colleagues, all from CSIRO:

Solar: 27 August 1955

The blue ribbon for the solar session should go to the Sydney observers. Pawsey presented corroborating evidence for differences of brightness distribution over the sun's surface depending upon the line along which this distribution is traced, with quite different results for east-west and north-south lines. With his "dynamic spectrum technique," J. P. Wild can follow a disturbance right from the lower chromosphere into the upper corona. Of special interest are the bursts that move outward with speeds as high as one-fifth the velocity of light [Type III]. These bursts often come in clusters, and they are apparently associated with the origin of solar cosmic rays. Helen W. Dodson, of the University of Michigan, has noted that generally the radio and cosmic ray phenomena are related to the ending of the corresponding visual phenomenon on the sun, a solar flare.

The second day began with an introductory lecture "Optical Investigations of Radio Sources" by Minkowski. We now moved to the unexplained phenomenon of unknown origin.

Discrete Radio Sources: 26 August 1955

Minkowski was the expert, along with Baade, in using the 200-inch telescope to identify radio sources. Their paper of 1954 was decisive in identifying Cygnus A, Cas A and Puppis A. In 1955 at Jodrell Bank, Minkowski outlined the dilemma. Very few identifications had been made. A number of shortcomings were recognised, the major problem being the poor accuracy of the radio positions, many with errors of some arc min. In addition, some of the radio sources (cf Cygnus A) showed radio structures considerably larger than the suggested optical identification.

Loose agreement of a radio position of low accuracy with that of some object listed in the NGC is not sufficient to provide the identification of a radio source. Even satisfactory coincidence of a precise position with that of an astronomical object requires supporting evidence . . . The radio spectrum, the optical spectrum, and the physical characteristics of the visual object also have to be taken into account. Observations of the radio spectrum should be particularly useful to support the identification of sources with H II regions which can be recognised from their thermal emission even if they are obscured and optically inaccessible.[19]

As we will point out in the theme Chaps. 35 and 36, the situation with optical identifications with radio sources didn't improve substantially until the 1960s and 1970s as arcsec radio positions were finally achieved.

---

[19]Minkowski also described the spectacular 200-inch optical data on the young supernova remnant Cassiopeia A; the high velocity filaments (velocities in excess of 1000 km/s) have been imaged in detail. Proper motions of these filaments with a time scale of only 3 years show systematic expansion.

During the afternoon of 26 August 1955, Hanbury Brown started the session, "Galactic Structure and Statistical Studies of Discrete Sources", with an introductory lecture "Galactic Radio Emission and the Distribution of Discrete Sources".[20] Brown stressed that for the emission at wavelengths greater than 1 m, the emission mechanism had to be non-thermal. The emission mechanism in the cm range for sources concentrated towards the galactic plane (Mills's Class I sources) was thermal emission from HII regions; in the metre range the mechanism was unknown. But Alfvén, Herlofson, Kiepenheuer and Ginzburg had all suggested "that the non-thermal radiation might be due to cosmic-ray electrons in interstellar magnetic fields." For the origin of the diffuse emission he concludes, "The generation of the energy . . . occurs in a very rarefied medium and is due to the deflexion [sic] of fast electrons in magnetic fields."[21] This is discussed in more detail in the Chap. 34.

Hanbury Brown agreed that the majority of the sources not concentrated in the galactic plane (Mills's Class II sources) may be extragalactic. "Recent work [the Halley lecture of Ryle] has shown that their distribution is remarkably isotropic, and it is difficult to associate them with any of these components of the background radiation which are clearly of galactic origin."

In *Sky and Telescope*, Bok reported on Hanbury Brown's introductory talk on the galactic background: "Towards the end of our second day, we came to the somewhat confused problem of the observed general galactic and extragalactic radio emission." Bok continued: ". . . Most of the earlier surveys of the sky were made with instruments of low angular resolving power that were not really capable of giving the full picture. Modern techniques and larger antennas now make possible high-resolution surveys . . ."

Bok had special praise for Shklovskii and Ginzburg, as these colleagues continued their advocacy of radiation from relativistic electrons (not yet called "synchrotron emission", Chap. 34:

> They have suggested that the radiation comes from relativistic (very fast-moving) electrons, the motions of which are determined by interstellar magnetic fields . . . The magnetic interpretation has received a terrific boost quite recently through the results of optical research on the [supernova remnant] Crab Nebula.

Both a Soviet group and the Leiden group had found strong optical polarisation from the Crab as expected with this new emission process. Oort and Walraven presented a paper on 26 August 1955 "Polarisation and the Radiating Mechanism of the Crab Nebula".

The next paper on the afternoon of 26 August was a short presentation by John Shakeshaft describing the "Cambridge Survey of Radio Sources". The fireworks of the day followed the next paper by Ryle, "The Spatial Distribution of Radio Stars". These discussions are summarised in Chaps. 35 and 36.

---

[20] Only Ryle used the term "radio star" at the conference.

[21] During the discussion of the Hanbury Brown paper, Geoff Burbidge used the new descriptive term "synchrotron radiation by relativistic electrons."

Fig. 26.2 Garden of University College Dublin. In the middle, left to right, Paul Wild, Elaine Wild (hat) and Joe Pawsey at the International Astronomical Union General Assembly in Dublin, 1955. Credit: IAU/Observatory of Paris

On day three, the morning discussions were on the quiet sun (introduction by Cla Allen) and then the active sun with the first talk by Paul Wild, "Spectral Observations of Solar Activity at Metre Wavelengths".

On 30 August 1955, Pawsey wrote Bowen with some highlights (and gossip) of the Jodrell Bank conference.[22] These are summarised in ESM 26.4, Pawsey to Bowen, gossip from the Jodrell Bank Symposium.

Ninth IAU General Assembly, Dublin, Ireland, 29 August– 5 September 1955

Immediately after the Jodrell Bank Symposium (IAU Symposium No 4), Pawsey went to Dublin for the IAU General Assembly. This was his first attendance at an IAU, having missed the IAU of 1948 in Zurich (see Chap. 17) and the IAU in Rome in September 1952, which had, of course, directly followed the URSI General Assembly in Sydney, and thus required astronomers to choose which event they would attend, since attending both was geographically impossible.

Frank Edmondson provided an enthusiastic description of the IAU in Dublin in a *Sky and Telescope* article, "Report from Dublin", December 1955. There were 800 attendees from 41 countries, the largest attendance to date at a General Assembly. The group photo of many of the participants was taken on the day of the opening in the garden of University College Dublin; see Fig. 26.2, an insert of the conference photo showing Paul and Elaine Wild along with Pawsey.

---

[22]NAA C3830 C25/5.

The assembly had numerous social events, such as an opening reception (Monday morning 29 August) given by the Irish Prime Minister at Dublin Castle. Two days later, there was a reception by the Irish President Sean T. O'Kelly at his residence at Phoenix Park, "a stately white mansion surrounded by green lawns and beautiful gardens." There were weekend excursions to Killarney or Connemara. On the day after the assembly ended, about 500 astronomers travelled to Belfast in Northern Ireland for a reception and tour. Later they visited the Armagh Observatory, Northern Ireland. Many of the astronomers then departed for home from Belfast or went to the Royal Astronomical Society meeting in Glasgow, Scotland.

Since the Jodrell Bank Symposium had just occurred, Commission 40 activities were somewhat limited. At one of the business meetings of the Commission, Pawsey presented his inaugural report as the President of Commission 40.[23] This report is likely the first presentation of a Commission 40 President's report. Pawsey took the opportunity to present his own view of the history of radio astronomy from 1945 to 1955. He included a thorough summary of the continuum from the Milky Way; he only hinted about the existence of the growing controversy of source counts (Ryle versus Mills).[24] Pawsey continued with summaries of the status of the remarkable progress made in HI work, the Galaxy and even the Magellanic Clouds. Both the quiet sun and the active sun were described; Pawsey was pleased with the recognition that the CSIRO groups (both solar and cosmic) had received at the earlier Jodrell Bank conference.

Frank Edmondson also mentioned that the radio astronomers had discussed a number of other topics during the Commission 40 meetings: (1) catalogue of radio sources, (2) terminology for radio astronomy, (3) galactic coordinate conversion tables, (4) standard sources for calibration, (5) publication issues, (6) international planning of programmes, (7) a list of radio observatories and (8) suggestions for future symposia.[25]

Edmondson made a striking plea on behalf of the radio astronomers as he reported his impression of the Dublin IAU:

> To an outsider, the most interesting and important subject was the question of radio frequency allocations. Many radio astronomers are finding it increasingly difficult to carry on their work because of interference from television and high-frequency radio stations. This type of interference is already a very serious matter in some of the more highly developed countries. A solution to the problem involves more than merely having certain frequency bands reserved for scientific purposes. Unwanted harmonics from stations broadcasting on other frequencies can cause trouble for the radio astronomers. They are measuring such exceedingly small quantities that any interference is troublesome. It is important that government authorities in all countries where there is television and high-frequency broadcasting be made to realize the serious nature of this problem and the importance of requiring

---

[23] See NRAO ONLINE.28.

[24] Bok reported that the discussions of the source counts continued with vigor at the IAU General Assembly in Dublin after the Jodrell Bank meeting.

[25] Several of these topics were discussed at length in Pawsey's report, see NRAO ONLINE.28. Bok reported that, due to the success of the Jodrell Bank meeting, Commission 40 had decided to hold a meeting on radio astronomy again in 1958 in Paris, to precede the Moscow IAU General Assembly.

stations to provide their equipment with all necessary safeguards to avoid broadcasting harmonics. Government cooperation has already been secured in a few countries, and it is fervently hoped that others will join in.

IAU Symposium 5 was held during the Dublin General Assembly starting 2 September 1955, "Comparison of Large-Scale Structure in the Galactic System with that of other Stellar Systems".[26] Edmondson complained that the time allocated for this meeting was far too short. Highlights were Magellanic Cloud results by Feast and Thackeray and the new optical techniques pioneered by W.W. Morgan of Yerkes Observatory. "HC van de Hulst's discussion of the distribution of hydrogen in the galactic system presented the latest Dutch results on the spiral structure in our galaxy. So much has been accomplished that it is hard to realize that the very beginnings of the subject were reported as recently as 1952 at the Rome meeting of the IAU." There were three radio astronomy papers in the publication of the IAU Symposium No. 5, including the van de Hulst presentation, "The Distribution of Atomic Hydrogen in the Galaxy". Additional papers were "Galactic and Extra-Galactic Radio Frequency Radiation due to Sources other than the Thermal and 21-cm Emission of the Interstellar Gas", by R. Hanbury Brown and "The Radio Emission from the Galaxy and the Andromeda Nebula", by J.E. Baldwin. All of these texts were similar to the papers presented by these three at the Jodrell Bank Symposium.

Other commissions also were concerned with radio astronomy at Dublin.

Proposal for Revised Galactic Coordinate System

A key decision was made at the Dublin General Assembly that would have major ramifications. IAU Commission 33 "Structure and Dynamics of the Galactic System" established a sub-Commission 33b "to investigate the desirability of a revision of the galactic pole and of the zero of galactic longitude."

Blaauw wrote Bok on 8 November 1955 (Blaauw archive Groningen): "As you will remember from the Dublin meeting a sub-commission consisting of Pawsey, Westerhout and myself was formed to examine the question of a revision of the galactic coordinate system (galactic pole/zero of longitude). It was generally felt that current observational programs might justify such a revision to be made within the next few years." In the years 1955–1960, Pawsey led a group of astronomers from Australia (Frank Kerr and Colin Gum) and the Netherlands (Adriaan Blaauw and Gart Westerhout) who worked on a revision of the old system of galactic coordinates. This complex and well executed project is presented in ESM 26.5, New Galactic Coordinates, which also includes references to four NRAO ONLINE (NRAO ONLINE.57.1, 57.2, 57.3, 57.4) texts that provide further background and

---

[26]The publication of the slim volume of 16 papers *Comparison of the Large-Scale Structure of the Galactic System with that of Other Stellar Systems,* 1958, with Nancy G Roman as editor.

supplemental information. (Extensive references for this material are included in Footnote 1 of ESM 26.5, New Galactic Coordinates).

The task of defining galactic coordinates is an example of important scientific work that is rarely accorded the recognition and status of discovery or a key publication. In the terms that Merton used to characterise the "reward system" in science, this sort of task often does not receive the attention that is commensurate with either the usefulness of the result, or the time and laboriousness of the work involved.[27] It is also a challenging, and not trivial, scientific task—a fact of which astronomers who do not study the Milky Way may not be aware. Undertaking this task could be considered an example of the "scientific ethos" described by Merton and often exemplified by Pawsey, in the form of (a) "communalism", ie, contributing an intellectual resource for the use of the scientific community as a whole, and (b) "disinterestedness", ie, undertaking a task for its scientific value rather than with consideration to personal benefit.

A few notable events were the May 1957 conference at Mt Stromlo Observatory in Canberra organised by the recently arrived (in Australia) Bok family and Pawsey with participants from Sydney and Mt Stromlo and the November 1957 Groningen (the Netherlands) meeting of Frank Kerr (on a long visit to Leiden and Dwingeloo) and Blaauw (Groningen) plus Westerhout (Leiden). The new coordinate system was discussed in detail at both events. After these discussions, the new system was discussed, modified and approved at the IAU General Assembly of 1958 in Moscow. In 1960, five publications describing the new system appeared in the *Monthly Notices of the Royal Astronomical Society* and the *Astrophysical Journal*. The authors were: Blaauw, Pawsey, Gum and Westerhout along with an additional paper which described recent Dwingeloo 25 m telescope observations of HI in the direction of the galactic centre by Oort and Rougoor.

---

[27] See Merton (1968) and Strevens (2006).

Part VII
Towards a Bigger Science

Chapter 27
Pawsey and the Giant Radio Telescope, 1951–1956

[A 100-metre diameter aerial] should lead to a better understanding of many phenomena of which we now have partial knowledge but it also opens up the possibility of entirely new discoveries.[1]

By 1955, a decade after the first exciting observing projects, the landscape of radio astronomy had shifted considerably. With many more research groups now starting up, and the lines of research growing both more diverse and more specialised, remaining a global research leader required making riskier, more strategic decisions. In retrospect it can be seen that this was largely accomplished in Australia by the building of the now-iconic large "Dish" radio telescope at Parkes, NSW, and by the success of the research projects led by John Bolton through the 1960s. The funding and construction of the large dish took nearly a decade to achieve and was by no means certain through much of that time. Robertson has provided a thorough description of these events in his two books from 1992, *Beyond Southern Skies— Radio Astronomy and the Parkes Telescope* and in 2017, *Radio Astronomer—John Bolton and a New Window on the Universe*.

In this section, we look at the complex funding possibilities pursued by Bowen and we explore Pawsey's role in the planning, design and early operational models for the Parkes radio telescope. This covers the period from 1951 to 1956.[2] From 1952 up to the time of the inauguration in Oct 1961 the telescope was called the Giant Radio Telescope (GRT) and we will refer to it by that name in the rest of this

---

Supplementary Information The online version contains supplementary material available at [https://doi.org/10.1007/978-3-031-07916-0_27].

---

[1]Pawsey in August 1952, "Notes on Applications in Radio Astronomy for a 100-Metre Diameter Telescope". NAA C3830 A1/3/11/1, Part 1.

[2]Ten supplementary online reference texts (NRAO ONLINE.38–NRAO ONLINE.47) provide a detailed year-by-year account of the GRT story from 1951 to 1961.

section. The GRT success was achieved only as a result of the complementary skills, values, and scientific styles of Pawsey and Bowen. Neither could have achieved a successful GRT alone. Nonetheless, the GRT also became the main source of increasing division between the two and would in the end result in complete schism at RPL.

Interest in a very large aerial of some sort had occurred very early, not least because of the building of what was then the largest radio telescope in the world, the Transit Telescope (a 218 ft (66 m) parabolic reflector zenith telescope) at Jodrell Bank in 1947. This instrument set a bar for what might be required to remain a leading research group in radio astronomy. But the expense as well as the design of such an instrument made choosing, and acquiring, the new aerial, a complex matter. In this chapter, we describe the early phase of these discussions and funding negotiations, taking place over the period 1948–1955.

The Emergence of the Big Dish Concept

Some differing views on the key factors that triggered the emergence of the big dish concept have been expressed. Peter Robertson speculates that it was the proposal by Woolley, the Commonwealth Astronomer and Director of the Mt Stromlo Observatory, to erect a radio telescope "of large resolving power" to investigate "the galactic structure of radio noise" that triggered the CSIRO into activity. Robertson (1992, p. 112):

> The possibility that Mt Stromlo might establish its own radio astronomy group or, going further, that radio astronomy might be wrenched from CSIRO and relocated in a national facility controlled directly from Canberra set the alarm bells ringing at RP. The challenge emphasised the need for RP to break with the free-wheeling approach of the 1940s and to formulate long-term objectives, in particular to decide the types of radio telescopes required for future work.

In Chap. 19, we described the tense discussions in early 1951 between Oliphant and Pawsey on one hand, and Bowen and White on the other, over this proposal. But Woolley personally had little to no interest in radio astronomy. Soon after the pressure from the Commonwealth Observatory Board and from Oliphant at ANU had eased, Woolley ceased to pursue such a project, and we do not think this proposal played a significant role in the decision to build a big dish.

Bowen has provided a succinct summary in "The Origins of Radio Astronomy in Australia" (in Sullivan, 1984, p. 298) of the increased interest:

> As in optical astronomy, steerable parabolic antennas are a basic part of the instrumentation for radio astronomy; they [modest size dishes] played a prominent part in early galactic research, particularly in investigations of [HI] line radiation. As in other establishments, there was an urge to increase the aperture of such instruments to the largest possible dimensions.
>
> Among the first options to be explored was a collaborative effort with our friends in the RAAF, with whom we had maintained a post-war connection. As early as 1949, we

discussed with them the possibility of building a really large air-warning antenna, with linear dimensions of several hundred feet. Several designs were roughed out and costed, and at one stage there even seemed to be a possibility of going to a horizontal dimension of 500 feet. Our interest in the project was based on the real hope that, if built for defence purposes, we would have the use of the instrument for radio astronomy.

RAAF decided against continuing instrumental development projects in Australia; thus this initiative had no impact on the big dish project. Another radar project being discussed in 1948 and 1949 was the detection of echoes from the moon and the sun. This would have required a very large aerial and was never funded as the scientific uncertainty was too great for the large expenditure in capital investment and manpower.[3]

In our view the primary trigger for the construction of the GRT was clearly the development of the large parabolic dishes at Jodrell Bank. In his interview with Sullivan on 24 Dec 1973 Bowen recounted:

> Now the big antenna, the Parkes antenna—the concept of a big dish was, as a matter of fact, certainly 1950—1949 even. To some extent it was indigenous—we thought this is the way to go but it was stimulated by the fact Lovell's beating this bandwagon in England. And it certainly looked a very good idea to me … a choice had to be made. Big money was involved—millions of dollars. And you just can't go around raising money of that kind without a very specific project in mind … As far as the Division of Radiophysics was concerned the choice was clear cut. That we wanted an instrument which for that kind of money would go on for 25 years or more. We'd still be a force in the radio astronomy field. And secondly, which could be used more or less simultaneously by a large number of people.

As we will see in Chap. 32 the great success of the GRT was largely a result of its flexibility and broad user base as the instrument continued to do leading research for another 60 years! Bowen had learned many of these lessons during the long fund-raising process. However when talking to Sullivan 20 years later, he implied he had known what was required from the beginning.

RPL Planning for the Future in 1951–1952

As early as 1948 White had noted that RPL had been pursuing large aerial projects for some time: "In 1948 RP put up to the Executive [of CSIR] proposals for the consideration of a large aerial system for solar and cosmic noise. The Executive agree that this was obviously the next step in the development of radio astronomy at the RP Laboratory."[4]

---

[3] These projects are discussed in more detail in NRAO ONLINE.38.

[4] CSIRO KE12/11, correspondence between RPL and the Commonwealth Observatory (Woolley and Martyn), also NAA C3830 A1/1/4.

On 18 February 1951 at a meeting of the RPL "Sub-committee on galactic work",[5] Pawsey and colleagues discussed plans for new improved radio astronomy instrumentation.[6] Proposals included a high frequency dish (50–60 ft), a very low frequency (18 MHz) array and a 100 MHz source survey interferometer with small aerials. The most ambitious proposal was due to John Bolton, two 80-ft steerable dishes which could be used as an interferometer; although this was supported by Pawsey the project was never funded. When asked by Sullivan why it was not funded, Bowen replied:[7] "Money. Let's say that at that time there were a multitude of such proposals going around ... whenever money was involved, it was out the window ..."

It is interesting to note that after Bolton moved to Caltech in 1955, as discussed later in this chapter, he implemented this concept as the Owen's Valley Radio Observatory—two 90-ft steerable dishes operated as a variable baseline interferometer.

The detection of the HI at Harvard by Ewen and Purcell on 25 March 1951 (Ewen & Purcell, 1951a, 1951b) provided the incentive to plan for a larger aerial with good efficiency at the higher frequency of 1.4 GHz (21 cm). Such an instrument would be used for high resolution imaging of the galaxy in HI with a resolution of 0.5–1°. Discussions with the Dutch astronomers began considering a collaborative project with a firm in the Netherlands, Werkspoor (a firm specialising in railway equipment) which was involved in the construction of the Dutch 25 m Dwingeloo antenna.

In March 1951, Carter, of RPL, designed a 60-ft antenna; this was described in a letter from McCready (in charge of engineering services at RPL) to Frank Kerr of RPL, visiting Harvard in mid-1951.[8] This was a proposed instrument for HI observations to be built at Potts Hill, ie a transit dish, only movable in elevation.[9] In 1952 a smaller telescope was constructed at Potts Hill, likely inspired by the proposed larger instrument. The 36-ft transit telescope at Potts Hill was completed in early 1953 and used extensively for HI observations of the galaxy and the Magellanic Clouds.

From our current perspective it might be assumed that the decision to build a much bigger dish would have been based on a scientific evaluation of the various proposals to follow-up on the exciting new radio astronomy discoveries. However, this was not the case as can be seen from Frank Kerr's comments in his letter to Sullivan:[10] "No, there was no way in which a consensus developed, or could have

---

[5]NAA C3830 A1/1/7.

[6]See NRAO ONLINE.38 for details of this meeting.

[7]Sullivan interview with Bowen 22 June 1978. Papers of Woodruff T. Sullivan III, "Interview with E.G. Bowen," *NRAO Archives*, accessed December 14, 2020, https://www.nrao.edu/archives/items/show/906.

[8]NAA C3830 A1/3/1(H).

[9]The design is shown in NRAO ONLINE.38, Fig. 1.

[10]Letter from Frank Kerr to Woody Sullivan (6 April 1987) providing comments on an early draft of "Cosmic Noise" (Sullivan Archive).

developed, on the proper way to evaluate competing proposals. It was all in Taffy's [Bowen's] mind. So, the basic arguments were not over technical points as such, but over Taffy's version of the future of RP."

In his interview with Sullivan[11] Bowen recollected that it would have been difficult to raise the funds to build highly specialised instruments (such as the Mills Cross), he needed an attractive "all-embracing" project, the big steerable dish. Bowen was already acutely aware of the difficulty of raising such funds in Australia: "The whole organisation was very well set up for salaries and operating ... However, the concept of giving large lump sums for scientific research, whether it's for a ship or a piece of equipment, just hadn't penetrated. And it wasn't at all easy to get anything like that passed in the political process."

Bowen—ambitious and attracted to the power and prestige of "big science"—then pursued possible GRT funding opportunities from both Australia and from overseas. Bowen embarked on no less than four different funding initiatives over the next few years. He utilised the extensive network of enormously influential scientists and science-funders in the US and the UK with whom he had regularly interacted since his wartime participation in the Tizard mission. These included Vannevar Bush, former vice president of MIT and now president of the Carnegie Institution in Washington,[12] Lee DuBridge, president of Caltech, and Alfred Loomis, multimillionaire physicist and at this time trustee of the Carnegie Corporation in the US. In the UK there were Sir Henry Tizard, formerly Chief Scientific Advisor to the Ministry of Defence and Sir Edward Appleton, Nobel prize winner. From 1951, big telescopes featured in Bowen's visits to these and other scientific leaders as he travelled regularly to the US and the UK. These initiatives were partly driven by Bowen's enthusiasm to build a big steerable dish, whether in Australia or the US. He took advantage of the growing interest from these scientific entrepreneurs in establishing radio astronomy as "big science" in the US. As summarised by Kellermann et al. (2020, p. 90): "Bowen was pursuing a two-pronged approach to support his ambitious radio telescope project. Either he would get American backing to finance the building of a radio telescope in Australia, or at least convince the Americans to build one in the US that he, along with 'his boys' would come and help run."

DuBridge, President of Caltech, hoped to create a world-leading radio astronomy program, with telescopes comparable to the recently constructed (1949) 200-inch optical telescope at Mt Palomar. These new instruments might well out-compete the fully steerable and very large Lovell Telescope, then planned for construction at Jodrell Bank. Vannevar Bush and Alfred Loomis had concurred with DuBridge and they agreed to ask Bowen to make a proposal on how to establish such a radio astronomy group and build a big dish. DuBridge wrote, "I hope you will let your imagination run wild ... I am sure this idea will catch fire and I hope you will find it

---

[11] Sullivan interview with Bowen 22 June 1978. See Footnote 7.

[12] A comprehensive biography of Vannevar Bush was written by G. Pascal Zachary: *Endless Frontier- Vannever Bush, Engineer of the American Century,* 1997.

possible to help us in preparing something to light the match".[13] Following a further exchange of letters and ideas, Bowen responded on 27 May 1952 with a "Draft Proposal for Radio Observatory".[14] The proposed aerial was 200–250 ft (61–76 m) in diameter, with the intention to use the entire aperture in the wavelength range 1–10 m (300–30 MHz), but the central 100 ft would have a higher surface accuracy so the higher frequency HI line at 1.4 GHz (21 cm) could be observed with optimal sensitivity.[15] The mounting was to be alt-az, with a sketchy proposal for a master equatorial to perform the coordinate transformation. The cost of the project (1952) was to be about US$1 million and running costs of $80,000 per year for a staff of 13. Bowen recalled in interview with Robertson that he anticipated taking John Bolton with him to Caltech as his second in command.[16]

On 20 June 1952, in reply to a question from DuBridge about competition with the planned 250-ft instrument at Jodrell Bank, Bowen pointed out the differences. His proposal had no plans for planetary radar but envisioned use of the aerial at higher frequencies, above 300 MHz. His letter discussed "the use of a Giant Radio Telescope", one of the earliest uses of this nomenclature.[17]

While Bowen was contemplating the excitement of a move to Caltech as director of a large dish, Fred White was worried about Australia losing its pre-eminent position due to competition with the 250-ft dish being built at Jodrell Bank. White was unaware of Bowen's interactions with Caltech at this time since he was not informed about this by Bowen until November 1952.[18] If White had known about these discussions, he would have been even more worried about the potential "brain drain" of scientific talent that was becoming an issue of national concern. Fred White wrote Bowen on 4 June 1952[19] indicating that Appleton, the Nobel Laureate and President of URSI, who was planning to visit Australia in August for the URSI conference "would be willing to stimulate an interest in the Government here [Australia] providing money for a large radio telescope". Details of possible encounters that Appleton had with government ministers are not known, but Appleton did mention the large radio telescope at his opening Presidential address at the URSI General Assembly on 11 August 1952 at the University of Sydney, "... those of us who follow the subject would much like to see in due course a similar instrument

---

[13] DuBridge to Bowen 21 Feb 1952 C3830 Z1/14.

[14] A document of 11 pages, NAA A1/3/11/1 part I.

[15] This innovative design feature was incorporated into the GRT and the Parkes radio telescope still has a higher precision central area.

[16] Robertson (2017, p. 118) states that Bolton had "already agreed" to go to Caltech; however we note that Caltech's offer of a position for John Bolton to start a radio astronomy group at Caltech was not made until 1954, Robertson (2017, p. 148).

[17] At the end of 1952 (10 November 1952) the name of the "Large Radio Telescope" committee at RPL was changed to "Giant Radio Telescope"—the GRT.

[18] NAA, C3830, Z1/7/B Part.

[19] NAA C3830 A1/3/11/3 Part 1.

[as the Jodrell Bank telescope] at the disposal of your radio astronomers here in the Southern Hemisphere."[20]

Bowen followed up on Appleton's comment and White's prodding with an exchange of letters with Sir Henry Tizard, exploring opportunities for funding from the UK. Bowen's letter of 15 July 1952[21] included " ... with the announcement of the Manchester project, local interest, or perhaps local pride, has revived and there is now just a possibility that funds for a similar project can be raised in Australia". The success of the project was dependent on "the possibility of obtaining some part of the finance elsewhere ... I am writing you to ask if you know of any philanthropic bodies who might be approached ...". Tizard replied on 11 and 20 August noting that he was not very hopeful that the Nuffield Foundation[22] would be able to provide funds, but he re-iterated his earlier suggestion that the best bet was the British Dominions and Colonies Fund of the Carnegie Corporation of New York.

In 1951, on one of Bowen's visits to the US, Vannevar Bush had also told Bowen about the British Dominions and Colonies Fund of the Carnegie Corporation of New York which had $0.25 M earmarked for expenditure in Commonwealth countries. Clearly this fund might provide support for an Australian GRT. This opportunity was to open up Bowen's third line of approach; on 17 June 1952, Bowen asked White for permission to inquire about potential funding from the Carnegie Corporation. Bowen then wrote to Vannevar Bush (22 August 1952) and received a reply on 3 September 1952. But the reply was less enthusiastic than had been hoped. Vannevar Bush wrote: "... radio astronomy has an extraordinary future, nor do I doubt that it is going to be an exciting field in which participation will be highly stimulating ... but it is not easy [for me] to see what form the actual construction should take to best advantage.[23]"

At this stage there were options on the table for a Caltech funded dish in California, a dish in Australia funded by US philanthropic organizations and an Australian government funded endeavour. Bowen immediately informed DuBridge that he had "a foot in both camps" and broadened the Caltech proposal to include a possible collaboration between Caltech and CSIRO with two large telescopes giving access to both the Northern and Southern skies (Robertson, 1992, p. 118).

However, the proposal was now becoming quite complex with multiple options. Not surprisingly, on 6 August 1952, Bowen received a neutral response[24] from DuBridge suggesting that Bowen should pursue the Australian funding, noting that competition with Australia might stimulate US funding! He concluded "we are

---

[20] In June and August 1952, there were also exchanges of letters with Mark Oliphant about the proposed GRT. He had visited Jodrell Bank in the UK. He hoped a similar telescope could be constructed in Australia. "Under prevailing conditions, it can succeed only as a national undertaking and as a matter of national prestige" [his emphasis].

[21] NAA, C3830, A1/3/11/3, Part 1.

[22] UK funding organisation for the Jodrell Bank telescope.

[23] NAA C3830 A1/3/11/3 Part 1.

[24] DuBridge to Bowen, 6 Aug 1952 NAA C3830 A1/3/11/3 Part 1.

exploring possibilities ... and will keep you informed." And with that the Caltech initiative died.

Bowen had not, in fact, offered a compelling science case for the GRT in his proposal to Caltech. However, Pawsey developed a scientifically motivated proposal in a separate document, also written around the time of the URSI conference of August 1952. In "Notes on Applications in Radio Astronomy for a 100-Metre Diameter Telescope" Pawsey set out a scientific justification more extensive than the document submitted earlier by Bowen to DuBridge. Pawsey suggested that the optimum wavelength would be 2 m, 150 MHz. However, a major use would be 21 cm imaging of galactic HI as well as the determination of source spectra in the range 100–1000 MHz. Finally, a major future facility would be suitable for radar detections of the Sun and Venus. With such an instrument, Pawsey anticipated that the number of discrete radio sources known would increase from about 100 to 10,000. Of course, the use of interferometers would still be required to determine the size and accurate positions of the stronger radio sources.

At least three meetings of the "Large Radio Telescope" committee were held in 1952 after the URSI conference in Sydney[25] with Bowen, Pawsey, Carter, McAlister and McCready in attendance. These meetings covered a number of important design issues.[26] It is noteworthy that at the time of the third meeting of this group on 10 November 1952, the name of the committee was changed from "Large Radio Telescope" to "Giant Radio Telescope"—the GRT.[27]

With Pawsey's report at hand, Bowen wrote again to Vannevar Bush (on 23 October 1952). Within a month, a negative answer was received from the Director of the British Dominions and Colonies Fund. The Carnegie Corporation reported that the Australian project was outside the scope of the Fund.

Frustrated with the lack of financial support and despite the goodwill expressed earlier, Bowen wrote to White on 22 October 1952[28] with a back-up proposal— some way had to be found to squeeze something out of the existing Radiophysics budget. He suggested that an even larger aerial, but one that was not fully steerable, could be constructed, a cylindrical paraboloid that could be built in increments. This could still retain Sydney's leadership in the field as the Jodrell Bank telescope was under construction. A wide range of ideas were being discussed at RPL.[29]

---

[25] C3830 A1/3/11/2, from Papers of Woodruff T. Sullivan III, NRAO Archives. The more formal "Radio Telescope Planning Committee" was formed in May 1954. These more formal meetings were held in the years 1954–1955 with at least 14 meetings from May 1954 to November 1955. A successor committee, the Technical Advisory Committee, mainly consisting of outside experts, began work in July 1955, with eight meetings up to June 1959.

[26] These are discussed in detail in NRAO ONLINE.38.

[27] The GRT terminology became the common designation in the last months of 1952; this was used up to the time of the inauguration of the Parkes telescope on 3 October 1961. Later the term Giant Radio Telescope became simply the "Parkes radio telescope".

[28] NAA C3830 A1/3/11/1 Part 1.

[29] Details provided in NRAO ONLINE.38.

1953–1954 Events at RPL

A style that explored cheaper (and clever) antenna designs likely suited Pawsey. He prepared an additional proposal on 11 March 1953, optimised for low-cost aerials. Pawsey's "Notes on Big Aerial" included a "tilting-barrel" transit aerial and a "pre-Arecibo" fixed dish.

However Pawsey's enthusiasm to find lower cost solutions as well and realistic evaluation of the science case exacerbated the slowly deteriorating relationship with Bowen. On 26 March 1953, Bowen wrote a letter of complaint to White about Pawsey, the first instance of open conflict between the two that has been located in the archive. Bowen was not pleased with Pawsey's lukewarm support of the full scale GRT plans.

> As you know [implying that this complaint had been expressed earlier], I have had a tough time with Joe on the question of a big aerial. He knows all the reasons why we should not have one. This ... is exactly the way to put our feet in the grave as far as radio astronomy is concerned. If we cannot find the money for a big aerial, that is an entirely different matter. But to produce arguments against it will not get us very far.[30]

In 1953, a major goal of the CSIRO Division of Radiophysics was to restart the dialogue with the Carnegie Corporation of New York after their December 1952 rejection of financial support for the GRT. In 1953, Fred White continued the contacts with the Carnegie Institution of Washington and the Carnegie Corporation of New York. By April and May 1954, these efforts were successful with the official announcement on 20 May 1954 when the latter organisation would award US$250,000 for the GRT, at the time assumed to be one-quarter of the required expenditures. Vannevar Bush of the sister organisation in Washington played a major role in the successful turnaround. He wrote Bowen in May 1954: "Nothing ... would bring our two countries closer together more efficiently than for Australia to lead the way in an important area of fundamental research."[31] Robertson has summed up: "As it turned out, the proposal to build a large telescope staffed partly by Australians had changed, two years later, to an Australian telescope funded partly by American money!" (Robertson, 1992, p. 120).

An important meeting of the radio astronomy group at RPL took place on 8 July 1954, a few days before Pawsey's departure for Europe and the US. Pawsey clearly saw that the small semi-independent research group paradigm was coming to an end and he proposed "An Observatory", as a future model for the Radiophysics Laboratory: "In the past, projects have been planned on the basis of a small group building apparatus and using it to get all the information possible. We are moving towards the **observatory procedure** [our emphasis], where complex equipment is used by a succession of observers to investigate explicit problems."

---

[30]NAA C3830 E2/2 Part 2.

[31]NAA C3830, A1/3/11/3 Part 1.

Pawsey began the 6.5-page report with a discussion of his philosophy of planning (see Chap. 33) and then provided a detailed summary of all the existing research environment at RPL—see ESM 27.1, RPL radio astronomy. In 1954, radio astronomy at RPL was in a state of transition as plans for the future were discussed. The small group model (with instruments constructed by the user) would disappear in half a decade. The observatory paradigm would replace this as the GRT came on line in 1961.

In 1954, Pawsey was in Europe and the US from July to October. His contacts with Tizard, Barnes Wallis and Freeman Fox and Partners (FFP) were important events in the GRT planning in the 1950s. On 29 July 1954, while in London, Pawsey asked Sir Henry Tizard, the WWII radar pioneer, for assistance in finding the "best engineers in England from whom to ask advice … " Tizard was uncertain and proposed to "consult a friend of his, B.M. Wallis, who is … one of the best engineers in Vickers [the aircraft manufacturer] …" Wallis was the well-known designer of the R100 Airship, the Wellington bomber of WWII and the "Dam Buster" bombs of May 1943. In August 1954, Pawsey met him at the Vickers factory (Morpurgo, 1972). Wallis had a number of ideas that would be incorporated in the final design of the 1961 GRT (later the Parkes telescope). However, in the end his relations with FFP soured, and he withdrew from active consideration in the period 1956–1957.[32]

In the period 9–12 September 1954, Pawsey visited Freeman Fox and Partners, who were to become the designers of the GRT and the managing consulting engineers. This firm was an obvious choice since they had carried out the detailed design of the Sydney Harbour Bridge, which opened in 1932. Likely, Pawsey's visit was the first personal contact between CSIRO and Freeman Fox and Partners.[33]

Following the news of the Carnegie grant, the official Planning Committee-GRT chaired by Pawsey began meeting. From May 1954 to mid-1955 there were about 14 meetings of this committee.[34]

Two years after the conclusion of discussions with Bowen, DuBridge was finally able to establish a radio astronomy group at Caltech. Kellermann et al. (2020, p. 90) describe the build-up of the radio astronomy groups in the US, including the formation of the Caltech group and the January 1954 Washington Symposium on the future prospects for radio astronomy (attended by Bowen and Mills). A hugely important outcome was the decision to support the idea of a National Radio Astronomy Observatory (NRAO), which would build and operate radio telescopes, making them available to all University based astronomers (Kellermann et al., 2020, Chap. 3). This was a bigger vision than maintaining the success of a single group such as RPL. It was not until the construction of the Australia Telescope Compact

---

[32]Full details provided in NRAO Online.44 additional note 1 "Wallis Disaffection with FFP-1956 and 1957". Also NAA C3830 A1/3/11/1 and A1/3/11/32.

[33]The successful and frequently turbulent relationship between FFP and CSIRO (especially the chaos between Gilbert Roberts of FFP and Bowen of RPL) is outlined in the NRAO ONLINE.41–47.

[34]Planning Committee documents, Papers of Woodruff T. Sullivan III, NRAO Archives, NAA McGee archive—C4632/4. Further details can be found in NRAO ONLINE.40.

array in 1988 that CSIRO agreed to manage all of their radio telescopes as a National Facility (ATNF—Australia Telescope National Facility) under Australian Government guidelines.[35]

DuBridge invited John Bolton to take charge of the new research program, including designing a new telescope. Bowen wrote a recommendation of Bolton in glowing terms, and Bolton accepted the job at Caltech (Robertson, 2017, p. 148–151). He arrived in February 1955 and in March appointed Gordon Stanley from RPL as a Senior Research Associate. The new radio telescope was two 90-ft steerable dishes operated as a variable baseline interferometer, essentially the same as Bolton's unfunded proposal to CSIRO in 1951. The success of the Owens Valley interferometer was a strong influence on the eventual decision by the US to build a Very Large Array of big dishes (the VLA).[36]

1955–1956 Events

The year 1955 was a period of increased activity with visits of Bowen and Pawsey to the UK as well as the release of the publicity booklet on 6 May 1955. *A Proposal for a Giant Radio Telescope* was published under the nominal authorship of Bowen and Pawsey. The document was mainly written by Frank Kerr. The book was given various cynical titles by the RPL staff: "promotion", "sales" and even "propaganda" (a term used later by Pawsey). Bowen was impressed by a suggested title from Merle Tuve (6 March 1955) "glossy line-shoot", a term meaning "excessive bragging". The booklet had a distribution throughout the world.[37]

Earlier in the year, Bowen and Pawsey had a discouraging exchange of letters with Merle Tuve of the Carnegie Institution of Washington. On 3 March 1955, Tuve made a strong suggestion for an equatorial aerial with a diameter of 130–170 ft, rather than an alt-az dish with the problems of coordinate transformation. In retrospect we know this would have been a very bad choice—it is essentially the same as the seriously compromised NRAO 140-ft equatorial telescope (Kellermann et al., 2020, Chap. 4, Sect. 4).[38] On 10 March 1955, a frustrated Bowen responded with an equally discouraging text. In addition to the advice to construct a smaller aerial, the response to the attempts to raise funds in Australia had been disappointing:

> ... The Government [of Australia] is quite apathetic. Even in research circles there has been a disappointing tendency to say that sheep are more important, and that radio astronomy is

---

[35] *Guidelines for the Operation of National Research Facilities*, S, A report to the Prime Minister by the Australian Science and Technology Council (ASTEC), Australian Government Publishing Service, Canberra 1984.

[36] This array concept is now known as "small N—large D".

[37] RPL 94 and NAA C3830 A1/3/11.

[38] John Finley was quoted: "No one with hindsight will deny that the choice of an equatorial mount was idiotic." See Kellermann et al. (2020), Chap. 4, p. 195.

all right for other countries. Finally, purely from the point of view of constructing a large device I have been surprised to find how scared the rugged Australians are of going one better than anyone else in the world.[39]

A few days later Pawsey corresponded with Barnes Wallis in the UK. He would meet Wallis later (their second meeting) in 1955 when he was in London. On 17 March 1955, Pawsey wrote to Wallis, complaining about the lack of interest in the GRT from the Australian government and the lack of funds from private subscriptions. Pawsey ended his letter to Barnes Wallis with an attempt at optimism: "However, it is always darkest before the dawn."[40] Wallis responded on 14 April 1955 with an upbeat letter suggesting that a dish even larger than 250 ft could be constructed: "I believe that we can adopt methods of construction [based on the large rigid airships] which will be cheaper than in your book [the "publicity book"]".[41]

Within 2 months, the pessimism at RPL was slightly dispelled when Bowen visited the Rockefeller Foundation in New York in late May;[42] the news was "optimistic". In June, Bowen wrote Pawsey from London with instructions to get the "ball rolling" on a proposal for a design contract. Mills, Minnett, and McCready began work on a document "Specifications for a GRT for Which a Design Study is Required". The final version was completed on 23 November 1955. Pawsey replied from Sydney to Bowen in London on 30 June 1955 with an update on the activities in Sydney regarding the "specifications". The group in Sydney were redrafting the texts as suggested by Bowen and "we are relying on you to make the necessary preliminary approaches to appropriate engineers. Who these may be is not yet clear." Pawsey mentioned three possible options: Husband ("detailed experience") with the Jodrell Bank telescope), Wallis ("bright ideas") and FFP (consultants for the Sydney Harbour Bridge) with whom Pawsey had met in September 1954 in London.

Bowen met Wallis for the first time on 6 July 1955. He was favourably impressed but could see major problems with obtaining permission of Vickers (his employer) to allow Wallis to work on the GRT design. He then met FFP for the first time, likely on 8 July 1955. He wrote Pawsey 5 days later, "My first impressions were good and they seemed to get an excellent grasp of the problem."[43] As Bowen was leaving London on 28 July, he wrote Pawsey summarising Wallis's design and suggesting that they start work with one of three consultants to carry out the detail work: (1) Freeman Fox and Partners (FFP), or (2) Sir William Halcrow and Partners or (3) Head, Wrightson and Company. On 20 September 1955, Pawsey, now in the UK in his turn, was impatient to wrap up the discussions in London, and introduced

---

[39] NAA, C3830 A1/3/11/3 Part 3.

[40] NAA C3830 A1/3/11/1 Part 3.

[41] NAA C3830 A1/3/11/1 Part 3.

[42] Bowen was to be in the UK from early June to mid-August 1955 and Pawsey was in the UK from late August through early October 1955. The two would overlap for a few days in Sydney at RPL in late August.

[43] NAA C3830 A1/3/11/1 and A1/3/11/32.

Barnes Wallis to Ralph Freeman of FFP. Pawsey wrote Bowen in Sydney on this date:

> My assessment of the position, I think, agrees with yours. It is that Wallis's design could be outstanding, and it is up to us to find out. Of the consultants we have thought of I think Freeman, Fox and Partners are probably outstanding and since they are thoroughly interested there is an excellent opportunity for getting them. The present position is that I arranged a joint discussion between Wallis and Mr Freeman at which Arthur Wills [consultant for RPL from the Aeronautical Research Laboratory in Melbourne] and I were present. The atmosphere was first-rate as Wallis outlined his ideas. These, of course, had gone far beyond anything Freeman had had time to think of and he seemed impressed with a number of bright thoughts. As far as I can judge we could get a high degree of co-operation between the two ... [I think] that he [Freeman] will draft a letter from his firm to the RP Division setting out the terms under which his firm would undertake a design study ... [including] an assessment of the general feasibility of the Wallis type of design and an approximate estimate of the cost size relationship ... It seems to me that the outstanding point for me to get cleared up is this one, to get a good firm of consultants lined up for checking and developing Wallis's ideas and I hope you agree with the sort of arrangements which appears to be coming out.[44]

Bowen sent a cable to Pawsey in London agreeing with his suggestions. Pawsey wrote to Bowen on 28 September 1955 as he was about to return to Sydney. He had achieved a major milestone in the design of the GRT as he summarised his final meeting with Freeman and Gilbert Roberts:

> [Freeman and Roberts] agree in principle to understanding the design study along the lines we wish ... They are quite agreeable to collaborating with Wallis. The position as I see it is that the designing engineers are employed by us, but are the responsible people in producing the design. Wallis would be in an advisory capacity. The current relations between the two look quite good and I think that the responsibility should be set fairly on [Freeman Fox and Partners] ... The way they handle a design study is to start from what appears to be the most promising design, in this case it would be the general scheme put forward by Wallis, and to investigate this along with other ideas. This means that they would attempt to assess the relative merits of steel and light alloys, of a rigid as opposed to a compensated structure, of alt-azimuth as opposed to an equatorial one etc. Since this is their procedure, they do not think it desirable that an independent parallel study should be undertaken ... the study would take about six months [an estimate that turned out to be vastly in error]. Liaison with [CSIRO] RPL looks to be very difficult and requires consideration.[45]

In the last months of 1955, two decisive GRT events occurred: the launching of the FFP design study and the announcement of the Rockefeller grant ($250,000) to the Australians in a letter to the Minister for the CSIRO, R.G. Casey, on 8 December 1955. In addition, Barnes Wallis's study "Giant Radio Telescopes" was sent to Pawsey in Australia on 14 October 1955.

Sadly, Wallis had a major fallout with FFP in the years 1956–1957.[46] The details are not known. On 20 April 1959, Bowen wrote Wallis[47] a letter of gratitude and

---

[44]NAA C3830 A1/3/11/1 Part 4. Pawsey to Bowen in Sydney, 20 September 1955.

[45]*Ibid.*

[46]Described in NRAO ONLINE.44, Additional Note 1.

[47]NAA C3830 A1/3/11/1 Part 10. Bowen was about to leave for the US the following day, after meeting Wallis in person on 17 April. On 15 May 1959, Bowen wrote White complaining about the

apology: "I would like to emphasise again how grateful we are for the effort you put into our radio telescope project . . . I deeply regret that difficulties have occurred with Freeman Fox and Partners, but we have run into similar troubles ourselves and can quite understand your [Wallis] point of view."

On 16 November 1955, FFP sent a proposal for the design study to Bowen. The agreement from CSIRO was sent back on 28 November 1955; a second version was sent to White on 23 December 1955 and agreed to by the Australians in January 1956. The design study proposed by CSIRO was entitled "Specifications of a Giant Radio Telescope for which a Design Study is Required". The FFP report was entitled "Proposed Radio Telescope Design Study".

A contentious point foreshadowed future conflicts. FFP would not accept a 6-month time scale for the report: "We shall do our utmost to complete our report in six months and if we find it will take appreciably longer we shall let you have an interim report at the end of June 1956 indicating the stage of our investigation has reached and the conclusion that may be drawn from it."

During 1956, CSIRO and RPL remained frustrated with the continual delays in the reception of the design report. The CSIRO had asked for a time scale of 6 months; in the end they had to wait almost 2 years for the first report. At last, the interim report was obtained in October 1956. A by-product of these delays was an abundant level of acrimony between Roberts at FFP and Bowen at CSIRO. Fortunately for CSIRO, the final choice of the GRT was an uncompensated (i.e. the panels could not be moved in real-time), alt-azimuth aerial that was constructed in a time frame of less than 2 years.

1956: Appleton[48]

On 10 January 1956, Sir Edward Appleton, Vice-Chancellor of Edinburgh University, wrote Pawsey, thanking him for a copy of the IAU Commission 40 Radio Astronomy report. (Pawsey was President of the Commission from 1952 to 1958.) "This is international co-operation at its very best, and it is very cheering to see it. Clear you are making great progress—a bit beyond me nowadays."

Pawsey responded to Appleton in detail on 22 February 1956:

As you remark, radio astronomy has gone a long way since the big advances at the end of the war. But it is still a tantalising subject; we have a wealth of factual material which is still loath to fit into a physical picture. This is particularly so in respect to the sun . . . where the physical understanding seems to diminish with the increase of facts.

---

treatment of Barnes Wallis by FFP: "We have the unfortunate business of Barnes Wallis; he is still well disposed to us, but very outspoken about Roberts and his ways." No mention of this break with FFP appeared in the *Biographical Memoirs of Fellows of the Royal Society* obituary for Barnes Wallis in 1981.

[48] NAA C3830 Z1/3/VI.

It is interesting also to see the parallel advances from large equipment [big science] and from simple things [little science, the small independent group]. Our "Mills Cross" ... is the current outstanding example of the former [large equipment]. It is giving a wonderful series of results showing galactic structure plus thousands of discrete sources. The combination is essential for understanding.[49]

Pawsey concluded the letter to Appleton with a description of the next "large equipment", the GRT.

We are now finally launched on the GRT project which you did such a lot to stimulate when you were here [URSI, 1952]. Bowen has done a remarkable job on money raising and has a quarter of a million dollars each from Carnegie and Rockefeller. We have now arranged for an outstanding London firm of consulting engineers, FFP, to undertake a design study.

In 1956, a major controversy concerned the uncertainty in the realisation of the pointing of an alt-az aerial in the pre-computer era. The simplicity of pointing an equatorial mount had to be balanced with the resultant complex changing gravitational forces with elevation. FFP spent little of their design effort on an equatorial telescope; their emphasis was on the alt-az mount concept in their final design study. Merle Tuve remained a vocal and frequent critic of the alt-az design. The advent of the Barnes Wallis concept of a ME (master equatorial) positioned at the intersection of the altitude and azimuth axes did become a major success, even though this was far from certain in 1955–1957. The Parkes telescope eventually achieved a surprising pointing accuracy, about five times better than the arc min specification envisioned in the late 1950s. In hindsight there is no question that the alt-az decision was correct and the strongly worded advice from the "expert", Merle Tuve, would have been the wrong decision.

During Tuve's round the world trip starting in July 1956, he spent a day with Freeman, Mike Jeffery[50] and Harry Minnett.[51] Though his visit was characterised by confused plans for the joint discussions (only 3½ h were available for discussions and Roberts was away for the day!) the visit was viewed as a success by Bowen. He felt that Tuve had acted "as a very useful catalyst" while there were decisions to be made. Roberts read about the visit and wrote to the Australians (on 19 July 1956) with warnings:

[Tuve] seems to be keen on the Polar Axis [equatorial] type of mounting, but the advantages of this, even if found practical for this size of dish ... are doubtful. For instance, deflections due to dead load can be readily compensated in the alt-azimuth mounting, but in the Polar Axis mounting present difficulties. With the Polar Axis mounting, deflections of the

---

[49] Pawsey described the "complete disagreement with Ryle" about the number counts, i.e. the excess of faint sources in the 2C survey compared to Mills. See Mills and Slee (1957) and Mills (2006). See Chap. 35.

[50] M.J. Jeffery, key member of the staff at FFP, who would spend a number of years in Australia until his death in 1969. See NRAO ONLINE.45 and 47. Robert Hayward has provided additional information, ESM 27.2, Biographical Sketch.

[51] A CSIRO (RPL) scientific staff member who had joined RPL in 1940. He was a CSIRO consultant in London who played a key liaison role for five plus years in London starting in October 1955.

structure supporting the axis bearing could not be corrected, whereas with our type of mounting any deflection of the axis of rotation is automatically taken up by the ME (master equatorial) . . . I feel that our system must be the better solution.

The interim report from FFP arrived on 13 October 1956. The best news was that a **rigid** dish of 325 ft in diameter could be constructed before the deflections reached 0.8 cm, a vast improvement compared to earlier expectations. The expected upper frequency for the telescope would then be about 3 GHz. But even Fred White was still apprehensive about the delay that had been experienced in 1955–1956. On 11 October 1956, White suggested that Pawsey might intervene (letter to Bowen): "However, the main point seems to me to be where some senior person should not access just where FFP have got to and if necessary crystalise their thinking in a particular direction. Probably, Joe Pawsey should be the person to do it."[52]

A major disappointment of the interim report was that no estimate of the costs was possible. There were too many uncertainties in the ME and servo control systems. But the good news was that an uncompensated structure with sufficient surface accuracy would be possible "up to the largest size for which the available sum would suffice."

As 1956 ended there was still a major uncertainty on the choice of alt-az (proponents: Freeman Fox and Partners, Wallis) versus equatorial (proponents: Tuve, Bruce Rule and John Bolton of Caltech). This conundrum would be resolved soon.

Recruitment of Bart Bok as Mt Stromlo Observatory Director

The successful recruitment of astronomer Bart Bok to replace Richard Woolley as Commonwealth Astronomer at Mt Stromlo Observatory must have been one of the most satisfying achievements for Pawsey in 1955–1956. When Woolley resigned to take up his appointment as Astronomer Royal in the United Kingdom in December 1955, Arthur Hogg (1903–1966) became Acting Director in his place. As will be seen, Hogg, who had joined the Commonwealth Solar Observatory in 1929 and remained there throughout his career, quietly contributing to many projects in Australian astronomy—among them, to the long process in which the GRT became reality.[53] In the meantime, Pawsey used the opportunity of Woolley's resignation to drive a search for a replacement who could provide the acutely-needed collaboration for radio astronomers (that Bolton was now so productively utilising at Caltech). Bok had studied at Leiden and Groningen, so had longstanding connections to Dutch astronomers (Oort, Kapteyn, van de Hulst and their colleagues). He then completed his PhD studies at Harvard (1932), focused on understanding the Milky Way, which,

---

[52] NAA C3830 A1/3/11/1 Part 7.

[53] https://www.science.org.au/fellowship/fellows/biographical-memoirs/arthur-robert-hogg-1903-1 966.

until 1926, had been hypothesised to be the whole Universe, rather than simply a galaxy. Bok became a Milky Way expert, publishing a famous work on the topic in 1941 with his wife, astronomer Priscilla Fairfield (later Bok). The two became a scientific team. Their mutual advisor was Harlow Shapley.

Bok was stimulated by the discovery of the HI line to become interested in radio astronomy and to begin building radio telescopes. In the late 1940s (starting in 1946 and continuing into the next decade), there were recurring pressures applied on Shapley as a result of the growing anti-communism of this era. It was an ideal time to recruit Bok to Australia.

In the book *Mt Stromlo Observatory: From Bush Observatory to the Nobel Prize* by Bhathal et al. (2013, p. 103). Bok recalled:

> Mrs Bok and I were well known to have a great interest in the southern Milky Way. So during a General Assembly of the International Astronomical Union in Dublin in Ireland (1955), the then Director of the Observatory, now Sir Richard Woolley, and Dr J.L. Pawsey, the second-in-command of CSIRO's Radiophysics Division in Sydney, approached us both together with the request, would we be interested in considering leaving Harvard and coming to Australia?

Another similar account was reported in an earlier history of Mt Stromlo by Frame and Faulkner, *Stromlo: An Australian Observatory* (2003, p. 132 and notes on p. 316). They described Pawsey's role:

> It was Joe Pawsey who suggested the name of Bart Bok. He saw virtue in appointment of someone with experience in *both* (optical and radio) branches of the subject ... Bok's Harvard background fitted the bill. [The notes attribute this assertion to an interview with Olin Eggen in 1988. The note continued:] Pawsey in turn, was acting on a suggestion from Father Daniel O'Connell at the Dublin IAU of 1955 (attributed to Ben Gascoigne).

In both versions Pawsey played the most central role in Bok's successful recruitment to Australia. He was appointed in 1956 and took up his post (7 March 1957) just as Mt Stromlo Observatory was transferred to become part of the Australian National University, securing the long hoped for access to research students in radio astronomy. Due in part to an early appearance in Parliament when the Observatory took the first photograph of the Russian satellite Sputnik (the first ever launched) in 1957, Bok swiftly became a well-known public figure with easy access to political leaders, strengthening the profile of Australian astronomy (Bhathal et al., 2013, p. 109–111). The Bok era (1957–1966) at Stromlo would be characterised by increased collaboration between optical and radio astronomers in Australia.[54] The close friendship of the Bok and Pawsey families remains a memorable aspect of this period in Australia.

Bok visited Australia from 26 September to 4 October 1956 prior to his move from Harvard to take up the Directorship of Mt Stromlo in early 1957. Bok visited Canberra and Sydney to scope out the "lay of the land". The highlight in Sydney was a 2-day "Symposium on Radio Astronomy" with an introduction by White (An Insider's View of the History of Radio Astronomy In Australia from 1945 to

---

[54] See Frame and Faulkner (2003) Chap. 7, and Bhathal et al. (2013), Chap. 6, for evaluations of the impact Bok had in Australia from 1957 to 1966.

1955), Pawsey (Radio Astronomy at RPL) and Bok (Radio Astronomy in the US). Many staff members presented papers.

At the symposium in Sydney, White praised the pioneers of RPL, Bowen and Pawsey:

> In this Laboratory, under Dr Bowen and Dr Pawsey, you have good facilities and will be able to continue to spend a reasonably large sum of money on this activity. When the large telescope [GRT] is built, this, together with the other facilities which you have, will make you, I believe, one of the best equipped laboratories in the world ... With these material resources, all that is required now is initiative and resource in research to keep the Australian effort in the forefront of this advancing science ... We are fortunate our large telescope will perhaps for some years be the only one of its kind in the Southern Hemisphere so that you will have unique opportunities to examine these parts of the sky which cannot be seen by the several telescopes being erected in the Northern Hemisphere. This may provide us with the opportunity of having many visiting radio astronomers. [We interpret this as a clarion call for OPEN SKIES.]

When Bok returned to his home in Cambridge, Massachusetts, in late 1957, he wrote the Pawsey family in Sydney an effusive letter:[55]

> You and Lenore have certainly contributed greatly to having me feel at home in Australia ... The two-day symposium was a wonderful experience and I have told all who wished to listen that I could not think of a more powerful scientific meeting in the field of radio astronomy than the one you and your group put on ...

In mid-1956, Pawsey and his colleagues continued to evaluate "possible [astronomical] experiments with the GRT" and the impact of the astronomical goals on the form of the future instrument. As had been the case in 1955, the group were concerned that "enhanced directivity" or interferometry would be an additional option to GRT science. A year later on 22 July 1957, at a similar meeting, Pawsey introduced a new topic which would be decisive with the new Parkes radio telescope of the 1960s: polarisation of continuum and line radiation. In addition, the question of the site location continued to be discussed. The question of a site free of interference and a location with flat ground with a size up to 20 km for suitable interferometry were desired properties.

---

[55]NAA C3830 Z3/1/VII 23 October 1956.

Chapter 28
Brain Drain: Trip to US and Canada 1957–1959

Letter from Pawsey to his mother, from Princeton end 1957:

> I am three persons (or even four): (1) a visiting scientist interested in meeting [fellow scientists] and discussing [scientific issues], (2) a visitor interested in showing the family round, (3) the President of Commission 40 of the IAU [Radio Astronomy] with organising duties for international meetings in Paris and Moscow and (4) the assistant chief of Radiophysics with remaining responsibilities re-appointments of staff and the giant radio telescope [GRT]. No one job is at all arduous, taken together they get me down...

Pawsey's 8½-month visit to the US in 1957–1958 occurred during a key period of the GRT deliberations (FFP design study completion at the end of 1957 and the site selection in early 1958). It also occurred in the context of shifts in relations within RPL and in the field of radio astronomy as it grew around the world. There was growing awareness in Australia about the increasing capacity, especially in the USA, to attract first-rate scientists overseas to lead the new research programs being established. Meanwhile, at RPL, Bowen's frustrations with Pawsey were growing to such a degree that Pawsey was beginning to feel some disquiet about his position in CSIRO. An important outcome of Pawsey's visit to the US was an unofficial "audition" for a leadership role in US radio astronomy. At this point Pawsey would realise that he would have more to offer a US community with its multiple new radio astronomy groups (similar to the multiple groups he had nurtured in the beginning of radio astronomy research in Australia), than the Australian groups which had become strong and less dependent on his leadership. Pawsey's scientific interactions during this time were also important as he planned for the Paris Symposium of August 1958 in his role as chair of the IAU organising committee.

Supplementary Information The online version contains supplementary material available at [https://doi.org/10.1007/978-3-031-07916-0_28].

427
W. M. Goss et al., *Joe Pawsey and the Founding of Australian Radio Astronomy*,
Historical & Cultural Astronomy, https://doi.org/10.1007/978-3-031-07916-0_28

Planning the Trip: See ESM 28.1, Trip to US, for Full Details

From the NAA records we can piece together Pawsey's role based on numerous letters during the period. In addition, the Joe and Lenore Pawsey Family Collection has provided many details of the planning for the complex trip for the five members of the Pawsey family in 1957–1958. These include letters from Pawsey to his brother-in-law and his wife (Ted and Kate Nicoll in Princeton New Jersey, from 30 January 1957 and 15 April 1957).

The US invitation to Pawsey for the visit in 1957–1958 came from the National Science Foundation's Panel on Radio Astronomy. The visit was initiated following a letter from Bart Bok (at Harvard) who was assuming the directorship at Mt Stromlo in early March 1957. On 31 July 1956, Bok wrote:[1]

> Have you heard from Merle Tuve already about the new desire expressed by the NSF Panel on Radio Astronomy that you should visit the US in the not too distant future? At our last meeting (12 July 1956) the topic came up again and I believe that we set aside up to $8000 to help make possible such a visit this year or next. Official confirmation of this will of course [come] from Merle Tuve or from the NSF, but I thought that you would want to know that the desire continues to be expressed that **you might come to the US for a somewhat longer visit in the not too distant future.** [our emphasis] Our Panel is composed of Greenstein, Minkowski, Kraus, Purcell, Hagen and Tuve and myself and there seems to be a very unanimous expression of opinion. The only demure was expressed by B.J. Bok [himself!] who would like to see you stay in Australia.[2]

Bok continued his correspondence on 23 October 1956 after his September visit to Australia to sort out their impending move to Australia:[3]

> Funny things are happening: I may tell you in strictest confidence that last week in New York, Donald Menzel [Director of Harvard College Observatory] came to me as head of a small AUI appointed committee to offer me formally and officially the Director-ship of the NRAO ... Apparently I was the unanimous first choice of the whole committee. This is really crazy, for everybody on the committee knew darn well that I am fully committed to Australia and that I would not go back on my Australian assignment for anything, even if this could be done—which it certainly cannot—without disappointing anyone or disgracing myself and the reputation of the US. But it was a gesture of some sort and I might as well take it as a token of expression of a lot of good will ... [Priscilla and I] cannot think of anything nicer than to go [to Canberra] for keeps ... The committee is apparently going to select an **American** [our emphasis] to head the observatory in West Virginia. I do not know who it will be in the end, but the names that I have heard mentioned most prominently are those of Townes, Goldberg, Greenstein, Whitford and Hagen for the

---

[1] NAA C3830 Z3/1/VI 1956.

[2] Later in the 23 July 1956 letter Bok wrote Pawsey: "Wholly off the record, you might like to know that the National Science Foundation approached me 10 days ago to ask if I could not change my mind about going to Australia and come to West Virginia instead. The reply was 'NUTS.' Priscilla and I are very happy at the decision we have made to go to Australia and we would not want to change it for anything. You will now have to fire us to get rid of us."

[3] NAA C3830 Z1/3/VI.

top director's job; there is no restriction with regard to nationality for any of the jobs next to the director.[4]

Bok felt Pawsey would be perfect for the NRAO director's job, and that if the committee could see him in action, they would change their minds about limiting the search to Americans only, though no written evidence of this deliberation has been located. In the end, Frank Edmondson, the acting Director for Astronomy with NSF, made $3000 available to Pawsey for travel expenses, with an extra $5000 as a consultation fee to defray the cost of Pawsey's trip. Pawsey wrote Bok on 21 January 1957, after receiving Bok's letter of 4 January 1957:[5]

> Thanks for your comments on my own trip. It is going to be very exciting. I have just had welcoming letters from Leo Goldberg [Michigan] and Jesse Greenstein [Caltech] for extended visits . . . Washington is also a "must" for several weeks. My general plan was to visit other places, either as side trips from these or in between. I shall certainly plan to visit the people and places you mention. I am now in a position to try to fix up a definite schedule and shall go ahead.

By the end of January 1957, detailed plans for the US and Canadian visits were coming together. The full details of this trip including the many groups visited by Pawsey are provided in ESM 28.1, Trip to US. J.L. Pawsey, Lenore and their 12-year-old son, Hastings, left Sydney on 9 August 1957, flying to San Francisco and then on to New York.

Pawsey attended the ICSU (International Council for Scientific Unions, now International Council for Science) meeting "Mixed Commission on the Ionosphere" from 14 to 16 August 1957 in New York City. Then Pawsey went to the University of Illinois in Urbana from 18 to 21 August 1957 where he attended a meeting of the American Astronomical Society.

A major component of the American Astronomical Society was a special symposium "Radio Sources Outside our Galaxy: A Symposium". There were only four speakers. The lead-off paper was by Pawsey, "Sydney Investigations and Very Distant Radio Sources". This was followed by three presentations: "The Distribution of Radio Stars" by Hewish, "The Problem of the Identification of Extragalactic Radio Sources" by Minkowski and "Model Universes Derived from Counts of Very Distant Radio Sources" by one of the hosts of the AAS in Urbana, McVittie. In Chap. 36, we discuss these papers which for the first time indicated broad agreement

---

[4]In 1961–1962, this requirement for the Director to be a US citizen was also discussed; Rabi made a point of informing Pawsey that this restriction was no longer relevant in late 1961 (see Chap. 38).

[5]Bok had written with the plans for the Boks' arrival in Sydney on 4 March 1957 and a description of the NSF grant for Pawsey's visit to the US, to be administered by Merle Tuve. Bok had hoped that Blaauw or Weaver would succeed him at Harvard. He had heard that Tommy Gold would be the new professor at Harvard; he was apprehensive: ". . . [T]here will have to be changes in the whole approach to radio and optical Milky Way astronomy at Harvard Observatory!" Bok also assured Pawsey that the NSF was prepared to spend substantial funds to stimulate radio astronomy in the coming decade in the US.

on the issues related to the radio source surveys, the nature of the radio galaxies and the implications for cosmology.

Pawsey's letter to Bowen at RPL (posted the third of August 1957) reported on the results of the new 3C survey based on discussions with David Dewhirst from Cambridge. It is included in full in ESM 28.2, Pawsey to Bowen 1957, and included the following key points:

> When I came to the A.A.S. meeting at Urbana I met a "teaser". Dewhirst, the young astronomer who works with Ryle at Cambridge, is going to Pasadena... [where] he hopes to compare these positions with the Palomar Schmidt plates. He asked me if he could have also the Mills positions ..., so that by inter-comparison he could make the best of both surveys. I was naturally taken aback and far from co-operative.[6] But on the other side, the Cambridge survey is likely to be fairly precise for the big sources, probably better than Mills. And for these sources, information from 2 surveys is indeed likely to be better than from either Mills or Cambridge alone.
>
> . . . .
>
> You will be pleased to know that Hewish admits in public that the original 2C survey [same aerial at 81 MHz] was over-interpreted, but he does not say how much (10% or 90%?).

Next on the whirlwind agenda was the URSI meeting in Boulder, Colorado, that began on August 22 and continued to 5 September 1957. Pawsey's summary report (11 October 1957) on the URSI meeting was succinct:

> The general impression I had was that this was a stage where techniques are leading ahead and there will be a spate of new results soon in consequence. On solar work we have been building stuff for years and have had very little to report, but we should be in a strong position right now. On galactic and extra-galactic things we are in a reverse position. We have got in long tedious investigations, e.g. the 21-cm spiral structure [at Potts Hill]. Some of these are all right but they should be balanced by short imaginative ones. Mills's source-size investigations I think a good one of the large variety. It might touch off something good. What I had been hoping for was more physics from the large-scale work. For example, Bernard [Mills] and Eric Hill were chasing a first-class problem—is there emission from intergalactic space. Similarly, Alex Shain has a beauty in the gun: the background brightness due to very distant galaxies.

The next stop was the University of Michigan at Ann Arbor on 6 September 1957 where there was a chance meeting with Campbell Wade, a newly recruited staff member at CSIRO RPL. (See ESM 28.3, Campbell Wade, for a description of the late night interview with Wade and his role as a postdoctoral fellow in Sydney 1957–1959.)

During the period that Pawsey was in Ann Arbor, he participated in a significant event in the life of the National Radio Astronomy Observatory and Associated Universities: the official transfer of the Green Bank, West Virginia, site from the NSF to AUI on 17 October 1957 (Open Skies, p. 165; Kellermann et al., 2020;

---

[6]This statement plus a following declaration with the phrase "unco-operative" are uncharacteristic of Pawsey's style of open communications.

Lockman et al., 2007, p. 20). Pawsey participated as a NRAO "Consultant",[7] traveling on Monday 14 October with Richard Emberson (Assistant to the President of AUI, Lloyd Berkner) to Green Bank. There they spent time in discussions with David Heeschen and John Findlay, two of the first scientific staff of NRAO. Kochu Menon told Goss on 2 September 2011 that one of the reasons Pawsey was invited to the NRAO event in October 1957 was to convince him to move to the new NRAO.

On the next day (16 October 1957) a meeting of the newly established NRAO/ AUI Advisory Committee[8] was held at Warm Springs, Virginia. The purpose of the committee was to "discuss the research program for the Observatory ... and also for radio astronomy in the US because of the feedback of the activities at Green Bank into the radio astronomy programs of the various colleges and universities."

Graham DuShane (1957) has written a comprehensive article on the beginning of the National Radio Astronomy Observatory at Greenbank. See ESM 28.5, The National Radio Astronomy Observatory, for a summary.

Starting in late 1957, Bowen became frustrated with Pawsey's absence from Australia and there was increasing conflict about Pawsey's plans for travel in 1958. "It is a pity though that you are not around in person when some of the more important decisions (e.g. the site decision) effecting the radio astronomy group are being taken". The exchange of letters with Bowen are summarised in ESM 28.4, Conflict about Pawsey.

On 23 December 1957, Pawsey spent most of the day visiting in Princeton with Frank Kerr. They discussed in detail (see Chap. 29) users' aspects of the planned GRT.

On 24 December 1957, Pawsey went to the Princeton University Observatory to meet Jan Oort[9] visiting from Leiden and Lyman Spitzer, the leading astrophysicist at Princeton. Pawsey's association with both was decisive in connecting Australian astronomy to US and European astronomy.

1958

With the new year of 1958, Pawsey had numerous activities in Washington, DC, with return trips on the weekend to Princeton. In Washington he met with a number of scientists involved in US science policy: Lloyd Berkner, President of Associated Universities, I.I. Rabi from Columbia University and founder of AUI and Robert

---

[7] A contract between Pawsey and AUI was signed on 30 August 1957 to "provide consultation on design of radio telescope and research problems for Radio Astronomy Observatory". The contract was only valid until the end of 1957.

[8] Members Donald Menzel, Armin Deutsch, Bill Gordon, Fred Haddock, Ed McClain, G.C. McVittie and Jerome Wiesner.

[9] Oort visited Princeton frequently since his son, Abraham, was a prominent climatologist at the Geophysical Fluid Dynamics Laboratory of Princeton University and the National Oceanic and Atmosphere Administration.

Bacher, Chairman of the Division of Physics, Mathematics and Astronomy at Caltech, the latter two members of Eisenhower's Presidential Science Advisory Committee. Berkner was concerned about Freeman Fox Partner's cost estimates for the GRT. He was convinced that they were vastly underestimated, with the expectation that the production costs for the telescope itself would increase dramatically.

On one of the first days of 1958, Rabi met Pawsey for breakfast in Washington. Possibly this was when Rabi first suggested that Pawsey might take on a leadership role in US radio astronomy.[10] The remainder of the day was spent at Department of Terrestrial Magnetism (of Carnegie Institution of Washington) DTM[11] with Merle Tuve, Bernard Burke (1928–2018), John Firor (1927–2007) and Bill Erickson (1930–1994).[12] Weekends were also filled with discussions: on Saturday, 4 January 1958, Pawsey had breakfast with Ed Purcell of Harvard.[13]

During the second week of January 1958, Pawsey gave a colloquium at the physics department of Columbia University (January 10). While he was there, Rabi[14] scheduled a luncheon with 12 people attending, including T.D. Lee (recently awarded the Nobel Prize, Physics, in 1957). During the afternoon, Pawsey was given a tour of the department by Charles Townes (Nobel Prize, Physics, 1964), meeting Akira Okaya (working on solid-state physics, electronics and lasers) and Polykap Kusch (Nobel Prize, Physics, 1955). Townes provided a comprehensive summary of maser research being carried out in the US in 1958; Pawsey was, of course, fascinated since the expectation in 1958 was that maser radio astronomy receivers would provide an order of magnitude increase in sensitivity due to their low system temperatures.[15]

---

[10] Rabi was an AUI Trustee in 1958; three years later on 21 April 1961, he became AUI President, continuing up to 19 October 1962. (Berkner had resigned as AUI president on 30 November 1960, after being in the position since 1951). As we describe in Chap. 38, Rabi was active in late 1961 recruiting Pawsey to become the NRAO Director in 1962.

[11] In 1957–1958, Pawsey had been appointed a Fellow of the CIW, Carnegie Institution of Washington. "This means that the US tax is reasonable and it seems to be much the best arrangement to have the money paid and taxed in America." Letter, JLP to Jack Cummins, Chief Scientific Liaison Officer, ASLO DC (3 June 1957—C3830 Z3/1/Part 7). The honorarium consisted of $5000 with a travel provision of $3000. ($1000 in 1958 is equivalent to about $8700 in 2018.)

[12] Erickson had departed DTM a year earlier, 1957, moving to Convair in San Diego. In January 1958, he was visiting DTM colleagues in Washington for discussions. See ESM 39.3, William Erickson.

[13] Ed Purcell had been a prominent radar researcher at the Radiation Laboratory during WWII (K band radar), shared the Nobel Prize in Physics in 1952 and the co-discover of the HI line in the Milky Way (with Ewen) in 1951, March 25.

[14] Apparently, this was the third meeting between Rabi and Pawsey within a week.

[15] Radhakrishnan has written in an obituary for Mayer (1921–2005): "In 1959, Connie collaborated with Charles Townes and his students at Columbia in the first application of the maser to astronomy. When Townes received the 1964 Nobel Prize for the invention of the maser, he asserted that Connie's desire to improve receiver sensitivity was influential in his work and shared a portion of his prize money with him."

Pawsey then returned to Washington on 13 January 1958 to continue his busy schedule, visiting colleagues at the NSF, ASLO, DTM, the Derwood radio observatory, the Naval Research Laboratory. At Derwood, Pawsey visited John Firor and Bernard Burke. Burke was a radio astronomer at DTM from 1953 to 1965; his 1955 discovery of the decametric bursts from Jupiter, along with Kenneth Franklin, at 22 MHz is one of the major discoveries of radio astronomy in the first decade after WWII.

The next stop was Colorado, visiting Estes Park, Central City, and Boulder, beginning on 1 February 1958 with arrival in Boulder the next day. Their host was Walter Orr Roberts, Director of the High Altitude Observatory and a friend and advisor from 1948 to Pawsey's death in 1962. On 2 February 1958 (Sunday) the family went to the Roberts's home for dinner. Roberts was one of the closest confidants of Pawsey; the decision to accept the AUI offer of the Directorship of NRAO in 1961 was heavily influenced by Roberts's advice (see Chap. 38). During the evening, Pawsey had long conversations with his host about his troubled situation. Clearly, Pawsey already had some concerns about his future in Australia in the post-GRT era. During the evening, he discussed with Roberts his concerns including discussions about "his [future] security at RPL".[16]

Pawsey spent the following 3 days at Stanford (Wednesday and Friday with Ron Bracewell) and the Lick Observatory (Thursday). Pawsey visited the construction site of the Stanford Microwave Spectroheliograph (Fig. 28.1). In Fig. 28.2, Pawsey is shown carving his name (Pawsey) on one of the incomplete piers. By the 1970s, well over 200 scientists had left their signatures on the piers.[17]

While visiting Stanford Pawsey had discussions with Bracewell about a proposed second edition of *Radio Astronomy* by Pawsey and Bracewell. This would rectify the main problem with the first edition, which was how swiftly much of its content had become out of date. Pawsey laid out a scheme of new and revised chapters with a division of labour, e.g. Bracewell would rewrite the solar radio chapter before the Paris Symposium in August 1958, as well as the chapter on radio astronomy techniques. But this second edition would never come to fruition.[18]

At Stanford, Pawsey met with two prominent electrical engineering professors, Hubert Heffner (1924–1975) and Allen Peterson (1922–1994). Heffner had expertise in parametric amplifiers and predicted that noise temperatures of 20 K would be possible in the near future with bandwidths of tens of MHz.[19] With Peterson, they

---

[16] Pawsey 1957–1958 diary page 28. Also, they discussed the controversial claims of Bowen from 1957. Both men had their doubts about the reality of the claim in Bowen's article "Relation between Meteor Showers and the Rainfall of August, September and October" (Bowen, 1956).

[17] In 2012 the Friends of the Bracewell Observatory (Bob Lash, President) and NRAO moved ten of the piers to the Jansky Very Large Array site in New Mexico. The Bracewell Radio Sundial was opened at the September equinox in 2013. Pawsey's signature at the VLA is shown in Fig. 28.3.

[18] See NRAO ONLINE.53.

[19] Pawsey suggested that Alec Little (his young colleague) would come to Stanford for an extended visit; a few years later Alec received a Masters Degree working on a prototype paramp. He returned to join Mills's new group at the University of Sydney a few years later.

Fig. 28.1 Pawsey visit to the Stanford Microwave Spectroheliograph on 5 March 1958. The instrument was under construction, under the leadership of Ron Bracewell. Photo by Pawsey with the Rolleicord camera. Credit: Joe and Lenore Pawsey Family Collection

Fig. 28.2 At the Stanford Microwave Spectroheliograph, 5 March 1958, Pawsey carved his name on the pier of the still incomplete instrument. Photo taken with Bracewell's camera. Standing Bracewell and Swarup (left to right). Credit: *National Radio Astronomy Observatory/ Associated Universities, Inc. Archives, Papers of Ronald N. Bracewell, Photographs*

discussed lunar radar research at 400 MHz with the surprising result of occasionally observed attenuation of the echoes of 10–15 db, of unknown cause.

Fig. 28.3 The noon pier of the Bracewell Radio Sundial at the Jansky Very Large Array of NRAO in New Mexico, inaugurated on September equinox 2013. Astronomers Pawsey, Jim Roberts, Bart and Priscilla Bok, Yutaka Uchida and Rudolf Minkowski have left their inscriptions on the pier in the era 1958–1970 above Pawsey's signature. Credit: © M Goss, Jansky Very Large Array, March 2018

On the afternoon of 18 April 1958 (Friday), Pawsey attended a colloquium at the Berkeley astronomy department given by Oort, one of multiple meetings Pawsey and Oort[20] were to have in the US in 1958. After the colloquium, a dinner for the guests (Oort and Pawsey plus faculty) was held at Harold Weaver's house, the founder of radio astronomy Berkeley. Weaver told Pawsey that he had plans for a big aerial for HI research, in particular for the investigation of HI in stellar associations.[21]

On Monday 21 April 1958, Struve (chair of the department and soon to be the NRAO Director) met Pawsey in San Francisco for lunch with Oort.

At last, on 23 April 1958, Pawsey boarded the plane for Sydney.

The 8½ months could be looked upon as a success. Amid discussions, lectures, field trips, colloquia, dinner meetings and personal conversations, Pawsey sought to connect the "new" field of radio astronomy with optical astronomy, solar astronomy,

---

[20]They met four times in 1957–1958, Princeton, Pasadena, Owens Valley, California, and Berkeley.

[21]By the time of the IAU in Berkeley in 1961 (Chap. 38), the 85-ft telescope at Hat Creek, California, had been brought into operation.

cosmology, ionosphere research and radar technology. His inexhaustible work ethic, planning and organizational skills, and capacity for connecting with others made a favourable impression. His networking skills were evident. And in 1961, Rabi would offer the Directorship of NRAO to Pawsey (Chap. 38); the US radio astronomy contingent was, as expected, enthusiastic. The 1957–1958 trip was a success.

Chapter 29
Driving the GRT, 1957–1959

It is obvious that FFP are a group of very eminent engineers of high calibre, and with a
tremendous reputation behind them. It is also clear that they are a bunch of old men who are
tired, over-worked and operate almost by an intuitive process. They give no responsibility of
any kind to their young engineers.—Bowen to White, 21 November 1958

The late 1950s were chiefly characterised by immense frustration and very difficult
relations with the British designers Freeman, Fox and Partners (FFP). A crucial
change forced by Bowen and initially resisted by FFP, was to call for competitive
bids. Despite the pressure to select a British firm for the construction contract it was
awarded to a German firm, Maschinenfabrik Augsburg-Nürnberg (MAN). This
dramatically illustrates the extent to which an ex-British colony was now truly
independent and capable of making what certainly transpired to be an excellent
solution. The combination of the innovative UK FFP mechanical design, the efficient MAN German construction and the Australian radio systems engineering was a
great success.

Neither Pawsey nor his close colleague, engineer Harry Minnett, who had been
placed in London specifically to oversee developments and maintain liaison with
FFP, managed the relationship with FFP effectively. In the end it fell to Bowen to
conduct two "tailtwisting" operations (see Footnote 13, Chap. 29) to drive the
protracted process through its barriers and difficulties and achieve the actual construction of the telescope. Throughout this process Bowen's frustrations with
Pawsey grew, but despite this, it was Pawsey who developed the science case and
operational procedures.

More details can be found in NRAO ONLINE.43 (1957), 44 (1958) and 45 (1959). Also in
Robertson (1992) Chap. 6 (p. 150–158) for design and Chap. 7 (p. 159–186) for siting and
construction.

Supplementary Information The online version contains supplementary material available at
[https://doi.org/10.1007/978-3-031-07916-0_29].

© The Author(s) 2023
W. M. Goss et al., *Joe Pawsey and the Founding of Australian Radio Astronomy*,
Historical & Cultural Astronomy, https://doi.org/10.1007/978-3-031-07916-0_29

1957–1958 Events

The major event of 1957 was the arrival of the belated Freeman, Fox and Partners draft design study of the GRT. In late October the draft proposal was sent to Sydney from London. Later a "preliminary" proposal was sent, followed in December by the "final" design study.

Throughout the year, Bowen was increasingly frustrated by the delays. In addition, during the year he became increasingly irritated by Harry Minnett, the RPL man at Freeman, Fox and Partners in London. Bowen wrote Frank Kerr, who was at Leiden for an extended visit, on 19 November 1957. Bowen praised Minnett while complaining that he had not:

> kept us properly informed of the magnitude of the job . . . As long ago as last July [1957] he was confidently predicting we would have the report in two or at worst three weeks' time. If we had been given a more accurate picture of what was going on we would still have had complaints about Freeman, Fox and Partners, but we would have been a good deal less critical of Harry [Minnett] . . . [1]

Bowen's harshest criticism was directed at Pawsey, probably exacerbated by the direct criticism he was receiving from Christiansen and, to a lesser extent, from Mills. In the same letter, Bowen wrote:

> What is needed then is for Joe [Pawsey] to put some enthusiasm behind steerable dishes and to express his enthusiasm loud, long and often. Alternatively, if he doesn't do this, then for the Radio Astronomy group to do it for him. The final decision on these things is always made by the [CSIRO] Executive, and if there are any dissenting voices or the dead hand of unenthusiasm [sic] around, they are sure to spot it and act accordingly.

Also, during the course of 1957, Bowen had become increasingly apprehensive about the increased cost estimates for the GRT. He realised that a likely third contribution from the US foundations might be required. In correspondence with Warren Weaver (1894–1978, the Director of the Rockefeller Foundation's Natural Science Division) on 18 October 1957 wrote to Bowen ("Dear Taffy"). Recently R.G. Casey, the Foreign Minister of Australia and the Minister for the CSIRO, had visited the foundation's president Dean Rusk as the minister reported on the CSIRO's funding shortfalls for the GRT. The funding for the 250-ft dish was not likely, with the suggestion that the size might be reduced to 220 ft. Bowen received a modest encouragement from Weaver: "I am not sure that any additional contribution would turn out to be at all possible. But, we would certainly wish to have all the facts before us."[2]

---

[1] John Deane archive—see Primary Sources, in Introduction. The archive was discovered in the 1990s by Deane in the rubbish collection at RPL; given to Goss in April 2011.

[2] NAA C3830 A1/3/11/3, Part 4. Bowen replied on 30 October 1957: "I always feel like a new man when I talk to you . . . The value of a radio telescope goes up roughly with size . . . [Due to mounting costs, the size might be reduced to 200 ft or even less.] This would limit the scientific value of the project in a serious way and it would be very sad if we had to follow this course." Bowen was fearful that the telescope would have a diameter of less than 200 ft. In the end, the Rockefeller Foundation

The conflicts between CSIRO and Freeman, Fox and Partners did not end with the receipt of the early design study in November 1957. After discussing the report with the experts in Sydney, Bowen sent a preliminary letter with harsh questions to Minnett, just as Minnett was preparing to return to Sydney on 27 November 1957. This letter contributed to the "troubled waters" between the two organisations. Bowen complained about a number of loose ends in the proposal, e.g. "in calculating the deflections of the dish, how is allowance made for the strength of the spiral members?"

Those at Freeman, Fox and Partners were not pleased and a strong letter was sent on 27 November 1957 by Ralph Freeman to Bowen, who pointed out that the questions posed to Minnett were:

> ... [o]utside his province and it would not be fair to expect him to stand up to cross-examination on such structural matters ... Judging by the elementary nature of these comments, and the fact that the answers to most of the questions implied are in fact to be found in the report, I can only suggest that the people from whom [i.e. not Bowen himself] the questions emanated had not studied the report very carefully ...

These exchanges set the stage for continued conflict in 1958 and 1959 as the design study led to contracts for the construction of the GRT.

The year 1957 did end on a very positive note, as Pawsey recognised that the design had a nice surprise for the Australians. Pawsey was in the US and wrote to Bowen with his appraisal of the Freeman, Fox and Partners study: "I should say that I think Freeman Fox have done an exceedingly good job. This applies particularly to the design of the dish which I regard as the heart of the problem." There was an excellent alt-az design (Freeman, Fox and Partners had not produced an equivalent equatorial dish) and both Bowen and Pawsey were struck by the fact that the dish would work well at 10 cm, much more favourable than the desired 21 cm. Pawsey even pointed out that the new dish would work well at 3 cm, with an improved mesh quality. Bowen had pointed out that the proposed ME was an improvement: "the idea of mounting the ME near the axes of rotation and then making the dish follow the ME ... is a clear cut advance which might go a very long way towards making the alt-azimuth type of mount acceptable." This idea conceived by Wallis was a most important design innovation to the Master Equatorial system.

Finally, Pawsey pointed out the importance of a provision in the design to adjust the setting of the reflector in the future, leading to the process followed in subsequent decades to carry out infrequent panel adjustments as the GRT was used at higher frequencies in the coming decades, such as 1.3 cm. Pawsey wrote: "I envisage the proper procedure as an original setting which is checked at the time of construction and then improved at a subsequent date when the behaviour of the dish is known."

In early 1958, the RPL group in Sydney awaited Gilbert Roberts's arrival from FFP in London in mid-January. Pawsey was still in the US for some months before his return to Sydney in April 1958. Bowen wrote to Pawsey on 13 January 1958 with

---

granted the CSIRO US$100,000 in December 1959, added to the earlier grant of US$250,000 in December 1955.

a summary of the current thinking about the GRT. RPL would likely (1) choose the alt-azimuth design, (2) agree on a diameter of around 210 ft and (3) use a finer mesh on the dish to optimise the use at high frequencies. Also, the expectation was that the Rockefeller Foundation might contribute additional funds. Bowen suggested that CSIRO would wait on a request until the estimated costs were more certain.

Then Bowen approached two controversial issues. Where would the GRT be located? Who would be the day-to-day leader of the team that supervised the contracts for construction? Bowen continued with a contentious question:

> It is the easiest thing in the world for a project of this kind to bog down unless someone is pushing hard the whole time. I am certainly not the one who will do this if the GRT goes to Canberra. Arthur [Higgs] and McCready have both said that they are not going to do it and no one in the radio astronomy group has yet volunteered. **This leaves you [Pawsey]. Are you prepared to take it on?** [our emphasis]

No reply has been located in the archives.

Are You Prepared to Take It on? Contracting and Construction of the GRT

On 16 January 1958, Bowen and other members of the Technical Advisory Committee were joined by Higgs, Mills and Wild in extensive deliberations with Roberts. The equatorial mount was clearly off the table; FFP had only provided a cursory look at this design and only because of "pressure from CSIRO" which in turn was generated by the bad advice being received from US "experts".

A major obstacle arose with the issue of the arrangements for construction. A prime contractor was not feasible "since no single firm would be supplying more than 30 per cent ..." CSIRO was wary of this management uncertainty. These concerns would increase in the following 2 years.

The CSIRO personnel White and Bowen accompanied Roberts to a meeting with the Minister for the CSIRO, R.G. Casey, on 6 February 1958 in Melbourne. By the end of February 1958, White began a negotiation with FFP for the "actual construction and erection of the radio telescope". FFP were to be the "Consulting Engineers in the matter of consideration of the tenders received and the supervision of the actual construction". White expected that the contract could be let in 6–9 months. This estimate turned out to be vastly overly optimistic.

The site selection process was a multi-year exercise with major roles played by Bowen, Pawsey, McCready, George Day, Mills, Christiansen and Kerr.[3] The three sites discussed were Cowra[4]/Parkes, Cliffdale (near Sydney) and Hoskinstown (close to Canberra). Hoskinstown was later chosen as the site for the Molonglo Cross designed by Mills and colleagues from the University of Sydney). Bowen

---

[3] see NRAO ONLINE.44, 44.2 and 45.

[4] Cowra is 104 km to the SE of Parkes and due west from Sydney, 300 km.

made a point of not attending[5] the March 1958 meeting, not feeling it necessary. Pawsey was still in the US. Although numerous letters were exchanged in the period February 1958 to early March 1958 between Pawsey and Bowen, confusion arose over their views on the sites. This led to further correspondence from Bowen to White complaining about Pawsey's indecision. For example, on 5 March 1958 Bowen wrote to White, "[Pawsey] thought the radio astronomy group had decided on the Cliffvale site last August [1957]. Nobody here seems to know about this decision, but he still thinks it is a good spot." Bowen continued:

> My own views are quite simple. In a country like Australia there are almost an infinite number of sites suitable for a giant radio telescope which have the necessary flat ground around them and offer an electrical noise level which is lower than can be achieved in the UK, in Europe or the USA. We would be foolish to throw away this natural advantage and put the device in a noisy area. This means Parkes or somewhere west of it.

> If, for other reasons, we have to get closer to civilisation, the choice lies between the outskirts of Canberra and the outskirts of Sydney. In this event ... I am resoundingly in favour of the outskirts of Sydney. In other words, I regard Parkes as the ideal site for the radio telescope. If, for any reason, we cannot go to Parkes then Cliffvale near Camden is the spot.[6]

The final site selection of Parkes was made at a meeting in Melbourne on 17 March 1958. The RPL staff who attended were Wild, McCready, Christiansen, Mills and Kerr. The CSIRO headquarters staff in attendance were Guy Gresford (research secretary Physical Science), Clunies Ross (Chairman), White (CEO) and possibly Bastow (member of CSRIO Executive). Presentations were given by Mills (technical requirements of site), Wild (administrative aspects of the site), Kerr (procedures used in overseas countries such as the US for radio telescope site selection), McCready (physical descriptions of the three sites), Christiansen (advantages and disadvantages of the three sites) and Wild (summing up). The summary of the meeting provided a recommendation of Parkes as the first choice.

In 2018 and 2019, a number of interesting details of the exact location of the GRT on the Parkes site have come to light due to the efforts of John Sarkissian who joined the Parkes Observatory in 1996.[7] The reader may be amused by "A Tale of Three Peg-Events—Locating the Parkes Telescope, 1958–1959", a collaboration with John Sarkissian—see ESM 29.1, Three Peg Events. In this text we describe a very 'Aussie' character, Australia "Austie" James Helm, the original owner of the sheep farm near the town of Parkes which would become the site of the GRT. Helm held the stake in March 1958 as Bernie Mills marked the first proposed site for the GRT; later in 1958, the final site was chosen at the south end of the Helm

---

[5] Bowen wrote at the time: "With this show of talent, there is not much need for me to come, too." Later White tried to convince him to change his mind, to no avail.

[6] In fact, Pawsey had written that he would prefer Hoskinstown, but if Bowen were opposed, he would respect Bowen's views. Bowen had given White the opinion that Pawsey was indecisive about the site decision. Was Bowen trying to undermine White's confidence in Pawsey?

[7] The site selection efforts of 1953–1957 are described in the NRAO ONLINE.44.

Fig. 29.1 Bowen's and McCready's re-enactment of the GRT site pegging event. Lindsay McCready is on the left and "Taffy" Bowen drives in the peg. This re-enactment event occurred in a period between late September to early October 1959, some weeks after the official peg was driven in by Sid and Murray Nash. As discussed in ESM 29.1, Three Peg-Events Parkes Telescope, 1958–1959, the re-enactment did not take place at the location of the telescope. Credit: CSIRO Radio Astronomy Image Archive B6586

property. Neither of these sites were at the location of the "famous" re-enactment of Bowen hammering in the peg (Fig. 29.1).

The relations between CSIRO and FFP remained troubled. Robertson (1992, p. 155) provides a succinct account of the situation in 1958:

> In view of this disappointing response in 1956,[8] Freeman Fox [in 1958] felt there would be little to be gained by throwing the project open to a competitive tender and, instead, decided to divide the project into three contracts and make its own selection of the contractors it believed best suited to the task. Early in 1958 these firms were chosen: Metropolitan-Vickers (Manchester) to act as the main contractor; Grubb Parsons (Newcastle) to develop the master equatorial system; and Sir William Arrols [*sic*, Sir William Arrol a Scottish civil engineer and bridge builder, 1839 to 1913] and Partners (Glasgow) to fabricate the heavy structural components and to construct the telescope at a site halfway around the globe. Metrovick and Grubb Parsons had been closely involved in the design study during 1957, so both firms seemed a logical choice.

Given the complexity of these arrangements, conflicts arose between CSIRO and FFP and even more troubling between FFP and the three firms they had chosen as contractors. There were too many players. An original agreement with

---

[8] FFP had approached a number of British firms who might have been able to contribute to the construction of the GRT. These firms expressed little interest in a project "involving so many untried engineering features and, not least, one which promised little financial reward". (Robertson, 1992, p. 154).

Metropolitan-Vickers and FFP fell apart in early May 1958 when Metrovick announced that they would only assume responsibility for the project if its share was 70% of the total. Their share was much less, and the situation reached an impasse. Roberts of FFP expressed his frustration in a letter to Bowen on 27 May 1958: "... [T]he commercial people [of Metrovick] are not at all keen on taking on what appears to them a disproportionate amount of the responsibility, but I hope we can sort this out ... The biggest difficulty may be to get them to take responsibility for the site erection work ..."

Roberts then ended the letter with a sentence that certainly angered Bowen and was likely ignored by him: "If you can think of any other difficulties in the project please **do not bother** (our emphasis) to write about them, because I am quite sure they have already been raised by some party or other here."

A few months later, Pawsey was in the UK during his trips overseas in 1957–1958 (Chap. 28). He was in London in July 1958 and visited FFP and Harry Minnett on 9 July 1958. He wrote Bowen on 11 July with "a clear picture of the developments on the GRT" after spending 2½ days at FFP.[9] On some of the issues he was clearly sceptical of the value of the information provided by FFP.

Pawsey described the design work at FFP of the "half-dozen engineers working on our jobs". The top priority for re-design in mid-1958 was the drive system, gear boxes etc. The second priority was the dish structure, prompted by comments from the Sydney conference of January 1958. The major concern was the proposed Metrovick contract: "... [T]here is known to be considerable diversion of opinion within the company itself and Roberts expresses himself as quite optimistic about settling the contract in principle within the next weeks." (The delay was to be many months, extending into 1959.) Pawsey reported that the senior management of FFP had left "the GRT in Roberts's hands". Pawsey said they would have to "wait in patience for a while until we see how the Metrovick situation works out and the essentials of the new design are clear. Just at the moment one sees a lack of progress. This situation could change overnight." Based on later events, Pawsey's mild optimism appears somewhat naïve.

On 24 July 1958,[10] Minnett wrote Bowen with a disconcerting description of a revealing visit to Jodrell Bank with Pawsey for 2 days during the week of 14 July 1958. The performance of the 250-ft telescope was quite discouraging, with low gain at 20 cm and severe pointing problems. Large deformation of the dish surface had appeared with use. This would seriously compromise the operation of the telescope at high frequencies [1.4 GHz]. This must have caused significant apprehension given

[9]NAA C3830 A1/3/11/1 Part 11.

[10]NAA C3830 Z1/14/A, Part 1 Minnett wrote: "Fairly long circumferential ridges or steps about 2 inches high have appeared here and there in the welded steel surface and are said to be due to twisting of the supporting purlins." It was difficult to obtain quantitative information on the shape of the dish. At 20 cm the beam was only 15 arc min with 25% side lobes at 45 arc min from the beam centre. The gain at 20 cm was only equivalent to a 100-ft antenna due to the surface errors. The pointing errors were in the range 2–5 arc min. "On the whole the drive and control system seems to be working as well as the specification required and is probably capable of improvement."

the engineering difficulties they might also encounter, but it may have also provided real optimism that the GRT would have a higher frequency niche without competition from Jodrell Bank.

On 7 November 1958, Bowen, Pawsey, Wild, Christiansen, Kerr, McCready and Arthur Higgs met to plan a course of action based on all the bleak news from London. The delays, uncertain delivery time and costs were driving incentives for a new proposal from Bowen. Item number one was to communicate directly with FFP with an expression of dissatisfaction about the current status. Item 2 was a proposed visit of Bowen to FFP for a personal confrontation; Bowen was impatient with Minnett, even suspecting that he was becoming a "member of the FFP" team.[11] From the 7 November 1958 Minutes of the GRT meeting Bowen offered his rationale for the trip; his frustrations were leading to possible consequences for the relationship with FFP:

> Dr Bowen then pointed out that it has been almost three years since negotiations with FFP began, but not a single item of hardware has yet been obtained, and not a single contract for the supply of equipment has yet been entered into. He felt we should have little hesitation in cancelling the present arrangements with FFP—and there were no contractual or other reason why this could not be done—if we were unable to obtain satisfactory answers on cost and delivery dates in the very near future …

End 1958: Tailtwister I, Bowen in New York and London, 13 November–23 December 1958

In January 1959, Jack Roderick, Professor of Civil Engineering at the University of Sydney and advisor to CSIRO for the GRT, proposed that Bowen's two trips to the US and the UK be called **Tailtwister.** Here we refer to the journey at the end 1958 as **Tailtwister I**, and the trip in early 1959 as **Tailtwister II**.[12]

On arrival in London on 18 November 1958, Bowen went to the FFP offices. Roberts was away visiting clients for a few days but by 21 November Bowen wrote White with initial impressions. The disaster with the dish design was serious. Bowen, as well as the junior FFP engineers, did not believe the claimed date of completion of the GRT by January 1961. Bowen wrote:[13]

> Roberts nor his engineers are prepared to talk about the dish, and freely admit that they have hardly thought about it the last six months. This is one of the clearest deficiencies at FFP. They are a small outfit [for] the work which they try to do …

---

[11] NAA C3830 Z1/14/A, Part 1.

[12] On 16 January 1959, Roderick wrote to Bowen: "Many thanks for letting me see the various letters concerned in your operation 'tailtwister': I think you have done an excellent job in the circumstances. One can offer explanations for the way in which these consultants [FFP] go about their business, but the important thing as you point out is to get things done quickly despite these shortcomings."

[13] NAA C3830 A1/3/11/1 Part 11.

[T]he dates and time in which things might or might not get done are kept in Roberts's head, [if he is absent] nothing seems to be done about it. Harry Minnett has told us of these things in a guarded way, but it sticks out when contact is made at first hand.

Bowen, Roberts and Minnett toured the north of England and Scotland, visiting Metrovick in Manchester, Grubb Parson at Newcastle and Sir William Arrol at Glasgow. The worst news was the "thoroughly depressing picture" of a delivery time of the master equatorial of 2.5 years! The predicted completion date was July 1962, 1.5 years later than the date given to Bowen by Roberts a week earlier.

Bowen met the managing partner of FFP (R.E. Fordham) just before he left for New York the first week of December 1958. Bowen told Fordham that he would return in February 1959 when the contract proposals for the GRT were due. Bowen wrote White: "I trust the project will have made a good deal of progress by that time and that we shall be able to proceed to the structural stage without too many difficulties."[14]

Tailtwister II: 1959

The lessons learned in Tailtwister I were utilised in numerous conferences in January 1959 as CSIRO and RPL faced the problems with FFP and the firms in the UK. Recriminatory correspondence, including from Minister Casey, was exchanged. By early 1959, relations between CSIRO and FFP had reached their nadir. Bowen now embarked on his second intervention.

Bowen left Sydney for the US (7 February 1959, a brief visit) and the UK, arriving on 22 February 1959. He travelled to Jodrell Bank, hosted by Lovell and Hanbury Brown. They learned about the low aperture efficiency at 21 cm and were told again about the necessity of having a single prime contractor, a firm responsible for the entire project. For the Australians, this was essential since they were displaced from firms who would likely be at distances greater than 10,000 miles, with resulting communication problems.

After a chaotic series of events,[15] the Metrovick contract arrived on 6 March 1959 with a total cost of £A 750,000 compared to the £A 500,000 earlier predicted by

---

[14] *Ibid.*

[15] Bowen had been assured that the Metrovick contract would be available in mid-February 1959. The chaotic series of events with the receipt of this contract continued (see NRAO ONLINE.45 for much more detail). The heated controversy that accompanied the public announcement of the MAN contract 5–6 months later in July and August 1959 is also described in detail in NRAO ONLINE.45. Given all the controversy over the past several years, this blunder was certain to raise a storm of ill will. The initial CSIRO press preleases in Melbourne on 16 July 1959 (the end of Tailtwister II as Bowen arrived back in Sydney) about the earlier completed contract with MAN, AEI and Askania made no mention of Freeman Fox and Partners. Bowen's initial text, shown in person to FFP in London, concerning the key role played by FFP as the consulting engineers for the GRT, was omitted from the submitted press release 2 weeks later. FFP sent angry letters to White and Bowen who immediately abjectly apologised (on 6 and 10 August). The key text had been omitted by

FFP, and a time to completion of about 4 years, implying operation only in late 1963! At this point, Bowen went to Ralph Freeman of FFP and insisted on going for competitive bids and a prime contractor, especially from US firms. Freeman agreed with this plan: "He seemed to agree that we were wasting time on the present negotiations [with Metrovick]".

Bowen wrote to White on 10 March 1959:

> After initial resistance, during which I was told about the glories of the British Empire and the ties which bind the Commonwealth together [hence the insistence that the contractor be a British firm], R.E. Fordham [the senior partner] finally conceded that we were taking the only possible course . . . and that FFP would cooperate [with competitive bidding].

The tender ("contracting") process began in March and April 1959. Bowen had informed FFP that Metrovick could provide the servo control system of the GRT "if this proved to be the fastest and most economical procedure".[16] Bowen was also in frequent contact with Pawsey about the tender process for the GRT.

By 17 April 1959, the tender documents were distributed to a number of British, German and US firms. Bowen had made the contacts with the three German firms MAN (Maschinenfabrik Augsburg-Nürnberg), Demag (Deutsche Maschinenfabrik in Duisburg) and Krupp,[17] all prominent steel companies. All tenders were based on the Metrovick servo system. MAN made substantial savings of £A 20,000, and a much shorter delivery time, by using the German optics firm Askania (a firm that had constructed optical telescopes) for the Master Equatorial. (The US bids were substantially higher. For example, D.S. Kennedy (Inc., Antenna Equipment, Cohasset, Mass., USA, 1947–1963) was £A 1,980,000. The high labour costs in the US were a likely cause.) The total costs would be about £A 650,000 after the FFP fee was included. The funds at hand were only £A 540,000. Both Pawsey and Bowen independently thought the only solution was to choose MAN and try to obtain additional funds of £A 100,000–200,000.

Pawsey now provided a valuable contribution as he and Bowen began detailed planning[18] for the acceptance tests that were to be negotiated with the prime contractor in early June 1959. The pointing accuracy tests as specified by FFP were straight forward. The surface accuracy evaluation was complex: "The question of acceptance tests for surface accuracy is one of several subjects on which we failed to reach agreement with Roberts, and nothing went into the technical specifications."

CSIRO in Melbourne. Additional damage followed on 6 August 1959 with an article about the new project ("German Radio Telescope for Australia") which appeared in the London-published *New Scientist* with no mention of FFP, AEI or Askania. Gilbert Roberts wrote a stern reply to the journal, published in the letters section on 20 August 1959. He corrected a number of points in the previous report, including a clarification of the role played by Dr Barnes Wallis.

[16] NAA C3830 A1/3/11/1, Part 12.

[17] Krupp withdrew from the tendering process in 1959.

[18] NAA C3830 A1/3/11/1 Part 13, Bowen to Pawsey, 4 June 1959. In a letter of 5 June, Pawsey congratulated Bowen on the MAN contract. "The first battle of your long campaign is now nearly over and I imagine that you are breathing a sigh of relief. If the money question is not too difficult [more funds required from the US foundations] we shall look forward to seeing you home soon."

Bowen asked Pawsey and the Technical Advisory Committee for comments on specifications and methods of testing, especially the impact of wind on the geometrical figure of the aerial. On 12 and 18 June 1959, Pawsey reported on progress with the surface accuracy acceptance tests after discussions with Roderick and Puttock (CSIRO Metrology). In the second letter, Pawsey enclosed three documents, two written by himself ("Some Geometrical Factors" and "Acceptance Tests on Shape of Reflector") and "Notes on a Desirable Method for Determining the Shape of the Reflector" by Arthur Higgs. Pawsey wrote in "Acceptance Tests . . .": "The errors in the shape of the reflector surface will be specified in terms of the departure of points on the actual surface from the nominal paraboloid (measured perpendicular to the surface of the latter)." The idea was to check the shape of the dish by determining the positions of 100–400 designated target points on the surface. For the acceptance tests, the dish was to be measured during the initial setup procedure. Then the process was to be repeated a week later and then a third time after 6 months. The expectation was then that the process would be repeated at regular intervals in the future, each measurement would lead to a readjustment of the panels.

On his way back to Australia at the end of June 1959, Bowen made a stop in New York to meet Dr Robert Morison (a neurophysiologist), Director of Medical and Natural Sciences at the Rockefeller Foundation on 2 and 3 July. The prospects for a final funding from this foundation were excellent, a one-fifth share or US$100,000 (A£ 45,000). On 8 December 1959, White and Casey were informed that a grant of US$107,000 had been awarded. During December, the Australian Government (Harold Holt, Treasurer) then made up the difference with an additional A£ 150,000.[19]

As Bowen returned to Australia on 16 July 1959, his gruelling 5-month trip to the US and Europe was at an end. The remarkable success was evident; the prospects for an economical antenna that would work at high frequencies were excellent.

As Bowen reflected on his success back in Sydney, he wrote Sir Henry Tizard in London feeling the need to explain to his staunch supporter the events of the last 6 months:[20]

---

[19]Robertson (1992, p. 169) had pointed out: "Although the shortfall anticipated in mid-1959 eventually blew out by a further £A 100,000 (giving a final total of £A 900,000 [US$1.8 million]), the excess was covered by CSIRO's capital works budget, without the need to again seek out sources of external funding."

[20]NAA C3830 Z1/14/A Part 2. 20 July 1959. Bowen pointed out in the attachment that Roberts had told the CSIRO in January 1958 in Sydney that: "We were unlikely to find a prime contractor to handle the job and would have to break it up into a number of sub-contracts. He advised against calling for competitive bids and gave reasons why FFP should negotiate contracts with a number of firms of their choice . . . In August and September 1958, it became clear that things were progressing badly in London, but the reasons were not apparent in Sydney. [On arrival in London in late 1958], I found a very muddled situation. The designs were not complete. There was disagreement with contractors on technical matters . . . and both the cost of the telescope and the completion date in Australia looked considerably worse than we were being led to believe . . . [The merits of competition soon became clear to CSIRO] and in March we decided to put the telescope up for

As discussed at the Athenaeum [Club, Pall Mall], I am sending you a summarised statement of the steps leading up to the award of the radio telescope contract to a German company. We tried hard to have it built by a British firm, but obtained a very disappointing response. What does not appear from the cold figures in the attachment [a chronological list of the tortuous events of 1957–58 with FFP] is the almost complete lack of interest we found among British firms. The Germans and Americans were wide awake, keen and anxious to do business, but this was conspicuously lacking among the British firms.

Tailtwister II was a major success for Bowen and the CSIRO. Bowen's achievements in this short period in late 1958 and up to mid-1959 represented one of the highlights of his career. His goal had clearly been to create a large radio telescope based more on his personal vision for the future than on any detailed science case. Certainly, the success of the GRT is based on his activities in coordinating the experience of the scientists and engineers at RPL in the post-war era. Bowen then moulded this environment with the capabilities and innovations that FFP represented, in spite of their inefficient and often counterproductive modus operandi. In the end, Bowen shaped an outstanding instrument, as the GRT came to fruition in 1959–1961. The success of this instrument after 60 years remains a remarkable achievement.

Scientific Plans: Pawsey's Views on Future Research Programs

After the MAN contract appeared likely in June–July 1959, a number of meetings were held at RPL in Sydney with discussions about planning for the GRT, expected to be operational in 1961–1962. Three aspects of the planning were coordinated by Pawsey in 1959: (1) receivers and backends which would be required, (2) the scientific programmes that would be carried out with the new GRT and (3) the future of low-frequency radio astronomy, both "cosmic" (Mills) and "solar" (Wild). The continuation of the Mills research would become controversial in 1959, a factor in the schisms of 1960–1962 at RPL as Christiansen, Mills and finally Pawsey were to leave CSIRO (Chap. 30).

Meetings were held on 3 March and 8 June 1959 at which staff replacement plans were discussed as well as plans for simple digital recording and data reduction schemes, under the leadership of Maston Beard (1917 or 1918–2000).

The meeting of 31 July 1959 dealt with operational requirements for the GRT site such as roads, buildings, receivers, computing and radio checks on the surface. Pawsey was not present. Bowen also presented his operational plans.

The major discussions organised by Pawsey occurred on 6 July, 11 August, 6 September and 13 November 1959. The meetings in August and September

---

competitive tender." CSIRO made a careful study of the eight US, three British and two German tenders. "We have no hesitation in awarding the contract to the lowest bidder, MAN."

were preceded by extensive written documents; detailed minutes were also produced.

The 6 July 1959 meeting was devoted to a discussion of Pawsey's own vision of the future of radio astronomy at RPL. The document "Radio Astronomy—Projected Programme"[21] with 14 pages and appendices was distributed earlier on 29 June 1959. The introduction set the scene:[22]

> The impending construction of the GRT implies a considerable re-orientation of the Laboratory programme and it is desirable to set down in general terms a plan for experiment and observations for the period ending a year or so after construction ... The plan involves the completion of current observations, the development of equipment and techniques for use with, and early observations on, the GRT and a decision as to which of our current lines of investigation, if any, should be pursued independently of the GRT.

Pawsey discussed the "world leading" status of the Australians in solar radio astronomy and in the radio source surveys in mid-1959. Pawsey's own Appendix 4, "Early GRT Experiments and Required Equipment", showed his insights for the future. His key experiments included HI line observations in the galaxy and especially extragalactic systems. He anticipated HI studies at high redshift. He foresaw a 10 cm all-sky survey carried out with the GRT, "complementary to that proposed by Mills with the super Mills cross." Pawsey's most significant suggestion was "a search for magnetic fields involving either circular (Zeeman effect) or linear (synchrotron mechanism) polarisation." His concluding suggestion concerned high angular resolution interferometry. "It seems that high resolution studies should not be taken on at Parkes until some time after erection. They form a natural second phase and involve a large constructional programme."

Pawsey, to Bowen's chagrin, also considered the future of the "continuing non-GRT projects":

1. Wild's radio studies of the outer atmosphere of the sun using a metre-wavelength spectroheliography [to become the circular Radio Heliograph at 80 MHz at Culgoora from 1967]
2. Mills's studies of very distant radio sources using a super Mills cross (continuum at a wavelength of about 1 m)

In the 29 June document, Pawsey insisted in continuing the support for Mills:

> ... [T]he field of study—galaxies at extreme distances—is perhaps **the most intriguing problem of all astronomy, our [Australian] position is currently right in the front rank** [our emphasis] and Mills is keen to continue at longer wavelengths as well as at the short wavelengths for which the GRT is suited.[23] A cross can be constructed at a wavelength of about a metre [300 MHz] which, for the restricted purpose of a survey at one wavelength

---

[21] NAA A1/1/7, also NAA A1/3/11/2 and Joe and Lenore Pawsey Family Collection.

[22] A more detailed description of the science programs can be found in NRAO ONLINE.45.

[23] As we will see in Chaps. 38 and 40, Pawsey was convinced that Mills should play a major role in the evolution of Australian radio astronomy after 1961 and the opening of the GRT.

should be both more sensitive and more directive than the GRT at its optimum wavelength. Observations with this cross would be exceedingly valuable in their own right; in combination with the GRT at short wavelengths we should have spectral information. This last aspect is likely to be important, since the main emission mechanisms differ in the two wavelength ranges: [mainly non-thermal synchrotron at 1 metre and partly thermal sources (HII regions) at 10 and 20 cm.].[24]

Mills wrote an appendix to the 29 June 1959 document with details. The Super-Cross was to be constructed in two stages, a 250 MHz instrument. The first stage was to be a cross with a total arm length of 2000 ft, later extended to 4000 ft. This would provide a resolution of 6 arc min. Mills asserted in the 1959 "Radio Astronomy-Projected Programme": "This programme, instituted very soon, would keep us in the forefront of work in the extragalactic field."[25]

In early September 1959, each member of the radio astronomy group was asked to write a proposal on their favoured GRT research topic. These were compiled in mid-September into a single document for a discussion with Pawsey on 16 September 1959. Pawsey had received seven proposals for GRT science[26] from Kerr, Mills, McGee, Shain, Piddington, Christiansen plus Pawsey, and Pawsey (alone). (Shain died at age 38 within 5 months on 11 February 1960.)

Frank Kerr was convinced that "one of the fields where most benefit can be obtained from the GRT will be 21 cm [HI line] studies of external galaxies ... The 12 arc min of the GRT at 21 cm would allow a determination of the resolved velocity field of the galaxy." Kerr even hoped that HI spiral structure would be observed in a few nearby galaxies. "The GRT will provide sufficient resolution for exploring the detailed structure of the Magellanic Clouds." Clearly the Large and the Small Clouds of Magellan would be important targets for the GRT. This research area was to lead to major GRT results in the next decades, observing both HI and continuum emission.

Kerr also discussed higher redshifted HI absorption line observations of distant galaxies, even though the claimed detection at the Naval Research Laboratory by Lilley and McClain (1956) of a line in the direction of Cygnus A had not been confirmed. Additional observations were certainly required of other high redshift systems since Cygnus A was not observable with the GRT. Also the attempt to detect intergalactic HI absorption was suggested; this experiment was tried by Brian Robinson, van Damme and Jim Koehler in 1963 in the early days of the Parkes telescope. However, the claimed detection of the HI absorption dip due to the

---

[24] A first draft of the "Radio Astronomy-Projected Programme" has been found with notes written in the margin in Pawsey's handwriting. See NRAO ONLINE.45.

[25] The final Super-Cross—the Molonglo Cross—was opened in 1967 after Mills moved to the University of Sydney (Chap. 38). This instrument was 5250 ft in extent (1.6 km) with a resolution of 2.8 arc min at 408 MHz.

[26] Also, three proposals for non-GRT science from Piddington (measure hydrogen in the interplanetary space using rockets), Shain I (measure low frequency radio emission from rockets) and Shain II (Jupiter decametric radio emission).

intergalactic HI in the direction of 3C 273 by Robinson et al. (1963) was never confirmed.

Based on Frank Kerr's own scientific interests, numerous galactic HI observations were also suggested: HI studies of the Milky Way central region (less than 3 kiloparsecs from the Milky Way centre), spiral arm structure, statistical studies of the HI emission distributions in selected regions of the galactic plane, and especially HI absorption studies against discrete sources. Statistics of the gas temperature, size and number of HI clouds and "new information about the distances of sources" were also topics for study with the high resolution of the GRT (e.g. Radhakrishnan & Goss, 1972). Kerr also suggested new spectral lines such as CH, OH, He^{3+}. The GRT could have discovered the OH lines near 1665 MHz (the frequency given in Kerr's table) in 1962; OH absorption lines were in fact discovered by Weinreb et al. (1963) and the strong, unexpected OH maser lines by Weaver et al. (1965). Kerr was aware of the fact that a major limitation for the detection of weak absorption lines was gain stability of the receiver, rather than sensitivity (see NRAO ONLINE.45).

Dick McGee wrote an extensive proposal with emphasis on high resolution studies of galactic HI with the GRT. Examples were the search for observational evidence for the boundary between HI and HII regions, and the determination of the cloud size spectrum of HI clouds in the interstellar medium of the Milky Way. McGee also presented an excellent proposal for studies of HI remote from the plane of the Milky Way: "a suggestion is that this work may lead to a detection of either the limits of HI gas in our Galaxy or of 'bridges' of intergalactic hydrogen." This observation was clearly related to the soon-to-be-detected high velocity HI clouds in the Milky Way by Dutch radio astronomers in late 1963 with the new Dwingeloo 25 m radio telescope.

Shain proposed a number of innovative high resolution observations of HII regions at 10 cm, as well as the galactic centre, ideally suited for Australian observations at a declination of -29°. Shain also realised the importance of 30 Dor, a prominent, massive HII region (in the nearby galaxy, the Large Magellanic Cloud) as a radio source, confirmed later by many investigators. Also, in a remarkable view of the future, he suggested that lunar occultations could be used to provide arc sec resolution. In Chap. 32, we describe how Cyril Hazard and colleagues identified the first quasar 3C 273 based on early GRT observations of a lunar occultation in 1962, published in 1963.

Pawsey provided a proposal (similar to his Appendix 3 in "Radio Astronomy-Projected Programme" of 29 June 1959) to continue the search for interstellar magnetic fields based on linear polarisation observations. Both the extended galactic non-thermal background (200–600 MHz) and compact galactic and extragalactic discrete sources were to be observed at 1.4–10 GHz. Parkes was to play a major role in the study of magnetic fields and Faraday rotation in the coming decades as Ron Bracewell, Brian Cooper, Marc Price, John Whiteoak and Frank Gardner pioneered these investigations in the early years of the Parkes telescope after 1962.[27]

---

[27] See Chap. 32.

A key meeting followed on 13 November 1959, with concrete plans for GRT receivers discussed earlier. Pawsey was chair of the meeting as priorities were established. Again, he emphasised that the list of instrumental projects from August 1959 provided special emphasis to the use of the GRT at 10 cm. Pawsey suggested that Christiansen supervise three projects: (1) the multi-channel spectral line backend for the GRT with John Murray as section leader, (2) the high resolution project at Fleurs (60-ft compound interferometer with the original Chris Cross) and (3) continuum observations with the Chris Cross. Frank Gardner was in charge of the maser front-end receiver project for the GRT with the assistance of Doug Milne.

Chapter 30
Schism at Radiophysics, 1960

> *My [Pawsey's] summary of the position is this: Mills's past work has been outstanding; his contribution has probably been the greatest single factor in giving Australia the high prestige it now enjoys ... [T]he simultaneous development of the metre wavelength studies proposed by Mills and the decimetre wavelength studies which will be undertaken at Parkes would provide Australian radio astronomy with complementary facilities in radio astronomy which no other country possesses.*
> *–Pawsey to Oliphant at ANU, late February 1960. Joe and Lenore Pawsey Collection.*

The year 1960 was pivotal for developments at RPL; the GRT construction was underway in the European factories and at Parkes in Australia. CSIRO made a decision to start the Paul Wild Radioheliograph project after the GRT was completed and the Super-Cross project of Mills was put on hold. Both Christiansen and Mills left for new positions at the University of Sydney in mid-year and Bowen and White also began the process that led to John Bolton's return to CSIRO in early 1961.

This new diversity led to an end of the CSIRO monopoly in radio astronomy in Australia. It reinvigorated the field as a new generation of young scientists were nurtured at the University of Sydney and the Australian National University in Canberra. Australian radio astronomy would flourish during the remainder of the twentieth century, but not in the directions that were envisaged in 1960.

After Christiansen and Mills departed from CSIRO, Pawsey's close knit and successful team had come to an end, leading to a highly stressed atmosphere at RPL. Pawsey, as a man of his time, was wrestling with a very human problem. The leadership style that brought him such success also had limitations when the context changed. His orientation to small autonomous teams who developed and used specialised instruments now became a negative constraint rather than an asset, in

---

More details for all the events described in this chapter can be found in NRAO ONLINE.46

Supplementary Information The online version contains supplementary material available at [https://doi.org/10.1007/978-3-031-07916-0_30].

W. M. Goss et al., *Joe Pawsey and the Founding of Australian Radio Astronomy*, Historical & Cultural Astronomy, https://doi.org/10.1007/978-3-031-07916-0_30

an era moving towards "bigger science" and larger, more general purpose, research facilities. The first phase of the successful era of post-war radio astronomy in Australia was coming to an end.

Delays to the Proposed Mills "Super-Cross"

In August 1959, the CSIRO executive agreed with the proposal to start Paul Wild's Radioheliograph project after the GRT was to be completed in 1961–1962. This meant postponing funding the instrument that Bernie Mills had proposed, the "Super-Cross". As a result, Mills began to look elsewhere for employment and funding. In early September 1959, Mills wrote his colleague Merle Tuve at the Carnegie Institution of Washington. He summarised the current status at RPL.[1] On 24 September 1959, Tuve responded ("Dear Bernie"):

> I have been somewhat unhappy to read your letter [of early September 1959] with the indication that you feel the overwhelming emphasis in the RPL division for the next few years will be on the 210-foot dish, to the neglect of important possibilities with large antenna arrays ... I think your work with large antenna arrays has been and is an outstanding example of accomplishments with modest investments. The interferometer technique with crossed aerials is not entirely beyond pitfalls ...[2]

A number of developments then quickly occurred. On 18 September 1959, David Heeschen, the Head of the Astronomy Department of the newly founded National Radio Astronomy Observatory at Green Bank, West Virginia, USA, offered Mills a one-year visiting appointment (on a leave of absence from CSIRO) to design a Super-Cross at Green Bank. Campbell Wade had just returned from a postdoctoral position at RPL (from 1957 to 1959). He knew Mills well, possibly suggesting that Heeschen contact Mills.[3] A few weeks later, a related proposal (7 October 1959) was made by Bowen (with the concurrence of Pawsey) to Lloyd Berkner (President of AUI, NRAO's parent organisation) that a joint AUI, RPL project might "co-operate

---

[1] Mills archive University of Sydney. P154-Series 8 File 2. Tuve's response of 24 September 1959, Mills to Tuve 13 October 1959, Mills to Oliphant 25 January 1960. The original Mills letter from early September 1959 has not been located. Also, additional correspondence from Heeschen and Wade at NRAO, Green Bank, Mills archive University of Sydney.

[2] Tuve was well known to be anti-"big science", see Kellermann et al. (2020) p 111–112. Thus, Tuve applauded Mills's accomplishment with small innovative investments, but was still apprehensive about some aspects of the cross, such as higher sidelobes.

[3] Mills archive University of Sydney P 154-Series 8 File 2. Heeschen thought that Mills might stay longer than one year. A handwritten postscript read: "[I hope] ... we can ultimately induce you to stay for considerably longer than a year". Mills considered the offer and turned it down in the course of early 1960 when the negotiations with the University of Sydney began. Wade wrote to Mills: "On the whole, I am very favorably impressed with Green Bank, and I'm glad to be here. Will you join in, sir?" In a turn of events, Mills offered Wade a position at the University of Sydney a year and a half later, 1961; see Chap. 32.

to build [Mills] crosses in both the North and South,"[4] similar to what had once been proposed for the US-Australia GRT collaboration.

On 13 October 1959, Mills wrote Tuve, discussing the rationale leading to the Super-Cross proposal:

> [The Super-Cross] did not originate purely from a chance to build a bigger and better instrument, but because the astronomical observations which, in my opinion, are crying out to be made, require such an instrument. I know of no other way of making them with comparatively limited funds. The scientific reasons for establishing a "Super-Cross" in Australia to work in conjunction with the 210-foot dish are so compelling that it makes me rather sad to see the large sums which are to be lavished on Northern Hemisphere instruments (at Leiden, Bologna and now perhaps at Green Bank).

On 25 January 1960, Mills wrote a frank letter to Oliphant (Director of School of Physical Sciences, ANU) about his version of the Super-Cross/GRT conflict.[5] Mills was aware that Pawsey and Oliphant had been in contact. Mills expressed his frustration:

> Now that RPL is getting the 210-foot dish it has become quite clear that this will absorb the major part of an unincreasable vote for [cosmic] radio astronomy and most of the remainder has been allocated for solar work. The result is that the possibility has been removed of significant increases in galactic and extragalactic problems at the lower frequencies, where the size of the dish is inadequate. This will in, my estimate, lead to a serious unbalance in the research programme and undermine the good position which Australian Radioastronomy has built up over the years. For this reason, I was particularly interested in the Melbourne Chair[6] as a possible means of filling the gap and further, in giving some stimulus to the subject which would help to overcome the **atmosphere of complacency and scientific stagnation** (our emphasis) which some of us detect here [RPL] in increasing amounts ...

There is little evidence of this "complacency and scientific stagnation" and we can now re-examine the arguments made for building a Super-Cross in relation to what actually happened in the following decade.

Both the GRT and what was effectively the Super-Cross (the Molongolo Radio Observatory, MRO[7]) made significant impact in astronomy but not in the ways that were being discussed in 1960. The low frequency surveys using the MRO made relatively little impact. However, surveys with the MRO to find new pulsars, which were still to be discovered at the time of these discussions, had a major impact. At Parkes the unexpected discovery of a quasar through a lunar occultation in 1962 (Chap. 32) not only opened a new field of research but showed that the radio

---

[4]Letter from Mills to Tuve 13 October. Robertson (1992) has described these events, including an instructive interview with Mills in 1984. The Bowen initiative did not lead to further action. Mills archive University of Sydney P 154-Series 8 File 2.

[5]University of Sydney Archive, P154-Series 8.

[6]University of Sydney Archive, letter from Oliphant to Mills, P154-Series 8, 9 October 1959. "I do believe that as head of the physics department in Melbourne you could develop your own work and play a very important part in the future of academic physics in this country. You can be sure I would press very strongly for adequate facilities for your research ..." Oliphant also suggested that it would be possible to hire a number of additional colleagues at Melbourne.

[7]Built later in the 60s by Mills at the University of Sydney with US funding.

emission immediately surrounding the black hole was stronger at higher frequencies and not weaker as had been assumed. We can now see that the arguments for the Super-Cross that were being made by both Mills and Pawsey, and which seemed justified at the time, were largely irrelevant and were being driven more by the personalities and relationships than by the science. Likewise, the GRT was driven more by Bowen's ambition than by any scientific rationale. On the longer term, the flexibility of the single dishes made them more effective.

On 10 February 1960, Bowen felt the need to explain the situation to White[8]: "I am afraid there are quite a few mis-statements and misconceptions going around, some of them in the Laboratory and it is just as well to get them straight."

Bowen claimed that he was "a strong supporter" of the Super-Cross. This new major instrument would not fit into the current budget of the RPL. "We have most definitely not made a decision against it. This, unfortunately, does not satisfy Mills, who is burning to go ahead and is exploring alternative ways of doing so." Bowen continued:

> [My] comment is that **we have the world's best radio astronomers, but they are the world's worst estimators.** [our emphasis] All told, the Chris Cross cost between £A 50,000 and £A 100,000, not the figure [sic] quoted [£A 20,000, attributed to Pawsey].[9] There is a tendency [at RPL] to think of a Mills Cross as something which is built in the background by two men and a few boys. The simple facts are that we have at the RPL perhaps the most cohesive group of radio astronomers in existence, with a budget which is larger than most. It was a major constructional effort on the part of this group to build the first Mills Cross and the Chris Cross over a period of four or five years ... The Super-Cross is a much larger project than any of these. If it were built at RPL, we would need to augment our present resources.

Bowen's comments had captured the essence of practical management issues which arose when scientists engaged in basic research were too protected from practical reality. Bowen went on to make a final point that was highly relevant: "There has not been a decision in this Laboratory to go ahead with the GRT in preference to a Super-Cross. The GRT was a going concern long before the Super-Cross came into the picture." The question for the group was the choice of a new large radio astronomy project, Wild's new imaging solar instrument or a Super-Cross. Only one could be built and the radio astronomy group chose the solar instrument, not the Super-Cross.

In January 1960, Ron Bracewell turned down an offer of the Chair of Electrical Engineering at Sydney University as he had just been offered a very attractive position at Stanford with the opportunity to lead a new Institute for Radio Astronomy. He told Mills, "it seems very likely that the Chair [of Electrical Engineering] will now be offered to you, but under difficult conditions for radio astronomy." In this letter, and in a second written on 16 February 1960, Bracewell offered Mills a position at Stanford in his new Institute of Radio Astronomy, with some expectation that the US Air Force funds might support the construction of a Super-Cross. But

---

[8]NAA C3830 Z1/7/B/2 Part 1.

[9]Bowen also claimed the astronomers at RPL habitually underestimated the manpower requirements.

Mills never followed up on the offer from Stanford.[10] Bracewell was to remain at Stanford in the US until his death in 2007.

Mills applied for a number of other positions in Australia (e.g. Adelaide and Melbourne), but his demands were too high and none of these possible appointments worked out.

Proposals to the Australian National University (ANU)

As will be recalled, Pawsey had long considered a closer relationship between ANU (where Mark Oliphant remained Director of the Research School of Physical Sciences and Engineering) desirable for RPL. Pawsey and Mills explored this as a possible means to resolve Mills's future at the end of 1959 and the beginning of 1960. Mills told Oliphant (25 January 1960) that: "Pawsey has ... suggested the possibility of ANU entering the field and tells me [Mills] that you have shown some interest in the idea." Mills also mentioned a "long-shot" proposal that had been discussed at RPL: "Another possibility which has been canvassed here, and which both Christiansen and I support, is the removal of the whole of Radioastronomy group from CSIRO to ANU." Mills thought "this may be too large a step for practical politics. Pawsey [also] thinks so and therefore has done nothing about it, but I know he would be in favour if the suggestion came from outside."

At this stage Pawsey only saw ANU (home of the Mt. Stromlo Observatory) as an alternate way forward. He did not trust Harry Messel (1922–2015) at the University of Sydney but had excellent relations with Bok and other astronomers at Mt. Stromlo. Pawsey wrote Oliphant in late February 1960.[11] He knew already that Oliphant was not enthusiastic: "Since [my last visit to Canberra], Bowen has spoken with you [Oliphant] and I subsequently gathered from him that ANU would

---

[10]University of Sydney Archive, letter from Bracewell to Mills, P154-Series 8.16 January 1960. Also Bracewell to Pawsey, NAA C3830 Z3/1/Part 10.16 January 1960. Bracewell wrote, "I do not know what to think about your working for Messel. However, I am sure that in the US you could not do better than here, taking into account ... supporting facilities of all kinds, and proximity to the California astronomers."

[11]An incomplete undated draft letter from Pawsey to Oliphant was found in the Joe and Lenore Pawsey Collection in 2010. Pawsey has written on the draft: "incomplete". Later in 2015, Hastings Pawsey found a ten-page undated document in Pawsey's handwriting: "Desirable Lines for Research in Radio Astronomy in Australia" (see ESM_30.1.pdf, Desirable Lines of Research in Radio Astronomy in Australia). Likely the latter document was an attachment to the earlier letter to Oliphant. (We are indebted to Harry Wendt for this suggestion.) In 2010, Mills wrote Goss (24 March 2010) with his response after reading the undated Pawsey letter to Oliphant: "Wow! This letter certainly stirred up some memories and explained why [Oliphant] had suggested me for the vacant Chair of Physics at Melbourne. However, I could not persuade them to support such a large project. The letter [Pawsey to Oliphant] would have been written in [late] 1959 ... or early 1960." Since the letter was written after Shain's death (11 February 1960), the likely date is late February 1960.

not be a likely starter in this matter [of taking over a new Super-Cross project]." But he was still attempting to persuade both Oliphant and Bowen of the desirability of a partnership between ANU and RPL that might solve the problem of how to fund the Cross. "However I think I should give my assessment of the situation on the chance that something useful might turn up." At this stage in discussing the reality of a new Super-Cross, Pawsey would have recognised that building a new instrument with Australian funds was unlikely. He appeared to be "grasping at straws". But within a few years, Messel (another Bowen style entrepreneur) had found the path to new funding with National Science Foundation (US) funds.

Here we summarise the substance of Pawsey's letter to Oliphant (late February 1960)[12] and in ESM 30.1, Desirable Lines, we summarise the accompanying document. Pawsey started with reference to his previous discussion about the future of Mills's low frequency research and possibilities of a University position.

> I myself would consider very favourably the possibility of the RPL supporting a joint project with another Australian institution, on the understanding that our contribution should begin only when the job of setting up the Parkes telescope is over, say in two years' time. Mills's proposal is to build a "Mills Cross", working at a wavelength of about a metre, which shall have considerably higher sensitivity [several times that of Parkes] and resolution [about 4 arc min, compared to 8 arc min for the Parkes dish at 10 cm] than any other instrument we have or expect to have . . .

Pawsey then described the scientific potential of a "Mills Cross", and a competing project, the more ambitious Benelux Cross. He acknowledged the funding difficulties: "These costs are very large by any standards, particularly so when we consider the other large investments in radio astronomy [in Australia]. I find it difficult to assess whether it is proper for Australia to invest so much in a single branch of science."

And Pawsey concluded with high praise for Mills who he sees as his successor, the future leader of Australian radio astronomy (see the epigraph Chap. 30).

Also, at the end of February 1960, Bowen and Oliphant had an exchange of letters. Bowen summarised the history of the Super-Cross in the context of the RPL programme of radio astronomy (24 February 1960): "I am afraid one or two aspects of our radio astronomy programme are being discussed in somewhat exaggerated terms and it is just as well to get them straight." Bowen pointed out to Oliphant that the policy in the past had been to execute one major project at a time. "The choice lay between a new solar spectrometer [sic] which Paul Wild is thinking about and a Mills Super-Cross. The decision of the Radio Astronomy group was to go for the Wild spectrometer. This unfortunately did not satisfy Mills, who is a dedicated man and wants to see a Super-Cross built at all costs . . ." His letter then repeats, with almost identical wording, the arguments included in his earlier letter (10 Feb) to White.[13]

---

[12] more details can be found in NRAO ONLINE.46.
[13] NAA C3830 Z1/7/B/2 Part 1.

In the Joe and Lenore Pawsey Collection copy of this letter from Bowen to Oliphant, three detailed comments are hand-written in the margins by Pawsey. In addition, an adjacent entry in the archive is a three-page handwritten letter [with no date, likely late February to early March 1960] to Bowen from Pawsey that began:

> There is a serious error in your letter to Mark Oliphant of February 24 [1960] with regard to the major features of a proposed Cross. [Bowen had asserted that the "sensitivity of a Cross is down by a factor of 10 or more on the GRT"...] The essential feature of the design of a Cross is that the sensitivity can be balanced to the resolution. [Pawsey provided a calculation showing that the sensitivity of a 6000 ft by 50 ft Cross would be a factor of at least three higher at a frequency of 300 MHz compared to the GRT.] The higher resolution of a Cross relative to a big dish demands a correspondingly higher sensitivity and it can have it ... I suggest you contact Oliphant and tell him of this considerable difference of opinion.[14]

Reconciliation of these noise estimates without knowing more about the assumed specifications remains difficult. In hindsight we can see that it was not the sensitivity but the versatility that made the biggest difference; a single dish was the superior performer. However, none of the proponents raised this point at the time. In discussions with one of the authors a decade later, John Bolton asserted that he was well aware of the advantage of flexibility of a system with few elements, thus his rationale for supporting the big dish.

The next day, 25 February 1960, Bowen replied again to Oliphant at ANU, continuing the discussion of their meeting the previous week, answering Oliphant's questions concerning the GRT: (1) completion plans, (2) scientific use and (3) staffing plans. "I am afraid there are some loose statements going around, many of them, I am sorry to say, having their origin in the Laboratory. The plain facts are that, as one would suspect with a project of this magnitude, there are some carefully laid plans for the telescope at Parkes." The construction was in good shape, for the first time ever at RPL, construction was being done by outside firms, and the time scale was ahead of schedule (in the end with delays of only 7 months). Three people were working on receivers overseas: Brian Robinson at Leiden, Alec Little at Stanford and Brian Cooper at Harvard. Bowen wrote that these staff would return to the lab in 1960 with high expectations. Pawsey had proposed a research programme for the GRT. (One half of the research staff was to be associated with the GRT, the remainder with solar work). Jim Roberts and Don Mathewson were overseas, at Caltech and Jodrell Bank, respectively. They "were burning to get to the 210-footer" and had many ideas for the use of the GRT. "Roberts is one of our best up-and-coming radio astronomers and already has first-rate work to his credit in this Laboratory." Bowen reported that Mathewson was

> one of our brightest hopes, at least as promising as John Bolton was at the equivalent stage in his development ... We have one or two people on the staff who are not keen ... anxious to go along their own lines. [Presumably Christiansen and Mills]. It is a pity that in making the very good case which exists for the construction of the Super-Cross it is accompanied by scare stories about the rest of our Radio Astronomy programme.

---

[14]Likely Bowen never followed Pawsey's suggestion to contact Oliphant with a "correction" of this difference of opinion.

On 29 February 1960, Oliphant replied to Bowen:

> It is very useful to have these clear statements of your policy and I will certainly use them in the future. I have not heard anything about the preparations for staffing at Parkes, loose or otherwise, but such may come my way later. I am sorry that you are facing these problems but such difficulties are bound to arise when a highly successful project breeds a set of **prima donnas** [our emphasis] whose ideas of what should be done next come into conflict. There is the point, too, that men who have been given fine facilities and protected from multitudinous tasks of administration etc., are apt to take such chores for granted and become very unrealistic in estimating costs and effort required.

Oliphant seems to have been well aligned with Bowen on this issue; he copied this letter on to Bart Bok so he could have an "authoritative statement about the position. We appreciate Mills's position, but the problem seems to be an internal one for you [Bowen] and your boys [at CSIRO] to work out between you."[15]

The implication from this letter is that Oliphant realised that bringing Mills to ANU would lead to controversy, given the issues and personalities involved. Oliphant did not follow-up any further on Pawsey's suggestion for a radio astronomy department at ANU; instead Oliphant handed the matter over to Bart Bok. Bok wrote Mills a personal letter:

> It is obviously not possible for Oliphant to make a positive move ... In order to have some chance of succeeding, such a proposal should have general support—or at least acquiescence—from Fred White [Chairman of the CSIRO] and Leonard Huxley [a member of the CSIRO Executive and soon to be Vice-Chancellor of ANU], and at the moment we just do not see how this may be done.

On 29 February 1960 (Monday), Bowen wrote to White. The letter was a summary of points discussed in person on the previous Thursday (25 February 1960). He had forwarded copies of the two letters to Oliphant on to White.[16] The main topic of the letter was the need for a thorough revamping of the structure of the radio astronomy group at RPL. Bowen's disaffection for Pawsey became obvious:

> I agree with your view that what we conspicuously lack in the Laboratory at present is the young man[17] who is going to run and make a success of the research programme on the GRT at Parkes. The three possibilities on the staff are Mills, Christiansen and Wild[18] and, of these Paul Wild is the ideal choice. Unfortunately for us, he wants to stay with the solar work for another five years or so. Neither Christiansen nor Mills are interested—in fact they tend to be hostile to the [GRT]. I would be happy for Mills to play a dominant role in the research programme of the telescope but unfortunately he has very firmly set himself on another course. As you know, the indications are that we shall lose both Christiansen and Mills. This will be a real loss to the Laboratory but I do not regard this quite as seriously as other people.

---

[15] Joe and Lenore Pawsey Collection.

[16] NAA C3830 Z1/7/B Part 2. Bowen also wrote on 29 February 1960 that: "The general administration of the RPL has been tightened up as of 10 am Friday, 26 Friday" [likely a memo sent to the RPL staff; the content of the memo was not described]. Also Bowen suggested that White attempt to invite HRH Prince Phillip to the opening of the Parkes telescope in 1961. This attempt did not succeed for the GRT opening on 31 October 1961.

[17] Clearly he is not referring to Pawsey who was 52 years old at this time.

[18] Bolton was not included in this list as he only joined RPL again in early 1961.

In the first place, it will make room for some of our bright younger men [Jim Roberts, Brian Robinson and Don Mathewson, all who had been overseas for some time].

Resignation of Christiansen, March 1960

Christiansen was likewise unhappy. He had always been critical of Bowen's push to build a large dish, "the last of the windjammers" being one of his favourite derogatory quotes.[19] Christiansen was more driven by innovative array technology and felt that new ideas were being supressed in the push to build the biggest dish which only involved innovation in mechanical structures, a topic of no interest to radio engineers. Christiansen had also written to Bok outlining his dissatisfaction with the direction Bowen was taking Radiophysics. Unfortunately for Christiansen, a copy of this letter found its way to Bowen via Oliphant.[20] Christiansen was summoned to Bowen's office where a fiery exchange took place and Christiansen tendered his resignation. Fortuitously, Christiansen had an offer for the position of Chair of Electrical Engineering at the University of Sydney, the position that had been declined by Ron Bracewell. Pawsey wrote him a glowing letter of recommendation.[21] Christiansen moved there in March–April 1960. He also had a long-standing invitation from Jan Oort of the University of Leiden to visit the Netherlands to work on the newly planned Benelux Cross and he took this up after he moved to the University of Sydney. Christiansen left in July 1960 for a 15-month visit to Leiden continuing the tradition of international collaboration championed by Pawsey.[22] After returning to Australia his group at the University of Sydney thrived and went on to produce some of Australia's most innovative radio engineers, including Bob Frater and both David Skellern and John O'Sullivan of WiFi fame (see Frater et al., 2017, Chap. 4).

---

[19] See Haynes et al., *Explorers of the Southern Sky: A history of Australian Astronomy,* 1996, p.233.

[20] Based on an interview that Roslynn and Raymond Haynes had with Christiansen in 1992 (Haynes et al., 1996, p.233). Christiansen had sent a letter to Bok in 1960 with critical comments about Bowen in which he complained about Bowen's interference with the radio astronomy group and was less than complimentary about Bowen's scientific achievements. We do not know the circumstances of the leak from Bok to Oliphant to Bowen.

[21] NAA C3830 Z3/1/X. On 4 March 1960 to the University of Sydney registrar: "I consider him an exceedingly good candidate, better than any of the previous applicants with the possible exception of B.Y. Mills, who may be a slightly more brilliant researcher but might not prove to be such a good teacher. Christiansen is a brilliant research physicist . . .".

[22] Christiansen and his family were to return in 1961, presumably just before the Parkes telescope opening on 31 October 1961. In Leiden, Christiansen had been the leader of the international design team for the Benelux Cross Project.

Resignation of Mills, May 1960

At about this time, Mills secured a position at the University of Sydney's School of Physics, only a few hundred metres from the RPL in 1960. On 25 May 1960, Mills met with Harry Messel to finalise the terms for his offer of appointment as a Reader in the School of Physics at University of Sydney and a funding grant of £A100,000 plus £10,000–£15,000 in annual running costs excluding staff, to establish a new radio astronomy section.[23] On 6 June 1960, Mills's appointment was officially approved by University Senate,[24] thus ending a 17-year career at the CSIRO Division of Radiophysics which had begun in 1942.

Within a few years, Messel had secured additional funding through the US National Science Foundation; the Molonglo Radio Observatory was built and fully completed in 1967. The operating frequency was 408 MHz with a resolution of 2.8 arc min. The MRO evolved into the MOST—Molonglo Observatory Synthesis Telescope—from 1978 to 1981, operating at a frequency of 843 MHz and a beam size of 43 arc sec. In practise, the Molonglo Radio Observatory was quite comparable to the envisioned Super-Cross. (see Frater et al., 2017, Chap. 3).

The Present Difficulties in Australian Radio Astronomy, Pawsey 31 March 1960

Within the next month, Pawsey prepared a major report, "The Present Difficulties in Australian Radio Astronomy", for the CSIRO Executive which was sent to Fred White on 31 March 1960.[25] Pawsey began with a call for action: "This report examines the present situation in radio astronomy in Australia. It strongly supports, from the point of view of the development of Australian science, a proposal by Mills to develop and construct a greatly improved metre-wavelength Mills Cross in Australia, but recognises that CSIRO cannot now finance it."

Two recommendations followed: (1) The CSIRO should support Christiansen and Electrical Engineering at University of Sydney in his continuing work using the ex-CSIRO site at Fleurs and (2) the CSIRO should communicate with ANU that metre-wave radio astronomy could not be pursued and would recommend that ANU might enter this field.

Pawsey began with a historical description of the origin of the schism within RPL. The inability to divert funds to metre-wave radio astronomy in competition with the GRT had consequences: "These diverse opportunities have led to

---

[23] Letter from Messel to Mills dated 25 May 1961, University of Sydney Archive – P154 – Series 8.

[24] Letter from Registrar to Mills dated 7 June 1960, University of Sydney Archive – P154 – Series 8.

[25] Joe and Lenore Pawsey Collection, located 2010. Also located in July 2014 in NAA C3830 Z1/20. Additional associated documents (e.g. draft letter to Oliphant) were only discovered in the Joe and Lenore Pawsey Collection.

difficulties in the radio astronomy group. On the one hand, the **completion of the Parkes radio telescope should provide Australians with the most effective paraboloid in existence** [our emphasis]." By 1959, RPL had five important observational programmes running simultaneously: (1) solar metre wavelength bursts with Paul Wild and colleagues, (2) solar decimetre wavelength observations by Christiansen et al., (3) cosmic 21 cm hydrogen line observations by Kerr, (4) cosmic 3.5 metre continuum by Mills and colleagues and (5) cosmic 15 metre continuum by Alex Shain. After the GRT completion, item (3) would accelerate as well as decimetre wave continuum observations. An ambitious new solar project proposed by Wild would lead to imaging the sun at 1 sec intervals. But RPL would have to drop or reduce items (2), (4) and (5). Christiansen was planning to move to non-solar work and the Fleurs Chris Cross would evolve into the Fleurs Synthesis Telescope in the 1970s. Item (5) tragically ended with the death of Shain in February 1960. Pawsey concluded "**But I regard the dropping of Mills's programme . . . as the most regrettable.**" [our emphasis].

Pawsey had consistently supported Mills throughout his career and had relied on his scientific advice during the period of controversy with Cambridge over the source counts. To Pawsey the departure of Mills was a huge loss. This event also led to the breakup of the Mills, Christiansen and Shain team, a major achievement created by Pawsey. On page two of the main report, he lists the five "stars" of RPL: Wild, Christiansen, Mills, Shain and Kerr; only two, Wild and Kerr, were to remain at RPL after mid-1960.

> On the staff side, we expect to lose Christiansen and Mills through resignations to take up university positions. We have lost Shain through his death. This means we have lost one and expect to lose two more out of five of our outstanding men. The achievements of these three have played a big part in building the reputation of Australian science. A substantial break-up of the RPL team now seems almost inevitable.
>   **These impending resignations are clear evidence of a crisis in Australian radio astronomy.**[26](our emphasis) It is imperative at this stage to recognise the significant factors. Proper moves now could lead to even greater achievements on an Australian-wide front; false moves could wreck the Australian effort.

Pawsey clearly was convinced that he needed Christiansen and Mills in the post-1960 era. His report on the "present difficulties" continued. Since the end of WWII, the Australian RPL group had developed numerous "special devices": radio telescopes with high resolution such as the Mills Cross, the grating interferometer, the Chris Cross, Michelson interferometry and dynamic solar spectrographs (rapid time and frequency response).

> The most important point is that the [scientists] who have been using them believe that such "devices" can be vastly improved, improved so far as to outdistance giant paraboloids in a large proportion of the fields of radio astronomy. No other country has an equivalent pool of scientific ability in this particular field. At the same time, [the GRT] will have a unique radio

---

[26] It is striking that Pawsey did not distinguish between "CSIRO" radio astronomy and "Australian" radio astronomy. At the time the two were the same in Pawsey's mind—see the following comments on "anchored beliefs".

telescope providing first-class opportunities in the several branches of radio astronomy to which it is suited.

At this time, no other country had such a pool of talented instrument builders. Pawsey's appraisal of the limited science potential for the GRT and his suggestion that the giant paraboloids would be outpaced by the low frequency arrays was well off the mark. This lack of impartiality may be indicative of the stress caused by the schism as well as his dependence on the views expressed by Mills. In reality it was the flexibility of the single dish, with the ability to respond to new discoveries, that was to be decisive in successive decades.

We now know that all humans use cognitive heuristics—shortcuts in thinking— that significantly bias judgement under certain conditions. One of the strongest is called "anchoring". Once we create a "mental model" of something, ie a kind of mental "gist" that sums up what we think about it, we "anchor" to this very strongly, and we fit each new piece of information to our pre-existing anchor, rather than adjusting our mental model (Kahneman, 2011). This effect is strongest when identity is threatened: when we are stressed, tired, and under pressure. Certainly, the situation Pawsey now experienced.

Pawsey saw the combination of the two types of complementary instruments as a way to reconcile the value of the two groups. But with the 50/50 division between the GRT and all other RPL projects in radio astronomy, the new solar project of Wild would take up the remaining resources. White, the CEO of CSIRO, had ruled out any substantial increase for funding a new large project in addition to the solar project.[27]

Clearly, Pawsey was worried about the overall future of radio astronomy in Australia. "I regard the discontinuance of Mills's line of investigation as the throwing away of quite outstanding possibility for Australian science." Mills had written a proposal for a new Super-Cross, to be built in two stages. An initial instrument working at 1 metre (resolution 5 arc min) with arms of 1370 by 12 metres would cost about £A 220,000 (not including the land) over a three-year period. A larger instrument with a 3 arc min resolution would consist of arms of 2130 by 15 metres with construction time of 4 years at a cost of £A 335,000. These expectations were to be compared with the planned Benelux Cross with a one arc min resolution at a cost of £ 1,000,000 sterling.[28] Pawsey continued: "I consider a project of this nature the most significant open to us in radio astronomy today. At the same time, because it operates at a longer wavelength than the really effective range of the Parkes aerial, it is not competitive with, but beautifully complementary to, the latter."

Pawsey thought a continued development of the Super-Cross would "provide in the future, as they have in the past, Australia's answer to the challenge of America's lavish expenditure." If ANU or another Australian university could participate with new funding, the project could proceed.

---

[27]Likely, Bowen would have agreed with this assessment.

[28]This was about A£ 1,250,000.

Bowen's Response to "The Difficulties in Australian Radio Astronomy" April 1960

On 4 and 5 April 1960,[29] Bowen sent two long letters to White with his response to Pawsey's "The Difficulties in Australian Radio Astronomy". The letters provided Bowen's response to: (1) Pawsey's assertions about the Super-Cross plans and (2) a response to general issues as well as his "personal views" on the points raised by Pawsey. Both letters were to be discussed later by the CSIRO Executive.

Bowen wrote on 4 April 1960:

> In my view, the core of the problem is very simple: whether it is desirable to go ahead with a Super-Cross right now or whether it should be deferred for a year or two. (I think it is quite clear that the merits of a Super-Cross as a device are not in question ...)

> The first attacks [the surveys] were carried out at about the same time by Mills using his original cross and Ryle using his interferometer at Cambridge ... It is generally agreed that the Mills Cross produced results which were more reliable and altogether of a better class than those of Cambridge, [see Chap. 35. After some confusion [not radio source confusion] and differences of opinion, it is also now generally agreed in astronomical circles that neither has solved the cosmological problem.

As discussed in Chaps. 35 and 36 on radio source surveys, the radio evidence strongly supported cosmological evolution, but the acrimonious debate between Cambridge and Sydney resulted in radio astronomy's loss of credibility in broader astronomical circles (Kragh, 1996). The irony of this situation was that Ryle had the correct interpretation (evolution of radio galaxies) even though the Mills catalogue was more reliable.

Bowen wrote that three additional instruments would be used to solve the problem:

> Professor Lovell has repeatedly indicated ... that the solution of this same problem is one of the main goals of the Jodrell Bank telescope ... and he intends solving it ... The 210-foot telescope at Parkes will give us a beam width of 10–16 min of arc [at 20 cm]. This is significantly better than [the Mills Cross and the Jodrell Bank antenna at 400 MHz 40 arc min] ... and it will be possible to take the matter substantially further than is possible [at Jodrell Bank] ... As a further step at Parkes, the 210-foot telescope will be used with the 60-footer as an interferometer to give a resolution of minutes of arc ... and will undoubtedly make important contributions to a variety of problems in radio astronomy.

In practice Lovell's 250-foot played only a minor and indirect role in source counts as no large-scale continuum survey was ever undertaken. The Parkes telescope 408 MHz survey of the early 1960s replaced the Mills Cross survey, but still had no significant cosmological impact. There was limited interest in more source counts, the Cambridge 3C had become a reliable catalogue and the addition of more sources from a southern survey would not be a significant step for cosmology. The Parkes catalogue also suffered from confusion bias (Chap. 36) at its survey frequency of 408 MHz so, although the high resolution 1.4 GHz follow-up

---

[29] NAA C3830 Z1/20.

observations made it an extremely reliable catalogue, the biased survey was of limited value for cosmology. The 60-foot interferometer was never used for a continuum survey.[30]

A summary of Bowen's views on big dishes was written for his colleague John R. Pierce (Bell Labs), Chair of the US Advisory Panel on Radio Telescopes.[31] Bowen was critical of the Cross concept and made a surprisingly astute projection that arrays of big dishes would be preferable. This concept was probably influenced by John Bolton's vision (eg *Stars and Stellar Systems Vol 1 Chapter* "Radio Telescopes", in *Telescopes,* ed. Kuiper & Middlehurst, 1960). In 2020, we now appreciate that the dish arrays built at Cambridge, Westerbork, VLA and the Australia Telescope Compact Array were indeed the most successful of the future paths.

Bowen continued his criticism of Pawsey in the 4 April 1960 letter to White:

> It is suggested [by Pawsey] that it is a disastrous thing that two senior members of our radio astronomy staff are considering leaving. On quite general grounds it could be argued that this event would be both reasonable and one which is to be expected.
>
> There is a marked absence of good candidates for the vacant Chairs of Physics and Electrical Engineering which exist in Australia today. At the RPL there is a notable concentration of people who might fill them. For a broad and balanced development of scientific effort in Australia, it could be argued that these are the people to fill them.
>
> In the same way, we have a number of very bright young people coming up in the RPL, for example Jim Roberts, Brian Robinson and Don Mathewson. They are at least as bright as anyone we have had in the past and it is going to be exceedingly difficult or impossible to give them proper scientific opportunities or advancement at the appropriate stage with the large proportion of senior staff we have . . .[32]

The tremendous team work that lead to repeated successes in the late 1940s and early 1950s was apparently evaporating.

The next letter from Bowen to White (5 April 1960) represented general comments and "personal views" about Pawsey's memo:

> Let me say how sorry I am to see the continuation of a process which has in fact been going on for a long time. One of the great strengths of the Radio Astronomy group of the RPL has been the way in which we have worked together as a team. The Radio Astronomy group has been, and probably still is, the largest and most cohesive group of its kind anywhere and this has been a large part of the reason for its success. Unfortunately, one or two sources [presumably Mills and Christiansen] of disaffection have been at work splitting the group for some time. **It is disappointing that Joe does nothing to stop it and occasionally seems to foster it.** [our emphasis]

---

[30] The Parkes telescope played an important role with the first of the surveys at a higher frequency (2.7 GHz – 11 cm) by Jasper Wall and colleagues. These surveys selected a different population of extragalactic radio sources, many with flatter radio spectra. (Shimmins et al., 1968, page 818). The 11 cm surveys were published in a series of papers the 1970s.

[31] see NRAO ONLINE.46

[32] Bowen advocated a steady flow of "scientific talent through [RPL], with bright young people coming in the bottom and some . . . of our best people going out near the top." He gave as examples Lehany, Burgmann, Bolton and Bracewell. "Some of our present worries come home to the fact this output of senior staff has not been taking place in recent years."

It is particularly unfortunate that the difficulties have reached a head at this stage in the construction of the giant radio telescope and I do not understand why the proposals for a Super-Cross are being pursued with such urgency at just this time. No doubt the reasons will emerge in due course ...

One reason why it has been possible to give so much encouragement to radio astronomy in the past is that it has taken place within the budget of the RPL. It has had to compete only with cloud physics and some of our activities and ... we have always given it lavish treatment. If the merits of radio astronomy had been argued out against nuclear and other branches of physics and against the merits of agricultural, biological and medical research in Australia, there is considerable doubt whether it would have obtained similar support. By ventilating more or less in public some of the present arguments and the expenditures involved, the group may be doing themselves a serious disservice.[33]

I am afraid that Joe's proposal does not do justice to the Executive or the Government generally, who have so generously supported radio astronomy in the past and are clearly committed to a large measure of support in the future. The present submission, while it lays claim to the cause of Australian science, does not pay any regard to the need for maintaining a proper balance between radio astronomy and the many other branches of scientific research which should be encouraged in Australia. I am all for more research and greater facilities for research in its many fields, but there are obviously some clear limits beyond which one cannot go in any one branch of science.

I have discussed all these points with Joe, but he is reluctant to see my point of view. Exactly what is done is clearly a matter for the Executive's decision. My own considered view is that this whole question of a Super-Cross is not one to rush into. It is one on which, for once, we should be cautious and see how the results of existing projects pan out.

Pawsey's document, "The Present Difficulties in Radio Astronomy", was discussed among members of the CSIRO Executive in April and May 1960. On 14 April 1960, Leonard Huxley (as a member of the CSIRO Executive) wrote Stewart Bastow (CEO of CSIRO from 1 January 1957 to 30 June 1959, then a member of the Executive until his death 23 January 1964[34]) describing the Pawsey document (and accompanying letters) plus two commentaries about the document by Bowen.[35] Huxley accepted Pawsey's analysis, especially the dangers of losing valuable scientific talent. He doubted the value of Bowen's commentary:

Bowen's documents do not add much in way of new or relevant arguments. The upshot is that [he does] not support Pawsey's proposal which in fact has been agreed to but he [Bowen] objects to chiefly on the grounds that the money has to come ultimately from the government and that we should not ask ANU to enter this field but should ourselves build a Super-Cross in a few years' time. In arguing thus it seems to me that he [Bowen] is not facing the facts: (a) Christiansen has gone [to Electrical Engineering, Sydney], (b) Mills is already negotiating with the University of Sydney. In this situation I do not see why ANU should not also approach Mills if it feels inclined to do so and can raise the money. I have not [summarised] Bowen's arguments [the contents of the letters of 4 and 5 April 1960], but I am not greatly impressed by them.

---

[33] See NRAO ONLINE.25.

[34] Bastow was a member of the CSIRO Executive from May 1949 to his death.

[35] CSIRO Archive KE20/2, Huxley document. The Bowen documents were most likely the letters to White on 4 and 5 April 1960.

Then, a month later, 2 May 1960,[36] the minutes of the 182nd Executive of CSIRO meeting included a report by Huxley on the radio astronomy programme of RPL. He described the current status of the Super-Cross, including the imminent departure of Christiansen and Mills from CSIRO: "It thus appeared that by staking the whole programme of the radio astronomy group on the GRT a major change of policy was evolving [to drop high resolution low frequency radio astronomy]." In effect the Executive made no substantial recommendation and the status was maintained.

At this time Lord Casey, who had been the Minister for the CSIRO from 1950 to 1960 and was a major supporter of the GRT, visited Parkes and the GRT site on 8 March 1960 (see ESM 30.2, Lord Casey).

Reactions to the Mills, Christiansen Resignations, May–July 1960

During the period May–July 1960, the impact of the two resignations (Mills and Christiansen) was discussed by Bok, Pawsey, Bowen, White and Bolton. On 23 May 1960, Bart Bok wrote Pawsey[37]: "Would you tell Ben [Gascoigne, on a visit to Sydney on behalf of ANU] about the latest developments with respect to Bernard Mills and yourself, for we continue to be very concerned about it all." Then on 6 June 1960, Pawsey wrote to Bok now with a sense of resignation: ". . . [T]he current situation re Bernard is that he is proposing to accept a Messel job which he rather expects to have offered to him today (or following a meeting today). He is keen . . . to get off on his own. So that is that."

Not surprisingly Bowen reacted to these events with a tone of relief, writing to White on 22 June 1960[38]:

> The news from the Lab [RPL] is encouraging. The recent excitement in the radio astronomy group has died down. Christiansen has settled in at [Electrical Engineering at the University of Sydney] and leaves in a month's time for seventeen months at Leiden. I am not sure how this ties up with the need for more teaching at the university to produce better graduates and more of them. Bernie has dickered successfully with the Universities of New South Wales and Adelaide and, leaving a trail of hard feelings behind him, has now fallen to the blandishments of Harry Messel. Although he has not finally made up his mind, it looks as if he will take a Readership at the School of Physics, plus £A1000 for himself and £A 100,000 to build a Cross. I wish him luck.

A month and a half later (31 July 1960), Pawsey expressed his ongoing concern over these events as his two colleagues left RPL. In a letter to Fred White, who was then visiting the UK, Pawsey wrote:

---

[36] CSIRO Archive KE8/3.

[37] NAA C3830 Z3/1/Part 10.

[38] NAA C3830 Z1/7/B Part 1. Bowen to White in London.

I presume that you have heard that Bernard Mills has now left us and is a member of Messel's staff and is planning to build a big Cross. His departure became inevitable about a year ago when the conflict in ideas between him and Taffey [sic] became apparent. But I had hoped rather he might have gone to ANU. With Chris and Bernard gone and also Alex Shain [deceased 11 February 1960] the old team is not what it used to be. However I hope that we can continue to work together.

Impact of the Schism on Australian Radio Astronomy

The immediate impact of the 1960 schism was the establishment of two separate radio astronomy groups at the University of Sydney. One was in Physics under Mills, which went on to build the Super-Cross (1960–1970), and to upgrade it to a synthesis telescope in the 1980s (MOST). It was upgraded again in 2015 (UTMOST) for FRB (Fast Radio Burst) searching. The other University of Sydney group in Electrical Engineering took over the Chris Cross at Fleurs in the mid 60s, and upgraded it to become Australia's first high angular resolution aperture synthesis telescope (FST) in the 1970s. These two groups at the University of Sydney were the primary training grounds for the next generation of Australian radio astronomers. See ESM 30.3, Goss and Shaver, pawns, for an account of how Mills used one of the authors (Goss) to help heal the schism and reconnect radio astronomy at Sydney University with that at CSIRO.

ANU never developed its own radio astronomy group, but the collaboration with CSIRO dreamed of by Pawsey and made a reality by Bok provided a stream of students, mostly using the Parkes telescope. By the 1980s the new Australia Telescope Compact Array together with the Parkes radio telescope were operated by CSIRO as a National Facility and triggered a huge expansion of radio astronomy research in many Australian universities.

In Tasmania, the University of Hobart continued a modest program in low frequency radio astronomy which had been started in 1954 by Reber and continued under Bill Ellis (Professor of Physics at the University of Tasmania) during the 60s. With the advent of Very Long Baseline Interferometry, the Hobart group became very active from the 1970s using an ex-NASA 26 m dish to provide a long baseline for the Australian array. In Western Australia, two groups in Perth (University of Western Australia and Curtin University, from 2007, now have the largest concentration of astronomers in Australia. The Curtin group finally revived the low frequency radio astronomy tradition started in CSIRO, by building the Murchison Widefield Array (MWA), a precursor to SKA low (Square Kilometre Array -low frequency). For more details see: "Radio astronomy in Australia: impact and the growth of a community", by Helen Sim, in the book *URSI at 100, celebrating the 100th anniversary of the International Union of Radio Science* (2021). In NRAO ONLINE.34 we include the draft text by Sim.

Leadership of the GRT Programme in 1960—John Bolton Returns to RPL from Caltech

As the end of the decade and as the completion of the GRT approached, John Bolton had reached a high point in his career with the very successful development of Caltech's Owens Valley Observatory. White and Bowen contemplated their visions for the evolution of Australian radio astronomy. Bowen considered the future organisation of the RPL. On 29 February 1960, he wrote Fred White[39]:

> [The departure of Mills and Christiansen] will provide an opportunity of putting our house in order in a very desirable direction, namely in the appointment of two people under Joe to take charge of the radio telescope programme and our solar programme, respectively. Paul Wild is clearly the man for the latter and there is an obvious place for us to look for the former. To invite someone to take charge of the GRT now would be to invite some kind of explosion. When Mills and Christiansen are finally settled, I think we will have no difficulty making a new arrangement smoothly and agreeable to all concerned.

Given the formal requirements for an open appointment process, Bowen would not have been free to name John Bolton at Caltech in a formal letter to White concerning re-structuring the group. White replied to Bowen on 2 March 1960 telling him that he (White) agreed with the suggested appointments: "I am quite willing to rest on your judgment about point 4 ... the younger man who is going to run and make a success of the research programme on the GRT." This may be the first time that Bowen had broached this sensitive subject in correspondence with White.

Apparently, Bowen also felt that he could not raise the Bolton issue with Pawsey, while the situation involving Mills and Christiansen was so sensitive. The relationship between Bowen and Pawsey had already deteriorated to the point of minimal mutual trust. Bowen may have presumed that Pawsey would interpret a move to employ Bolton as a further attempt to disenfranchise Mills—even though Mills had no interest in supporting the GRT. Regardless of speculation on Bowen's thinking, at the end of February 1960, it is likely that Pawsey knew nothing of this discussion between Bowen and White.

Bowen wrote to Bolton on 2 June 1960 suggesting he consider a position at RPL.[40] Bolton and Bowen had already discussed a possible return to Sydney as early as December 1958 during Bowen's visit to California. In the correspondence that ensued[41] Bowen also suggested that White visit Bolton in the USA, during White's travel there in August. We find a revealing comment in one of Bowen's letters to Bolton that indicates Bowen's perceptions of Pawsey at this time: "[Pawsey] blows hot and cold as far as the GRT is concerned. If it is a gigantic flop, he clearly wants to say 'I told you so.' If it is successful, he wants to be in it." This statement, not characteristic of Pawsey, provides more evidence of the deteriorating relationship

---

[39] NAA C3830 Z1/7/B/2 Part 1.

[40] NAA C4633/1 2 June 1960 Bowen to Bolton.

[41] See NRAO ONLINE.46 for details.

between the two men and the increased likelihood of each misunderstanding the other.

As the possibility of Bolton's return surfaced, unanimity at RPL remained elusive. As Don Mathewson discussed with Goss in 2011,[42] most of the RPL scientists in 1960 viewed, not Bolton's hire, but Bowen's handling of the transition, mistaken. Mathewson has pointed out that if the process had been above board with no secrecy, Bolton's recruitment would have been much less controversial, and misunderstandings might have been at least partially avoided.

Bolton wrote Fred White on 22 June 1960[43]:

> **I told him [Bowen, during the Owens Valley Radio Observatory opening in December 1958] that I did not intend to spend the rest of my life at Caltech** [our emphasis] ... I indicated that while I intended to stay here until the observatory I had built had got on its feet and was a scientific success, I would be glad of an opportunity to return to RP. Taffy recently asked me whether I was of the same opinion and whether I would like to go back to run the non-solar [sic] research on the GRT. As you know, several of the top staff have left RP recently. There has been for many years a conflict of opinion on whether the correct way to do galactic and **radio star work** [a surprising term for Bolton in 1960, since the term was not generally used outside of Cambridge] is with partially filled arrays at metre wavelengths or steerable dishes at the centimetre and short decimetre wavelengths. Those who have left RP are of the former school and Taffy and I of the latter ...[44]

On 22 June 1960, Bowen wrote White with details of his communication with Bolton in the previous weeks[45]:

> I have had in mind for some time—namely to advertise two senior positions under Joe to look after our solar work and the GRT, respectively ... They would carry an appropriate title and a good salary and might attract Paul Wild and John Bolton, respectively. If this came off, it would then compensate Paul for the very attractive job he has turned down recently [Cornell] and should be good enough to bring John Bolton back to us ... I have been in touch with him in recent weeks and he is most interested in the prospect of returning ... he has done magnificently at Caltech, so he deserves the best possible position we can give him ... I will wait until you get back to discuss the details.

On 1 August 1960, Bowen wrote to White with information "you require for a discussion with John Bolton in San Francisco"[46]:

> It is quite clear that he is excited by the prospect of coming back ... He did wonderfully well with us in the first place and still hankers after the Australian environment. Secondly, he is most decidedly in favour of big telescopes and looks forward to having under his charge the best of which is likely to exist for a year or so ... [My] intention is to advertise two positions

---

[42] Personal interview at Mt. Stromlo observatory.

[43] NAA PH/BOL/5 Part 2.

[44] This view is entirely consistent with Bolton's vision of the future of radio astronomy and was not a result of Bowen's influence. NAA C4633/3 On 4 July 1960, White replied to Bolton: "I was very interested in your idea of returning to RP. You probably know that Taffy Bowen would like this very much. Your opinions about the use of the dish interest me greatly, because of my anxiety to see this telescope project really go ahead with first class work."

[45] NAA C4633/3.

[46] NAA C3830 Z1/47/BOL

under Joe—one scientist in charge of the telescope at Parkes and the other in charge of solar work. We shall probably have a number of good applicants, but it would be hard to find anyone better than John Bolton for the former and Paul Wild for the latter. An appointment dating from January 1st 1961 is the one to shoot for ... It is fairly certain that, when it is known that John is coming back, he will get a number of attractive counter offers in the USA. We should, therefore, be as generous as we possibly can, both in the salary and the prospects we hold out to him ... I think we should also clearly recognise one fly in the ointment, **namely that Joe is not entirely happy about John coming back.** [our emphasis] The reasons are probably deep seated and of the kind we are slowly learning to live with. However, Joe has no good alternative to suggest, and the alternatives open to us, namely not having anyone in a senior position to look after the research programme of the telescope [GRT], is too awful to contemplate.

Even though the agreement between Bowen and Bolton had been discussed (with the blessing of White) by late June 1960, many additional details were to be settled. The formal job application procedure had to be fulfilled, as well as Bolton's resignation at Caltech. Bolton announced the resignation at Caltech in August 1960, to the consternation of senior Caltech administrators DuBridge, Greenstein, Bacher and others.[47] When Bowen visited Caltech at the end of September 1960, he wrote to Bolton after meeting his colleagues:

I spoke to Lee DuBridge, Bob Bacher and Jesse [Greenstein] about your impending departure from Caltech. They are still wondering what hit them and seem to be suffering from delayed reaction shock. They are understandably worried about a replacement and do not think they will be able to appoint anyone until at least June 1961. It would be ok for Jim Roberts to stay on a little longer, but in view of the above paragraph I would not be surprised to receive a request from Caltech for him to stay on for several months, or at least until they find a replacement for you.[48]

On 15 October 1960, John Bolton sent his "formal application" for the senior research position to Bowen.[49] This was a hand-written letter with abbreviated curriculum vitae, including his nominated references Robert Bacher (Caltech), Fred Hoyle (Cambridge), Rudolf Minkowski (Mt Wilson and Palomar) and R. Hanbury Brown (Jodrell Bank).[50]

On 14 November a very confused set of communications were transmitted to Bolton via the Australian Scientific Liaison Office in Washington, D.C., with the offer of a new position at CSIRO. The problem was that the ASLO personnel did not know any of the details of the new appointment. Bowen was especially frustrated by

---

[47] California Institute of Technology archive, L.A. DuBridge archive, Box 34. Letter 16 August 1960, DuBridge to Admiral Rawson Bennett, Chief of Naval Research.

[48] Roberts remained at Caltech until March 1961.

[49] A few weeks earlier, Bowen had pointed out to Bolton (6 October 1960, NAA C4633/2) that the application was a pro-forma exercise: "The wheels are turning slowly but inevitably in the direction I indicated when we spoke [in London and Pasadena]. An advertisement will be issued ... and I will send you a copy as soon as it appears. I might say that the Executive have already agreed to the appointment, but it is necessary for CSIRO to go through the motions so that there is no question about payment of fares and so forth."

[50] NAA PH/BOL/5 Part 2. It is unlikely that any of the referees were contacted by letter as there are no letters of reference in the Bolton personnel file at CSIRO (PH file).

the confused bureaucracy ("... the clerks [in Canberra] thought they would make some work for themselves, so they sent it first to the Gestapo,[51] then to Washington—who knew nothing about it—then to you! It looks as if we all have our private cross to bear!") By necessity, Bolton had already organised the trip back to Australia from Los Angeles before the formalities were complete: he, Letty and their younger son Peter left on the *SS Orcades* on 12 December 1960, arriving in Sydney on 30 December 1960.[52] The older son, Brian, had returned earlier.

Both Wild and John Bolton were appointed at the top of the Senior Principal Research Officer rank. By mid-1961, both Bolton and Wild had been approved by the Minister for CSIRO to be promoted to Chief Research Scientist. On Bolton's "commencement of duty" form, signed on 3 January 1961, an entry appears: "If not a new position indicate name of person replaced. [The name was] **B.Y. Mills**." This choice was remarkably ironic.

Pawsey, Disillusioned with CSIRO: Late 1960

Pawsey's concern for the future of RPL, as well as his own future role, increased in the course of 1960. The departure of his senior colleagues Christiansen and Mills, plus the Super-Cross demise, were major factors. Also, the management of RPL had been reorganised by Bowen with minimal input from the Assistant Chief, Pawsey.

In September 1960, Pawsey was at a crossroads in his career; after the founding of Australian radio astronomy in 1945–1950, he could see his influence dwindling. Pawsey's discouraged state of mind in September 1960 can be gauged by his personal correspondence (located in the Joe and Lenore Pawsey Collection) during this period. A few weeks before Bowen returned to Sydney on 27 September 1960, Pawsey wrote letters to Leonard Huxley (CSIRO Executive and soon to be the Vice-Chancellor of ANU), Fred White (Chair of CSIRO from 1 July 1959 to 22 May 1970, there was no CEO of CSIRO from July 1959 to December 1986) and Bowen with expressions of concern.

Two "personal and confidential" letters were written by Pawsey on 5 September 1960 to Huxley and White, the letter to Bowen on 8 September. Huxley replied on 16 September 1960; no response from White has been found in the archives. However, White did visit Pawsey a week later, Tuesday 13 September 1960. Based on the immediate contact with Bowen, clearly the two discussed Bolton's appointment at length.

Bowen responded from Washington on 22 September 1960. Both Pawsey letters from 5 September 1960 were typed (likely in the Pawsey home), the one to White on

---

[51] Bowen clearly intended this crude statement as a joke.

[52] NAA PH/BOL/5 Part 2 NAA C4633/3. The process involving the Australian Scientific Liaison Office in Washington, D. C was still standard practice decades later when Ekers moved from NRAO (VLA) back to Sydney to take on the position as the Australia Telescope Director at CSIRO.

a CSIRO RPL letterhead. The content of both letters is similar with additional elaboration in the Huxley letter. The letter to White (5 September 1960):

> I wrote you while you were away [until about 1 September] saying that I was anxious to see you to discuss the possible appointment of John Bolton. I had previously advised Taffey [sic] that his appointment was likely to cause considerable internal friction ... The essential point the boys [the radio astronomy scientific staff] are querying is whether I, or John through Taffey, will be in control of the research program. I am told that they unanimously want me ... [Then a] second development took place ... A rumour reached RPL that John had been appointed as from this November. I, as Assistant Chief of the Division and head of the radio astronomy section, have no word of this. If this rumour is true I am not going to quarrel with John whom I regard as a friend, but I want specific assurances from the Executive on certain points. Firstly on seniority, I regard Paul Wild as having definitely higher attainments than John and I want to be quite sure that John is not appointed over Paul's head.

The letter to Huxley (5 September 1960) was similar with a number of additional points. The letter may be viewed as a "cry for help", directed to the CSIRO Executive. Pawsey clearly felt he could be more frank with Huxley than White, his boss at CSIRO.

> I am worried about the way certain things are working out in the RP laboratory and I should like to have a talk with you as soon as convenient after your return [from Europe and the US on 21 September 1960]. The problem is the differences in outlook concerning radio astronomy and whether in the circumstances, I should hang on or get out. I think it will be useful if I set down the essential factors so that you may have a chance to think things over. These fall under three heads: (1) the RP situation, (2) pressure on me to take a State university chair, and (3) my own ideas on what I can do best. Firstly the RP situation. You are familiar with the general position. I shall add only the new developments. Taffey [sic] some time ago suggested a new senior appointment of a man to look after things at Parkes and suggested two possibilities: Hanbury Brown[53] and John Bolton. I told him I should welcome Hanbury warmly but that I advised against John on the grounds that his appointment would cause grave internal friction. I have since repeated this argument to Taffey ... [At the lab, among the scientific and receiver groups] I find that the mistrust of John is far more widespread than I realised. The essential point is whether I or John through Taffey will be in control of the research program, and my unofficial informants tell me that the unanimous view is that they want me. [Then] the rumour that John has been appointed as from a date of about next November [1960], and already handed in his resignation at Caltech. As Assistant Chief and nominal head of the radio astronomy section I am not informed of this. **In simple terms I have notice to quit on the one hand and an appeal to hang on from the other. Where do the Executive stand?** [our emphasis] [Pawsey next described the suggestion that he might take a chair of physics at Adelaide or Melbourne.] ... I am obviously at a turning point. This RP situation cannot go on indefinitely. I have to get out or establish a new independent basis with the Executive. The move should be based on what I can best do. It seems to me that what **strength I have lies in my ability to stimulate**

---

[53] In the letter from Pawsey to White of 31 July 1960 (see above, Joe and Lenore Pawsey Collection), Pawsey wrote: "Taffey [sic] tells me that he is hoping to get either John Bolton or Hanbury Brown to fill the gap. I, and I am quite sure I can also speak for others in the group, would be delighted to have Hanbury join us if he should be interested. John's appointment would raise some very tricky points and I wish you were here [in Australia, due to return in September 1960] so that I could discuss these with you. Taffey knows my doubts."

and develope [sic] scientists at the research level [our emphasis]. An exceptional proportion of those who work with me seem to reach top level whereas others I have thought to
have equal inate [sic] ability do not make the grade. If so my most useful contribution would
seem to be in a "research institute" of some sort. The RPL has served me very well, but apart
from the present difficulties, I should like to see some modification. I do not think we have
enough contact with the younger generation. What I should consider excellent would be a
research group with the facilities of RP but loosely tied in with a university and have a fair
proportion of research students in its [RPL's] circle ... I feel here that the potentialities of
ANU are very considerable; if only one could transfer RP or part of it to Canberra and fix up
some sort of amalgamation of resources.[54] I feel that I should like to play a part in an
organisation with a dual objective: research and the development of research scientists in
Australia. [If I were to move to a State University] I should wish to continue in radio
astronomy but I do not think it really feasible to build up another big radio astronomy group
in competition with RP. So I should have to retire from, or to the outskirts of, this branch of
science after having built it up in Australia. If I continued in radio astronomy, I feel that I
could still play a very useful part in keeping Australia in the forefront. I am intensely
interested in particular in the technical developments in our field ...

Huxley wrote from London (he had been at URSI) on 16 September 1960, a
reserved, neutral but realistic response:

... [M]ay I say that it seems to me that you may find yourself out on a limb. Bolton's return
has been mooted for some time and the new set-up will be entirely built around the GRT. The
Executive is not likely to sponsor any other group on the grounds of ceremony[55] ... I am
afraid this is a [quickly composed] letter but the external conditions are not helpful.

A more significant letter was handwritten[56] by Pawsey to Bowen from 8 September
1960. Pawsey:

Just what is happening about John Bolton? There have been two developments here:
(1) there is a rumour[57] to the effect that John is appointed from Nov. next and (2) there
have been a series of questions at the lab. For (1) I have no official information and I
definitely don't like being out on a limb. On (2) there were actually questions about
"conditions of work at Parkes" in a GRT committee meeting ... [T]his is simply the John

---

[54] The co-supervision of University students by CSIRO staff started just a few years later, ironically
lead by John Bolton and initially through an agreement with Bok at ANU. One of the authors
(Ekers) was an ANU student.

[55] Huxley also encouraged Pawsey to make up his mind quickly about the Adelaide University offer
since another candidate was about to be selected. Huxley also had some disparaging remarks about
the Physics Department at Melbourne, "which in any case is a rundown department".

[56] A handwritten copy of this letter was found in May 2010 in the Joe and Lenore Pawsey
Collection. The letter could only be deciphered with the assistance of Elizabeth Pawsey,
J.L. Pawsey's daughter-in-law. No copy has been located in either the NAA or the CSIRO archives.

[57] The rumour had also reached the US. University of Sydney Archives, P154- Series 2. Letters to
Mills- 9 September 1960. Campbell Wade (NRAO Green Bank) wrote Mills with a discussion of
the radio properties of M84 in the Virgo cluster. Wade had heard about Bolton's appointment: "The
imminent departure of the eminent JGB [Bolton] has occasioned a feeling of relief in certain
quarters [presumably NRAO which was in dispute with Caltech at this time] in this country. You
mentioned that certain people at RP are feeling sick over the matter, and that some recruiting might
be possible as a result. Since I do not know the precise feelings many of them bear towards [Bolton],
could you suggest the most likely prospective?" No response from Mills has been found in the
University of Sydney archives.

Bolton question and there is widespread distrust of John in RP. Much wider than I thought. You will remember that I warned you of this, but I find it extends to people I had not previously suspected—whom I had thought supporters.[58,59] On the other hand I agree with you that John is a good astronomer with the right range of interests and an excellent organiser. I find myself in a most awkward position. I myself have got on well with John in the past and definitely do not want this to build up into a personal quarrel with him. What then is the best to do? If this appointment is not yet made I definitely recommend a cooling period before appointment. I suggest the end of 1961 as suitable (if earlier is he not letting down Caltech?). He might well come here on a working visit ... If the appointment is made then I should like to see things more clearly defined. Firstly seniority with respect to Paul Wild. I regard Paul's scientific attainments over the last 10 years and his capacity for scientific leadership is definitely superior to John's and I shall strongly oppose an appointment over Paul's head—secondly the broad balance of the RP program. I want a firm assurance that John's coming will not block the GRT/non GRT balance that we have previously agreed on. The right way to proceed is the GRT will give us a blaze of glory in the near future. This will then become more of a routine instrument as bigger ones are completed elsewhere, and we must hope to score again on the "enhanced directivity" [i.e. interferometry] side. To be able to do this we must keep our hand in and back Paul's scheme when it comes good and in the nearer future keep going with Chris's Fleurs set up. Yours sincerely, Joe

Pawsey's perceptions about John Bolton seem at odds with Bolton's successful leadership of the Caltech radio astronomy group. In retrospect, we recognise that Bolton was an effective leader of the GRT science and technology developments after his return to Australia. We think the issue was that Bolton was not a team player in the style that Pawsey had cultivated. Bolton could be very dominating; he made it clear that he would lead the science. His support for the big dishes, or arrays of large dishes, was also certain to generate conflict with Mills's Super-Cross proposal.

The comment by Pawsey about big dishes is revealing. Apparently, the expectation was that a big dish such as the GRT would only be able to do a small range of projects well—hence the "blaze of glory" and also that it would be rapidly superseded by larger dishes. In reality the GRT continued to make scientific contributions for another 60 years. A decade later bigger dishes were constructed, such as the Effelsberg 100 m MPI antenna in Germany and in 2000 the Robert C. Byrd Greenbank Telescope (100 m–GBT) in West Virginia, USA. In the current era,

---

[58]This perspective was succinctly summarised by Frank Kerr in a letter to Sullivan 12 Aug 1986 "Bowen and Bolton got on very well together, as they were similar types in many ways. Neither of them could be trusted and they were both successful politicians." W.T.Sullivan, III, archive.

[59]This animosity felt by the staff towards Bolton was later addressed by Bowen. (NAA C4633/3 from 8 November 1960), Bowen wrote to Bolton: "As you know Fred White and I are all for [your appointment] and look forward to you being responsible for a lively and really active programme of research on the GRT. However, there have been difficulties down the line in the Laboratory and a few people are not reacting too well to your return. This is unfortunate and reflects only on the people concerned. It will, I am sure, sort itself out in the end, but we have to face up to the fact that these difficulties exist. This is not the kind of thing one can write about easily, and I will have a full and frank discussion with you as soon as you arrive [30 December 1960]." Bowen urged Bolton to be cautious in writing to the Laboratory, also suggesting addressing "all questions on the research programme to Joe and myself ..." and to be discrete in asking questions on aerial feeds and receivers. "I think you will quickly see the reason for this."

these instruments are matched in productivity by the "enhanced directivity" of the arrays such as Cambridge, WSRT, VLA, ATCA, and ALMA.

On 16 September 1960, Pawsey wrote a more tempered letter to Bowen at ASLO in Washington. The series of four cables (summarised in ESM 30.4, Bolton's Appointment) are referred to in the letter of 16 September 1960. These had been initiated by Fred White's visit to RPL in Sydney on Tuesday 13 September, Pawsey to Bowen:

> Fred White was here on Tuesday [13 September 1960] and we discussed John Bolton's prospective appointment ... He recommended and I agreed to, the offer described in the cable he sent you. I have just received a copy of your cable in reply. I take it this will almost lead to an appointment, but I shall not announce this to the Laboratory until I receive confirmation that the negotiations are concluded. At this time it is important, from the point of view of harmonious relations in the Laboratory that a fairly definite statement regarding working arrangements should be made. Both your cables [ESM 30.4, Bolton's Appointment] state that Bolton and Wild should be "appointed under Pawsey". I wish ... to make a statement on fairly general lines on how I propose to organise the research programme. [The main points were those discussed at the 4 October 1960 meeting, "The Radio Astronomy Research Programme", described below.[60]] ... In view of the queries in my last letter, I have promised to give a preliminary talk to the Lab on the program at Parkes. [This meeting was held on 4 October 1960.]

Bowen replied on 22 September 1960[61]:

> Your handwritten letter of 8 September 1960 [the angry message] has only just reached me here in Washington simultaneously with your letter of 16 September. In view of the interchange of cables with Dr White on the 13th, there is no point in discussing prior events, except perhaps to say that up to that time, it was my intention to recommend to the Executive that we advertise two positions in the Radioastronomy Group under your direction, and to invite applications in the usual way. It was also my intention to do this on my return to Sydney after a full discussion with you and Dr White. However, Fred's cable of 13 September [see ESM 30.4, Bolton's Appointment] forced the issue and I think you will agree that in view of its wording I had no option but to make a definite recommendation in reply.

Bowen seemed to have dismissed the angry handwritten letter of 8 September 1960; he implied that he had no interest in discussing "prior events".

Both the cable exchanges and Bowen's 22 September 1960 seem inconsistent with the offer made by Bowen to Bolton in June; even White had been involved in the June and July correspondence. Bolton surely viewed the offer as a "done-deal" since he had resigned from Caltech in August 1960. Thus, the claim by Bowen that

---

[60] As an example of the mechanism to set the research programme, Pawsey wrote to Bowen: "It will be up to John and me jointly to encourage the most worthwhile projects; to interest appropriate persons in ones we initiate, or to filter proposals put forward by others. They will lead to a limited number of proposals of varying merits. The individual proposals will then be put forward by the individuals responsible for the investigation (including objectives and procedure) to a meeting of research staff ... at which the acceptance of the projects and allocation of telescope time will be discussed. I should have the right of veto, which I would never have to exercise on a project on which agreement was reached, and should be the person to decide in cases of disagreement."

[61] NAA C3830 Z1/47/BOL

he had intended to wait until his return to Sydney to discuss the new appointment seems disingenuous. Perhaps both Bolton and Bowen had made the arrangements in June 1960 without the CSIRO Executive's formal approval; then Pawsey found out via "rumour" in August 1960 that Bolton had been appointed starting in November 1960. Bowen's claim "Fred's cable of 13 September forced the issue . . . and I had no option but to make a definite recommendation" is inconsistent with the events of the previous months, but it may indicate that White was forcing Bowen to make Bolton's appointment public and transparent.

Two of Pawsey's colleagues also had comments about the Bolton appointment. On 1 December 1960, Frank Kerr wrote Pawsey[62] on his way from Amsterdam to Lisbon.

> . . . [I]t is quite clear that John Bolton has been telling people all over Europe [during his visit to URSI in September 1960] that he is to be in charge of the GRT, and setting his plans in considerable detail . . . At URSI, he [Bolton] is reported to have proclaimed to a group which included Lovell and Hanbury Brown that he was going to be the "A.C.B. Lovell of the Southern Hemisphere". When John told Taffy this later apparently, Taffy said, "Oh no you aren't, I am going to be!". . . John says that he had been actually persuaded into going back, and that it has not been his doing . . . None of this seems to agree with the "official view" which was repeated at our last GRT meeting, and in particular **John gives no indication that he considers himself under your direction.** [our emphasis]

Frank Kerr's assessment could be partially correct since by this time Bolton would not be taking scientific directions from Pawsey, but he would have accepted Pawsey as an administrative manager.

A telling exchange occurred in October 1960 between Jim Roberts and Joe Pawsey.[63] Pawsey had heard from Paul Wild (who had visited Caltech, September 1960) that Roberts was quite concerned (due to the conflicts in Sydney) about returning to RPL from Caltech in March 1961 after his two-year visit to the US. On 1 October 1960, Pawsey wrote Roberts a letter from his home. He summarised the events of 1960 at RPL: Mills and Christiansen leaving RPL and Mills's new Super-Cross project at Sydney University. "It seems that metre wavelength work will continue in Australia." Then Pawsey put forward an overly optimistic picture of the newly agreed upon management structure at RPL: "In each case [GRT and solar] I propose a fairly careful committee type organization with me as chairman with a right of veto so that things have to be done very much in the open. There are quite clear difficulties but I hope I can make it work." Pawsey asked Roberts about his plans. Since he had experience in "theory, solar, interferometers, big dishes", Roberts had a bright future. Pawsey saw an interesting management challenge ahead: "For limited objectives it [Pawsey's non-dictatorial style] is not as efficient as a dictatorship with restricted objectives. I think you win in the end. It definitely worked in the past. May it again."

---

[62] Joe and Lenore Pawsey Collection.
[63] *Ibid.*

Jim Roberts wrote from Pasadena on 26 October 1960: "After reading [your letter of reassurance] I felt quite a little relieved—I just don't know how much was being done completely behind your back ... [but] I admire your optimism". Roberts did feel uncomfortable: "I hate any sort of political manoeuvring. I have told [Bolton and Bowen] that I do not want to be involved in any conflict." Thus, this was the reason to "shy off the big dish". But Roberts was keen to continue "the cosmic work" and Bolton's enthusiasm had been contagious at OVRO. Thus, he proposed to work on a mixture of solar and GRT projects. However, when he returned to RPL, he did spend most of his subsequent career working on cosmic radio astronomy topics such as Jupiter decametric radiation, polarisation and propagation problems based on GRT (Parkes telescope) observations.[64]

---

[64] A fascinating autobiographical, well-written text by J.A. Roberts "Have Gen Will Travel - Imperfect Images from the Life of a Radio Astronomer", privately published by Jim Roberts, July 2002. See NRAO ONLINE.49.

Chapter 31
John Bolton Returns, 1960–1961

While there have been differences between us in the past, I believe, on what kind of GRT to build, I do not believe we have had any differences on scientific aims ... As I have said to Taffy, I am sure the GRT will succeed and once we begin a chain of research success and discovery—personal animosities quickly heal over.—Bolton to Pawsey 19 December 1960

The GRT was nearing completion, the early frustrations were now in the past and progress exceeded expectations. Plans were being made for the scientific research with the GRT and an observing program selection committee was formed. This new major research facility was coming to life.

John Bolton had agreed to come back to Australia to lead the GRT group, but the RPL staff had not yet been informed, creating confusion and tension. There were conflicting plans for new receivers. En route to Australia Bolton informed Pawsey of the discovery of a "radio star" at high redshift–3C 48. The high redshift was later confirmed but, sadly, Bolton had already withdrawn his initial interpretation.[1]

A rapprochement between Pawsey and Bolton was underway, but as 1960 ended, Pawsey's future was clouded. His outstanding radio astronomy team was gone; what was his future role?

Supplementary Information The online version contains supplementary material available at [https://doi.org/10.1007/978-3-031-07916-0_31].

[1] As discussed below, the high redshift of 3C 48 was confirmed in 1963 (Greenstein and Matthews), 0.367, compared to the first recognized quasar 3C 273 at 0.16.

W. M. Goss et al., *Joe Pawsey and the Founding of Australian Radio Astronomy*, Historical & Cultural Astronomy, https://doi.org/10.1007/978-3-031-07916-0_31

The GRT Nears Completion

In early 1960, Harry Minnett made plans at Freeman, Fox and Partners (London) to return to Sydney by 21 May 1960. Bowen provided information on the situation in Australia on 6 January 1960. Bowen was especially concerned about the lack of detailed drawings of the mesh panels for the surface.[2] On 1 April 1960, he wrote Minnett with suggestions of possible scientific projects after his return to the RPL. Bowen thought that Minnett would only be occupied with AEI (servo contractor) and Askania (the master equatorial contractor) engineers "on the servo and control systems and in the acceptance testing of the whole telescope" during the period from late 1960 to roughly April or May [1961], "when the telescope comes into operation as a research instrument". For the future, Bowen suggested that Minnett might either join the receiver group of Brian Cooper or join the research programme of the GRT under the supervision of Pawsey. Bowen was keen that Minnett take a break from the telescope design and construction issues: "... You have a complete break from telescope problems and associate yourself, perhaps on the analysis side, with one of the projects which are a going concern in the Laboratory."

On 2 May 1960, Harry Minnett and his wife were on a ship on the way back to Sydney. He wrote from Aden with a report of his last visit to MAN in Germany on 24 and 25 March 1960. The telescope was taking shape; some design issues at FFP were sorted out at the last minute before their departure from Tilbury. "I rather doubt if it will be practical [in the interim before completion of the telescope in 1961] to divorce myself entirely from telescope matters ..." He would certainly be full time preparing for telescope testing. For example,

> ... [T]he arrangements for computing the paraboloids of best fit and accessing errors should be looked into and agreed in advance. I doubt if FFP will take initiative in such details ... I have produced a draft programme, but further discussion and elaboration is best done in Sydney as FFP have not shown much interest in the details. With the above in mind, I don't think it would be profitable to become involved in a specific research problem for the present. As you know it is almost ten years since I was actively concerned with radio astronomy research and I now feel very much out of touch with the overall picture.

Minnett would look around at RPL "before committing myself".

White's visit to Europe in June 1960 provided an opportunity to visit MAN in Gustavsburg to check on the progress of the GRT. Gilbert Roberts accompanied him to MAN Germany from London on 13 and 14 June 1960.[3] White wrote Bowen and Pawsey with a full report. The MAN staff were beginning trials of the turret and the cylindrical hub with the associated AEI servo gear; trials of the alt-azimuth control

---

[2]NAA C3830 A1/3/11/10 Part 50, 37.

[3]NAA C4633/3. Also White had reported to Pawsey (Joe and Lenore Pawsey Collection) in a handwritten aerogram from the UK on 15 May 1960: "Please tell Taffy lunched with FFP- Freeman, Roberts and the engineer who is coming out (liked him) [Mike Jeffery] ... I am going over to Germany with Roberts on 13 June ... Then they will be in midst of a test assembly of the turning [azimuth] gear. They have had troubles with the castings for the rails on the top of the tower ...".

were to be on 24 June 1960, a delay of 6–8 weeks compared to the 1959 schedule. (The master equatorial was not yet available.) The azimuth track had been levelled at the factory. The trials were to last 8 weeks; then the telescope components were to be packed, ready for shipment from West Germany to Sydney. Mike Jeffery of FFP was to be present at the trials. On 22 June 1960, Bowen wrote to White that FFP had also done a thorough job of keeping RPL abreast of progress at MAN: "We are infinitely better off than appeared possible about a year ago and I am well satisfied. The important thing is that MAN is going at it with plenty of push and enthusiasm." Feelings of frustration with FFP seemed to belong to the past. "Tailtwister" visits to FFP in London by Bowen were no longer required.

White provided Bolton a summary on MAN activities in his letter of 4 July 1960: "I have just recently been to Frankfurt to see the turret of the telescope in its final stage of trial erection ... You probably know that the concrete tower is already completed in Australia." White reported to Bowen on 22 July 1960 (again from London, after discussions with Roberts) that the detailed tests at MAN had begun, likely to last 2 weeks. AEI (formerly Metrovick) were participating with the servo tests with their own personnel present.

Bowen continued his report to Bolton on 29 July 1960:

> The telescope is at a very interesting stage. Weber, the MAN Director concerned with erection problems, arrived [in Sydney] last week, Mike Jeffery flew in this morning and some of MAN's erection gear is already en route from Germany. The turret and [cylindrical] hub will follow very shortly ...

MAN was still hoping for acceptance tests to occur in late April 1961.[4]

On 8 November 1960 (6 weeks before Bolton arrived in Sydney), Bowen wrote Bolton in California[5]:

> The more I see of the work MAN have done and are doing, the more I am convinced that we made the right choice ... [T]he derrick has been completed and the azimuth track is fitted to the top of the tower. The turret and hub structure have been off-loaded in Sydney and some of the parts are already on their way to Parkes. Things are really humming.

In ESM 31.1, NASA and the GRT, we summarise the negotiations with NASA in 1960 for possible use of the GRT for tracking deep space probes.

---

[4] Also, on 29 July 1960, Bowen wrote White with a few additional details. Weber (of MAN) had made most of the "arrangements for the arrival of his men and for the complete set of erection equipment which MAN are sending out. Jeffery ... is flat out getting the local arrangements [in Parkes] organised."

[5] NAA C4633/3.

Plans for the GRT: Sydney, October 1960

Soon after Bowen's return to Sydney on 27 September 1960 from the UK via the US, activities began at RPL for planning the administration of the GRT. Pawsey had initially suggested holding the planning meeting for operation of the telescope on the day of Bowen's return. Instead, Bowen's meeting on the administration of the GRT and Pawsey's operational planning meeting were both held a week later on 4 October 1960. Rumours, second-hand knowledge of the arrangements with John Bolton, were floating around the corridors of RPL. There had been no official announcements. For example, on 5 August 1960,[6] Brian Cooper had written John Bolton about details of the receiver plans for the GRT: "Taffy [Bowen] tells that you may soon have a direct interest in the GRT and has asked me to let you know how the receiver program is shaping up. The program, which was decided some time ago, and which I have taken on in recent weeks may be summarised as follows …"[7] (A detailed summary of five receiver projects going on at RPL followed.)

Even in early October 1960, Pawsey, the Assistant Chief of RPL, was still not sure of Bolton's status as the future leader of the GRT. The relationship of Bolton and Pawsey remained uncertain.

On Tuesday, 4 October 1960, a meeting of the radio astronomy group at RPL was held to discuss two topics: "General Arrangements Relating to GRT" lead by Bowen, followed by "The Radio Astronomy Research Programme, with Particular Reference to Parkes Radio Telescope Arrangements" organised by Pawsey.[8]

Bowen's document consisted of a block diagram, an organogram. At the top, Bowen (Chief) and Pawsey (Assistant Chief) are shown, followed by radio astronomy (Pawsey) and Cloud Physics (Bowen). Under rubric "radio astronomy" there were five boxes: solar work (Wild), GRT (blank), receivers (Cooper), digital processing (Beard) and Misc (blank).

> In the radio astronomy group, there will be two main research sections, one dealing with solar work under Paul Wild, the other dealing with researches [sic] on the GRT under a senior officer yet to be appointed. It is proposed to have an advertisement for this position to invite applications in the usual way and to appoint the best applicant. We have been in touch with several [quite likely none were contacted] likely candidates. It is already clear that John Bolton of Caltech will be an applicant and is highly likely to be appointed.[9]

As we have seen, Bolton had already accepted the position on an informal basis 3 months earlier, in June 1960, and had publicly submitted his resignation from Caltech. However, these statements by Bowen would be a necessary part of the required formal appointment procedure. The second page of Bowen's document described the role of Harry Minnett ("in charge of all mechanical and operating

---

[6] *Ibid.*

[7] Cooper implied that he was uncertain whether he would be working for Bolton in the near future.

[8] NAA C4633/3 and W.T. Sullivan archive.

[9] The files show no evidence for any other candidates.

details") and Lindsay McCready ("officer in charge of arrangements at Parkes field station").

The second meeting on 4 October 1960 was organised by Pawsey to refine the plans for the operation of the GRT, including equipment plans and major scientific programmes; for the first time, **a telescope observing program selection process** was proposed. In addition, the division of the scientific staff among the various GRT and solar programmes was to be determined. See Chaps. 27 and 33 for earlier discussions of an "observatory procedure" based on a report written on 9 July 1954: "An Observatory Model for a New Radiophysics Laboratory".

Pawsey's introduction set the stage:

> Our intention is (1) to progressively curtail pre-1960 projects, (2) to build up versatile facilities which will permit very flexible use of the Parkes telescope as soon as practicable, and (3) while observations are proceeding at Parkes, to undertake a new major solar development project under Wild ...[10] Subsequent plans must depend on the direction of development of radio astronomy. My guess is that there will be a demand for directivity greater than that given by the Parkes telescope itself, which will lead to our developing enhanced directivity systems [interferometry], possibly for the detailed mapping of restricted areas in the sky. I do not think we should commit ourselves on the desirable facilities (simple interferometer or more complex) until we are nearly ready to proceed.[11]

A number of receivers under development were described, e.g. a 21 cm line receiver, the Brian Robinson Leiden parametric amplifier at 21 cm, a multi-channel 21 cm line receiver (backend) with narrow channels for detecting HI absorption lines and a 400 MHz receiver for a sky survey. In addition, a series of test observations at 10 and 20 cm were suggested by Minnett to evaluate the high frequency performance of the dish. Scientific key programmes were also suggested: the galactic centre by Kerr, HI in external galaxies by Robinson and others, the detection of possible radio recombination lines at 1400 MHz by Murray, HI absorption by McGee and Murray, and the Magellanic Clouds (Mathewson) in HI and continuum. An innovative text followed, "selection of specific investigations" or a telescope observing proposal selection committee:

> The procedure will be for members of the Parkes research team to select problems of special interest ... from the above list or elsewhere. It will be the responsibility of the leader of the

---

[10]In a letter from Wild to Bolton of 21 October 1960 (NAA C4633/3), Wild explained his early thinking of the instrument "to obtain metre-wavelength pictures of the sun showing background and bursts". Wild wrote Bolton: "I am still thinking of the radio camera in terms of about 60 crude dishes [at frequencies below 200 MHz]. The instrument could go at Parkes, presumably close to the GRT facilities, or, if space does not permit some miles away." This instrument, the Culgoora Radioheliograph, was built in 1967and located 400 km north of the Parkes telescope near Narrabri, NSW.

[11]Bowen's interest in having an interferometer at Parkes is somewhat difficult to understand. He was always a strong supporter of moving the 60-foot Kennedy dish from Fleurs to Parkes and we speculate that this may have been Bowen's response to a request from John Bolton to have an interferometer at Parkes for radio position measurement to make optical identifications. As we point out below, the 60-ft interferometer (with the 210 aerial) was never used for accurate position observations.

Parkes section [Bolton] and myself [Pawsey] to stimulate the selection. We shall also help arrange for necessary collaboration. The person concerned will then submit a research proposal (giving observations and ways and means) to the Parkes steering committee. This committee will consist of Bowen (*ex officio*), leader of the Parkes section (convenor), Cooper and research officers of the Parkes group. The committee will accept a proposal if it sees fit and will allocate telescope time. I [Pawsey] shall retain the right to make decisions in case of disagreement ... [A] limited number of projects have been tentatively accepted. Research proposals should be prepared for these. I am anxious to arrange for the preparation of future proposals. These should involve a reasonably limited observing time. They will be reviewed later, but are likely to be accepted ... This proposal is experimental and exploratory ... The objective is to make a truly flexible arrangement where the versatility of the equipment can be exploited ...[12]

No mention was made of external members of this committee or of external users of the GRT. Bowen's discussion of an "open skies" policy with the US foundations in the early 1950s was, however, acknowledged in 1961 discussions between Bowen and the Rockefeller Foundation (see NRAO ONLINE.47, 19 April 1961). Bowen described requests of overseas astronomers to use the GRT: "... [I]t is one of our objectives to throw the instrument open to any competent astronomer who has a worthwhile problem to tackle." Bowen wrote the same message to Sir Walter Bassett after his visit in March 1961 concerning "open skies". However external users and external members of programme committees would only become a common practise from the 1980s.

GRT Receiver Plans

A major source of conflict in the previous months had been the choice of receivers for the GRT; discussions had already started at RPL in 1959. In a letter to Pawsey of 1 December 1960, Frank Kerr reported that Bolton had suggested that the committee be disbanded. This was clearly a bad idea since many decisions had already been made during the past year at RPL and the RPL staff still had not been informed of the role that Bolton was to play. The two meetings of early October 1960 had only partially cleared the air.

A confused discussion began on 5 August 1960, when Brian Cooper from RPL reported to John Bolton the receiver plans for the GRT. Bolton responded with a critical letter questioning many of the decisions which had been made at RPL. But Bolton already had direct experience at the Owens Valley Radio Observatory (OVRO). He dismissed the necessity of the 120 times 25 kHz spectral line backend: "I think this is a complete waste of effort." Based on current knowledge of the properties of the HI interstellar medium, we now recognise that 25 kHz (5.3 km/s at

---

[12]The proposed scheme was to schedule the GRT in one-month blocks, with the possibility of longer- term allocations. "Director's Discretionary Time" was also to be possible. The minimum observing team consisted of an "astronomer" and a "radio man", someone responsible for the equipment. The two were to share in the planning and interpretation of the observations.

the 21 cm line) was only marginally suitable for HI emission studies and completely unsuitable for the narrower HI absorption lines. Bolton favoured much narrower channels of 5 KHz (about 1 km/s) required for studies of the interstellar medium based on HI absorption. In addition, Bolton also discarded the idea of an initial use of the GRT at 10 cm (2.7 GHz) until efficiency measurements had been made. Since most of the RPL staff members would have had no perception that Bolton was to be in charge of the GRT in a few months, this criticism would have seemed unwarranted.

Based on the HI results obtained in 1959 by Radhakrishnan, Clark and Wilson at the OVRO, as well as contacts that he made at the URSI meeting in September 1960, Bolton wrote a thorough description of his plans on 3 October 1960: "I have arrived at some fairly definite plans for the GRT. I believe that we should roughly divide the effort between investigations of the [radio] sources and investigations of normal galaxies—the latter both in continuum and the hydrogen line. 21 cm absorption spectra of sources comes in both these categories ..." With HI absorption it was possible to derive the properties of the interstellar medium as well as information on the distance of the background object. Bolton also insisted on a sky survey at 440 MHz of the southern sky, including the Magellanic Clouds and a survey to find 500 extragalactic sources. This was the genesis of the very successful Parkes Catalogue of more than 1840 radio sources (at 408 MHz) and their optical identifications (Otrupcek & Wright, 1991).

Bolton was already considering

> ... projects involving the 210-foot as the principle element of more elaborate systems. One such system would be the addition of two smaller dishes for high precision positions where identification is suggested with very distant objects [referring to the identification of the radio galaxy 3C 295, the most distant galaxy known at the time (Minkowski, 1960, p. 908)]. Another major project would be the simultaneous combination of the 210-foot data with a number of close-by moveable small dishes and a number of radio-linked out-stations for synthesis of the brightness distributions of the sources.[13] This investigation in my opinion is one of the most promising fields and our own first attempts here [at OVRO] have been very rewarding. It promises both the physics of the source mechanism and a method of continuing radio observations of the [distant] universe when we have to leave the optical correlation behind. For this sort of program, I believe we should use the highest frequency at which we can use the full aperture of the [GRT] ...

The suggested program involving the additional small dishes was implemented by moving the 60-foot Kennedy dish from Fleurs to Parkes in 1963.[14] The primary intention had been to use it as a connected interferometer to determine accurate positions in order to make optical identifications as Bolton had suggested. A

---

[13] This proposal is similar to the AST, Australian Synthesis Telescope, proposal of mid-1977 made by RPL staff, consisting of movable antennas at Parkes that formed a compound interferometer with the 64 m Parkes telescope. The AST evolved into the ATCA, Australia Telescope Compact Array at Narrabri of 1988. (Frater et al., 1992, p. iv).

[14] Orchiston (2012). The 60-foot Kennedy antenna had been in operation at Fleurs since May 1961. The Parkes interferometer (the 210-ft dish and the 60-foot movable antenna) was to begin operation in October 1965.

secondary objective was to extend the successful OVRO program to measure the brightness distribution of southern radio sources. For the later project, Bolton had conceived an innovative modification of the OVRO interferometer by implementing a continuously variable baseline. This unique and highly flexible arrangement was successfully used to determine the structure of southern radio sources (Ekers, 1967) and for spectral line absorption measurements.[15] However, the trailing cable used to connect the moving 60-foot telescope to the GRT did not have adequate phase stability to derive accurate positions. Fortunately, the pointing accuracy of the GRT alone exceeded expectations and was adequate for making optical identifications. The interferometer was never used for accurate position determination (see below).

Brian Cooper responded on 25 October and 9 November 1960. He was aware at this time of the conflicts: "With regard to your proposals for an observing program, I feel that this is properly a matter for committee discussion, and I believe that Taffy, in consultation with Joe, will be writing to you about this." Then in the second letter: "You are certainly coming with some interesting results in your extragalactic source work and I can appreciate that you are most anxious to press on with this on the 210-footer. However, the whole question of observing programs is a rather intricate one and we had best leave further discussion until you come over."[16]

Pawsey-Bolton GRT Discussion, Late 1960

In late 1960, as John Bolton was preparing to leave the US (12 December 1960), he wrote Pawsey on 16 November 1960. The letter contained a comparison of the Mills, Slee and Hill (MSH) 80 MHz catalogues with the newly obtained OVRO data at 960 MHz. Many of the extended sources in the MSH catalogue were not confirmed at OVRO or were smaller than 2 arc min; also, the position errors in MSH were often underestimated. This may have been the trigger for Bolton to start a new discrete source survey using the Parkes dish.

The most striking paragraph in that letter was about 3C 48. This letter (an aerogram) was written a month before the Bolton family departed from California by ship in mid-December 1960:

> A couple of weeks ago I wrote to Taffy and said I thought we had a star [that is, detected a star in the radio]. It is not a star. Measurements [by Jess Greenstein] on a high dispersion spectrum [from the 200-inch, was obtained by Allan Sandage] suggest [to Bolton] the lines are those of Neon [V], Argon [III] and Argon [IV] and that the red shift is 0.367. (See footnote 1). The absolute photographic magnitude is then–24 which is two magnitudes greater than anything known. The continuum is still going up towards the blue and may well be synchrotron. I think this must be the early stage of a radio galaxy, probably short-lived and so very rare ... The source is 3C 48 and can be seen clearly on the 48-inch Schmidt

---

[15]Goss et al., 1970, and Radhakrishnan in Goddard and Milne (ed.), 1994 (see Footnote 16, Chap. 32).

[16]NAA C4633/3.

plates ... I don't know how rare these things are going to be but one thing is quite clear—we can't afford to dismiss a position in the future because there is nothing but stars.[17]

Pawsey replied[18] on 7 December 1960 with a frank discussion of their growing conflicts. The letter was posted in care of the P and O Steamship Company, *SS Orcades* in Los Angeles and also to an address in Honolulu, the first stop of the *SS Orcades* on the way to Australia. John, Letty and Peter Bolton were to depart on 12 December 1960. Pawsey discussed the "pseudo-star which turns out to be the brightest known astronomical object" (3C 48). He congratulated Bolton on his achievements in collaboration with Caltech optical astronomers.

> You will find less facilities in this line available in Australia, but even more goodwill. Bart Bok [who served as Director of ANU's Mt Stromlo until 1966] is extremely interested and anxious to be cooperative. Until now, cooperation with regard to source identification has been unsuccessful [due to radio source position accuracies of at best a few arc min].

Pawsey described, and enclosed the minutes of, the meeting of 4 October 1960, establishing the division of responsibilities at RPL. Pawsey tried to "clear the air":

> As you remarked in a recent letter to Taffy [not found in the archives], program discussions at a distance are too tricky. One picks on the wrong things. For example, you are critical of what you have heard [clearly Pawsey had seen the letters from Bolton to Brian Cooper] of our plans for a new multi-channel receiver. What I want is a receiver with facilities for narrow bands ... [for HI absorption] ... My guess is that you will agree with my policy. [This was the case; Bolton had suggested this to Cooper on 9 August] ... similarly I have heard gossip [via Cooper on 9 August 1960?] that you are very much against using 10 cm on the big dish ... [I]t appears imperative to me to get going soon at the shortest feasible wavelength [10 and 20 cm] ... The real point is that these programs require discussion here in Sydney and what goes on 10,000 miles away creates difficulties.

> [Pawsey then provided advice about Bolton's role at RPL.] Programs involve both things and people and your role here should be a dual one: partly scientific leadership and partly individual research. It is terribly important that you should gain the confidence of the people under you, **both as regards scientific judgement and integrity and also from a personal point of view**. [our emphasis] You are familiar with this from your work in directing the Owens Valley Lab.[19]

Immediately, Bolton wrote (handwritten on *SS Orcades* letterhead) from Honolulu on 19 December 1960[20]:

> I am very much looking forward to my return to Sydney at RPL. I hope I can contribute to getting the big dish and decimetre observations under way ... and later perhaps on the optical side from what I have gained in the last five years. The last few months have been

---

[17] Bolton has described the 3C 48 experience in 1990 (Bolton, 1990, page 381).

[18] NAA C3830 Z3/1/X.

[19] Pawsey ended the letter with a surprising message: Gilbert Roberts of FFP was in Parkes for a fortnight's holiday with his wife. Pawsey asked "Will he last the distance?" Roberts did witness the completion of the GRT; he died 1 January 1978. (Kerensky, 1979). See Fig. 30.2.3, Roberts at Parkes, Christmas 1960, in ESM_30.2.pdf, Lord Casey.

[20] Bolton informed Pawsey that he had been quite ill since leaving California. He had been in bed with fever, treated with antibiotic injections every 4 hours.

somewhat difficult. A number of people from RPL have been through and there has been certain correspondence. Answers have been given to questions which when removed from context inevitably succeed in irritating someone ... [Bolton explained that he was against 10 cm use due to his doubts about the quality and over-size of the individual surface panels.] I have deliberately avoided writing to you as I felt it was foolish to run the risk of any pre-Sydney arguments. While five min of discussions across a table can reach amicable agreement, ten times the amount of letter writing can lead to all sorts of cross purposes. While there have been differences between us in the past, I believe, on what kind of GRT to build, I do not believe we have had had any differences on scientific aims. I would like to assure you of my complete cooperation. I shall do my best to avoid any unnecessary dissension within the group—or escalating any that exists. As I have said to Taffy, I am sure the GRT will succeed and once we begin a chain of research success and discovery— personal animosities quickly heal over. We will have to work a lot closer together than in the past. [Since the users of the GRT will be a single group] ... It is always easier to dislike someone you don't know very well!

Bolton ended his letter with a postscript on 3C 48, a disappointing conclusion that turned out to be incorrect: "The last news on 3C 48 as I left Caltech was—it is most likely a star—all astrophysicists had admitted defeat on identification of the lines and had agreed to publish same for open competition."

But as is now well known, the redshift of 3C 48 was confirmed at 0.367 (the value originally suggested by Bolton in 1960 but then withdrawn) in 1963 by Greenstein and Matthews after the redshift of 3C 273 was determined by Schmidt based on the 1962 lunar occultations of the radio source observed with the Parkes dish.[21] This discovery is also discussed in Chap. 32.

As 1960 ended, Pawsey could foresee a clouded future; he had lost three of his most trusted colleagues, he could see that his role as the leader of radio astronomy in Australia was diminished and he had perhaps lost the confidence of the CSIRO Executive. Was there a sense of betrayal? Was his management style obsolete? Should he "quit" or "hang on"? As Frank Kerr told Woody Sullivan in 1986: "Competition had to grow when radio astronomy became "Big Science" ... There were too many entrepreneurs for RPL to hold them anymore. It was inevitable that some would move [on]. It was a great pity that many of the people concerned didn't have a sense for this to happen without acrimony."[22]

Pawsey's outlook can be summarised by a reflective exchange he had with one of his most trusted protégés Ron Bracewell. On 16 July 1960,[23] Pawsey sent a confidential letter to Ron (typed at the Pawsey home), as he described his state of mind in mid-1960: "Another intangible is my general feeling of unrest. A year ago we had what I was sufficiently egotistical to consider the outstanding radio astronomy group in the world. Now Alec [sic Alex] Shain is dead and Chris and Bernie gone. I no longer see the future clearly."

Ron Bracewell wrote back immediately from Stanford (27 July 1960):

---

[21] See Hazard, Jauncey, Goss, and Herald, Hazard et al., 2018, page 6.
[22] W.T. Sullivan archive.
[23] NAA C3830 Z3/2 Part 3.

All radio astronomers are puzzled as to the future in Sydney, and it is often discussed. The fragmentation that has taken place is an inevitable concomitant of maturity. I look forward to a rearrangement of the pieces that will favour continued successful development of radio astronomy in Australia and send my best wishes for your part in it.

Chapter 32
Reflections on GRT Science, post 1961

*Sited as it is in the heart of New South Wales grazing country,
this magnificent scientific instrument emerged to become, not
only the envy of many research groups all over the world, but
also an enduring source of pride to Australians from farm and
factory alike. In a subtle way, it had constituted a visible
symbol of Australia's intellectual "coming of age". It finally
shattered the outworn image of the broad-hatted sheep-
grazing Australian, which thoughtful Australians had been
trying unsuccessfully to blur for some time.*
W.F. Evans, "History of Radiophysics Advisory Board
1939–1945" (1970)

Planning the First Months of the GRT: The Parkes Telescope[1]

As 1961 began, John Bolton re-joined CSIRO in early January. On 24 February
1961, a meeting of the 210-foot Radio Telescope Committee (aka the GRT) was held
with Bowen (chair), Minnett (secretary), Pawsey, Beard, Higgs, Bolton, Cooper,
Day, Mathewson, McCready and McGee.[2] Pawsey had prepared a document 2 days
earlier that set out a plan of action for the GRT; the construction was expected to be
completed by August 1961. Pawsey described Phase 1 tests of the new aerial,
consisting of determinations of the shape of the dish, pointing accuracy as a function
of position, wind and temperature. An example was the determination of the "...

Supplementary Information The online version contains supplementary material available at
[https://doi.org/10.1007/978-3-031-07916-0_32].

---

[1]In 1955, the formal proposal was titled "A Proposal for a Giant Radio Telescope". But during
1961, the GRT name disappeared. At the opening on 31 October 1961 the official programme
was titled "Inauguration of the Australian National Radio Astronomy Observatory
and the Commissioning of the 210 Ft Radio Telescope". Soon the term "Parkes radio telescope"
or "Parkes Telescope" was in common use. In 2020 the common name is "Parkes Observatory".
[2]NAA C3830 A1/1/7.

W. M. Goss et al., *Joe Pawsey and the Founding of Australian Radio Astronomy*,
Historical & Cultural Astronomy, https://doi.org/10.1007/978-3-031-07916-0_32

directional patterns at various polarisations to access dish accuracy, the determination of the radio axis of the dish ... The measurements will require a series of radio sources of known positions distributed over the coverage of the telescope."

The 22 February 1961 document described Phase 2: initial observations with existing receivers at 20 and 75 cm in the continuum and at 21 cm for the hydrogen line (HI). The major continuum research programmes proposed by Bolton were discussed: positions and spectra of 400 catalogued sources at 75 and 20 cm with the goal of optical identifications. Frank Kerr was to start galactic HI observations, concentrating on the galactic centre, while Dick McGee was to start a 21 cm line survey of both Clouds of Magellan. The planning for a promising programme of planetary research (Jupiter, Venus and Saturn) was to await the return during the next month of Jim Roberts who had been seconded to Bolton's group at Caltech.[3]

At the meeting of 24 February 1961, the status of GRT construction was described:

> The access tower on the turret has been erected and all welding work on the turret and hub is complete ... all the ribs have been bolted together in groups and painted ready for erection. The spiral purlins have been bent to shape and are being painted ... It is now reported from London (by Frank Kerr) that the control desk (Askania) may be shipped on 15 March 1961—2 weeks earlier than previously expected.

In the section concerning new spectroscopy, HI observations of external galaxies by Brian Robinson, using the "Leiden receiver" (21 cm parametric amplifier), were discussed along with "high order 21 cm spectral lines", clearly the H 158α hydrogen recombination lines of ionised hydrogen from HII regions, a project of John Murray.[4]

In Phase 2, John Bolton had prepared an observing scheme that occupied 3 months, planned to begin 1 October 1961: (1) Jupiter observations of the intensity and polarisation at 10, 20 and 75 cm for 3 days, (2) "First radio star finding programme" (in 1961 Bolton still used the old terminology of "radio star") for 10 days, (3) Continuum and HI line survey of the Magellanic Clouds for 20 days, (4) Selected regions at right ascension 03–05 h and declinations −20 to −35 degrees (the northern declination limit of the GRT was +27 degrees, due to the 60 deg. zenith angle limit).

A discussion followed on the topic "operating arrangements at Parkes" with two possible models: (1) the users themselves would operate the telescope or (2) telescope "controllers" (the Jodrell Bank terminology) or operators who would control the movement of the telescope.

The minutes of the meeting continued:

---

[3] See Chap. 19 and Roberts, 2002 unpublished memoir *Have Gen, will Travel*. This text is presented in NRAO ONLINE.49.

[4] Apparently, Murray's proposed observations were never carried out.

1. Bolton proposed that, in the initial stages, responsibility for operating the tele-scope should be taken by the small group of people who would become familiar with the instrument during the commissioning period. For the longer term, these people would train and "license" other suitable members of the research staff. Thus a typical research team might consist of a Research Officer, Experimental Officer and Technical Assistant, and they would operate the telescope during the 12-h observing period on a rostered basis. [Initially the telescope was only used for astronomical observations during a 12-h period at night]

2. The other scheme is that a number of men [No women!] should be recruited and trained for the special job of operating the telescope. One of these "controllers"[5] would be attached to the observing team and would carry out the movements required by the observer in charge. The safety of the telescope would be the responsibility of the "controller". This system provides continuity of operating skill and safety procedure, whatever the observing team composition.

> The first scheme has been used successfully with the somewhat simpler control system of the Caltech telescopes [at Owens Valley Radio Observatory]. The second is similar to that used at Jodrell Bank, and Dr Mathewson reported that the skill of the operators there allowed the astronomer to concentrate on the observations. Dr Bowen said this relation was similar to that between pilot and scientist in rain-physics experimental flights. The scheme finally adopted might well combine some of the features from each proposal.[6]

The next meeting of the radio astronomy group (organised again by Pawsey) occurred on 10 March 1961 with the title "Radio Astronomy Group to Discuss (1) Technical Manpower Questions and (2) Technical Liaison in Laboratory". At this time, two groups from Fleurs were present: the compound interferometer group of Krishnan, Harting, and Payten using the 60-foot Kennedy antenna with the Chris Cross, and the radio link interferometer group of Peter Scheuer (visitor from the Cavendish Laboratory), Bruce Slee, Higgins and Fryar using the Mills Cross at Fleurs as the main element at 85.5 MHz. A major decision at this time was the setting of a firm time scale for the GRT receivers, e.g. the 75 cm receiver was to be ready by 1 September 1961.

Progress with the GRT Construction: 1961

On 15 February 1961, Bowen wrote White with a comprehensive report concerning GRT status. A major problem was in the erroneous calculation of the desired amount of counterweight for the GRT, a remarkably serious oversight.

> FFP have been secretive about it, but final instructions have now been given and the counterweight material is going in. When this is completed, the ribs, which were finished

---

[5] A term used at Jodrell Bank; the word never caught on at Parkes.

[6] The scheme finally adopted was a variation on proposal 1, with a member of the Parkes technical or maintenance personnel who became the "telescope driver" at night or during the weekend.

some time ago, can be lifted into place. [The amount of counterweight would remain a problem well into 1962.] There has been further fiddling around with the mesh panels ... Roberts has always been petulant and a bit shy about this one, and this is ... having its effect on the fabrication.[7]

The main problem was with Askania and the ME (Master Equatorial) and the control desk; the delays were to extend for at least 2 months.[8]

On 19 April 1961, Bowen provided a detailed update to Dr. Robert Morison, Director of Medical and Natural Sciences, Rockefeller Foundation, on the status of the GRT.[9] This Foundation was the most recent and largest of the overseas benefactors.

The erection of the structure on the site at Parkes has been progressing steadily for the past eight or nine months and a few days ago the last rib was lifted into place on the dish ... The remaining structural work will be completed about mid-June and the electrical control gear shortly afterwards. Some of the precision components are ... behind schedule but these should allow completion of the device in July or early August [too optimistic by 2–3 months]. For this [time scale] we have to thank the excellence of our contractors, the MAN Company, and our design engineers, Freeman, Fox and Partners.

Bowen then described the interest shown by US agencies in the progress of the GRT: NASA, Office of Naval Research and the Air Force's Lincoln Laboratory of the Massachusetts Institute of Technology. They were interested in the scientific use of a southern hemisphere instrument but "I suspect they are interested more ... in the refinements which have been built into the device and in the economy with which it has been constructed ... [W]e are anxious to give them all the assistance we can in return for the generous help we have always received from our friends in the USA".

Finally, the question of "open skies" was raised by Bowen in the letter to the Rockefeller Foundation; as had occurred earlier,[10] the issue of "repayment of the debt" of the overseas support from the two US foundations for the GRT:

[W]e have informal requests from around a dozen radio astronomers [from outside Australia] for permission to work with us on the telescope ... This makes us exceedingly happy and it is one of our objectives to throw the instrument open to any competent astronomer who has a worthwhile problem to tackle.[11]

---

[7] In spite of the rapid progress made with the construction, Roberts continued to irritate Bowen in 1961.

[8] NAA C3830 Z1/14/A Part 2. Also Bowen told White that he hoped that Casey, Walter Bassett and himself could visit Parkes in mid-March 1961; the visit did occur on 13 March 1961, almost exactly a year after the visit described in NRAO ONLINE.46.

[9] NAA C3830 A1/3/11/3 Part 5. The first Rockefeller grant had been made in December 1955 followed by the second grant in December 1959.

[10] Discussed in 1955 by Bowen in correspondence with the Rockefeller Foundation.

[11] NAA C3830 Z1/14/A Part 2. In a letter to Sir Walter Bassett on 19 April 1961 (after the visit of the previous month), Bowen was proud of their achievements: "On the whole we seem to have a powerful combination of British design, German fabrication and Australian get up and go." Bowen also mentioned, as he had to the Rockefeller Foundation, the desirability of an "open skies" policy with respect to overseas visiting observers. Even the Parkes *Champion Post* newspaper of 30 October 1961 contained a story entitled "Available to World Scientists": [text] "The Giant

CSIRO paid only lip-service to this commitment until 1988, when the Parkes radio telescope and the newly completed Australia Telescope Compact Array were combined into a single National Facility. These instruments were then open to the best scientific proposals from all over the world. Prior to this period the facilities were operated by CSIRO for its own scientific staff. While collaboration with outside users was encouraged, there was no open access process. In the modern era, this "open skies policy" is supported by the international community of astronomers. Thus, there is a sharing of facilities, providing all Australians with reciprocal access to many observational facilities worldwide.

In April–May 1961, Bowen had an exchange of letters with Merle Tuve[12] concerning the Carnegie Corporation of New York grant. He had given an up-to-date report to the Carnegie Corporation of New York (Stephen Stackpole); the Corporation had given the first overseas funds in 1954 that provided the essential kick-start to the GRT. "It has been a long haul and we are happy to see this phase of it drawing to a close. We would never have got so far without the encouragement and continued interest of our friends overseas."

On 20 July 1961,[13] Pawsey sent a memo to all "Research Officers and Experimental Officers in Radio Astronomy Group" at RPL. This was just before he departed on 26 July 1961 for a visit to the US and Europe for the IAU in Berkeley in August 1961 (Chap. 38), returning only just before the opening of the GRT at the end of October 1961.[14] Pawsey wrote:

> Last October (colloquium 4 October 1960) when arrangements for operating the Parkes radio telescope were discussed it was stated that there would be three basic sections in the radio astronomy group: GRT, Solar and Receivers under Bolton, Wild and Cooper, respectively. The time has now come to implement this arrangement and I now wish to allocate responsibility as shown on the attached page [an organogram].
>
> As I stated last October I wish free interchange between observing and technical groups and, as a receiver, for example is completed, the appropriate individuals should transfer to Parkes. Such transfers should normally be arranged between heads of section.[15]
>
> I should also like to take this opportunity of informing you that Bolton and Wild have been promoted to the grade of Chief Research Officer. CRO's who are not actually Chiefs of

---

Radio Telescope will be available to overseas scientists to assist them in studies for which the instrument power and capabilities are essential."

[12] NAA C3830 Z1/14/A Part 2. Tuve was a member of the scientific staff of the Carnegie Institution of Washington.

[13] NAA C3830 A1/1/1 Part 14.

[14] On 19 July 1961, White and Bowen exchanged letters about Pawsey's reluctance to be present for the opening of the GRT in late October 1961. Pawsey was anxious to be in London at this time for the presentation of the Hughes Medal of the Royal Society. He wrote to White on 21 July 1961, agreeing to be at the opening. As we discuss in Chap. 40, Fred Hoyle presented the Hughes Medal to Pawsey at his hospital 4 weeks before his death on 30 November 1962.

[15] Pawsey informed the staff that during Paul Wild's absence (at the IAU) Kevin Sheridan would be his deputy; Bolton was to coordinate the Parkes, receivers and data processing groups. The Parkes group included Bolton, Kerr, Roberts, Mathewson, Hill and McGee. The receiver group consisted of Cooper, Gardner, Robinson, van Damme, Gruner and Mackey.

Fig. 32.1 In June 1961, the tripod was positioned with the aerial cabin mounted. This image, on a stormy day, from the Austie Helm collection (via his son Denis Helm and grandson Scott) shows the final moments of the lift. Note the man on the right most feed leg of the tripod. The erection crane was used by MAN for the construction of the telescope starting in late 1960. Credit: Austie Helm Collection, all rights reserved

Divisions are very rare in CSIRO and we heartily congratulate them on this recognition of their work.

Opening of the GRT, 31 October 1961

The contractors (MAN) had arrived in Parkes in September 1960, 13 months before the scheduled opening on 31 October 1961. In June and October 1961, memorable events occurred, "dramatic events" according to Minnett ("The Construction of the Parkes 210 -ft Radio Telescope" in Goddard and Milne (1994, p. 16).[16] In Fig. 32.1 we show the aerial cabin being lifted to the three feed legs, the tripod (about June 1961). Minnett was a participant in this heroic event. He wrote in the 1994 volume: "When the aerial cabin had been secured in position on the tripod, the telescope had reached its highest point. A photograph from the erection crane was suggested to mark the occasion with some figures in it to add human touch."

---

[16] *Parkes 30 Years of Radio Astronomy* (1994), Goddard and Milne, CSIRO Australia has contributions from many of the key players who were still alive and provide lively first-hand discussions of the commissioning and early science results.

Fig. 32.2 Harry Minnett (left) and Dennis Gill in June 1961 on top of the Parkes telescope aerial cabin, as photographed by George Day from the erection crane. Credit: CSIRO Radio Astronomy Image Archive JP32–2

George Day took the photograph shown in Fig. 32.2 (from the George Day collection) from the erection crane (height 230 feet; see Fig. 32.1). Indeed, the position occupied by Dennis Gill (to the right) and Minnett was precarious. Minnett confessed in 1994 that he would not like to "repeat the event all these years later." Day was at an even higher elevation than the aerial cabin when the photo was made.

The second memorable event was the "tilting the dish from the zenith for the first time", on 7 October 1961, 3 weeks before the opening. The *Parkes Champion Post* wrote on Monday, 30 October 1961 (the special edition for the opening on the following day):

> On October 7 engineers performed the first trial tilt of the Big Dish and brought to life an idea first mapped out on the drawing boards of a London firm of engineering consultants 6 years earlier.
>
> A CSIRO photographer [likely Ken Nash] aboard a low flying aircraft watched the 1000 tons of mesh and metal bow towards the horizon. This is possibly the first photograph [Fig. 32.3, 7 October 1961] published showing the actual instrument in the tilt position.[17]

The "Inauguration of the Australian National Radio Astronomy Observatory and the Commissioning of the 210-foot Radio Telescope" occurred on Tuesday

---

[17] Confirmed by Minnett ("The Construction of the Parkes 210 Ft Radio Telescope" in Goddard and Milne, 1994). An additional image of this event on 7 October 1961 from ground level was presented by Minnett (in Goddard and Milne, 1994, p. 17, Fig. 3).

Fig. 32.3 The 7 October 1961 "first tilt" image, taken from a low flying aircraft. Credit: CSIRO Radio Astronomy Image Archive B6573-11b. In Minnett (Goddard and Milne, 1994), an image of this event taken from the ground is shown

afternoon 31 October 1961. A copy of the three-page programme is show in Fig. 32.4. The cover image is the 7 October 1961 aerial photo of the first tilt of the dish, shown in Fig. 32.3. The official opening was carried out by the Governor-General of Australia, His Excellency the Rt Hon Viscount De L'Isle, Victoria Cross (WWII). Other speakers would be F.W.G. White, the Chairman of CSIRO, Dr. D.A. Cameron, the Minister of the CSIRO and Lord Casey, former Minister (and long-time supporter of the GRT). After the official commissioning by the Governor-General, E.G. Bowen, Chief of RPL, concluded the proceedings.

In Fig. 32.5 we show the photo of the Parkes dish on 31 October 1961 taken by Pawsey. The cars of the early arrival VIP guests are shown on the right as well as the special grandstand built for the speakers. On this day, Pawsey's duty was to escort the VIP guests through the telescope tower. The arrival of the guest of honour, Viscount De L' Isle and his wife is shown in Fig. 32.6, greeted by Fred White. In Fig. 32.7, the departure later in the day is shown with the long-term CSIRO staff member Cliff Smith standing to the reader's right of the door.

During the presentations, both George Day and Austie Helm took coloured photos, shown in Fig. 32.8 (George Day photo of Lord Casey) and Fig. 32.9 (Austie Helm, of De L' Isle). In Fig. 32.10, an image of the assembled audience is shown; the photo was taken from the telescope backup structure.

In the 1994 volume (Goddard and Milne, 1994), John Masterson, photographer at RPL for almost 40 years, published an amusing essay "The Parkes Radio Telescope—A 30-Year Photographic History" with 37 images from 1964 to the 1990s.

Fig. 32.4 The Programme of the Opening Ceremony of the Parkes telescope on 31 October 1961. The image on the front cover is from Fig. 32.3. Credit: CSIRO Radio Astronomy Image Archive JP32–4

Apparently, on the day of the opening, Masterson began a tradition at the GRT: photographing famous and not-so-famous people as they popped up into the dish via a manhole cover. Likely the first and prototype subject was the Governor-General Viscount Lord de L'Isle on 31 October 1961, Fig. 32.11. Masterson: "While touring the telescope, Lord De L' Isle [sic] was photographed entering the dish surface through a hatch. I have since photographed many visitors and staff climbing through this hatch, and find that some enter with a certain amount of style, some don't." In Fig. 32.11, Bowen is at the far right, while one of the Governor-General's staff appears to be offering a hand while the other seems to be pressing the Governor-General back into the inner dish!

An important aspect of the opening day on 31 October 1961 was the recognition of the local Parkes and Goobang Shire citizens who had played a major role in the

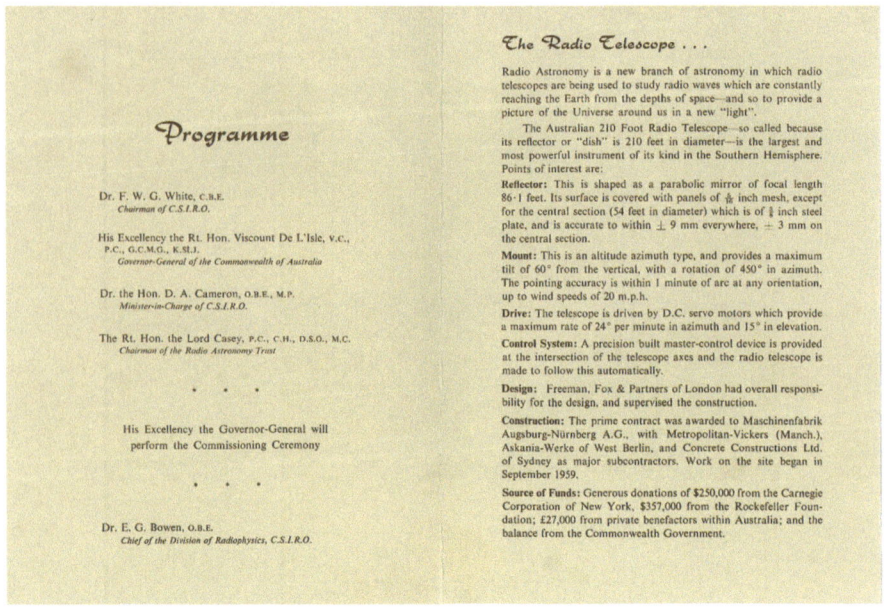

Fig. 32.4 (continued)

choice of the site and the preparation of the landscaping. Both the Helm and Jelbart families were invited as VIPS to the event. The authors have seen the official guest badges of both groups, the Jelbart family invitation and passes are shown in Fig. 32.12.

Events After the Opening: RPL Scientists, Publicity in the International Press

On 23 November 1961, about a month after the GRT opening on 31 October, the radio astronomy group had a meeting organised by Pawsey to discuss plans for the next months. Bolton began with a summary of the status of the GRT. A major point of the discussion was the status of the receivers for the telescope. Other groups at RP gave their reports, including Paul Wild on the "design study for equipment for metre-wavelength pictures of the sun" (the future Culgoora Radioheliograph). Scheuer (visiting from Cambridge see Chap. 34) and Slee reported on their 85.5 MHz long baseline interferometry plans. This was a survey initiated by Mills and Scheuer to

Fig. 32.5 Pawsey's image of the opening day, 31 October 1961. The early arriving VIP's cars are shown as well as the temporary grandstand. Credit: Joe and Lenore Pawsey Family Collection

settle the remaining disagreement between Sydney and Cambridge on the effect of source angular size on the source counts[18] (discussed extensively in Chap. 34).

On 16 November 1961, Bowen wrote Morison (Director of Medical and Natural Sciences, Rockefeller Foundation) with news of the opening of the 210-foot telescope on 31 October 1961, including a number of photographs. Morison replied on 15 December 1961[19]:

> In light of our experience here, it is simply fantastic the way you have succeeded in meeting all of the specifications in the allotted time. *The New York Times* [see below- 5 December 1961] carried a very good story during the week of our Trustees' Meeting, and several of the Trustees commented with satisfaction on the modest part the Foundation played in making the project a success. The comments of some of those who know something about our American efforts in the same field were, I'm afraid, mingled with a certain degree of

---

[18] About a 1000 radio sources were observed in the southern sky; the baselines ranged from 6 km to 32 km providing partial angular size information in the range to 3.5 arc min to 20 arc sec. A preliminary publication appeared in the 1963 issue of *Proceedings of the Institution of Radio Engineers Australia*, edited by Pawsey shortly before his death, "Apparatus for Investigating the Angular Structure of Radio Sources" by Scheuer, Slee and Fryar, p. 185, vol 24, but the main survey results were never published. Bruce Slee, hoping to publish the unique data set, tried for some years to obtain the data from Scheuer, with no response. After Peter Scheuer's death in 2001, Bruce Slee asked John Baldwin to look for the records at the Cavendish Lab and Lord's Bridge, with no success (interview with Baldwin by Goss, Cambridge, August 2010).

[19] NAA C3830 Z1/14/A Part 2.

Fig. 32.6 The Governor-General Viscount De L'Isle and his wife arrived at Parkes, welcomed by Fred White, CSIRO Chairman. Credit: CSIRO Radio Astronomy Image Archive B6607-1

Fig. 32.7 The Governor-General Viscount De L'Isle's departure later in the day. Credit: CSIRO Radio Astronomy Image Archive B6607-14

Fig. 32.8 George Day's photograph of Lord Casey at the Parkes opening, 31 October 1961. Credit: CSIRO Radio Astronomy Image Archive JP32–4

Fig. 32.9 Austie Helm's photo of Viscount De L'Isle during his opening of the Parkes telescope on 31 October 1961. Credit: Austie Helm, all rights reserved

embarrassment. Please accept our warmest congratulations to you and the entire group that worked with you.

On 19 December 1961, a thoughtful report on the GRT appeared in the US newspaper *The Christian Science Monitor* written by Albert E. Norman,

Fig. 32.10 The audience at Parkes, 31 October 1961. Photo taken from the telescope backup structure. Credit: CSIRO Radio Astronomy Image Archive B6607–13

"Radiotelescope, Deep Probe into Universe". He provided an update on the radio telescope located

> within 15 miles of this quiet rural town [Parkes] where sheep graze among wheatfields . . .
> The Australian telescope has the advantage of being located on the level plain of the
> Goobang Valley, a region free from industrial radio "noise" generated by factory machinery
> and where temperature variations are not generally great.

Norman described the 1945–46 pioneering solar research of Pawsey and the discovery of "radio stars" by the CSIRO group of Bolton and Stanley in 1946 at Dover Heights. A picture of Bolton at the new GRT was included, but with no identification; however the text described both Pawsey's and Bowen's role. Norman continued:

> In 1945, an Australian research team under Dr J.L. Pawsey demonstrated with radio
> equipment that the atmosphere of the sun was very much hotter than was supposed and
> that sunspots were the sources of strong radio emissions.
>   . . . It was international admiration of the pioneering work of Australian astronomers in
> this field under the leadership of Dr. Edward G. Bowen . . . that led the Carnegie Corporation
> to offer $250,000 toward an Australian radiotelescope. This was followed by donation of a
> similar sum from the Rockefeller Foundation and the balance needed to provide the $2
> million came from private Australian donors [sic] and the Australian Government.

Fig. 32.11 John Masterson's (1994) caption of this first "popping up into the dish" iconic photos reads "Governor-General Lord De L'Isle pops up and into the dish. 31 October 1961". Bowen at the extreme right. Credit: CSIRO Radio Astronomy Image Archive B6607–8

The precedence of the Jodrell Bank radio telescope was emphasised:

> The Australian telescope owes a good deal to the British one since discoveries made after the instrument was begun revealed directions in which the Australian telescope would need to be improved if it was to exceed the performance of its British partner, the only two in the world of this type of such giant size.

The article also foreshadowed an important contribution the GRT would play with the next decade in the Apollo programme of NASA: "It is expected that Australia's debt of gratitude for the very generous American gifts will be a little repaid when the United States begins launching a truly deep space probe such as the projected moon shot."[20]

This return to NASA was soon to be realised. Following the commissioning, Parkes was used in the Mariner space tracking mission as an extension of the NASA

---

[20] The *New York Times* article of 5 December by Harold M. Schmeck, Jr. (Schmeck, 1961) was a brief account that appeared before *The Christian Science Monitor* report: "The only larger instrument of its kind in the world is the 240 [sic]-foot dish at Jodrell Bank in Britain. Nothing in the Southern Hemisphere approaches the Parkes telescope in size. Specialists believe that its design, more refined in important respects than that at Jodrell Bank, should give the Australian radio telescope capabilities unequaled, at present, anywhere in the world."

The Chairman and Members of the Executive of the
Commonwealth Scientific and Industrial Research Organization
request the pleasure of the company of

Mr. and Mrs. P. I. Jelbart

at the inauguration of the
Australian National Radio Astronomy Observatory
and the Commissioning of the 210 ft. Radio Telescope by His Excellency
The Rt. Hon. Viscount De L'Isle, V.C., P.C., G.C.M.G., K.St.J.
Governor-General of the Commonwealth of Australia
at Parkes, N.S.W., on Tuesday, 31st October, 1961, at 2.45 p.m.

R.S.V.P. by enclosed card before 9th October.
Admission cards will be sent on receipt of acceptance.

OFFICIAL GUEST

INAUGURATION
of
AUSTRALIAN NATIONAL
RADIO
ASTRONOMY OBSERVATORY
PARKES, N.S.W.
31st OCTOBER, 1961

Fig. 32.12 The Jelbart family invitation and passes. Credit: CSIRO Radio Astronomy Image Archive JP32–4 provided by the Jelbart family

contract to use Parkes as part of the design study of NASA's deep space network of 64-metre antennas.[21] Then, in the late 60s, CSIRO agreed that the Parkes dish be used to support the Apollo mission. Most of the televised video of the first moon landing (Apollo 11 in July 1969), including Armstrong's and Aldrin's walk, was

---

[21] The 64-m telescope at Parkes was to play a major role in the Deep Space Network (DSN) design of the three 64 m and later 70 m NASA tracking antennas around the world (California- from 1966, Tidbinbilla near Canberra, Australia–from 1973, and Madrid in Spain (from–1974).

received by the Parkes telescope. The Parkes Telescope also played a major role in the Apollo 13 rescue mission in April 1970.[22] Further contracts with NASA and later with ESA provided Parkes a significant role in the Voyager II, Giotto and Galileo space missions.

The interactions between FFP and NASA are described in ESM 32.1, NASA–Freeman, Fox.

1962: First Full Year of the Parkes Radio Telescope

As 1962 arrived, Bowen was honoured in the New Year's Honours' list: CBE, Commander of the British Empire. On 2 January 1962,[23] Ralph Freeman (also a CBE) wrote a letter of congratulations and thanked Bowen for pictures of the opening at Parkes from 31 October 1961. The Parkes telescope was still having teething troubles, but progress was reported with the Master Equatorial (ME). Bowen gave White a thorough report of the status of the telescope on 2 February 1962; he had been on a visit to the US, Canada, Chile and London (January to March, 1962).[24] The Bowen to White letter was an exuberant and breezy letter, as Bowen reported on a number of items from his trip[25]:

> Many thanks for your nice note about the CBE. It feels good to be raised to these giddy heights.
> The NASA Grant is as good as fixed, but I will send you a cable when it is fixed. [The grant was to use the Parkes telescope as a test bed for the new 64 m NASA Deep Space Tracking instrument. Harry Minnett was at JPL in Pasadena from 15 to 28 February 1962 for discussions.][26] The Grant could easily have been for double that amount of money in the first year. Incidentally, it is preferable that there should be no publicity about this Grant.

Bowen described the status of the Ford Foundation request for funds for the "Paul Wild solar instrument". On 25 January 1962 at the Ford Foundation, Bowen met Borgmann (Pawsey's friend, see Chaps. 38 and 40) and lunched with John McCloy (Assistant Secretary of War in WWII, High Commissioner for Germany in the post-war, chair of the Ford Foundation, 1958–1965) as he continued the lobbying for the new Australian initiative. Bowen was optimistic that CSIRO would come in "for a half share. If we are lucky and tread the primrose path carefully enough, they may go the whole way." A major concern was the worry that the Foundation had to "resolve ... how the [CSIRO project] relates to their other activities in the social and medical

---

[22] One of the authors, Goss, observing at Parkes with Radhakrishnan and Brooks, was a witness to the events of 13 April 1970 as the GRT rapidly was converted to a tracking station at 11 cm.

[23] NAA C3830 Z1/14/A Part 2.

[24] NAA C3830 Z1/91962.

[25] NAA C3830 Z1/14/A Part 2.

[26] Bolton has provided a three-page summary of the Parkes role "Parkes and the Apollo Missions" in Goddard and Milne, 1994, page 134.

sciences which the Foundation had favoured so much in the past." Later in 1962 RPL was successful (Chaps. 38 and 40); the Ford Foundation provided US $630,000 for the solar instrument. As we have pointed out earlier, Pawsey also played a major role in the negotiations with the Ford Foundation and communicated frequently with Borgmann.

Also, Bowen had just experienced an extensive lecture tour in North America beginning at MIT; he described the new Parkes radio telescope. Additional lectures were given at Montreal, Pittsburgh, Washington and Bell Labs. Bowen's lecture at MIT had an impressive audience of over 1000 people.

> The [audience] seemed to enjoy what they heard. Our stocks are certainly high at the moment. Poor old Rabi is getting desperate about the 140-footer at Greenbank [sic], [as the project ran over budget]. They have spent $7M and expected to spend another $6M on the redesign. However, the bids are coming in and they are nearer $9M, so he is thinking seriously of doubling up with the Canadians on a 150-footer based on our [Parkes] design.

Bowen also bragged to White as he concluded the letter of 2 February 1962:

> Money is being poured into scientific work—good and bad—in this country as never before. The standard offer for boys like Harry Minnett and Paul Wild is $15,000 and $20,000 per annum [salary], and for the like of you and I, $30,000 and up. It is hard to believe, but true. You will have to talk hard to Bob Menzies [Australian Prime Minister] and his boys about this or else the losses from the ranks are going to be heavy.

A few months later (16 April 1962), Bowen again wrote White with a more detailed report about the scientific achievements of the Parkes telescope[27]:

> The telescope is performing like a dream and it is clear that, except for some of the Askania components, the performance is even better than we expected a few months ago. The research programme is running on a regular basis from about 6 pm to 4 am every night, most of the receivers and telescope maintenance being done during the day.
>
> The morale of the troops is sky high, which is in marked contrast to a year or so ago. They are all clamouring to get their programmes on the dish and this is being done in a very orderly fashion. [Bowen enclosed a summary of the Parkes telescope observing schedule programmes in early 1962.]
>
> ... [A] quick survey of the Magellanic region on 408 MHz [75 cm] is also attached. This was taken by Mathewson and Healy ... and illustrates the wonderful detail which shows up even on this low frequency. [Bowen asserted that the quality was much better than the old 3.5 metre observations with the original Mills Cross, with comparable resolution of 50 arc min.] Mathewson now has similar plots on 20 cm [15 arc min] and will shortly have them at 10 cm [8 arc min resolution]. It is really wonderful stuff.

On 27 February 1962,[28] Bolton wrote to Bowen with a detailed report of the status at Parkes. Major problems with the power supply of the ME had been temporarily fixed by a new power supply suggested by John Shimmins of the RPL staff at Parkes. The deformation of the dish from zenith to the 60 deg. zenith angle limit (of the dish motion) was quite favourable with only 2 arc min pointing changes and a slight loss

---

[27]NAA C3830 Z1/7/B Part 2.
[28]*Ibid.*

of gain (less than 20 per cent) at large zenith angles (tipped closer to the horizon) at 10 cm. Herr Putz from MAN was back installing the 20 tons of missing counter-weight. The stability of the tracking was remarkable, about 7 arc sec. Additional good news was that a total of eight of the "galactic group" were learning to drive the telescope (including Frank Kerr, Jim Roberts, Don Mathewson and Dick McGee).

On 2 March 1962,[29] Bowen wrote White a long letter from London with more news about the status of the joint project with JPL for a detailed evaluation of the new Parkes telescope as it related to planned NASA tracking instruments. As Hanbury Brown, Minnett and White wrote (1992, p. 42) for the biographical memoir of Bowen:

> The Parkes Telescope also proved timely for the US space programme. Bowen received a NASA grant for Minnett to participate in studies at the Jet Propulsion Laboratory ... for the design of a 210 ft instrument [in the end three of these were constructed] for communicating with very distant space probes. Many of the Parkes features, including the drive and control concepts, were adopted.

Minnett had two visits to Pasadena in 1962 as a consultant. "NASA was greatly impressed by the performance achieved in the Parkes radio telescope, compared with any US design ... One aspect of the contract [with JPL] was the supply of engineering data to NASA on the performance of the Parkes antenna." (Thomas and Robinson, 2005, p. 199).

Parkes Telescope Scientific Programmes, 1962

As the new GRT began operation in 1962, the scientific programmes can be categorised in two groups: (1) projects that were anticipated based on the telescope specifications which had been achieved and (2) a small but significant number of projects that had not been anticipated but arose unexpectedly due to unforeseen discoveries.

First, we will examine anticipated science. In 1962, the new telescope made significant contributions due to advanced planning of the scientific staff throughout the 1950s. In 1962 the staff possessed a telescope that was easy to use and even surpassed many of the design specifications.

Planned Observations

As we have discussed,[30] in 1959 Pawsey had summarised the types of observations to be carried out with the GRT:

---

[29] *Ibid.*
[30] NRAO ONLINE.45.

These observations will include hydrogen line observations of external galaxies and selected galactic objects, continuum surveys of the sky with special reference to discrete sources at about 10 and 20 cm and less exacting continuum surveys at longer (50 and 100 cm) wavelengths. In addition, observations concerned with items of current interest, e.g. polarisation, will form an important part of the programme which cannot well be predicted.

The polarisation observations produced immediate successes. Polarised radio emission had been detected in the Crab nebula in 1957 (Mayer et al., 1957, p. 468) following the predictions of the new synchrotron radiation theory (see Chap. 34). In 1960 Pawsey wrote (Pawsey and Harting, 1960, p. 740):

Current theories of the mode of origin of galactic radio-frequency radiation assume the main components to be due to "synchrotron" emission by relativistic electrons in interstellar magnetic fields. Such emission is almost completely linearly polarised at the point of origin. The received radiation could, however, be substantially depolarised owing to its origin in extended regions of inhomogeneous magnetic fields, or to effects associated with the rotation of the plane of polarisation in ionised regions with magnetic fields along the line of sight [i.e. Faraday rotation]. The detection polarisation is a most important observation, which could substantiate the synchrotron emission hypothesis and provide direct evidence on magnetic fields in interstellar space.

The Parkes telescope was a circularly symmetric dish with the receiver mounted on-axis at the focus on a rotating platform. Thus measurements of polarisation were straight forward with minimal instrumental polarisation. During Easter 1962 Ron Bracewell, while visiting Sydney University on sabbatical leave from Stanford in California, observed the strong radio galaxy Centaurus A. He detected strong polarised signals at 20 cm immediately (Bracewell et al., 1962, p. 1289). Marcus "Mark" Price, a Fullbright student from the US, based at Parkes at the time, has provided a lively account of this unexpected sequence of events.[31] The detection of linear polarisation in Centaurus A was rapidly followed by the first observation of Faraday rotation at radio wavelengths. Again, the ease of changing the observing frequency with the single dish was decisive. Frater et al. (2017, p. 142–143, NRAO ONLINE.2) have described the controversial role played by Bowen in the publication of these two papers in *Nature* later in 1962. The wavelength squared dependence of the position angle of polarisation was easily detected in Centaurus A (Cooper and Price, 1962, p. 1084) and other sources (Gardner and Whiteoak, 1963, p. 1162). One of Pawsey's key science drivers for the Parkes telescope had been the measurement of polarisation and the extension of these measurements to higher frequencies. This direct observation of Faraday rotation in Centaurus A was made in the months just before Pawsey died in late November 1962 (see Chap. 40) and Pawsey told a number of colleagues at RPL that he was particularly proud of this major achievement at Parkes. In addition, Pawsey wrote to both Gerald "Jerry" Tape (1915–2005) on 5 September 1962 and I.I. Rabi (25 September 1962) of Associated Universities in the US (see Chap. 40) about his excitement. He told Rabi (exactly 2 months before his death):

---

[31]Price, M. (2012). In: *Parkes @ 50 Years Young*, eprint arXiv: 1210.0986.

[The new data from Parkes] seem to be almost certainly due to Faraday rotation in an ionised interstellar region between us and [Centaurus A] ... The plane of polarisation rotates with changing frequency in the precise manner which would be expected from passage of a wave through an ionised medium containing a magnetic field ... This effect gives the first real chance of measuring magnetic fields in interstellar space, which I think is an overwhelmingly important objective in astronomy ... This is one of the very good examples of a subject which requires study from both northern and southern hemispheres, so that work undertaken at Green Bank could be indeed complementary to that which is being done here.

By 1962 linear polarisation from Cygnus A had been detected (Mayer et al., 1962, p. 581A) and polarisation had at last been measured in the galactic plane (Wielebinski and Shakeshaft, 1962, p. 982). This field of research prospered and was expanding rapidly. For many years from this time Parkes dominated observations of polarisation and Faraday rotation in many different classes of radio sources: radio galaxies, diffuse galactic emission, supernova remnants, pulsars, spectral lines and galactic Faraday rotation. Whiteoak and Milne have both provided excellent reviews of these early Parkes polarisation observations (Goddard and Milne, 1994).

The radio continuum surveys were another well planned and executed programme for which the Parkes telescope was ideally suited. A team now led by John Bolton was able to produce a catalogue of about 2000 southern sources. This was a low frequency survey compiled from 408 MHz scan records, but with sources confirmed or rejected by scans at 1410 MHz where the resolution and positional accuracy were superior. The source list resulting from the 408 MHz survey became the first version of the highly respected *Parkes Catalogue of Radio Sources*. The survey was published in four papers between 1964 and 1966 and a convenient version of the combined catalogue was put together by Jennifer Ekers (1969a).

The combination of the low frequency survey and higher frequency follow-up made it difficult to assign a finding frequency (the effective frequency on which the source was initially detected) to this survey—effectively a frequency between 408 and 1410 MHz. Thus, the surveys were not suited for statistical or cosmological studies. However, the process had a fortunate side effect. A large number of previously unknown flat or inverted-spectrum sources were detected; these had been missed at low frequencies. A new industry of investigating different populations of extragalactic radio sources was opened (Savage and Wall, "Identifications, Confirmations and Tribulations", in Goddard and Milne, 1994).

The discovery of the peaked spectrum source PKS 1934–63 highlights the unexpected value of the move to a higher frequency. This object was the first of the class of Gigahertz Peaked Spectra radio sources and provided the evidence for the synchrotron self-absorption model of radio sources (Kellermann, 1966, p. 195). This particular source became the primary calibrator for most Southern Hemisphere radio telescopes; with its inverted spectrum, the radio source was too weak at low frequencies to have been included in previous catalogues such as the Mills, Slee and Hills survey at 85.5 MHz.

Another development, which had been planned by John Bolton, was the interferometer using the 60-foot Fleurs dish as one element along with the 210-foot telescope. Bolton was aware of the resolution limitations of the single dish and wanted to continue the highly successful projects using the two 90-foot dishes at Owens Valley as an interferometer. By 1965 a unique, variable baseline interferometer using the 210-foot with the 60-foot dish relocated from Fleurs was operational. Radhakrishnan, (in "The Parkes Interferometer", Goddard and Milne, 1994) provides an insightful description of this unusual instrument and has drawn attention to the fact that Ekers and Goss, both authors of this book, cut their teeth using this modest interferometer. The absolute phase stability of the exposed trailing cables was inadequate to determine the precision positions which were required to identify extragalactic radio sources. However, the interferometer was very useful in determining source structure. The first detailed observations of the structure of the southern radio sources was carried out by Ekers (1969a, b). Radhakrishnan, Goss and colleagues used this interferometer to determine the HI absorption profiles of weak extragalactic and galactic radio sources, ("A Personal View of Parkes Spectroscopy in 1967-1974" by Goss in Goddard and Milne, 1994).

Surveys of atomic hydrogen gas (HI) had always been an obvious target for the big dishes due to the required brightness sensitivity and the observations at Parkes were very successful. The 12 arcmin beam ensured that the galactic HI features could be studied at a resolution of tens of parsecs. Both the galaxy and our nearest neighbours, the Magellanic Clouds, provided high quality images. These surveys continued for the next decade and were particularly effective because the centre of the Milky Way, Sgr A, passed overhead at Parkes. Then in 1972, Don Mathewson discovered the Magellanic Stream, an immense cloud of HI gas emanating from the Magellanic Clouds, arcing more than 120 degrees across the sky. "Parkes and the Magellanic System" (Mathewson in Goddard and Milne, 1994). Another planned project involved the study of radio emission from the planets. Thermal emission from the planetary disks is stronger at higher frequencies due to the Rayleigh Jeans law. Kellermann detected Mercury, Venus, Mars, Saturn and Uranus at 11 cm in 1964 and 1965.[32]

Unexpected Science with the Parkes Telescope

The Parkes dish embarked on many productive but often routine observing programmes that had been anticipated by Pawsey and his team in the 1950s. However the agility of this single dish led to many other unexpected and surprising results.

The most significant was the lunar occultation of the radio source 3C 273 in 1962. To measure lunar occultations it was essential that the telescope could be pointed to any position on the sky where an occultation was expected. Pawsey had invited Cyril Hazard from Manchester University, then working at the University of Sydney in Hanbury Brown's intensity interferometer group, to make lunar occultation

---

[32]Kellermann, K.I. (2012) In: *Parkes @ 50 Years Young*, eprint arXiv: 1210.0986.

observations with the Parkes telescope. The timing of the occultation of 3C273 provided a precise position at the sub-arc sec level, leading to a possible identification with a stellar object with a jet. A spectrum of the bright star-like object was obtained by Maarten Schmidt (1929–2022) using the Mt. Palomar 200-inch telescope. The remarkable result was that the redshift was found to be z = 0.158. (Schmidt, 1963). Thus the first quasar was discovered. Details of the occultation and events leading up to this discovery are described by Hazard (2018). Observations of quasars continued at Parkes for another two decades and PKS 2200–330 held the record for the most distant objects in the universe for 25 years.

Soon after observations started with the Parkes dish, a new spectral line emitted by the OH radical in interstellar space was discovered by Sander Weinreb and colleagues at 1667 MHz. The Parkes engineers were able to quickly retune an existing receiver to this new frequency and confirmed the existence of the absorption line. In addition, they were able to detect the strong OH absorption in the galactic centre of the Milky Way ("Spectral Line Astronomy at Parkes" by Robinson in Goddard and Milne, 1994).

One of the greatest discoveries in astronomy during the 1960s was the pulsating radio sources (pulsars) which were found serendipitously at Cambridge by Jocelyn Bell in 1967. Parkes was quickly adapted to observe pulsars and within a month was able to take advantage of having long observing tracks to determine a more precise pulse period. The favoured model for the emission of the radio pulses was the lighthouse effect from a rotating neutron star containing a narrow radio beam. In December 1968, Radhakrishnan, Cooke, Komesaroff and Morris used the Parkes polarimetry to observe the position angle of the linear polarisation of the strong Vela pulsar during the pulse. They found a regular sweep of the position angle, providing direct evidence for the rotating neutron star model (Radhakrishnan et al., 1969, p 44), establishing the slightly oblique magnetic field model which has remained a key component of pulsar emission theory to this day. Later Radhakrishnan and Cooke (1969, p 225) wrote: "Comparison of the polarisation structure of PSR 0833-45 at different frequencies leads to the conclusion that pulsar radiation must emanate from the neighbourhood of magnetic poles."

Later Developments

Further advances in astronomy and new technology have enabled the Parkes radio telescope to maintain world class status for a period exceeding 60 years. The basic telescope structure has remained unchanged; however there have been a series of other changes that have resulted in renewed bursts of activity every decade or so since it began operation. In the 1970s the surface was upgraded to work at millimetre wavelengths (for the central 60-foot diameter); this allowed Parkes to search for other molecular lines. In this era, the Green Bank 140-foot telescope had been a pioneer in the detection of new molecular lines at cm frequencies. New technology has revolutionised the receivers first, by lowering receiver noise to levels not anticipated in the 1950s. Then, the receiver bandwidths have increased by two orders of magnitude. The data processing advanced with fast digital signal processing. Computing power has increased in both the areas of computer control and data analysis.

In 1967–1968 the technique of Very Long Baseline Interferometry was developed in Canada and the US. When VLBI observations were started in 1982 in the south, the Parkes telescope became the dominant element of the southern hemisphere VLBI network.

In 1997 a new 13 beam receiver was installed at the focus of the Parkes telescope. This multibeam receiver revolutionised searches for HI in galaxies and made it the most productive pulsar telescope in the world, doubling the number of known pulsars. In 2007 this multibeam receiver on the Parkes telescope discovered a completely new and enigmatic class of radio source with bursts lasting only one thousandth of a second (FRBs) arising from the distant universe.

The decision to build a Giant Radio Telescope has certainly been vindicated. Many generations of astronomers have benefited from this legacy of the two radio astronomy pioneers: Bowen, the visionary entrepreneur and Pawsey, the science leader.

Why Was the Parkes Design so Successful Compared to Contemporary Radio Telescopes?

The flexibility of the single large dish at Parkes ensured that the instrument could be adapted to new, unexpected discoveries. During the 1960s, the rate of discoveries in this new field of radio astronomy reached a peak, see Fig. 32.13 (Ekers, 2010). The scientists using the new Parkes telescope responded to a number of new events: (1) the lunar occultation of 3C 273 leading to the discovery of quasars, (2) studies of the newly detected OH line in absorption and emission, (3) polarisation measurements of pulsars and (4) rapid response in supporting space missions, including the Apollo 11 moon landing coverage and the Apollo 13 rescue mission.

The technical performance of the Parkes antenna was superior to the other big telescopes of that era such as Jodrell Bank and the Green Bank 140-foot.[33] The key difference in the antenna design, construction and commissioning was the integration of the broad range of disciplines involved. The *Parkes: 30 Years of Radio Astronomy* (Goddard and Milne, 1994) symposium included excellent reviews of the construction and commissioning of the GRT by those involved during the 1950s. Harry Minnett asserted that the skills of both the antenna and servo engineers had been combined effectively. Several participants at the symposium provided

---

[33] However, it is interesting to note that the 140-foot telescope was scientifically more productive by the mid-1960s with its superior receivers and spectrometers and a large astronomy community in the US taking advantage of NRAO's open skies policy.

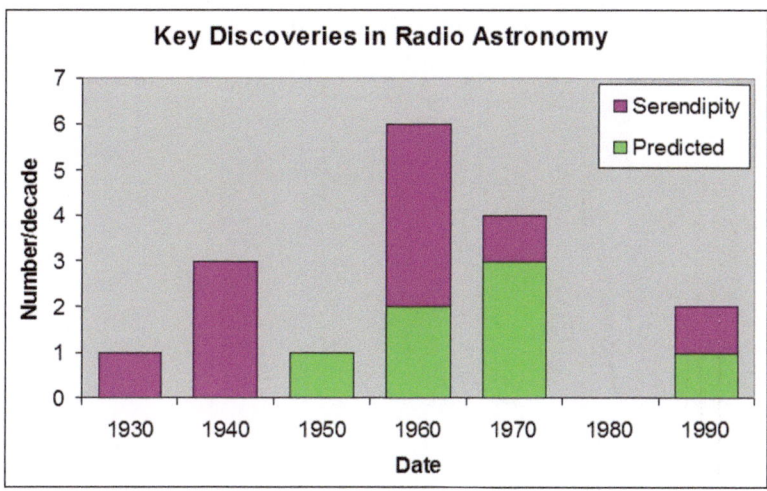

Fig. 32.13 Distribution in time of the key serendipitous and predicted discoveries in radio astronomy based on those listed in Wilkinson et al. (2004). Credit: R Ekers

examples of astronomers and receiver builders collaborating in the early days of the Parkes telescope. In addition, close collaboration existed among the software personnel, receiver experts and astronomers, producing innovative science. These connections were especially important in the remote environment of Australia, where industrial expertise was lacking. This cooperation and integration of skills, part of Australian culture, made major contributions to the rapid success of the new Parkes telescope in the following decades (Ekers, 1993).

Part VIII
The Development of Understanding

Chapter 33
Pawsey and Philosophy of Science

I defined the scientific method of today as a willingness to question (and investigate systematically) any question which comes up ... One of the most blatant applications [of this process] is the questioning and re-examination of old established ideas...

Science has had amazing successes in the conquest of the physical world through a reasoning, questioning process. [Science] fundamentally mistrusts emotional conclusions ... There is thus a clash between religion and science.

–Pawsey letter to his mother 14 March 1932 from Cambridge.

Scientists are mostly not especially reflective on the nature of science itself—or at least not in public. The history, philosophy and sociology (HPS) of science may be of interest, but science is complex and time for contemplation of the enterprise is limited. Pawsey, from the records that remain, was little different. But he had grown up in a family where ideas and values were taken seriously, and in an era that questioned established truths concerning religion, morality, the relations of Empire, the rights of working men (and women), and of course, after WWII, unquestioned loyalty to authorities.

This section is devoted to an analysis of the major conceptual debates and achievements of the 1950s: new ideas about radiation and the development of synchrotron theory; the co-development of survey telescopes and statistical survey methods with the rejection of Hoyle, Bondi and Gold's "Steady-state" model of the cosmos; and the development of aperture synthesis. RPL scientists were central to each of these significant shifts in understanding, and Pawsey was engaged with, and an indirect contributor to, them all. These chapters offer a richer and more detailed "history of ideas" of these signature achievements of 1950s radio astronomy, analysing the conceptual and evidential circumstances that made new thoughts and steps possible, and untangling the many occasions where scientists struck on what would turn out to be the correct idea on the basis of erroneous assumptions or misinterpretations of evidence. The detail of these episodes pulls into view the complexity of science: where knowledge is—as so many HPS scholars write about—"underdetermined" by method and evidence alone, and multiple social

© The Author(s) 2023, corrected publication 2024
W. M. Goss et al., *Joe Pawsey and the Founding of Australian Radio Astronomy*,
Historical & Cultural Astronomy, https://doi.org/10.1007/978-3-031-07916-0_33

factors shape its development. Not the least, by determining through practice, training or sometimes through dispute, just what is truly "evidence", or otherwise.

Before we embark on this analysis, we present Pawsey's *own* view of science, from the few records that we have across the years. The reader can then better imagine how Pawsey might have reacted to the excitements, the challenging conceptual issues, and the disagreements that are discussed in the following four chapters. First we explore Pawsey's reflections on science, then we discuss the core values that, developed in reaction to the two world wars that shaped the first part of his life, informed both his philosophy of science and his philosophy of science management. The last section of the chapter discusses a detailed example, his views on "planning research along 'theoretical lines' or 'exploratory lines' and on conditions of success". We identify three independent contemporaneous analyses (one Pawsey's own) of the role of serendipity in scientific discovery, whose agreement indicates a robust sociological theory of serendipity. We also classify the discoveries of the period according to his framework, identifying both those successfully made on the basis he set out, and those that were missed.

Early Thoughts on Science

Throughout Pawsey's scientific life, starting during his student days, he often expressed his opinions concerning the meaning of science to his family, friends and colleagues. These beliefs became the basis of his activities and success during the years of his prominent leadership of the radio astronomy group at the CSIRO Division of Radiophysics, 1941–1962.

We are fortunate that his first written record concerning the philosophy of science is in a remarkable letter of about 600 words written on 14 March 1932 to his parents in Victoria (Australia); he had only been at Cambridge in the UK for about five months. The 24-year-old Pawsey was exposed there to a greatly expanded intellectual atmosphere, with an abundance of new ideas and experiences.

We do not know what, if any, ideas about the nature of science specifically, he might have explored at Cambridge. In this period, the philosophy of science was in its infancy as a field of scholarship, and was devoted to empiricism, that is, to the notion of neutral and unbiased observation and to understanding the methodological basis for scientific truth. The dominant research came from German "logical-positivists" (that is, those who argued that logic plus scientific method was the basis for reliable knowledge), and it was not until later in the 1930s that many would come to the UK in flight from Germany and Austria. But prior to the involvement of philosophers, excitement about what new forms of empiricism had to offer had been widely celebrated by field and bench scientists across disciplines for more than a century, with the revolutionary advances in medical, agricultural, industrial—and military—sciences offering their own practical justification.

Regardless of exposure, Pawsey, wondering about his future and settling his framework of values in place, discussed ideas about science in his correspondence with his parents, as he did his various reflections on politics and marriage. He valued questioning authorities—he mentioned the fallacies of Isaac Newton—and was clearly not strongly attracted by religion.[1] Pawsey wrote to his parents on 14 March 1932:

> In my last letter, I touched on the spirit of science and the effect on the world. I wish to emphasise one point. I defined the scientific method of today as a willingness to question (and investigate systematically) any question which comes up … One of the most blatant applications [of this process] is the questioning and re-examination of old established ideas. These ideas have been upheld by men we look to, men to whom we do not claim superior intellect. May we then question their conclusions? … The old school of thought held … that to question one tenant [sic] of the teaching of a great man was to despise him and to reject one tenant was to reject his whole philosophy. My idea is to accept no man as infallible and to base my conclusion not on those of greater men than myself, but in those cases when it is feasible, on my own reasoning from the facts at my disposal.
>
> … You will point out it is impossible to personally investigate every question, and actually most things must still be accepted "from the wisdom (or ignorance) of the ages". [This] point of view … should have an important secondary effect. It teaches tolerance. Most questions you conscientiously investigate have two sides to them … Those conclusions which you obtain second hand are almost always one sided …
>
> Science has had amazing successes in the conquest of the physical world through a reasoning, questioning process. [Science] fundamentally mistrusts emotional conclusions … There is thus a clash between religion and science. Religion says unequivocally "Do this!" and gives not reasons. Science says: "Here are the reasons but do[es] not say—Do this! — with any great [conviction]" …

Pawsey stressed that science must be separated from emotions and made a strong plea for tolerance. These twin views were indeed characteristic of his style as a research leader. He pointed out that most questions have multiples sides to them, again suggesting caution in the handling of controversial situations. While his philosophy of keeping all options open was clearly very successful for the development of the new field of radio astronomy, we have seen that it could result in indecisiveness and, towards the end, produced bitter disagreements with Bowen.

Pawsey's central orientation to the values of tolerance and objective decision making can also be interpreted as the result of his reactions to the political turmoil of the twentieth century, beginning from his schoolboy visit to WWI battlegrounds and solidified in the strong commitments that grounded his reaction to WWII. His generation was, indeed, characterised by strong liberal, international and equity-oriented views across a variety of public issues (Martin Ryle, for example, was like Pawsey, concerned about nuclear power and weapons and became increasingly publicly focused on advocating for science for public good (Smith, 1986)). The connection between Pawsey's core commitments and his philosophy of science was perfectly captured by the sociologist of science who was almost his exact

---

[1] Pawsey's sons, Stuart and Hastings, verify this lack of any strong Christian belief in both their parents. They did not attend a church on either Christmas or Easter. The Pawsey parents did attend the Congregational Church in Vaucluse, Sydney, occasionally when Stuart sang in the church choir. The children attended Sunday school and other activities at this church. Hastings has strong memories of his father's knowledge of numerous biblical verses by heart.

contemporary in age: Robert Merton. Merton's famous 1942 essay arguing that science was both productive and crucial for democracy because it was structured by ethical principles (the "ethos of science" structured by "norms" in Merton's formulation), provides a fitting analysis of what we know of Pawsey's own views about science (Merton & Barber, 2004).

These views and commitments from scientists in general would play a major role in resistance to McCarthyism in the USA and to anti-Communist sentiment in Australia (Buckley-Moran in Martin et al., 1986, pp. 11–23). At RPL, the scientific staff varied in the degree to which they considered themselves "political", with most preferring to eschew such debates, as is the case today. But the general norms were progressive and strongly oriented to internationalism (David Martyn, for example, went to great lengths to generate professional connections to Japanese radio researchers, despite lingering post-war ill will). Several, including the inimitable Ruby Payne Scott and, in a quieter way, Chris Christiansen, held very strong pro-labour union, pro-Communist political views, and these were in many ways little changed from the 1930s, when the Depression was perceived by many to underscore the importance of Unions and of safeguarding workers' rights. Many were active in the Australian Association of Scientific Workers (AASW), which played an important role in Cold War policy issues in the 1950s.

Pawsey and Public Policy[2]

In the immediate post-war years, Pawsey was involved in debates over atomic policy and the role of secrecy in military research. For example, on 18 November 1945, Pawsey sent a letter[3] to David Rivett, then Chief Executive of CSIR, reporting on the complex and divisive Sydney AASW deliberations about the desired security policy in the era of nuclear weapons. There were three possible choices: (1) Complete secrecy with each nation. This could lead to world war within six or more years and would likely cause a "serious break on general progress of science". (2) A worldwide abolition of secrecy without international agreements and (3) International control of all major weapons. Option 1 was held by Pawsey to be a "dangerous" policy, while they all noted that Einstein had publicised his support for option 3. Pawsey declared:

> My own idea at present is that the most important action which could be taken . . . is to obtain agreement between scientists of all countries on general policy, and then make our pronouncements . . . I think it is good to have the general question discussed in Australian scientific societies . . . and approaches made to overseas bodies with the expressed objective of attempting to form that united front.

---

[2]Full details of Pawsey's participation in these debates are provided in NRAO ONLINE.36.
[3]NAA A9874/85.

In notes made in preparation for a 22 November 1945 ABC broadcast on atomic weapons, he offered a perfectly Mertonian view that the "scientific attitude" would prevail:

> We do not admit a thing is impossible because it has not been done before. We have a method of thinking, just plain common sense guided by the utmost care in expressing our exact ideas which has been astoundingly successful in certain branches of human endeavour. Can humanity not achieve some success in this vastly important matter of preventing war? This thing is essentially practicable if the people of the world detest war. And I for one think this is so.

A year later, he spoke publicly on the moral and social obligations of scientists, explicitly referring to his experience as a developer of defensive weapons[4]:

> [I]t's an obligation on scientists to study the implications of their work, and then they have to advise governments on request, and at the same time keep on this question of informing public opinion. Now this has a tacit assumption in it that the advice of scientists is worth something and I'd like to back this idea very strongly.
>
> In Australia, [knowledge of nuclear weapons is] limited, but it's probably still more than is available to members of the public. Next, we have a knowledge of scientific method. Now, by that I mean the way in which these physical developments are achieved, and it's known to the people who are working in scientific laboratories, and it's only the outstanding scientists who really are masters of that, and those are very few. **I think that the application of the clear methods of thinking is the main hope of the world to get out of the difficulty of her present social morass.** [our emphasis] If we 're to do that, the most obvious way is to somehow or other involve those really first class scientists, those few first class scientists who really know the method, and one other point is that we've got something in the nature of a world organisation or world contact already set up. This is a very proud boast of scientists, that we already have, or we had before the war, a world-wide freedom of the press and a great number of personal contacts.
>
> I'd like to finish on one very important thing. This question of secrecy again ... Now secrecy, national secrecy, in science, is just a form of armament. As that it has three effects. It can give tactical surprise in the use of new weapons. You hoped it would give a monopoly. As a matter of fact, in a lot of cases it didn't give a monopoly. **Radar was a very high secret and it was not a monopoly. It did that, it causes distrust and fear among nations because of not knowing the developments in other nations, and lastly it hinders science in its application to industry and to the good of humanity. The net result is that each case must be treated on its merits, but as a principle, international secrecy is a thoroughly wrong principle.** [our emphasis].

---

[4] ABC programme "The Nation's Forum of the Air: Has the Atomic Bomb Created a Moral Dilemma for Scientists?" The complete text was published by the ABC on 1 May 1946, at a cost of threepence. The participants were Richard (R.E.B.) Makinson, .Pawsey, C.E.W. Bean and W.E.H. Stanner. Makinson was a physicist at the University of Sydney who had been a war-time colleague of Pawsey working on radar related research, Bean was a World War I correspondent who had been at ANZAC cove at Gallipoli, wounded in the latter part of the campaign. After the war he was the editor of the 12-volume *Official History of Australia in the War of 1914–1918*. Bean was extensively quoted in the *Sydney Morning Herald* obituary for Kurt Offenburg of 16 May 1946. Stanner was a well-known anthropologist at ANU; he is credited with championing Australian Aboriginal people in the post-war era. In WWII, he organised "Stanner's Bush Commandos" (called Nackeroos) in northern Australia as an officer in the Australian army.

Pawsey later participated in debates within the AASW concerning British nuclear testing at Maralinga in remote South Australia, typically expressing moderate views. In 2020, what is perhaps most worthy of comment is that neither he nor anyone else from RPL expressed awareness of or concern for the Maralinga Tjarutja of the Pitjantjatjara people, the deeply affected traditional owners of the land (Parkinson and the ABC, 2007).

"On Planning Research Along 'Theoretical Lines' or 'Exploratory Lines' and on Conditions of Success", 1948

That Pawsey's philosophy of *science* was inextricably bound up with his philosophy of *science management* was set out in a letter of advice sent to the RPL scientific staff during his lengthy absence in 1947–1948. As noted in Chap. 17, we discuss what this letter reveals about Pawsey's philosophy of science here.

During Pawsey's 13-month absence, both Kevin Westfold and Lindsay McCready wrote to him with concerns regarding the solar noise programme. On 20 April 1948, Westfold wrote a comprehensive summary of his research activities since Pawsey's departure the previous year: "The problem I am trying to solve with these equations is the ionosphere problem of propagation in a plane-stratified medium under a uniform magnetic field. We would then be in a better position to attack the problem of propagation in the solar atmosphere."

The crux of the uncertainty in the solar noise group was described by Westfold:

> As you are probably aware, the theoretical knowledge of conditions in the solar atmosphere and in particular of the interactions between the radiation and the high temperature medium has always been far behind the experimental knowledge we have gained [by observations]. We need to know more about the dynamical state of the corona, the collision processes involving absorption and emission of radio frequency energy, why the apparent temperature discontinuity between chromosphere and corona etc. etc. Radio frequency observations have not discovered more about the physical state of the solar atmosphere than the astrophysicists already knew [i.e., using radio observations to determine local density, filling factor, temperature, velocity, and magnetic field had raised a number of questions]. Christiansen [having taken Fred Lehany's place], [Steven] Smerd and I have been discussing the present solar noise **experimental** programme. So far it has been directed along **exploratory** lines with little success [our emphasis]. We were wondering whether . . . you had found out anything which would enable the programme to enter a new phase of making some critical experiments which would help to decide some astrophysical question or to find out something to decide the next step to take in attacking some astrophysical problems. Since you left us we have missed very much the help of your stimulating criticisms and suggestions. We would therefore be grateful to hear what you think of the present programme and what future steps it should take, as well as your impressions of similar work you have now seen in the US and UK.

As we discuss in the following Chap. 34, it is clear that the generation of solar bursts was the combination of many complex processes and a simplistic model with theoretical predictions and experimental tests wasn't going to be fruitful. It took decades of observations and classification of the different types of activity to make progress. What could not be foreseen by the solar theory group was that the experimentalists like Payne-Scott and Christiansen were developing the techniques that *did* enable Paul Wild to make huge advances a decade later.

On 23 (or possibly 30) April 1948, Lindsay McCready wrote Pawsey with the message:

> Westphall [sic] has written to you. He let off some steam at a recent ... [solar?] meeting on the subject of us carrying out a lot of experiments without an adequate theoretical basis. He has been, I think ... influenced by outside criticism by Smerd [following the latter's "carpeting" by Bowen for low productivity, see Chap. 17 and ESM 17.5, Difficulties at RPL]. I've had the suspicion that other people are a little envious of what you and your group have done and can see prospects of other good papers coming out. It may be in some people's psychological make up to seize on a [minor] point of criticism and exaggerate it grossly ... We've always realised that a good theoretician would relieve you of a lot of detail and assist in devising crucial experiments etc. Are you having any luck enticing someone out to RP to [fill] the gap?

On 5 May 1948, Pawsey sent McCready a handwritten letter that was the precursor of his official response to Westfold. "Westfold's point of experimentation following the theory is a good one ... I shall work soon on this ..." Two weeks later on 18 May 1948, Pawsey sent out an (official) detailed response to all members of the solar group with the addition of a personal letter to Westfold. "I started to reply to your [Westfold's] remarks about basing experiments on theory and then decided to widen my remarks and have a hit at a lot of people in the laboratory. Consequently, I have let off a lot of steam in the note to the solar noise group accompanying this letter."

The three-page document (18 May 1948, see footnote 16, Chap. 34), titled "On planning research along 'theoretical lines' or 'exploratory lines' and on conditions of success", was addressed to "Mr Westfold, Mr McCready—Members of the Radio Astronomy Group".[5] (This was one of the first occasions in which the group was described as "Radio Astronomers"). The report consisted of (1) general principles, (2) specific examples of past experience, (3) criticism of activities of the radio astronomy group, (4) list of required future experiments and (5) suggestions for increased collegial behaviour among members of the group.[6]

Pawsey began by questioning the use of "exploratory" and "theoretical" research and instead suggested a better classification as "discoveries" exemplified by a "primary discovery" such as Jansky's discovery of the galactic background in the 1930s, and "verifications" of an idea derived from theory or common sense. His point was to use the classification to illustrate the different processes required.

The primary discoveries were:

> made essentially by accident,[7] the observation of a new effect or a discrepancy with existing ideas. It is found through a combination of two things. First keen powers of observation, the

---

[5] This is one of the first times the new term "radio astronomy" was used by Pawsey, a term invented by him and others in 1948 (See Chap. 17, and EMS 17.1, Pawsey's Co-Invention of the Name "Radio Astronomy", January 1948).

[6] The organisation of the report is chaotic with numerous repetitions and items discussed out of order.

[7] Today the term "serendipitous discovery" would be applied to an accidental discovery.

ability to recognise something significant; and secondly choice of work [the instrument] which increased the probability of encountering a new effect.

In this respect good presentation is an outstanding factor [i.e. an instrumental output that can be readily interpreted].[8]

Pawsey gave an example: the newly planned Penrith swept frequency instrument with intensity modulated frequency versus time display.

The verification discoveries: "depend on having a good imaginative picture on what is likely to be occurring and then being able to choose the easy experiments which give a lot of information."[9]

It is interesting that Pawsey's classifications bear remarkable resemblance to the serendipity pattern identified by Merton, which we have discussed in various other chapters in this book (Merton & Barber, 2004). (We note that Nobel prize winning scientist Irving Langmuir (1881–1957) has more comprehensive discussions of the requirements for serendipity, that are put forth by Merton.) Table 33.1 summarises Pawsey's classification compared to the examples suggested by Merton.

Pawsey compared "Primary discoveries" with his "Verification discoveries" which are the process of normal analytical experimentation and were not discussed by Merton in this context; Merton did differentiate between planned and unplanned research. Pawsey made this classification "to illustrate the different mental process required" and included the requirements needed to make such discoveries to help his theory group improve their research output.

Pawsey was not to know that these same themes would be debated twenty years after his death. A workshop on "Serendipitous Discoveries in Radio Astronomy" was held at Green Bank radio Observatory in 1983.[10] The participants included many of the scientists who had been involved in primary discoveries as they discussed the enabling factors. Summarising this workshop, a common pattern was identified in all the discoveries: technology driven, unanticipated, understanding the instrument, curiosity and persistence, a flash of insight, and the time had to be right. There was no evidence that the work of Merton was known to any of the participants and the Pawsey classification concepts had not been recovered from the archives at the time of this meeting. We thus have three independent classifications of the nature of discoveries; in the case of Merton there is no overlap in the scientific examples used. Our conclusion is that this classification is quite fundamental to scientific progress and Pawsey's ideas would have effectively shaped the research environment in the Sydney group.

We can explore these ideas further for research planning. They were based on Pawsey's own experience, so they certainly apply to the early research in the group; but we can also then ask how well *later* discoveries matched these ideas. Table 33.2 includes a list of radio astronomy discoveries made in the Pawsey era (1940–1963).

---

[8]We note that by "good presentation", Pawsey did *not* mean as in a talk or a paper. He meant something more fundamental. The importance of this instrumental output would have been clear from his ionospheric research for his PhD, in which the development of reliable "inscription devices" was so important (Chap. 5). See also Yeang (2012).

[9]Most scientific observations are in this category and involve measuring and refining explanations of known phenomena.

[10]Kellermann and Sheets (1984).

Table 33.1 Comparison of Pawsey and Merton's classification of Discoveries

Pawsey[a]		Merton[b]	
Primary discovery definition	Accidental	Serendipity pattern	Unanticipated
	Discrepancy with existing ideas		Anomalous—Inconsistent with prevailing theory or other established facts
	New effect		Strategic—Implications which bear upon general theory
Verification discovery definition	Good imaginative picture		
	Choose easy [suitable] experiment		
Requirements for recognizing discoveries	Perseverance	Curiosity	
	Keen power of observation		
	Ability to recognise a significant result	Scientist is prepared to take advantage of an unexpected occurrence	
	Good presentation		
Requirements to make verifications	Good imaginative picture		
	Ability to design suitable experiment	Planned research	

[a]Pawsey letter to Westfold and McCready, 8 May 1948
[b]Merton and Barber (2004, p. 196)

The table is based on a more extensive list of all radio astronomy discoveries compiled by Wilkinson et al. (2004). We have added Pawsey's categories to illustrate how well his criteria were met. Highlighted entries are Pawsey's "primary discoveries" or Merton's "serendipity pattern".

It is also interesting to look at discoveries that were missed, even though the CSIRO Radiophysics group had the relevant technology and knowledge. This can provide further insight into the attributes which are needed to make discoveries. Obviously, the one missing factor in a missed discovery is serendipity, but serendipity alone is not enough to make a discovery—and in any case, most of the following list were not missed because of bad luck. Serendipity might influence *when* a discovery is made, but not *whether* it is made.

Missed discoveries (that is, discoveries made by others, that would have been possible at RPL in this era):

1948

Ionospheric scintillation—The earlier Australian result was pre-empted by joint Cambridge and Manchester publications in *Nature*, 19 March 1950. (See Chap. 18)

1951

First HI line detection—RPL had technical capability but were too busy doing other projects and had no astronomers. (See Chap. 20)

Table 33.2 Radio astronomy discoveries made in the Pawsey era (1940–1963)

Discovery	Date	Telescope	Reference	Accidental	New effect	Discrepant	good model	Critical experiment
Radio burst - sunspot correlation	1945	Collaroy	Pawsey (1946)	Yes	Yes	Yes	No	Yes
Hot corona	1945	Collaroy	Pawsey (1946)	Yes	Yes	Yes	Yes?	Yes
Sunspots identified as the location of the radio bursts	1946	Dover Heights	McCready et al (1947)	No	Yes	Yes	Yes	Yes
Identification of the Crab Nebula	1949	Dover Heights and NZ	Bolton et al (1949)	Yes	Yes	Yes	No	
Identification of radio galaxies Vir A and Cen A	1949	Dover Heights and NZ	Bolton et al (1949)	Yes	Yes	Yes	No	
Solar bursts type III - moving 0.1c	1946	Dover Heights & Hornsby	Payne-Scott et al (1947)	No	Yes	Yes	Yes	Yes
taxonaomy of Solar bursts type I,II, III	1950	Penrith	Wild & McCready (1952)	No	Yes	No	Yes	Yes
Radio continuum from the LMC	1951	Potts Hill Mills cross	Mills & Little (1953)	No	Yes	No	Yes	Yes
Milky Way a spiral galaxy	1952	Potts Hill & Kootwijk		No	Yes	No	Yes	Yes
Galactic Centre	1954	Hole in the ground dish	McGee & Bolton (1954)	No	Yes	Yes	No	
Aperture synthesis	1955	Potts Hill Chris Cross	many	No	Yes	No	yes	Yes
Solar limb brightening	1953	Potts Hill Chris Cross	Christiansen & Warburton (1953)	No	Yes	No	yes	Yes
Faraday rotation	1962	Parkes 210'	Cooper & Price (1962)	Yes	Yes	No	yes	No
Quasars (3C273)	1963	Parkes 210', Palomar 200"	Schmidt et al (1963), Hazard et al (1963)	Yes	Yes	Yes	No	Yes

1951

Decametric radio bursts from Jupiter—Observed by Shain (1951) but dismissed as radio frequency interference; the Jupiter connection was not recognised.

1953

Evolution of radio sources—Bolton, Stanley and Slee (1954a, p. 110).Using the cliff interferometer, this group correctly suggested that their radio source counts could be explained by evolution in the Universe, but in the end this correct interpretation was discarded by the authors due to a conceptual error. (See Chap. 35)

1953

Polarisation of galactic emission—the radio group tried and failed because of depolarisation at the lower frequencies used. At the time the depolarisation by Faraday rotation was not understood. (See Chap. 34)

1953

Synchrotron mechanism—Mills dismissed this option because of the lack of polarisation. (See Chap. 34) This was a case of knowing too much; the incorrect observational limit on polarisation was not known to those who proposed this mechanism.

Pawsey's main concern was to solve the "solar noise" problems in two main areas: (1) the thermal component of the solar emission and (2) the non-thermal emission processes (the classification of Type I, II and III bursts that would be carried out by Wild and McCready from 1950). He pointed out that the steps required to understand the non-thermal solar radiation in 1948 were:

(1) the enumeration of a detailed theory, (2) a study of the consequences susceptible to experimental verification, and (3) verification of such consequences. What were the crucial observational checks? Until the nature of the radiation was established its study was unlikely to yield worthwhile evidence about the sun.

Pawsey completed the summary by listing four projects, which could be decisive in understanding the nature of solar radio emission. Projects 1 and 2 concerned the statistics of solar emission: "What is the waveform (envelope) of bursts?"; then (2) "Is the waveform that of random noise?" This project had been suggested to Pawsey by the well-known English radio scientist Ron Burgess (see Sullivan, 2009, p. 114 for a description of Burgess's contributions.) Payne-Scott published her results on projects 1 and 2 in 1949 ("The Noise-like Character of Solar Radiation at Metre Wavelengths", Payne-Scott, 1949a, p. 228).

Project 3 was: "Can we observe the atomic hydrogen spectral lines or others?" A missed opportunity (Chap. 20).

Project 4 was an all sky continuum image of the southern sky. Already in 1948, a major limitation in understanding the galactic background and the "radio stars" was the lack of a known non-thermal mechanism.

Pawsey wrote:

In the cosmic noise field, progress is similarly held up by lack of knowledge of mechanism (see Chap. 34) . . . Exploration is obviously profitable . . . Another serious obstacle [is] lack of adequate technique. Bolton's last letter points this out all too plainly.[11] There is obviously room for further work along present lines but I want to know where to go from here. I shall make sure of getting [more] information on Lovell's 200-foot [fixed] mirror but I do not like the idea of movable huge mirrors.[12] There is room for invention corresponding to our original application of fixed distance interferometry [sea-cliff interferometer]. *How about variable distance interferometry*? [our emphasis].

In mid-1948, Pawsey was thinking of beginning a Michelson interferometer project. Before his departure from Sydney in late 1947, the three-element lobe-switching interferometer for solar and Cygnus A observations had been proposed. By 1949, Payne-Scott and Mills were observing the sun and Cygnus A with the three-element 97 MHz instrument at Potts Hill.

The letter combined these observations with some philosophy of scientific management—familiar, doubtless, to subsequent generations in academia—that pushed for "grit" and productivity, but also allowed for exploratory experiments:

> Procrastination is the enemy of enthusiasm and kills research. What has been done? If nothing has been done, then who is responsible? I am getting only one answer. It is up to the originator of an idea either to work it out himself or sell it to someone else to do for him ... The place most of us fail ... is not in conception of experiments but in lack of perseverance. I consider it reasonable to spend a short time in exploratory measurements which I am prepared to scrap. It is a major error to carry out a series of measurements and then drop them before the stage of a paper or report. Work which is not written up is essentially wasted ... Remember that if work does not go through all stages (1) experiment ... (2) writing (3) publication either in journal or issue of laboratory report, it is still-born.

Caution and Risk

As we've discussed, caution was one of Pawsey's characteristics, a feature for which he was criticised by both Bowen and John Bolton. He was indeed conservative with respect to the generation of new astrophysical concepts. But at the same time, he equally consistently saw a place for the quick testing of conjectures with experiments that might be risky. This was the approach advocated by philosopher of science Karl Popper (who, fleeing his native Austria, arrived at Canterbury College, University of New Zealand, Christchurch, in the same year as Fred White, 1937, where Popper wrote the influential *The Open Society and Its Enemies*). Popper suggested that science progresses, not by confirming discoveries, but by correctly identifying which ideas and observations are *incorrect*, ie, by refutations; and advocated for advancement of science through strong tests of "risky conjectures".[13]

---

[11] Bolton's paper was published in *Nature* of July 1948. The claimed errors in the determined positions were of the order of one degree, but these were vastly underestimated (Sullivan, 2009, pp. 142–143).

[12] Pawsey was planning to visit Jodrell Bank later during his trip to the UK. See Chap. 17.

[13] https://plato.stanford.edu/entries/popper/.

Most scientists probably intuitively favour some mix of likely/confirmatory work with exploratory/riskier exploration. Pawsey was explicit about the importance of such a mix. For example, he wrote a report titled "An Observatory", 9 July 1954. This report was written at the same time as the "Observatory" concept—of building large facilities that could be used by multiple groups—was being operationalised as the National Radio Astronomy Observatory in the USA. Pawsey began the report with a discussion of his philosophy of planning:

> Planning involves an assessment of the lines of investigation which, because of the current state of the science, or of our abilities or facilities, are likely to be the most fruitful scientifically. Two general principles involved are: (1) There should be a reasonable compromise between long-term projects which are thought to be in main lines of development and (2) short-term experiments which, if successful, might happen to reveal unexpected or new phenomena (**wildcat experiments** [our emphasis] the extreme case). In the case of the long-term projects there are obvious advantages in following up work initiated here. The large projects also have the feature that they are obvious ones, carried out in competition with the world, and, when undertaken, it is imperative that they be completed and put into operation quickly (e.g. the Fleurs aerial [the Mills Cross at 80 MHz]). Another factor is the desirability of undertaking work which places us in a position to profitably follow new lines when they appear.

Pawsey had several terms for trials of conjectures. As we pointed out in Chap. 20, and ESM 20.1, A Review of Recollections, Paul Wild wrote Sullivan in 1986 with a description of Pawsey's affinity for "long shots" or "wildcats" or of projects carried out "on the wrong job number".[14] Christiansen and Mills, two of Pawsey's closest colleagues at RPL, commented on this, among other features of his approach science in their obituary for him (Christiansen & Mills, 1964):

> Most of Pawsey's time was spent, however, in the affairs of his research group. The team of radio astronomers which he built up was unusual in many ways. There were, and are, few scientific groups of comparable size where the head of the group had such a detailed knowledge of the work of each member and where every paper was criticised in detail by him. Yet this intense scrutiny and discussion of the scientific work of each member did not lead to an authoritarian regime. Pawsey's criticisms were usually accepted not only because they were sound but because they were so clearly and intelligibly expressed that acceptance was inevitable. But Pawsey never forced his opinions on a younger colleague: if the matter was open to doubt he was willing to leave it to experiment. He was in fact, the arch-empiricist. **"Suck it and see"** was one of his favourite expressions. Brought up in the school of Rutherford he had little faith in theoretical predictions. In his view, the predictions of the theoreticians became really interesting only after experiment had shown that they were correct. He did not in general accept theoretical predictions as a guide to experiment; he preferred to investigate the questions that arose from previous experiment. "**Following his nose**" was how he described the process.
>
> He was cautious in undertaking new experiments and subjected all suggestions for them to a highly critical examination: however, he felt that any large research organisation could afford to have one **"wildcat"** experiment in progress. One of his own "wildcats" was an investigation of the effect of electromagnetic waves on the growth of plants.

---

[14] In the ABC Interview of 11 Nov 1960 (see NRAO ONLINE.56), Pawsey referred to the "wrong job number" projects.

... He insisted on treating any problem in its simplest terms, and was a master of the rapid "order of magnitude" calculation. This was one of the main factors in his success as a scientist and as the head of a scientific group.

Chapter 34
The Development of a Theory for Radio Emission

I at once believed that the synchrotron mechanism was responsible for non-thermal cosmic radio emission. I ascribe this not to any keen insight, but to the fact that I was closer to physics and rather far from classical astronomy. In this situation the synchrotron mechanism seemed clear and realistic, whereas hypothetical, strange "radio stars" remained purely speculative.
–Ginzburg recollection from Sullivan (1984, p. 295)

Introduction

The discovery of the synchrotron radiation mechanism as the explanation of galactic and extragalactic sources of radio emission marked a major development in radio astronomy, providing much needed coherence to so many unexpected observations from the late 1940s. Through the decade 1948–1958 Joe Pawsey was keenly aware of how hampered his researchers were by the absence of theory—both of mathematical, and of big-picture conceptual, understanding of the phenomena they were observing. This chapter explores the technical and social difficulties that had to be overcome for synchrotron radiation to be understood and accepted as the primary non-thermal emission mechanism in non-solar radio astronomy; it extends, with additional detail and commentary, Sullivan's discussion of this question.[1]

As discussed in Chap. 5, in the period up to the mid-1930s ionospheric research dominated the development of long wavelength radio communications. The

---

Supplementary Information The online version contains supplementary material available at [https://doi.org/10.1007/978-3-031-07916-0_34].

---

[1] The development of the radio emission theory up to 1953 is treated very thoroughly by Sullivan (2009). This section was originally written independently without direct influence from Sullivan's interpretation of the history and the links to Sullivan (2009, *Cosmic Noise*, Chap. 15, p. 366) have been added later. In this way we present an independent view but completely consistent story.

W. M. Goss et al., *Joe Pawsey and the Founding of Australian Radio Astronomy*, Historical & Cultural Astronomy, https://doi.org/10.1007/978-3-031-07916-0_34

reflecting layer in the ionosphere was directly observed through the effects of multi-path interference (Appleton, 1924). Pawsey's PhD research (Chap. 7) was part of this era of ionospheric experiments and contributed to the scientific interpretation of what turned out to be a challengingly complex phenomenon: the effects of a turbulent moving multilayered ionosphere on radio wave propagation. In some ways the development of early radio astronomy, including the dramatic series of discoveries in radio astronomy that occurred after the war from mid 1940s until the late 1950s during the time Pawsey led the Radio Physics research group, similarly saw shifts in scientific understanding as the first and second generation of radio astronomers began to grasp something of the complexity of the phenomena they were investigating. During the period of Pawsey's leadership, the origins of the extra-terrestrial radio sources were understood and theories for the radio emission mechanism were unravelled. These developments in radio astronomy have been discussed chronologically in the previous chapters but now we depart from the strict chronological sequence and discuss the development of radio emission theory as a theme that shaped the field.[2]

Discovery of Non-thermal Radio Emission

In 1937 an amateur radio ham, Grote Reber, undertook the challenge to understand the nature of the radio emission that Jansky had discovered four years earlier. Reber built a home-made $2000 31-foot (9 metre) parabolic dish in his mother's backyard in Wheaton, Illinois, and started trying to detect those radio signals. At first, he observed at much shorter wavelengths (9 and 32 cm) in contrast to Jansky's observations at 20.5 MHz (14.6 metres). The only type of continuous natural radio emission known at that time was thermal radiation, which was predicted to be stronger at shorter wavelengths for optically thick emission. But nothing was detected at the shorter wavelengths of 3300 MHz (9 cm) and nothing was detected later at 910 MHz (32 cm). Reber then modified his telescope receiver to work at even longer wavelengths until he finally detected a signal from the galaxy at a frequency of 160 MHz (1.9 m) which was similar to the signal originally seen by Karl Jansky (Reber, 1940). Contrary to what had been predicted, this radio emission was stronger at longer wavelengths, so some other non-thermal process was required to generate it. In the 1940s, the only other known cause of radio emission was the coherent motion of charged particles, either naturally as in the radio noise from lightning, or man-made from electrical circuits and radiators. These are known as non-thermal radiation processes since they are unrelated to the thermal radiation generated by the random motion of charged particles in thermal equilibrium at a well-defined physical

---

[2]In 1962 one of the authors (RDE) started his career in radio astronomy by reviewing the two competing theories for the powerful extra-galactic radio sources: the popular colliding galaxy model and the then unpopular (but now accepted) active galactic nuclei alternative.

temperature which is a measure of the kinetic energy per particle. No non-thermal emission processes were known in astronomy, so this was a major puzzle and some even thought that Reber's observations must have been wrong.

At the same time, Pawsey was carrying out ionospheric propagation research at much lower frequencies in the 1 MHz range (Chap. 6) at the Cavendish Laboratory in Cambridge. We have no evidence that Pawsey, or anyone else at the Cavendish at that time, realised that the non-thermal radiation discovered by Reber was to become a dominant part of their future research a decade later.

What Was the Radio Emission Mechanism?

In 1946 the CSIR[3] Future Programme of the Division of Radiophysics included a summary of research being done by Pawsey and Ruby Payne-Scott during the first year of post-war radio astronomy. After summarising the early solar radio observations which included establishing the correlation between the solar radio bursts and sunspot activity,[4] the report went on: "This work is being extended to the reception of noise[5] from other parts of the universe, certain galaxies being prolific sources of radiation similar to that from the sun. It is thought to be of the same character and to have its origin in similar effects." At the time, such a statement indicates a remarkable vision of a future radio astronomy that includes not only the sun but also other galaxies that are presumed to have radio emission similar to that detected from our own galaxy by Jansky and confirmed by Reber. The concluding paragraph of McCready, Pawsey and Payne-Scott (1947) also touched on this topic when they noted the problems encountered when explaining the galactic radio emission as thermal radiation from hot gas. The issues could be resolved if instead there was a process similar to the non-thermal solar radio emission occurring in stars and if the galactic emission was then the sum of all the stars in the galaxy. The fact that the solar radio emission was bursty would not have been an issue because the galactic emission would have been averaged over millions of stars. However, this conjecture also contained the seeds of a misconception about the similarity of all types of non-terrestrial radio emission that misled the interpretation of the radio sources and the emission process for many years.

Should we see this assumption by Pawsey and Payne-Scott—that all astronomical sources of radio emission that they had found would be the same basic (though as yet

---

[3] NAA C3830 D1/1 Submitted by Bowen to the CSIR on 3 July 1946.

[4] These early observations are discussed in Chap. 12 and for a broader precis of the solar work, see NRAO ONLINE.20 and 23.

[5] When Jansky did his experiment there were various sources of "noise"—so the term cosmic noise got connected to radio astronomy emission.

unknown) phenomena—as a natural use of (so-called) Occam's razor,[6] that is, the law of parsimony? Occam's razor is an attractive philosophical proposition that suggests that if there are multiple explanations, the simplest is more likely to be correct. We question here what constitutes "simplicity": the numbers of parameters of a causal explanation? The number of different ontological entities (real things) an explanation proposes? These are debated by philosophers of science.[7] Others have also argued that there are very few instances in which scientists have successfully chosen between explanations on the basis of simplicity. In this case, the simplest explanation of the many newly discovered phenomena was that solar radiation, the diffuse galactic radiation and the discrete sources must all be the same thing. Sullivan (2009, his Chap. 15) has described this very well. The mental model ran something like this: we know that the sun emits radiation at radio wavelengths; the sun is a star; the galaxy also emits radiation; the galaxy is full of stars; therefore, the radio emission observed must simply be the sum of that emitted by all the stars. We refer to this as the "radio star" model of emission. This model, and its assumption that researchers were dealing with a single phenomenon, was not only understandable, but likely also unavoidable at that time— although Fred Hoyle provocatively questioned the closed mindedness of these assumptions (with minimal evidence) in the early 1950s. Nonetheless, it led the early radio astronomy researchers badly astray for the next 6 years!

In hindsight we now know that similar radio wavelength emission was being produced in some very different processes by many very different types of astronomical objects, so it is of interest to speculate on why so many in the community took the wrong path. Was it because of a deeply embedded preference—an epistemic value (Ruse, 2012)—for simplicity? Simplicity is particularly valued in experimental physics,[8] but is that value sometimes misplaced in radio astronomy? Francis Crick (1988) noted "While Occam's razor is a useful tool in the physical sciences, it can be a very dangerous implement in biology. It is thus very rash to use simplicity and elegance as a guide in biological research." Other biologists have made the same point, since in biology things can get very messy.[9] This biologist's view may also be relevant in astronomy, which in some ways is closer to biology than to the physical sciences. In the traditional physical sciences, the scientist can design experiments in the laboratory to purposefully achieve greater simplicity by isolating particular factors. In astronomy, however, the primary task is one of only observing the

---

[6] In fact the concept of simplicity in explanation vastly predates William of Ockham, to whom many versions of "Occam's razor" are misattributed.

[7] http://www.iep.utm.edu/simplici/#SH3d.

[8] "The Tyranny of Simple Explanations" https://www.theatlantic.com/science/archive/2016/08/occams-razor/495332/.

[9] "Simplicity", Stanford Encyclopedia of philosophy, https://plato.stanford.edu/entries/simplicity/.

behaviour of complex systems.[10] Astronomers can only interpret what they observe, and so the explanations often must be complex, as is the Universe being observed.

In any case, the development of a theoretical basis to interpret the non-thermal radiation unfolded as a very confusing story, and this assumption that all radio emission sources would have the same basic origin certainly added to the confusion. Hey et al. (1946) found that the intensity of the strongest radio source, Cygnus A, was fluctuating on time scales of less than a minute, so inferred that it had to have small angular diameter, like a radio emitting star. This was the first of the discrete radio sources. Bolton (1948) went on to find six more radio sources (one of the six was never confirmed, No. 5.47) using the cliff interferometer at Dover Heights (Chap. 16). At this time most radio astronomers assumed that the explanation of the diffuse galactic emission was the collection of all the discrete sources which were also assumed to have small angular diameter with similar properties as the sun (e.g. Bolton (1948), and Ryle (1949)).[11] They were called *radio stars*. Some astronomers had questioned this model (eg Greenstein et al., 1946) because of the filling factor issue—the density of stars as bright as the sun was much too low to explain the brightness of the diffuse galactic emission. Even when the first radio sources were identified with supernovae remnants and two external galaxies in 1949 (see following section) these identified sources were considered to be anomalous and the radio star model was not abandoned. A spirited debate between Ryle, advocating a galactic radio star model, and Gold and Hoyle suggesting an extragalactic location for the discrete radio sources, took place at a meeting at University College London in 1951.[12] In his summary Hoyle questioned the dogmatic assertion that the discrete radio sources were galactic and argued for keeping an open mind, especially in view of the identifications already made with extragalactic nebulae which were being discounted. The shift in thinking away from the radio star model required extraordinary evidence, and it was not until after Cygnus A and Cassiopeia A, the two strongest discrete radio sources in the sky, were identified in 1952 that it was generally accepted that the radio sources were a mixture of galactic nebulae and extra galactic sources. None were "radio stars"! Radio emission from supernovae remnants and the intense radio emission from external galaxies was a complete surprise, as neither type of object had been predicted to be a source of radio emissions.

We will describe the development of the radio emission theories following Pawsey's involvement and discuss these developments as seen through Pawsey's eyes at the time. In this way we can see how this field evolved from the perspective of the early radio astronomers as they were making their new discoveries in the era 1946 to 1955.

---

[10] See Chap. 5 for Yeang's discussion of observing and intervening in ionospheric research in the 1930s.

[11] Sullivan (2009, p. 169) discusses this view of the Ryle group. Ryle (1949) summarises this model but mainly discusses these radio stars as the origin of cosmic rays.

[12] This exchange is described in detail in Sullivan (2009, pp. 375–376).

The Hot Solar Corona

As discussed in Chap. 14, Million Degree Corona, in 1946 Pawsey identified a component of the solar radio emission which corresponded to a temperature of 10^6 K, not the 6000 K expected for steady thermal emission from the photosphere of the sun (Pawsey, 1946). David Martyn had already realized that the solar corona would be optically thick at 200 MHz and hence predicted that the radio emission must be arising from an optically thin corona at a much higher temperature (Martyn, 1946).[13] Martyn had access to the latest observation results from Pawsey. Similar predictions had been made on theoretical grounds by Ginzburg (1946) and Townes, unpublished[14] and there was optical evidence for the high temperature (see discussion in Chap. 14).

The success of this thermal emission model in explaining the stronger emission at low frequencies as a changing optical depth effect combined with the huge change in temperature with depth followed the classical scientific methodology of theory, prediction and confirmation. Some attempts were made to explain the spectrum of the galactic emission which was also stronger at low frequencies in the same way. As Jim Roberts pointed out,[15] by the early 50s the radio astronomers realised that a thermal model was completely inconsistent with the observed non-thermal radio spectra of the galaxy and many other radio sources.

The Active Sun

Pawsey and others had also clearly identified the radio bursts from the sun which occur over periods of hours to days and which were associated with sunspot activity, clearly the result of a non-thermal process. Pawsey et al. (1946) referred to this as follows: "it seems improbable that the radiation should originate in atomic or molecular processes but suggests an origin in gross electrical disturbances". Already in 1948 Pawsey had a clear understanding that it would be necessary to have a theory for the radio emission mechanism before further advancement in the field could be made. In a memo to his theory group in CSIR,[16] Pawsey expressed his frustration at the lack of progress. He acknowledged the role of serendipity in the initial discovery of solar radio bursts but insisted that a theoretical understanding was required. At this time his attention was still focussed on explaining solar radiation since it was assumed that all the other sources of radio emission would be the same process,

---

[13] Both papers can be found in "Classics in Radio Astronomy", Sullivan (1982).

[14] See footnote 44 on page 135 of Sullivan (2009) *Cosmic Radio Noise*.

[15] This is a summary of a discussion between Jim Roberts and Ron Ekers at Marsfield in February and March 2013.

[16] NAA C3830 F1/4/PAW/1 Part 2. Memo from Pawsey to the CSIR Radio Astronomy Theory Group 18 May 1948 (see Chap. 33).

i.e. the galactic emission and the point sources were all thought to be radio emitting stars like the sun. The steady thermal component of the solar radiation had been explained, but then Pawsey continued:

> On the other hand, the non-thermal radiation is waiting on theory. We have collected a lot of important facts and not one has been predicted in advance in any worthwhile manner. The steps one hopes for are: (1) the enumeration of a detailed theory, (2) a study of the consequences susceptible to experimental verification and (3) verification of such consequences ... What are the crucial observational checks? Until the nature of the radiation is established its study is unlikely to yield worthwhile evidence about the sun.

Paul Wild (1985) summarised this early period in a much more positive way:

> The situation around 1948, when I joined the Sydney group of investigators led by Joe Pawsey, was one characterised by mystery, incredulity and intense interest. A whole new field of research lay ahead with obvious objectives: to disentangle the confused conglomeration of phenomena; to interpret and understand them; and to put the results to use in the mainstream of research for solar physics, astronomy and physics.

Following Wild (1985) we see that the basic theoretical foundations for the physical processes which were occurring in the solar atmosphere were already well-developed in the early 1900s using Maxwell's electromagnetic theory. Key concepts included the gyro frequency (Heaviside, 1904), the dispersion relation for electromagnetic waves in a plasma (Lorentz, 1909), the harmonic electromagnetic emission from electrons gyrating at the gyro frequency (Schott, 1912) and the concept of plasma oscillation (Tonks & Langmuir, 1929). In 1945 Schwinger had described the emission from electrons accelerated to high energy (later called synchrotron radiation). This result was not published until Schwinger (1949), possibly due to the security classification of this research area in the West.

Paul Wild then continued: "Before the 'origin of species' could be identified there had to be an exercise in taxonomy", and he proceeded to classify the different types of radio emission, labelling them type I, II, III. Much later Boischot (1958) defined type IV and Wild added type V. Different mechanisms were invoked mostly involving a source of excitation moving out through different plasma conditions in the solar atmosphere, sometimes moving at high velocity. The details of the physical processes involving electron streams, plasma oscillations, scattering, high energy particle acceleration, etc. were complex and occupied theorists for many years. Wild (1967) went on to build the solar radioheliograph, an instrument designed to test solar theory. Discussions of these early years of solar radio astronomy are well covered elsewhere and we have provided a summary of these developments in NRAO ONLINE.20. See reviews by Pawsey and collaborators from the post-war era, e.g. Pawsey (1953) and Pawsey and Smerd (1953). There is also a good discussion in Roberts's autobiography (2002).[17]

---

[17] Autobiography by J.A. Roberts, 2002 *Have Gen Will Travel, Imperfect Images from the Life of a Radio Astronomer.* NRAO ONLINE.49.

Non-thermal Radio Sources

Pawsey extended the scope of research in his group to include **all** the non-thermal sources of emission; he was suggesting that an all-sky radio continuum survey of the southern sky be carried out, again noting that a major limitation in understanding the galactic background and the "radio stars" was the lack of a known non-thermal mechanism. "In the cosmic noise[18] field, progress is similarly held up by a lack of knowledge of the mechanism." The first Australian survey was conducted using the cliff interferometer at Dover Heights (see Chaps. 13, 16 and 37) and resulted in the discovery of six more (one source was not confirmed, see above) discrete sources of radio emission (Bolton, 1948). But the positions of these sources had very large uncertainty, because at these long radio wavelengths, even with the height of the cliff above the sea, the angular resolution of the instruments was still very poor. Sometimes even the constellation was wrong. At this time the two brightest discrete sources in the sky, Cygnus A and Cassiopeia A, had not been identified because of these large position errors. A collaborative effort was started between radio astronomers at Cambridge and Sydney to obtain better positions (Chaps. 18 and 21). In particular, Ryle and Bolton exchanged information, but disagreements in their position estimates were so large that no joint paper was ever written. They even discussed the possibility that the sources were moving to explain their differing results. Cassiopeia A could not be observed from the South and Cygnus A was at such low elevation that the large corrections for the ionospheric refraction was a major issue. However, for the other sources, Bolton's New Zealand expedition in May 1948 took advantage of the higher cliffs that were both east and west facing and used a sea-cliff interferometer to measure much more accurate positions. Bolton, Stanley and Slee (1949) published these positions and suggested identifications for three of the radio sources. These were Taurus A which was identified with the Crab Nebula (a very well-known supernova remnant), Virgo A identified with M87 (NGC 4486), and Centaurus A identified with NGC 5128. The dominance of the existing galactic radio star model at the time was such that even though both M87 and NGC5128 were classified as extragalactic nebulae, Bolton, Stanley and Slee questioned this classification in their paper as being unlikely for radio sources, which were assumed to be galactic. We return to this topic below.

From October 1950, we have discovered a fascinating correspondence between Joe Pawsey and Jim Roberts.[19] Jim Roberts, then about 23 years old, was a Sydney University physics graduate who had received a scholarship the previous year to study abroad for 2 years. Because of the conflicts that had arisen between Cambridge and Sydney (see Chaps. 18 and 21) Australians were not welcome in Ryle's radio astronomy group. Jim Roberts decided to do his PhD with Fred Hoyle. Roberts attended the 1950 URSI General Assembly in Zurich where he met Pawsey. Pawsey remained concerned about the problem of the non-thermal sources. What was the

---

[18]The term "cosmic noise" is used to refer to all non-solar radio astronomy.

[19]NAA C3830 Z3/1, C3830 F1/4/PAW/2 and C3930 F1/4/ROBE/1.

mechanism? He viewed the young Jim Roberts as a gifted colleague who might be able to solve this dilemma. Fortunately, Roberts was now working with Fred Hoyle, one of the top theorists in this area.

Pawsey asked Jim Roberts to provide a report on non-thermal radiation mechanisms that could be incorporated in the proposed book on radio astronomy he was planning to write with Bracewell (see NRAO ONLINE.53 on the Pawsey and Bracewell book). Roberts replied with the following suggestion: "The report was to be titled 'Possibilities of Non-thermal Processes as the Origin of Extra-terrestrial Radio Noise'". There would be four sections: (1) Radiation from a Medium Capable of Coherent Oscillation, (2) Radiation from Accelerated Electrons not in Thermal Equilibrium, (3) Amplification of Waves in the Presence of Ordered Electron Motions and (4) Mechanism for Producing Non-thermal Electron Assemblies. Pawsey had doubts about the four main points in Roberts's synopsis. He suggested an alternative division of the four chapter headings: (1) introduction, (2) production of oscillations in gases, (3) escape of radiation from oscillating regions and (4) conclusions. Pawsey also had definite ideas about the proposed style of the report:

> I should aim at giving an account which is readily intelligible to physicists. Present the problem clearly and when the solution is known with certainty say so. When the answer is, in your opinion, uncertain, say so ... Use examples freely to help the reader ... [It is] essential [to point out] the differences between thermal and non-thermal processes, and between ordered and disordered radiation—the separation of generation and escape ... I should be very happy [if you handled the incoherent radiation case also].

In retrospect we can see that even the range of topics considered for inclusion in this report in October 1950 were very constrained by the contemporary thinking based on solar radiation theory. There was no hint yet of the synchrotron radiation caused by the acceleration of very high energy charged particles; this was the missing but essential step. Reading the papers written at this time reinforces the degree to which this misconception about the possible radiation mechanisms and also the feeling that sources would be inconceivably powerful if they were at extra-galactic distances had influenced thinking. Even the seminal Bolton, Stanley and Slee (1949) paper which included the identification of Centaurus A and Virgo A with two external galaxies was titled "Positions of three discrete sources of **galactic** radio-frequency radiation" [our emphasis], and in the paper they had noted:

> NGC 5128 and NGC 4486 (M87) have not been resolved into stars, so there is little direct evidence that they are true galaxies. If the identification of these objects (M87 and NGC5128) with discrete sources of radio frequency energy can be accepted, it would tend to favour that they are diffuse nebulosities in the galaxy, for the possibility of an unusual object in our own galaxy seems greater than a large accumulation of such objects at a great distance.

In his autobiography Hoyle (1994)[20] describes this common practice in astronomy to pick the more trivial when two alternatives are available—he calls it the principal of maximum trivialization. Within a short period, Minkowski, a famous optical

---

[20] *Home is where the Wind Blows*, 1994, University Science Books.

astronomer from Caltech, had written to Bolton assuring him that both objects were most certainly extragalactic and that he must have discovered a new class of very powerful, very distant sources. Bolton never again mentioned the galactic interpretation and when one of the authors, (RDE) who was a student of John Bolton, asked about why he called them galactic in his paper, John replied that they had to say they were galactic to satisfy the *Nature* referee. However, we have been able to examine the manuscript and correspondence with *Nature* and the paper had been accepted without refereeing; creative remembering is common to all humans.

This seminal paper by Bolton, Stanley and Slee not only showed that there were more classes of radio sources, it was also the first link between radio astronomy and traditional fields of non-solar astronomy. The extra-galactic identifications, if correct, added a new facet to the emission mechanism problem which now had to also explain the huge energy requirements for such powerful sources.

By this time in Sydney, the existence of a significant extragalactic population had already been accepted. Bolton had identified M87 and Centaurus A in 1949 and Mills (1952b) had divided the discrete radio sources into two populations based on his survey statistics: Class I (distributed in the galactic plane) and Class II (isotropic distribution, possibly extragalactic). In early 1952 other questions were arising about the radio star model. Bracewell (1952) summarised the increasing evidence that many of the discrete sources had extended structure that was much too large to be stars. He included discussion of new observations by Mills in Australia who was using two Michelson interferometers with different baseline lengths which showed that some sources were more than half a degree in angular size— much larger than the size of any stars. Mills (1953) published a detailed description of these new measurements of the angular sizes of four discrete radio sources (Cygnus A, Taurus A, Virgo A and Centaurus A), taken over the previous few years. He compared the radio structures with optical features described by Baade and Minkowski (1954a, b) and found that there were certain similarities between the radio and optical structures for all sources. For the radio source Taurus A, the Crab Nebula supernova remnant, Mills pointed out that the radio shape was very similar to the shape of the optical so the central brightness temperature in the radio was "too high to be explained in terms of thermal radiation from the mass of such gas … It seems that a non-thermal process is operative in the gaseous mass of the nebula as also appears likely in the case of the Cygnus source."

The year 1952 was a turning point for the radio astronomers because it heralded the demise of the radio star model; it finally became clear that the majority of discrete radio sources were not "radio stars" but extended objects, some of which could be identified with external galaxies and others with remnants of supernovae explosions in our galaxy. Suddenly there was a dramatic change in Ryle's view on the nature of the "radio stars", which is discussed in detail in Sullivan (1990) and in Chap. 35. This started with the identification in 1952 of Cygnus A with a distant galaxy using a more accurate position measured at Cambridge. Ryle had dismissed the earlier Bolton et al. (1949) identification of the radio source Virgo A with the galaxy M87 and the radio source Centaurus A with the galaxy NGC5128 as not certain and less interesting since M87 was a fairly normal galaxy. However, contrary to

Ryle's opinion at the time it was M87's very unusual jet and nucleus that eventually led to the concept that the nucleus of a galaxy is the source of energy that powered the radio galaxies, and not the collision of two galaxies as is discussed at the end of this chapter.

Ryle's shift in thinking was also influenced by the first observations from the new Cambridge survey of radio stars. There were far more fainter sources than expected in a finite disk of stars in our galaxy and the increasing numbers of fainter sources were seen in all directions around the sky. This strongly suggested an extragalactic population. Ryle's, and therefore the Cambridge group's, acceptance of extragalactic radio sources was the beginning of the realisation that radio sources could be used as cosmological probes of the distant extragalactic Universe, a topic also discussed in detail in Chaps. 35 and 36.

Discarding the radio star model, however, brought radio astronomers no closer to understanding the mechanism(s) that produced extragalactic radio emission. In a letter to Ratcliffe (19 June 1953),[21] Pawsey discussed possible projects in Cambridge for a new young Australian student, Brian Robinson (Chap. 24). Pawsey suggested a speculative topic of increasing relevance as radio astronomers still struggled to understand the radiation mechanism in the galaxy and in the discrete sources: "If we could get a lead on any connection between cosmic rays and cosmic noise it would be first rate. However, I shall leave you to think it over."

Cosmic Rays and Synchrotron Radiation

Hess (1912) made high altitude balloon flights with Geiger counters and found that the ionising radiation level increased with altitude. This was the opposite to expectations based on the idea that the ionising radiation was coming from radioactive material in the earth. He correctly interpreted this surprising result as high energy ionising particles coming from outside the earth's atmosphere. These became known as *cosmic rays*. Outside the radio astronomy community, Fermi (1949) had explained how the charged relativistic cosmic-ray particles were accelerated to high energy in the interstellar medium. Langmuir had observed radiation from accelerating high energy particles in the General Electric synchrotron accelerator (Elder et al., 1947). The theory explaining this electromagnetic emission from high energy accelerated electrons was published by Schwinger (1949), becoming known as *synchrotron radiation*. But the particle physicists and the radio astronomers were from very different fields of research; thus, the connection between high energy charged particles in the interstellar medium and the cosmic radio emission was not made.

When Pawsey wrote to Ratcliffe in June 1953, did he already have some inkling of the role cosmic rays were soon to play? Had he been influenced by Messel, a

---

[21]NAA C3830 Z3/1/V.

particle physicist who had just arrived at Sydney University? This seems unlikely as Messel saw the CSIRO radio astronomy as competition and relations were frosty.[22] It seems more likely that Pawsey was referring to the suggestions that the radio emitting stars were either a source of cosmic rays (Ryle, 1949) or stars trapping cosmic rays which then radiated in the strong stellar magnetic field (Alfvén & Herlofson, 1950). Ratcliffe did not respond to Pawsey's suggestion, and Brian Robinson carried out a PhD thesis on an unrelated topic involving the motions of the ionosphere.

Although looking for what might connect "cosmic rays" with "cosmic noise" in the way Pawsey imagined was speculative and did not inspire productive research in either Sydney or Cambridge, this oversight was soon to change.

Synchrotron Model for Radio Emission

In 1950 the first connections across these different branches of physics began to be made. While Unsöld (1949) had interpreted the anomalous non-thermal radio emission from sunspots as an effect of plasma oscillations, Alfvén and Herlofson (1950) suggested that this emission was more likely to be synchrotron radiation from stars that had trapped cosmic ray electrons in their magnetic fields, i.e. the stars were behaving like synchrotron accelerators. Kiepenheuer (1950) took this a crucial step further by suggesting, in a small note in *Physical Review*, that the galactic radio emission could also be generated by the synchrotron process, but in the interstellar medium (ISM) rather than in the stars. The optical polarisation of starlight caused by aligned dust particles required an interstellar magnetic field of 10^{-6} gauss. Kiepenheuer used this estimate of the magnetic field strength and assumed cosmic ray density to suggest at least 1% of the relativistic electrons could be produced by the interactions of the cosmic ray primaries with the interstellar medium. He estimated the intensity of the resulting synchrotron radiation might well agree with the observed galactic radio emission.

Although Kiepenheuer's note was published during a productive 18 months of working in US Observatories, this interpretation was mostly ignored in the West; there were only three citations by 1957—two Russians and one particle physicist! Kiepenheuer then returned to his solar-focussed research back in Germany. The idea was, however, enthusiastically embraced by Vitaly Ginzburg in Russia who argued that the Kiepenheuer concept was a very natural explanation given the clear evidence

---

[22] In his book *Boffin* Hanbury Brown (1991) relates the following discussion with Joe Pawsey in relation to his own decision on whether to join the Messel group in Sydney: "There had been considerable friction between Harry [Messel] and the Government Laboratory (Radiophysics) of which Joe was a leading light and Joe had no love for Harry; roughly speaking Joe regarded Harry as the last man in Australia with whom he would choose to work, but he was scrupulously fair-minded and hated to speak ill of anyone. After he had studied my ceiling for some time he said, 'Well Hanbury, there's only one good thing I can say about Harry, he always keeps his word'."

for both the magnetic field and the cosmic-ray particles (Ginzburg 1953). Later Ginzburg recalled:

> I at once believed that the synchrotron mechanism was responsible for non-thermal cosmic radio emission. I ascribe this not to any keen insight, but to the fact that I was closer to physics and rather far from classical astronomy. In this situation the synchrotron mechanism seemed clear and realistic, whereas hypothetical, strange "radio stars" remained purely speculative (epigraph, this Chapter)... The reaction of astronomers was quite the opposite, i.e., the synchrotron mechanism seemed mysterious and speculative, whereas "radio stars", although posing riddles, were more acceptable—for what kinds of stars cannot exist? (Sullivan, 1984, pp. 295–296).

Here we again have Hoyle's principle of maximum trivialization at work! Ginzburg (1951) made some corrections to the Kiepenheuer (1950) paper and together with his student Getmantsev further developed this new theory.[23] Shortly after (in 1953), another Russian theorist, Ioseph Shklovskii, who had initially rejected the synchrotron model, changed his view and published his seminal paper explaining the radiation from the Crab nebula as radio **and** optical synchrotron emission (Shklovskii, 1953).

But most radio astronomers in the West were still unaware of the importance of cosmic-rays for their field. Jim Roberts recalls knowing about Schwinger's theory for radiation from relativistic electrons and Korchak's theory for relativistic protons (Korchak and Terletsky, 1952) at the time of their publication in 1952: "I studied this paper of Schwinger's in some detail. As this was before the days of the photocopier, (and much before the days of down-loading a copy from the internet!), I made a handwritten copy of substantial parts of the paper and used this as a reference for many years" (Roberts, 2002, p. 43). While Roberts had calculated the intensity and polarisation of barely relativistic electrons, he has commented that radio astronomers were not familiar with high energy particle physics, so they did not think about a link to cosmic rays, except in relation to the sun. While real theoretical physicists were completely familiar with quantum and relativity physics, Roberts recalled that his knowledge (and probably other astronomers in the West) was limited to classical radiation theory, with little knowledge of quantum or relativistic effects because these were not in the curriculum. He himself (and probably others) were focussing on non-relativistic effects of electron streams in ionised plasma and magnetic fields.

This view is corroborated by Sullivan (2009) from a discussion with another young researcher of the period, Peter Scheuer, who suggested that synchrotron theory was extremely difficult and would require an effort beyond what most astronomers were willing to invest for a speculative theory. Roberts also noted that because he had no big picture view of astronomy, he could not have made the links between the interstellar medium, cosmic rays, and the presence of magnetic fields which had been deduced from observations of optical polarisation caused by aligned dust grains. Here we see the clear advantage of the multi-disciplinary connections.

---

[23] Sullivan (1982) conveniently provides all these key papers in his book *Classics in Radio Astronomy*.

Had Roberts Interacted with particle physicists and been aware of the evidence for interstellar magnetic fields, he may well have made the same connection as Kiepenheuer. Hoyle (1954) made a sociological comment that in physics there were the "plasma crowd" and the "high energy crowd"—he favoured the high energy synchrotron model and he had even suggested this possibility for solar radio emission much earlier (Hoyle, 1949). However, he never followed this up.

Bernie Mills Discussing the Radiation Mechanism and the Polarisation Prediction

On 30 September 1953 Mills wrote to Pawsey[24] from the US; he remarked that in his search for possible mechanisms for the production of radiation he was led back to the old idea of Schwinger radiation from cosmic ray particles moving in the galactic magnetic field. He then went on to note that this could be tested by looking for the longitude dependence of intensity and polarisation assuming a fairly regular field following the shape of the spiral arms and suggested a method for making this test. This is one of the first documented suggestions of a direct observational test for this new theory. (See ESM 34.1, September 1953 letter, for a detailed description]. Pawsey replied in a letter to Mills of 27 October 1953: "With respect to polarisation I spoke with John Bolton since writing you and they have made various attempts, all with negative results. The latest is at 400 MHz [75 cm] where no plane polarisation was detected down to a limit of about 1%". In Australia the synchrotron radiation idea was dropped because of these low limits on the predicted galactic polarisation. However, this limit was not published. The result was likely not known outside Australia. As we will discuss later, the polarisation of the emission at these low radio frequencies would have been destroyed by Faraday rotation gradients in the interstellar medium, but this was not appreciated at the time. Bernie Mills's argument for what was to become known as *synchrotron emission* was dismissed for the wrong reason.

We can gain an interesting perspective from a note Mills wrote 35 years later in 1988 at the time of the 40th anniversary of the construction of the Dover Heights cliff interferometer[25]:

> My own conversion [from the radio star model] would have been quicker and more complete except that I could think of no possible physical mechanism for the non-thermal radiation from interstellar matter. I completely missed the letter to *Phys Rev* by Kiepenheuer (1950). Although familiar with Alfvén and Herlofson on similar suggestions for radio stars I didn't 'connect it up—conditions [in the interstellar medium] so different. Neither did anyone else I knew! However, in 1954 I noticed what seemed an interesting paper by Shklovskii (1953) and I had it translated—everything fell into place. I disagreed with details, but this was the mechanism I had been looking for. In retrospect one thing stands out—the enormous

---

[24]NAA C3830 F1/4/MIL/1 and A1/1/1 Part 81953.

[25]unpublished document from the Mills's family archive.

resistance to change in conventional wisdom—even in the face of strong evidence. It is late 50s before everyone had abandoned radio stars. I guess this is a fact of life and should be accepted as such.

Indeed, we can make an analogy with the long period of time (roughly 15 years) that surface diffraction, rather than ionospheric reflection and refraction, remained the preferred theoretical explanation for how low frequency radio waves propagated around the globe in the early twentieth century (despite its inability to explain practical radio phenomena such as fading and skip zones), as discussed in Chap. 5. Are these examples related to the Kuhnian "paradigm shifts", albeit on a small scale rather than on the larger scale revolution that radio astronomy represented within astronomy in general at this time? We suggest that these dramatic shifts which only occur after the evidence is overwhelming, are indeed the way science progresses and there is a pervasive continuum of such shifts from minor to revolutionary.

By 1953, then, the idea of an alternative radiation mechanism was beginning to be taken up in the West. The next step was to follow Pawsey's sense of the scientific method and subject the idea to tests by verifying or refuting the theory/concept's prediction.

Roberts, who had completed his PhD with Fred Hoyle in Cambridge on radiation theory and returned to Australia in 1952, wrote a review article which provides a good indication of why it took so long for synchrotron radiation to gain widespread acceptance. In his review article (Roberts, 1954) he discussed the synchrotron mechanism. Without any detection of the predicted radio polarisation and with no direct evidence for relativistic electrons in the ISM,[26] he considered the link to synchrotron radiation theory very speculative and to be treated with considerable caution.[27] Roberts (1954) continued:

> The idea [of synchrotron emission] has been developed in various forms by other authors and, to date, is the only theory able to explain the intensity and spectrum of the observed radiation. Unfortunately, no sensitive test of the theory has yet been proposed, and until one is forthcoming the theory must be treated with some caution . . .

Thus, while the Russians had a clear sense of a theory that made intuitive sense and were discussing other implications, the issue remained unsettled in the West for another three, perhaps arguably six, years. (Historians of science refer to this as a period of "stabilisation", suggesting that "facts" only become facts by being "stabilised" as a network of methods, observations and social and institutional agreements (Chalmers, 2013).) All the early radio observations of the diffuse galactic emission were at a sufficiently low frequency that they would have been completely depolarised. Mills and others in Australia had already predicted that synchrotron emission would be polarised, and the lack of any radio polarisation

---

[26]The cosmic ray electrons were not detected until 1961 by Earl (1961), but Kiepenheuer had already noted that relativistic electrons would be produced when cosmic ray primary particles interacted with the ISM.

[27]Sullivan (2009, p. 386), also discusses Roberts's dismissal of the synchrotron radiation theory.

detection was considered strong evidence against the synchrotron radiation theory by the Australians. Nonetheless over these years the theory was being tested and debated within the astronomy community. For example, as early as 1956, Oort and Walraven (1956) had estimated that the polarisation in the Crab nebula would be wiped out by Faraday effects for wavelengths longer than about 3 cm, which would explain the absence of confirmatory observations.[28] The earlier Australian observations had been made at 75 cm where the Faraday effect is many hundreds of times larger. Mayer et al. (1957) were aware of this and searched for and detected polarisation at a shorter wavelength of 3 cm. However, the effect of Faraday depolarisation was *also* not generally appreciated in the broader radio astronomy community until after the Parkes telescope (GRT) made a direct measurement of Faraday rotation in Centaurus A five years later in 1962 (see Chap. 32).

One of Pawsey's key science drivers for the Parkes Telescope was the measurement of polarisation and the extension of these measurements to shorter wavelengths. The Parkes direct observation of Faraday rotation in Centaurus A was made just before Pawsey died, and as he was dying, he said he was particularly proud of this achievement (Chap. 40).

IAU Symposium on Radio Astronomy, Manchester 1955

The August 1955 Manchester IAU symposium #4 on radio astronomy (van de Hulst, 1957), was the key turning point for the stabilisation of the concept of the synchrotron radiation mechanism in the West. In an opening talk, Hanbury Brown summarised the situation: "the emission mechanism for the Class I sources [sources concentrated in the galactic plane] is now known to be thermal emission from HII regions for observations in the cm range; in the metre range the mechanism is unknown. But Alfvén, Herlofson, Kiepenheuer and Ginzburg had all suggested that the non-thermal radiation might be due to cosmic-ray electrons in interstellar magnetic fields".

Oort visited Russia to attend the official opening of the reconstructed Pulkovo Observatory on 21 May 1954 and became aware of Ginzburg and Shklovskii's enthusiasm for the synchrotron mechanism as a source of radio emission. With Oort's broad understanding of all aspects of astronomy he was quickly able to pursue this idea. By the time of the Manchester meeting Oort and Walraven (1956) were able to present a paper based on their measurements of strong optical polarisation in the Crab nebula. They had already been making detailed optical observations of the brightness of the Crab nebula and modified their equipment to confirm the earlier indications of polarised emission by Vashakidze (1954) and then later by Dombrosky (1954). Although Shklovskii had not himself predicted the linear

---

[28] Oort and Walraven attribute knowledge of this depolarisation effect to Unsold and Seeger but they provide no reference.

polarisation, Gordon and Ginzburg had recognised this possible test (see discussion in Sullivan, 1982, p. 384).[29] For Oort and Walraven (1956), the observation of strong polarisation provided overwhelming evidence for the Shklovskii (1953) radiation theory and they were the first to coin the term "Synchrotron Radiation".[30] They agreed with Shklovskii's suggestion that the radio emission from the Crab nebula is also synchrotron radiation and went on to speculate that the jet and the larger radio lobes in the radio galaxy M87 could also be synchrotron radiation. At this time the M87 "jet" was usually called the "wisp" and the term "jet" only replaced "wisp" when the concept of an ejection of a stream of high energy particles from an active nucleus became acceptable more than a decade later.

The importance, and the difficulties, of Russian contributions was very marked at this meeting in 1955. Neither Shklovskii nor Ginzburg were able to attend; we can only speculate as to what might have resulted had radio astronomers from the West been able to discuss their findings and perspectives in person.[31] Both submitted papers to the meeting. Ginzburg's paper, "The nature of cosmic radio-emission and the origin of cosmic rays", was provided in advance for the meeting and it would have discussed the synchrotron radiation interpretation. However, it was not included in the Symposium publication edited by Henk van de Hulst.[32] The paper by Shklovskii (1957)[33] discussed the nature of M87: "it has a striking peculiarity, a small and very bright jet" and by analogy with the Crab he proposed that both optical and radio are relativistic electrons in magnetic fields and that relativistic particles are supplied by the nucleus. The optical jet would have a short lifetime, it would disappear, but the radio synchrotron would last for 10^8 years. After the jet disappeared there would be an apparently normal elliptical galaxy with a strong radio source. He speculated on how the nucleus could create this energy, but it would

---

[29] In his autobiography (Roberts, 2002, NRAO ONLINE.49) Roberts noted: "Contrary to some later reports, Shklovskii (1953) did not suggest searching for linear polarisation in the light of the Crab Nebula as a means of testing his theory. His paper was in Russian, but Frank Kerr had arranged for such papers to be translated for the Radiophysics Library. When I saw the translation it immediately occurred to me that a search for optical polarisation should be made and I suggested this to a meeting held at Mt. Stromlo Observatory at about this time. In the event the first detection [of optical polarisation] was made by Dombrovsky (1954) and this was quickly followed by other measurements."

[30] Previously this had been called "magnetobremsstrahlung" and the accelerator community just called it radiation from accelerated particles in the synchrotron.

[31] Leonid Gurvits (private communication) notes that in the 1950s, attendance at scientific meetings abroad was subject to decision at the very top of the Central Committee of the Communist Party. In the opening phrases of his Paris paper Ginzburg referred to the Manchester paper "presented by S.B. Pikel'ner". But Pikel'ner himself did not attend the Manchester symposium.

[32] Ginzburg's (1959) "Radio astronomy and the origin of cosmic rays", which was included in the IAU/URSI symposium in Paris in 1958 includes the statement: "A paper sent to the Manchester Symposium on Radio Astronomy in 1955 described the views developed earlier (Ginzburg, 1953) concerning a magnetobremsstrahlung (synchrotron) origin of non-thermal cosmic radio emission. Unfortunately, for unknown reasons, it was not included in the Symposium volume."

[33] This paper is included in the proceedings but there is no attached discussion so this paper may not have been presented orally at the meeting.

be another decade before the accretion of material into a black hole would be generally recognised as the source of energy in the nuclei of galaxies.

Following this meeting, we can trace the widespread acceptance of synchrotron radio emission theory and its application to a wide range of radio phenomena. For example, Burbidge (1956), following up on a suggestion by Baade, also recognised that the optical jet in the radio galaxy M87 could also be explained by synchrotron radiation at both radio and optical wavelengths, and he calculated the required magnetic field strength and particle energetics. The pieces of the non-thermal radio synchrotron puzzle were finally all falling into place.

Symposium on Radio Astronomy: Paris 1958

By the time of the Paris Symposium on radio astronomy in 1958, the issues of the radiation mechanism had been resolved and the key role of the synchrotron mechanism recognised. In his contribution to the symposium, "Radio astronomy and the origin of cosmic rays", Ginzburg provided a critical commentary on the now discredited theory that the galactic radio emission is the sum of a great number of radio stars. In the same paper he then went on to establish much of the basis for the use of radio emission to study the origin, acceleration, lifetime, propagation and losses of cosmic ray electrons in the galaxy.

In his opening presentation in the Fourth Session, "The Large-Scale Structure of Galaxies", Pawsey emphasised the importance of the recent acceptance of the synchrotron emission process by the astronomy community and Hoyle, in his concluding lecture at the Paris Symposium's Sixth Session ("Mechanisms of Solar and Cosmic Emissions"), echoed this opinion: "Undoubtedly the outstanding advance of the last few years in theoretical astronomy has been the widespread recognition of the importance of synchrotron emission by relativistic electrons."

Pawsey began his summary:

> The continuum observations, on the other hand, refer to at least two quite distinct sub-systems in a galaxy: to ionised interstellar gas (HII regions), and to regions emitting a non-thermal component. HII regions emit radio waves by the well-known thermal process involving free-free transitions, and absorption in these regions can also be important. The non-thermal component is the dominant one in the metre-wavelength range. Following Shklovskii's suggestion,[34] the mechanism of emission is now believed to be the synchrotron mechanism, in which radiation is emitted by relativistic electrons spiralling around lines of magnetic force in interstellar space. The main evidence for this mechanism is the observed linear polarisation of the radio and optical emissions from a very few discrete radio sources. Workers are currently trying to extend this evidence by studying the polarisation of other regions. If this hypothesis is correct then theory suggests that sources of the non-thermal

---

[34] Many papers in the West attribute the recognition of the synchrotron mechanism as the explanation of the non-thermal radio emission from discrete sources to Shklovskii without appreciation of the earlier and seminal contributions by Ginzburg. This may be a result of the much stronger interactions between Shklovskii and the astronomy community in the West.

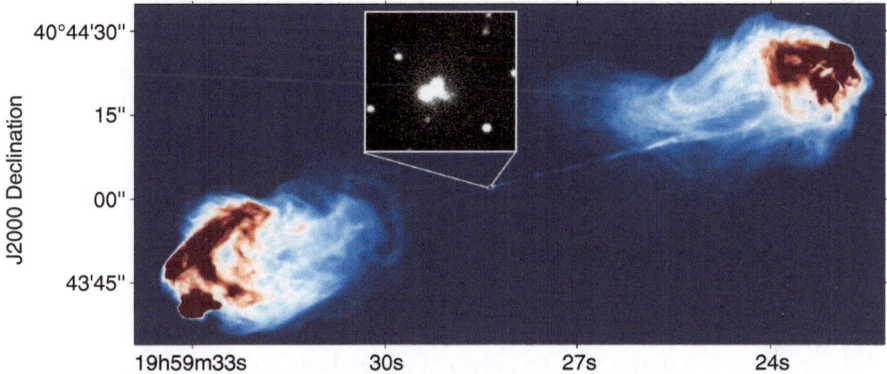

Fig. 34.1 VLA radio image of Cygnus A showing the jet powering the hot spots and the two lobes from the central black hole. Insert: First optical image of Cygnus A by Walter Baade with 200" Hale reflector. Credit: NRAO/AUI/NSF

> radio emission (i.e., regions characterized by magnetic fields and high-energy electrons) are likely also to be regions favourable to the production of cosmic rays. Hence, studies of this component are related to those of the origins of cosmic rays in the Galaxy ...

The new issue now was not the radiation mechanism but the prodigious energy requirements to power these radio galaxies and a new debate emerged on whether this was provided by colliding galaxies or active galactic nuclei. The colliding galaxy model was proposed by Baade in 1951 to provide a source of energy for the extragalactic radio sources (see Sullivan, 2009, p. 345 for a lively account). Following the identification of Cygnus A with what appeared to be two galaxies in collision (Fig. 34.1) the case for the colliding galaxies seemed overwhelming (Baade & Minkowski, 1954).

The colliding galaxy theory so dominated thinking that observations of galaxy pairs became an additional criterion for choosing the "correct" identification from the many candidates when position errors were large. In this way the preconceived ideas about the role of colliding galaxies were influencing the observational evidence for the theory even though there was never a satisfactory explanation of how the kinetic energy in a collision could generate the radio emission. At the same time the alternative notion that the nucleus of the galaxy was causing the radio emission (Shklovskii, 1957; Ambartzumian, 1958) had little support in the West.[35]

The matter was finally resolved with the discovery of quasars by Schmidt (1963) and the associated direct evidence that the active region was so small that only gravity of a massive object in the nucleus of a galaxy could provide the energy. Since this time, the explanation of the radio galaxy phenomena has been a black hole in the

---

[35] One author (RDE) recalls a public lecture by Bernard Mills in the early 1960s which included a detailed description of the colliding galaxy theory but was accompanied by derisory comments about the alternative Russian theory that radio galaxies were caused by explosions in the nucleus of a galaxy.

nucleus of a galaxy (AGN) surrounded by an accretion disc which forms high energy plasma jets powering the radio galaxy lobes (see Fig. 34.1). Ironically, the interacting galaxy model is again current but now the interaction disturbs the gas so it can fall into the black hole, providing the energy required, a totally different concept than that envisaged by Baade in the 1950s.

How Isolation Impacted Radio Astronomy Research in Australia

We have noted that throughout this period Pawsey had continued to emphasise the importance of understanding the radio emission mechanism, yet little progress on the theory was made in Australia despite the huge progress in observational radio astronomy. We now discuss why this was the case for galactic and extra-galactic radio emission, but not for solar radio astronomy.

For solar radio astronomy, the basic physics was all known but understanding details of the specific mechanisms required clever observations, taxonomy and interpretation. These were all accomplished by the Ruby Payne-Scott and Paul Wild teams established by Joe Pawsey. Sullivan (2009, Sect. 13.2) notes "progress in the field was dominated to an extraordinary degree by the Radiophysics Laboratory in Sydney." Australia's isolation from other fields of astronomy and high energy particle physics was not an issue for the solar research. But for extragalactic and galactic non-thermal source interpretation, the applicable basic synchrotron theory was not recognised by the Australian radio astronomers; this recognition required links to other fields to take this step. The slow adoption of the new theory was not confined to Australia. Sullivan (2009) has an excellent discussion (Sect. 15.4.3) of why the synchrotron mechanism, which was well recognised in Russia by 1950, remained so unpopular in the West until 1955 and even later.

Kevin Westfold had worked as a theoretician in the Pawsey group in the late 1940s, but it was not until he spent some time with John Bolton at Caltech in the late 1950s that his theoretical worked flourished. While at Caltech he published the definitive paper on the polarisation of synchrotron emission at radio wavelengths (Westfold, 1959). He later returned to the new Monash University in Australia but had little further involvement in radio astronomy.

Australian radio astronomers had made all the key observations required to understand what was happening. RPL staff had made the majority of the first quantitative measurements of radio spectra confirming the non-thermal nature of many sources. Bolton had discovered that some discrete sources were extra-galactic. Mills had separated the galactic and extragalactic populations through his extensive surveys, and Mills had also put the nail in the coffin of the radio stars with his observations of extended structure in many discrete sources showing they could not be stars. So, what else would have been needed for the Australian radio astronomers to recognise the importance of the synchrotron theory model to explain these observations?

The Australian astronomers had made all the key observational steps but were tripped up by their own observation of no polarisation, ironically a case of knowing too much—the lack of the predicted polarisation would not have been known in Russia! Just as the predicted Crab optical polarisation, detected by the Russians and confirmed by the Dutch, provided the convincing evidence for optical synchrotron radiation from the Crab supernovae, the detection of polarised galactic radio emission as predicted by Mills would have confirmed the synchrotron mechanism years earlier if only the observation had been made at a slightly higher frequency where the emission was not depolarised by differential Faraday rotation. Strong polarisation in the galactic emission was later detected at higher frequencies.

Also, the Australian group had other handicaps. They would have needed a much broader background in physics as a discipline than existed in CSIR, plus a theorist of the calibre of the outstanding David Martyn, who was isolated from the radio astronomy community in Australia (see NRAO ONLINE.7, an indirect result of a wartime indiscretion). We see the "tyranny of distance" at work isolating the Australians from the traditional "centres" of research, continuing to create a constraint on Australian scientific achievement. This isolation meant that Pawsey's group were not embedded in a community with enough breadth to recognise the links to the interstellar medium and its magnetic field. There were no particle physicists until Messel arrived in 1953, and there were almost no extragalactic optical astronomers. As Jim Roberts has noted[36], relativistic effects were not part of physics taught in Australia (or the UK). This aspect of theory was well outside Pawsey's sphere of knowledge so he could not have pointed his theorists in the right direction. He could, and did, make them continually question their understanding of the observed phenomena. Pawsey was clearly frustrated by the lack of progress. As discussed above, in his document of 18 May 1948 (exploratory science and theory), Pawsey wrote to RPL colleagues: "We have collected a lot of important facts and not one has been predicted in advance in any worthwhile manner."

It is also significant that most of the radio astronomy group in Australia had Engineering backgrounds, so their primary methodology involved the application of known theory to build better instruments and interpret observations. Any knowledge of astronomy was as amateurs. With this background, theoretical speculations involving far-out ideas of relativistic particles and active galactic nuclei did not emerge.

Pawsey's strong emphasis on extensive international travel and his drive to build broad collaborations across engineering, physics and astronomy, was probably the most effective strategy possible for the geographically isolated radio astronomers in Australia.

---

[36] Roberts (private communication 2013).

Chapter 35
Radio Source Survey: Disputes, 1948–1957

*Perhaps the greatest discontinuity [in my career] was with
the identification of Cygnus A. That showed that [in 1953] we
were in the cosmology game ... To me that was the point
where one said, "Well now, if other things like Cygnus exist,
here is something which we can likely see, even with our little
instrument, as far away as the [Mt Palomar] 200-inch can
see. This is something much more interesting than it might
have been—much more interesting than [the radio sources]
being galactic objects ...*
*–Martin Ryle (Interview with W. Sullivan, 1976, cited in
Sullivan, 1990, 2009)*

Introduction

The decade of the 1950s became one of the most eventful in the history of not only
radio, but of all astronomy, as a result of the first significant surveys of radio sources.
During this decade it was realised that the radio sources included an isotropic
population of extragalactic sources which could be observed in the radio at greater
distances than the deepest optical observations. Radio astronomy suddenly became a
part of extragalactic astronomy, and most significantly, a completely new and
unexpected probe of cosmology. The story of the development and interpretation
of these surveys is a complex one, involving technically challenging instrumental
issues, one controversy involving the different results obtained by the observational
groups located in Sydney and in Cambridge, and another controversy between the
cosmologists and the observers. The situation was further exacerbated by the intense
emotional interactions between some of the key personalities involved.

The facts of the story are now fairly well known, with substantial reviews and
interpretation in Edge and Mulkay (1976), Bertotti, Balbinot, Bergia, and Messina
(1990), Kragh (1996), and Sullivan (2009). Here we summarise these events and
present the controversy from the Sydney perspective, which is less well known.

Supplementary Information The online version contains supplementary material available at
[https://doi.org/10.1007/978-3-031-07916-0_35].

557
W. M. Goss et al., *Joe Pawsey and the Founding of Australian Radio Astronomy*,
Historical & Cultural Astronomy, https://doi.org/10.1007/978-3-031-07916-0_35

Conflicts between the Sydney and Cambridge groups had already started over failed collaboration attempts, especially in relation to the early optical identifications proposed for some radio sources.[1] New, more serious, conflicts were now arising due to major discrepancies between the early survey catalogues and disagreements about how to interpret the results, and these differences are explored more thoroughly in this theme chapter.

Exploring the disagreements and controversies about the surveys of the 1950s offers many enticing opportunities for gaining insight into how the field of radio astronomy developed in this period. We can also delve into how and why some scientists missed the key insights that might have delivered swifter progress, and how and why others reversed their opinions. This was a period of fast-changing conceptions about what radio observations could accomplish, the nature of the phenomena observed and the implications of these observations for not only astronomy, but also cosmology. These are the details that we have attempted to elucidate in this chapter.

Early Radio Source Surveys

In 1948 Pawsey suggested a radio continuum survey of the southern sky to complement the solar work.[2] The Cambridge group under Martin Ryle also started a survey in 1948. The initial objective of these surveys and those that followed throughout the 1950s was to extend the short list of discrete radio sources then known in order to understand what they were. It was naturally assumed that the brightest radio sources in the sky would be associated with the brightest optical stars, thus making a catalogue to identify and classify individual radio sources was the obvious next step. Initially there were no obvious identifications with bright stars apart from the sun and the Milky Way. The groups needed to measure more accurate positions so the radio sources could be identified with known astronomical objects. This initial work started at Sydney and Cambridge at about the same time. As the 1950s progressed, additional objectives of understanding the distribution of the *population* of radio sources became a topic of intense interest. How were they distributed across the sky—isotropic or concentrated along the Milky Way? How were they distributed in intensity—were there more faint sources farther away? By 1950 the radio astronomy research group at Jodrell Bank[3] also began to conduct surveys using their large transit dish.[4]

---

[1] This was discussed in Chap. 18. See also NRAO ONLINE.21 for a description of pre-1951 conflicts. See Chap. 16 for the first Bolton source identifications and relevant work before 1950.

[2] As early as Dec 1945 (see NRAO ONLINE.20) an internal report by Ruby Payne-Scott mentioned the need for an all sky survey.

[3] See Chap. 16

[4] No other comparable research organisations had yet emerged at this time (See Sullivan, 2009).

Table 35.1 Radio source surveys

Date	Survey	Location	Number sources	Freq (MHz)	Reference	Telescope
1950	1C survey	Cambridge	50	81	Ryle, Smith, and Elsmore (1950)	Long Michelson interferometer
1951	Badgery Creek	Sydney	77	101	Mills (1952a)	3 element interferometer
1953	Dover Heights	Sydney	104	100	Bolton, Stanley, and Slee (1954)	Cliff interferometer
1953	Jodrell Bank	Manchester	23	158	Hanbury Brown and Hazard (1953)	218' transit dish
1954	2C survey	Cambridge	1936	81	Shakeshaft, Ryle, Baldwin, Elsmore, and Thomson (1955)	4 element array
1955	MSH preliminary	Sydney	368	85	Mills and Slee (1957)	Mills cross
1956	3C	Cambridge	471	159	Edge, Scheuer, and Shakeshaft (1958)	2C array at double freq
1958	MSH +10 to −20	Sydney	1159	85	Mills, Slee, and Hill (1958)	Mills cross
1959	3C—final	Cambridge	471	159	Edge, Shakeshaft, McAdam, Baldwin, and Archer (1959)	3C array
1960	MSH −20 to −50	Sydney	892	85	Mills, Slee, and Hill (1960)	Mills cross
1961	4C preliminary	Cambridge	910	178	Scott and Ryle (1961)	4C aperture synthesis array
1961	MSH −50 to −80	Sydney	219	85	Mills, Slee, and Hill (1961)	Mills cross
1962	3CR	Cambridge	328	178	Bennett and Simth (1962)	3C & 4C arrays

The main radio source surveys conducted in the 1950s and early 1960s are summarised in Table 35.1. We discuss some of the more critical results here; more technical details can be found in the ESM_35.1.pdf, Radio Source Surveys.

For their survey, the Cambridge group built a two-element interferometer called the "Long Michelson" (see ESM_35.1.pdf, Radio Source Surveys). Fifty radio sources were detected and a "Preliminary survey of the radio stars in the Northern Hemisphere" was published by Ryle et al. (1950), later to be known as the 1C (First Cambridge) survey. Based on the lack of any correspondence with the brightest galaxies, they concluded that these sources were "radio stars" in our galaxy although the positions were still too poor to make any associations with specific stars. There were too many possibilities.

In Sydney, Bernard Mills used three antennas configured as two phase-switched 101 MHz interferometers with baselines of 60 and 270 m at a site at Badgery's

Creek, southwest of Sydney (Fig 35.1a).[5] This was first introduced in Chap. 21 and more details are provided in ESM_35.1.pdf, Radio Source Surveys.

The survey yielded 77 sources (Mills, 1952b) and an important result was that Mills divided the radio sources he found into two classes: Class I sources were concentrated along the galactic plane with galactic latitudes <12 deg. tending to be the more intense sources. On the other hand the Class II sources were displaced from the galactic plane and showed a different log N- log S behaviour (Fig. 35.1b).

They used the famous log N- log S description of the counts of the number of radio sources of different intensity, a powerful diagnostic for the way the radio sources are distributed in space. N is the number of sources per unit area with a flux density S or greater. This presentation is often characterised by a slope, α, where $N \propto S^{\alpha}$ or Log (N) $\propto \alpha$ Log (S). In a 3D Euclidian Universe, the volume of space and hence the number of sources, N, increases with the cube of the distance and the flux density of the sources decreases as the square of the distance so $\alpha = -3/2 = -1.5$. However, if the sources are confined to a finite volume there will be fewer fainter sources at larger distance so $\alpha > -1.5$.[6] In the Mills survey the galactic plane sources of Type I showed a slope (α) of -0.75 while the Class II sources indicated a slope of -1.5. Class I sources were characteristic of a population of Milky Way objects in a finite depth disk as observed from the sun. Mills pointed out that the Class II sources could arise either from a nearby local population (e.g. stars) in the galaxy or from extragalactic objects.

At the time many astronomers, and in particular the Cambridge group, held the view that the discrete radio sources were stars in our galaxy because if these brightest radio sources in the sky were outside our galaxy they would have to be extraordinarily luminous (see Chap. 16). Therefore, Mills's data, one of the first suggestions of an extragalactic population of radio sources which were not just a few abnormal sources such as Virgo A and Centaurus A (identified by Bolton, Stanley, & Slee, 1949, as external galaxies) was potentially revolutionary. Mills was cautious, but he wrote to Bowen, who was in the UK to give a presentation to the RAS in London at the time: "I can say, but not for publication, that the evidence favours the extragalactic hypothesis ...".

Mills needed to be cautious since, already in 1951, there were discrepancies between the first survey results, as everyone involved was keenly aware. The Ryle et al. (1950) 1C survey had 50 sources fewer than detected by Mills; they claimed that it showed no evidence for the two classes. We note that Sullivan (2009, p. 365) repeated the analysis of the source counts from the 1C list and obtained results very

---

[5] On 30 April 1951, Mills wrote Bowen in London with information concerning a presentation that Bowen was to make at the Royal Astronomical Society on 11 May, "Recent Developments in Radio Astronomy in Australia". Bowen (1951) described the southern all sky survey being made by Mills at Badgerys Creek. NAA C3830 Z1/9. Bowen personal correspondence.

[6] In its simplest form, both the distance and the intrinsic luminosity of the sources cancel so the source counts provide information about the distribution of sources in the volume of space surveyed without detailed knowledge of the distance or luminosity of the sources. We return to this point later.

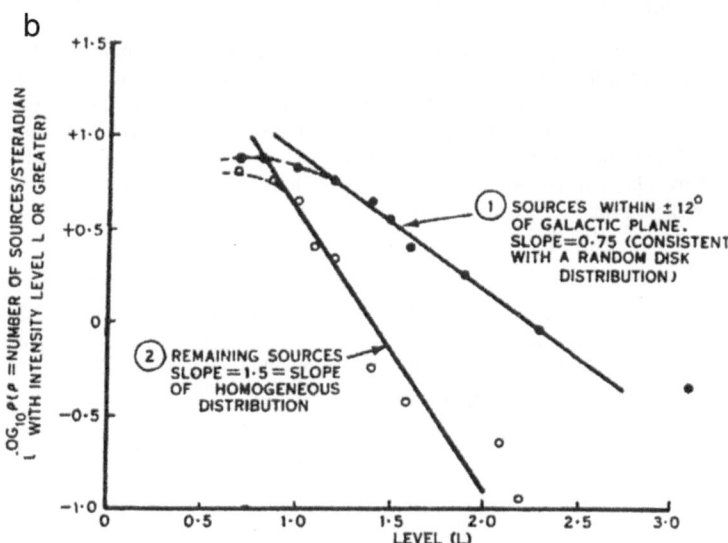

Fig. 35.1 (a) One of the three elements used for the 101 MHz all sky survey made by Mills in 1950–1951. Mills stands next to one of the aerials (See also Chap. 21, Fig. 21.1). The primary beam was 14 by 24 deg. The survey was conducted in 10 deg. steps, i.e. the aerial was fixed at a certain elevation (0 to 90 deg) and the sky drifted past. The declination ranged covered was +50 to −90

similar to Mills: "I find a log N-log S slope of −2.1 with a demarcation similar to Mills's (1952a) for sources at low and high galactic latitude. The 1C sample also shows an excess of strong sources near the galactic plane (six of seven of the strongest sources are at latitudes ≤12°)." Mills wrote to Bowen that he thought a major problem in the northern galactic plane for Ryle and Smith was the effect of sidelobes from the two strongest sources in the north which were both near the galactic plane, Cassiopeia A and Cygnus A.[7] The former was circumpolar (i.e. always above the horizon as observed from Cambridge). Mills wrote: "Since Ryle has to deal with two powerful sources right on the galactic plane, I think ours is more likely to be correct (it may not be a good idea, however, to claim any superiority, particularly if Ryle should be present)." Ryle did attend the 11 May 1951 presentation by Bowen at the RAS in London.[8]

In Sydney Bolton, Stanley, and Slee (1954) published the results of the final survey made with the Dover Heights sea-cliff interferometer (see ESM_35.1.pdf, Radio Source Surveys). The 104 discrete sources detected were compared with other radio source catalogues made up to that time. As already noted by others there was good agreement for the brighter sources but significant disagreements for many of the fainter sources. Bolton et al. stated that some disagreements were to be expected since, in addition to the differing responses to extended sources, there was the issue of confusion. For cases where the source density was high enough to have two sources in the primary beam, both interferometers and single dishes will give incorrect [but different] results. They emphasised the value of multiple surveys with different types of antenna and noted that some source confusion was removed with the sea-cliff interferometer due to the sharp edge of the earth's shadow. Remarkably Bolton, Stanley, and Slee (1954) concluded their paper with a most significant result.

> ... There seem to be too many faint sources compared with an isotropic distribution of objects ... A plausible explanation is that the Sun (if these sources are galactic) or the Galaxy (if these sources are extragalactic) is in a local region of low source density and that somewhere towards the limit of the survey we reach a region of much higher density ... however, there is not much point in speculating too far on this result as it could also be produced by a large dispersion in absolute magnitude [intrinsic luminosities] amongst the sources of the survey.

Fig. 35.1 (continued) degrees. Credit: CSIRO Radio Astronomy Image Archive B2594-4. (**b**) Shows the different behaviour of sources near ① or away ② from the galactic plane. Credit: Fig. 7, "The Distribution of the Discrete Sources of Cosmic Radio Radiation", Mills, B. Y. (1952b). *Australian Journal of Scientific Research* A Physical Sciences 5: 266

---

[7] This problem remains in the current era. The southern hemisphere observers using low frequency arrays such as the Murchison Widefield Array in Western Australia still have an advantage because there are no sources as powerful as Cygnus A or Cassiopeia A in the Southern Hemisphere.

[8] Independently, Pawsey wrote Bowen (NAA C3830 Z1/9) on 1 May 1951. He stressed to Bowen that "Mills's results are much more likely to be right but I don't think we can back this objectively."

This is the first published evidence for an evolving population of radio sources, but unfortunately, Bolton and his colleagues did not realise that their concern about the effect of a large dispersion in luminosity was misguided. This was understandable. Fainter sources could either be further away, or intrinsically less luminous; one would naturally think that these two effects could not be disentangled. But this argument for an evolving population of sources was far stronger than Bolton, Stanley, and Slee (1954) realised. The next year Ryle and Scheuer (1955) showed that while a large dispersion in intrinsic luminosity could decrease the slope of the source counts, it can never increase the slope! Unfortunately, Bolton was not in a position to use this information. Due to the decision to fund the Mills Cross telescope rather than a new instrument at Dover Heights (see Chap. 22), Bolton had left radio astronomy in Aug 1953 to work on cloud physics (rain making). He had no further involvement in radio astronomy until after he moved to Caltech to set up the Owens Valley Radio Observatory in 1955. But, as we will see at the end of this chapter, Bolton eventually had the last word on the surveys.

Sullivan (1990) has presented a striking image that showed that for the four main surveys up to 1953 (see Table 35.1) few optical identifications had been made and the surveys "profoundly disagree with one other" (Sullivan, 1990, p. 318 in *Modern Cosmology in Retrospect*). Some of the disagreement related to an issue which was to continue to be controversial for many years: that of the nature, as well as the position, of many of the radio sources. Were these (sun-like) stars or other kinds of objects, and were they inside, or beyond, our galaxy? In this period, Ryle was quite committed to the assumption that the radio sources were stars in the galaxy, which would have extremely small angular size.

All three radio astronomy groups were strongly influenced by their understanding of the instruments they were using, which were very different. Many of the discrete sources (especially in the galactic plane) were later found to be extended, so the responses were quite different for single dishes—as were used particularly at Jodrell Bank—which detected all the radio emission from an extended source, and interferometers which are insensitive to extended structure.

In his thorough analysis of the underlying reasons for the survey controversies of 1948–1953, Sullivan (2009, p. 361) discussed the complexity of disputes centred on instrumentation. He pointed out that Ryle's use of long interferometer baselines was well motivated, since it was the result of his conviction that the radio sources were stars which have very small angular size. His success in using an interferometer with its phase switch (invented at Cambridge, see Chap. 37) to remove the strong diffuse galactic emission also made this the optimum instrumental design for a survey of point sources. Ryle saw this as a major advantage over the single dish, which would be confused by structure in the diffuse galactic emission. Sullivan (2009, p. 361) has described the importance of using an interferometer to eliminate confusion caused by small scale variations in a letter that Ryle wrote to Tandberg-Hanseen on 10 February 1950:

As Ryle explained at the time[9]:

> It is very easy [with large dishes] to obtain maxima in the received flux which appear to be due to a source of small angular diameter, but which in reality are due to the angular variations of the general structure of the Galaxy. For this reason, we have always used interference systems of considerable resolving power to discriminate between "point" sources (i.e., sources having a diameter of less than 5–10 minutes of arc) and the general background radiation from the Galaxy.

This is a nice example of how a concept that turned out to be mostly incorrect—the notion of radio "stars"—can be central to instrument design.[10] As Sullivan points out, in this case it was an example of an effect discussed by Pickering (1981) and Galison (1987) where a researcher shaped his instrument to obtain the anticipated result. While this is often necessary for success, there was also the obvious danger of prolonged misleading research results. So it is interesting to see how these preconceived ideas affected the eventual outcome. As we will see by the end of this chapter, Ryle was in one sense correct—the interferometers and arrays were far superior to the single dishes for surveys of the extragalactic discrete source population. But this was not obvious at all in the early 1950s, and the conflicting results from the single dishes were a huge distraction that greatly confused the discussions at the time.

After their first survey, the Cambridge group began planning for a new enlarged instrument, designed in 1950–1951 and constructed in 1952. (See Fig. 35.2 and ESM_35.1.pdf, Radio Source Surveys, for a description of the 2C aerial.) But since Ryle only expected point sources which would have the same amplitude on any interferometer baseline, he only used a single baseline. This produced a disastrously incorrect survey. Multiple point sources within his beam, or even within the sidelobes of his beam, were incorrectly catalogued as single sources in the wrong position if the interferometer fringes added, or were missed altogether if the fringes cancelled.

Disagreements about instruments in this period were worsened by Ryle's personality and difficulty in collaborating. Ryle's disdain for other instruments made it harder, not easier, for a collective scientific assessment of the affordances and limitations of different instruments and of the consequent issues of confusion and potential identifications of sources with astronomical objects. For example, Hanbury Brown reflected later on the first Jodrell Bank survey (see ESM_35.1.pdf, Radio Source Surveys) which he and Cyril Hazard had conducted with a 218-foot transit

---

[9] Graham Smith in Sullivan (1984, p. 244) has also discussed the dilemma of radio stars in the era 1951–1952.

[10] See also Chap. 5; for example, the Austin-Cohen formula became a design rule for antennas in the early twentieth century.

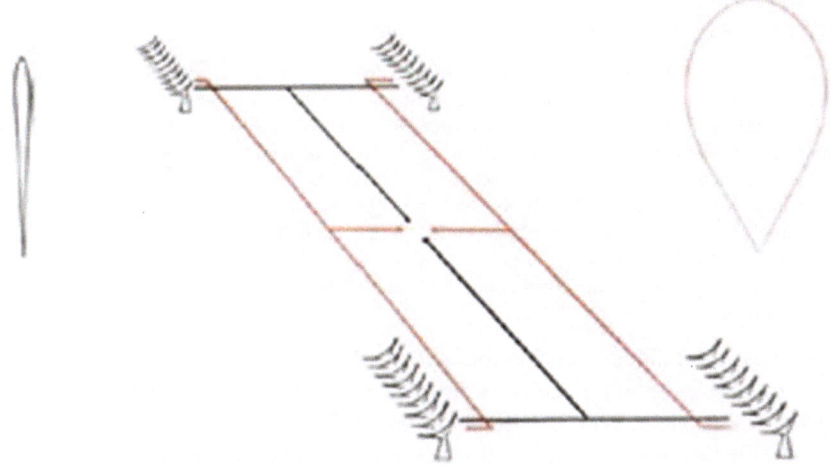

CAMBRIDGE INTERFEROMETER, shown in photograph on page 210, is diagrammed here. The circuit indicated by red lines sets up a high-resolution reception pattern (*left*). The circuit indicated by the black lines produces a pattern with lower resolution (*right*).

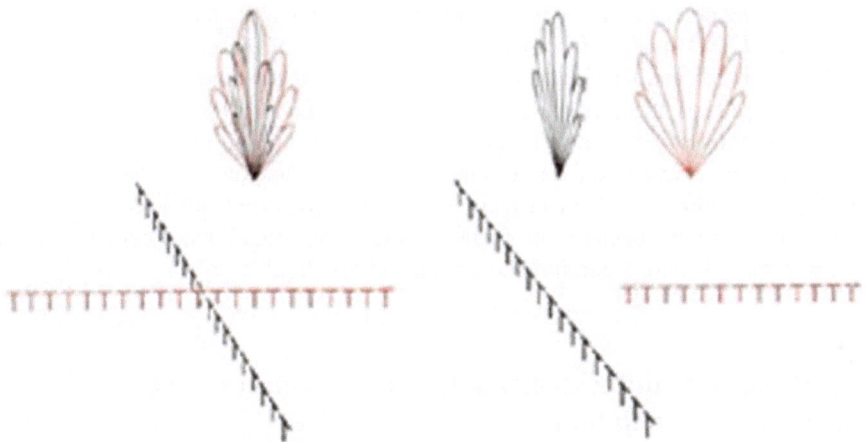

ORIENTATION OF ANTENNAS at right angles increases the resolution of radio interferometers. Australian "Mills cross" is diagrammed at left; another Cambridge installation, at right. The two systems register only sources that are picked up by both of their antennas.

Fig. 35.2 Comparison of the Cambridge 2C array (top) and the Mills Cross (bottom). See ESM 35.1.pdf, Radio Source Surveys, for more details. Diagrams based on sketches by Martin Ryle. Courtesy of and copyright: Cavendish Laboratory, University of Cambridge, all rights reserved

parabolic dish (Hanbury Brown & Hazard, 1953b) and compared with the Cambridge 1C interferometer survey of Ryle et al. (1950)[11]:

> A comparison of the two surveys showed that of the 13 most intense sources detected with our pencil-beam (10 of which lay within 5° of the galactic plane) only 4 appeared in the [Cambridge 1C] survey. I shall never forget the strenuous arguments we had with those dedicated interferometrophiles in our efforts to convince them that our sources in the galactic plane were real and not side-lobes of our pencil-beam. It was part of the conventional wisdom at Jodrell Bank that Cambridge had only three standard reactions to our work: (1) "it is wrong", (2) "we have done it before", or (3) "it is irrelevant". Indeed, as we later showed, at least 6 of the 10 sources in the galactic plane had angular sizes exceeding 1° and were either partially or totally resolved, and therefore largely undetected, by their interferometer.[12]

The issue of the influence of possible extended extragalactic sources took another decade to settle. Although these were real, and were missing in the Cambridge survey, it was a small effect that had not influenced the conclusions from Cambridge. However, the fact that some of the strongest sources in the southern hemisphere were very extended (eg Centaurus A) led Mills astray. He assumed there were similar sources in his survey, and he overcorrected for this effect. So, while Pickering and Galison were correct in noting a potential bias from instrument design at Cambridge, this bias did not affect the final outcome. Even though Ryle started with strong pre-conceptions and a personality that hampered the capacity to acknowledge error, his own evidence eventually convinced him to dramatically change his view.

Sullivan (2009) ends his comprehensive coverage of the early history of radio astronomy at the end of 1953 (though reference is made to some post-1953 events in his discussions of the repercussions of some of the pre-1953 radio astronomy history). Technology was developing at a rapid rate and an entirely new field of astronomy was being opened. Completely new and unexpected classes of radio sources were being found (or at least, tentatively suggested) all over the sky; the radio source surveys became the focus of future investigations, especially for the groups in the UK and Australia who dominated this field.

The 2C survey and Extragalactic Radio Sources: Radio Sources as a Population

During 1953, plans for the next generation of survey instruments were being developed in Sydney and in Cambridge. These included the conceptual design of the Mills Cross (as a survey telescope) in Australia, while Smith at Cambridge

---

[11]Two of the authors (Ekers and Goss) were present at the IAU in Patras, Greece, when Hanbury Brown gave this talk in August 1982. As the famous "three reactions" statement was made, the audience of 50 or so broke out in extended laughter, though Goss observed that Cambridge colleagues present (and thus in a complex position) did not join in the hilarity.

[12]NAA C3830 A1/1/1 Part 5.

emphasised the advantages of the planned 2C antenna, at 3.7 m. These instruments could not and did not resolve the issue of confusion, i.e. the probability of multiple sources simultaneously in the telescope beam, and discussions continued about how to interpret their findings.

Thus perhaps the greatest impact of the 2C survey was its effect on Martin Ryle, who would suddenly reverse his position that radio "point sources" were new radio-emitting forms of stars that occurred within our galaxy. Ryle's reversal of position—as he well understood—had huge cosmological implications.

In Sydney, this extragalactic population had been long recognised following the identification Bolton et al. (1949) of Virgo A (M87) and Centaurus A (NGC 5128)[13] and Mills's 1951 division of the discrete radio sources into two classes: Class I (galactic) and Class II (possibly extragalactic), as we discussed above. Those cosmologists—Fred Hoyle (1915–2001), Thomas Gold (1920–2004) and Hermann Bondi (1919–2005)—and astronomers—Baade and Minkowski (see Chap. 16)—who were aware of radio astronomical investigation in this period were also very willing to consider alternative explanations and extra-galactic distances for radio sources.

But as is well known, before 1953 Ryle had inflexibly rejected speculation that most sources were extragalactic. Why was this? The strongest source in the northern sky had no bright nearby galaxy identification; from Ryle's perspective, if the emission was from a very distant extragalactic source, it would be so powerful as to stretch credibility. He had dismissed the Bolton et al. (1949) identifications of Centaurus A as "not certain" and the case of Virgo A identified with the galaxy M87 as "less interesting", just a normal galaxy like the nearby Andromeda galaxy, which had been detected as a very weak radio source at Jodrell Bank. Extragalactic emission was also not compatible with his view that the radio emission from the Milky Way was the sum of all the radio stars in the galaxy (Boyd, 1951).

Hoyle's autobiography (Hoyle, 1994) includes a colorful description of an exchange between Ryle and Gold following talks by Gold and Ryle at the Massey meeting (Boyd, 1951) in April 1951.

By 1951, about half a dozen radio sources had been definitively related to astronomical objects, not one of which had turned out to be a star. There had been identifications at Jodrell Bank with weakly emitting nearby galaxies, and there had been the two cases identified by John Bolton, the Crab Nebula and the galaxy NGC 5128. So Gold said that perhaps the other possibility for explaining the isotropic distribution of radio sources—namely, that most radio sources were very distant—should be taken seriously. He said nothing more than that, and he expressed it temperately. It was in these circumstances that Ryle began an attack that was to persist for almost two decades ... he [Ryle] began, "What *theoreticians* have failed to understand ... " (with the word theoreticians implying some inferior and detestable species)
...

I do not think it unreasonable to say that Ryle's motivation in developing a program of counting radio sources, a program that was to occupy a major fraction of his group over the next ten years, was to exact revenge for his humiliation over the radio-star affair. This was to

---

[13]The initial misclassification of these two galaxies as galactic is discussed in detail in Chap. 34.

be done by knocking out the new form of cosmology with which Gold, Bondi, and I were associated.

Ryle could hold onto his misconception that the radio sources were mostly galactic stars, even though Minkowski had identified Cygnus A with a distant galaxy using the Cambridge position,[14] by assuming Cygnus A was an abnormality, an outlier which, while interesting, could still be ignored. But in late 1953 he suddenly saw how all the data could make sense, if this "outlier" was typical for the population.

Sullivan (1990, p. 321) has analysed Ryle's notes from the early 1950s to find out when this dramatic change of view occurred. Before 1953 Ryle was an adamant supporter of the galactic radio star model. But in October 1953 he presented a talk to the Cavendish Physical Society:

> Now it so happens that there *is* a large isotropic component of radiation which cannot be explained in terms of emission from galactic sources. If indeed it is extragalactic, it offers the possibility of being able to distinguish between some of the cosmological theories.

> Whether the observations will ever be sufficiently accurate one cannot say, but it is nice to think that the cosmologists may one day [lose their] complete freedom of choice of the conditions beyond the optical limit.[15]

The first direct archival evidence related to the new (2C) survey found by Sullivan is from 30 July 1954 when Ryle made a note: "Study of extragalactic sources— cosmological application—little hope of much identification ... Main object will be to see further—even if area of sky has to be restricted ...".

Sullivan goes on to document Ryle's later recollection of this period:

> Perhaps the greatest discontinuity [in my career] was with the identification of Cygnus A. That showed that we were in the cosmology game ... To me that was the point where one said, "Well now, if other things like Cygnus exist, here is something which we can likely see, even with our little instrument, as far away as the 200-inch can see. This is something much more interesting than it might have been—much more interesting than [the radio sources] being galactic objects, much more interesting than M87's [fairly normal galaxies]" ... (Interview with W. Sullivan, 1976, cited in Sullivan, 1990, 2009)

It was the interpretation by Shakeshaft et al. (1955) of the 2C catalogue of radio sources paper, called "The Spatial Distribution and the Nature of Radio Stars" by Ryle and Scheuer (1955), that caused a commotion. In the logN-logS plot in Fig. 35.3, N is the number of sources per steradian (I) with intensity greater than (I) in units of 10 Jy. Ryle and Scheuer pointed out that the steep slope (< -1.5 which is the expected slope for an isotropic distribution in a static Euclidean universe) could only be explained if the sources were extragalactic, of similar luminosity as Cygnus A, and "of much greater number density at larger distances than nearby" (giving more faint sources than expected and hence the steep negative slope at faint flux density levels). (What seems surprising to us now is that they were not concerned by the very rapid flattening of the counts for faint sources. This effect

---

[14] See Chaps. 16 and 22.

[15] From Sullivan (1990)—14 October 1953, notes for a talk to the Cavendish Physical Society, p, 17, file 8 (uncat), RYL

Fig. 35.3 2C survey results from Ryle and Scheuer (1955) showing the log of the number of sources versus the log of the source intensity. This is referred to as the log N- log S source counts. Credit: Fig. 1, "The Spatial Distribution and the Nature of Radio Stars", Ryle, M., and Scheuer, P.A.G. (1955). *Proceedings of the Royal Society of London* Series A 230, pp. 448–462

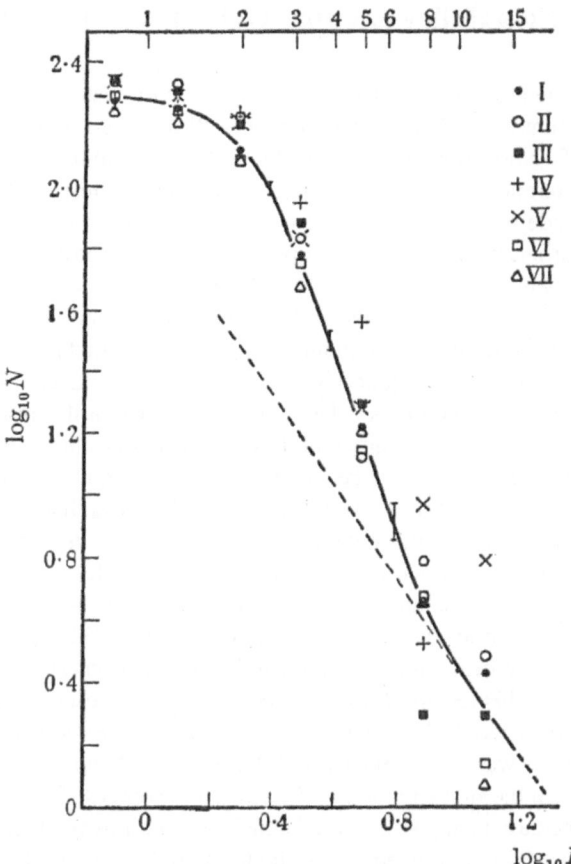

would only be possible in a finite universe or if there was a mistake in measuring amplitudes for faint sources; it should have alerted the group to instrumental error.)

Scheuer (Scheuer & Bertotti, 1990, p. 331) included a brief historical timeline. He asserted that Ryle was already convinced in late 1953 that the numbers "were not thinning out with distance, as they should, sooner or later, if they were local galactic stars ... At this point—it must have been in the winter of 1953-54 or the early spring of 1954—Martin Ryle's attitude on 'radio stars' changed almost overnight. They had to be extragalactic." And the population of sources had to increase with distance. This was evidence that contradicted the Steady-state theory recently developed by Hoyle, Gold and Bondi, which predicted a constant density in the universe. This effect could not be explained by a local distribution of galactic stars unless the sun were located in the centre of a spherical hole in the nearby universe.

Ryle's Halley Lecture, Oxford, 6 May 1955

Ryle (1955) presented his new view on the nature of the radio sources in his Halley Lecture. He described a series of well-focussed steps leading to the need for a survey of radio sources to understand their distribution in the universe. He made no reference to similar, earlier suggestions of just such an extragalactic population made by the Dutch, and by the Australians who had made this suggestion 4 years earlier. It had also been proposed by his adversarial theoreticians Gold and Hoyle, all of whom he had previously refuted.

He gave no hint that he had just made a complete reversal of his previous galactic radio star interpretation and instead he presented his new world view as if it were all the logical consequence of the Cambridge work. This was a very Cambridge-centric version of the history. Nor should we necessarily find this surprising. Memory, we now know, does not withdraw information about the past from storage. Rather, memory is an active reconstructive process that generates and validates a particular conception of the self; indeed, research shows that when we describe our memories differently to different audiences, the memory itself (and not just the message) alters in what is known as the "audience tuning effect". Ryle's version of history can be understood as an example of this effect.[16] Ryle became very assertive, even dogmatic, about his new views; interestingly, Mills, who had already made this case long before, remained characteristically conservative.

In this lecture, Ryle proceeded to clearly describe the nature of the source counts which would have $N(S) \propto S^{-1.5}$ in a static Euclidean Universe. He noted that the shape was not altered even if the sources all had different luminosities.[17] The results from the first catalogue (2C with 1936 sources, Shakeshaft et al., 1955) required an increase in apparent spatial density or luminosity of "radio stars" with distance. Note that Ryle continued to use the term "radio star" to describe both galactic and extragalactic sources; clearly the terminology no longer implied that the objects were "stars", but by keeping the old terminology the Cambridge group had disguised their previous misconceptions. Most other radio astronomers of this era had ceased using the terminology "radio stars".

Ryle proposed that many of these radio sources could have luminosities comparable to Cygnus A, but be located more distantly in the universe. They would likely not be identifiable even with the 200-inch telescope. This point was very important—and with this step forward he had leapfrogged all the other groups. It explained why most radio sources could not be identified with galaxies—a problem which

---

[16]John Bolton and Taffy Bowen also showed examples of this effect, published in later decades in personal retrospective publications. Also, Schacter (2012), Brown, Kouri, and Hirst (2012), Echterhoff, Higgins, and Groll (2005).

[17]As described later, this insightful analysis is only true if the source counts are described by a single power law on all flux density scales. This is not correct for any real universe but this notion that the radio luminosity function did not matter persisted in some camps for more than another decade.

many observers of the survey controversy took to imply a poor-quality radio catalogue and a lack of progress.

A reason why this step may have been more obvious to the Cambridge group is the asymmetry between the Northern and Southern Hemispheres. In the South there were several relatively nearby bright radio galaxies: NGC 5128 (Centaurus A), NGC 1316 (Fornax A) and M87 (Virgo A) while the brightest source in the Northern Hemisphere is the far more distant Cygnus A. It had been implicitly assumed that the brightest sources would be associated with the brightest and closest galaxies, but this was not the case.

Ryle had been explicit in his conclusion of the Halley Lecture: "This is a most remarkable and important result, but if we accept the conclusion that most of the radio stars are external to the galaxy, and this conclusion seems hard to avoid, then there seems no way in which the observations can be explained in terms of a Steady-state theory."

As noted by Sullivan (1990, p. 23), the Steady-state group was quick to respond.

This Steady-state theory had been invented in the late 1940s at Cambridge by Bondi, Gold and Hoyle and provided an attractive and testable alternative to the various older "big bang" models developed in the decade following the introduction of general relativity. Ryle's sweeping disproof of Steady-state theory naturally caught the imagination and attention of the general public and the scientific community. Here was no less than a major theory of the universe being overthrown.

Within a week, the Steady-state proponents had a chance to respond at the Royal Astronomical Society meeting the following Friday (13 May 1955). An image of the 2C survey and a newspaper billboard from London on 13 May 1955 are also shown in Fig. 35.4.

Shakeshaft and Ryle spoke on behalf of the radio astronomers, while Gold and Bondi responded from the Steady-state camp. Gold praised the quality of the survey but did not miss the chance to revive the earlier controversy with Ryle about whether the radio sources were extragalactic, as well as questioning the current cosmological evidence. Gold wrote:

I have been greatly impressed by this magnificent survey, which has exceeded all expectations of a few years ago. I am also glad to see that there is now agreement that many of these sources are likely to be extragalactic, as I suggested here and elsewhere, with much opposition, four years ago. Mr Ryle then considered that such a suggestion must be based on a misunderstanding of the evidence.[18]

If the interpretation which has now been adopted is correct, then it is true that radio observations may make a direct contribution to cosmology. The great distance of the identified Cygnus source is an indication that some other sources may also be far, and some even further than any optically recognizable objects. But on present evidence it is very rash to regard the great majority of weak sources as extremely distant. Yet this is implied in attributing cosmological significance to the curve of the number-intensity relation. A wide

---

[18] Gold's comment was entirely correct, and his earlier suggestion that the radio sources might be extra-galactic had also been strongly opposed by Ryle who had now changed his interpretation.

Radio astronomy enters cosmology

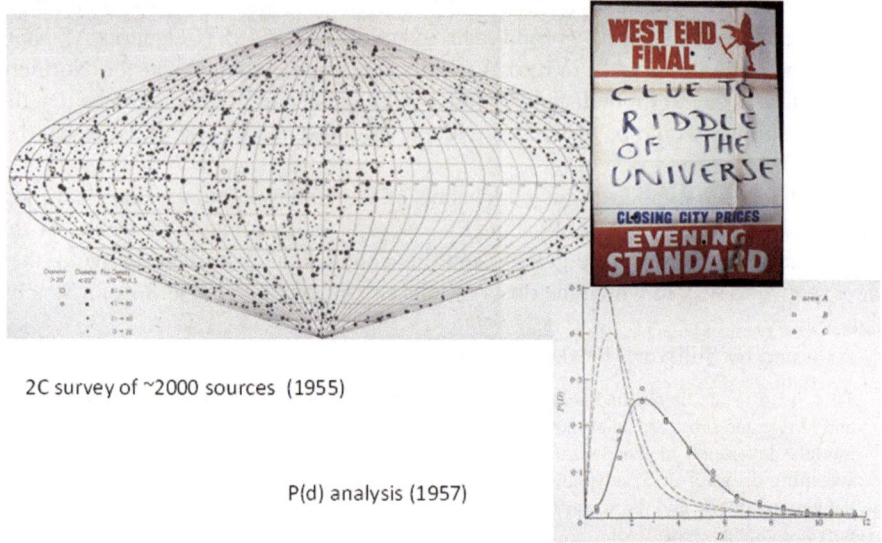

2C survey of ~2000 sources (1955)

P(d) analysis (1957)

Fig. 35.4 2C survey (Shakeshaft et al., 1955) and P(d) analysis (Scheuer, 1957) with a newspaper billboard from London on 13 May 1955. Credits: Fig. 3, "A survey of radio sources between declinations −38° and+ 83°", Shakeshaft, J. R., Ryle, M., Baldwin, J. E., Elsmore, B., & Thomson, J. H. (1955). *Memoirs of the Royal Astronomical Society, 67*, 106; "A statistical method for analysing observations of faint radio stars", Scheuer, P. A. G. (1957). *Proceedings of the Cambridge Philosophical Society* 53, pp. 764–773

spread in the intrinsic intensity of sources would imply a dilution of far with near in the count of the weak sources.[19]

At Jodrell Bank, Hanbury Brown and Hazard had detected faint radio emission from nearby "normal" galaxies using the 218-ft fixed reflector. These were faint and close by, but the prevailing view was that fainter radio sources were simply more distant. At the same RAS meeting, Hanbury Brown asked the very good question about the role of the large variation in intrinsic luminosity of different sources based on the cosmological interpretation. Ryle in his Halley lecture had argued that the luminosity of the sources does not matter because sources of different luminosity will all have the same source count slope of −1.5 so the combination of sources of different luminosity will still have a slope of −1.5. However, this argument is only correct if the distribution of intensities is a power law. This would be the case for a Euclidian universe with no boundary, but it does not apply in a real universe at large distances

---

[19] As already noted a wide spread in the intrinsic intensity of the radio sources, which is also called a broad radio luminosity function, can decrease the slope of the radio counts but an increase is not possible; this aspect of Gold's conclusion was incorrect.

where the curvature of space and time dilation break the power law assumption. Ryle and Scheuer (1955) presented a more sophisticated version of this argument and showed that the luminosity function could not increase the slope, however, a reduction was possible.

In fact, we can now see that the effect of the shape and width of the luminosity function would continue to be either ignored, or blamed for unexpected effects, until it was fully included in the models by Longair (1966) and explained by Von Hoerner (1973). They both showed that it was still the case that either the luminosity or the number of sources had to increase in the distant universe. Von Hoerner (1973) also demonstrated that for some luminosity functions a population of sources need not have an inverse relation between *average* flux density and distance, i.e. the weaker sources in a survey are not necessarily more distant as was often naively assumed.

Gold continued: "It is a fortunate fact that the Steady-state theory of cosmology is very definite in its observational implications. If that theory is correct, then it must be expected that any error in an observation or its interpretation will lead to discord."

It is interesting to note that Gold, and some other cosmologists, used the predictive power of the Steady-state theory to assert that the observations were flawed, rather than contemplate abandoning a theory when predictions are not confirmed.[20] Proponents of other theories with less clear predictions would not engage in such contentious arguments.

Sullivan (1990, p. 325) reports that in 1976, Ryle told him that he had been unprepared for the intensity of the opposition from the Steady-state adherents. As Ryle recalled to Sullivan:

> [The intense controversy] was a considerable shock, because of course the trouble with cosmology up till then was that it had been a playground for mathematicians—"Is space curved this way or that way?"—and all these things. It was nothing very much to do with the real world and observations had never, and apparently would never, make any effect on it. It was a game mathematicians could play, safe from all possible attack.

> But the development of the Steady-state model was an important break-through. Here was something that made specific predictions in wide range of not-necessarily-thought-of possible measurements. It said that the universe was in a state that could remain the same through time as well as space ... And as soon as you know you can detect sources at redshifts large enough for things to happen on other cosmologies, then you can detect a difference ... It was remarkable what an absolute storm it provoked. Well of course, it wasn't helped by the fact that the press got hold of the story.

---

[20] These events occurred 4 years before philosopher of science Karl Popper (who was at Canterbury in New Zealand for 4 years at the same time as Fred White) published his work *The Logic of Scientific Discovery* in English. In this book, he set out the then-revolutionary argument that science progressed as a result, not of confirming correct theories, but from falsifying incorrect ones. He argued that a good theory ought to be falsifiable, that is, empirically testable, as the Steady-state theory was; but the corollary was of course that when contradictory evidence was found, the falsified theory ought to be instantly discarded. In reality, of course, it requires many observations and experiments to determine whether an apparently falsifying observation is an anomaly, an error or malfunction in the experiment (as Gold was suggesting) or an incorrect theory; nor are these the only problems with Popper's model of scientific progress (see Chalmers, 2013).

The Australian group were taken by surprise as Ryle promoted his new evolutionary paradigm based on the 2C Survey statistics. From May to July 1955, leading up to the conflicts during the August IAU conference at Jodrell Bank, Ryle and Pawsey corresponded about discrepancies in their results and their interpretation; Pawsey in a characteristically cautious, but open-minded style.[21]

On 14 June 1955 Pawsey wrote Ryle after having read *The Times* newspaper from London:

> I recently saw in a cutting from *The Times* that you have completed all, or at any rate a major part, of the job of analysis of your 81 MHz source records and had described your results at the Halley Lecture [at Oxford on 6 May 1955]. It must have been a tough job with this large number of sources.

> *The Times* article then went on to say that conclusions of major importance followed from the analysis, but I was unable to gather any essential points from the article. So, we in Australia are left highly intrigued by things which are now common gossip in England, and I wondered if you could let us in on the secret. Perhaps you have a spare copy of the lecture ...

> I shall look forward to seeing you at Jodrell Bank [IAU Symposium on Radio Astronomy] and Dublin [IAU General Assembly].

On 18 June 1955, Ryle answered an earlier letter of 24 May 1955 from Pawsey [which has not been located in the NAA archive]:

> Many thanks for your letter of 24 May enclosing a preprint of the catalogue of sources [likely a list of the new Mills Cross sources] ...

> As soon as I have some copies of the figures, I shall be sending you a manuscript of a paper which has gone to the Royal [Publications of the Royal Society].

On 27 June 1955, Ryle wrote a handwritten letter to Pawsey, now responding to the letter of 14 June:

> There was, of course, certainly no intention of keeping Australia out of the picture! In fact, we have intended a more or less simultaneous "release" of the new data coincident with the Halley Lecture. Before that, I am afraid we had been somewhat cagey because of the large number of hungry cosmologists we have prowling around [probably Hoyle, Gold and Bondi— the proponents of the Steady-state theory of the Universe] —we wanted to be able to collect our thoughts a little on the next stage before they pounced![22]

> Rather than send you the simplified account given in in the Halley lecture (which you might have punched holes in!), I thought it best to send you the real paper—the latter is not yet common gossip in England.

> When you have had a chance to read this, we should very much appreciate your comments— both on what you think of the arguments and on the observational side. I imagine that Mills's

---

[21] NAA C3830, A1/1/1 Part 9 (1954–1955).

[22] Sullivan (1990, p. 321) has pointed out that Ryle had previously made a sarcastic remark about cosmologists in course notes written 2 years earlier: "Cosmologists have always lived in a happy state of being able to postulate theories which had no chance of being disproved—all that was necessary was that they should work in the observable Universe out to regions where the velocity is about ½ c ... Even if we never actually succeed in measurements with sufficient accuracy to disprove any cosmological theory, the threat may discourage too great a sense of irresponsibility."

survey is now in a position where we could make a similar analysis—he has in fact probably already done so and it will be most interesting to see what his results show ... We look forward to seeing you and Paul Wild [at Jodrell Bank and the IAU].

In mid-May to mid-June 1955, Pawsey continued his conversations with Ryle, sending some early results from the Mills Cross. Independently, Fred Hoyle wrote Mills asking if his survey was complete at a level that could provide a check on the 2C survey statistics.

Pawsey replied to Ryle once more on 6 July 1955 before his arrival in London on 21 August.

Thank you for your two letters and for the copy of your paper, which arrived today.

The content of the paper is revolutionary. If the data are correct, I think your explanation the most plausible (i.e. the best) and the implications in cosmogony[23] immense.

I immediately checked the current state of the 80 MHz Mills Cross results. The present position is this. We have not yet begun a systematic survey but have a lot of observations of areas chosen on a peculiar basis. In these areas discrete sources have been noted as the records became available. There are 500 or 600 on the list and when these are plotted, log I versus log N they fall on a slightly irregular line of slope of -1.5 which flattens off (sources getting small) [weaker?] at an intensity of about 5 [Jansky]. Thus, the observations appear not to agree with yours, but we must investigate the situation much more carefully before there is any certainty.

Another count of a small number of sources away from the Milky Way showed a slope steeper than -1.5 but at an intensity less than where you found this.[24]

This is all mysterious. We will do our best to clarify the issue as soon as possible, I hope before I come over, and shall let you know what eventuates. In the meantime, I don't need to do more than say our results constitute a case for careful investigation. In this context it will be most helpful to have the full paper which you sent me so that we can compare the really crucial points. Thank you very much for sending it.

The next stage was to occur when Pawsey was at Jodrell Bank in late August 1955. As additional checks were made in Sydney and a larger sample of the southern radio sky was available, the statistical disagreement between the two surveys had increased as the papers from the conference were presented.

---

[23] Pawsey used the word "cosmogony" not "cosmology". Cosmogony (the origin of the Universe) is more appropriate than cosmology (large scale structure in the Universe) but cosmogony has gone out of common use.

[24] "a slope steeper than -1.5" was a scarcely noticed critical point. Even a slope as steep as -1.5 already requires evolution. It was the disagreement with Ryle's much steeper 2C value of -3 that triggered the continuing dispute. Both results required evolution.

1955 IAU Symposium no 4 on Radio Astronomy[25]

This 3-day symposium was held on 25–27 August 1955 at Jodrell Bank. The second day of the programme included a discussion of the statistics and optical identifications of discrete radio sources.[26]

After some introductory papers on radio source identifications, and a series of papers on the emission mechanism (see Chaps. 26 and 34), there was a short presentation by John Shakeshaft describing the "Cambridge Survey of Radio Sources". The main fireworks of the morning followed with Ryle's presentation "The Spatial Distribution of Radio Stars". This was followed by Pawsey, who presented the competing point of view, on behalf of Mills, from Sydney, "Preliminary Statistics of Discrete Sources Obtained with the 'Mills Cross'". Ryle's presentation was similar to his Halley Lecture[27] and also included the P(D) curve (see Figs. 35.3 and 35.4). From the data shown, Ryle asserted again that the steepening of the log N- log S at faint intensities was a real effect, with the spatial density or the luminosity of the sources showing a progressive increase with distance.

Pawsey began his presentation with a stirring declaration: "The statistics of the discrete sources observed in Cambridge and the interpretation given by Ryle and his colleagues constitute one of the most interesting items of recent astronomy. It is therefore of great importance to check the observational data and this can be done [by using the new Mills Cross at 85 MHz]".

Pawsey explained to the audience that he had received, some months previously, a pre-publication copy of the 2C survey. At that time, the Mills survey had detected 550 sources over a solid angle of about one steradian at 85 MHz; 180 of the sources were well displaced from the galactic plane. In Fig. 35.5, we show his comparison with the Ryle and Scheuer distributions; in Fig. 35.6 we show the additional sources detected (total 1030 sources) just before the Jodrell Bank Symposium; the three Sydney curves showed no deviation from the −3/2 law "which we can be sure is significant". Pawsey wrote:

> There is thus a substantial disagreement between the Cambridge and the preliminary Sydney results, and it seems best to withhold judgement on the more interesting interpretation put forward by Ryle and Scheuer until the Sydney observations are complete. At that stage quite definite conclusions should be reached because the pencil-beam technique used is substantially free from confusion

---

[25] See Chap. 26 and NRAO ONLINE.51 for details on the organisation of this first IAU sponsored symposium in the new field of radio astronomy.

[26] Other topics discussed in this symposium are covered in Chapters 26 and 34. On 30 August 1955, Pawsey wrote Bowen in Sydney (from the IAU in Dublin) with a report of the Jodrell Bank Symposium. The symposium volume edited by van de Hulst was published in 1957 and includes the papers presented and the discussions and introductory comments. Surprisingly, Pawsey referred to this session on the Cambridge source statistics as the "fizzer" of the conference. We interpret this as Pawsey's dissatisfaction that no resolution had been reached rather than being a dull event.

[27] published in *Observatory* in 1955 (vol. 75, p. 137).

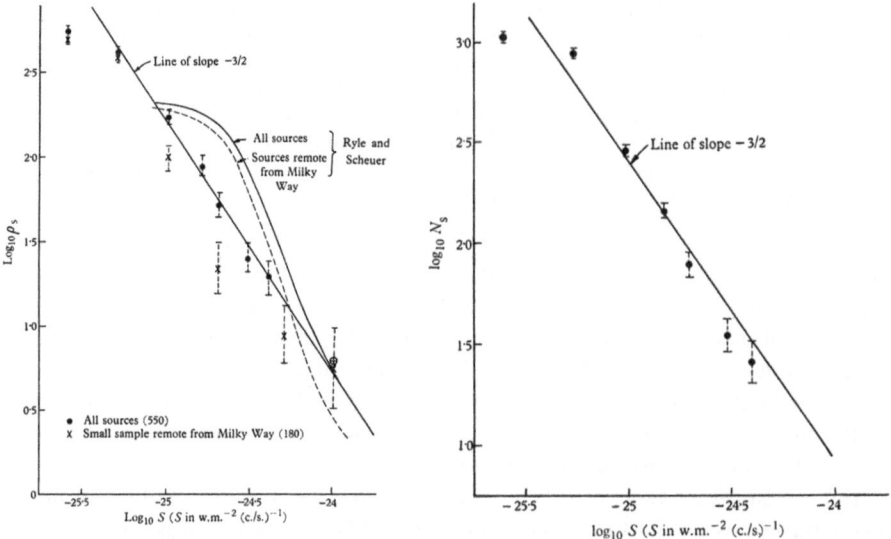

Figs. 35.5 and 35.6 Radio source counts. Left: Mills Cross compared to Ryle and Scheuer. Right: additional 1030 Mills Cross sources. Credit: Fig. 1 & 2 from "Preliminary Statistics of Discrete Sources obtained with the 'Mills Cross'", Pawsey, J. L. (1957). IAU Symposium No 4 Radio Astronomy, ed. van der Hulst, H.C. pp. 228

During the discussion there were heated exchanges between Ryle and both Gold and Bondi, the latter two suggesting that confusion might be a problem with the 2C survey. Gold stated after Pawsey's talk:

> Another way in which a steepened curve could be brought about is by the erroneous judgment of intensity of some of the faint sources. When there are several sources in the beam, it might frequently occur that one is recorded of greater than correct intensity. This would produce an increase in the number in one range of the curve at the expense of a proportionally much smaller decrease in a higher section of the curve.[28] An interpretation of that sort would imply that the Cambridge survey is much more liable to such an error, and already at a higher intensity than the Australian one.

Bok (1955, p. 21) reviewing the Jodrell Bank Symposium on Radio Astronomy, was cautious as he summarised his opinions about the conflicts. He was worried about the need for a detailed comparison of the two surveys and the failure of the Cambridge group to secure optical identifications: "the Australian observers were somewhat luckier, with Pawsey reporting 10 normal galaxies, one pair of apparently colliding galaxies, one additional supernova remnant and several [HII regions] . . . as having marked radio sources at their optical positions." Bok concluded:

---

[28] This analysis by Gold is correct and was the reason for Ryle's very steep slope. This problem had also been recognised by Scheuer and was his motivation for introducing the P(D) analysis discussed in detail in the following "Probability of a Deflection" section.

There was considerable discussion—especially later in Dublin—of the possible cosmological consequences of the surprisingly large numbers of faint sources in the Cavendish
statistics, but in view of the rather different Australian results and the lack of positive
identifications with optical objects, the time does not seem ripe for such speculations. We
are still very far from a real understanding of the nature of the faint sources. Ryle considers
that they are mostly galaxies (possibly like the faint colliding pair 200 million light-years[29]
from the sun that is responsible for the very strong Cygnus A source), and practically all of
them beyond the reach of even the 200-inch telescope.

In retrospect we can see that contrary to Bok's, and indeed everyone's, assumptions,
identifications with optically known objects did not help the early radio astronomers
understand their sources better—instead, these identifications led Mills astray! With
typical position errors of fractions of degrees in 1955, the only secure identifications
were with nearby galaxies. By chance this meant that a few Southern sources were
able to be identified with nearby galaxies. Unlike Ryle, Mills assumed that most
radio sources were relatively nearby, so he expected a Euclidian source count with
index −1.5, and interpreted this as no evolution.

But we can see now that Ryle's assumption that the dominant radio population
had luminosities similar to Cygnus A and were at large redshifts was correct; hence
comments about the lack of identifications, like those from Bok, were misleading.
Lots of identifications with bright optical galaxies would not be expected, and Bok's
concluding comment that "... practically all of them beyond the reach of even the
200-inch telescope" was correct and explains the low identification rate since faint
galaxies could not be identified with the poor positions available.[30] We can now see
that Ryle was fortunate in his choice of Cygnus A as the prototype since this was the
correct assumption about the average distance of the radio source population. With
this large distance even a source count index of −1.5 required evolution. This
assumption of a large average distance and not the bogus steep slope of the 2C
source counts, was the key to Cambridge case for evolution.

Pawsey Correspondence with Southworth, Oort and Appleton

Pawsey and Ryle were similar insofar as both were strongly influenced in their
interpretation of the data by their knowledge of the reliability of their own instruments (and presumably by human factors such as loyalty to, and defense of, their
own group). Thus Pawsey clearly backed Mills and Mills's results in his letters to
other astronomers at the time. For example, on 26 November 1955, George

---

[29] 760 million light-years on the new distance scale of the twenty-first century.

[30] To "improve" identifications probability, colliding galaxies were also preferentially included, and
this was another assumption (later shown to be wrong) that distorted the statistics.

Southworth (see Chap. 11 and NRAO ONLINE.20) had written to Pawsey. Pawsey's reply included an update on the Mills Cross survey that had begun in 1954:

Our major current program is a survey of the sky with the 85 MHz Mills Cross. [see ESM_ 35.1.pdf, Radio Source Surveys] . . . It is working excellently. Incidentally, we are engaged in a hot controversy with Ryle on results. He deduces thoroughly interesting cosmological ideas from results from the recent Cambridge survey with an interferometer, for example, that most of the observed faint radio stars are beyond the limits of the 200-inch telescope. We think his interpretation interesting, but his observations phoney.

A letter from Pawsey to Oort was sent a few weeks later (16 December 1955) with a similar message:

I think the principal gossip from here is that the Cambridge-Sydney controversy over the statistics of discrete sources is reflected in similar discrepancies in comparisons between individual sources. So, the Sydney guess as to the cause of the discrepancies is still inadequate resolution for the Cambridge observations.

On 22 February 1956, Pawsey wrote Appleton[31]:

Incidentally, as you may know, we are in complete disagreement with Ryle as to the discrete sources. Ryle found from his statistics many more faint sources than would have been expected in a Euclidean universe with a uniform (average) density of radio sources. He drew very pretty cosmological inferences about the days when the universe was young. The disagreement between Mills's and Ryle's statistics is paralleled in a sample comparison of individual sources.

On 19 September 1956, Appleton asked Pawsey for advice as he prepared to give the Reith Lecture later in 1956.[32] Appleton wanted to tell the story of the 2C survey, writing:

As a matter of mere sentiment, I would like the radio people to be right and the quite arrogant theorists wrong. But that merely indicates the need to steel oneself to the discipline of science.

Can you, then, please give me the Australian view, with advice as to what can be said and can't. You have different parts of the universe to examine, but Ryle says he gets the same result in all directions. (Appleton's emphasis)

Pawsey replied on 28 September 1956 with a decisive letter.[33] He noted that Appleton was dealing with an "intensely controversial question and I should like to suggest that you move with great caution." Pawsey pointed out that the results of the previous year from Cambridge had major implications but only if the data were correct. However, "the disagreement [of the two surveys] is appalling. One or other of these surveys is completely haywire." Pawsey then pointed out that Ryle's data was in error due to confusion with only two beamwidths per source compared to tens

[31] NAA C3830 Z3/1/VI.

[32] BBC radio flagship lecture series.

[33] Also in this letter, Pawsey replied to Appleton in a whimsical mood: "[You have pointed out in your address to UK scientists] that science should be fun. I too often get overwhelmed and forget that."

of beamwidths per source in the Mills survey from Sydney.[34] The Mills survey showed no excess of sources at the low flux density level over that expected from a "Euclidian universe with a constant space density". Pawsey showed all the reasons for remaining cautious.

1955–1957: Increased Controversy Between Sydney and Cambridge

After the heated debates of 1955, the Sydney group (Mills, Slee and Hill) continued their southern sky survey, counting sources down to 7 Jansky; 368 sources were detected over an area of about one steradian. The region of declination of +10 to −20 deg. and RA range 0 h to 8 h was chosen since it overlapped with the 2C survey preliminary catalogue sent by Ryle to Mills "for the purpose of checking".

Publishing a paper in 1957 detailing these results, Mills et al. were even more vehement than at the Jodrell Bank 1955 conference. The source lists were discrepant: "Simple inspection of the maps reveals that the two catalogues are almost completely discordant. The conclusion follows that instrumental effects play a decisive part in determining the positions and intensities of sources in at least one of the surveys."

The authors have carried out an analysis of the two surveys, concluding that the major limitation was caused by instrumental errors in the 2C survey. The claimed culprit was an effect of the low resolution of the 2C interferometer.

> There are two important factors which tend to increase the apparent number of sources with flux densities just above the survey limit ... Confusion or blending effects in which sources below the limit cause a random variation in the output ... Large chance excursions are then counted as single sources. [Also] the effect of observational selection in the presence of noise; the rapid increase in numbers with decreasing flux density provides many more sources ...

The Mills's and colleagues' logN-logS curve was almost identical to the one shown in the 1955 Jodrell Bank conference proceedings. The message was clear:

> We have shown that in the sample area, which is included in the recent Cambridge catalogue of radio sources, there is a striking disagreement between the two catalogues. Reasons are advanced for supposing that the Cambridge survey is very seriously affected by instrumental effects which have a trivial influence on the Sydney results. We therefore conclude that discrepancies, in the main, reflect errors in the Cambridge catalogue, and accordingly deductions of cosmological interest derived from its analysis are without foundation.

> An analysis of our results shows that there is no clear evidence for any effect of cosmological importance in the source counts ...

---

[34]The number of beamwidths per source was a critical design criteria to avoid confusion and is discussed more extensively in the Paris Symposium section (Chap. 34).

Scientific American Interchange Sept 1956 Between Ryle and Mills

In 1956, Ryle wrote an article, "Radio Galaxies", for *Scientific American* in what was probably the first time the scientifically minded public were made aware of this remarkable new window upon the Universe. His focus was, of course, on the great distance of radio galaxies "beyond the range of the 200-inch [Palomar] telescope!". He presented an account of the 2C survey including the claim that the slope of the log N- log S was very steep, providing evidence for an expanding universe. Ryle noted:

> The Cambridge conclusion about the distribution of radio stars in space far beyond our galaxy has been questioned by workers in Australia. A survey with the Mills-cross radio telescope has failed to show a marked excess of faint sources such as was found by the Cambridge group. The Australian survey, however, has not yet covered a large area of the sky and it does indicate that radio stars are not distributed uniformly with distance.[35]

The article concluded with a preference for an evolutionary cosmology:

> If these surveys verify that the density of radio sources in space does indeed increase with distance, they should help to make possible a decision between the evolutionary and Steady-state theories of the universe ... Thus, our present conclusions from the radio work at Cambridge support the evolutionary view.

This article—perhaps not surprisingly—made the Sydney-Cambridge disagreements both public and (hence) more acrimonious. Mills, though cautious in interpreting data in general, was as inflexible as Ryle about the validity of his own survey. He also had a strong value for exactitude in undertaking surveys, and this was more important to him than cosmological speculation. The group in Sydney at CSIRO were offended by having what they considered to be superior results publicly dismissed. Mills wrote a two-column rejoinder. Pawsey and Bart Bok, who was still at Harvard but was preparing to move to Australia to the Mt. Stromlo Observatory in January 1957, helped to arrange its publication. Bok wrote to Pawsey on 23 October 1956[36] that he had just been in New York at the *Scientific American* offices and had met the editor, Dennis Flanagan.[37] Flanagan was "honoured and delighted to be able to print the Mills letter. They promised to put it in a good conspicuous spot, so that the reader couldn't possibly miss it." Mills's rejoinder was published in the letter section in December 1956 (page 8); Ryle followed with three plus columns and a complex figure on page 10.

Mills's letter—rather like Gold's earlier critique of Ryle at the 1955 IAU symposium—accurately questioned the accuracy of Ryle's observations but then tried to justify his own results with an erroneous assumption. Beginning with a precise

---

[35] Before Mills made his erroneous correction for extended sources his source counts did show a small excess of faint sources.

[36] NAA C3830 Z3/1/VII.

[37] Flanagan was the well-known editor from 1947 to 1984.

description of the large discrepancies between the two catalogues ("it is obvious that at least one of the catalogues is hopelessly wrong"), he explained how "difficulties might be expected to arise when two or more radio sources are sufficiently close together for a radio telescope to respond to each simultaneously" and noted that this will occur frequently with the Cambridge instrument. After this persuasive argument that it was almost certainly the 2C catalogue that was in error (which was true), he destroyed his own case with a circular argument that the Mills catalogue must be right because it was consistent with a static Euclidean Universe!

Ryle's rejoinder in the *Scientific American* began mildly:

> In a subject which is advancing as rapidly as radio astronomy it is natural that there should be differences of opinion from time to time; in some cases these differences may be finally resolved only by more detailed observations. It is possible that in the present case a definite conclusion may have to await the completion of the new survey of greater resolving power which is now in progress at Cambridge on a frequency of 160 MHz.[38]

Ryle stated that preliminary results from the 3C survey agreed fairly well with the 2C sources. But then he made the point that was key to the whole dispute: "Errors in the positions of individual sources have no effect on the estimated number of sources falling within a given intensity range."

In other words, Ryle was saying that it didn't matter if the sources in the 2C catalogue were in the wrong place because incorrect positions did not affect the slope of the source counts—a statistical result, and one that requires evolution. Ryle was able to make this argument because of Scheuer's P(D) analysis (see next section), something that would have been too difficult to explain in the *Scientific American*. Mills had been making a much more straightforward assertion that if the catalogues don't agree, you can't trust the results; Ryle was changing the question from a focus on understanding individual sources primarily from their position, to what can be learned from *the distribution of amplitudes* of a population of sources.[39]

Ryle also asserted that the flatter curve from Sydney was influenced by a contamination of flatter spectrum galactic sources; subsequent data from Sydney showed that this was not the case.

Ryle's concluding remarks: "We look forward to seeing further details of Mills's survey, but in the meantime, we feel that our conclusions concerning the spatial distribution of sources of small angular size are substantially correct."

---

[38]The new 3C survey, made with the 2C antenna but at twice the frequency. Thus the beam solid angle would be four times smaller and the effect of source confusion much less. See next chapter and ESM_35.1.pdf, Radio Source Surveys.

[39]Similar insights were only just being developed in other areas of science. 1950–1955, famously, saw the first publications by Sir Richard Doll and Sir Austin (Richard) Bradford Hill linking tobacco smoking with death from lung cancer, which ushered in entirely new—and hotly disputed, particularly by tobacco companies—statistical methods to understand distributions of disease risk in populations (rather than sick individuals) in medicine. (Parascandola, 2004).

In later reflections, Mills told the author (WMG) that he now regretted that he had entered the *Scientific American* exchange which had further inflamed the dispute and had resolved nothing.[40]

Probability of a Deflection—P(D) Analysis

While Ryle consistently used the erroneous steep slope of the 2C catalogue to argue for evolution, his young Cambridge collaborator, Peter Scheuer (see Chap. 39), had realised in 1954 that an analysis of the population based on the statistics of the observed interferometer deflections would be more effective than using the sources listed in the 2C catalogue. Scheuer was well aware that confusion from multiple sources was resulting in serious errors in the 2C catalogue; but he also realised that his statistical analysis of the raw interferometer deflections could provide information on the population which allowed for the effects of overlapping sources and eliminated the subjective element when identifying individual sources.

Ryle was distrustful of theory and still preferred to use direct counts of sources in the catalogue, but Scheuer's P(D) statistical analysis supported his case for an evolving population. When reflecting on the controversy over source counts, Scheuer (1990) provided a succinct description of his invention and utilisation of this "Probability of a Deflection" P(D) method of analysis and also his assessment of the view of the Cambridge group at the time:

> One models the probability distribution **P(D)** of deflection amplitudes D on the interferometer records to be expected for sources with a given **N(>S) - S** relation sprinkled in random positions over the sky and compares this directly with the histogram of observed interferometer amplitudes. This eliminates the subjective element in extracting individual sources from the records ... On the one hand, the P(D) analysis confirmed that the observations required a log N(>S)-log S relation with a slope steeper than −1.5, and that was very important in the long controversy that followed. **On the other hand, it indicated a slope much less steep than the slope of −3 that came from the [2C catalogue] source counts. The second conclusion was almost as unwelcome to the rest of the Cambridge group as the first was reassuring [our emphasis].**

The statistical P(D) approach was underlying Ryle's comment in his *Scientific American* letter when he argued that a determination of the statistics of the source intensities was more important than having the correct positions of the sources in a catalogue. However, the P(D) analysis was always inconsistent with the very steep slope initially claimed by Ryle; but this detail was omitted in most presentations Ryle made in this era. While Ryle included use of the P(D) analysis in the 1955 paper interpreting the 2C survey results, he had no enthusiasm for the more theoretical approach and preferred to base his argument on the properties of the individual

---

[40] The full set of Mills Survey catalogues are summarised in ESM_35.1.pdf, Radio Source Surveys: The final log N- log S slope based on 1159 sources over 3.24 steradians was −1.8, a value in excellent agreement with current observations.

sources in the catalogue. As we have indicated, the positions of many of these sources were wrong.

In retrospect we can now see that if Ryle had only admitted that his original 2C catalogue and the resulting steep source counts was seriously flawed, much of the acrimony in the ensuing controversy between Sydney and Cambridge would have been avoided. The case against the Steady-state cosmology would have still been valid.

Due to undertaking military service, Scheuer did not publish his P(D) methods until 1957, at the time of the first publication of the preliminary Mills Cross catalogue of the first 383 sources. Therefore, Mills and Slee were only aware of the assertions based on use of the P(D) analysis from the 1955 Ryle and Scheuer publication (above) but not the methodology used. Two footnotes in Mills and Slee (1957) dealt with the statistical analysis method:

> Ryle and Scheuer (1955) give curves which they have derived from this probability distribution with various types of source distribution, but no flux density scales are appended, and no details of the calculations are given, so that we are unable to check their correctness.

> Note added in Proof: We have just received from PAG Scheuer a copy of his paper (1957 in the *Proceedings of the Cambridge Philosophical Society*) giving the theoretical derivations on which the probability distributions curves are based; however, we have not been able to compare the curves directly with our own results because of the lack of essential numerical data. [No details of the nature of the missing data were given.[41]]

Scheuer's publication of the P(D) method did not enable Mills to improve the analysis of his own survey. We now know that the process would have been ineffective since it only works for surveys which are close to, or at, the confusion limit. The 2C survey was close to the confusion limit in contrast to the Mills Cross survey. Mills himself summarises the reasons for his continued support for a non-evolving (Euclidean) interpretation of his source counts in his review paper in *Annual Review of Astronomy and Astrophysics* (2006). Mills gives a positive view of the P(D) analysis process and goes on to explain why his interpretation of the source counts was flawed:

> Although the Cambridge 2C catalogue was largely useless, an important result had been obtained from an analysis of the statistics of the interferometer output (from Scheuer, 1957). This showed that the observed radio emission could not have originated in a population of unresolved discrete radio sources randomly distributed throughout a nonevolving universe. My view expressed at the time was that many of the stronger sources would have been resolved by the interferometer [this turned out to not be the case], producing smaller output deflections, and it seemed likely that the distribution of the sources was not random, so that nothing could be said directly about evolution.

We now know that the Cambridge assumptions, both about the large distances and the minor effects of extended sources, were correct. Thus Mills's remaining

---

[41] To make a P(D) analysis it is necessary to work from the raw survey data and not the catalogue made from this data. We assume that Mills did not have easy access to this raw data which would have required reprocessing very large quantities of chart records.

reservations about the use of P(D) were unfounded. This was perhaps more a result of good luck than an incisive understanding of the radio population.

Pawsey's Matthew Flinders Lecture

Pawsey was aware of the importance of the bigger picture, while Mills remained focused on defending the accuracy of his catalogue. In 1957, Pawsey was asked to present the prestigious first Matthew Flinders Lecture at the Australian Academy of Science[42] which was a tribute to his standing in the scientific community. In the talk, he reviewed all of radio astronomy at that time, and his summary picked up on Ryle's interest in the possibilities for radio research, as we can see from the following comment on the radio continuum surveys and cosmology:

> Radio studies of "discrete sources" [not radio stars!] have been reported which appeared to show an 'edge of the universe' effect but we now consider the observations to be at fault ... In the field of cosmology, then, radio astronomy is in a tantalising position. There is every reason to suppose that some galaxies which are far beyond the current optical limits are visible to radio telescopes. But so far we have been unable to make use of this potential source of information because of lack of detail in radio observations. This position has been aggravated by the disagreement between Cambridge and Sydney observations after the raising of high hopes by the Cambridge work. But there is every reason to suppose that improved radio telescopes and more critical analysis will take the effective radio horizon out to distances where the recession velocity approaches the velocity of light and decisive results on "world models" can be obtained.[43]

Although Pawsey remained a strong critic of the 2C catalogue and sceptical of Ryle's cosmological inferences, he was certainly not blind to the far reaching possibilities raised by Cambridge. By 1957 these ideas were influencing his planning for instruments and research directions, particularly in relation to the question of investment in a large dish "Giant Radio Telescope" (see Chap. 27). His analysis of the performance of radio telescopes for observations of extremely distant objects is included in ESM_35.2.pdf, Surveys with arrays.

AAS Symposium on "Radio Sources Outside Our Galaxy"

The recognition of the increasingly important role being played by the radio astronomers captured the attention of the optical astronomers and cosmologists in the US. After World War II the US had fallen behind Australia and the UK in observational radio astronomy and that was now starting to change. The group at Harvard

---

[42] 2 May 1957.

[43] Compared to discussions with fellow radio astronomers, Pawsey was more assertive and optimistic in this public forum.

had detected the 21 cm radio emission line from neutral hydrogen, the Naval Research Laboratory was operating a 60-foot dish and going to higher radio frequencies than other groups. The US had also recognised the extra-ordinary opportunities to embark on radio astronomy research as "big science" with the genesis of the National Radio Astronomy Observatory (Kellermann, Bouton, & Brandt, 2020).

In this environment the US astronomy community held a symposium at the Urbana, Illinois, meeting of the American Astronomical Society on 20 Aug 1957. A remarkable collection of papers, which are not at all well-known today, were included in this small symposium volume published by the Astronomical Society of the Pacific in 1958. This symposium heralded the beginning of the era of rapprochement between the observational groups, although the acrimonious public debate was to continue, culminating the infamous Paris IAU symposium the following year, which we discuss in the next chapter.

In Illinois, Pawsey presented the status of the "Sydney Investigations and Very Distant Radio Sources" based on the report discussed in the last section. Remarkably, this paper also included one of the earliest numerical simulations of the effects of radio source confusion using realistic beam models. The simulations were carried out by Mullaly and T. Pearcey (unpublished) using Australia's CSIRAC digital electronic computer (McCann & Thorne, 2000), one of the few astronomical projects for which CSIRAC was used. For the simulation they generated an artificial sky with a random distribution of point sources in a static Euclidean universe. Fig. 1 from Pawsey (1958) is shown below (Fig. 35.7). It clearly demonstrates the effects of confusion which can cause sources to be lost or have the wrong amplitude and position. Below about 10 or 20 beam areas per source was a "dangerous level" where peaks could disagree seriously with the real sources. Pawsey compared this to the Sydney and Cambridge (2C) surveys and concluded that the Sydney survey just avoided this limit (beams per source), but the Cambridge survey was well below it, possibly by a factor of 10.

Hewish, from the Cavendish Laboratory in Cambridge reported on "The Distribution of Radio Stars" based on the new 3C catalog. Hewish would have been privy to discussions (eg by Scheuer) of the problems with the 2C survey, and now he admitted the serious confusion errors in that catalogue. Minkowski, from the Mount Wilson and Palomar Observatories, discussed "The Problem of the Identification of Extragalactic Radio Sources" and McVittie, a cosmologist from the University of Illinois, discussed the cosmological implications.

By the end of the meeting it was clear that the major disagreements between the surveys had been resolved. It was clear, and accepted by Hewish, that the 2C Survey was limited by source confusion but Hewish now had the much superior 3C Survey results at hand; the evidence for evolution had persisted but at a much more moderate level. This harmony may have only been possible due to the fact that two surrogates for the strong personalities were involved. Hewish represented Ryle and Pawsey represented Mills.

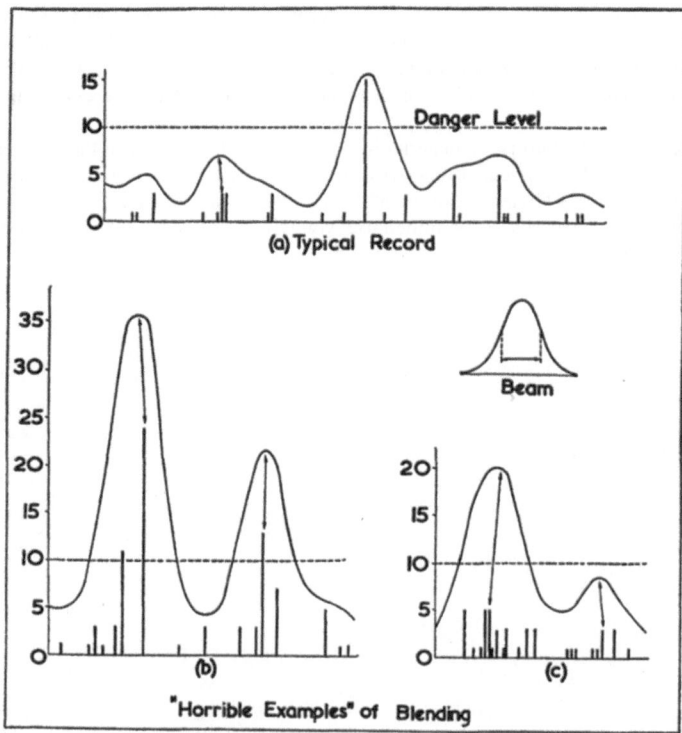

Fig. 35.7 Examples of the trace resulting from the passage of the ideal beam shown over randomly distributed point sources. The dotted line indicates a level below which serious blending effects occur. Credit: Fig. 1, "Sydney Investigations and Very Distant Radio Sources", Pawsey, J. L. (1958). Publications of the Astronomical Society of the Pacific 70, pp. 133–140. All rights reserved

Following this meeting Pawsey wrote Bowen a letter from the US (received in Sydney 28 August 1957)[44]:

> The third [Cambridge] survey, called 3C, agrees much with the Mills survey. In fact, there are about 30 per cent coincidences . . . According to Mills and me [Pawsey], confusion due to background sources becomes very serious at the level corresponding to 10 or 20 beam areas per recorded source for a pencil beam, and probably about the same for an interferometer. I therefore conclude that the 3C survey probably contains about 100 sources per steradian which are reasonable, the rest are probably phoney . . . You will be pleased to know that Hewish admits in public that the original 2C survey was over-interpreted.

---

[44] NAA C3830 F1/4/Pawsey/5.

Chapter 36
Radio Source Survey: Reconciliation, 1958–1962

> We are now [1958] at the time of maximum uncertainty and
> confusion in the history of work on radio sources. Agreement
> between the two major groups engaged in survey work is
> minimal, and the status of many of the observations is
> radically in doubt ...
> – Edge and Mulkay (1976)

It was one matter for Hewish and Pawsey to come to a quiet agreement in Illinois—and quite another to achieve a change in view across a scientific community. This community was now more diverse and included astronomers, cosmologists, and newly retrained radio physicists and engineers, with varying predilections, experiences and assumptions about what information about the cosmos was important or reliable.

This chapter retells the story of an event that had such strong social and emotional impact in the radio astronomy community that leading scientists recalled and referred to it for decades afterwards—the Paris Symposium of Radio Astronomy in 1958. This event was notable for the ferocity of the debate between disagreeing scientists and research groups—a ferocity the more shocking in that it was over issues for which the science case was *already resolved*. In this chapter we explain this apparent paradox. We also explore how disagreements get resolved. Given the ongoing controversy over the accuracy of surveys and their cosmological implications, Edge and Mulkay noted that by late 1958, the hope of reaching consensus might have appeared remote—"But this is what the radio astronomers actually did and they achieved consensus in the early sixties, with startling ease."

We now will show how this remarkable transition occurred; in the end, even the sceptical observers in Sydney were reluctantly convinced. However, the more difficult transition was convincing the cosmologists; the dispute between radio experts had destroyed confidence on the part of the remainder of the astronomical community.

Supplementary Information The online version contains supplementary material available at [https://doi.org/10.1007/978-3-031-07916-0_36].

W. M. Goss et al., *Joe Pawsey and the Founding of Australian Radio Astronomy*,
Historical & Cultural Astronomy, https://doi.org/10.1007/978-3-031-07916-0_36

The 3C Survey—Early Results 1957 and 1958

The year 1958 was a challenging year for the Cambridge radio astronomers as they presented the preliminary results of the new 3C catalogue. They had to admit in public (at the Paris Symposium of Radio Astronomy and the IAU General Assembly in Moscow) that the 2C catalogue was largely incorrect. In parallel, Scheuer and colleagues at Cambridge continued to try and convince astronomers that, nonetheless, the P(D) analysis (of 2C) remained valid.

Observations for the 3C catalogue began in 1956, using the 2C antenna at double the frequency (see ESM_35.1.pdf, Radio Source Surveys). Their beam area was now about four times smaller and, based on lessons learned from 2C, they also reduced the number of sources—only the brightest 471 sources were in the published 3C catalogue (Edge, Shakeshaft, McAdam, Baldwin, & Archer, 1959) which was a decrease in the number of sources per beam by a factor of 20! Only about 25 per cent of the 2C sources were confirmed by the 3C catalogue. The problem with the 2C design was that the instrument had been planned to detect a single source per beam area; the modern criterion is one source per 20–30 beam areas. With a rms noise of only 1 Jy, even the 3C survey suffered from some confusion at the lowest levels. The Mills/Slee survey 1957–1961 (see ESM 35.1 Radio Source Surveys) was noise limited with a 50 arc min beam and a typical noise of 3.5 Jy. This survey had a similar effective detection limit as the 3C survey.

A preliminary publication[1] of a small fraction of the sky from the 3C survey was sent to *Monthly Notices of the Royal Astronomical Society* on Christmas Eve 1957 by David Edge.[2] The full publication, Edge, Scheuer, and Shakeshaft (1958), occurred later in 1958.

The writing of the 3C survey paper towards the end of 1957 was quite contentious (Scheuer later termed it "one of the most gruesome papers I've ever had to write").[3] Scheuer had just returned from a three-year stint in the British Army, his National Service, starting in October 1954. The issue was the impact of extended sources. With the double interferometer, sources larger than 7 arc min were resolved out by the interference fringes, and sources larger than 3 arc min were recorded with a reduced intensity. The effect was to remove or reduce the intensity of some of the larger and hence closer and more intense sources. This effect would artificially increase the slope of the radio sources counts in a way which could not be corrected by the P(D) analysis.

---

[1] Hewish had also presented a preliminary account of the 3C survey at the August 1957 AAS meeting in Illinois, attended by Pawsey as noted in the previous chapter.

[2] David Edge was the lead author of a milestone book in the sociology of science: "Astronomy Transformed: The emergence of radio astronomy in Britain" written with sociologist Michael Mulkay (1976).

[3] Details of this conflict were reported by Peter Scheuer in an extensive interview with Woody Sullivan in 1976. In ESM 36.1 Scheuer Interactions, we summarise the issues involved, including those related to the effect of extended sources on the P(D) analysis.

Peter Scheuer was in strenuous conflict with his co-authors and Ryle about the potential severity of this issue. In 1958 the most striking evidence for an important effect due to angular resolution was Mills's and Slee's (1957) detection of a large number of extended sources (i.e. sources comparable in size with their 50 arc-minute beam) even at high galactic latitudes. In the end, ironically, most of these turned out to be chance groupings of weaker sources. Scheuer commented much later, "I was inclined to be very much more cautious about what one could assert. Largely for the wrong reasons. Because what I said was that we didn't know what the angular diameters were, and therefore we couldn't be sure we weren't cutting out a whole lot of bright sources for that reason." After a period of dissension, Ryle and Scheuer's co-authors persuaded Scheuer to go along with the publication, entitled "Evidence on the Spatial Distribution of Radio Sources Derived from a Survey at a Frequency of 159 MHz" (Edge et al., 1958). The source counts for this preliminary 3C survey provided a logN-logS slope in the range −2.2 to −2.7.

Within a year the final 3C catalogue of 471 sources was completed for the sky in the declination range −22 degrees to +71 degrees, (see Edge et al. (1959)). This final 3C survey (3.2 steradians and flux densities from 8 to 200 Jy) provided a slope of −2.0 for the logN-logS source counts, now approaching the value accepted at the present day, consistent with the Mills Cross survey slope. The P(D) statistical analysis was also found to indicate that the distribution of deflections could not be explained by a uniform distribution of sources with depth in space, just as had been found from the P(D) analysis of the previous 2C survey.

Royal Astronomical Society (RAS), 14 March 1958: Scheuer, Ryle and Bondi

On Friday 14 March 1958, Peter Scheuer gave a long presentation at the RAS in which he described the paper by Edge, Scheuer and Shakeshaft soon to appear in MNRAS. The main goal was to illuminate the now-admitted deficiencies of the 2C survey, by describing—and keeping attention on—the successful 3C survey. He concluded by asserting that for both the 2C and 3C surveys there is a deficit of strong sources which provides the evidence for evolution and: "[i]t can also be shown that it is not possible to give a consistent explanation of the statistics of deflections from both the 2C and 3C records by invoking angular diameters or clustering of sources alone."

Bondi, one of the two RAS Secretaries, immediately responded as he had done in 1955 when the 2C survey was introduced—by focusing on the weakness of the empirical case against a Steady-state Universe:

This is clearly work of the greatest interest and importance. It is gratifying to me personally [that it] confirms the doubts about the first survey which my colleagues [Gold and Hoyle] and I expressed three years ago, and we are looking forward to more and more definite results. As for the statistical method of interpretation, I have little confidence in it as long as the unexplained discrepancies in the location of the sources is so great, for then a non-random source of error can hardly be excluded.

Bondi could not trust that the group who had produced such an incorrect catalogue would get the more subtle analysis (ie, P(D)) correct. Ryle invoked the reliability of the much better 3C survey in his reply:

> There now exists a large number of completely reliable radio sources and it is quite clear that we can expect only a very small proportion of these to be identified with visual objects. Any kind of evidence which can be obtained is therefore of importance. We have all been disappointed by the lack of agreement between the different surveys and further observations are clearly necessary. I cannot, however, agree with Prof Bondi's summary dismissal of the statistical method. This powerful technique was introduced solely to avoid the type of error which arises when attempting to isolate a source in the presence of confusion from neighbouring ones. When the results are treated in this way both the 2C and 3C surveys agree quite precisely in their indication of a divergence from a uniform distribution.

However, although this exchange shows Ryle accepting (with disappointment) the lack of agreement between catalogues, we note that he never publicly explicitly admitted that his original 2C catalogue was flawed and his inferences from that catalogue, unjustified. He had drawn his first cosmological implications from source counts using the 2C catalogue and shown these plots in many presentations. It was only the P(D) analysis of the original observations that were used to make the 2C catalogue of radio sources which *did* justify an evolutionary cosmology.[4] To the cosmologists, who were not radio engineers and not familiar with interferometry, the empirical data remained inaccurate, and its producers were not to be trusted!

Bakerian Lecture, June 1958

On 12 June 1958, Martin Ryle presented the Bakerian Lecture of the Royal Society of London.[5] Ryle (1958) summarised the tentative results from the new 3C survey, and in addition he anticipated the construction of the new 4C aerial at the new Lords Bridge site (work on the new site had begun in July 1956):

> A new radio telescope having a greater resolution and sensitivity than any previous instrument will soon be in use at Cambridge, and should provide reliable observations of weaker sources; the number-flux density relationship may then be tested over a greater range, and the angular distribution of the sources may be examined to greater distances.

The increased reliability of the new 3C survey compared to the 2C survey was described; Ryle continued to assert that it was not compatible with the Steady-state

---

[4] Albeit a much weaker effect than Ryle found from the catalogue but still requiring evolution of the population.

[5] Founded in 1775 by a grant of £100 by Henry Baker. Other famous lectures were given by Humphry Davy, M. Faraday, J.C. Maxwell, W.H. Bragg, E. Rutherford, N. Mott, W.L. Bragg, E. Appleton, M. Oliphant, M. Rees, J. Silk and A. Ghez (among others) during the last two centuries.

cosmology: "Although the comparison of the theoretical predictions with the obser-vations is limited by the small number of intense [nearby] sources ... there appears to be a real discrepancy between the observations and the predictions of the Steady-state model."

Ryle summarised by asserting that the key fact was that the majority of radio sources must indeed be at considerable distances outside the galaxy, and that these results were entirely independent of any optical observations. They only required the simplest of radio observations, ie. something about the numbers of sources of different intensities (the source counts), the integrated brightness of the radio sky and the isotropy of the radio sources.

Ryle understood very well that the really important point was not whether the survey data would refute the Steady-state model of the Universe, but that the use of radio techniques was about to utterly transform astronomy as a field of science. He continued:

> Whether or not the present arguments are regarded as conclusive evidence in favour of an evolving cosmological model, I believe that what they have shown is that most radio stars are very powerful sources at very great distances. The ability to observe to distances at which large redshifts occur, without the restrictions set in optical astronomy by the form of the spectral distribution, means that the predictions of different cosmological models can be tested with a directness which has been impossible photographically [with optical telescopes].

Ryle had recognised the fact that even the brightest radio sources were at great distances. He felt this was the real reason radio astronomy could have a much bigger impact on cosmology than optical astronomy. And in fact, there was no disagree-ment on these issues. But instead, the radio astronomers continued to argue about details of the surveys, even though they did not impact this conclusion. Jasper Wall (2016) has made this point very well: "The course of cosmology/galaxy formation should have been altered in 1955."

But it took years. Why? The scientists involved were busy defending issues related to an important epistemic value, that of empirical exactitude. They were also responding defensively to Ryle's failure to admit he was wrong, which can be interpreted as a violation of the Mertonian norm of disinterestedness. He was, all too obviously, worried that admitting an error would result in loss of credibility, when in fact the opposite was the case.[6]

---

[6]In Chap. 14 we include a brief summary of social theories that science is organised around particular 'norms' of behaviour, something first proposed by Robert Merton in 1942. In relation to this comment on Ryle's behaviour, readers may be interested in: Mitroff, "Norms and Counter-Norms in a Select Group of the Apollo Moon Scientists: A Case Study of the Ambivalence of Scientists" (Mitroff, 1974).

IAU Paris Symposium No. 9, 30 July-6 Aug 1958

The Paris Symposium was jointly sponsored by the International Astronomical Union (IAU) and the International Radio Science Union (URSI). This joint sponsorship by the primary international unions representing astronomers and radio engineers was a result of Pawsey's vision to bring these two communities closer together—see NRAO ONLINE.51. Pawsey chaired the organising committee which included radio and optical astronomers, cosmologists and theorists.

At Paris, the radio astronomers were to hear a lot about source counts and cosmology. The Paris Symposium was a remarkable example of how the events surrounding the source count controversy drove many in the scientific community to behave in an unprofessional and emotional manner. The symposium presentations and discussions were published by Bracewell (1959).

Numerous authors have remembered the heated discussions at this meeting.[7] Even 52 years later, conference attendee John Baldwin (1931–2010) clearly recalled vigorous and acrimonious debates. These occurred particularly during the fifth session, "Discrete Sources and the Universe" (below), which was chaired by Pawsey, in which there was "a storm of vituperation on all sides". Baldwin summarised: "At the Paris Symposium the participants were firing at targets that no longer existed. As I look at the pages of the book now, I can smell the smoke and even imagine flames arising from the pages! The irony is that the problems of the surveys were mainly resolved by 1958."[8]

Peter Scheuer (1990)[9] also had "strong memories of the mood at Paris": "The arguments came to a head at the Paris Symposium in August 1958 (Bracewell, 1959). Dr. Pawsey (Sydney) chaired the session on 'Discrete Sources and Cosmology'. The mostly rather carefully written version of the discussion gives only a pale reminder of the heat of the argument."

It is clear that the participants in this symposium were not listening to each other, often arguing at cross purposes and settling old disputes. Heated exchanges about previous errors occurred. Many of the discussions focussed on irrelevant topics such as big dishes versus interferometers. Yet as Baldwin commented, all the key technical issues related to the survey controversy had been understood and mostly resolved by the experts before this meeting.

The topic of the Fifth Session of the Conference was "Discrete Sources and the Universe". R. Hanbury Brown of Jodrell Bank provided a clear overview of the

---

[7]Edge and Mulkay (1976, p. 167): "The ensuing discussion revealed all the passions and misunderstandings kindled by this topic—as will be obvious to those who read both on and between the lines of these extracts." Bracewell, in the preface to the conference publication: "Some roughnesses of expression which remain serve to remind us of its spontaneous and unrevised character".

[8]Personal communication to M. Goss on 2 July 2010 (only 5 months before his death on 7 December 2010). He had never experienced in his career such heated exchanges.

[9]See ESM_36.2.pdf, Peter Scheuer Source Count.

territory in his inaugural lecture titled "The Distribution and Identification of the Sources":

> The reason for the particular interest in the distribution of the sources in depth is that it might be used to test cosmological theories, and we are to hear a paper on this subject. I believe, however, that the papers in this session will show that the **radio astronomers must make considerable progress before they can offer the cosmologists anything of value.** (our emphasis) Take first the method of counting the sources: this technique is subject to serious errors when the number of sources is too great in relation to the resolving power of the aerial system. We are to hear two papers on this subject. The limitations of the second method, namely measuring the probability distribution, are less well understood and perhaps we can touch on this subject in discussion.

Here, Hanbury Brown was being less dismissive than some others on the value of the statistical method, P(D), but he was correctly questioning the assumptions about the range of apparent angular diameters. This was the same question that plagued Peter Scheuer in the Edge et al. (1958) paper.

Hanbury Brown emphasised the success that he and Cyril Hazard had in finding weak radio emission from nearby galaxies such as M31 and M33 at luminosities much less than the "colliding galaxies" Cygnus A and Perseus A. While, for any individual source, the assumption that a closer source would be brighter, was reasonable, this effect need not apply to a population of sources. This had earlier caused misunderstandings throughout this period. Hanbury Brown was concerned a thorough understanding of the extragalactic population required a knowledge of the radio source luminosity function.[10] The luminosity function describes the relative numbers of radio sources of different luminosity going from the most luminous (abnormal) galaxies down to the weak nearby galaxies. Even though the abnormally luminous galaxies are rare, they can be detected to much greater distances: thus they were more numerous in the catalogues. The galaxies such as Andromeda, which have similar radio luminosity as the Milky Way, were such weak radio sources that few had been included in the catalogues.

Hanbury Brown concluded his introduction by introducing another, new, issue that would play an important role in research at Jodrell Bank: determination of angular sizes of the extragalactic radio sources. "[M]easuring diameters is a more promising way of testing cosmological theories than counting sources in different [flux density] ranges."[11] Hanbury Brown also hinted that the determination of source sizes might contribute in the determination of the distances of the objects. Later in

---

[10] This occurred slowly during the 70s. The problem was more observational than conceptual. It required many identifications with distant optical galaxies and optical measurements of the distances. The field was dominated by Minkowski and his successors in the USA since the largest optical telescopes were required.

[11] Cosmology can also be done by counting sources of different angular size: very analogous to counting the number of people in a population by size rather than age. Measuring angular sizes was a different cosmological test as well as being useful to correct for the extended sources. To continue this analogy—the extended sources were distorting the population statistics. Thus the situation was similar to neglecting all the tall people who could not pass through the door (that is, the interferometer) in order to be counted.

the symposium Mills made a similar suggestion for the next generation surveys, and in his summary of the Paris symposium Minkowski made the same point. This idea did have impact in the following years, although not directly related to cosmology. Following up small diameter (and presumably more distant) sources led to the identification of some of the most distant radio galaxies, and eventually to the discovery of quasars (see Chap. 33).

Ryle's Contribution to the Session

The next paper by Ryle was "The Problems of Confusion in Surveys of Sources", a two-page summary of source counts and the statistical method, P(D), of Scheuer. The first sentence indicated the influence of Pawsey as chair of the symposium:

> Dr Pawsey has asked me to make some general remarks on the problem of confusion when the traces of adjacent radio sources overlap. This is a question that affects all surveys in which the observations are limited by resolving power and not by sensitivity; there is perhaps some misunderstanding about the relative performance of interferometers and pencil-beam systems of similar sizes, and there seems to be an impression that the former are more seriously affected than the latter. I think this impression may have arisen from the fact that a number of pencil-beam systems have been limited in the detection of weak sources less seriously by confusion than by the use of smaller collecting areas, by the difficulty of distinguishing between sources and irregularities in the general galactic emission, and, in some cases, by the greater influence of man-made interference.[12]

Ryle's paper showed that the Cavendish group had learned an important lesson from the problems with the 2C survey. He asserted that a very precise survey required one source per 100 beamwidths while a less accurate (flux densities with errors of 25 percent) could work to a level of one source per ten beams. He then asserted that the new Cambridge method of analysing P(D) (of which he provided a brief description): "if we are only interested with the distribution of sources in space and are not interested in their individual positions ... allows us to work down to about one source per beamwidth."

Hazard-Walsh Presentation: Subsequent Heated Discussion

The Hazard-Walsh paper followed Ryle's (Dennis Walsh, 1933–2005). Titled "A Comparison of an Interferometer and Total-Power Survey of Discrete Sources of Radio-Frequency Radiation,".[13] The presentation generated lengthy and heated

---

[12] Eg, radio waves emitted from electric trams and motors, which had a more pronounced impact on a single dish than on an interferometer.

[13] Paris Symposium, Bracewell (1959) page 477.

discussion. The authors were young staff members at Jodrell Bank (see Chap. 16) and their talk referred to the much earlier, and now withdrawn, Cambridge view:

> A fundamental limitation to the number of radio sources observable with a given aerial system is set by the finite solid angle of the aerial beam's reception ... With an interferometer, beating will occur between the fringe patterns of each source and the record's appearance will depend on their relative phases. Thus, the interpretation of the record in a confused region will in general be different from the interpretation that would be placed on a total-power record covering the same region. It is therefore to be expected that an interferometer survey of a given region of sky will give rise to different results from a survey made with total-power equipment of the same resolving power at the intensity level at which the confusion effects become serious.[14] A comparison of the results obtained from a total-power and an interferometer survey of similar resolving power should enable us to make an estimate of the reliability of a survey that is resolution limited.

At Jodrell Bank, Hazard and Walsh had carried out two 92 MHz surveys of the same region, one with the 218-foot fixed parabolic dish (beamwidth of 3 degrees), and the other with an interferometer made with the dish and an array of dipoles with a spacing of 500 metres (a lobe spacing of 22 arcmin). Two sets of records were obtained, resulting in two source lists which could be compared. Sources much larger than 22 arc min were not detected with the interferometer (resolved out). In the common region of sky, the total sources numbers were 81 from the single dish records and 202 from the interferometer. However, 40 sources on the single dish list and 60 sources in the interferometer list did not coincide with the sources on the other list. Figure 36.1 gives an example. The clearly defined interferometer source at 12 h 30 m RA has no counterpart in the single dish record.

Hazard and Walsh concluded their presentation:

> It has sometimes been said [referring indirectly to Ryle's claim made 8 years earlier in 1950!] that the maximum number of sources that can be resolved in a radio survey is equal to the number of beamwidths in the sky. The observations presented in this paper show that the number of sources that can in fact be reliably catalogued is very much less than this ... This limitation must be borne in mind in drawing conclusions from source counts, and it places a severe restriction on the number of sources that can be reliably observed by existing aerials operating on metre wavelengths where most surveys have been carried out. In order to extend the investigation of the spatial distribution of the localised sources to greater distances, surveys with much greater resolving power must be used, or more reliable methods of analysis must be devised.[15]

The remarkable discussion of the Hazard and Walsh paper is presented in full in ESM 36.3 Discussion after Hazard. This was the longest discussion of any

---

[14] As discussed in the previous chapter Bolton, Stanley, and Slee (1954) had made exactly this point based on their sea-cliff interferometer survey made 5 years earlier.

[15] In a follow-up paper published in late 1959 in MNRAS ("An Experimental Investigation of the Effects of Confusion in a Survey of Localized Radio Sources"), Hazard and Walsh (1959b) were more specific, pointing out that substantial errors would result if the source survey had a source density of more than one source per 25 beam widths. They pointed out that the cut-off for the 2C survey should have been 56 Jy and not the published cut-off of 25 Jy. Confusion errors were expected in both single dish and interferometer surveys but "it is to be expected that they are more serious in the interferometer surveys because of its larger solid angle of reception."

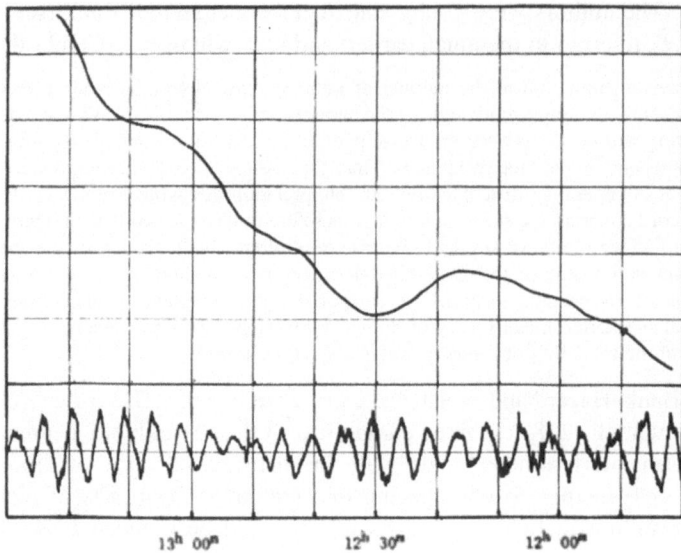

13ʰ 00ᵐ　　　　　12ʰ 30ᵐ　　　　　12ʰ 00ᵐ

Fig. 36.1 Total power and interferometer records over the same region. Credit: Fig. 2, "A comparison of an interferometer and total-power survey of discrete sources of radio-frequency radiation," Hazard, C., and Walsh, D. (1959). URSI Symp. 1, Paris Symposium on Radio Astronomy 9, pp. 477

presentation at the conference, with a text of 5.5 pages and discussions running to 4.5 pages. Here we comment on a few highlights.

Several participants are noteworthy in the discussion: Scheuer, the Chairman Pawsey, Ryle and Gold. Scheuer spoke twice, recognising at the outset one issue, the lack of familiarity with the P(D). "Since there is not time to explain the statistical method, P(D), for analysing confused records, I think it would be best if I asked for questions, which I shall try to answer."[16] While Scheuer's answers were intended to clarify and thus substantiate claims about the utility of the P(D) analysis and hence the reliability of the 3C catalogue, Scheuer's other remarks in this discussion session were prompted by Mills, who again asserted that a number of the sources he and Bruce Slee had found were extended. This led Scheuer to defend the Cambridge catalogues' reliability by pointing out that only a few of the 2C sources (and now 3C) were >5 arc min in size, thus minimising any errors in the source counts. While agreeing with Mills that to resolve this issue, more complete information about source sizes was required, Scheuer pointed out that this was not feasible in 1958. This issue was only resolved some years later. The arguments about the effect of extended sources were correct in principle, but the number of sources was later found too small to have any appreciable effect.

---

[16]This was in response to Hanbury Brown's question about the angular size of the radio sources impact on P(D).

The intersections between cosmological implications and the accuracies of catalogues and instruments soon became emotive. Indeed, many years later, Scheuer wrote in 1990 (*Modern Cosmology in Retrospect*, p. 338)[17]: "Though disputes about catalogues were irrelevant to the P(D) analysis, it still bore guilt by association. In one discussion I was allowed to come to the front to answer questions about it, but the real attacks came after I had sat down again, and I was not allowed to reply to any of them."[18]

One such concern was voiced by Thomas Gold. Gold had picked up on the single dish vs interferometer argument, noting that this was an issue that had to be settled. He remarked that there was no unique relation between the output of the receiving system and the actual radiation pattern on the sky, pointing out that if more information was measured, the range of possible distributions on the sky would be reduced. He then went on to equate the size of the dish with the information content and ended with a strong critique of the P(D) analysis, which he claimed could not overcome a lack of information. He concluded: "We therefore are left with a clear case for larger aperture antennas and we could not get the same information **'on the cheap' [our emphasis]** by any device of analysis." (see ESM_36.3.pdf, Discussion after Hazard, for the full text). This argument was seriously flawed, leading to confusion in this discussion about dishes and interferometers. The single dish only generates more information by scanning, so it is not a simple comparison. While a larger dish responds to more Fourier components, they are combined into a single output so no more information was obtained. Gold's bias against the Cambridge view—due to his unfailing support for the Steady-state cosmology and his long-term strained relations with Ryle—was undoubtedly inflaming the debate.

Roger Jennison from Jodrell Bank continued the attack along the same lines, arguing that the only satisfactory solution would be a single dish or completely filled aperture synthesis array. "No applications of statistical analysis can fill in the gaps in the Fourier components." The development of radio telescopes over the next few decades showed that Jennison's conservative view—only a completely filled aperture would suffice—was quite wrong. All of the most powerful survey telescopes now in existence such as Westerbork, VLA, GMRT and ATCA are interferometer arrays, as are the next generation telescopes such as MeerKAT, ASKAP and the SKA. As Pawsey had argued earlier, the single dishes only became more powerful survey telescopes at high frequency (see ESM 35.2.pdf, Surveys with arrays).

Ryle and Pawsey had the last words, stimulated by the Hazard and Walsh presentation.[19] Ryle said, "As Dr. Scheuer has pointed out, we have never claimed that the statistical method could supply Fourier components not present in the

---

[17] See also ESM 36.2 Peter Scheuer Source Count.

[18] Scheuer had also mentioned the conflict with Gold during his 1976 interview at the Grenoble IAU with Woody Sullivan: "We had a session in which there was a rather tense debate about the source count business. And at the end of which I felt, I think, somewhat resentful at not having been allowed to reply to various criticisms, which were just allowed to stand."

[19] George Field and Graham Smith also had brief comments and questions at the end of the discussion.

aerial!" He pointed out that the surveys now included a range of interferometer spacings.

Pawsey's contribution to the discussion:

> It should be recognised that the Cambridge conclusions depend on the application of the inductive method. The observations are in themselves incomplete, cf. the restrictions of the Fourier components of the distribution of brightness as discussed by Mills. But a plausible hypothesis is advanced relating source distribution in space, source sizes, degree of clustering, and so forth, and from it and the observations a conclusion is drawn. It is the essence of the inductive method that the hypothesis must conform with all available observations, and the present controversies strongly suggest that this is not true in this case. Hence it is imperative to resort to the most direct means for determining the distribution of radio brightness in the sky. This implies the use of pencil-beam techniques of a resolution adequate to resolve the existing uncertainties.

By the "inductive method" we surmise that Pawsey was referring to a transition from particular observations to hypothesised general principles, thus the creation of a hypothesis or theory from a synthesis of known data. Was he referring to the fact that the P(D) analysis did not provide the actual distribution of radio source intensities but could only be used to compare observations with a model of the expected distribution of source intensities? In this situation it is indeed very dependent on any assumptions which are being made. The one confusing aspect of Pawsey's summary was to state that a pencil-beam technique was required to fully sample all the Fourier components needed to determine the size distribution of the sources in the survey. Pawsey was well aware that if all Fourier components were measured by an interferometer the distribution of radio brightness could be recovered. (see Chap. 37).

Mills: "A Survey of Radio Sources at 3.5 m Wavelength"

Mills presented his paper as the seventh of 13 presentations in this session on discrete sources. His overall tone was conservative as he discussed the cosmological implications of the Sydney survey (see ESM 35.1.pdf, Radio Source Surveys).[20] He described the survey in qualified terms: "only in the 'promising' stage since the data are not yet sufficiently conclusive ... [The slope of the logN-logS was] not conclusive ... the errors are large ...".

By 1958, Mills and colleagues no longer believed that the Class II (non-galactic) radio sources displayed large-scale clustering as they had earlier suggested (Mills & Slee, 1957). After 1958, similar tests on a much larger area showed no significant clustering. This meant that one of the previous objections to the use of the P (D) method had been removed.

---

[20]Mills only discussed the region of sky +10 to −20 deg. declination. The total number of sources was 1159; the final MSH catalogue would contain 2270 sources in the declination range +10 to −80 deg.

Mills and colleagues continued to find a few optical identifications, still the main goal of survey research.[21] With the positional errors (one sigma) of 10–13 arc min, only 14 possible identifications had been made. These included NGC 1068, a well-known Seyfert galaxy, and NGC 4038/49, a now famous interacting galaxy pair known as "the antennae". The number of new radio galaxy identifications was discouraging. The conclusion of the Sydney group in 1958 was:

> ... the suggestion is clear that a substantial proportion of galaxies may have a slightly abnormal radio emission ... about one galaxy in thirty emits between two and five magnitudes more than a normal galaxy[22] ... It is necessary to increase the statistical reliability of the identifications by extending the catalogue to a larger area and, if possible, by increasing the sensitivity ...

Mills's line of argument was that the small number of radio galaxy identifications was a result of the powerful radio sources being so distant that the associated galaxy was not visible to the optical telescope. Mills considered these powerful radio sources as an abnormality; as a result he attempted to search for more of the assumed "normal galaxies", which would require more sensitivity and a larger survey area. He had missed the key point that the majority of radio sources in a radio survey are "abnormal", and this is just what made them interesting and visible to large distances, so they become powerful cosmological probes. This classification and subsequent rejection of the "abnormal" is a common thread seen throughout science.

At this point in time the radio astronomers were caught in a difficult situation. Positions were not good enough to identify the fainter, more distant galaxies, because chance alignments were too frequent, so only the closer galaxies with lower radio luminosity could be identified. The argument turned on the assumptions that different astronomers were making about the characteristics of a "normal" galaxy. Ryle used the one possible identification with a distant galaxy, Cygnus A, as representative of the whole class of radio sources, and then emphasised this point as the key to the use of radio sources for cosmology. To some extent these differing views might have resulted from the North–South asymmetry discussed earlier. The prominent object in the North, Cygnus A, was a very distant object (230 Mpc), while the prominent object in the South, Centaurus A, was quite close (4 Mpc). It is also likely that Ryle could more easily take this leap, because it was part of the much larger transition he experienced as galactic radio stars became extragalactic sources. Likely he never found the objects that Mills called "normal" galaxies to be of major significance. Ryle assumed Cygnus A would be a typical object in the distant universe, similar to most of the sources detected in the early surveys.

---

[21] See Sullivan (2009) for a discussion of why optical identifications—identifications of radio sources with known, real, visible objects—remained so powerful a goal.

[22] In 1958, the term "normal galaxy" referred to what we would, in the present epoch, consider a low luminosity radio galaxy such as Centaurus A, Virgo A and M84. The term "normal galaxy" in the present era signifies objects such as the Milky Way or Andromeda (M31) with no prominent AGN activity.

It is also of interest to speculate whether Mills's perspective would have been different if Minkowski had accepted Mills's earlier (and correct) proposed identification of Cygnus A instead of waiting for the more accurate Cambridge position [see Chaps. 16, 18, 21 and 22].

During the discussion after Mills's presentation at Paris, Ryle commented that for both the 2C and 3C survey, the Cavendish group had looked for "extended sources" at frequencies of 81.5 and 159 MHz with sizes up to 1–2 deg. They found a much smaller fraction than found with the Mills Cross. "We also have available the observations taken with the interferometer of intermediate resolving power and the total-power observations made with a single element. The purpose of these observations has been to search for sources that might affect the number-intensity counts and the corresponding statistical method of analysis."[23]

The logN-logS for the 1000 sources in the Mills catalogue resulted in a slope of -1.8. Mills then corrected this for the resolution effect of the aerial and for the noise in the weaker sources as he obtained a slope of -1.65. At the time he did not consider this to be significantly different from the -1.5 value expected for a static Euclidian universe.[24] He considered the resolution of the closer more extended sources to be the "principal uncertainty", the same issue that was of concern to Scheuer.

After the Paris Symposium, the next step for Mills was to construct an interferometer with a baseline of 10.2 km to determine the angular size with an interferometer of baseline 2920 wavelengths (lobe spacing of 12 arc min). This instrument was constructed using the Mills Cross as one element and a simple distant aerial connected by a radio link. Angular sizes in the range 20 arc sec to 1 arc min were determined (see Goddard, Watkinson, and Mills (1960)). The only publication based on this instrument was by Mills, Slee, and Hill (1960), "On the Identification of Extragalactic Radio Sources". The sizes were used in making the identifications. Forty-six possible identifications with galaxies were proposed, with 38 determined angular sizes. Nine of these were <30 arc sec. In the paper, a radio luminosity function was derived. After all this effort, Mills was disappointed with the results. As he has written in his superb autobiographical text in *Annual Reviews of Astronomy and Astrophysics*, "An Engineer Becomes an Astronomer" (2006), page 1: "However, the first program [luminosity function determination] was not very significant with the yet to be discovered quasars and I had left CSIRO [for Sydney University] before a useful amount of angular size information had been obtained in the [angular size] program."

---

[23] In his article of 1990 ("Radio Source Counts", page 338), Scheuer refers to the Cambridge observations in an off-handed manner: "Some not very rigorous attempts of this sort were indeed made, using parts of existing aerials, and a remark made by Ryle [after Mills's presentation at Paris] probably refers to one of these."

[24] However in his Mills et al. (1960) paper he realised that this difference was very significant when a real cosmological model was used instead of the static Euclidean case which never applies in an expanding universe.

Impressions of Many Participants after Paris

As the Paris Symposium concluded, we can surmise that the majority of the attendees agreed with Mills, Gold, Bondi and Hoyle, not with Ryle and Scheuer. Pawsey's doubt likely exerted a strong influence: disagreements between the surveys made skepticism and caution the preferred epistemic values of the day. But in hindsight we can surmise that Pawsey allowed himself to be too strongly influenced by Mills—partly through his own preference for caution—his usual strength for impartiality and keeping the peace was not evident.

For example, Bart Bok's (1958, p. 620) summary of the Paris conference in *Sky and Telescope* was likely typical (see ESM 36.4 Bart Bok Summary). Bok commented on the lack of identifications and the poor agreement between the surveys, questioned the P(D) statistical approach, supported the big dish arguments, and equivocated on cosmological importance.

Rudolf Minkowski provided a more balanced and extremely perceptive view in his summary lecture for the "Discrete Sources and the Universe" portion of the Paris Symposium. Based on his lecture, he understood the pros and cons of the source count controversy. He was even-handed in his treatment of the two sides as he began his summary of the heated discussions of the previous session; he realised that both sides [Cambridge/Sydney surveys] had major potential. He could see that progress was just visible on the horizon:

> The lively and extended discussions in this session have given to some extent an unduly pessimistic picture of the present state of the investigation of faint radio sources. It seems necessary to emphasise the fact that progress has been made, even if relatively few results are definitive and few problems have been solved.

> ... The discussion of the relative merits of the pencil-beam instruments and the interferometers for surveys of radio sources has perhaps tended to obscure the merits of the interferometer, which under favourable circumstances can give us the most accurate positions ... Actually, as Ryle has emphasised, there is no basic difference between pencil-beam and interferometer as regards the effects of confusion. The main difference between the Sydney and the Cambridge surveys is the fact that the Mills cross is sensitivity limited, while the Cambridge interferometer is confusion limited. The inability of the interferometer to record sources beyond a certain size is an inherent difference, however.

> It is now generally recognised that the Cambridge survey at 81.5 MHz was very severely affected by confusion. The new survey at 159 MHz is obviously much improved, but it is not free of the effects of confusion. Also, sources larger than 5 minutes of arc are not recorded by the Cambridge interferometer. The Mills cross on the other hand is not affected by confusion and will record all sources regardless of size. Side-lobe effects, however, may produce spurious sources, which should be relatively more frequent among the fainter sources.

Minkowski pointed out that that the recent comparison of the 3C and Mills et al. surveys had produced a list of 43 radio sources in common to both surveys; this was a major improvement compared to the disappointing number of well-determined radio sources in the Pawsey (1955) catalogue. But the accompanying optical identifications of radio sources was disappointing, "All these attempts have led to the same result: **not more than a small fraction of the sources can be identified with**

optical object." (our emphasis) Minkowski also foresaw that determinations of angular sizes of many sources was essential. In addition, he had a nuanced view of the P(D) method of Scheuer:

> Scheuer has attempted to avoid the effects of confusion by investigating the statistics of the deflections recorded with the Cambridge interferometer. This seems to be a powerful method, but it does not remove the effects of the finite sizes of sources. The result of this attack on the problem is that the observed frequency distribution of small deflections is in agreement with a uniform distribution of sources, but that there is a deficiency of large deflections. No agreement has been reached on the question of whether this discordance can be understood as an effect of the finite sizes of sources.[25]

> All results can be explained, as Shakeshaft has mentioned, by a deficit of intense sources. Whether this interpretation is correct can only be decided by deeper surveys that reach substantially larger numbers of sources. This is the only possible way to remove the influence of a deficiency of intense sources that, however unlikely, may exist as a statistical fluctuation in our neighbourhood.

The question raised here about a small deficiency of intense sources due to a statistical fluctuation in the Milky Way had not disappeared, because all the bright sources in the sky were now known, and the statistics could never be improved. However, it was now clear that both the Steady-state and static Euclidean universes were strongly excluded without including the effect of this small deficit of very bright sources.

Minkowski ended his summary with a cautious note:

> At this moment the available data are obviously not a sound basis for cosmological discussions . . . The problem is indeed not basically different from that in optical astronomy, where counts of galaxies do not seem to provide a manageable way to attack the cosmological problem.

> The existence of a minimum apparent size in certain cosmological models puts additional emphasis on the importance of measurements of angular sizes. **But it seems clear that considerable time will elapse before the study of radio sources has reached a stage in which the results may be used with confidence to attack cosmological problems.**[26] (our emphasis).

But as Helge Kragh (1996, p. 323) has emphasised, Minkowski was far too pessimistic: "It turned out that the considerable time was less than three years." The speed with which a consensus was reached will be described below.

---

[25] This was indeed the key remaining issue, as was recognised by Scheuer, but in the end further observations showed that it was not an issue.

[26] A similar sentiment was expressed by Hanbury Brown in the first presentation of the session on "Discrete Sources and the Universe".

IAU MOSCOW, August 12–20, 1958

After the Paris symposium concluded, many of those present continued to the International Astronomical Union meeting in Moscow, which was held less than a week later. Pawsey attended, partly because he had led the establishment of Commission 40, the IAU Commission on Radio Astronomy (see NARO ONLINE.51). Pawsey had in fact prepared his Commission 40 Report well before the meetings of July and August 1958.

Pawsey's view of where matters stood after Paris was clear in his report, which stated:

> ... In the field of cosmology, there has been a halt. Previously announced radio results which appeared to invalidate the Steady-state continuous-creation hypothesis concerning the universe are now believed to be invalid and the point is now non-proven. More certain knowledge must await the construction of more powerful tools.

The Cavendish group had asked Pawsey for permission to present a codicil to this report. The justification of adding the codicil was the extreme rate of change in the field and the view from Cambridge that their observations were misrepresented in the already outdated report by Pawsey. In particular, while the Cavendish group had openly admitted that the usefulness of the 2C survey was severely limited by confusion, the cosmological evidence for evolution depended not so much on this catalogue as on the P(D) statistical analysis by Scheuer. The Cavendish group asserted that the conclusions about evolution of the radio source population were still valid. This codicil by F.G. Smith on "Aspects of Recent Cambridge Work" is included in full in ESM_36.5.pdf, The Graham Smith Codicil.

The minutes of the business session of Commission 40 were ironically written by the Secretary of Commission 40, B.Y. Mills. His item number one read:

> A statement was read by F.G. Smith covering some aspects of the recent work of the Cambridge group not included in the President's report. It was resolved that the statement should be included in the report of the meeting. (The reader is referred to the proceedings of the Paris Symposium for a general discussion on the highly controversial questions raised by Dr Smith.)

As Secretary of the Commission, Mills had added the parenthetical clause to ensure that the reader did not miss the controversial points of the Cambridge-Sydney rivalry.

We note that Hermann Bondi also described the events at Paris in the second edition of his book *Cosmology* (published 1952 and revised second edition 1960); this section must have been written before March 1959. It is quite refreshing to read the account by someone very close to the scene in the 1950s but not too emotionally involved. Bondi did support the Steady-state Cosmology, as one of the co-originators, but not in the evangelical manner of Hoyle and Gold. Some of his statements were so similar to Pawsey's that they likely had some level of communication. His assertion that "the first Cambridge survey was quite unreliable owing to instrumental limitations and [that] no valid conclusion could be drawn from it" echoes Pawsey—but the summaries of both were a little too strong, since the arguments against the Steady-state hypothesis based on P(D) were in fact correct.

In fact, and despite Mills's conservativism in the face of limited identifications of sources, the same argument against the Steady-state hypothesis could also have been made using the Mills survey, as Mills himself realised (discussion with Miller Goss in 2006).

In NRAO ONLINE.35, there are a number of photographs from the 1958 Paris Symposium of Radio Astronomy, the Moscow IAU and a trip to the Crimean Astrophysical Observatory. Pawsey was accompanied by Professor Cecilia Payne-Gaposchkin of Harvard on the trip to the Crimea and additional sites in the southern USSR.

Summary of Radio Cosmology 1960

In 1960, Pawsey and Hill (1961) wrote a review paper of about 45 pages: "Cosmic Radio Waves and Their Interpretation". Edge and Mulkay (1976, p. 184) described this publication:

> This review article, although not particularly influential, since it was somewhat dated by the time it reached print, was a child of its time and has considerable historical interest. It reflected the current negative valuation of the status of radio source surveys in general, and of the Cambridge work in particular. The 2C survey was referred to as a:

> false start ... the Cambridge group drew unwarranted conclusions from the survey of radio sources in which they had catalogued about 2,000 sources ... The reliability of the observations was challenged by the Sydney group and the objections appear now to be accepted by the Cambridge group (3C).

Jasper Wall[27] makes the following insightful comments:

> Pawsey and Hill (1961) base their discussion on the commonly-held view amongst radio astronomers at the time that static Euclidean was a good enough approximation. This, from the start in 1955, was unfortunate ... two other cosmologists prior to 1960 were into radio astronomy, Hoyle and McVittie. Hoyle knew of course that static Euclidean was useless, but it served his purpose to keep it in the frame continuously. He knew that his frame of the Universe showed even more drastic curvature effects than standard Friedman models, but he never said so. McVittie (1959) said that for powerful sources in a proper universe, you'd have to have evolution. He had then weakened, forcing the counts to -1.5; i.e. increasing numbers per co-moving vol of weaker radio sources, i.e. evolution of the population. Right on. Perhaps Pawsey and Hill did not make enough of the intervention of a true and impartial cosmologist. But they were reflecting and reviewing and capturing, quite accurately I suspect, the feeling of the times.

Pawsey and Hill did reference McVittie (1959) and note that for the intense sources at large distance, like Cygnus A, a large increase in space density would be needed just to maintain a slope of -1.5. But their review only concluded that more information on the luminosity would be required to make progress. Some progress in this direction was made by Mills et al. (1960) as they finally concluded that the

---

[27] J.V. Wall private communication 2018–2019.

revised much greater average distances for the radio sources and their (more reliable) Sydney catalogue were not consistent with the Steady-state cosmology. However, Jasper Wall noticed an astounding bias in the preceding paper in the same journal, Mills et al. (1960). He stated that the identifications of radio sources, which are needed in order to determine the average distance, he states that a large number of "abnormally" luminous radio galaxies would be very unlikely because if correct they would drastically alter the source counts!

Pawsey and Hill concluded with an optimistic tone:

> What then of the future? A pertinent factor from the radio astronomy viewpoint is that equipment improved in performance by several orders of magnitude should be available soon. With such equipment, and the better understanding of the problem, which is emerging, there is a high probability of a **substantial advance** [our emphasis]. Whether this should take the form of discrimination between existing theories or of a completely unexpected discovery is an open question.

It is interesting to note how just such an unexpected discovery provided direct evidence for evolution. As discussed in Chap. 32, the discovery of quasars in 1963 provided a new class of objects which could still be detected in the optical even at distances just as great as the radio sources. The optical observations of the redshifts of quasars by Maarten Schmidt (1968) provided the direct evidence for evolution— and this evidence showed the same rate of evolution that was required to explain the radio source counts.

1960 And Beyond—Radio Astronomy's Source Counts Achieve Consensus

The speed with which the radio astronomers achieved a common ground in the field of radio counts in the 1960s was remarkable; Pawsey's hint of optimism was justified. Kragh (1996, pp. 318–324) wrote:

> At the end of the 1950s, after a decade of controversial existence, the Steady-state theory was as alive as ever. At that time the more emotional and philosophical resistance to the theory had weakened and no longer played a significant role . . . Interest shifted to observational and experimental issues, a change that, to a large degree, was induced by new observations that seemed to allow for new possibilities of discriminating between the two main rival models of the universe.

Indeed, we might suggest that the field of astronomy as a whole was reluctant to make any cosmological pronouncements on the basis of new radio observations, since the majority, optical astronomers, still had little appreciation or understanding of radio approaches. For example, prominent optical astronomer V.C. Reddish[28] quoted aptly from the book he was reviewing "Radio Astronomy" by Graham Smith (1960, p. 158):

---

[28] University of Edinburgh and the University of Manchester, later Astronomer Royal of Scotland.

> It may be wondered how it comes about that radio astronomy, with quite limited observational results, should so soon have grown into a position where it is presuming to answer deep cosmological questions, while in comparison optical observations carried out in such detail by telescopes of great sensitivity and resolving power should have failed to penetrate so far towards the edges of the observable Universe.

Reddish commented, "Perhaps the answer really is that the optical astronomers are more aware of the inadequacies of existing data, more appreciative of the pitfalls to be met in analysing and interpreting it, and in their long dealings with the Universe have learned that it is unwise to be presumptuous. On this point, however, the reviewer may be regarded as biased."[29]

The final clarification of the source counts occurred in the early 1960s at four key conferences. The first was the URSI General Assembly of September 1960 in London. In his introductory talk at the session on discrete sources, John Bolton (1960) presented impressive new results from his Owens Valley Radio Observatory using his new 960 MHz observations with the twin 90-foot interferometer of Caltech to check the reliability of sources in the Cambridge and Sydney surveys. Edge and Mulkay (1976, p. 192) have pointed out the significance of these new data taken at a much higher frequency:

> This [new effort] constituted the first serious attempt at empirical mediation by a third party in the dispute between Cambridge and Sydney. Bolton was the more critical of the results of his former colleagues [in Sydney]. Of a sample of 100 3C sources checked by the new Owens Valley instrument, the existence of only three could be seriously doubted. Of 216 of Mills's sources however only 75 per cent were found and, of his "extended" (which were, of course, at the heart of the controversy), roughly half could not be detected, and many of the remainder appeared to be complex blends of point sources. Bolton summed up:

> In general, rather better confirmation was obtained of the sources in the 3C catalogue and their relative intensity than of the sources in the Sydney catalogue. The extended sources of the Sydney catalogue in some cases do not appear to be extended and in others are complex, perhaps blends. I must stress that this conclusion is not very firm, but at the moment I find it difficult to believe that the slope of the Log N Log S relation can be forced down to -1.5. Whether the interpretation of a greater slope proposed by Ryle and Scheuer is correct ... is another question.

This was indeed game changing, and it was the first indication that the Mills survey also had problems, although by no means as severe as those affecting the original 2C catalogue. The primary evidence for a population of extended sources, which was the remaining concern for the Cambridge surveys, had been based on the Mills catalogue and this was clearly now in doubt, probably a result of the smaller but still significant effects of confusion. At this time Pawsey was pre-occupied with many other policy issues in Australia (see Chaps. 31, 32 and 38) and there is no indication of any interaction between Pawsey and Bolton on this topic.

The next year was to be a watershed. Kragh (1996, p. 324) has provided a graphic description of the year 1961:

---

[29]Reddish and Hill, review of the 1960 book *Radio Astronomy* by Graham Smith (Pelican Books), *Observatory* vol 81, 1962 p. 207.

[This year] marked the point of no return and the beginning of the end of dissension as far as the radio astronomical data were concerned. Based on extended counts of sources and an improved statistical method due to Scheuer, Ryle's group had ready new and definite conclusions in the beginning of 1961. These were published in papers by Ryle, in collaboration with Clarke and Scott, respectively, and presented on 10 February [1961] at a meeting of the Royal Astronomical Society. Two weeks later Ryle gave a Royal Institution lecture on the new Cambridge work, where he discussed the implications of the work for a broader audience.

The paper by Scott and Ryle (1961) provided new evidence using the 4C aerial at 178 MHz that the slope of the LogN-LogS was a straight line of -1.80 with permissible limits of -1.68 to -1.93.[30] The excess of weak sources or the deficit of more intense ones was confirmed. Suggestions that this effect was "caused by the presence of extended sources or by a strong clustering tendency have been investigated and it has been shown conclusively that neither effect would be sufficient to produce a [LogN-LogS curve that had been observed.]" Ryle and Clarke (1961) derived a LogN-LogS for the Steady-state model using luminosity functions based on identified sources. All models produced a logN-LogS that were appreciably flatter than -1.5. An important change had been the use of better luminosity functions based on the radio source identifications at that time. In effect this was making the old "abnormally strong" radio galaxies the main population in the catalogue instead of treating them as outliers. Mills et al. (1960) constructed a new radio luminosity function based on optical identifications and obtained predicted LogN-LogS produced slopes of -1.3 for a Steady-state universe in full agreement with Ryle and Clarke (1961).

At the Royal Astronomical Society meeting on 10 February 1961, a famous exchange between Bondi and Ryle followed the latter's presentation:

Bondi: With regard to the historical development of Ryle's work we should remember that six years ago he gave the slope of the logN-logS relation as - 3 and now it has been reduced to -1.8. Perhaps there is still a residual error which might account for the relatively small discrepancy between -1.8 and -1.5. There is no doubt, however, that Ryle and his team have carried out a most painstaking and careful investigation and I should like to congratulate them upon it.

Ryle: Prof Bondi has not given a very accurate picture of the past work since he appears to ignore the statistical analysis of Scheuer which gave results entirely consistent with our present findings. I should also point out that the disagreement with the Steady-state theory amounts to a factor about 10 when expressed in numbers of radio sources. It is difficult to regard this as a small discrepancy.

The extreme reaction was not recorded in the *Observatory* proceedings. Kragh (1996, p. 337) has, however, filled in the missing pieces: "Ryle did not appreciate Bondi's sarcasm and, according to a witness of the debate, 'flew into a rage, which

---

[30]The new 4C antenna allowed direct counts to sources as weak as 2 Jy; the P(D) analysis provided information for sources in the range .05 to 2 Jy.

resulted in the nastiest public display of tempers between scientists that I [Martin Harwit] have seen in more than 30 years as a professional astrophysicist.'"[31]

In the second edition of his book on cosmology Bondi (1960) made an interesting and quite subtle comment which was a significant insight into scientific methodology. He points out the quite well-known fact that Steady-state theory made strong predictions while evolutionary models can fit any observations, and he then went on to argue that at the observational limit when errors are large (e.g. the Cambridge 2C Catalogue), highly predictive theories like Steady-state will be preferentially excluded. This was all he was trying to express at the RAS meeting when he was attacked by Ryle.

Kragh (1996, p. 325) reported on the end of the debate that began at the URSI in Tokyo in 1963:

> Contrary to the 2C survey [in the era 1955–1958], the new Cambridge results [in 1963] remained stable, soon receiving support from other radio astronomers. At a meeting of the International Scientific Radio Union in Tokyo in 1963, Ryle reported an improved figure for the logN-logS slope. It was now settled to -1.8 ± 0.1. Finally, the Sydney group announced the result of new measurements of frequency 408 MHz in 1964,[32] ending up with a revised figure of -1.85 ± 0.1, evidently "in very good agreement" with the Cambridge value. It was further confirmed by results from the 4C survey, carried out during 1958–64 at frequency 178 MHz. From this time, the controversy over the observational value of the slope definitely came to an end, and already in 1961 most specialists realised that, whatever the correct value of the slope, it could not be -1.5. However, the immediate impact of the consensus on the cosmological controversy was limited. Even though advocates of the Steady-state theory had to accept the observational data, they could still avoid giving up the theory, either by questioning the significance of the data or by modifying the theory (or by a combination of the two strategies). This was just what happened.

Edge and Mulkay summarised their conclusion of 1976 with a droll statement:

> On this sly note [at the Tokyo URSI] the controversy died. Nobody any longer considered that the previously apparent inconsistencies defined a problem worthy of further attention. Advances in the technologies had made it all seem rather trivial; there were other, obviously more important problems to be tackled. Consensus led to intellectual migration.[33]

---

[31] Also the public read about this controversy: "The report of 10 February [1961] was widely discussed in the press, both in England and abroad. According to McCrea, news about Ryle's conclusion had leaked out to the press, so that 'astronomers on their way home after the meeting were able to read all about the final overthrow of Steady-state theory!'" Kragh (1996, p. 324).

[32] This was a new group now led by Bolton and using the new Parkes radio telescope, the GRT. This Parkes survey which included one of the authors (RDE) also failed to confirm many of the weaker MSH sources.

[33] Additional convergence occurred in the mid-1960s. By 1964–1967, optical identifications of the revised 3C catalogue were virtually complete and the optical and radio universe "were brought into broad alignment". (Edge & Mulkay, 1976, p. 279).

Chapter 37
The Evolution of Aperture Synthesis Imaging

> The importance of the Cambridge work is well known, and perhaps this is why the more humble first use of the technique in a distant land tends to be ignored. Christiansen (1989)

Introduction

The theme of interference between radio waves played a key unifying role throughout Pawsey's career. Pawsey used radio-wave interference to study the structure of the ionosphere for his PhD research (Chap. 7), and it was Pawsey who first realised that radio images of the sky could be made from measurements of radio interference. Since these observations are made in the aperture plane and not the image plane, this is referred to as "indirect imaging". When electromagnetic waves from the same source combine, they can either reinforce or cancel depending on the path difference. This makes the classical beating interference patterns often referred to as "fringes". The first interference patterns in the radio were seen by Hertz between 1886 and 1889 during the course of his experiments to prove that the radio waves he had detected had the interference properties predicted by Maxwell's electromagnetic theory (Pierce, 1910).

Radio wave interference with a very practical application arose in WWII when the Chain Home (UK) and Royal Australian Air Force (RAAF) coastal radar systems were used for air and seaborne defence. Interference between the direct transmission and its reflection from the sea generated a pattern of maxima and minima, and these could be used to estimate an aircraft's elevation as it flew through this pattern. Fred Hoyle and others in the US and Australia had calculated charts for this purpose (e.g. Jaeger, 1943; Mitton, 2011; Domb, 2003). Note that the term "under the radar" refers to the practice of flying at low elevation where the interference pattern created a null so the aircraft remained invisible.

The development of aperture synthesis in radio astronomy has its origins in the measurement of these interference patterns, and the interpretation of these

© The Author(s) 2023, corrected publication 2024
W. M. Goss et al., *Joe Pawsey and the Founding of Australian Radio Astronomy*,
Historical & Cultural Astronomy, https://doi.org/10.1007/978-3-031-07916-0_37

observations is based on an understanding of Fourier synthesis. Because of the need to calculate the Fourier transforms, progress in this field was tightly linked to the dramatic evolution of electronic computers during the same period.

In this chapter we first compare the evolution of indirect imaging in different disciplines: optical, X-ray crystallography, medical and radio. Then we will follow in detail the radio astronomy development which started after WWII simultaneously in Ryle's group in Cambridge, UK, and in Pawsey's group in Sydney, Australia. (We will summarise a few salient points made at more length earlier in the book, so that the reader can easily follow these developments). The existence of these two groups in Cambridge and Sydney provides a remarkable opportunity to explore the development of this completely new imaging technology in two independent groups. The events occurred in different countries on opposite sides of the globe, that started with essentially identical shared academic background and identical equipment from the WWII radar research in the UK and Australia. By comparing the parallel developments in these two quite different and independent research environments, we can elucidate some of the factors, sociological as well as technical, which influenced the discovery of aperture synthesis imaging in radio astronomy, a discovery for which Sir Martin Ryle was awarded a Nobel Prize in 1974. Edge and Mulkay (1976) also discuss these independent developments with particular emphasis on the different leadership styles, but largely ignoring the many other factors separating the groups: in a highly developed research environment in Cambridge and the other from an isolated fledgling research group in the colonies.

Early Development of Indirect Imaging in some Other Disciplines

Optical Astronomy

Michelson (1890) discussed the possibility of using an optical interferometer to measure the diameter of stars. He defined fringe visibility and gave the Fourier equations but did not refer to this formulation as a Fourier transform. In Michelson's application to starlight it is assumed that the stars are symmetrical circular disks so the diameter can be estimated from the visibility amplitude alone. There was no need to consider the phase of the fringes, which would have been quite impossible as the flickering interference patterns were recognised by eye. Binary star separations could also be determined without the need to measure the fringe phase by looking for the minimum in the fringe visibility amplitude when plotted as a function of baseline length. Michelson and Pease (1921) measured the diameter of Betelgeuse using a 20-foot interferometer mounted on the Mt. Wilson 100-inch telescope. The radio equivalent of Michelson's stellar interferometer has been referred to as a Michelson interferometer since the analogy was noted by Ryle and Vonberg (1946).

Fourier Theory

Zernike (1938) published a simple derivation of the van Cittert-Zernike theorem: *The spatial coherence over a space illuminated by an incoherent extended source is described by the Fourier transform of the intensity distribution over the source.* This theorem is the mathematical description of the relationship between the image of a source formed in the focal plane of a telescope and the interference pattern observed by an interferometer. Although the theorem is now considered the basis of Fourier synthesis imaging, it was developed in the field of optics and played no role in the early developments of aperture synthesis imaging in radio astronomy either in the Netherlands or elsewhere, even though Zernike had been an assistant to Kapteyn at the astronomical laboratory of Groningen University in the Netherlands in 1913. The theorem appears in the radio literature only after the publication of the Born and Wolf *Principles of Optics* (1959).[1] In 1953 Zernike received the Nobel Prize in physics for his invention of the phase contrast optical microscope (Zernike, 1955). The phase contrast microscope is another example of indirect imaging where changes in the propagation of light through a transparent sample are converted to changes in brightness using techniques closely related to those used in radio astronomy imaging.

X-Ray imaging in Medicine

Two-dimensional X-ray images are normally exposed on a photographic plate, and any structure in the third dimension is integrated along the line of sight from the X-ray source to each pixel in the image. The 2D images are called projections. As early as 1896, stereo X-ray image pairs were being used in medicine to give depth estimates for features in the 2D images. Between 1920 and 1970, 3D X-ray tomography was developed using ingenious analogue devices to do 3D image slice reconstruction from multiple 2D projections at different angles. This reconstruction step takes medical imaging into the indirect imaging domain, and the reconstruction of a 3D X-ray image is identical to the process of reconstructing a 2D radio image from a one-dimensional array of receivers. We discuss this in detail in the following sections. This reconstruction technique is referred to as "back projection" since it is the reverse of the process of projecting an image into a space with one less dimension.

The medical practitioners generally had no background in physics or mathematics, and these images were recorded on photographic plates with far too much detail for any practical numerical imaging approaches at the time. Consequently, there was

---

[1] Jim Moran has pointed out that the small angle approximation which leads to the Fourier transform relation first appears in Born and Wolf Principles of Optics (1959) and is not explicitly included in the van Cittert-Zernike papers.

no application of Fourier methods or any use of computers until the 1970s, e.g. *From the Watching of Shadows* (Webb, 1990). The first link to radio imaging techniques occurred in 1967 when Pawsey's protégé Ron Bracewell[2] (Bracewell & Riddle, 1967) published a paper that provided a mathematical solution to a problem involving back projection in medical imaging.[3]

X-Ray Crystallography

X-ray Crystallography needs special attention because it played a key role in the development of aperture synthesis in radio astronomy and is a fine example of the value of cross-disciplinary interactions. In 1912 X-ray diffraction in crystals was discovered by von Laue (1914). W. Lawrence Bragg (1929) suggested the use of Fourier methods for determining crystal structure. By 1936, Fourier synthesis calculations were routine in X-ray crystallography using Lipson-Beevers strips (Beevers & Lipson, 1936) and making hand calculations. A review of the history of Fourier methods in crystal structure determination is published in Beevers and Lipson (1985). In 1939, Bragg's X-ray crystallography group was flourishing at the famous Cavendish Laboratory in Cambridge. The key problems at this time were the 2D Fourier analysis and the impossibility of making a direct measurement of the phase from the detected X-ray pattern which remained one of the biggest issues in X-ray crystallography for many decades. At the same laboratory in Cambridge, Martin Ryle's group of radio astronomers was exploring indirect imaging techniques. The following quote is from Ryle's 1974 Nobel Prize autobiography: ". . . both Ratcliffe and Bragg [W. Lawrence] gave enormous support and encouragement. Bragg's work on X-ray crystallography involved techniques very similar to those we were developing for 'aperture synthesis'".

Ratcliffe and Pawsey: The Cambridge—Sydney Connection

The Sydney and Cambridge radio astronomy groups were of course very similar due to their use of WWII hardware and radar expertise in early radio observations of the sun and cosmos (Sullivan, 2009, p. 170). The reader will also recall (Chaps. 6, 7 and 8) that Cambridge and Sydney were connected by Pawsey, in ways that influenced developments for the next two decades. Pawsey had gone to Cambridge to do his PhD on the ionosphere, working with Ratcliffe as his advisor from 1931 to 1934

---

[2] Ron Bracewell had moved to the Stanford University from his position in Sydney where he had earlier developed much of the theoretical framework used for radio astronomy imaging. See *Four Pillars of Radio Astronomy*, Frater et al. (2017).

[3] See below, Section "Impact of computers on medical imaging".

Fig. 37.1 54RS (Collaroy Radar Station 54) during WWII, site of first Australian Radio Astronomy in October 1945. COL Mark V British Radar, 200 MHz (Chain Overseas Low Flying radar). Credit: Photo obtained from Warringah Council Libraries in 2015 (Michelle Richmond, historian and librarians Rose Cullen and Gaynor Cotter), photo number 41/WAR41678

(Chap. 7). As part of his PhD thesis, Pawsey (1934) used the interference between the direct reception of BBC radio broadcast signals and the signal reflected from the ionosphere to show that the fading effect in low frequency broadcasts was primarily an effect of interference and not absorption. Once he had moved to the CSIR Radiophysics Laboratory in Sydney, Pawsey continued to maintain his strong links with Ratcliffe in Cambridge. This close relationship played an important role in the radio astronomy developments in Australia.

Soon after arriving in Australia, Pawsey (1940) wrote a classified document proposing an improved radar system to measure aircraft elevation based on the use of a Michelson interferometer.[4] This system was never built, but it is clear that Pawsey understood the physics of the two-element interferometer and was well aware that it could be used to determine directions. At the end of the war Pawsey's group started investigation of the newly discovered radio emission from the sun using the Airforce radar station at Collaroy, a hilltop 15 miles north of Sydney (Fig. 37.1). To further investigate the nature of the unexpectedly strong bursts of radio emission that they observed from the sun and perhaps gain insight into their mysterious origin, both Pawsey's group in Sydney and Ryle's group in Cambridge

---

[4]20 Dec 1940 "Phase Dependent Elevation Aerial", RP58/1. CSIRO Division of Radiophysics archive.

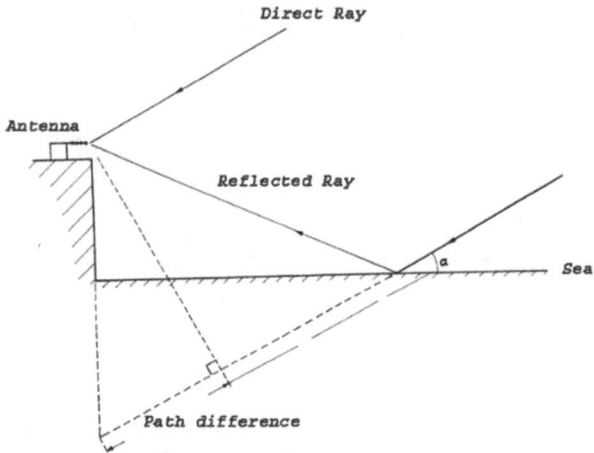

Fig. 37.2 Sea-cliff interferometer (Lloyd's mirror). Credit: CSIRO Radio Astronomy Image Archive B1639-4

started a series of experiments designed to determine the size and location of the solar radio bursts (see Chaps. 12 and 16).[5]

At about this time Pawsey introduced Bracewell to the duality of physical and mathematical descriptions following the style he had learned from Ratcliffe. In 1946 Bracewell left Sydney to do his PhD with Ratcliffe at Cambridge. He returned to Australia in late 1949 and was to play a major role in the formal development of aperture synthesis theory. He became well known for his book on the Fourier transform (Bracewell, 1965) (see below, "Fourier Synthesis").

The Sea-Cliff Interferometers

The Sydney group's early research used sea-cliff interferometers, the radio equivalent of a Lloyd's Mirror, combining the direct signal with the signal reflected from the ocean (see Fig. 37.2). All those involved in WWII radar systems would have been very well aware that the radar installations along the coastline near Sydney could be used as sea-cliff interferometers to determine the location of solar or extrasolar radio sources. (During the war, the Australian 200 MHz radars had also detected fringes from incoming aircraft using the interference between the direct and sea-reflected echoes.) The higher the cliff, the better the angular resolution. The receiver at Collaroy had a height of 122 m above the sea and was used at 200 MHz. A second station was set up on top of the 85 m cliff at Dover Heights using a 100 MHz Yagi antenna (Fig. 37.3). Pawsey led a team including Lindsay McCready and Ruby Payne-Scott in 1947, as they measured the phase of the sea-cliff interferometer fringes (called lobes by the radar researchers and radio astronomers) to

---

[5]See also NRAO ONLINE.20.

Fig. 37.3 Dover Heights–1952. Radio astronomy instruments on the site of the WWII radar station. Credit: CSIRO Radio Astronomy Image Archive B2763-6

determine the position and angular size of the solar emission (see Chaps. 11, 12 and 13). This procedure required a sophisticated analysis of the effects of atmospheric refraction at the low elevations needed for the sea-cliff interferometer.[6]

In their seminal paper published in the *Proceedings of the Royal Society*, McCready et al. (1947) showed that the compact radio bursts are at the positions of the sunspots and a size of about 6.5 arc min (compared to the solar diameter of about 32 arc min) of the sunspots. This paper also included the first published statement of the aperture synthesis concept in radio astronomy. They noted that it is possible in principle to determine the angular distribution of the emission by Fourier synthesis using the phase and amplitude of the interference fringes measured at a range of separations. They then pointed out that this could be achieved either by having a range of cliff heights or a range of radio wavelengths. They considered the use of wavelength as a suitable variable as unwise, since the solar bursts are likely to have frequency-dependent structure. They also noted that getting a range of cliff

---

[6]For more details see *Under the Radar*, Goss and McGee (2009, p. 276).

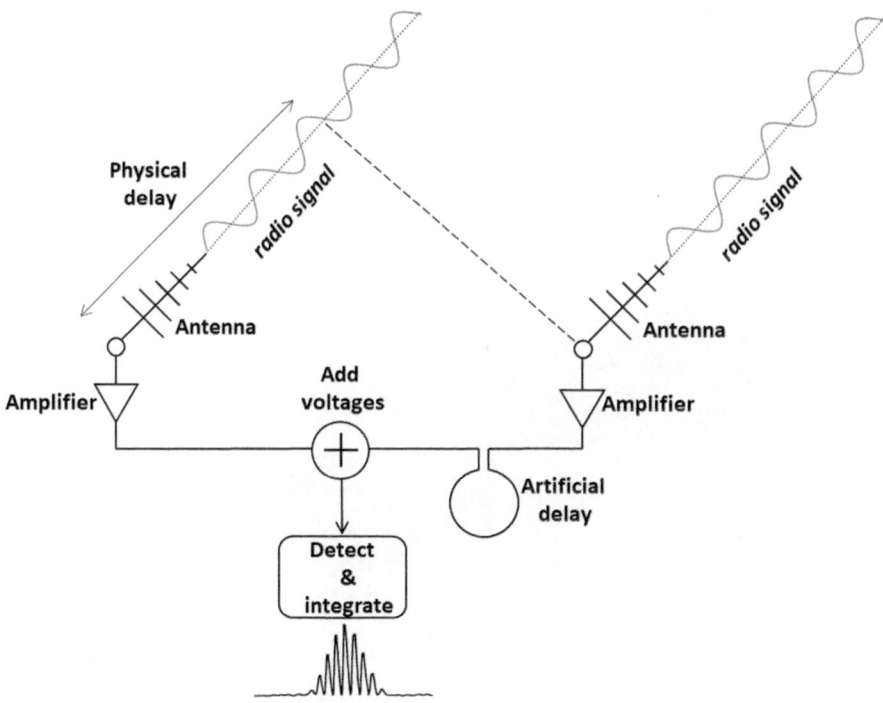

Fig. 37.4 Michelson or horizontal interferometer. Credit Ron Ekers

heights is clumsy and suggested a different interference method would be more practical. Based on Pawsey's earlier proposal to build a Michelson interferometer for radar determination of elevation, we assume that this comment referred to a Michelson interferometer (Fig. 37.4) but they did not make this explicit in the paper. We also note that the concept of using wavelength to vary the effective baseline length is an obvious extension of the Appleton frequency scanning technique used to determine the height of the ionosphere (see Chap. 7). In Bracewell's interview with Woody Sullivan,[7] he recalled that the first Fourier Synthesis concepts emerged in 1947; the idea of restoring the source distribution from measurements of the Fourier components was being discussed at CSIR Radiophysics Laboratory in Sydney at this time. For example, in order to measure the position of a burst, they realised that they should wait for a single unresolved solar burst with unit visibility and then measure the phase. They knew that with a single phase measurement they could not uniquely determine the position of any radio source more complex than a point source.

The paper was submitted to the Royal Society in London on 22 July 1946 by David Rivett who was visiting the UK. However, it did not appear in the *Proceedings of the Royal Society* until 12 August 1947—a 13-month delay which had

---

[7]Sullivan interview of Bracewell, 8 January 1980. W.T. Sullivan III, Papers. Archives, National Radio Astronomy Observatory / Associated Universities Inc.

some significant consequences.[8] During this period Ryle and Vonberg (1946) published their paper in *Nature* (submitted 22 August and published 7 September 1946) associating radio bursts with sunspots. As discussed at the end of Chap. 13, a 13-month delay was not exceptional at this time, but this paper also included the first statement of the Fourier synthesis concept in radio astronomy and the timing of this idea is noteworthy.

Sullivan (2009, Sect. 8.5.3) found that the first reference to the Fourier Synthesis concept in Cambridge is an entry in Ryle's (unpublished) notebook, from 23 August 1946, describing the harmonic analysis of fringe amplitudes from a large number of aerial spacings to determine the distribution of sunspots. We note that Pawsey had sent a copy of his paper directly to Appleton on 5 August 1946 so it would have been available in the UK a few weeks before Ryle's notebook entry. Given the rivalry over solar work between the two groups it is most likely that Appleton would have shown Ryle a copy of the paper and Ryle would have recognised the significance of the Fourier synthesis concept. There is no indication that Ryle ever claimed precedence for this concept; he did not publish it himself for another five years, and when he did (Ryle, 1952), and in his other early Cambridge papers, he gave credit to McCready et al. (1947) for the Fourier Synthesis concept. But in striking contrast to Ryle's publications, the Sydney work was not cited in any of the papers by Ryle's students, nor was it mentioned again in any of the Cambridge papers published after 1955 (see discussion section at the end of Chap. 37).

Pawsey was first author on the McCready et al. paper when submitted to the *Proceedings of the Royal Society*, but it was alphabetised following Royal Society policy—hence published as McCready, Pawsey and Payne-Scott. Joe Pawsey was clearly the leader, with Payne-Scott observing and deriving the mathematical quantities while McCready was responsible for the design of the equipment. This paper clearly includes the first published description of Fourier Synthesis in radio astronomy, but the subject of the paper was locating solar radio emission in sunspots. Many later citations to the paper refer to the pioneering solar research with no mention of the aperture synthesis proposal. Thus, it is not surprising that recognition for this new proposal was lost to a broader radio astronomy community.

Fourier Synthesis at Cambridge

Back in Cambridge in 1946, Ryle started to use the radio equivalent of the Michelson interferometer to resolve the strong diffuse galactic radio emission that made it difficult to separate out the weaker signals from the sun. This technique worked well; his group often detected fringes from the sun. In July 1946 a giant sunspot appeared that had sufficient signal strength for determining the angular diameter of

---

[8]The consequences of this delay for the solar burst sunspot association has been discussed in detail in *Under the Radar* (Goss and McGee 2009, Chap. 7).

the source of the radio emission using the formalism developed by Michelson to describe his interferometric measurements of stellar diameters (Michelson, 1890). These pioneering solar observations were published in *Nature* by Ryle and Vonberg (1946). The association of the radio emission with the sunsport was based on the similarity of the radio size with the prominent sunspot. No positions were determined in these initial Cambridge observations.

Following Sullivan (2009), we now trace the development of the Fourier Synthesis concepts that evolved from the sequence of student experiments led by Ryle at the Cavendish Laboratory in Cambridge. These experiments all involved the use of Michelson interferometers and were designed to better understand the solar radio emission. Many of these experiments involved observations taken over many days by physically moving the aerials to different positions. If the structure of the solar radio emission changed during this time, as it did when the sun was in an active state, the observations would be spoiled. In retrospect we can see that these experiments did not contribute much to the understanding of solar radio emission, but they were very important steps towards the practical implementation of the aperture synthesis technique in radio astronomy.

Ryle, with his student H.M. Stanier (1950), used the original pair of Würzburg dishes and a quantity of low-loss coaxial cable captured from the Germans as a two-element radio interferometer operating at 60 cm wavelength (500 MHz). They measured solar visibility for 17 different positions of the interferometer dishes, giving a range of east-west spacings out to a maximum baseline of 220 m. Despite attempts to wait for times when the sun was quiet, the results taken over many days were still somewhat confused by solar variability caused by sunspot activity. They computed the 2D radial profile of the quiet sun using Lipson-Beevers strips and a Hollerith punched card machine to calculate the Fourier integral. They had only observed with interferometers along an east-west line. Thus, the assumption was made that the solar radio emission had a circular symmetric structure (like the optical disk of the sun). This assumption was later found to be wrong; consequently both the equatorial bulge and the predicted limb brightening of the radio emission were missed.

Ryle's next student, K. E. Machin (1951), used an array of four fixed and two moveable elements at a longer wavelength of 3.7 m (81 MHz). This gave many more simultaneous interferometer measurements. Again he had to wait for periods when the sun was quiet, even less frequent at these longer wavelengths. Given the assumption of circular symmetry, Machin realized that he could simplify the very time-consuming calculations of the 2D Fourier Transforms by fitting the data to Bessel functions, which are an analytic form for the Fourier Transform of circular functions.

Michelson interferometers add the signals from each antenna and receiver. This includes the signal from the source which produces the interference fringes, but it also includes large signals from the receiver noise and the diffuse sky background. This large DC signal makes the detection of weak interference fringes difficult, especially with the instability of the receiver gain at that time. To remove this unwanted signal, Martin Ryle invented the phase switch, published by

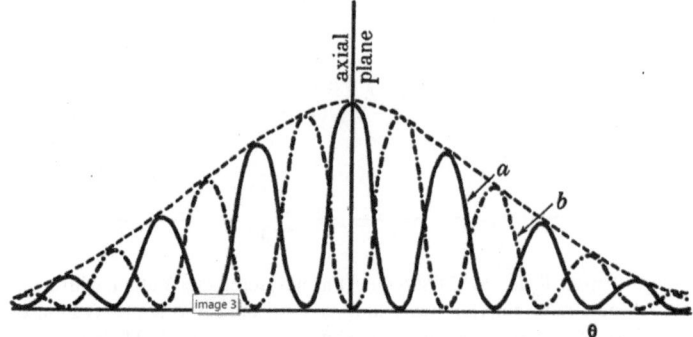

Fig. 37.5 Figure 1 from Ryle (1952) illustrating the principle of the phase switch, (**a**) connected in phase and (**b**) connected in anti-phase.Credit: Fig. 1, "A new radio interferometer and its application to the observation of weak radio stars", Ryle, M. *Proceedings of the Royal Society of London*. Series A. Mathematical and Physical Sciences 211, no. 1106

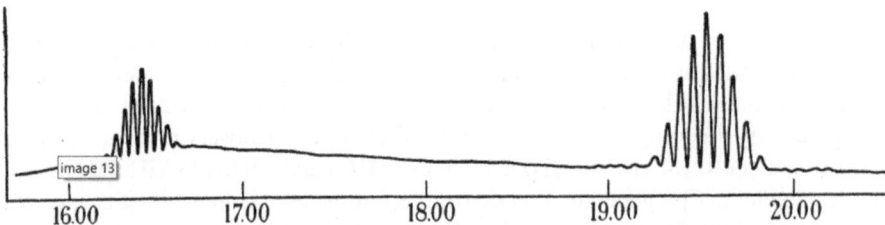

Fig. 37.6 Figure 4 from Ryle (1952) showing a total power output without phase switching. Credit: Fig. 4, "A new radio interferometer and its application to the observation of weak radio stars", Ryle, M. *Proceedings of the Royal Society of London*. Series A. Mathematical and Physical Sciences 211, no. 1106

Ryle (1952).[9] By changing the phase of the signal from one antenna, the interference pattern will change, but not the DC signal. Consider two antennas A and B: two voltage outputs can be formed: first A + B, then by switching the phase of B by 180 degrees, A-B is measured (see Fig. 37.5). If the phase is switched continuously, the difference between the two power outputs can be synchronously detected:

$$(A + B)^2 - (A\text{-}B)^2 \rightarrow A.B$$

In this way the product of the two voltages (A.B) can be measured without the contribution of the two much larger total power signals (A^2 and B^2).

Figure 37.6 shows the interferometer output achieved in Cambridge including the total power. Figure 37.7 is the output after the phase switch removed the total power

---

[9] See Sullivan, page 114, in *Classics in Radio Astronomy* (1982) and *Four Pillars of Radio Astronomy* (2017) for a discussion of the almost independent invention of the phase switch in Sydney by B.Y. Mills.

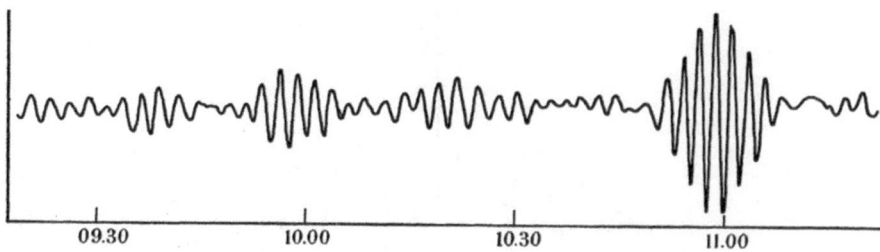

Fig. 37.7 Figure 9 from Ryle (1952) showing the interferometer output with phase switching. Credit: Fig. 9, "A new radio interferometer and its application to the observation of weak radio stars", Ryle, M. *Proceedings of the Royal Society of London*. Series A. Mathematical and Physical Sciences 211, no. 1106

signal. Many weaker radio sources could now be detected. This was the first correlation (rather than adding) interferometer.[10] This technological innovation was to have major implications for the research directions at Cambridge (see later section).

Two more key developments were made when Ryle's next student, P.A. O'Brien (1953), used the earth's rotation and published the first solar images using a full 2D Fourier Synthesis calculation. As we discuss in the next section, the Australians were independently pursuing a similar path at this time. O'Brien observed the quiet sun at three frequencies: 1.4 m (215 MHz), 3.7 m (81 MHz) and 7.9 m (38 MHz). He used a moveable element to generate a range of interferometer spacings along a line, and, for the first time, he let the rotation of the earth change the orientation of these spacings. This was the first published use of earth rotation to measure different Fourier components. Ryle's notebook for 22 July 1954 contains a passing reference to "earth rotation synthesis" (e.g. Scheuer, 1984). O'Brien explained the procedure: "What we were doing was really extremely primitive—picking up an aerial, walking along and then plonking it down somewhere else. It was the string-and-sealing-wax tradition, except ours was iron and concrete."

On his busiest day O'Brien measured 43 spacings at five different baseline orientations in nine hours. It was essential to work fast while the sun remained quiet, since any change in the angular distribution of radiation would invalidate this process. O'Brien's 1.4 m data was the most interesting. He now had measurements in two dimensions and only had to assume reflection symmetry (not circular symmetry) to process his data. O'Brien (1953) concluded that the sun was 25% larger in the equatorial direction. This observation showed that the simplifying circular symmetry assumptions made in the previous Cambridge observations were incorrect. This had resulted in the limb brightening being washed out in the Cambridge experiments,

---

[10]In a modern radio telescope, the voltages would be sampled and digitised directly and the product calculated in a very fast special purpose computer (a correlator). This process is quite easy when working at radio frequencies but nearly impossible at optical or shorter wavelengths where it is only possible to record the detected photons.

which led to the subsequent controversy when the limb brightening was reported by the Sydney group.[11]

Other Radio Interferometry Developments in the UK

Soon after joining the Jodrell Bank group, Hanbury Brown expressed his opinions of his Cavendish colleagues to his former fellow WWII radar researcher, Taffy Bowen (in Sydney). Hanbury Brown wrote to Bowen on 30 April 1950:

> Ryle is doing some interesting work at Cambridge. He has a large collection of point sources (50-1C) which he is publishing. I haven't even got a list myself as, of course, it is harder to find things out in Cambridge than in Sydney. He is concentrating on interferometers and has some very elegant stuff. I have a whole colony of schemes on interferometers and am bursting to try them.[12]

Hanbury Brown's comment (above) about "a whole colony of schemes on interferometers" is very interesting and an ironic twist as one of his schemes was one of the most significant discoveries in interferometry. This scheme was not developed by the "interferometrophiles" at Cambridge.[13] This letter to Bowen was written in 1950 when Hanbury Brown (in Kellermann & Sheets, 1984) recalls having a vision of the same noise being detected in two spatially separated receivers observing a strong radio source. He realised that this would be the case if the predominant noise was the source and if the source was unresolved. Two years later, in 1952, Hanbury Brown and his students Jennison and Das Gupta had built the intensity interferometer and demonstrated that the scheme worked. They successfully used this correlation of the intensity fluctuations to measure the diameter of some of the stronger discrete radio sources (Hanbury Brown, 1974). Richard Twiss worked out the theory of the intensity interferometer; this was published by Hanbury Brown and Twiss (1954), known as the Brown-Twiss effect. Although only useful for very strong sources (e.g. the sun, Cas A and Cygnus A), it was still a valuable technique when working with very long baselines. The procedure used the correlation between detected intensity rather than the wave amplitude, with no necessity to preserve the phase coherence between the two aerials. Furthermore, the correlation was not destroyed by atmospheric turbulence. Later they realised that the effect would also occur at optical wavelengths, seemingly in contradiction to quantum mechanics. This produced some controversy.: Brannen and Ferguson (1956) argued that it would require a revision of quantum theory. However, Ed Purcell (1956) showed that, on the contrary, it was an effect that could only be explained by quantum theory, since it depends on the clumping of bosons, which is a quantum effect and would not occur with classical particles. A full quantum optics description of light, which includes

---

[11] See Chap. 16 (Sect. 5) and NRAO ONLINE.20.
[12] NAA C3830 A1/1/1 Part 5.
[13] See Chap. 35.

this Hanbury Brown-Twiss effect, was provided by Glauber (1963), resulting in his Nobel Prize in physics in 2005.

A stellar intensity interferometer based on this Brown-Twiss effect was built at Narrabri, Australia, in 1963 and was able to measure the diameters of stars, extending the endeavours of Michelson after his coherent optical interferometry reached its limit in 1937. Remarkably, the Michelson interferometer, when applied to the radio, triggered the development of a new method to carry out optical interferometer measurements of stars (Hanbury Brown, 1974).

While working on the intensity interferometer, Jennison (1958), at the Jodrell Bank radio astronomy group, derived the phase closure relations between interferometers with three or more antennas. This rather simple procedure was to have a very profound influence on the field—well beyond anything that could have been anticipated at the time. An interferometer measures the difference in the phase between the radio signals received at two different locations. If these phases are summed around a closed loop of three or more receivers, all the instrumental and atmospheric errors are cancelled out and what remains is called a phase closure. A simple analogy would be to consider walking around a path in a hilly countryside. When a person gets back to the start, all the elevation changes along the path will have cancelled out and he will be back at the same elevation. If one station on the path were higher (like an instrumental error in one receiver), this will cancel out when the loop is completed. However, if the person is observing a source with structure it will look slightly different from each different location, and this information about the structure of the source will not cancel out when the phases around the loop are summed. The use of phase closure and the closely related self-calibration techniques and triple correlations would have a very big impact on Fourier synthesis imaging in radio astronomy a few decades later.[14]

Imaging Arrays at the Sydney Field Stations

Now we return to the development of the Australian arrays, which followed a different course to that taken in Cambridge. Initially the Sydney group used the sea-cliff interferometer where both direct and reflected radio signals are combined in a single receiver. This makes it easier to recover the fringe phase because it avoids the difficult technical problem of maintaining phase stability over the long lengths of coaxial cable needed to connect the two separate aerials in a Michelson interferometer. (Compare Figs. 37.2 and Fig. 37.4.) Of course, it was also a great advantage to have suitable cliffs, such as Collaroy and Dover Heights, quite nearby in the Sydney area. The first use of a Michelson interferometer in Australia was led by Ruby Payne-Scott. She had observed the sequence of burst events passing from higher to lower

---

[14]The phase closure relations are also the underlying principle for the closely related structural invariance in X-ray crystallography for which Hauptman (1986) received the Nobel Prize.

Fig. 37.8 Ruby Payne-Scott, Alec Little and Chris Christiansen at Potts Hill Reservoir in late 1948. Observing the sun with the swept lobe interferometer. Credit: CSIRO Radio Astronomy Image Archive B14315

frequency. There were various models that required motion of the "physical agency" causing the radiation. To measure this motion her group needed an interferometer, but they also needed to measure changes on a one-second time scale, which was much faster than the slow (three minute) fringes generated by the rotation of the earth. To do this they invented the "swept lobe interferometer" that produced rapidly changing fringes by offsetting the frequency of one of the two receivers by 25 cycles per second. This could not be done using the single-receiver sea-cliff interferometer; thus, a Michelson interferometer was built at the new Potts Hill reservoir site near Sydney (Fig. 37.8). The main advantage was that the sun could now be observed continuously from two hours before noon to two hours after noon, not just for an hour at sun rise as with the sea-cliff interferometer. A third antenna was added at a different separation to remove ambiguities in the position determination.

In 1951, John Bolton considered the design of a radio telescope that would be suitable to follow up on the identification of the discrete radio sources (Bolton Stanley & Slee, 1949). He made a proposal for a Michelson interferometer using two much larger 70-foot dishes on moveable equatorial mounts.[15] Bolton argued that a pair of large dishes would provide flexibility to modify receiving equipment while still having the angular resolution and sensitivity needed for radio source identifications. Pawsey suggested that Bolton make a detailed proposal, but the idea was not pursued further in Australia at that time. However, this concept was to emerge later as the basis for the Owens Valley Radio Observatory (OVRO) interferometer built in the US and operating in 1960. This, in turn, strongly influenced the eventual design of the Jansky Very Large Array in the US, which is the largest aperture synthesis telescope built to this day and is only now being matched by the SKA precursors 40 years later.

Other members of the Australian group led by Pawsey continued to work on solar radio emission. The solar emission was already known to be highly variable in time, so to determine the changing structure, it was necessary to have instantaneous measurement of all the Fourier components. Hence neither the variable spacing Michelson interferometers used in Cambridge nor a single cliff were suitable. In 1951 Christiansen[16] started building the Potts Hill grating array with 32 small steerable parabolic dishes so he could simultaneously measure many Fourier components (Fig. 37.9). From this point the paths taken by the Australian group started to diverge from what was happening in Cambridge.

Christiansen had added a north-south array at right angles to the east-west array (Fig. 37.10), enabling the determination of positions of the emission regions in both coordinates. This arrangement gave Bernie Mills[17] the idea of a cross with outputs from each linear array multiplied to form a single narrow beam instead of two orthogonal fan beams. He built the famous Mills Cross, which used analogue beam formation to make images without any need for computing Fourier transforms. An image can be built up by scanning the beams with analogue phase gradients applied along each arm. Mills was completely confident that his idea would work, but Bowen still forced him to first build a small 36 m prototype at Potts Hill. This was a success, and he detected the continuum radio emission from our closest galaxy, the Large Magellanic Cloud, for the first time. He then proceeded to build a full-scale version, completed in 1953 at Fleurs near Sydney—now the site of the in-development Badgerys Creek International Airport. The Mills Cross concept was a huge technical success, providing high resolution and high sensitivity radio images with no need to compute the Fourier transforms. The concept was quickly adopted around the world, with Mills Cross-type telescopes built in the US (Carnegie Institute DTM—Washington), Italy (CNR—Bologna) and Russia (Puschino). In Australia a very low frequency version was built (19.7 MHz, Shain Cross) and in 1953 a second-generation grating array, the Chris Cross, was also built at Fleurs. A much bigger cross, the "Super Cross" (which was competing for funding in Australia

---

[15]CSIRO Division of Radiophysics, Minutes of the Meeting of the Radio Astronomy Sub-Committee on Galactic Work, 12 Feb 1951.

[16]For more information see *Four Pillars of Radio Astronomy* (Frater et al., 2017).

[17]*Ibid.*

Fig. 37.9 (Chris) Christiansen and his 32-element solar array at Potts Hill Reservoir 1953. Credit: CSIRO Radio Astronomy Image Archive 2976-1

Fig. 37.10 Aerial view of the Potts Hill grating array in 1954 showing the N-S and E-W arms. NS arm is in the foreground (16 dishes) and the EW arm in the centre has 32 dishes. Credit: CSIRO Radio Astronomy Image Archive 3475-4

with the Giant Radio Telescope—GRT, Chaps. 27, 31) was finally built at Molongolo in 1965. However, the analogue beam forming cross type telescopes eventually became a dead end. As the flexibility of the aperture synthesis telescopes combined with the power of the new digital computers, their ability to image individual objects at high sensitivity and the ease with which they could go to higher frequencies prevailed. Some of the crosses were later converted to synthesis telescopes, including the Chris Cross, which became the Fleurs Synthesis Telescope (FST) in 1973, and the Molongolo Cross, which became the Molongolo Observatory Synthesis Telescope (MOST) in 1978.

Fourier Synthesis

In 1949, Ron Bracewell returned to Australia from Cambridge, where he had done his PhD with Ratcliffe. At this time, Bracewell started applying mathematical rigor to Fourier synthesis ideas. The radio astronomers' understanding of the principles of Fourier synthesis imaging was developing rapidly. Bracewell and Roberts (1954) published the seminal paper "Aerial Smoothing", clearly laying out the underlying principles of aperture synthesis for the first time. This paper introduced the concept of invisible distributions, which are structures that cannot be recovered if only a limited number of Fourier components are measured. They also defined the principal solution that is obtained by setting all un-measured Fourier components to zero. The process of image generation (principal solution, or dirty images, in current terminology) was now separated from the deconvolution problem, which estimates the un-measured Fourier components. This separation was critical for the future development of synthesis imaging algorithms. Pawsey had asked Jim Roberts to write a paper with Ron Bracewell after Bracewell gave a colloquium on this topic (Bracewell & Roberts, 1954). In Cambridge, at almost exactly the same time, Peter Scheuer included "Theory of interferometer methods", as Chap. 5 in his PhD thesis (1954). The chapter contained a full analysis of Fourier synthesis, including what Scheuer called 'indeterminate structures', which are the same as the invisible distributions of Bracewell and Roberts. Scheuer also included multi-frequency synthesis and minimum-redundancy arrays, which are now two important concepts in modern aperture synthesis techniques. However, Scheuer's thesis Chap. 5 was never published. Scheuer (1984) later noted that, "Martin Ryle took the severe line, that on engineering topics you shouldn't write mere theory, you should jolly well build the thing first." When Bracewell was at the Cavendish, Scheuer was still an undergraduate and Bracewell had left for Sydney in 1949 before Scheuer started his PhD in Ryle's group. Hence these parallel developments in Sydney and Cambridge must have been completely independent, although both groups acknowledge the influence of Ratcliffe's unpublished lectures on the Fourier Transform. In 1961 Roger Jennison from the University of Manchester published Ratcliffe's lecture course on Fourier Transforms (Jennison, 1961). In the preface Jennison notes, "I am most fortunate to acquire in 1951 a set of notes compiled by the late Dr. I C

Browne from a series of lectures given by Mr. J A Ratcliffe of the Cavendish Laboratory. I most gratefully acknowledge my debt to Mr. Ratcliffe for this valuable introduction to the subject." In the preface to Bracewell's book, *The Fourier Transform and its Applications* (Bracewell 1965), Bracewell noted:

> My interest in the subject was fired when I was studying from Carslaw's *Fourier Series and Integrals* at the University of Sydney in 1939 . . . in solving these problems I benefited from the physical approach to the Fourier transformation that I learned from J A Ratcliffe at the Cavendish Laboratory, Cambridge.

We note that Pawsey was already familiar with Fourier techniques before he went to Cambridge as he included Fourier analysis in his MSc thesis from the University of Melbourne in 1931 (Chap. 6).

The First Earth-Rotation 2D Aperture Synthesis Image

In 1953, Christiansen completed the construction of the Potts Hill east-west array of 32 6-foot dishes with a 700-foot baseline along the side of a Sydney water reservoir (Fig. 37.10; see Chap. 25.)[18]

This array of equally-spaced dishes is analogous to a diffraction grating with a 3-arc min fan beam and grating responses separated by 1.7 deg.; one beam at a time was centred on the disk of the sun. This fan beam integrated the 2D solar image along a line which rotated on the sun as the earth rotated during the day. Christiansen noted that, "The way in which a 2D radio brightness distribution may be derived from a number of 1D scans is not obvious. However, rather similar 2D problems have arisen in crystallography and solutions for these problems, using methods of Fourier synthesis have been found." (Christiansen & Warburton, 1955). Following the crystallographers, Christiansen calculated the 1D Fourier transform of each scan, filled in the radial slice at the appropriate position angle in the aperture plane, and then drew contours of visibility amplitude. The contour mapping avoided the over-weighting of the short spacings which would occur if the visibilities were just added into the 2D aperture.[19] Christiansen then calculated the 2D Fourier transform of the visibilities read off the 2D contour map of visibility amplitude. After doing tests, his group assumed symmetry for each cut, which only required reflection symmetry and not the assumption of circular symmetry for the sun that had been made by Stanier (1950) and Machin (1951) at Cambridge. When the sun was active, as was observed on many other occasions, bright asymmetrically positioned sunspots were observed, and the symmetry assumption would then have been incorrect. The enhanced emission was ignored. Govind Swarup, a Columbo Plan Fellow from India who

---

[18] See also NRAO ONLINE.23.

[19] This is an interesting technical problem related to the correction of the weights in back projections which causes the "fog" in medical images made using analogue back projection summation (see earlier section: X-ray imaging in medicine, and Webb, 1990).

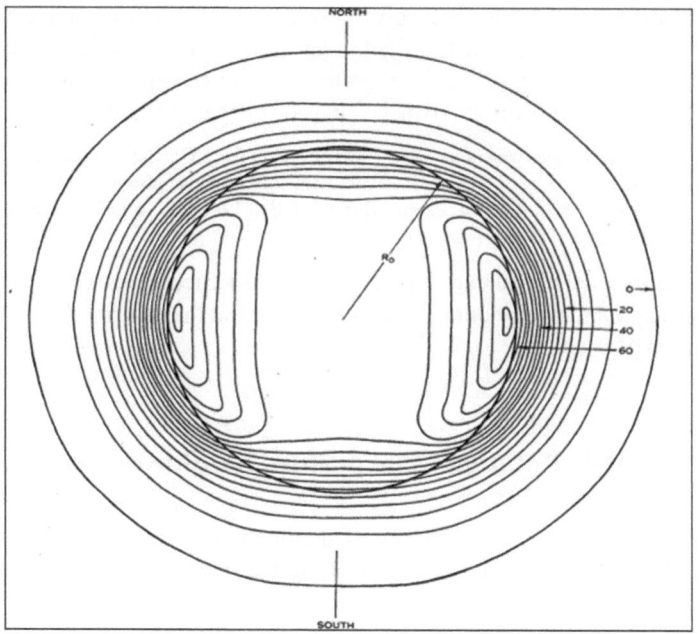

Fig. 37.11 First earth rotation aperture synthesis image of the sun at 21 cm. Credit: Fig. 37.10, "The Distribution of Radio Brightness over the Solar Disk at a Wavelength of 21 Centimetres. III. The Quiet Sun? Two-Dimensional Observations", Christiansen, W. N., and Warburton, J. A. (1955). *Australian Journal of Physics* 8, no. 4: 474–486

was then working in Christiansen's group, calculated the Fourier transforms, taking more than a month with an electric calculator and using Lipson-Beevers strips.[20] The result was the first earth rotation 2D aperture synthesis image, an image of the quiet sun at 21 cm (Fig. 37.11).

The enhanced brightening of the limb in the equatorial direction was observed for the first time (Christiansen & Warburton, 1955). Christiansen and Warburton referenced O'Brien (1953) who had first used the earth rotation in Cambridge to create a 2D image; this image did not have enough angular resolution to detect the limb brightening. (The active sun cannot be imaged in 2D using this earth rotation method because the structure changes during the observation.) Ten years would pass before Paul Wild (1967) made the first 2D images of an evolving solar burst using the Culgoora Radioheliograph.

Chris Christiansen had made the first 2D aperture synthesis image. More than 30 years later (Christiansen, 1989), he wrote:

---

[20] Govind Swarup, who later became head of radio astronomy in India, moved to Stanford to work with Ron Bracewell after Ron left Australia. See NRAO ONLINE.32 and 33 for more details of Swarup's contribution.

Fig. 37.12 Peter Scheuer recreating the use of Lipson-Beevers strips from the 1950s. Credit: Woody Sullivan, 1988

> The further development of this technique required a faster way to make the Fourier Transforms, and this discouraged use of this technique in Australia. The turning point took place in Cambridge when digital computers revolutionised the speed of computation. The importance of the Cambridge work is well known, and perhaps this is why the more humble first use of the technique in a distant land tends to be ignored (see chapter epigraph).

Christiansen's comments are accurate on both counts: the pioneering work in Australia is largely ignored, and his first image was also the last aperture synthesis image computed in Australia for nearly two decades!

The Role of Electronic Computers: Calculating the Fourier Transforms

The technology used for the calculation of the Fourier transforms has played a critical role throughout the development of aperture synthesis. We will now go back and see how these methods have evolved so dramatically and how they influenced the radio astronomy developments.

In the 1940s Lipson-Beevers strips were the preferred method for calculating Fourier transforms, a technique which had been developed by the crystallographers (see Beevers and Lipson's review, 1985). A 2D Fourier transform of a 25 × 25 array with two-digit accuracy could be calculated by one person in 24 hours (Fig. 37.12).

By the late 1940s, punched card tabulators (Fig. 37.13) were providing an alternate option, and these were also used for the Michelson interferometers in Cambridge. Pearcey (1953) from CSIRO, Sydney, describes the method of calculating 2D Fourier transforms and estimates that a state-of-the-art punched card sorter and tabulator could do a 2D Fourier transform of a 25 × 25 array to three digits in 14 hours. While this was more accurate, it was a more expensive operation requiring four operators!

Fig. 37.13 1957 IBM082 Hollerith punched card sorting machine From https://commons.wikimedia. org/wiki/File:Punch_card_ sorter.JPG. Credit: waelder, CC BY-SA 3.0 <http:// creativecommons.org/ licenses/by-sa/3.0/>, via Wikimedia Commons

Fig. 37.14 1962 Ann Neville feeding paper tapes from Cambridge observations for the North Pole Survey into EDSAC II (CavMag, issue 11, p. 11). Courtesy of and copyright: Cavendish Laboratory, University of Cambridge. All rights reserved

The role of computers and signal processing was now changing dramatically. This was to have a huge impact on the development of aperture synthesis in the UK. By 1949 the EDSAC (Electronic Delay Storage Automatic Calculator) I electronic digital computer (Fig. 37.14) had been programmed by Wilkes at Cambridge to do a 1D Fourier transform. John Blythe, a student in Ryle's group at Cambridge, used EDSAC I, and this was the first use of an electronic computer for aperture synthesis in radio astronomy. He had built an east-west line of dipoles with an additional single moveable dipole which he moved step by step in the north-south direction. Blythe used this array to observe the galactic plane at 38 MHz. The interferometer outputs (Fourier components) were transformed into an image using EDSAC I. The 360 38-point 1D transforms took 15 hours (Blythe, 1957). The output of the computations were the locations of the image contours, not the intensities of the pixels in the image! We might now find this quite unusual, but in 1957 there would have been no way to either store or list all the pixel intensities for the more than 20,000 pixels needed for a fully sampled image (Blythe had mapped half the celestial sphere at 2.2 degrees angular resolution). Only the contours would be published- the final results of the computations. There was no concept of a digital image at this time. Blythe's telescope synthesised the equivalent of a Mills Cross but had substituted an electronic computer for analogue image formation. As surmised by Scheuer (1984, p. 251), Ryle realised that using the digital computer and moveable elements would be a less formidable undertaking than building a beam forming array.

In 1958 the much more powerful EDSAC II computer (Fig. 37.14) was completed and applied to Fourier inversion problems using an implementation of the Fast Fourier Transform (FFT) by Wheeler at the Cavendish Mathematical Laboratory. The FFT had been known since Gauss (1805) and was reinvented many times but only came into common usage after Cooley and Tukey (1965) published a convenient implementation of the algorithm. The importance of this step was recognised in Ryle's Nobel lecture (Hargrave & Ryle, 1974). From this time the calculation of the Fourier transforms was no longer a dominant issue in aperture synthesis imaging in radio astronomy.

The fact that Australia did not take advantage of the electronic digital computer held back progress in Australia, while Cambridge went on to exploit the aperture synthesis technique to its fullest. It is particularly interesting to note that Australia's first digital electronic computer, which was designed by Trevor Pearcey at CSIR in 1948, was operating between 1951 and 1955. This computer was physically located in the same building as the radio astronomy group at the time that Christiansen and his colleagues were calculating Fourier transforms by hand. When questioned later, radio astronomers who were present at the time said that it never occurred to them that the arcane art of machine language programming and punching holes in paper tape would have any relevance for the radio telescopes.[21]

Even though the Australian computer (later called CSIRAC) was of comparable performance to EDSAC I in Cambridge (McCann & Thorne, 2000), there was an important difference in the way it was run. In Trevor Pearcey's CSIR department in Sydney, the computer was part of an electronic computer research project and was in a state of continual development. In Cambridge, the EDSAC was operating as part of the Cavendish Mathematical Laboratory with mathematicians, including David Wheeler (Campbell-Kelly, 2006), who were implementing key algorithms on the new digital computers. In his autobiography Mills (2006) summarised his view of the situation as follows: "The only way to perform Fourier transforms in a reasonable time was to use the Radiophysics computer (CSIRAC), which was still under development and hardly to be considered as an essential part of an observing program." He was also concerned about the length of time required for an aperture synthesis observation and admitted that Fourier synthesis had no appeal for him and that he did not consider it seriously. Bracewell (in private communication) said it was too big a jump in tradition for the Australian group who were accustomed to

---

[21] Ron Ekers interviewing Ron Bracewell and Jim Roberts in 1980.

Fig. 37.15 Pawsey Supercomputer Centre at Kensington, Western Australia. Credit: Pawsey Supercomputing Centre

building and experimenting with aerials rather than coding computers in machine language. Clearly it was the influence of the X-ray crystallographers and the mathematicians who implemented the Fourier transform algorithms at the Cavendish that had made the jump possible there. Writing a reference for Pearcey on 10 June 1955, Pawsey said: "The [CSIR] machine was in due course completed and, because this Division does not itself have an adequate requirement for such a machine, it is being sent to Melbourne to continue work."[22] The development of electronic computing in CSIRO was discontinued in 1955; the computer was dismantled and transferred to the University of Melbourne (Willis and Deane, 2006). This CSIRO decision was supported by Bowen and Pawsey in order to concentrate resources on radio astronomy and cloud-physics (rain making). However, in 1957, CSIRAC was used by Pawsey (1958) to make one of the first simulations of the effect of confusion in radio surveys (Chap. 35).

In a complete reversal of the 1950s situation, Australia's largest super computer complex (Fig. 37.15) has now been named the Pawsey Supercomputing Centre, although neither Joe Pawsey nor anyone else in his radio astronomy group had anticipated the impact that the digital computer would have. The Pawsey Supercomputing Centre computers are now essential for processing the prodigious data output from the new generation of aperture synthesis radio telescopes—the MWA and ASKAP, which are SKA precursors.[23]

Impact of Computers on Medical Imaging

Developments in medical imaging were happening in the same time frame but were taking a quite different path. Initially the large 2D X-ray images on photographic plates made any computed imaging technique totally impractical. Only analogue

---

[22] 10 June 1955, letter of recommendation to the Australian Atomic Energy Commission who were thinking of hiring T. Pearcey, NAA C3830 Z3/1/VI.

[23] https://www.pawsey.org.au/research/the-square-kilometre-array/.

solutions to determine the 3D structures were used until 1972. These analogue procedures had no way to adjust the weights of the Fourier components in the image. Thus, the over-weighted low spatial frequencies resulted in the problem of the "fog" in the images (see Webb, 1990).

The medical imaging community had a very different culture from radio astronomy and X-ray crystallography. Mathematical treatment was not part of the medical culture and there was a justifiably conservative attitude on the part of physicians and instrument manufacturers. The Radon transform, which does the correct inversion, had been known since 1917 but was not applied to medical imaging until 1963.[24]

At Stanford in the mid-1960s, discussions between Swarup and Bracewell[25] led to the idea of applying a convolution to the projected distribution which would correct the radial weighting error in back projection imaging (Bracewell and Riddle, 1967, see above in this chapter). This method had limited application in radio astronomy but made a significant impact on medical imaging as it was more practical than previous approaches.

In 1979 the Nobel Prize in medicine was awarded to Cormack and Hounsfield for the discovery of Computer Assisted Tomography (CAT). Bracewell and Cormack were both students at Cambridge. We have no indication that they had collaborated. Cormack had found a mathematical solution in 1963 based on the Radon transform. In 1972 Hounsfield solved the formidable technical problems and made the first image. At the time, Hounsfield was not aware of the work that Cormack had done on the theoretical mathematics for such a device.[26]

Aperture Synthesis Developments at Cambridge

The invention of the phase switch at Cambridge in 1952 (see earlier section and Chaps. 16 and 18) had a major influence on future research directions there. With both the strong DC component from the receivers and the diffuse galactic background removed, the much fainter signal from the discrete radio sources could be observed. Without the phase switch it was only possible to observe intense radio sources such as the sun. These discrete radio sources were not strongly variable like the sun, so it was much more practical to make observations which built up Fourier components over a period of time, by moving aerials to different separations. The Cambridge group would have become very frustrated making solar observations which were so often spoiled by changing solar activity and had to be discarded. By the late 1950s they ceased observing the sun. They left it to the Australians, who had

---

[24] Ambartzumian, an Armenian astrophysicist, had already solved an identical problem in 1936 when he deduced the 3D velocity distribution in a star cluster from the observed radial velocities (Ambartzumian, 1936) but this connection was not recognised until much later.

[25] Private communication from Swarup.

[26] https://en.wikipedia.org/wiki/Godfrey_Hounsfield.

invested more resources in solar observations. During the following years, the CSIRO continued to dominate the field with little competition. Denisse and colleagues did initiate a major project to observe the sun at 169 MHz at Nançay in France (starting in 1953) with frequent collaborations with the Australians.

Thus, the Cambridge group concentrated on the fainter "radio stars". This situation was summarised in Ryle's 1957 paper on the proposed move to the Mullard Radio Observatory (Ryle, 1957) where he described the new types of radio telescopes for the new site. A key decision was made at this time which affected all future research at the Cavendish Laboratory. They would not continue the solar observations but focus on the study of galactic radio emission and the discrete sources, which they still referred to as *radio stars*. This decision meant that the instruments must have high angular resolution and high sensitivity to detect radio sources which are orders of magnitude weaker than the sun. But since these signals are not time variable, the large aperture required could be synthesized by combining Fourier components measured sequentially by moving the array elements. Ryle argued that such an instrument would have important advantages over the Mills Cross type of telescope, primarily because "scanning the beam" is accomplished by computation with exactly the same input data. By comparison the Mills Cross would have to form separate beams in real time by applying an appropriate phase gradient along the array for each beam. Ryle noted that this method of observing was only possible because the new EDSAC II computer was sufficiently powerful to provide the mathematical computation needed, as had been recently demonstrated by Blythe (1957).

Ryle and Hewish (1960) published the key Cambridge paper describing aperture synthesis, "The Synthesis of Large Radio Telescopes". Surprisingly, there are no references to the original McCready, Pawsey and Payne-Scott paper. Ryle and Hewish did, however, describe the Mills Cross concept as a less practical and more complex system.

When we try to trace the first recognition of the possibility of using the earth's rotation to synthesise the aperture, the picture becomes rather murky. Scheuer (1984) recognised this tangled path when he described the development of aperture synthesis at Cambridge. He did not include the parallel developments taking place in Australia. Jan Högbom, who was a PhD student at the Cavendish Laboratory, recalled that in 1958 he realised that the earth's rotation could be used to change the interferometer baselines. He claimed that he described earth rotation synthesis to Ryle, but Ryle never responded (Högbom, 2003). Högbom had run the calculations on EDSAC II to confirm that this would work, but he did not consider earth rotation synthesis to be a very useful technique because he had not considered the possibility of using steerable antennas to track the sources across the sky as the earth rotates. Högbom later realized that Ryle already understood the principle and probably had been keeping it to himself. In retrospect, Scheuer (1984) considered that Högbom should have known that O'Brien had already used this method five years earlier. But, presumably due to the closed culture then prevalent in the group,[27] Högbom had not been aware of this. At this point in our reconstruction of past events using people's hazy recollections, we note that oral history is confounded, not only by changes in memory, but also by memory's partiality. That is, what is reported can only be the personal (and partial) view of the individual. We will continue this analysis in the discussion section at the end of Chap. 37.

Christiansen and Högbom (1969) wrote the definitive text on the design of imaging radio telescopes in *Radio Telescopes*. A new word, "supersynthesis", had been introduced by the Cambridge group to describe the combination of both movable aerials and earth rotation. It is defined on page 183 of Christiansen and Högbom first edition (1969): "This double type of synthesis has sometimes been called *supersynthesis*". However, the word went out of common use after a few years; in the second edition of Christiansen and Högbom in 1985, the term had disappeared. The word was used in discussions at Cambridge, and occasionally used in the literature, but never formally used in any of the Cambridge publications. With the advent of 2D arrays and the need to incorporate the effect of earth rotation in any aperture synthesis observation of finite duration, there was no need to distinguish between the two types of synthesis. In the present era, *supersynthesis* has disappeared completely from the aperture synthesis lexicon.

In the 1960s, Jan Högbom went on to introduce the earth rotation synthesis concept to the Dutch and persuaded them to modify the design of the beam forming Benelux cross as this instrument evolved into the Westerbork 1D earth rotation synthesis array. However, Högbom was more prominently recognised for his later invention of the first practical deconvolution algorithm in radio astronomy, CLEAN: image artefacts (sidelobes in radio astronomy terminology) were removed by the estimation of the values of the missing Fourier components (Högbom, 1974).

The first Cambridge earth rotation synthesis image was published by Ryle and Neville (1962). Observations were taken in June 1961 using 4C aerials at 178 MHz and 2D Fourier transforms were computed using EDSAC II (see co-author Ann Neville feeding in paper tapes in Fig. 37.14) They imaged an 8-degree by 8-degree area around the North Pole. By choosing the North Pole, which does not move in the sky as the earth rotates, they were able to use the 4C aerials, which could only observe in a fixed direction. To observe any other region using earth rotation would require mechanically more complex aerials which could track a region of sky rising in the east and then setting in the west. This observation was made seven years after Christiansen had made his earth rotation aperture synthesis image of the sun; the Ryle and Neville observation is the first earth rotation synthesis image of any region

---

[27] Quote from Harry van der Laan (Ryle's PhD student) re the scientific environment at the Cavendish: "We have an outsider visiting (Mort Roberts)—if he asks an intelligent question you must answer intelligently and honestly, but don't tell him anything he hasn't asked about. They have so much more money that we can't compete." Papers of Woodruff T. Sullivan III, "Interview with Kenneth I. Kellermann (with Harry van der Laan)," NRAO Archives, accessed August 6, 2021. https://www.nrao.edu/archives/items/show/14994. Interview 19 March 1975 credit NRAO/AUI/NSF.

Fig. 37.16 Ryle & Neville (1962) North Pole synthesis image at 178 MHz. Credit: Fig. 37.6 from "A radio survey of the North Polar region with a 4.5 minute of arc pencil-beam system", Ryle, M., and Neville, A. C., *Monthly Notices of the Royal Astronomical Society* 125, no. 1

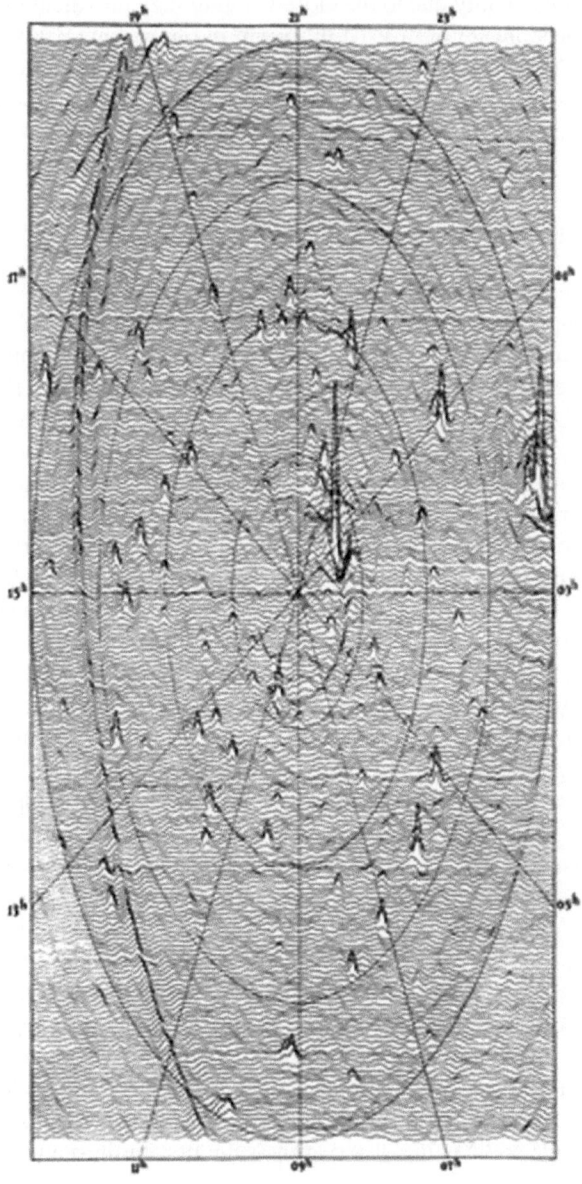

of the sky other than the sun (Fig. 37.16). This is an impressive image, one of the largest single steps in sensitivity and resolution ever made in radio astronomy.[28]

---

[28]This image strongly influenced Jan Oort in the Netherlands and triggered the redesign of the beam forming Benelux Cross (private communication to Ekers). The Benelux cross later became the

Elizabeth Waldram provides an interesting example of the close links between the Cambridge radio astronomers and the developments in X-ray crystallography. She used EDSAC II to generate the very impressive ruled surface display used in this paper. This was the first realistic 2D radio image, and it made a much greater impact than the conventional contours which became unwieldy for an image of this complexity. Waldram had been transferred to Ryle's group from the X-ray crystallography group at the Cavendish Laboratory in 1960 because of anxiety related to her exposure to X-ray radiation. She was already very familiar with the use of the electronic computer and its software when she joined Ryle's group.[29]

The Cambridge Earth Rotation Aperture Synthesis Telescopes

In 1962 Ryle published the design of the One Mile Telescope (frequencies 408 and 1407 MHz). This publication included many aperture synthesis ideas that Ryle had been considering for the previous decade. In Graham Smith's *Biographical Memoirs of Martin Ryle*, page 508 (1986), he noted that Ryle was worried:

> that their simple and elegant techniques could easily be adopted elsewhere, particularly in the USA where much larger resources could be assembled, and the next set of observation pre-empted. The result was a strict policy: nothing was said about new results until the occasion arose, or until a paper was accepted.

The final design included large steerable dishes that could be moved along an east-west track, and these could track a source as the earth rotated, measuring all the Fourier components along concentric ellipses in the aperture plane. As described above, earth rotation combined with moveable elements was referred to as "supersynthesis" at the time. The array, consisting of three 60-foot dishes and a one-mile east-west track, was completed in 1964 (Fig. 37.17). Ryle made no reference in this paper to the earlier Australian work on the theory of aperture synthesis!

Construction of the Cambridge 5-km Radio Telescope was started in 1967 and completed in 1971 (Fig. 37.18). This telescope had four moveable and four fixed antennas forming 16 simultaneous non-redundant interferometer spacings (frequency 15 GHz). The 5-km array achieved one-arcsecond resolution which was the first time radio astronomers had produced images of the sky comparable to optical resolution. The Cambridge group used back projection[30] to compute the 2D images in real time, and the output from the array was at this time the digital images.

Westerbork (Aperture) Synthesis Radio Telescope (WSRT) (Högbom and Brouw, 1974). The initial primary goal was the study of cosmology using the extragalactic radio source population.

[29] Ron Ekers interview with Elizabeth Waldram, Cambridge June 2010, and *CavMag*, Feb 2014, issue 11.

[30] See discussion of "back projection" in the section on X-ray imaging.

Fig. 37.17 The One Mile
Telescope at the Lord's
Bridge Observatory,
Cambridge. P2026.
Courtesy of and copyright:
Cavendish Laboratory,
University of Cambridge.

This telescope also marks the completion of a transition from survey telescopes to telescopes used to image individual objects.[31] The statistics of the distribution of radio sources as a function of position and flux density had finally been settled after the acrimonious disputes between Cambridge and Sydney (see Chaps. 35 and 36). As outlined by Scheuer (1984), it was time to move on and understand the inner workings of the radio sources through detailed observations of individual objects, a vision pursued by Peter Scheuer (and numerous colleagues) throughout his career in Cambridge.

A composite of historical images of Cygnus A (Fig. 37.19) illustrates the dramatic advances which were made in the imaging of individual sources. Jennison and Das Gupta (1953) combined visibility observations made at Cambridge, Jodrell Bank and Sydney and showed, for the first time, two distinct centres of radio emission. The minimum in the visibility amplitude and changing position of the minimum with hour angle could only be modelled with a symmetric double source. A decade later

---

[31]The One Mile telescope was used for both deep surveys (eg the 5C survey) as well as the investigation of many individual radio sources, both galactic and extragalactic.

Fig. 37.18 The 5 km telescope at the Lord's Bridge Observatory. Cambridge. P2027. Courtesy of and copyright: Cavendish Laboratory, University of Cambridge.

Ryle et al. (1965) published an aperture synthesis image made with the Cambridge one-mile aperture synthesis telescopes at 1.4 GHz (Fig. 37.17) confirming the elongated double structure. Then Hargrave and Ryle (1974) obtained higher angular resolution aperture synthesis observations with the new Cambridge 5-km telescope at 5 GHz (Fig. 37.18). The nuclear emission was detected as well at intense hot spots at the edges of the lobes.

Figure 37.20 is a modern image of Cygnus A obtained with the VLA by L. Sebokolodi,R. Perley and O. Smirnov in preparation (2021). The VLA images show the jet powering the hot spots and the two lobes from the central black hole.

Nobel Prize

In 1974 the Nobel Prize was jointly awarded to Sir Martin Ryle "for his observations and inventions and, in particular, of the aperture synthesis technique"; and to Tony Hewish "for his decisive role in the discovery of pulsars". The award presentation for Ryle included: "The radio-astronomical instruments invented and developed by Martin Ryle, and utilised so successfully by him and his collaborators in their observations, have been one of the most important elements of the latest discoveries in Astrophysics."

While the Nobel committee may have overlooked the role of the Australian group in the invention of aperture synthesis, Ryle and his Cambridge group had clearly made much more effective use of the aperture synthesis technique for the study of extra-solar radio sources.In addition, their research had a major impact in the

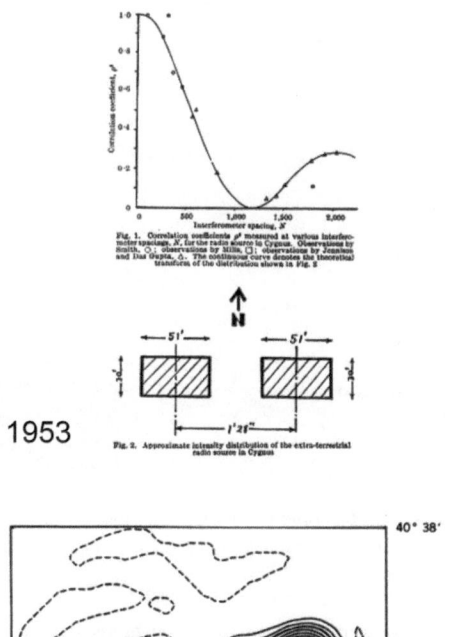

1953

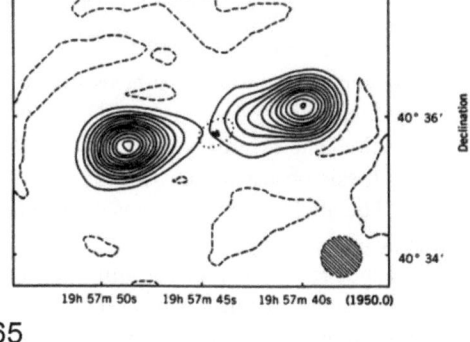

1965

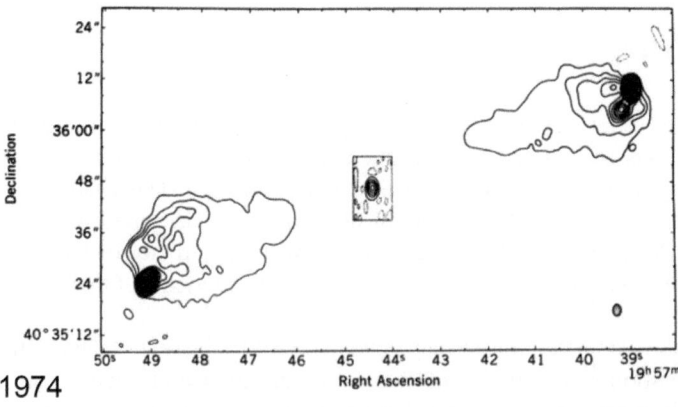

1974

Fig. 37.19 History of Cygnus A Imaging: Top; visibilities and model (Credit: Figs. 37.1 and 37.3, "Fine structure of the extra-terrestrial radio source Cygnus I", Jennison, R. C., and Das Gupta, M.K. (1953). *Nature* 172, no. 4387: 996–997.) Middle: Aperture synthesis observations with the Cambridge one-mile telescopes at 1.4 GHz. (Credit: Fig. 37.1, "High-Resolution Observations of

understanding of the astrophysics of these objects. Ryle's Nobel Prize lecture was very Cambridge-centric and made almost no mention of the prior Australian work. His lecture did, however, strongly emphasise the important role of the electronic computer in making aperture synthesis a practical method. This step, more than anything else, resulted in the much greater impact of the Cambridge group.

Further Developments of Aperture Synthesis Telescopes in the US

When the US started discussing a National Radio Astronomy Observatory (NRAO) in 1954, Robert H. Dicke wrote a memo to the ad hoc AUI committee on radio astronomy in which he outlined a proposal for a synthesis radio telescope for Green Bank, West Virginia.[32] His concept (Fig. 37.21) was based on summation of interferometer responses and was completely independent of the work being done in Cambridge and in Sydney at that time. However, the NRAO was advised by a committee to build a 140-foot equatorially mounted single dish. Thus, the US may have lost an early opportunity to become a world leader in aperture synthesis radio astronomy!

In 1955 John Bolton from Sydney was asked to start a new radio astronomy group at Caltech in the US (see Chap. 27). Bolton brought in his Australian colleague Gordon Stanley, and they completed the Owens Valley Radio Observatory (OVRO) two-element interferometer which commenced operating in 1960. Bolton's emphasis was to have a small number of large moveable elements in an interferometer to provide maximum flexibility at a higher frequency. This idea can be traced back to his 1951 proposal to Pawsey's Sydney group to build a Michelson interferometer using two 70-foot dishes on moveable equatorial mounts; the idea was not pursued further in Australia at that time. (See earlier in this chapter "Imaging Arrays at the Sydney Field Stations".) After 25 years, a telescope of this type (The Australia Telescope Compact Array) would be built in Australia, inaugurated during the Australian Bicentenary Year of 1988.

The 1960 Pierce report to the National Science Foundation (Keller, 1961), based on Bolton's proposal, recommended that the US build an array of large dishes. In

← ──

Fig. 37.19 (continued) the Radio Sources in Cygnus and Cassiopeia", Ryle, M., Elsmore, B. and Neville, A.C. (1965). *Nature* 205, pp. 1259–1262.) Bottom: Aperture synthesis observations with the Cambridge 5-km telescope at 5 GHz. (Credit: Fig. 37.3. From "Observations of Cygnus A with the 5-km radio telescope", Hargrave, P. J., and Ryle, M., *Monthly Notices of the Royal Astronomical Society* 166, no. 2.)

─────────────────────────────

[32] Memo from Robert Dicke to the AUI ad hoc committee on radio astronomy, 7 June 1954.

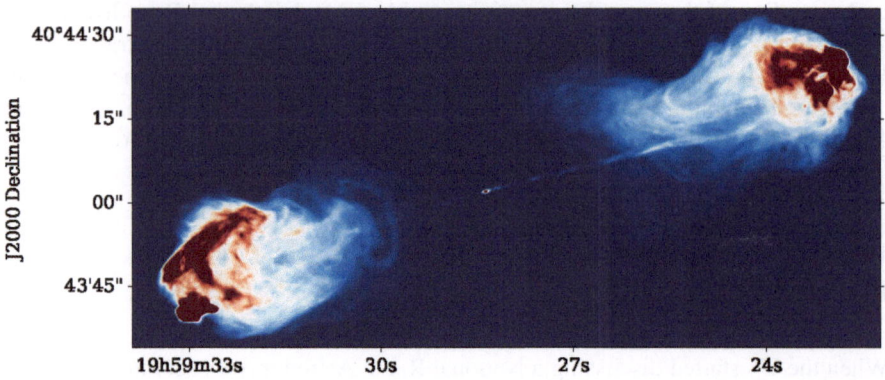

Fig. 37.20   VLA image of Cygnus A Credit: Rick Perley/NRAO/AUI/NSF

Bolton's chapter on Radio Telescopes in *Stars and Stellar Systems* (Kuiper and Middlehurst, 1960), he noted that, *"If experience proves Ryle's aperture synthesis idea successful in practice, this will greatly reduce difficulties encountered with telescopes of the Mills Cross type."*

Joe Pawsey was appointed as NRAO Director in 1961 (see Chaps. 38 and 39) and realised that radio astronomy needed a high angular resolution radio telescope array. He started promoting plans for an imaging array during 1961–1962 but he died before the VLA concept emerged (see Kellerman et al., 2020), a telescope that would fulfill Pawsey's vision well beyond his expectations (Chap. 40).

Discussion

One of the greatest discoveries in radio astronomy in this post WWII period was the technique of aperture synthesis. This ability to make sharp and impressive radio images of exotic objects never previously observed has revolutionised knowledge of the Universe. It was possible to achieve this angular resolution by synthesising a large aperture even though the radio waves are millions of times longer than optical waves. This technology has directly influenced almost all the major radio telescopes built since then: WSRT, VLA, ATCA, MWA, LOFAR, MeerKAT, ASKAP, and one day the SKA.

But just when did this discovery occur and who should get the credit? In this chapter we have traced in some detail the evolution of the technology and the numerous complex and intertwined developments that led to this discovery. We find a multiple step process with many players moving along this path. Joe Pawsey and Ruby Payne-Scott were first to recognise the possibility of measuring and synthesising an image indirectly from its Fourier components after observing the sun with an interferometer formed by a sea-cliff. Martin Ryle led a group of young scientists through an incremental series of experiments using Michelson

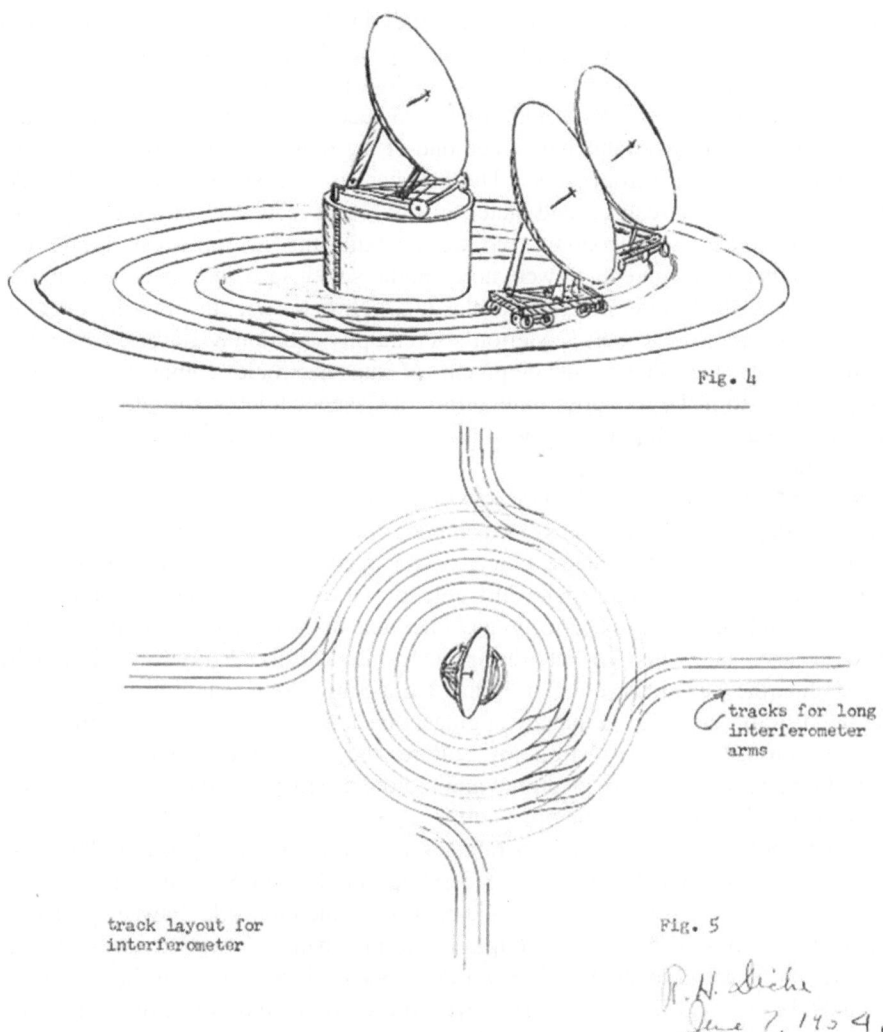

Fig. 37.21 Bob Dicke three dish interferometer (aperture synthesis array) design 1954. Credit: Figs. 37.4 and 37.5 in AUI committee on radioastronomy report, 7 June 1954

interferometers to measure the Fourier components a few at a time. The procedure consisting of moving aerials over the available terrain, allowing the rotation of the earth to change the angle on the sky. Frustrated by the continually fluctuating regions of solar emission, Chris Christiansen built an array of many small dishes. He transformed this process into an image of the sun using laborious hand calculations of the 2D Fourier Transform. It is not possible to clearly credit any one step in this process. As the Nobel committee awarded the Nobel Prize to Martin Ryle in 1974, they not only singled out one of the many individuals involved, but also may well have provided an answer to identifying the key step in this discovery.

The Nobel prize citation states: "The radio-astronomical instruments invented and developed by Martin Ryle and utilised so successfully by him and his collaborators in their observations, have been one of the most important elements of the latest discoveries in astrophysics." The key here is *"utilised so successfully"*. As Martin Ryle clearly realised and stated in his Nobel lecture, "Why then, with its obvious simplicity and economy, did we not build this instrument in 1954? The answer is that at this time there were no computers with sufficient speed and storage capacity to do the Fourier inversion of the data." Now we can view this discovery as a striking example of Robert K Merton's serendipity (Merton & Barber, 2004). If it were not for the chance development of an electronic computer, EDSAC, programmed to make a Fourier transform in a Cavendish laboratory next door, the aperture synthesis technique may well have fallen on barren ground, as was the case in Australia.

In retrospect we can see the key role played by technology along this path. The need to adapt to new and changing technology, especially electronic computers is clear. We also note that these changes can occur on time scales shorter than construction time scales for major instruments. Thus it is essential to have funding and construction models that encourage rather than discourage the need to adapt. We also observe the value of interdisciplinary interactions. X-ray crystallography research at the Cavendish Laboratory had a huge impact on radio imaging.

In Australia the radio frequency engineers became the astronomers. Unimpeded by any strong traditional astronomy culture, they were more innovative: extragalactic radio sources were identified and radio emission from the sun was imaged, classified and understood. The open culture in Sydney, fostered by Joe Pawsey, can be contrasted with the closed and rather secretive environment at the Cavendish under Sir Martin Ryle. Pawsey travelled extensively in the US, Canada and Europe, sharing the ideas developed in Australia openly around the world. Pawsey promoted open discussions between the groups, triggering many innovative ideas. A new generation of world leading, influential telescope designers arose in the US, Netherlands, France and later China. However, these groups did not have such a clear focus as the Cambridge group, who made outstanding contributions to the pioneering studies of extra-solar radio sources.

It is interesting to ask why, when all the early Cambridge papers by Martin Ryle do credit the Australian McCready et al. (1947) paper for the Fourier Synthesis concept, no Sydney work is referenced in any of the papers by Cambridge students or in any publications from Cambridge after 1955? This was a period of rivalry and animosity related to disagreements between Ryle in Cambridge and Mills in Sydney on the interpretation of the radio source counts (see Chaps. 35 and 36). These disagreements were intensified by the arguments between Ryle, who interpreted his counts as evidence for evolution in the universe, and Hoyle who was using the Mills results to discredit the case for evolution. Furthermore Ratcliffe, who had kept in close contact with Pawsey in Australia, left the Cavendish in 1960,[33] breaking the strongest link with Australia. Graham Smith summarised the situation on page 508 of his *Biographical Memoirs of Martin Ryle* (1986):

> We took very little notice of any publications, either in journals or textbooks, and relied on Ryle's insight. We were indeed guilty of underestimating, for example, Bolton's work on identification of four radio sources, and Pawsey, McCready and Payne-Scott's work on Fourier analysis of the brightness distribution across the sun. But we were in the full flood of discovery, and we were self-propelled.

Throughout this development we can see how the different paths taken were influenced by a number of factors: firstly, by the nature of the radio sources being imaged—the sun was bright and time-variable. Thus, many small elements were the best way to build an array. Secondly, the availability and acceptance of electronic computing were key ingredients. Finally and more subtly, the pioneer radio astronomers had to make the conceptual leap required to accept that Fourier components observed at different times could be coherently combined into a synthesised image.

After the Cambridge group changed direction in 1957 as solar work was de-emphasised, they had a greater impact using the aperture synthesis telescope to explore extra-solar sources in the universe. The initial key to understanding these sources clearly still came from the Sydney group with John Bolton's identification of the Crab Nebula and the extragalactic sources (Cen A and M 87). However, the Australian group did not follow-up with high resolution radio observations of these sources until decades later. Yet, the Australians continued to dominate the field of solar radio astronomy, initially by elucidating the detailed evolution of the radio outbursts which led to the first taxonomy of radio events in the sun.[34] This work eventually culminated in Paul Wild's solar radioheliograph built at Culgoora, NSW, Australia in 1967 (Fig. 37.22). The radioheliograph made 2D movies of the time-resolved images of the sun using a new synthesis method which involved the electronic summation of Bessel functions. This enabled Wild's Australian group to image the spatial motions of Type II, III and IV bursts.[35]

It is perhaps ironic that the Australian solar work lead by Pawsey had a far greater global impact than the much more specialised and internally focussed work at Cambridge. On the other hand, this Cambridge research, empowered by the adoption of the digital computer, was the work justifiably recognised with a Nobel Prize.

---

[33] To become the Director of the Radio and Space Research Station at Datton Park, retiring in 1966.

[34] See NRAO ONLINE.20 and Chaps. 12, 13, 16 and 25.

[35] For a more detailed overview of Paul Wild's accomplishments see his biographical memoirs (Frater & Ekers, 2012) and *Four Pillars of Radio Astronomy* (Frater et al., 2017).

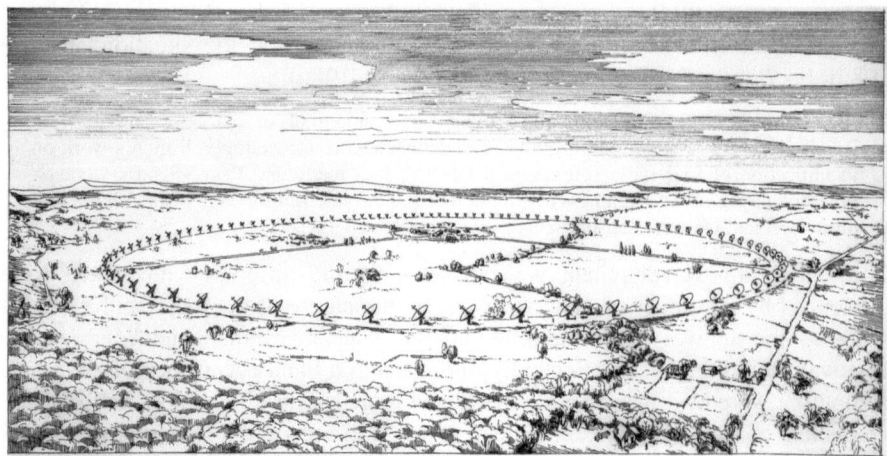

Fig. 37.22 Artist's impression for the Radio Heliograph built at Culgoora near Narrabri in north-west NSW in 1967. Credit: CSIRO Radio Astronomy Image Archive B6636-22A

Part IX
Death and Legacy

Chapter 38
To the US National Radio Astronomy Observatory, 1961

Rabi to Pawsey 31 October 1961:

> I submitted the nomination I have for you to be Director of the National Radio Astronomy Observatory to the Board of AUI at this last meeting (26–27 October 1961). I am happy to say that the Board unanimously empowered me to offer you the post. You will also be pleased to hear that I had a meeting with the Director [Struve] and the principal scientific staff of the Observatory in which I very frankly discussed the whole matter of the directorship of the Observatory and told them of our discussions in London. I questioned each man in turn, and I am very pleased to be able to tell you that the enthusiasm was unanimous. They join me in urging you most strongly to come to Green Bank.[1]

By 1960, radio astronomy was flourishing in the USA. The pace of development had greatly intensified from 1955 to 1960. Radio astronomy had developed in multiple groups spread across the country, a very different pattern from the single Australian group and the two groups in the UK. By 1957 the universities of California (Berkeley), Cornell, Harvard, Ohio State and Stanford all had active radio astronomy programs. At the Carnegie Institute Department of Terrestrial Magnetism (DTM), Franklin and Burke had discovered the intense bursts of radio emission from Jupiter, and the Naval Research Laboratory (NRL) had made detailed studies of the thermal emission from the moon and the planets using their 50-foot dish. Following the detection of the 21 cm hydrogen line by Ewen and Purcell at Harvard, Bart Bok (Harvard astronomy department) had built up a group of astronomers focussed on the interpretation of observations of the 21 cm hydrogen line. As noted by Kellermann et al. (*Open Skies*, 2020, p. 54), the Harvard project was managed by astronomers and not by radio scientists as in Australia and the UK. Many of these Harvard graduates were to become members of the NRAO scientific staff, a very

---

Supplementary Information The online version contains supplementary material available at [https://doi.org/10.1007/978-3-031-07916-0_38].

---

[1] Joe and Lenore Pawsey Family Collection.

W. M. Goss et al., *Joe Pawsey and the Founding of Australian Radio Astronomy*, Historical & Cultural Astronomy, https://doi.org/10.1007/978-3-031-07916-0_38

different team composition than the instrumentally based groups of radio scientists and engineers that dominated the Australian groups.

In addition to having the appetite—and budget—for building big facilities, by 1953 the leading scientific entrepreneurs of the 1950s, including Lloyd Berkner[2] (1905–1967), had begun envisaging how radio astronomy might develop on a national scale: by building an observatory to which all astronomers would have access. Such a national observatory would provide far more scientific capacity than would be possible with resources, whether financial or human, spread over multiple smaller groups. This move to a national big science focus and the ensuing controversies within the US astronomy community is described in *Open Skies,* Chap. 3. In May 1955 the US National Science Board (NSB) issued a policy statement (see Kellermann et al., 2020, p. 79) with a recommendation for government support for large-scale basic research facilities, also including a national astronomical observatory. The concept that this might take the form of a big dish was triggered by Bowen's visits to the US and his promotion of a vision to build an even bigger dish than that planned by Bernard Lovell for the UK. At a 1954 meeting of an advisory committee to discuss options for the new observatory, Robert Dicke (1916–1997)[3] from Princeton University, pointed out that an aperture synthesis radio telescope (see Chap. 37, "Further developments of Aperture Synthesis Telescopes in the US") would be a cheaper and more effective option.[4]

In 1955, the National Science Foundation had agreed to fund Associated Universities Incorporated (AUI)[5] with a grant of $140,500 for the establishment of a radio astronomy facility. By the end of that year, Green Bank, West Virginia, had been chosen as the site. This site was officially opened in late 1957 (with Pawsey in attendance, see Chap. 28). In the summer of 1958, observations began with the 12-foot antenna, while construction started on the 85 and 140-foot dishes. In 1960, as it became increasingly clear that the 140-foot dish would be seriously delayed, a relatively inexpensive 300-foot transit telescope was planned to provide interim observational capability for the new observatory.

Personnel was, of course, the key ingredient for fulfilling the vision of a broad, successful research program in radio astronomy. In July 1959, Otto Struve, an outstanding astrophysicist, but with no radio experience, became the NRAO's first Director. Success required recruiting radio experts globally. At this time, within the US, a common procedure was to attract leading scientists internationally. John

---

[2] Head of the Navy radar section in WWII and became a leading national and international science policy advisor. See Dramatis Personae. Biography by A.A. Needell in 2000, *Science, Cold War and the American State.*

[3] Princeton University professor and inventor of the Dicke switch—see Chap. 16.

[4] This was independent of the work being done in Cambridge and Sydney and predates any published description of aperture synthesis from Cambridge. The significance of Dicke's suggestion was completely missed by the US astronomy community at that time,

[5] At this time, Associated Universities Incorporated was based upon nine eastern US universities and in the late 1950s managed the Brookhaven National Laboratory for high energy particle physics.

Fig. 38.1 16 May 1961 Honorary Degree from the Australian National University. Lenore and J.L. Pawsey. Joe and Lenore Pawsey Family Collection

Bolton had been successfully recruited to Caltech in 1955, and Ron Bracewell had been similarly recruited to establish the solar radio astronomy program at Stanford in the same year. Six years later Bolton left the US and returned to Australia to lead research using the GRT at Parkes.

Among those of interest to NRAO, was the clearly very successful leader of radio astronomy research, Joe Pawsey.

Additional Honours and Questions

Pawsey was at the legacy-building stage of his career. The honours were accumulating. For example, he was awarded the first Matthew Flinders Medal and Lectureship by the Australian Academy of Science in 1957.[6] J.L. Pawsey received an honorary Doctor of Science degree (honoris causa) from the Australian National University (ANU) on 12 May 1961 (see Fig. 38.1). This award had been suggested by R.A. Hohnen, the Registrar of ANU, on 19 December 1960. Pawsey's response on 29 December 1960 included his customary recognition of younger colleagues: "I have been most fortunate in having such excellent scientists in my group—men like Mills, Christiansen, Wild, Shain, Kerr and Bolton—also in the encouragement and material support given us over the years by the Chief of this Division and the Executive of the CSIRO." Likely, Bart Bok, professor at ANU, had initiated the honoris causa degree process.[7]

---

[6] See NRAO ONLINE.51.

[7] NAA C3830 Z3/1/X. Correspondence with ANU.

But Pawsey was also the leader of an increasingly fractured group at RPL amid a swiftly expanding and changing field, with poor and deteriorating relations with Bowen.

The NRAO Directorship

On 16 June 1961, Bart Bok (Director of Mount Stromlo Observatory at ANU) sent a handwritten letter to Pawsey.[8] Bok was visiting Green Bank-NRAO for an extended period; he was well known in AUI and NRAO circles since several of his former PhD students were on the scientific staff (Dave Heeschen, Frank Drake, Kochu Menon and Cam Wade). Bok had served on AUI and NRAO advisory committees starting in July 1954, continuing into 1956.[9] Bok wrote critically of the state of NRAO:

> [In my IAU report of radio astronomy[10]], I have to stay away from saying too much about the difficulties of [Green Bank]. [I will describe these when we meet later in Santa Barbara—at the IAU Symposium and Berkeley—at the IAU in August 1961.] It is an amazing story of mismanagement of the steel construction of the 140-foot reflector—which is really in the soup. Struve seems most unhappy as a Director. On several occasions there have been hints (from the sides of NSF, AUI and the staff here at NRAO) that they wish they knew if you consider yourself at all available. If you can find the time to write me a few lines on the subject I would appreciate this. [Possible salaries were then described.] No one has said anything definite as to whether or not they can (politically) offer the Directorship to a non-American—but I thought it would be well to know if, in principle, you would be interested. I am well impressed with everything here except the progress of the 140-foot. The 85-footer is [working well] and very fine work is rolling off the assembly line. The 300-foot transit telescope for 21 cm work looks good ... The NSF very much want to see the place run right and Otto Struve is frankly not the man for that job ....[11]

Pawsey's immediate response (21 June 1961) to Bok, written by hand at his home[12]:

> Thanks for your letter from Greenbank [sic]. You have certainly given me a teaser to think about. As you know it is a bit difficult to see into the future here but things are going reasonably with the Parkes dish [no longer called the GRT] coming along well. In the near future it would give us a new look at 21 cm [HI] features in the galactic centre and in the not too distant future we can look at the Ryle problem [source counts]. As to the solar, work is going well with Paul Wild's big project looming up if we can get support. We want outside

---

[8] Joe and Lenore Pawsey Family Collection. Written on NRAO letterhead.

[9] NRAO Archives.

[10] Being prepared for the IAU General Assembly, Berkeley, California in August 1961.

[11] Bok also mentioned that he was optimistic that the NSF would provide support for the Mills Cross at Sydney University; this was to occur in early 1962 (NRAO ONLINE.8 "Mills Cross versus Parkes Dish" in *The Bulletin* of 31 March 1962- an Australian news magazine).

[12] The Joe and Lenore Pawsey Family Collection included a carbon copy of the original written with a pencil. This text was initially essentially unreadable. Rob Birtles of CSIRO archive provided a Photoshop rendition after much trial and error; Liz Pawsey, J.L. Pawsey's daughter-in-law, was able to decipher this indistinct text.

help. CSIRO is overwhelmed by one dish. Greenbank looks a difficult one to me. It is a place committed to big paraboloids and with major trouble looking with its major investment. My special interest is in techniques which are trying to beat the big steerable dishes, Paul Wild's solar initiative, the big Mills Cross [often called the Super-Cross by Pawsey], the Illinois cylindrical paraboloid [George Swenson's project]. I have little doubt that what a newer radio astronomy needs is a blending of these. The big dishes are fine but the next step is something different. But you cannot start at a place like Greenbank with a big dish vying for competition. So, though I would be quite interested in trying to start something worthwhile in the States, I don't think I am the Greenbank man. I also want to see Paul get started [on the radio heliograph]. I hope the world is treating you well.

Pawsey Leaves for the US and Europe, 1961

The planned departure for the trip to the US and Europe was 26 July 1961 with a full schedule for the months of August and September. In Europe, Pawsey had made commitments in October, hoping to attend an Organisation for Economic Co-operation and Development OECD sponsored conference on Large Telescopes which had been organised by J.H. Oort from 23 to 24 October 1961 in Paris.[13] In addition, he hoped that it might be possible to extend the visit into November to meet colleagues and visit his daughter and son-in-law (Margaret and Don McLean[14] and their new baby Ann) in Paris. Fred White and the CSIRO Executive, however, set the date of the GRT opening for 31 October 1961. Pawsey was to be presented with the Hughes Medal of the Royal Society of London at that time. White was frustrated. He wrote to Bowen on 26 June 1961: "I think we would all like him [Pawsey] to be there [at the opening]. He cannot expect to have everything his own way. I am sure we must have the opening to take place in Oct . . . I would think it wise of him not to be overseas then. May I write to him?"[15]

On 19 July 1961 (a week before departure from Sydney), White wrote Pawsey after talking to him on the phone about the opening:

> I can quite see that you are anxious to be in attendance at the meeting of the Royal Society [to receive the Hughes Medal]. On the other hand, I think you now know that we cannot delay the opening ceremony into November because of the conflicts with the Elections [Federal Election of 9 December 1961]. I would very much regret, and I am sure you would too now, postponing the opening until next year. It seems to me that MAN has made such remarkable progress that we should announce to the world the fact that the telescope is complete as soon as possible.

Two days later, Pawsey confirmed that he would in fact attend: ". . . there will only be one Opening Ceremony". He could receive the Royal Society Hughes Medal at a

---

[13]Later in the year the meeting was postponed to 12–14 December 1961, well after Pawsey had returned to Sydney on 27 October.

[14]Working on a PhD degree in Paris with Denisse.

[15]NAA C3830 Z1/7/B/2 Part 2.

later date.[16] He did suggest that he hoped that the Parkes event would be delayed as much as possible and would need to avoid the Melbourne Cup (Australia's major thoroughbred horse race, often called "the race that stops a nation"), which would be on 7 November 1961.

On 26 July 1961 (Wednesday), Pawsey, Lenore and her mother, Mabel Nicoll, left Sydney for San Francisco. Lenore and her mother stayed in California with relatives while Pawsey continued to New York. The family would eventually visit Ted Nicoll, Lenore's brother, in Princeton, New Jersey, for an extended visit.

In New York Pawsey visited his friends and colleagues at Bell Labs and met with a number of well-known physicists and engineers: Arthur B. Crawford, Bill Jakes, David C. Hogg (of the "Hogg horn" antenna), J.R. Pierce and Arno Penzias.[17]

Next, he proceeded to Blacksburg, Virginia, to attend the Virginia Polytechnic Institute conference titled "Physics of the Solar System and Re-entry Dynamics", 31 July–8 August 1961.

Numerous experts were present: Grant Athay-solar atmosphere, Kinsey Anderson-solar cosmic ray, Sydney Chapman-van Allen belt properties, Fred Singer-interplanetary content, especially dust, and Ludwig Biermann-state of interplanetary plasma, including comet tails.[18]

During the weekend of 5 August 1961, David P. Stern, a post-doc from Goddard Space Flight Center[19] went with Pawsey and Ludwig Biermann on a trip to see the Navy's Sugar Grove 600-foot paraboloid construction site[20]; the amusing details of this (including some surprising photos of Pawsey "spying" on the closed site after climbing a tree) are described and shown in ESM 38.1, Sugar Grove. The chair of the physics department at VPI had organised that Stern take Pawsey, Biermann and a fellow graduate student, on a weekend excursion west of Blacksburg, Virginia, in the

---

[16]Finally presented by Fred Hoyle to Pawsey on 3 November 1962 in the hospital, four weeks before his death. See Chap. 40.

[17]Arno Penzias, a future Nobel Laureate in Physics, with Bob Wilson, in 1978 for the discovery of the cosmic microwave background. In late August 1961, Penzias had just arrived at Crawford Hill.

[18]Pawsey's notebook includes a single page of calculations of the frequencies of the higher order hydrogen recombination lines. He calculated the frequency of the lines at 20 cm and 90 cm. He also derived the equation for the large radius of the hydrogen atom that would emit in the radio (proportional to the square of the principle quantum number, n). For example, the n = 166 line would emit at 1425 MHz with a radius of 1.5 microns. This line is observed with many radio telescopes in the modern era.

[19]Graduate of Hebrew University and the Israel Institute of Technology. He came to the US in 1959 to the University of Maryland (Fred Singer group), moving to Goddard Space Flight Center in 1962. Stern had a prolific career at NASA. He retired in 2001 having worked on the mapping and physics of the magnetosphere. He also had major interests in science education and the history of science. Goss read a note about him in Pawsey's notebook of 1961. After Goss contacted Stern in October 2015, Stern sent Goss a number of the images reproduced here, complementary to the images found in the Joe and Lenore Pawsey Family Collection. In addition, during the Blacksburg meeting of 1961, Stern wrote letters to his young bride, Audrey, that contained information about Pawsey. Some of these are summarised in ESM 38.1, Sugar Grove. Thank you to David, his wife Audrey and their son Allon Stern for assistance.

[20]See Kellermann et al. (2020, p. 469), for a lively description of the Sugar Grove fiasco.

neighbouring state of West Virginia. They had intended to visit the National Radio Astronomy Observatory at Green Bank, West Virginia (some 75 miles distant), but ran out of time.

Toward the end of the conference, Pawsey explained to Bowen[21] that "the conference got dull" so he left early and went to Washington to visit NSF on 8 August 1961. He met with Geoffrey Keller[22] (Program Director for Astronomy), Randall Robertson (Assistant Director for Mathematics, Physics and Engineering Science) and Gerard F. Mulders (Assistant Program Director for Astronomy).[23] The main discussions were about possible NSF support for the radioheliograph at Culgoora. The NSF colleagues encouraged Pawsey to collaborate with a US group, who could then ask for NSF support for "Paul's [Wild] project".[24]

After the Washington visit, Pawsey flew to Pasadena on the evening of 8 August 1961 to visit the radio group at Caltech, under the leadership of Gordon Stanley. Pawsey and one of the group members, V. Radhakrishnan, talked about the use of maser amplifiers, a new low noise "front end" receiving device that led to a remarkable increase in sensitivity in radio astronomy. In the 11 August 1961 letter to Bowen, Pawsey wrote:

> An important point came up. Rada Krishnan [sic, actually V. Radhakrishnan, a prominent radio astronomer at the end of the twentieth century] has written you that he wants to stay until September-December 1962 at Caltech [before joining RPL].[25] The reason is that Gordon [Stanley] wants him to go to Bell Telephone Lab to take delivery of 2 (repeat 2!) 21 cm TW [traveling wave] masers and put them into service at Bishop [the Caltech Observatory, Owens Valley Radio Observatory]. Gordon would be [quite sad] if he lost Radha now. I think we have to agree but let us cash in on it. The point is that BTL [Bell Telephone Lab] appears to be ready to give masers away. We want one. What [is] the best policy? I have asked Gordon to sound them out ... So reply to Radha in a friendly way and get him working for this in the maser business. I have told him we won't oppose his delay, so write him and Gordon.

On 17 August 1961, Bowen wrote Pawsey with the news that Radhakrishnan could delay by six months (in fact four years). Also, Bowen told Pawsey that he would ask

---

[21] NAA C3830 F1/4/Paw/7.

[22] Keller also attempted to convince Pawsey to consider accepting the Directorship of NRAO, three months before the formal offer of 31 October 1961.

[23] On 13 November 1961 Keller was promoted to the position held previously by Robertson, who became Associate Director for Research. At this time, Mulders became Program Director for Astronomy.

[24] Letter from Pawsey to Bowen from Santa Barbara 11 August 1961. NAA C3830 F1/4/paw/7.

[25] Radhakrishnan had had earlier contact with CSIRO Division of Radiophysics (likely both Bolton and Pawsey) about moving from Caltech to Sydney to work with the GRT. In fact, he was to arrive in Sydney after a sailing voyage across the Atlantic and Pacific Oceans in a small yacht—the Cygnus A—four years later (1964–1965, arrival 24 December 1965). See Goss et al. (2016): *Radio Astronomers at Sea: Martin Ryle and V. Radhakrishnan Correspondence, 1963–1966- the Voyage of the 'Cygnus A' from the UK to Australia.* Contact Goss for a copy of this volume.

Jim Fisk of Bell Telephone Laboratory for a maser for the Parkes telescope[26]: "He is a very old friend of mine who has never refused anything yet." The transaction never occurred.

The Santa Barbara Conference IAU Symposium no 15

After the Caltech visit, Pawsey attended IAU Symposium Number 15, "Problems of Extragalactic Research", 10–12 August 1961 in Santa Barbara, California, (see the volume edited by G.C. McVittie, 1962). Most of the prominent astronomers in the field gave presentations: Bertil Lindblad (spiral structure), Oort (masses of elliptical galaxies), Woltjer (magnetic fields in galaxies), Maarten Schmidt (evolution of galaxies and stars), Rudolf Minkowski (radio sources), George Abell (clusters of galaxies), Henry Palmer (angular diameters of radio sources), Martin Ryle (counts of radio sources), Fritz Zwicky[27] (clusters), Sandage (distance scale of the universe) and Hoyle (Steady-state cosmology). G.C. McVittie (1904–1988) also contributed the conference summary "Galaxies as Members of the Universe".

The Santa Barbara conference exhibited a friendlier atmosphere than had occurred during the heated discussions of the 1958 "Paris Symposium on Radio Astronomy" (see Chap. 36). The young researcher John Baldwin (1931–2010)[28] remarked that the Santa Barbara meeting was far less of a "vituperative storm" than the earlier conference during which the source counts of the Australians (Mills et al) were compared with the 2C counts presented by Ryle and colleagues. In 1961, an example of the calmer approach was shown by Mills in the discussion after the Ryle paper:

> There is a tremendous amount which could be said about this most interesting and provocative paper by Prof Ryle, but since the discussion [must be brief], I would like to make just a few remarks about the number flux density relation. First, it is gratifying to see that the Cambridge *observational* results have now approached the Sydney ones so closely.

Mills followed with praise for the new 4C aerial at Lord's Bridge: "The different philosophy of observation underlying the two surveys may be the cause of difference rather than errors in actual measurements."

---

[26]Fisk (1910–1983) President of Bell Telephone Laboratories 1959 to 1973. Bowen knew him from the Rad Lab at MIT in WWII, where Fisk had worked on high frequency radar research.

[27]Zwicky ("Observations of Importance to Cosmology") provided a comprehensive summary of the missing mass problem in clusters as perceived in 1961. Based on observations and comparison of the total visible light of galaxies and the total mass derived from their velocity dispersion, a large unaccounted mass was implied. Zwicky favoured intergalactic stars or even pygmy galaxies. The recognition of the solution in the form of dark matter would occur over the next several decades.

[28]Interview with Goss, Cambridge, July 2010.

The 11th General Assembly of the IAU 13–24 August 1961, Berkeley, University of California

On 15 August the opening ceremony was held as an open-air event on the Plaza at the University of California in front of Dwinelle Hall. More than 10,000 people attended the event, the highlight and the drawing card being a long address by UN Ambassador Adlai Stevenson (1900–1965). He had been a previous Democratic Candidate for President, who lost to Dwight D. Eisenhower in 1952 and 1956.[29] The ceremony began with numerous short speeches, with Prof Leo Goldberg of Harvard as the master of ceremonies; a telegram of welcome by President John F. Kennedy, followed by an address of welcome by the IAU President, Prof J. H. Oort of Leiden University, first in English, then French. Contrasting accounts of the opening appear in a US publication and in a Soviet publication:

1. "Politics Haunts Astronomers at Berkeley" by Gene Marine in the *Bulletin of Atomic Scientists* Oct 19, 1961, vol 17, no. 8, p. 345.
2. "The Eleventh General Assembly of the International Astronomical Union in Berkeley" in *Soviet Astronomy*, vol. 6, no2 September-October 1962. Translated *Astronomicheskii Zhurnal*, Vol 39, no 2, Page 376, March–April 1962.

The nature of the Stevenson opening address is summarised in ESM 38.2, Controversy IAU, with some details of the "China Controversy".

During the IAU, Pawsey participated in numerous discussions including a meeting concerning the US Project Westford, carried out by MIT Lincoln Laboratory for the US Military. The idea was to create an artificial ionosphere above the earth in order to enhance US military communications. This project consisted of an array of 350 million copper dipole antennas ("needles"—1.8 cm by about 20 microns, resonant at a radio wavelength of about 3.6 cm), placed in earth orbit at 3500 to 3800 km. The 1961 launch failed, while most of the dipoles from the 1963 launch were dispersed by solar radiation pressure. During the IAU in 1961, both optical and radio astronomers were keen to coordinate the impact on astronomical research. Light pollution (due to reflected sunlight) and problems with cm astronomy were of concern. There was nearly unanimous objection to the proposal. With the launch of communications satellites in 1966, the needle system was obsolete. Pawsey attended a session at the IAU in which the 8 GHz properties of the radiation belt were discussed as well as the optical properties (about two 10 mag stars per square degree); the session was led by E.F. McClain of the Naval Research Laboratory.

On 17 August 1961, a number of the IAU guests in Berkeley went to Palo Alto to visit the Stanford radio telescope, organised by Ron Bracewell. See Figs. 38.2 and 38.3.

Friday afternoon, 18 August 1961, Pawsey attended a joint discussion on solar magnetic fields with talks by H.W. Babcock, Alfvén, Leighton, Zirin and others. Of

---

[29]David Halberstam, a prominent US journalist, wrote: "Stevenson's gift to the nation was his language, elegant and well-crafted, thoughtful and calming." (Halberstam, 1994).

Fig. 38.2 At the IAU 1961 in Berkeley, many participants visited the Stanford Radioheliograph during a weekend excursion. Joe and Lenore Pawsey Family Collection

Fig. 38.3 Delegates from the IAU visit the 150-foot telescope at Stanford in August 1961—under construction. Martin Ryle is at the left. Joe and Lenore Pawsey Family Collection

Fig. 38.4 Sunburned astronomers in the Napa Valley vineyard tour—Mills has his head covered with his red handkerchief. IAU 1961 , August. Joe and Lenore Pawsey Family Collection

special interest to Pawsey was the review talk by Tony Hewish, "The Sun's Magnetic Field from Radio Observations". Hewish presented results of radio observations at 38 and 80 MHz of the annual June occultations of the radio source Taurus A (Crab Nebula) by the sun. The presentation described the Cambridge observations from 1953 to 1958. The long series of occultation studies of Taurus A were shown to provide important constraints on the physical conditions in the far outer corona of the sun (Hewish, 1958, and Högbom, 1960).

The weekend was a time to socialise with the other delegates. Joe and Lenore Pawsey attended a dinner at Rudolf and Luise Minkowski's house with fellow guests Martin and Rowena Ryle, Victor Ambartzumian and wife (from Armenia), Madam Alla Massevitch (see NRAO ONLINE.35 of a photo of her taken by Pawsey at the previous IAU of 1958) of the USSR and Thornton Page of Wesleyan University, Middletown, Connecticut. On Sunday a tour of the Napa Valley and a vineyard was characterised, according to Pawsey, by "dinner, wine and sun". The sunburnt astronomers (see Fig. 38.4) were in a jovial mood.

Joint Discussion on Problems Requiring Radio-Astronomical Observations of High Sensitivity and Resolution: Berkeley IAU- Monday 21 August 1961

At short notice, a joint discussion between Commission 28 (Galaxies) and Commission 40 (Radio Astronomy) was organised for Monday, 21 August. The emphasis was on astronomical questions that would be elucidated by high angular resolution at radio wavelengths. Seven papers were presented. Oort presented a summary paper, "Review of Problems Requiring a Resolution of One Minute of Arc", providing a description of the type of science the proposed Benelux Cross (400 MHz) would enable, such as observations of galactic HII regions and, of course, radio cosmology. Three papers were given about HI in nearby galaxies: Frank Kerr discussed the Magellenic Cloud observations from CSIRO in Sydney while Louise Volders described new Dwingeloo (the Netherlands) and Jodrell Bank (the new 250-foot antenna) observations of 25 galaxies. Mort Roberts provided a summary of the new Harvard data obtained with the maser constructed by visitors Brian Cooper (CSIRO) and John Jelly (from the Harwell Nuclear Labs, whose wife was the prominent nuclear physicist Joan Freeman, Pawsey's former radar colleague from the wartime years at RPL), which enabled a number of higher redshift galaxies to be observed. Tom Matthews presented new Caltech OVRO observations "Radio Sources: Identifications, Structures and Diameter of 175 Sources". Minkowski and Ryle presented papers similar in nature to their presentations at the Santa Barbara meeting of two weeks earlier. In the discussion Mills described the improved state of relations between Sydney and Cambridge: "We are happy Cambridge agrees with the Sydney observations. But we still differ over interpretation." Minkowski provided a poignant description of the problem: "Lack of distance measures hinders progress. At present we must first make an optical identification, then estimate the distance by redshift." A luminosity function based on 45 identified sources could be derived which was in agreement with the source counts obtained by the Sydney group (Mills and colleagues).[30]

On the last afternoon of the IAU, Thursday 24 August (after the closing session had ended at noon), the Western Development Laboratory of Philco (an electronics Company in Palo Alto—the contractors for the University of California 85-foot antenna at Hat Creek, California) sponsored a special symposium, "Future Trends in Radio Telescopes of Very High [angular] Resolution" with Pawsey as chair. The IAU *News Bulletin* described the meeting: "Ways and means of achieving [high] resolution will be featured at the Philco Symposium … The meeting is open to all participants in the General Assembly." The speakers were: Martin Ryle, "The

---

[30] A few days earlier (16 August), Pawsey had written in his notebook that he was impressed by the disagreements in radio source evolution models between the Cambridge group (Martin Ryle) and Jodrell Bank (Henry Palmer). Also, there were major disagreements between the newly obtained polarisation observations of the radio galactic background as observed at Cambridge (Ryle, Smith and Hewish) and at Dwingeloo (Westerhout).

synthesis of large radio telescopes"; Paul Wild, "Circular aerial arrays and the possibility of image formation"; W.N. Christiansen, "The Mills Cross and its modifications"; Frank Drake (speaking for Sebastian van Hoerner), "Comments on the construction of high-performance telescopes"; and Ron Bracewell, "Guiding principles for the design of future large telescopes".

After catching his breath with a tour of Yosemite National Park, Pawsey flew to El Paso, Texas, to attend IAU Symposium No. 16 (his fourth conference in 1961!) while Lenore went to Regina, Saskatchewan (Canada), to visit relatives.

The symposium was titled "The Solar Corona at the Sac Peak Solar Observatory in Cloud Croft, New Mexico", organised by W.O. Roberts and colleagues (28--30 August 1961). The preface to the conference proceedings stressed the importance of the solar corona as a unique astronomical phenomenon:

> The solar corona is one of the simplest and most interesting objects of astrophysical research. To some extent it serves as a testing ground for the methods of analysis which must be used in the more complicated problems of the chromosphere and the photosphere, and ultimately, stellar atmospheres in general.

Pawsey met many of his friends from the solar community such as Clabon "Cla" W. Allen (1904–1987). Pawsey also met the prominent scientists in rocket X-ray and rocket UV solar work: Richard Tousey (1908–1997), pioneer in solar UV investigations of the sun and Herbert Friedman (1916–2000), NRL pioneer in solar X-ray astronomy. During the conference Pawsey also helped organise informal discussions of the properties of Type IV bursts and possible radar observations of the sun at low radio frequencies. An excursion to White Sands National Monument near Alamogordo, New Mexico, was an additional attraction.

Pawsey took a number of impressive coloured photos of prominent astronomers at the conference at Sac Peak in New Mexico. These photos are included in NRAO ONLINE.52.1 to 52.6. Numerous prominent scientists appeared in these photos: Vaniu Bappu, L Biermann, A, Covington, Tommy Gold, S Chapman, John ("Jack") Evans, Paul Wild, J.F. Denisse, Monique Pick, John Firor, Einar Tandberg-Hanssen, Mrs.Zirin and Marshall Cohen.

Post IAU in the US, 1961

During the period 31 August to 6 September 1961, Pawsey visited G.C. "Mac" McVittie at the University of Illinois in Urbana. Pawsey had a number of discussions with George Swenson.[31] On both 1 and 4 September, he visited the new radio

---

[31] During this period Pawsey and George Swenson discussed the second edition of the textbook *Radio Astronomy* by Pawsey and Bracewell. George Swenson suggested in particular that the book should not increase substantially in size. In addition, Swenson had a number of suggestions for improvement. For example, he urged Pawsey to give a detailed derivation of the radiometer equation (fluctuation proportional to the inverse square root of the product of integration time

observatory (a cylindrical parabola) at the Vermillion River Observatory near Danville, Illinois (see frontispiece), the first day with McVittie and Swenson and the second visit with E.L. Jordan, the Chair of Electrical Engineering at the University of Illinois. He took advantage of the long Labor Day weekend to write six short reports to groups of RPL scientists (and two individuals) with ideas for future research topics stemming from discussions at the two IAU symposia and the IAU General Assembly. He suggested that Alan Weiss, Dick Mullaly and Paul Wild (who had been at the Sac Peak solar corona conference) investigate "what, in a solar flare, is most closely related to X-ray production. It could be the cm-wave burst; it seems unlikely to be the metre-wave Type IV burst. I enclose a list of observed X-ray enhancements observed by satellite. Many of these are in our observing time [in Sydney]." The most extensive report was written for John Bolton, Eric Hill, Bruce Slee and Peter Scheuer under the rubric "Sources". This report dealt with the controversial source counts (log N- log S), comparing the Mills counts with the Cambridge results. By 1961, the 4C source counts were becoming available; Ryle had presented these results at Santa Barbara and the IAU. In addition, a luminosity function of radio sources had been produced by the Cambridge group (e.g. Ryle & Clarke, 1961, see Chap. 36). Pawsey emphasised the differences between their results and the luminosity function proposed by Minkowski, Bolton and Mills. Pawsey especially praised the Caltech and Bolton results[32]:

> Excellent papers were given by the Caltech boys (various ones) on source sizes. You know the story. Also, a beauty by Palmer on Jodrell Bank results. Their story is that they have statistical information on a very adequate sample ... Incidentally, the new Cambridge survey seems to include only the 200 or 300 sources published by [Scott et al., 1961, using the new 4C aerials]. [These authors'] emphasis is on theory, not catalogues.

Pawsey was worried by "the lack of coincidence between radio parts of a source and the associated galaxy. [The double nature of many radio sources] ... sets a serious handicap to identifications ... [T]he demand for source-shape information is accentuated."

On 5 September 1961, Pawsey went to Ann Arbor, Michigan, to meet Fred Haddock at the University of Michigan to attempt to form a collaboration with the solar astronomers there as participants in the solar project of Paul Wild. Pawsey met a number of the electrical engineering faculty as well as the solar astronomer Helen Dodson. He wrote in his diary that he would remain in contact with Haddock, but there were "no real conclusions" concerning a realistic collaboration. On 8 September 1961, Pawsey continued his lobbying with Carl Borgmann of the Ford Foundation[33] who was visiting the Ford Motor Company head office in nearby

---

multiplied by bandwidth). Swenson said that this relation "deserved a better proof." The first edition of Pawsey and Bracewell Sect. 2.4.2 was titled: "The limit to sensitivity is set by fluctuations", followed by a cursory proof. See NRAO ONLINE.53.

[32] Bolton had been back in Sydney since early 1961.

[33] RPL had earlier begun discussions with the Ford Foundation about possible funding support for the Paul Wild Radioheliograph project. See also Chap. 17.

Fig. 38.5 Aerial view of the Arecibo construction site, September 1961. Joe and Lenore Pawsey Family Collection

Detroit. Borgmann planned to contact W.O. Roberts and Jack Evans for advice about the feasibility of the CSIRO solar proposal. Pawsey also was following another avenue of possible funding: Leo Goldberg (having moved from Michigan to Harvard) had suggested contacting John Lindsay at NASA—Goddard.[34]

By mid-September 1961, Pawsey was in the tropics in Puerto Rico visiting the new Arecibo Observatory, run by Cornell University. The 1000-foot fixed spherical dish was under construction; the opening was to be on 1 November 1963.[35] Pawsey met William E. "Bill" Gordon (founder and Director of Arecibo, 1918–2010), Gordon Pettingill (1926–2021), well-known planetary radar astronomer, and Don Yabsley (1923–2003), a CSIRO colleague on leave from RPL at Arecibo. He toured the construction site of the new telescope Fig. 38.5) and presented a talk, "Astronomy in Australia". He also toured the University of Puerto Rico at Mayaguez, departing for New York on 15 September 1961.

---

[34] John C. Lindsay (1916–1965, Associate Chief of the Space Sciences Division of NASA) was a major contributor to the early exploration of the sun via satellite and rocket-borne instruments. He originated the Orbiting Solar Observatory, with a first launch of OSO 1 in March 1962.

[35] On 1 Dec 2020, as this book was going to print, we received the news that the Arecibo Telescope had collapsed as a result of a support cable failure. This catastrophic event represented the end of one of the world's leading astronomical facilities for the last 57 years.

Pawsey, Recruited in Earnest

Pawsey returned to his in-laws' house in Princeton and then took a late night trip to the Newark airport to meet Walter O. Roberts. Likely Rabi had asked Roberts to discuss Pawsey's becoming the NRAO Director.[36] Pawsey's notebook reads: "Greenbank—strongly urged to accept job". As Goldberg had suggested, Roberts advised that the best place for help with the Wild solar project was NASA, John Lindsay.

During his last month in the US, Pawsey was in Princeton on three separate occasions. He went to Rabi's house on Sunday, 17 September 1961 and afterwards wrote in his notebook, "Asked me about Green Bank—said he thought me the best man, salary [would be] 2 times CSIRO [salary]. Super [annuation] to be discussed." Pawsey may well have recognised by September 1961 that his future at RPL was at best uncertain. Conflicts were becoming more frequent with Bowen. He feared that he was losing the continued support of Fred White, Chairman of CSIRO. He must have realised that his role within the CSIRO had decreasing importance, that he could be marginalised in the future. Clearly, he needed a new environment. A Green Bank position at NRAO offered a remarkable challenge: he was being asked by Rabi and Keller to provide new leadership for a troubled institution. The report to Rabi from Pawsey from 5 October 1961, summarised in ESM 38.3, Confidential Report, was the beginning of an active involvement with the future of NRAO.[37]

On 17 September 1961, Pawsey departed to Green Bank where he began an intensive three-day visit with the NRAO staff. He carried out an "inspection" of Green Bank, listing the astronomy, engineering and business management staff with whom he met. He learned of construction plans: a possible multi-feed system for the 300-foot transit telescope, a tour of the 140-foot telescope. Individuals discussed their research: Beverly Lynds and her catalogue of dark clouds in the Milky Way;

---

[36] On 31 January 1962, Pawsey wrote Roberts (Joe and Lenore Pawsey Family Collection): ". . . [Y] our visit to Newark last September bore fruit and I am accepting the position of Director at Green Bank."

[37] Ken Kellermann found a single page document in Box 36 of the Rabi papers in the Library of Congress (LOC) in April 2012: "Manuscript Division, Classified Items Removed, The following item has been removed from the collection because it contains security classified information: Letter of October 5, 1961, Located in the Columbia University Office File, Greenbank (National Radio Astronomy Observatory), J.L. Pawsey to Rabi." A handwritten note on the document was: "Confidential (Australia), 1992?" On the advice of the LOC, Kellermann sent a Freedom of Information Request to the US Department of State in 2012; he received a reply from the LOC on 13 March 2014, with a copy of the Pawsey letter, the report to Rabi about the status of NRAO in 1961. Pawsey had written "Confidential" on the letter due to the sensitive nature of the report; the material was clearly not related to military "security". Likely the LOC personnel saw this classification on the letter from an individual in a foreign country and withdrew the letter from circulation in about 1992. After a two-year delay, and additional pressure from a member of the Virginia delegation to the US Senate, the letter was made available in March 2014 after being cleared by the US Department of State on 14 February 2014. Ironically, Goss also found the **same** letter in July 2014 in Australia in the Joe and Lenore Pawsey Family Collection, with no "confidential" restrictions!

Sander Weinreb[38] (see ESM 39.2, Sander Weinreb), Dave Heeschen and the radio continuum spectra of normal and radio galaxies; the evolutionary state of the Orion system with Kochu Menon[39]; Frank Drake's high frequency imaging with the 85-foot telescope; Hein Hvatum's plans for a 21 cm HI multi-channel receiver for the 300-foot, as well as dual frequency feeds for 700 and 1200 MHz.

At the end of a long day, Pawsey took the night train to Washington.

One of his objectives here was to meet Geoff Keller, Program Director for Astronomy at NSF. Keller urged Pawsey to accept the offer of the Directorship of NRAO at Green Bank with the comment "see Rabi for further discussion". He promised Pawsey that the NSF could provide one million dollars ($seven million in 2015 dollars) for the high-resolution project. In the afternoon, Pawsey visited Merle Tuve, Director of the Department of Terrestrial Magnetism of the Carnegie Institution of Washington.[40] Tuve was apprehensive about Pawsey's possibly becoming the NRAO Director with major concerns about his independence with respect to the AUI management and the fact that "no new blood [for NRAO] was available".[41] Tuve (like Pawsey in 1962) hoped to recruit Bill Erickson to work in his group at Carnegie Institution of Washington later in 1962.

The following day, 22 September 1961, John Lindsay, head of Solar Physics at the Goddard Space Flight Center, met with Pawsey to discuss the "Paul Wild" project and possible connections with solar X-ray missions. Although NASA had an interest in establishing a scientific connection with the radio heliograph project, direct funding opportunities seemed unlikely. It was not NASA's policy to provide financial support for ground-based observatories.

That night (22 September 1961), Pawsey returned to Princeton, where he met Lenore at her brother's (Ted Nicoll) house. He rang Rabi to arrange a meeting for Sunday afternoon, 24 September. In a letter of 31 October,[42] Rabi wrote: "I want to thank you for the thoroughly splendid afternoon and evening which I had with you and Mrs. Pawsey and your brother and sister-in-law [in fact brother-in-law and his wife]." The following day (Monday 25 September) Pawsey visited the head office of the Ford Foundation in New York, following up on a contact he had made when visiting Ford Headquarters in early September in Michigan.

---

[38] Pawsey wrote, "He hopes [in the future] to apply several years to Zeeman [HI] experiment [at NRAO]. If not successful might try Sydney or Jodrell."

[39] In the 5 October 1961 report to Rabi (ESM 38.3, Confidential Report), Pawsey was quite critical of the level of science he found at Green Bank: "... [S]everal members of the staff tend to be not sufficiently careful in drawing conclusions from observations of limited accuracy . . . I exempt from criticism two research items currently in progress: (1) an observational study of several emission nebulae by [Menon and Kahn] and (2) an attempt to observe the deuterium line (327 MHz) . . . by Weinreb . . .".

[40] Tuve also expressed to Pawsey his continual fear of a possible "atomic war", during a discussion of the political state of the world in late 1961.

[41] See Kellermann et al. (2020, pp. 104–108) for an in depth analysis of the animosity between Tuve and AUI management.

[42] Reported in the offer letter (Rabi to Pawsey) for the NRAO Directorship of 31 October 1961.

Although he had scrutinised the NRAO work at Green Bank extensively, he had not yet made up his mind on the Directorship offer.

To Europe

The Pawseys left New York on 26 September 1961, flying to London. In his one-month visit to Europe, Pawsey would visit the Netherlands, Jodrell Bank, Cambridge, London, Paris and Bologna before leaving from Frankfurt for Sydney on 26 October 1961.

In Leiden, Jan Oort and Chris Christiansen discussed plans for the Benelux Cross with Pawsey. A number of the key personnel in Leiden were Australians: the project leader Christiansen, along with Cyril Murray and John Murray. Jan Högbom, Jean Casse (from Belgium), N.G.V. Sarma (from India) and L.H. Sondaar were also involved in the project. Visits to Utrecht, Dwingeloo and Groningen (Kapteyn Laboratory) followed.

On 5 October 1961, Pawsey travelled to England to Jodrell Bank where he met Henry Palmer who described the possible construction of three movable antennas to be used in the long baseline campaigns (each with a diameter of about 8 m). These would be used at different stations up to about 100 km at 160 MHz, providing a resolution of a few arc sec. Pawsey talked to Rod Davies,[43] who had new HI data obtained with the 250-foot telescope. He had a weak Zeeman effect detection (2.5 sigma) on Taurus A with an implied magnetic field of about 12 micro Gauss (not confirmed in subsequent observations). Davies was testing a non-switched receiver with some success; the total power changed by only one percent in a half hour.[44]

On 12 and 13 (Thursday and Friday) October 1961, Pawsey gave two colloquiums at the Cavendish Laboratory at Cambridge. The title was "Radio Astronomy in Sydney". He included seven topics:

1. The Scheuer experiment—Long baseline interferometry at 80 MHz 1800 to 9200 wavelengths using the Mills Cross EW arm as the main element;
2. Shain map of the sky at 19.7 MHz, continued by Max Komesaroff after Shain's death in February 1960;
3. HI clouds—Hindman, Kerr and McGee;
4. HI flow pattern in the Milky Way, McGee and Murray;

---

[43] Davies was a former staff member at RPL, after a degree at the University of Adelaide. He moved to RPL in 1951 working on solar data from Potts Hill. He moved to Jodrell Bank for a PhD in 1953. He was the Director of Jodrell Bank 1988 to 1997.

[44] Pawsey also had discussions with Davies of a possible sabbatical visit to Sydney; within a few years this visit did occur. Also Pawsey had discussions with Michael Large about a position at RPL after the completion of his PhD. Later, Large did move to Australia, but with a position in the School of Physics at the University of Sydney, joining Mills's group in early 1963. Large played a key role in the Molonglo Cross and the MOST until his death in 2001.

5. Type IV solar bursts, Mullaly and Krishna;
6. Corona occultation of Taurus A at 80 MHz, Slee;
7. Solar flare associated phenomena.

Pawey and Martin Ryle discussed the completed MSH catalogue of Mills and colleagues. Pawsey was impressed with the aperture synthesis results obtained with the new 4C radio telescope. The errors in determinations of the positions of radio sources represented a vast improvement compared to previous radio survey instruments (for the weakest sources of a few Jy—0.5 arc min in right ascension, and in declination 3 arc min).[45] During the visit to Cambridge, Pawsey also had discussions with Fred Hoyle about a 3-4 month visit to Australia in late 1962, including Sydney University, RPL and MSO.[46]

I.I. Rabi was in London, and he and Pawsey continued their discussions about the NRAO Directorship. The conversation was the last personal conversation Rabi and Pawsey had before the formal offer of the Directorship was made at the end of October 1961.

On 18 October 1961 (Wednesday), Joe and Lenore Pawsey went to see their daughter Margaret, their son-in-law Don McLean and their new granddaughter, Margaret Ann (later known as Ann) born 17 February 1961 in Sydney. The McLean family had arrived in Paris in June 1961 where Don began his PhD working with J.F. Denisse in the field of solar radio astronomy.[47] While Lenore spent time with their daughter and granddaughter, Joe visited observatories: the radio observatory at Nançay; the partially completed "Grand Radiotelescope de Nançay"; the 169 MHz solar interferometer, the "Grand Interféromètre", used by Boischot and Denisse in 1957 for the discovery of Type IV bursts.[48]

On 24 October 1961, the Pawseys went to Bologna, and then on to Frankfurt, where they departed for Sydney for the official opening of the Parkes radio telescope (GRT) on 31 October 1961.

---

[45] Also at Cambridge, Pawsey talked with three advanced graduate students about positions at RPL. He was most impressed with Ivan Pauliny-Toth, who was to have a long career at NRAO and Bonn. Pawsey was less impressed with Wielebinski (who would join Christiansen in a few years at Electrical Engineering at the University of Sydney and then later at Bonn) whom he found "over confident; he is weak in physics; he is an excellent electronics man." He also had doubts about Turtle (who would join Mills in 1963, remaining at the University of Sydney until 1998.) He "lacks enthusiasm".

[46] Hoyle did visit Australia in late 1962 (Chap. 40), visiting Pawsey in Sydney on 3 November to present the Hughes Medal of the Royal Society, four weeks before Pawsey's death.

[47] Interview with Don McLean 20 September 2008. Margaret Pawsey was born in 1937; she died unexpectedly in London in late December 1977.

[48] The timing of Pawsey's visit to Paris had originally been set by a conference planned by Oort: the OECD Conference on Large Telescopes on 23–24 October. This meeting was postponed to 12–14 December 1961, at which point Pawsey was back home in Australia after attending the Parkes telescope opening on 31 October 1961.

GRT Opening 31 October 1961, Simultaneously with the NRAO Directorship Offer by Rabi

On 31 October 1961, Pawsey participated in the GRT opening. Despite how much had been made of the importance of his presence, Pawsey did not play a major role.[49] Coincidently, on the day of the opening of Parkes, Rabi sent Pawsey a formal offer for the Directorship of NRAO for three to five years.[50] The offer had been ratified by the Board of Trustees of AUI on 26 October. Rabi wrote:

> You will be pleased to hear that I had a meeting with the Director [Struve] and the scientific staff of the Observatory in which I very frankly discussed the whole matter of the director-ship . . . and told them of our discussions in London. I questioned each man in turn, and I am very pleased to be able to tell you that the enthusiasm was unanimous. They join me in urging you most strongly to come to Green Bank.

The salary was $25,000 per annum (roughly equivalent to $215,000 in 2019 with a house provided at nominal rent). Rabi also suggested that Pawsey might bring some of the Australian staff with him to NRAO: "There may very well be certain individuals whom you wish to bring with you. These people would be most welcome, if appropriate budgetary arrangements can be made." Rabi had discussed the offer with Alan Waterman (1892–1967, Inaugural Director of the NSF from 1950 to 1963), "who was enthusiastic at the prospect of your coming. One small formal step [the issue of Pawsey's non-US citizenship] has yet to be taken, which has no relation to any legal problem, since you are not a citizen of the United States. This move is now underway." Rabi hoped that Pawsey could come by 1 October 1962, as Struve was stepping down then. In fact, Struve planned to retire 1 December 1961.[51]

---

[49] Pawsey was invited to the opening and was the escort for the VIP guests (see Chap. 32). The implication in Robertson (2017), *Radio Astronomer* p. 226, that he had been shunned is incorrect.

[50] Surprisingly this important letter has not been found in NRAO, AUI files or in the Rabi collection at the Library of Congress. The letter was located by Hastings Pawsey in the family archive in Sydney on 25 May 2010. Also in the Joe and Lenore Pawsey Family Collection, a letter from Pawsey to Struve from 1 December 1961 has been found ("Dear Otto") which gives further details of his negotiations with Rabi: "This whole business has come as a surprise to me. I arranged my last [three days during week of 18 September 1961] visit to Green Bank before I had any thought of living there. When I came [to Green Bank] Rabi had mentioned the possibility to me but I felt so uncertain that I thought it best not to mention it to you. Thank you very much for your very effective efforts to show me round and introduce me to all the people. This was most helpful . . . It would be nice to see into the future but we can only see darkly. Instead I shall remind you of an episode which you may well have forgotten. When you visited Australia [see Chap. 23] you helped us radio astronomers very much by giving our work such good recognition. Praise from an eminent astronomer meant a lot both in morale and in helping gain support."

[51] University of Sydney Archive. P154-Series 2. Campbell Wade to Mills, 2 November 1961 (two letters, one "personal", other "official"). Wade wrote Mills turning down the offer of a position at Sydney University in Mills's new radio astronomy group. "It has not been easy to choose between Green Bank and Sydney. Each offers powerful attractions, and it has taken two months of inquiry and considerations to reach a decision. I have decided to remain at Green Bank, although this may not prove to be the wisest course in the long run. A major factor in our decision to stay here has been

(Additional details of Struve's troubled departure are summarised in ESM 38.4, Struve and appointment).

On 9 November 1961, Pawsey wrote back to Rabi with a brief reply. He explained that he was very interested in the offer and had started discussions with his boss, Bowen. He needed to meet the Chairman of CSIRO, Fred White, before he would be able to make a decision. Rabi responded on 17 November that he hoped Pawsey could arrive in Green Bank earlier than October 1962.

Pawsey and White met on 21 November 1961. White seems to have realised that there might be a chance that the Australian government and perhaps the Australian public would suspect that White/Bowen were possibly pushing Pawsey out of CSIRO, due to the conflicts that had begun in 1959–1960.[52] White tried to prevent this perception by telling Pawsey that the AUI offer was a compliment and that CSIRO would not oppose the US offer. White wrote to Pawsey:

> If you do decide to accept for 3 years I would like the event to be publicly noted as an invitation to you of merit. This would be more than justified. I could take this up with Rabi and Waterman, perhaps more easily than you. To do so I have in mind that either AUI or NSF or both should write officially to CSIRO saying the right things about your position and stressing the advantage to the USA. Either the Minister [for the CSIRO] or the PM [Prime Minister, Sir Robert Menzies] might announce this in Australia. This would put the whole matter on the right plane. Would you therefore let me know before you write Rabi that you are about to do so ... and tell me you agree with this?

On 26 November 1961, Pawsey accepted the position for three years. He hoped he could return to CSIRO at the end of the period. White had suggested that Pawsey could take leave without pay from CSIRO. The details of the arrangement were to be worked out between White and Rabi/Waterman.

On 26 November 1961, Pawsey responded to Rabi:

> I would like to thank you and the members of your Board for the honour you have done me in inviting me to take this position. My ... acceptance has been very much influenced by the expressions of confidence conveyed in your letter from yourself and the Green Bank folk and I shall, if appointed, do my best to justify them ... I am deeply interested in the future of radio astronomy in the US generally and I hope, in conjunction with the other radio astronomers in the country and the NSF, to be able to play some part in endeavouring to stimulate radio astronomy in America ... I shall retain very close ties and interests with Australia and I hope that my presence in the United States may lead to cooperation between the two countries which could be most helpful to both.

---

the excellent performance of I.I. Rabi as president of AUI. He is providing overall leadership which is both wise and firm, and this has given us a new feeling of confidence in the future of NRAO. Also, an important unknown quantity is the successor to Dr. Struve, who retires with the year. At a meeting of the staff last week, Dr. Rabi indicated that this matter is progressing well, and we have every reason to expect a favorable result." If Wade knew about Pawsey's appointment, he did not let on. But it seems most likely that he had some inkling of the offer.

[52] Clearly the US colleagues were completely unaware of this aspect. The gentle "test of the waters" by Bok in June had represented an unofficial signal from AUI to Pawsey about a possible move to NRAO; there was no indication that Bok had been influenced by Bowen or White.

Pawsey acknowledged that Rabi wanted him to start the Directorship earlier in 1962, but his son Hastings would be finishing high school at the end of 1962; Pawsey wanted him to do so in Sydney. He suggested that, as a compromise, he alone could spend a month at Green Bank in either March or April 1962, an orientation visit.

Also, on 26 November 1961, Pawsey wrote Fred White: "I am in the process of burning my boats [similar to the US expression "burning my bridges"] and accepting the Green Bank position."

On Tuesday 28 November 1961, White wrote Rabi, their first correspondence concerning the Pawsey appointment. White framed Pawsey's appointment as "repaying the US for all the financial help given to Australian institutes", such as GRT funding by Carnegie and Rockefeller, Super-Cross by NSF (pending, to be announced in March 1962) and the Radioheliograph by the Ford Foundation (pending, to be announced in March 1962):

> We value very highly the considerable help that has been given to us by the USA to help to develop the subject of radio astronomy in the Division of Radiophysics in Sydney. Now that Pawsey has made up his mind to accept your offer I have come to regard this move as allowing Australia, through him, to repay to some degree the generosity of the US to us here in Australia. Pawsey has, I understand, accepted your offer for a period of three years and has asked me if he might be given a leave of absence from CSIRO for that period, while retaining his superannuation privileges with us. It would make my task in this regard easier if you, as President of AUI Inc., would write me officially asking that Dr. Pawsey be given leave of absence for this period. You will, I hope share with me the wish that the right things be said publicly about this arrangement. Dr. Pawsey is a very valued scientist, both in CSIRO and Australia, **whose personal contributions have been largely responsible for the advancement of radio astronomy in this country**. [our emphasis] To some it may seem rather odd that we are willing to release him so readily. I wonder if you and Dr. Waterman [Director of NSF] could give some thought to **expressing to the press at the appropriate moment your reaction to our agreement** [our emphasis], as well as your satisfaction at the possibility of having Pawsey as Head of Green Bank for this period. I hope that during Pawsey's period in the US we will be able mutually to encourage further the already close ties between our two countries that exist in this exciting field of science.

In the US, Rabi was elated with Pawsey's acceptance. On 2 December 1961, he sent Bowen a handwritten letter ("Dear Taffy ... Sincerely Rabi")[53]:

> Congratulations piled on congratulations on the 210-foot [opened 31 October 1961]. It puts us to shame as purse-proud[54] plodders, which we are. Count your blessings, amongst them moderate poverty. My radio astronomy interest is strong but incidental to my succeeding Berkner [at the end of 1960] as President of AUI. Pawsey is saving my life by coming as director of the NRAO next year. The present incumbent [Struve was not named!] although a great optical astronomer had no administrative talent and no knowledge of the techniques of radio astronomy. I hope we are off to a better start. We are changing contractor of our 140-foot[55] and hope it will now go through. It will cost 6 times as much as you [sic] gimmick and

---

[53] C3830 Z1/14/A/II.

[54] Definition: "wealthy in a showy manner".

[55] The contract with Bliss was terminated at this time with AUI becoming the prime contractor. In addition, Stone Webster was to provide design, engineering and construction supervision.

I hope it will in some way pay for itself ... How did you do it anyway for two million [dollars, the GRT cost]?

Three days later, Rabi wrote Pawsey, again a handwritten aerogram ("Dear Pawsey, Sincerely I.I. Rabi")[56]:

> Your decision leaves us all rejoicing. Waterman [NSF Director], Scherer [Associate Director NSF for Administration], myself and the whole board of AUI, the staff of the observatory etc. We are glad that you will be able to come for a month in March or April and hope that you will be able to assume your post earlier than October. My best wishes for a successful career at NRAO. We have (at his request) released Dr. Struve of active responsibility as of Dec 1st. He stays on to settle his affairs till March 1st. However, Dr. Heeschen has taken over as acting director. I think you will find him an extremely reliable and conscientious interim. He is looking forward to seeing you and preparing the ground for your ultimate direction ... Again I can assure you complete support of all concerned.

On 11 December 1961, Struve wrote Pawsey (sent to his home address— "Dear Joe, Very sincerely yours, Otto Struve").[57] He provided an update for his ultimate retirement date, 1 February, with a leave of absence in December and January:

> Many thanks for your letter of December 1. I am very pleased that you are planning to come to Green Bank, and I wish you the very best of luck in this new undertaking. I am quite certain that all the staff members of the NRAO will greatly welcome your decision and will fully collaborate with you.
>
> In the meantime, you may have learned that my personal plans have been somewhat changed since your visit to GB [September 1961]. David Heeschen has been appointed Acting Director as of December 1 and will presumably function in this capacity until your arrival. I am officially resigning as of February 1, 1962, and am now on leave of absence during the months of December and January.
>
> You will find many difficult problems, scientific and non-scientific, in connection with the Observatory but there is no doubt that the potential strength of the Observatory is enormous because of the desire of the Science Foundation [NSF] to provide all necessary funds for the most effective work in radio astronomy. I also believe that in connection with the AUI is a source of added strength, partly because of the fact that the AUI operated not only the Observatory but also the Brookhaven National Laboratory and partly because AUI is based upon the long experience and success of nine eastern universities.[58]

A few days later (14 December 1961), Pawsey wrote David Heeschen, the Acting Director of NRAO. He assumed Heeschen knew that he was accepting Rabi's invitation to take the Director position.

> When I made my plans I expected that Struve would have continued until about the time I came to stay [October 1962]. Struve's [immediate] resignation [on 1 December 1961] means that things do not look so tidy but I think it may be an excellent thing [since Heeschen was Acting Director]. I well appreciated the leadership problem [with Struve] when I visited you [in September 1961]. I have been looking forward to working with you when I arrive ... I

---

[56] Joe and Lenore Pawsey Family Collection.

[57] Joe and Lenore Pawsey Family Collection and NAA C3830 Z3/9. Pawsey had initially met Struve in Sydney in early January 1954, while the American astronomer was visiting Mt. Stromlo.

[58] Joe and Lenore Pawsey Family Collection and NAA C3830 Z3/1/IX. Pawsey had met Struve in Australia in January 1954, when he discussed the Sgr A 75 cm image with Pawsey.

have every confidence in your ability to steer a properly balanced course ... [T]he course
will be fairly straightforward over the first few months until we can get together.

On 18 December 1961, NSF and AUI provided a press release in the US and
Australia announcing the appointment of Pawsey as NRAO Director.[59] The language was the customary wording of appointment news releases: the quotes were
clearly written by PR staff and attributed to Rabi and Waterman. The accomplishments of former Director Otto Struve were praised, and the new Director, Pawsey,
was introduced. Following White's suggestions to Rabi, the connections between the
US and Australia were emphasised. "[T]his will greatly improve interchange
between Australia radio astronomers, who in many respects lead in this important
scientific field, and our own [US] radio astronomy community." CSIRO was thanked
for the "arrangement" which allowed Pawsey to come to AUI on a leave of absence.
There was no explicit mention of the gesture of "repaying the US" for the generosity
of the two private foundations whose support had been decisive in funding the
GRT.[60]

On 18 December 1961, the *New York Times* ran the press release: "Dr. Pawsey is
credited with being one of the guiding spirits of radio astronomy research in
Australia, one of the nation's most active and effective in this field ... At Green
Bank, he will head a staff of about sixty that includes a dozen senior scientists."

The *Sydney Morning Herald* followed on 19 December 1961:

Canadian-born [sic-his wife Lenore was born in Canada!] Dr. Pawsey has been with the
CSIRO for 20 years and in May this year was awarded an honorary Doctor of Science degree
for distinguished research work ... He took up his post in 1951 [sic, it was 1939]. His work,
using the sea as a mirror for radio waves, produced some of the first evidence of the existence
of radio stars. Dr. Pawsey's research, with the help of a colleague in Holland, first
established the spiral nature of the earth's own galaxy. Last year he was awarded the
Royal Society's Hughes Medal for his work in this field with the radio physics laboratory
in Sydney.

The year ended with a few letters of congratulation to Pawsey from US colleagues.
H.F. Weaver (1917–2017, AUI Trustee) sent a telegram from Berkeley on
28 December 1961. C.L. Seeger (1912–2002, older brother of the famous folk
singer, Pete Seeger, 1919–2014), an astronomer at Stanford, wrote a frank letter
on 20 December 1961:

---

[59] On 17 January 1962, White wrote Rabi thanking him for the "excellent statement" put out by AUI
concerning the Pawsey appointment.

[60] The *New York Times* had an article about Pawsey on 18 December 1961; in Australia numerous
papers (Joe and Lenore Pawsey Family Collection) had reports: *Sydney Morning Herald* on
18 December 1961, *The Telegraph* (Sydney) and *The Age* from Melbourne on the next day.
Some of the Australian papers referred to Dr. Allan [sic] Waterman, the Director of the West
Virginia [sic] National Science Foundation. Pawsey's hometown newspaper the *Cobdon Times* of
10 January 1962 also mentioned the new appointment in the US: "In the Christmas break, [Pawsey]
visited old friends in the Camperdown district. He was accompanied by his wife and his two sons
and by his mother Mrs. M. Pawsey."

I heard the announcement [of your new position] with mixed feelings. Your leaving CSIRO is sad. On the other hand, their loss is our gain. There is a real job to be done at Green Bank and I can imagine no one better to undertake it . . . I feel certain that when I welcome [you], I am speaking not only for myself but for most of the other radio astronomers. Come to think of it, it's only fair that if Australia gets Bart Bok, it should pay the penalty of sending you to us.[61]

Additional letters of congratulations followed. These are summarised in ESM 38.5, Letters of Congratulation.

Two letters of congratulation criticised the state of NRAO. Gart Westerhout wrote from Leiden on 9 January 1962; he was planning to move to the University of Maryland on 1 April 1962. Westerhout:

[I would like] to express the hope that under your direction this observatory will at last become the major undertaking that it was meant to be. I am looking forward to very close contacts with you and hope to profit greatly from your experience and advice. We shall be close neighbours: on 1 April 1962 I am starting at the University of Maryland as professor of astronomy . . . I have taken on the task of establishing an astronomical institute . . . There is a great need for academic training in this region, where optical and radio observations are so densely distributed. For research, I hope to be able to draw heavily on the facilities of these existing observatories, [such as NRAO] . . . .[62]

On 8 March, Tuve wrote a more critical letter about AUI:

I was delighted by your Christmas card and by the advantages to all of us in your acceptance of the post at Green Bank. I hope that my warnings about previous organizational mistakes which crippled the director there resulted in giving you a firm hand and full control. The fact that I feel our country is on an irrational binge, partly military and partly screwball, induced by perfectly logical fears, is not altered by your coming here. I hope that the USA is a reasonable compromise, although many of us are unhappy about it. There are instabilities in our national picture of a different sort from the stresses in Australia.

Since you made the decision with some knowledge of realities here, all I can say is that we rejoice in having you and in finally having a firm and intelligent hand directing things at Green Bank for NSF. . .

Westerhout and Bill Erickson will be professors of radio astronomy here at Maryland. Maybe the Washington area can help provide you with some collaborators.[63]

Accepting the role of NRAO Director was a big change in terms of the scope of a US national observatory that Pawsey would now direct, and a huge step in terms of the administrative and political complexity of the organisation and its interactions with the US community. With the benefit of hindsight, and from a purely organisational

---

[61] Possibly Seeger did not realize the irony of his statement. Pawsey played a major role in convincing Bok to move to Australia in 1957. Likewise, Bok (June 1961) made an initial gesture to Pawsey to gauge his interest in coming to NRAO.

[62] Joe and Lenore Pawsey Family Collection. On 9 January 1962, Westerhout made a plea for "open skies": ". . . [F]or some time I have been advocating the principle that no new radio [or optical] observatory should be built in the US, and that it should be easily possible for others to use existing equipment. In the coming years I shall [attempt] to prove this."

[63] Joe and Lenore Pawsey Family Collection. Tuve was pleased that two outstanding young radio astronomers were moving to the Washington area at the University of Maryland, Erickson and Westerhout.

perspective, it was certainly time, both for the benefit of RPL and for Pawsey's own professional growth, for him to move on to a new role. New horizons were opening.

Chapter 39
Visions for NRAO, 1962

Edge and Mulkay (1976, p. 453):

> It is possible that Pawsey's influence has been "hidden" in this way. One American [unnamed] correspondent wrote to us with this comment:
>
> A remarkable number of the big ideas in radio astronomy can be traced to Pawsey, and this has been little recognised. I think every new idea in Australia actually came from him. You can find in his early papers the first clear understanding of what an interferometer does, and quite clearly the first exposition of the principle of aperture synthesis ... But, he was such ... a quiet guy that he didn't sell the ideas, and few people took notice.
>
> The citation figures certainly do not reveal any such central influence.

Introduction

Pawsey was quite active in early to mid-1962 as he discussed plans for his expected directorship of NRAO. He had anticipated some of these issues in his report to Rabi of 5 October 1961 (see ESM 38.3, Confidential Report) when he outlined a preliminary version of his vision for the evolution of NRAO. Of particular importance he pointed out the necessity of training of scientists and planning for sub-arc min imaging:

> ... [We must emphasise] concentration on developing men rather than building very ambitious equipment ... The major challenge in instrumental radio astronomy is the problem of producing "pictures" of appropriate areas in the sky in the light of radio waves ... I suspect that for higher resolution, techniques yielding detailed pictures of quite limited selected areas will prove more economical ... I should also like to encourage small-scale experiments leading to high-resolution techniques, both at Green Bank and elsewhere, with

Supplementary Information The online version contains supplementary material available at [https://doi.org/10.1007/978-3-031-07916-0_39].

the understanding that the future large-scale development should pick the best from all these experiments.[1]

Even after the serious cancer appeared in March 1962, Pawsey continued to engage in discussions with colleagues about his plans including possible recruitment of new colleagues at NRAO for both long-term and short-term positions. He prepared a major document while a patient in Massachusetts General Hospital (see Chap. 40), completed on 17 July 1962, "Notes on Future Program at Green Bank".[2] Ironically, on the next day (18 July 1962), Pawsey and AUI personnel (Tape and Heeschen) agreed that Pawsey could not be the NRAO Director due to the severity of his illness.[3]

"Notes on Future Program at Green Bank" Report Prepared at Massachusetts General Hospital 17 July 1962[4]

Introduction: The purpose of this report is to set down the general ideas concerning prospective activities at Green Bank which I have been able to formulate during my short visit to Green Bank [in March] and my time in hospital [in Boston]. They should form the starting point for a general statement on policy which I should like to prepare in co-operation with the Green Bank staff and AUI executive as soon as I am able.

(1) The development of a first-class scientific team at Green Bank.
(2) The provision of extremely powerful radio astronomy equipment for the use of American radio astronomers. These may be members of the [NRAO] organization, of universities or of other research institutions. **The use of the equipment should not be restricted to American radio astronomers**, but it is essential, since America is paying the bill, that Americans with appropriate abilities should have access to the equipment. **The essential qualification is that the person concerned be likely to be able to produce results of high scientific value.**[5] (our

---

[1] Pawsey had already discussed some of these ideas with the NRAO astronomers: Wade, Menon, von Hoerner and Drake at the IAU in Berkeley in 1961.

[2] Joe and Lenore Pawsey Family Collection, NRAO archive. On the title page, Lenore Pawsey has a handwritten note: "By J.L. Pawsey July 17, 1962", the date of distribution to AUI and NRAO.

[3] NRAO archive, "Notes on Pawsey for Discussion with Wilde" [AUI was not aware that the correct spelling was J.P. Wild, sic]. 25 July 1962. Unsigned document stamped private. The probable author was Gerald Tape, the Vice-president of AUI, written for the President of AUI, I. Rabi.

[4] The top copy of the report was located in the Joe and Lenore Pawsey Family Collection, also NRAO archive.

[5] These statements indicate a clear intention that Green Bank would have an "Open Skies" policy for access to observing time on the telescopes (i.e., observations would be open to all users, even those outside the US).

emphasis) I think the allocation of facilities should be the responsibility of the director [of NRAO].

(3) The stimulation of research in radio astronomy generally in the US. This, again, is inter-dependent with objectives (1) and (2) because while there are so few good and experienced radio astronomers in the USA, it is extremely difficult for Green Bank to acquire the staff necessary to realise the first two objectives.

Two Major Projects

(1) 140-Foot Radio Telescope. In 1962, the completion of the 140-ft radio telescope was essential (see ESM 39.1, Pawsey–Rabi 1961, for details of Pawsey's involvement with Rabi about the 140-ft project in 1961–1962).

> Plans for the completion of this instrument are already well under way and I would propose to continue existing arrangements and to maintain close liaison with the engineers in the hope of anticipating snags before they become serious or permitting the possibility of changes in design which might facilitate the construction without appreciably impairing the performance of the instrument.
>
> I think it is important that Green Bank should have a quite large fully steerable paraboloid like the 140-ft telescope. Such an instrument is extremely versatile and permits the undertaking at short notice of numerous types of observations which current results may suggest as being important—e.g. observations of linear polarisation.[6]
>
> It will be desirable to provide special receivers for the 140-ft radio telescope to be ready immediately, as soon as the erection is complete ...[7]

(2) High Resolution Project[8]

> One of the greatest challenges to instrumental radio astronomy has been the design of radio telescopes with sufficient angular resolution to show the significant physical features of objects of interest in the sky ... There are three basic approaches to the provision of greatly improved angular resolution: (a) making the structure very large, (b) drastically reducing the wavelength, (c) using one or other of the interference techniques which have been developed in radio astronomy for this purpose. With respect to (a) at Green

---

[6]The latter three words are written in Lenore Pawsey's handwriting.

[7]Pawsey suggested that the shortest wavelength that might be optimised was in the range 3–5 cm.

[8]The main text is partially quoted here. The table of contents indicated that the original documents contained an appendix, "Notes on possible method of realising the high-resolution project". Unfortunately, the NRAO archive as well as the document found in the Pawsey family documents do not contain this missing text (appendix).

Bank I should be reluctant to embark on a program of building an extremely large paraboloid. I think the 300-ft [soon to become operational at the end of 1962] and the 140-ft when completed will supply quite outstanding facilities of this type.

With respect to (b) we are planning in the millimetre wave project to make just such a step but it should be pointed out that if successful the gross change in wavelength may well mean a considerable modification in the relative emission from different parts of radio sources and indeed if the spectral trends in the known wavelength range [metre and cm] can be extrapolated to the millimetre wave region, the signals from most of the known radio galaxies will be too weak to be observable ... This project is pioneering work in a new spectral range rather than an improvement of resolution in the radio frequency range.

The third approach (c), the application of interference principles, **involves the combination of signals from a limited number of moderately large antennas and in my opinion is much the most promising for the provision of greatly improved angular resolution** (our emphasis).

Pawsey discussed three methods of achieving high angular resolution: Mills Cross or the Chris Cross which had been constructed at RPL at an expense of some tens of thousands of dollars, and the California Institute of Technology interferometer (two 90-ft fully steerable antennas, cost about US$1 million) at a wavelength of about 30 cm with angular resolution of some arc minutes. Pawsey explains that the results for a few dozen of the most intense "discrete sources" that had been observed greatly strengthen the case for having high enough angular resolution to make an image of the radio sources.

> I consider the specific objective of the Green Bank high resolution project should be the provision of equipment furthering the study of radio galaxies. The discovery of radio galaxies ... has provided some of the most interesting problems of present day astronomy. It is found that certain galaxies, and we cannot tell from optical observations which ones, show a radio emission vastly in excess of most other galaxies. We refer to these high radio emission galaxies as radio galaxies. **An extraordinary feature of these galaxies is that the radio emission appears to come from areas in the sky which are adjacent to the optical nebula but which [often] do not even overlap them** (our emphasis) ... The radio emission is currently believed to originate in regions ... in which high energy electrons associated with cosmic rays move in magnetic fields but the detailed nature of these regions and the reason for their occurrence at places apparently outside the optical galaxy are still [open] for speculation. It would seem that one of the most powerful methods of attempting to solve this mystery would be the study ... [of the] radio brightness distributions over a reasonable sample of radio galaxies.

Pawsey pointed out that the radio luminosity of the radio galaxies was so enormous "that they can be observed at distances far beyond the limit of observations" of most optical telescopes. "Observations of radio galaxies thus provide the possibility of finding out something of the nature of the universe beyond our present horizon and so contributing information on cosmology." He acknowledged the controversial source counts and the implications this had for the Steady-state model of the universe.[9]

After a discussion of the Benelux Cross (1 arc min resolution at 1420 MHz) and the planned Mills Super Cross (3 arc min at 400 MHz), Pawsey suggested that higher resolution was also required.

> I think of these [Benelux Cross and Super Cross] as analogous to the 40-inch Schmidt at Mt Palomar and I should like at Green Bank to aim at the design of an instrument somewhat equivalent to the 200-inch in that it would have considerably higher resolution so as to be able to show the physically significant details of an adequate sample of radio galaxies. I should be willing to permit much slower operation that is feasible with a whole sky instrument, i.e. to trade observing time for resolution.
>
> I do not at this stage have a clear idea of the best method of realising this objective, but wish to approach it from the point of view of a design study. There are at present time three or four methods which seem theoretically possible. These include (1) 2-antenna aperture synthesis, probably using a number of antennae instead of simply two moving ones, (2) a big Mills Cross or (3) the "ring" of Paul Wild.
>
> I think that the existence of a multiplicity of methods means that the project will prove to be technically feasible but I do not like any of the methods in their existing form. However, I think that there is a high probability of some simplifying ideas arising in the course of the design study which could make all the difference to the practicability of the project. I have not until now mentioned the resolution and sensitivity required in this project because I am not fully conversant with existing work, but it is clear that a resolution of a fraction of a minute of arc will be required. I think the first step on this project should be a careful evaluation of the existing observations to help determine the specifications to be aimed at. Observations highly relevant to this objective have been made in Jodrell Bank (Palmer) and Sydney (Scheuer) and at the California Institute of Technology.

Pawsey proposed that the project should have three phases, and he had hoped to have had an informal conference at Green Bank with Palmer, Scheuer and Erickson[10] to discuss the various proposals and methods. Phase 1 would be a careful study of the existing observations. "I should like to see Scheuer and Palmer come to Green Bank to make the best possible statistical estimate of the distribution of Fourier components for the radio sources ... I have written to Scheuer asking if he would be interested (no reply yet). Heeschen has spoken with Palmer who is reasonably interested." Phase 2 was to make a list of all methods with an additional list of the basic limitations. Pawsey was especially interested in the "sensitivity-time aspect and also the technical difficulties. At this stage we should be able to make some generalisations, e.g. is the aperture synthesis method fundamentally more sensitive than other methods?" Phase 3 was to use the previous stages to provide knowledge of the most suitable sources for future observations plus the technical difficulties anticipated. At that point, the optimal method would be chosen.

---

[9] However, by 1962 the Cambridge and Sydney source counts disagreements were largely resolved (see Chap. 36).

[10] Pawsey also wrote: "We should attempt to appoint a leader for the project. W.C. Erickson is my choice and I have already approached him ... (no reply yet). When the leader is appointed, we should attempt to appoint several new members of staff who are skilled in these high resolution fields."

I should then like to see this put into operation on a relatively small scale in order to have practical experience before embarking on any large scale venture. We should then be in a position to assess the whole project and make reasonable compromises. It is probably at this stage or [earlier] that a number of subsidiary experiments may be required. With the results of these at hand, the final decision whether to proceed or not should be taken.

A major question must be raised based on Pawsey's 17 July report. He did not mention the 5 March 1962 memo by David Heeschen, a draft of a "development program for the very large telescope",[11] written a few weeks before Pawsey visited Green Bank on 20 March. The project was a request for $3M to be included in the FY1964 budget for "the first phase in the development of a very large telescope." The idea was to establish system requirements, followed by design of antennas plus electronics, as well as expected phase behaviour. The instrument would have a resolution of better than 1 arc min at 21 cm and be usable up to a higher frequency, "probably 10 cm and a configuration which will allow later expansion to give still higher resolution. At present a cross array of large antennas is considered to be the most promising telescope configuration." A prototype instrument was proposed (phase one), possibly at a site not at Green Bank. A time scale of 1965 or 1966 was suggested for the beginning of phase two: "construct full telescope by expanding the portion built in phase one." The cost would "certainly be tens of millions".[12]

Kellermann et al. (2020) have written: "It is curious that, while Pawsey's report ["Notes on Future Program at Green Bank", from 17 July 1962] acknowledged his discussions with Drake and Heeschen about their proposed millimetre initiative, he makes no reference to any discussions about an imaging array."

One possible explanation may be that, due to the onset of the brain tumour only 4 or 5 days after arrival in Green Bank on 20 March, Pawsey did not have time for an extended discussion with Heeschen. However, Heeschen does recall[13] having this discussion so it is also possible that this proposal was so similar to his own concept that it may not have been clearly separated in his mind.

Other Projects   Pawsey was quite keen to exploit the new 300-ft transit telescope at 21 cm, where the beam would be about 10 arc min. "This should give most interesting results both on the hydrogen line and the continuum." In addition, he was interested in supporting the planned millimetre wave project.[14] "It is an attempt to open up a new region of the electromagnetic spectrum for astronomy."

---

[11] NRAO Archive, the same memo in two different locations https://www.nrao.edu/archives/items/show/9019 or https://www.nrao.edu/archives/items/show/16227

[12] Heeschen wrote Pawsey (in Sydney) on the same day describing possible future plans that would be discussed later in the month, including "the first phase of a very large telescope", with no further details.

[13] See "Pawsey's interactions with David S. Heeschen" at the end of this chapter.

[14] The proposed NRAO project of Frank Low, 1962 (see Kellermann et al., 2020, Chap. 10).

Pawsey also mentioned a few astronomical fields of investigation which would be important for understanding various astrophysical phenomena. He was especially interested in the determination of the properties of the interstellar magnetic field; he understood the importance of magnetism in the galaxy. He foresaw that the Green Bank telescopes could play a major role in this research. "Recent thinking in astrophysics is permeated by concepts involving the existence of magnetic fields in interstellar space. Direct evidence for the existence of such fields, however, is very hard to obtain." He realised the importance of the detection of linear polarisation of the radio emission from both the galactic background and individual radio sources.

> Definite radio evidence at present is restricted to the observations of plane polarisation in the radio waves from the Crab Nebula and ... other discrete sources. In these cases the mechanism of emission of radio waves [is due] to the synchrotron mechanism and the observations of linear polarisation supplies strong confirmatory evidence ... Accepting this hypothesis, the polarisation gives evidence of the existence and direction of magnetic fields ...

An additional new tool had been proposed by Bolton and Wild (1957, p. 296) to determine the magnetic field in the hydrogen gas in the Milky Way, using the Zeeman effect in the 21 cm hydrogen line. A group including V. Radhakrishnan at Owens Valley Radio Observatory had used HI absorption lines to set a limit of some tens of micro Gauss. Based on a conversation with the young MIT graduate student Sander Weinreb at Green Bank in September 1961, Pawsey suggested that accurate spectroscopy in the HI line at Green Bank might lead to a Zeeman detection (See ESM 39.2, Sander Weinreb).[15]

Pawsey continued his detailed description of Weinreb's work on the Zeeman effect:

> [T]here is another method of obtaining radio evidence of the existence of magnetic fields in interstellar space. This is through observations of Zeeman splitting of the 21 cm hydrogen line. Various attempts at such observations have been made including one by Sadow [sic], a Massachusetts Institute of Technology Ph.D. student who worked for some time at Green Bank. The results in all cases so far have been negative. I should like to see Weinreb extend his observations under Green Bank auspices if he thinks that there are reasonable chances of success.[16]

The detection of the Zeeman effect in HI turned out to be quite difficult. The group working with Rod Davies at Jodrell Bank attempted this challenging observation in the early 1960s. The first successful observation was reported by Gerrit Verschuur (1968, p. 774) with fields in the range 10–20 micro G in the Cas A HI absorption

---

[15] Pawsey was keen to recruit Weinreb at NRAO. A few years later, Weinreb was recruited to work at NRAO, where he worked for 24 years. His career was characterised by ground-breaking achievements at NRAO for the Green Bank telescopes, the 36-ft mm telescope at Kitt Peak, the VLA and the VLBA. In 1999, Weinreb moved to the Jet Propulsion Laboratory in Pasadena, where he has played a major role in the instrumentation developments for the Deep Space Network.

[16] Weinreb reported these results in 1962 (*ApJ*, vol 136, p 1149) with upper limits on the magnetic field in both cases. The 85-ft telescope at Green Bank was used.

lines for the Perseus arm, based on HI absorption data obtained with the 140-ft telescope at Green Bank.

A final project was discussed by Pawsey: a university sponsored solar endeavour.

This new information, in conjunction with the optical, has begun to yield a physical understanding of what is going on in the solar atmosphere which promises to transcend all that was previously known. Among these new data, observations of the radio waves emitted by the sun have played an outstanding part, and studies, pioneered in the Radiophysics Laboratory of the spectra of short duration bursts of emission at times of solar disturbances have been particularly significant. These, and associated studies, have led to hypotheses of giant explosions on the sun which require for their verification ... more powerful observations than any yet available.

Pawsey suggested that detailed imaging at short time intervals would be required of solar flare phenomena. He suggested a metre wave instrument and a 10 cm instrument would be required. The former was clearly related to the Culgoora Radioheliograph which was to be opened in 1967 in Australia; the 10 cm instrument was perhaps an upgraded version of the Stanford Microwave Spectroheliograph that had become operational in 1960.

Pawsey concluded:

I do not think that we have the staff and facilities to undertake such a project at Green Bank, but I should very much like to see a university or research institute with adequate technical facilities undertake it. From the Green Bank and general US point of view, it would have the merit that it would train a number of Americans in the art of high-resolution radio work.

Thus, this endeavour would serve as a training facility for US radio astronomers.

Pawsey's Notebook with Additional Plans for NRAO in 1962

Astronomical topics

In a series of notes written in Pawsey's notebooks[17] from 1962, we have discovered that he also had plans for additional astronomical projects that could be carried out at NRAO; these items were not described in the "Notes on the Future Program at Green Bank". This list consisted of a number of relevant topics: (1) Imaging of individual galactic objects such as supernovae remnants and HII regions, (2) galactic wide-scale imaging of the Milky Way corona and disk, (3) HI line studies of the Milky Way spiral structure (with the 300-ft telescope) and the galactic centre (140-ft telescope), (4) planetary observations especially at short wavelengths, (5) normal galaxies in continuum and especially in the HI line and finally (6) other spectral

---

[17] Joe and Lenore Pawsey Family Collection.

lines. In parenthesis he added after the final item "Sandy?". (Finally Pawsey spelled Weinreb's first name correctly! See ESM 39.2, Sander Weinreb.)[18]

Policy Issues for NRAO in 1962

A list of important policy questions for the new observatory was located in the notebook:

(1) How to recruit new staff to Green Bank?
(2) What were the amenities at Green Bank for the new staff, especially the schools?
(3) What was the future of the troubled project, the 140-ft telescope?
(4) Would a new headquarters of NRAO be considered? Pawsey wrote: "Would the NRAO headquarters be shifted to civilisation? Possible sites could be Princeton, Harvard, Charlottesville, Virginia, or a site in Maryland."[19]
(5) What was the best method to encourage University cooperation? Perhaps joint projects with the university groups might be efficient. But since Pawsey anticipated a major recruitment effort in Green Bank in order to enhance the scientific and engineering staff, the university collaboration might not be important.
(6) What was the policy of international collaborations?[20]
(7) Should a multi-feed system be constructed for the 140- and 300-ft telescopes? What were the consequences for the off-axis performance? What type of receiver system was required?

Pawsey's Attempted Recruitment of Erickson and Scheuer, 1962

Bill Erickson

In parallel with the discussion of the document "Notes on the Future Program at Green Bank", Pawsey wrote Bill Erickson (then working on the Benelux Cross) a 2½ page letter on 27 June from the hospital in Boston.[21] Many of the points in the

---

[18] The possible solar program at a potential US university appeared twice in these notebooks.

[19] The move to Charlottesville occurred in 1966. The Joe and Lenore Pawsey Family Collection contained a page of the *New York Times* from 29 January 1966 "Astronomers Find Hillbilly Life Too Limited- Scientists Will Go 110 Miles to West Virginia Facility" (describing a possible commute).

[20] The "Notes on Future Program at Green Bank" had stated "The use of the equipment should not be restricted to American radio astronomers . . .".

[21] We have copies of this letter from both the NRAO archive (records of the NRAO New Operations series) and the Joe and Lenore Pawsey Family Collection. The former copy was the copy received by David Heeschen in 1962. J.L. Pawsey wrote to Heeschen (per Lenore Pawsey): "This is a copy

"Notes on the Future Program at Green Bank" were duplicated in the letters to both Erickson and Scheuer.

Pawsey began by explaining to Erickson the details of his illness. On 27 June 1962, he was unrealistically optimistic about his recovery.

> Radio isotope test shows that the cause of my trouble was a brain tumour which has been removed. (see Chap. 40).
>
> The present position is that I am substantially recovered from the effects of the operation, but I am left with the serious weakness in the left arm and leg. However, Dr Sweet [William H. Sweet, 1910–2001, Chap. 40] expects me to recover sufficiently to lead a normal life and does not expect any residual effects on my intelligence, thank goodness. On this basis AUI plans to go ahead with my appointment as Director ... My purpose in trying to see you in Paris was to try to interest you in a project which I regard as the Number One project at Green Bank, and to see if you might be interested in taking a senior position at Green Bank where you would be the person in charge of the [high-resolution] project. I think the project really stimulating and I should love to have you working on it.

Pawsey then described the high-resolution project (the interferometer proposal), using almost the same language in the "Notes on Future Program at Green Bank". He described to Erickson the phases of the process. But his main emphasis was on the suitability of AUI as the managing organisation for the project:

> I should also like to comment on the suitability of the AUI set-up for achieving an ambitious objective like this. Firstly, I accepted the directorship because of my faith in Dr Rabi, the president of AUI, as a scientific leader. This faith has been strengthened through my association with him since my arrival in the US, and I am sure he can provide both inspiration and very strong support. Secondly, the remainder of the management of AUI appears to me to be able to provide absolutely first-class scientific management. They have a strong desire to make a success of things and a remarkable freedom from red tape. Thirdly, NSF is solidly behind Green Bank with tremendous financial support. Fourthly, the existing scientific staff at Green Bank I think are quite good. They are very keen and quite able though inexperienced and suffering from the effects of inappropriate leadership in the past [Struve]. The outstanding weakness at present is, of course, their lack of experiences in all the phases of radio astronomy excepting those utilising single parabola antennae. This weakness, of course, runs right through the US. We have vacancies, and could appoint promising scientists readily. What we want is one or two like Jan Högbom or Peter Scheuer.
>
> You might be interested to hear of a comment of Rabi's on overcoming the technical difficulties. He remarked that three times in his lifetime he has seen a similar situation of apparently unsurmountable technical complexity conquered by the tremendous technical resources of America. [Pawsey then described his proposal to stimulate solar radio research at US universities in order to stimulate training in radio astronomy.]

---

of the letter of which I spoke to you yesterday [presumably by telephone]. The original has not yet been posted. I have sent it to Jerry Tape [Gerald Tape, AUI Vice-President] to look over prior to dispatch. In any case, it gives my current views on this project and I think these [remarks] can be helpful to you. I am sending a copy to Rabi [AUI President] provided it passes Jerry Tape with the remark that it is highly provisional and could well be modified after further discussions with you people at Green Bank [that is the NRAO staff]". See ESM 39.3, William Erickson, for additional details of the complex correspondence between Pawsey and Erickson (and Tape) during the year following October 1961; ESM 39.4, Scheuer-Pawsey Interactions, contains similar details about Peter Scheuer, a visiting scientist in Sydney for 3 years, beginning in 1960.

... I should very much like to know your reactions to this project, and if they are favourable, whether you would be interested in taking part. You told me that you were committed to taking a position at the University of Maryland. I should be interested to know the details. If you are definitely to go to Maryland it is possible we might arrange some plan of co-operation, but as you can see, the project will be an exciting one and will demand quite a lot of work.

Peter Scheuer

The letter to Peter Scheuer in Sydney at RPL was sent on 4 July 1962. The content was an abbreviated version of the letter sent to Erickson a week earlier. Pawsey knew Scheuer well since he had been a staff member at RPL since 1960. Pawsey was more frank in the description of his illness than he had been with Erickson:

The second phase of my illness has been a devasting business, but I am pleased to say that a silver lining is now shining through. For quite a while I was very depressed, partly as the direct effects of the operation, partly from sheer hospital life, and also from the fact that I have been substantially crippled over this period. However, Dr Sweet's opinion is that in a reasonably short period I shall be able to live a reasonably normal life,[22] and if this eventuates, the plan is that I should go ahead with the directorship of Green Bank as originally planned. [After a short convalescence period], I should be able to return to Australia ... and wind up things there, returning to Green Bank in the northern autumn ...

Pawsey asked Scheuer to join the discussions about the new high-resolution project along with Erickson and Henry Palmer of Jodrell Bank. Pawsey and Scheuer had started this discussion earlier in the year in Sydney. Pawsey outlined the circumstances of a possible long-term visit to the US by Scheuer:

Now I remember that you do not wish to leave England permanently because of family considerations, but what I should suggest is that you come over to Green Bank for a period of something like four or five months with Palmer if I can arrange this, and that you work together on this ... If you came to the States you would have the opportunity to visit various radio astronomy centres in addition to Green Bank, and I think that you will find it a very worthwhile experience. I do not know whether it will conflict with your English commitments, assuming you get the job at Cambridge [see ESM 39.4, Scheuer-Pawsey Interactions], but I would suggest that it might well be compatible with them. The only snag I can see is that Martin Ryle might be on just the same track and wants to be secretive about it. However, if Martin Ryle is on this track, the information I am asking for is right up his street in any case. The only trouble would be if he has detailed schemes of how to achieve this, he doesn't want to divulge at this stage. [Pawsey suggested that the long-term visit could be the end of 1962.] ... I would very much like to have you visit Green Bank just so as to mix with the boys and stimulate the discussions generally on the sorts of topics in which you are interested.[23]

---

[22] As we will show in Chap. 40, Sweet was far too optimistic in his discussions with J.L. Pawsey about the prognosis of recovery of his patient.

[23] In the letter, Pawsey had asked Scheuer to report on his improved condition to the staff at RPL in Sydney; in a pencilled note as a PS, Lenore Pawsey wrote that she had written Lindsay McCready at RPL with an update about the release from the hospital. "Please excuse a few irregularities in the

Scheuer's response was apparently never posted. Goss searched for some years for Peter Scheuer's response in various archives in Australia, including the Joe and Lenore Pawsey Family Collection. It is probable that Peter Scheuer did not answer in July 1962 since the rumour had reached RPL that Pawsey was unlikely to remain the NRAO Director due to his terminal illness. Another possibility is that Pawsey and Scheuer had spoken personally about the offer after Pawsey's return to Sydney on 29 July 1962.[24]

In March 2012, Jane Scheuer invited Goss to her home near Cambridge. With her encouragement, Goss looked at the Peter Scheuer archive, finding two draft copies of his response to Pawsey, both apparently never posted. These were likely written in July 1962 in Sydney; Scheuer left Australia for the UK by ship in early 1963, a few months after Pawsey's death.

Scheuer's draft handwritten letter (with many corrections such as crossed out words and insertions), never posted, read:

> It was indeed a great pleasure to receive your letter today, with the good news that you are improving and due to leave the hospital soon. Naturally everyone here has been very anxious about you over the past months and it is a great relief to know that you will gradually be able to pick up the threads again and that you are already in the mood for making plans.
>
> I have rather a bad conscience about [producing] preliminary interferometer results which I should have sent you. [Interferometry at RPL at 85.5 MHz with baselines of 6 to 32 km with resolutions of 2 arc min to 22 arc sec.] These are only just getting into reasonably presentable form but I shall send then over separately ... There is not yet enough fully analysed material to permit even tentative statistical conclusion; the one notable result which seems to emerge so far is that quite a considerable proportion of the extended sources (resolved by the Caltech observations) have cores of smaller dimensions in them, apart from Centaurus A and Hydra A which were [already known to have a compact core]. These [now] include Pictor A.[25]
>
> Many thanks for your invitation to visit Green Bank together with Henry Palmer. This is something I would like to do. The relevant facts are as follows: the Cambridge job is now settled. I shall be back with Ryle next year [1963]. The first thing I had to do was to ask for a postponement of the starting date so that I could [finish] the interferometer programme here, and the university will give me a [one] term extension; but they could hardly be expected to look kindly upon a further absence of several months afterwards. However, it might be possible to arrange something in the long vacation next summer [1963], and I imagine that this might be the best time for Palmer too, since he presently has lecturing duties at Manchester to consider.
>
> On the other hand, I do not think that Ryle will have any objections on the grounds of trade secrets. [Thus answering a major query in the Pawsey letter of 4 July 1962]. Since the beginning of [this year], he has published a short account of his new machine in *Nature* and the construction phase [of the One Mile Telescope] is just beginning. It is essentially an

---

composition of this letter which was dictated from a hospital bed." Likely the Pawsey family had organised for secretarial assistance to take dictation. The spelling is US rather than Australian.

[24] Peter Scheuer also signed the Sydney radio astronomy group letter to Pawsey from 25 October 1962 (see Chap. 40).

[25] The only known publication of these data was by Scheuer, Slee and Fryar in the Pawsey volume of *Proc IRE Australia* of February 1963. The data from baselines of 6, 10, 17 and 32 km are shown in Figure 5 of this *Proc* paper, which shows that Pictor A has a compact core with 10% of the total flux density of 570 Jy (at 85.5 MHz) in a source of less than 10 arc sec in size.

interferometer of two 50-foot steerable dishes on an East-West baseline. The rotation of the earth rotates the baseline, while the individual dish follows one point in the sky. By aperture synthesis one builds up the equivalent of a large aerial at the North pole. The third dish is only there to halve the observation time. Essentially the device will work on 400 MHz with 1.5 arc min resolution and if the phase stability is good enough it may be pushed up to 1400 MHz. This is a device of the same general type as the one you contemplate in that it will examine small patches of sky in some detail, though I imagine you would want considerably more detail on smaller patches of sky. For the moment, the most relevant point is that the design is published and any further schemes there may be are presumably very remote.

I look forward to hearing more on your progress when Paul Wild and Kevin Sheridan [return] from their trip overseas and on seeing you when you come to Australia.

Yours truly, Peter

Pawsey's Concept of an OBSERVATORY, 1962

Pawsey had a vision of a new model for the use of radio astronomical instruments. The transition in thinking from "small science" to "big science" had already begun nearly a decade earlier as we discuss at the end of Chap. 33. In his "Notes on Future Program at Green Bank" (section two of this chapter, 17 July 1962) Pawsey wrote: **"The essential qualification is that the person concerned [the prospective NRAO telescope user] be likely to be able to produce results of high scientific value.** I think the allocation of facilities should be the responsibility of the director [of NRAO]."

In 1954 Pawsey described this concept in the minutes of the meeting held in July 1954 as he prepared for the URSI conference in late August in The Hague).[26] The minutes were dated 8 July, "Radio Astronomy Group—Current Program" and were discussed in Chap. 27. At this time Pawsey saw that the small semi-independent research group paradigm was coming to an end as he viewed the future in 1954: "In the past, projects have been planned on the basis of a small group building apparatus and using it to get all the information possible. We are moving towards the **observatory procedure** [our emphasis], where complex equipment is used by a succession of observers to investigate explicit problems."

The 6.5-page report contained numerous details of the research environment at RPL at the time, and this is summarised in ESM 27.1, RPL Radio Astronomy. Pawsey began the report with a discussion of his philosophy of planning which is included at the end of Chap. 33. In 1954, radio astronomy at RPL was in a state of transition as plans for the future were discussed. The small group model (with instruments constructed by the user) would disappear in half a decade. The observatory paradigm would replace this model as the GRT came on line in 1961.

As the GRT became a reality, Pawsey reconsidered the question of the observatory model as discussed in Chap. 31. In October 1960 Pawsey described the

---

[26] see Chap. 24, also Pawsey family and Sullivan archives.

arrangements for a research programme with the Parkes radio telescope.[27] This now included a new topic, "selection of specific investigations", involving the formation of a telescope observing proposal committee. Details are given in Chap. 31, including the procedure for scheduling the observations selected.[28]

Likely Pawsey envisioned that the Green Bank telescope would be operated in a similar fashion. The concept of a national observatory began to become a reality.[29]

J.L. Pawsey's Interactions with David S. Heeschen and Frank D. Drake, 1962

In 1962, both Heeschen and Drake met Pawsey at Green Bank in the last week of March 1962. As we will discuss in Chap. 40, Pawsey and Jerry Tape arrived in Green Bank on the morning of 20 March (Tuesday). The paralysis became pronounced sometime during the following weekend (23–25 March); on Tuesday 27 March 1962, Pawsey was taken to the hospital in Washington, D.C. He only interacted with the NRAO staff for 4–6 days at Green Bank. Additional interactions occurred in June and July based on correspondence between Pawsey (at Massachusetts General Hospital in Boston) and NRAO/AUI staff.

Heeschen wrote Goss on 28 November 2007, describing the short visit by Pawsey to Green Bank at the end of March 1962:

> It was a good visit [by Pawsey in 1962]. We told him all about our plans and hopes, including the VLA.[30] We had not yet settled on any of the details of the VLA of course, but

---

[27] NAA C4633/3 and Sullivan archive.

[28] The proposed scheme was intended to provide a schedule for the GRT in 1-month blocks, with the possibility of longer-term allocations. "Director's Discretionary" was also to be possible. The minimum observing team was to consist of an "astronomer" and a "radio man", someone responsible for the equipment. The two were to share in the planning and interpretation of the data.

[29] In the GRT discussions of 1960 there was no mention of "outsiders"; only in the mid-1980s did this change occur at Parkes.

[30] Memo from Dave Heeschen to Green Bank colleagues (including Findlay, Hvatum, Hogg, Drake, Wade, Vinokur, Walton and Callender) from 5 March 1962, "Very Large Radio Telescope". This proposal was for $3 million from the NSF to start an eight-step program for phase one of a large array. The funds were never allocated; but phase one became the Green Bank Interferometer, the prototype of the future VLA. "Three million dollars is requested for the first phase in the development of a very large telescope". Phase I was "to establish performance requirements for the telescope and to select the telescope construction which will meet the performance requirements". The envisioned instrument was to work at 21 cm and possibly 10 cm with a resolution of 1 arc min or less. The array would be in the form of a cross. The site selection was described: "Green Bank may not be suitable. Green Bank and other possible sites must be investigated." Phase II was to be the full telescope based on phase I experience. "No estimate can be given now for the cost of Phase II, but it will certainly be tens of millions." For historical discussions of the VLA see "Radio Astronomy: A Large Antenna Array", and "Reminiscences of Early Days of the VLA" by DS Heeschen in Cornwell and Perley (1991, p. 150) and "The Very Large Array", by DS Heeschen, in Burbridge and Hewitt (1981, p. 1).

were sure we wanted a general purpose instrument that would do high resolution imaging, have compatible resolution and sensitivity limits, and minimal instrumental effects. We wanted the instrumental characteristics of an optical telescope, but with higher resolution. He listened to all we had to say. Shortly after his visit [in fact in July 1962] he wrote ["Notes on Future Program at Green Bank", section two of this chapter and in it] he encouraged us to go on with our plans for a big instrument. He also described his own idea which was to bring together a few experts to review what was then known and unknown about discrete radio sources and try to decide specifically what sort of instrument might best address the unknowns.[31]

The next day, 29 November 2007, Heeschen told Goss:

Well, we planned to create an instrument to solve many unforeseen problems ... We were overwhelmed by the concept of the problem of designing an [array with a wide frequency coverage with a wide range of angular resolutions]. We were guided by the mismatch between sensitivity and resolution of all the existing instruments [in Australia, UK and France]. One of them would have great sensitivity and no resolution and the other one would have high angular resolution and no sensitivity. This was affecting the dynamic range of problems that could be studied. And then the instrumental effects of the high resolution ones were causing all kinds of problems [such as the 2C catalogue] ...

Goss interviewed Frank Drake on 14 December 2007 concerning the Pawsey visit of 1962 to Green Bank. His upbeat remarks follow:

I met him in passing at the 1958 Paris Symposium on Radio Astronomy, where the shoot-out between Mills and Ryle occurred. But when I met him at Green Bank [1962] that was the first time I ever had any real contact ... He was a real breath of fresh air at Green Bank the time he was there ... We were back to doing serious radio astronomy instead of struggling with a depressed director like Struve. That was not good for morale ... So he came [to Green Bank] and he was to stay a month [actually less than a week]. Then he was to go back to Australia, pack up his stuff and come right back. So that's the circumstances under which he arrived, and he quickly met with all of us. And you probably heard it from lots of people, but he was just a wonderful guy. He was interested in what people were doing. He was a gentleman. He was not pompous; he had no ego. He was very interested in scientific work. It fascinated him; he liked to talk about it. And that's what he was doing during those first few weeks [less than a week] he was there, and he had ... one general meeting at which he said he wanted to put together plans for future instruments at Green Bank, and on his list was an interferometer system because he sensed the importance of aperture synthesis as the wave of the future. And I guess you have discovered—you surely have discovered he was really the originator of that idea ... Yes, Pawsey was the one who recognised you could describe the sky in terms of spatial frequencies and this could be measured by interferometry—that was all his idea.[32] Then you know, later it was seized upon by Ryle and company. But Pawsey was the originator, although he wasn't the first to really push it hard. I guess Ryle was the one to do that ... Anyway he was already well aware of the potentialities there, and was talking about all of us thinking through what we should build in the way of a multi-element interferometer to start doing imaging of the sky. And he was very enthusiastic. He was getting up to date on what everybody was doing at Green Bank, and then encouraging them

---

[31] It is surprising that Pawsey's "Notes on Future Program at Green Bank" (17 July 1962) did not mention the 5 March 1962 memo by Heeschen nor does it refer to this conversation with Heeschen. We discussed this omission in the beginning of this chapter.

[32] Pawsey's key role in the early development of aperture synthesis is discussed extensively in Chap. 37.

to think big, and that's what was happening up to the moment he got hit by the tumour ... It wasn't the first day by any means. He had been there for a while ... because we were all familiar with him and we were talking on the same wavelength. I believe we had a one-on-one discussion about my plans for the future ...

Goss asked Drake what would have happened at NRAO if Pawsey had survived in 1962:

It would still look the way, it is, because his presence did introduce into NRAO the idea of building a bigger array. That would've happened, but he disappeared before he could do that. But very soon after that, that idea arose within the NRAO on its own without the influence of Joe, and that, of course, led to the proposal for the VLA.[33]

Viewed from the year 2021, it is impossible to say with certainty what the VLA would have become if Pawsey had lived. Pawsey's style, scientific drive and instrumental knowledge would clearly have had a huge positive impact on NRAO at that time. He had clearly recognised the need to have staff with more instrumental experience; eventually some of that experience came from John Bolton's Caltech students. But would Pawsey have been as bold as Heeschen and colleagues to create the VLA telescope with its variable baseline of 1–30 km and a frequency range of 80 MHz to 50 GHz? Would he have been able to provide the focus and real-world management skills that Dave Heeschen brought to the VLA project?

In 2007, David Heeschen told Goss facetiously: "We were so bold with our design of the VLA because of our inexperience." Campbell Wade reiterated this statement on 16 October 2018: "Our group just did not realise that the VLA was an 'impossible' concept. We were saved by our lack of experience."

Since the VLA opened in 1980, this instrument has become one of the most successful ground-based astronomy facilities ever built. The VLA has certainly provided the radio source imaging capability that Pawsey had foreseen.

---

[33] As pointed out above, the proposal of 5 March 1962 by Heeschen was made a few weeks **before** Pawsey's arrival in Green Bank.

Chapter 40
The Final Year, 1962

I.I. Rabi, President of Associated Universities, Inc in New York to Wild on 20 August 1962:

> We were glad to hear that Joe is holding his own [on return to Australia 30 July 1962] and possibly improving. If character were all, Joe would now be a well man.

On 26 January 1962, Pawsey began organising a 1-month stay in Green Bank for March, an orientation visit. After the US, he planned to go to Paris to visit his daughter Margaret and her family, then continue back to Sydney via India. He hoped to arrive back in Sydney by 24 April 1962, Hastings Pawsey's 17th birthday. As expected, Pawsey arranged his trip to the US and Europe in order to meet colleagues. In Pasadena on the way to New York, Green Bank and Washington, he planned to meet Gordon Stanley and Otto Struve.[1] Later he was to meet Rabi in Boston in early March, just before his visit to Green Bank, which was to start about 20 March. These plans were outlined in a letter to Rabi on 26 January 1962, explaining that his home base in the US would his brother-in-law's (Ted Nicoll) house in Princeton, New Jersey. Pawsey wrote:

> I am looking forward very much to getting into the job at Green Bank. I must say that your letter welcoming my decision to join you was most cheering and stimulating. We can make Green Bank really good, but I want to get over there and get clear ideas of circumstances and people before I make my firm suggestions.

On 31 January 1962, Rabi replied, again with a handwritten letter: "I have set the machinery in train to help with your visit to make you an [NRAO] employee during the period [March-April] . . . Your visit will come at a crucial time [for the 140-foot

---

Supplementary Information The online version contains supplementary material available at [https://doi.org/10.1007/978-3-031-07916-0_40].

---

[1] Joe and Lenore Pawsey Family Collection and letter to Rabi from Pawsey 26 January 1962, Rabi collection Library of Congress.

telescope[2]]." On the same day, Charles Dunbar, Secretary of AUI, wrote Pawsey with the details of the temporary visa process, the filing of a petition for permission to import a non-immigrant alien.[3] Since Pawsey was "an alien of distinguished merit and ability", the expectation was that the process would go smoothly. In fact, within 2 weeks, the NSF reported to AUI (15 February 1961, Dunbar to Pawsey) that "our petition to import you from Australia has been granted by [Immigration and Naturalisation Service]". Dunbar was surprised: "I had no idea that such rapid action could be secured."

On 16 February 1962, Pawsey wrote Gordon Stanley[4] about his short visit planned to Caltech on 15 March; he would split the day with visits to Stanley and Struve. Pawsey recognised the unique success that John Bolton and Gordon Stanley had created with the founding of the new Owens Valley Radio Observatory of Caltech.

> I am anxious to see you to get the general picture of progress at Caltech. I should like to take an intelligent interest in radio astronomy throughout the US in addition to that in our own corner [i.e. NRAO] and I regard your territory as a particularly important corner. I hope things are continuing in the very excellent way I found them [at Caltech in 1961]. You [have] done a phenomenal job.

On 26 February 1962, Fred White wrote a striking letter[5] to Alan T. Waterman, Director of the NSF. Clearly White wanted to provide Waterman with an overall view of Australian radio astronomy. The letter was likely inspired by the December 1961 visit by Geoffrey Keller (Assistant Director of the NSF). White wrote:

> It is obviously very difficult for you and senior colleagues to keep closely in touch with what is happening in countries so far away as ours. This may be of very considerable interest to you in view of the generosity of the US institutions in helping us maintain the standard of this science in Australia. You will be fully aware ... of the fact that it was the Division of Radiophysics of the CSIRO that made the initial break through. [This was made possible in this Laboratory under Dr Bowen and Dr Pawsey], men who had the ability to make such great advances in this science in its early stages. We in CSIRO are determined to maintain our position in this science, but have, of course, every intention of encouraging others in Australia to come into the field as well. I personally am very happy indeed that the University of Sydney has chosen to embark on research in this field and that two of our best men—Mills and Christiansen—have seen fit to leave us to join the staff of that University. This can do nothing to hinder radio astronomy in Australia; in fact, the opposite

---

[2]The years 1961–1962 were turbulent for the 140-ft telescope project (David Heeschen "The First Ten Years, 1955–1965", in Lockman et al. 2007, p 277–279). The contract with Bliss was ended and Max Small became construction manager; the Bliss design was modified considerably as Stone and Webster Engineering Company carried out a thorough design review. Struve had resigned in November 1961 (Chap. 38) during this turmoil.

[3]A complexity was that AUI would pay Pawsey's salary during the short visit in March–April 1962; the process for an immigrant visa would have lasted at least a year.

[4]Caltech Archive.

[5]C3830 Z1/7/B2. Bowen-White correspondence.

may well be the case, that through them younger men in physics will enter this very exciting field of research.[6]

The above text, containing a double negative (e.g. *can do nothing to hinder)*, may well indicate that White was not certain that the two new University of Sydney radio astronomy groups would achieve *immediate* success.

White also had a message to Waterman about Pawsey's new role in US radio astronomy:

> When I heard from Rabi that he wanted Pawsey to go to Green Bank my first reaction was to ask Pawsey whether he personally wished to undertake this assignment. When I found he did, my next reaction was to say that this was a very good thing. We in Australia have received very considerable help from the United States, and if by Pawsey going there we can make some return [sic] we certainly should try to do so. However, I must in all honesty say that I would hate to think that Pawsey would remain in the United States for the rest of his career. As far as we are concerned this is really up to him, but I would certainly like to see him back in Australia and in the Division of Radiophysics. In a sense, of course, the fact that we have had so many younger men coming to prominence in this science has presented us with a financial embarrassment. CSIRO has quite deliberately decided to support its Radiophysics group. We were extremely grateful to Carnegie and Rockefeller for the very generous donations which when matched by private donations here [the private donations claim was a clear exaggeration] and by our Government, have enabled us to give to that Laboratory the large radio telescope. However, you will, I think, appreciate that, within a short space of time and particularly when faced with other competing claims of equal merit from equally good scientists in other fields, it is not always easy to persuade one's Government to donate one large sum after another. Hence I must record the very consider-able interest I take in the approach that has been made by the University group to your Foundation[7] for assistance in going ahead with the project that Mills has put up to you. I told Keller this when I saw him in Australia (mid-December 1961) and I was very pleased that you sent him out here so that you and your Foundation would know at first hand something of the situation.

Bowen had visited a number of US colleagues at NASA, the Ford Foundation and the NSF in January and February 1962. The CSIRO lobbying at the Ford Foundation continued when Bowen visited Borgmann on 25 January 1962. Bowen was very positive about the chances of the Wild project being funded as he wrote White on 2 February 1962:

---

[6]This letter was likely a pre-emptive attempt to dampen the perception at the NSF of Australian conflicts between the University of Sydney and CSIRO. This conflict was exposed a month later on 31 March 1962 with a controversial publication in the Australian weekly *The Bulletin* in their series "Science at Work", entitled "Mills Cross versus Parkes Dish" (by an unnamed science correspon-dent). See NRAO ONLINE.8 and Chap. 38. The assertion was made that CSIRO had attempted to block NSF funding of the Mills Cross. "The split between the University groups . . . and the CSIRO group is so marked that there have indeed been efforts to influence the US National Science Foundation against financing the Mills Cross project and rather devote funds to the CSIRO group." Not surprising this article created ill will in Australia (NRAO ONLINE.8).

[7]This letter from White to Waterman may have been in reaction to the letter from Bowen to Waterman from 5 October 1961, with criticism of the proposal from Mills for support of the Super Cross project at the University of Sydney.

[T]hey are being very friendly. I had a long talk with Carl Borgmann and then John McCloy, Chairman of the Board. We [have] many mutual friends, like Alfred Loomis and Jay Stratton [a trustee]. Nothing at all was said about how the decision might go. If I were asked to guess the outcome, I would say they will come in for a half share. If we are lucky and tread the primrose path carefully enough, they may go the whole way.[8]

Pawsey Returns to the US

On 13 March 1962,[9] Pawsey left Sydney for San Francisco. During a layover in Honolulu, he met with Bowen at the Reef Hotel. The main topic of conversation was the "Wild" project and the likelihood that the Ford Foundation would fund it.[10] He flew on to Los Angeles, and on 14 March 1962, he visited Caltech with a full day of conferences. In the morning, he met Gordon Stanley.[11] Stanley warned him about his new position as NRAO Director. Pawsey noted in his diary: "Greenbank (sic)—very difficult. Gordon hostile to the idea [of Pawsey being director] because Greenbank gets all money and universities none. No suggestion or basis for cooperation." In addition, Pawsey heard from Stanley that the Caltech group was quite "distrustful" of the Stanford group of Bracewell: "Ron has stirred up trouble. Charles [Seeger] will do his best to suppress funding [of other university groups]." Stanley must have anticipated that Pawsey would do everything in his power to improve cooperation within US radio astronomy. Pawsey also met Bruce Rule who continued to be quite negative about the likelihood that the 140-ft telescope would ever be a success: "strong suspicion that [it would be] cheaper to start again". The long meeting with Otto Struve was also fruitful; Struve gave Pawsey his opinion about the AUI officers, the character of Rabi and Reynolds as well as the AUI trustees, Weaver, Goldberg, Whitford and Backer. Struve also gave his assessment of the NRAO staff.

Pawsey travelled to New York for the AUI Executive Committee meeting, joining a number of colleagues at the Harvard Club: Dunbar (Secretary of AUI), Gerald "Jerry" Tape (Vice President), Edward Reynolds (from Harvard University)

---

[8]C3830 Z1/17.

[9]The trip was reorganised with a departure 2 days earlier than originally planned due to insistence of the AUI board that Pawsey attend their meeting of the Executive Committee in New York on Friday 16 March 1962. Pawsey had to forgo a visit to Stanford with this tight schedule.

[10]They also discussed the NASA project to carry out studies on antenna performance using the new 210-ft radio telescope (see Chap. 33).

[11]The visit to Caltech was cut shorter than had been initially planned, due to the desire to reach New York on the evening of 15 March. Dinner, hosted by Helen and Gordon Stanley, included Mukul Kundu and his wife (then at the University of Michigan) and Charles Seeger (from Stanford). Stanley (1994) reported that Pawsey stayed "the night with my family … He asked me why things had happened that way [apparently the departure of Bolton in December 1960 from Caltech to return to CSIRO in Sydney], to which I had no answer."

and David Heeschen (from Green Bank) for discussions before the AUI Executive Committee meeting the next day.

Since Rabi was in Boston to receive the Compton Medal at MIT, Tape introduced Pawsey at the Executive Committee meeting. He told the committee that Pawsey would assume fulltime duties on 1 October 1962. After the meeting, Pawsey had a private conversation with Dave Heeschen, acting NRAO Director. Then Pawsey and Tape travelled to Boston where they met with Rabi on Saturday, 17 March 1962. Strikingly, the "high resolution" [i.e. the interferometer] project planned for NRAO by Pawsey was discussed in some detail among the three.[12]

Later, Pawsey flew to Newark, New Jersey, from Boston (18 March 1962), where he was met by his brother-in-law, Ted Nicoll. On Sunday, he enjoyed some family time. He, Ted and Kate Nicoll all discussed the suggestion that 17-year-old Hastings would go to Cornell after the Pawsey family (only Joe, Lenore and Hastings) moved to Green Bank in October. Pawsey thought he would also ask Jerry Tape for advice about Hastings's university prospects.

On Monday, 19 March 1962, Pawsey met his friend Carl Borgmann of the Ford Foundation (Program Director for Science and Engineering) to talk about possible support for the "Paul Wild" project—the newly planned radioheliograph[13]—and soon thereafter Borgmann rang to say the Ford Foundation Board of Trustees had approved the proposal for $550,000. There would be an additional $80,000 in March 1966, with no requirement for an additional US collaborator.

On the next Friday, 23 March, Pawsey wrote[14] a letter on NRAO stationary to Bowen and Wild with the good news: "So all is well, I am delighted." He had already convinced Borgmann that CSIRO would provide continued operations support for the radioheliograph if the Ford Foundation made the grant.

Tragedy at Green Bank, March 1962

On 19 March 1962 (Monday evening), Pawsey and Tape travelled together by train from New York arriving at Green Bank the next morning. Tape soon departed; they planned to meet again on 27 March in Washington for discussions with the NSF. The purpose of Pawsey's visit to Green Bank during the following week was to orient

---

[12]Letter from Pawsey to Rabi from the hospital in Boston on 27 June 1962. Joe and Lenore Pawsey Family Collection.

[13]C3830 Z3/1/X. On 27 September 1961, Borgmann took Pawsey to Idlewild (the international airport in NY in 1961). They discussed the proposed grant. Pawsey: "What are the prospects on the solar side? I wonder? When we parted at Idlewild you were going to do a little informal exploration as to the prospects of getting us some financial backing for Paul Wild's project for radio pictures of the sun. I suggested at the time that I could prepare a better statement on the subject than verbal story and I should be very happy to do this now if it would be of service." During the meeting, Pawsey also discussed the idea of Hastings attending Cornell, an idea supported by Borgmann.

[14]C3830 F1/4/PAW8.

himself in the new position and to meet the NRAO staff. The 300-ft and the 140-ft telescopes were under construction.

At some point on Friday afternoon or during the weekend, Campbell Wade, David Hogg and Frank Drake (Drake and Sobel, 1992) reported that Pawsey was observed walking with difficulty, dragging his left leg.[15] Drake quoted a conversation in which Pawsey said: "I am asymmetrical." He experienced pronounced paralysis on his left side, both arm and leg.

Based on four letters from Lenore to Joe and one letter from his mother,[16] written in the period 17–26 March 1962, it is apparent that Pawsey was already unwell before his departure from Sydney on 13 March. On 17 March, Lenore wrote: "I wonder if your trip has been going according to schedule **and if you are feeling strong again**" (our emphasis). On 18 March 1962, his mother wrote: "It is distressing that you were not 100 percent in health when you had to go. After the meeting [in New York of the AUI Board] I am guessing that your plans will be your own and you could rest if necessary." Then a few days later (21 March) Lenore wrote a letter to Pawsey in care of her brother in Princeton: "I hope you have no more 'incidents', it didn't make the [newspapers in Sydney] the next day . . . Hope you are feeling well and regaining your strength. You'll need it all for the next 6 weeks [until his expected return to Sydney in mid-April] to say nothing of the next 6 months." On 24 March 1962, Lenore wrote again: "Hope you are getting over your tiredness. Get as much rest as you can for the gruelling week at the end [expected visits to Washington] and with a trip to Europe and then to Canberra [for the expected Australian Academy of Science meeting in late April]."[17] On 26 March, Lenore wrote again to Pawsey: "Do try to get as much rest as possible, [Dr Ian L. Thompson, the family doctor] had said that the wog [minor illness] you had will leave you tired for weeks. He was really surprised how quickly you recovered that you were able to get off [for Green Bank from Princeton] on Tuesday [20 March]." This statement implies that Pawsey had written Lenore from either Princeton or Green Bank around 20 or 21 March 1962 with news of the illness. Clearly, Pawsey had some type of unspecified illness consisting of "tiredness" before and during the trip from Sydney to the US, and that his family was quite concerned.

The local Green Bank and NRAO doctor, Dr Martin, examined Pawsey and said he was fit to travel on Tuesday 27 March 1962. The plan had been that Pawsey would return to Washington to meet Jerry Tape and the NSF staff.[18] After

---

[15] The letter to Bowen and Wild from Friday 23 March about the Ford Foundation is likely the last correspondence he completed before the signs of the brain tumour appeared in late March 1962.

[16] Joe and Lenore Pawsey Family Collection. (Also see ESM 40.1, Correspondence, Lenore).

[17] Pawsey had planned to visit his daughter Margaret (and the new granddaughter Ann) in Paris and then to be back in Sydney for Hastings's 17th birthday on 24 April 1962.

[18] JLP's calendar contains a listing of appointments for 27 March 1962. He was to see Tape at 7:30 and Keller of the NSF at 9:30. Later in the day he was to have dinner with Keller. The next day (28 March) he was to visit the Naval Research Laboratory, Ed McClain. None of these appointments occurred due to the illness.

consultation with Rabi and Tape, Pawsey was driven[19] to Washington, where Tape contacted Pawsey's friend Merle Tuve for advice about the paralysis. Fortunately Tuve's wife, Dr Winifred Whitman (1901–1993), MD—a psychoanalyst—could advise as she had personal experience with paralysis. Dr Whitman recommended Dr Hugo V. Rizzoli (1916–2014), a well-known doctor and Chief of Neurosurgery at the Washington Hospital Center (WHC). Pawsey was admitted into WHC on Tuesday evening, 27 March. Polio was excluded immediately.

Two days later on 29 March 1962 (Thursday),[20] Campbell Wade from NRAO came to WHC to assist. Wade had been asked by David Heeschen to be a "gofer" (Wade's description of his role, as an "errand boy") for Pawsey. He was to run errands, make phone calls and look after the post[21] for Pawsey. For 3 of the 5 weeks that Pawsey was in WHC, Wade spent the week days in Washington and returned home for the weekends, looking after his young family in Green Bank.

In addition Wade brought Pawsey some news of Green Bank. On 5 April 1962, he brought a fine photograph (Fig. 40.1) of the Green Bank observatory site in the valley from Allegheny Mountain looking west. Wade wrote that this was a "glimpse of the outside world . . . You can make out our buildings and telescope if you squint closely enough!"

On 29 March 1962 Pawsey dictated a key letter to Wade,[22] which was to be sent to the family in Sydney. On 2 April, Lenore replied:

> The letter you dictated to Campbell Wade came this afternoon with the news of your illness. You don't need me to tell you how much I will be thinking of you. I will be waiting anxiously for more news . . . I am so glad Campbell is with you. It is such a help to have someone right on hand to do things for you—like write letters. I do hope you are not in actual pain. Campbell says you have a very nice room, Joe, for which I am glad and apparently good medical attention. We must just hope that everything will turn out all right—your normal philosophy. I have always admired your philosophy and now we must both use all of our "powers of positive thinking" and think only of a quick and complete recovery. Margaret [their daughter in Paris] will be so disappointed if she can't see you. Perhaps you will still go

---

[19] The circumstances of Pawsey's trip to Washington remain unclear. Neither Wade nor Heeschen took him to DC. Wade arrived in DC 2 days later. Heeschen wrote a letter on 28 March (Wednesday): "We were all most sorry to hear that you have been laid up. I hope it will not prove to be unduly serious and that you will be up and around again soon," implying that Heeschen had not been in Washington at this time.

[20] Wade's visit is based on a 2 April 1961 letter from Lenore to Pawsey, a letter from Lenore's mother to Pawsey in Washington and his appointments calendar. Mabel Nicoll, Lenore's mother, was at her son's house in Princeton, soon to leave on 14 April for the UK. In the original itinerary proposed by Pawsey, he expected to travel from Washington to Princeton on 29 March. Mabel wrote: "When you phoned that your trouble was in the leg, I immediately thought of the trouble you once had with varicose veins. I hope soon it will leave as suddenly as it came and that you will be resuming your trip as planned."

[21] A copy of a letter that Wade wrote on Pawsey's behalf to colleagues in Hobart (Tasmania, Australia) has been found in the Joe and Lenore Pawsey Family Collection. Pawsey had to decline giving an invited lecture at the University of Tasmania.

[22] Mabel Nicoll also reported that Campbell Wade "has come to your aid with helping you with your correspondence".

Fig. 40.1 Site of the Green Bank Observatory, 1962. Photo provided by Campbell Wade to Pawsey. Joe and Lenore Pawsey Family Collection

across. The AUI is certainly very generous to look after all your medical bills so you don't have to worry. Don and Thelma [McLean, parents of their son-in-law, Margaret's husband Don McLean] send their best wishes and said that they will do anything to help—take the boys [Stuart and Hastings] if I had to go across [to the US] etc … Have you told your Mother yet?

(Additional correspondence between the Pawsey family in Australia and Pawsey in the US is summarised in ESM 40.1, Correspondence, Lenore).

In addition to Merle and Winifred (Whitman) Tuve, two support groups in Washington were mobilised:

(1) The Nicoll family. Ted was already in Washington during the weekend of 31 March.
(2) The Australian Scientific Liaison Office of the Australian Embassy in Washington.[23] William Hartley, the ASLO Officer was assisted on a daily basis by Margaret Pennington, his secretary. Both Hartley and Pennington spent many hours at the hospital, where numerous letters were dictated to Margaret Pennington over the phone or in person. They informed CSIRO headquarters (Guy Gresford, the CSIRO Secretary) and the Division of Radiophysics in Sydney by cable. On 5 April, Hartley wrote Gresford a detailed letter with a summary of the history of the illness up to this time. He emphasised

---

[23] In 2014, Goss discovered this invaluable collection in the NAA C3830 Z1/52, Correspondence re Dr Pawsey from 5 April to 31 July 1962. This collection had not been listed in the online description of C3830. Apparently, CSIRO considered this personal correspondence when it was classified in 1962.

the key role played by Dr Winifred Whitman in organising the WHC care and pointed out that originally the symptoms were not thought to be serious but "it was thought desirable to put him into hospital for observations and checks. At this stage the condition **developed** (our emphasis) into a partial paralysis of one arm and leg." After 27 March the paralysis had increased in severity; many tests were run. Heart conditions, brain injury and spinal troubles were eliminated. The formal diagnosis was "toxic sensitivity", in modern terminology MCS, multiple chemical sensitivity.[24] Pawsey was naturally quite frustrated by the "absence of any definite diagnosis."

Of course, the family back in Sydney (Lenore, Stuart and Hastings) and in Paris (Margaret Pawsey McLean, Don McLean) were concerned. Finally, on 7 April, Pawsey phoned from the US to Australia, an unusual experience in 1962, to reassure the family. The call was a great relief for the Pawsey family. On 11 April, Lenore wrote Joe: "This is mainly a business letter, but if you don't feel like taking any action, don't bother, as I think everything is under control ..." She had sent in Hastings's application to Cornell and had received a:

> very nice letter from Campbell Wade yesterday telling me how things were going and that he and Mary Jane [Wade's wife] were going to drive over [from Green Bank] to see you the next day ... Try to make use of the enforced rest and build yourself up. It will be much better when you can sit up. I am hoping you are improving every day ... If all the good wishes of your friends are effective, you should soon be on the mend. I hope you continue to feel well yourself. That must be important in your recovery, also your optimism must be a great help.[25]

On Friday 13 April, another key letter[26] was sent by Pawsey to his family in Sydney with copies to his mother in Melbourne and to Margaret in Paris. The letter had been dictated to Margaret Pennington at the hospital. He explained that when he arrived in the US, he was **tired** [our emphasis] but no serious symptoms were evident. "Towards the end of that time [in Green Bank] some rather curious symptoms developed. I found that my left arm and left leg were becoming very weak." Dr. Martin recommended that he seek expert advice. When he arrived on 27 March in Washington: "I was starting to become crippled." He recounted the role of Merle Tuve and Winifred Whitman as well as Dr Rizzoli. After entering the hospital: "For the first few days I went downhill and I think things were quite serious at that stage, but [by 30–31 March] I began to recuperate. [Many tests were run, none with any positive results.] As Dr Rizzoli put it—no one knows why the lesion developed in my nervous system."

---

[24] A chronic medical condition characterised by symptoms attributed to low level exposure to toxic substances, such as solvents. Paralysis due to MCS was rare.

[25] Contact with all of the family became frequent. For example, Stuart wrote his father on 2 and 18 April with an addendum written by his younger brother, Hastings, on 2 April.

[26] Joe and Lenore Pawsey Family Collection. During this period, his recovery was mixed. On his calendar, Pawsey wrote for 12 April: "PT [physical therapy]" and "Horror night". The next day he wrote the word "Wrecked" in his calendar.

The phone call to Australia on 7 April 1962 early morning (late evening 7 April in Australia):

> ... gave me a tremendous stimulant. It was a sort of claim for recovery and I have been getting better steadily ever since. The present situation is that I have partial use of my hand and leg ... The outstanding feature of my stay here has been the way in which my friends have looked after me. Merle Tuve has been my "sheet anchor." I have known him for many years as an outstanding scientist, but he has also a tremendous power of installing confidence into me and I owe a lot to his help and moral support.

AUI had also been extremely generous, paying the hospital and medical bills. Jerry Tape of AUI and the ASLO people had visited him, bringing flowers and running errands. "The doctors' report is very favourable, so I don't think you should worry. I shall be with you fairly soon and completely recovered before long."

About the same time (11 April 1962), Bill Hartley sent a positive update to Bowen: "He got out of bed on 11 April, was very alert and was much more cheerful than formerly and feels that he is now set on the way to recovery, [even though there was no definite diagnosis]. All things considered Pawsey is quite comfortable in hospital and is surveying the American scene as revealed by TV and radio." He had frequent visitors including several by Ted Nicoll. Bowen replied on 13 April to Hartley, thanking him and Tuve as well as revealing good news about funding: "We are delighted to hear the wonderful news from the Ford Foundation, together with the earlier grant from NASA [for GRT development work with applications to the Deep Space Network, $172,000]."

Numerous letters of concern were sent to Pawsey: Bok,[27] Bowen, White,[28] Wild, Mathewson, Bracewell and others. Wild's letter of 15 April 1962: "We have rung Lenore several times—she seems to be bearing up well with the anxiety, which is magnified a hundred times by the distance between you ... The news about you rather took the sugar off the Ford Foundation cake, but we were very pleased to get the official letter last week." He ended on a high note: "... [G]et well quickly—we can't afford your wasting your time away in hospital!"

On 19 April 1962 the AUI minutes of the AUI ExComm reported:

> [Pawsey's] recent progress has been encouraging; there is every reason to suppose that his recovery will be sufficient to permit him to assume full-time duty at the Observatory starting October 1, 1962, as planned ... At [Rabi's] request, Dr William H. Sweet [1910–2001, AUI trustee and Professor of Medicine at Harvard], who talked with [Dr Rizzoli], explained that although there has been no precise diagnosis of Dr Pawsey's malady, his recovery is going forward briskly.

---

[27] Bok wrote on 18 April from Sydney (on a lecture tour). "All your friends are terribly relieved to learn that you are getting back the use of your left arm and leg and that there is not anything very serious back of the sudden illness and partial paralysis. But do take it easy before you head back for home, for we would rather see you a bit later and in good shape than to have you take to bed again right away!"

[28] "It must be a very great disappointment to you to have taken ill during what was to have been no doubt a very important and interesting visit. However, the diagnosis seems now to be fairly certain and I do hope you will rapidly recover."

Fig. 40.2 Ted Nicoll and his emaciated brother-in-law J.L. Pawsey, date end April early May 1962. Princeton New Jersey (photo Ted Nicoll, Joe and Lenore Pawsey Family Collection)

Dr Sweet expected that Pawsey would recover and be able to function effectively as NRAO Director.

On 23 April 1962 (Monday), there was progress. The calendar has the word "GOT DRESSED"; the next day he remarked that "Mrs B [presumably a nurse[29]] took me for walk outside". Two days later (Wednesday), Pawsey wrote "got brace. Rizzoli says I can go Sunday. Take few steps". Thursday, his optimism continued: "Exchanged crutch for stick. Got up off ground [perhaps he fell]."

On 27 April (Friday), Bill Hartley had good news for Bowen. Pawsey would be taken on Sunday 29 April to Princeton by Ted Nicoll (Fig. 40.2). He was to spend 1–2 weeks before returning to Sydney (via Europe to see Margaret and his granddaughter). Hartley was fearful that he was being released too soon from WHC, since his mobility was limited. "[H]e is now able to use his left arm, and especially the left hand. However, his left leg is in a rather poor state and he can only hobble around with difficulty using a crutch for any larger distance."

Pawsey wrote in the calendar: "Final discussion with Rizzoli. He said I could take the [NRAO] job tomorrow."

Pawsey was discharged 29 April 1962. He wrote Fred White thanking him for his get-well letter. Pawsey was frank about his illness: "It has been a bad illness; I think at one stage I did not have much margin, but I am pleased to be able to say that I am

---

[29]David Hogg remarked in a letter from 24 April that he heard that Pawsey had a good long walk, assisted by a friend nurse. Joe and Lenore Pawsey Family Collection.

well on the mend." He expected to make a full recovery, with perhaps some "traces of weakness". He also told of his meeting with Borgmann of the Ford Foundation in March:[30]

> ... I could reassure [Borgmann] that you and Bowen were completely behind Paul [Wild] in the project and would do everything necessary to make it a success even though there could be changes in the organisation of the Radiophysics Division itself [that is, Pawsey was to leave CSIRO later in 1962]. I would like to discuss this a little bit further when I see you on my return. With the Ford grant assured it meant that the three big projects in which I had been interested in seeing established—the GRT and Bernard's [Super Cross] and Paul's [Radioheliograph] projects—were all successfully launched, and in the dark days of my illness, this knowledge caused me very considerable satisfaction. Each of these three projects is now a joint American-Australian effort and this introduces a quite new phase in Australian radio astronomy. The two countries are now involved in large-scale co-operative science. I cannot see the implications, but it is pretty clear that I am very closely involved [as the new NRAO Director].

On 30 April 1962, Pawsey continued his optimistic assessment in a letter to Bowen (dictated on the phone to Margaret Pennington): "I am approaching the end of this period of illness. I am improving mightily from day to day and I expect that in a few months' time I shall be completely recovered".

The *NRAO Recreational Association Newsletter* reported on 30 April 1962:

> Dr Wade visited Dr Pawsey in the hospital in Washington last week, and reports that Dr Pawsey was to be discharged from the hospital on Sunday, April 30 [in fact it was 29 April], much improved. He will spend a couple of months with relatives in New Jersey recuperating before going back to his home in Australia. Dr Pawsey's recovery has been much better than his doctors expected, and he should be able to return to NRAO in September as planned.

A noteworthy factor at this point was that the paralysis had never been diagnosed at Washington Hospital Center. On 2 May 1962 while in Princeton (going to physical therapy every day), he could walk with a cane. Two days later he could not walk at all. A serious setback had occurred.

On Sunday 6 May 1962 Pawsey's brother-in-law took him to Massachusetts General Hospital (MGH) in Boston. I.I. Rabi, the AUI President and resident of Princeton,[31] arranged for William H. Sweet to be in charge of the MGH treatment.[32] Pawsey was admitted to MGH on Sunday evening 6 May 1962. In addition to the deterioration in the paralysis (thought at first to be atherosclerosis), the MGH medical staff described a condition of latent diabetes and prostrate problems;[33] further tests were carried out. On 8 May, Ted rang his sister Lenore back in Sydney

---

[30] Joe and Lenore Pawsey Family Collection.

[31] Rabi visited the Nicoll house on 4 or 5 May. Ted Nicoll's daughter Ruth Bronzan told Goss in September 2010 (she was home from her studies in Philadelphia) that her mother had told her: "A small man will call at the door. He is very famous—a Nobel Prize winner. He is here to organise the next step of Joe Pawsey's health care." Ruth Bronzan reports that he came in the house and "took charge immediately."

[32] The 150-page medical report from MGH obtained in 2011.

[33] Pawsey had been to a doctor in Sydney in March; a prostatectomy had been planned for July back in Sydney. In fact, it was found to be benign and was removed on 6 June 1962 at MGH.

with the news of Pawsey's rapid decline, suggesting that she travel immediately to Boston.

Dr Sweet had begun to suspect a possible brain tumour. A WAIS (Wechsler Adult Intelligence Scale) was administered to determine a baseline of Pawsey's cognitive functions before any possible surgery. The test was carried out on 10 May 1962. The results were:

> The patient is functioning at a very superior level and dull normal non-verbal. Test results are indicative of right-sided organic involvement. Dr Pawsey is a 53-year-old right-handed physicist from Australia. During the examination session the patient was cooperative, and rapport is considered good. On the WAIS, general intelligence is bright normal with a striking discrepancy between a very superior verbal level and dull normal non-verbal. Memory for designs is average and that for words better. Information, verbal abstraction, and mental calculations are exceptional, and comprehension, reasoning, and vocabulary very good as is auditory span (7 digits forward and 6 backward). Detection of details missing from pictures, analysis of block designs, and arranging pictures to make sensible stories are average. Coding and object assembly are markedly poor. There appears to be impaired ability to recognise the whole from its parts and to organise spatially.[34]

By Saturday, 12 May 1962, Pawsey was deteriorating with increased paralysis, as well as gradual mental decline. Dr Maurice Victor, a neurologist, wrote in the medical report: "There is no question about the progression. There is virtually no voluntary movement in the left arm and none in the left foot ... He thinks very slowly, but clearly, is fully oriented ... He is easily moved to tears, when talking of his wife and children ..." Then the paralysis spread to the right hand. Dr Victor thought that Pawsey had atherosclerosis due to the diabetic condition but admitted that carotid artery disease was also possible. He wanted to determine the process in the right cerebral hemisphere. By 13 May, communication between Pawsey and Ted and Kate was almost impossible. Dr Sweet was not optimistic: "Even eating a banana this morning wore him out."

On 14 May (Monday), a major tumour was found with a right common carotid angiogram. The results were not a surprise. The radiologist reported: "The appearance would be consistent with a glioma [a tumour that arises from the brain] or possibly a metastasis."

Lenore arrived in Boston 15 May 1962 (Tuesday). The operation to remove the tumour—temporo parietal craniotomy[35]—was scheduled for the next day. Dr Sweet was the surgeon, removing a $4 \times 3 \times 2$ cm tumour in a 6.5-h operation. Pathology immediately confirmed that this was a Glioblastoma Multiform (GBM), an aggressive malignant primary brain tumour. Even in the present era, half the patients with

---

[34] Almost 2 months later, the post-operative WAIS was repeated. Fortunately there was little change: "The general level is again bright normal, very superior, and non-verbal dull normal ... There is no confusion verbally on the WAIS. He responds without impairment in accuracy or quality. Non-verbal remains same as before [with ability to learn design pairs less able] ..." Thus the mental ability of Pawsey was still unimpaired.

[35] *Craniotomy* is the surgical term for the removal of part of the bone from the skull to expose the brain.

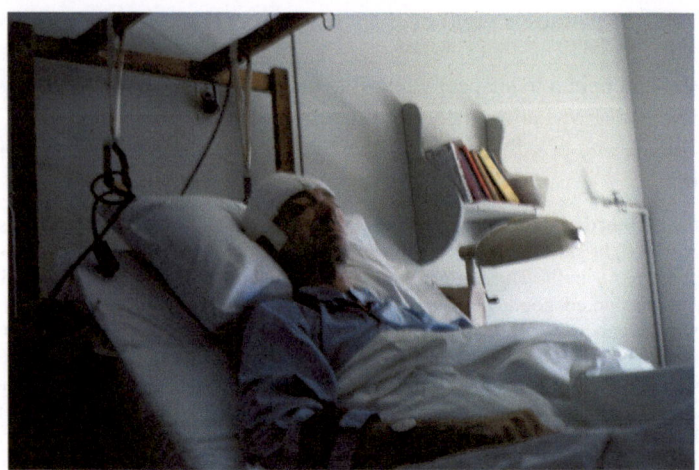

Fig. 40.3 Pawsey at the Massachusetts General Hospital shortly after the surgery, date 17 or 18 May 1962 (photo by Ted Nicoll, Joe and Lenore Pawsey Family Collection)

GBM die within 1 year, while 90% are dead within 3 years. In 1962 the life expectancy was only 9–12 months. The outlook on 17 May 1962 was bleak.

The day after the surgery Dr Sweet visited Pawsey. Pawsey could speak and repeat test phrases (Fig. 40.3). Dr Field, the internist concerned with the diabetes, wrote: "Perhaps more alert than immediately before the operation. Recognises me, recalls contents of our conversation [previously] re Fred Hoyle's theory of cosmology. Can lift left leg ... Making good recovery from surgery." He felt the diabetes should pose "no great problems."

The final report of the surgery was written by Dr William Sweet on 28 May 1962:

> This man has a highly malignant glioblastoma with surprising little mental deterioration although considerable motor and sensory deficit. I would agree that radiotherapy certainly might be tried and some benefit hoped for. Being highly anaplastic and invasive, I would think that probably the major portion of the hemisphere is going to have to be included in the treatment field ... We will start [radiation] treatment as soon [as possible].

This radiation treatment started on 29 May 1962 by Dr Schulz who wrote at the end on 27 June: "Treatment went exceptionally well ... [T]here has been a rather pronounced improvement in this man's condition. He will return to Dr Sweet for follow-up." On 29 May, Pawsey went outside for the first time and the news on 30 May was good: "Patient psychologically doing better and visit of physiotherapy, outside during day in wheel chair." Then on 29 June 1962 (a month later), additional improvement was indicated when the medical record states that he could walk down stairs holding onto the bannister.

On 2 July 1962, Pawsey dictated a letter to Lindsay McCready back in Sydney. He said he could walk with a stick and would likely be discharged from MGH in 1–2 weeks. The plan was to go to Princeton and then visit Green Bank in order to show the site to Lenore, before returning to Australia for recuperation. The Pawseys would return to Green Bank in October as originally planned. He would be discharged from MGH on 12 July 1962.

Clearly there was a disconnect—a disparity between reality and Pawsey's perception. Pawsey's viewpoint concerning his prognosis was far too optimistic. As often can occur, the doctors may have de-emphasised the long-term prospects. This misplaced perception was shown by Pawsey the next day (3 July 1962) in a letter[36] of congratulations to Don Mathewson on his upcoming marriage to Xenie Federoff. "I am still in hospital . . . but I'm getting along fairly well and the doctor expects that I shall live a normal life in a few months' time. I shall always be left with a residual weakness in the left arm, I'm sure, but I think I shall be able to walk normally. At present I walk with a stick, but considerable instability still."

Two CSIRO RPL colleagues, Arthur Higgs and Paul Wild, visited the US in June and July 1962. Arthur Higgs, Division Secretary RPL, was in the US visiting family and US institutions as part of a world tour that later included Europe, Israel and Japan, before he was to return to Australia. Higgs provided the CSIRO Executive and Bowen with a realistic assessment of the status of Pawsey's health, and discussed Pawsey's future with AUI, meeting with Tape and Rabi at the AUI office on 15 June 1962. Higgs pointed out that there had been far too much false optimism about Pawsey's recovery.

At the AUI meeting that day (16 June 1962), Dr Sweet presented a negative report.[37] The minutes show:

> Dr Sweet considers it highly unlikely that Dr Pawsey will be able to serve as Director of the National Radio Astronomy Observatory for the full three-year term of his appointment . . . The chance of permanent recovery in the case of a brain tumor of the type which Pawsey suffered is remote. Dr Sweet believes that Dr Pawsey's own wish is to spend a relatively short period of time at Green Bank when he is in a position to leave the hospital [three to five weeks]. He then wishes to return to Australia to complete some unfinished business and hopes to be able to come back to the United States and Green Bank when his work there is completed. Dr Sweet emphasised the numerous uncertainties involved in the situation . . . [There was little sign of optimism.]

He reported that the median life expectancy for this condition was about 9 months, and that the chance of Pawsey completing the 3-year appointment to NRAO was 1%. "There is no doubt that the tumour was malignant and that the recurrence in some form is extremely likely. In the meantime he'll have some paralysis but should be mentally fit."

On 16 June 1962, Higgs wrote to Bowen explaining that Rabi insisted he could not terminate the directorship appointment at this time without the concurrence of Pawsey. He continued:

> Neither Joe nor Lenore have yet been fully informed of his medical situation though I think he guesses. In view of Sweet's report, AUI are obviously very doubtful as to whether they should proceed with the appointment as originally offered, but Rabi feels that he couldn't possibly even suggest such a thing to Joe at the present time, and they all feel much the best

---

[36] Joe and Lenore Pawsey Family Collection.

[37] AUI archive.

thing for Joe would be to return to Australia to convalesce as soon as he is fit to travel. They will probably write to you or [White] privately to ask you to try and persuade Joe to do this—on the pretext of getting thoroughly well first before beginning to shoulder the man-sized worries of Green Bank. This afternoon Joe talked of wanting to show Green Bank to Lenore and getting a few of his ideas for their programme under way. Mid-July was suggested as a date when he might be able to leave here [Boston].

Pawsey remained unrealistically optimistic about his recovery.

We can trace the dramatic events of July in a series of files located by Ellen Bouton and Ken Kellermann in the AUI archives. The documents are four unsigned memos. All are dated 25 July 1962.[38]

(1) Note on Pawsey for Discussion with Wilde
(2) Notes on Future Work for Pawsey—for Discussion with Wilde [sic]
(3) Notes on NRAO Directorship—for I.I. Rabi; Summary of Ideas from Conversation with Trustees and Discussions at the AUI Executive Committee Meeting on July 20, 1962
(4) Notes on Dave Heeschen as NRAO Director

Based on the fact that Rabi was not present for the 19 and 20 July 1962 AUI Board, Jerry Tape ran the meeting in his absence. The likelihood is that Tape wrote the four documents for Rabi's benefit. During a closed Executive Session on 20 July, we may surmise that these documents were discussed in detail as they were likely introduced by Tape. (No report of this session was presented in the minutes.)

On 7 July Dr Sweet and Tape met with Lenore Pawsey and Ted Nicoll (brother of Lenore) at the MGH. Sweet provided the statistics of survival for glioblastoma patients with major surgery and post-op radiation treatment. The typical survival rate was 35% after 1 year, 17% after 2 years and only 5% after 3 years. In the document No. 1, Sweet is quoted:

> This is survival; the statistics for useful life are unknown but clearly more severe. We convinced Mrs Pawsey and Dr Nicoll that it would be unfair to Dr Pawsey and to NRAO to have Joe the Director in view of his recent medical experience, his difficulty in getting around and the mental pressures associated with the Directorship. Tape made the suggestion that Joe be "retained" on the NRAO program in some way because his future is tied up in its program ... [T]o sever all connection with NRAO would cause [Pawsey] to "give up".

The suggestion was made that Joe might be a consultant for a vague "large antenna system".[39] This project was a stimulus to prevent his "giving up" in the next few months after returning to Australia.

---

[38] NRAO/AUI Archives, NRAO, Founding and Organisation, Antenna Planning (Range 2B Box 2, Joseph L. Pawsey, 1961–1962). No 1 and 2 have a note in pencil "not shared with Dr. Wilde", in Charles Dunbar's handwriting. (Secretary of the AUI Board).

[39] Tape wrote that Wild was building a large array in Australia which was expected to be a superior instrument in a few years. "The US needs to study similar instruments for North America." Clearly Tape did not realise that the Australian array was a solar instrument with limited use for non-solar radio astronomy.

A major concern was the continuance of health and disability benefits which could only be insured if he returned to CSIRO employment; the time of only a few months as an AUI employee provided no disability benefits and minimal sick leave.

The following day (Sunday, 8 July 1962) Ted Nicoll took his brother-in-law on a long drive around Boston. The purpose was a frank discussion of Joe's future. By the end of the drive, Nicoll reported that Joe was finally reconciled to his bleak future: "... [He] was probably resigned to the point of view that he could not be Director."

The next meeting reported in the AUI memo of 25 July 1962 was at the Nicoll house in Princeton on 18 July. Dave Heeschen and Tape met with Lenore, Ted Nicoll and this time Joe. From Memo no. 1:

> I [Tape] told Dr Pawsey that he should not be Director of NRAO, but that we would like his help with the program if possible. I stated that under no circumstances should he sever his CSIRO affiliation, not even taking leave if that leave jeopardised his eligibility for benefits. He agreed with my arguments.
>
> He said he would return to Australia July 27 [1962] and spend the first couple of months on rest and therapy. He would like to keep a hand in NRAO programs. When Mrs Pawsey suggested that he might be the "Scientific Director" for Green Bank, I suggested that I thought work on a specific project, e.g. a large array, would be better. [Tape was reluctant to be too negative.]
>
> We left with the understanding that we would first work on Pawsey's health and then explore future work with CSIRO staff. I explained the possibility of a "cooperative program" on large arrays with CSIRO. Joe could stay on their payroll and we could transfer funds to CSIRO if necessary to assist in the work there.

The memo to Paul Wild (called Wilde) about Pawsey and his future work outlined the cooperative study proposal suggesting that NRAO personnel might visit Australia and CSIRO staff might go for short visits to Green Bank. (In the end nothing came of this suggestion.) Tape wrote in the memo No. 2:

> Such a study would capitalise on Pawsey's contributions over the next year by which time his own physical condition and future productivity can be re-evaluated.
>
> I had planned to write White along these lines so that they could have more flexibility in helping Joe with a work program. **I gathered from Pawsey that he would prefer communication with White to Bowen!** (our emphasis) Wilde [sic] should be able to carry these ideas back; however, check with him re desirability of a letter to White.

On 9 July 1962, White wrote Pawsey,[40] following the suggestion of Higgs:

> ... I sincerely hope that you will soon be well enough to return home. It does seem to me that the next proper step is for you to get back here so that you may recover your health and strength in your own home. I know it will be a very severe disappointment to you to have had, by force of circumstances, to embarrass AUI by not being able to go on with the Green Bank project as they planned. However, your future health is much more important than that, both to you and to us. No doubt Rabi will discuss with you whatever future arrangement might be possible before you leave. You may rest assured that we will do all we can to facilitate anything that may be arranged.

---

[40] Joe and Lenore Pawsey Family Collection.

Joe Pawsey was released from MGH on 12 July 1962 (Thursday). On 20 June, Pawsey had been placed on the CSIRO payroll, retroactive to 1 May 1962. This step solved US visa and tax issues, as well as allowing a reversion to Australian superannuation [retirement benefits]. On 8 July, AUI agreed to pay all of Lenore's expenses to and from the US. A few days later, the decision was made by CSIRO that they would pay all medical and hospital expenses for Pawsey after 1 May. On 12 July, the CSIRO Executive declared that "we would of course welcome his return to his post at Radiophysics ... [H]e will be welcome back in the laboratory as soon as he is fit again."

The official AUI minutes of 20 July 1962 reflected the final decision:

> Dr Tape reported that Dr Pawsey had been discharged from Massachusetts General Hospital on July 12 ... Arrangements have been made for him to return to Australia on July 27, and he and Mrs Pawsey will be accompanied by Dr Paul Wilde [sic] ... Dr Pawsey is now reconciled to the decision that he should not undertake the Directorship. Dr Pawsey has indicated his wish to continue some participation in the work at Green Bank, and the possibility of a cooperative program of some kind will be explored ... The future of the Directorship is still uncertain. In the meantime, Dr Heeschen will continue to function as Acting Director.

After his release from MGH on 12 July 1962, Pawsey and Lenore went to the Nicoll home in Princeton.

In Fig. 40.4, we show Pawsey in Princeton shortly before his departure to Australia on 27 July 1962; his appearance shows deterioration compared to Fig. 40.2. The casts on his left arm supported his paralysed arm.

Pawsey Returns to Australia

Pawsey posted a handwritten letter to Bowen[41] on 20 July 1962; the quality of the writing showed no sign of increased deterioration. Possibly this was the last letter that he was able to write on his own:

> I am at Lenore's brother's in Princeton, having come out of the hospital a week ago. [He then described the details of their arrival in Sydney with a request that Stuart Pawsey bring their car to the Sydney airport.] ... Paul Wild[42] has delayed his return [by a week] to come on the same plane and he will be a great help. I am still pretty much of a cripple and shall have to spend a long time convalescing but conditions should be good. In general health I am making rapid strides—in fact this last week I appear to have put on 10 pounds in weight ...

The Pawseys and Wild departed from New York City on 27 July 1962 on QANTAS (First Class) to Sydney via San Francisco and Fiji. At the intermediate stops, Pawsey got off the plane for short walks; in Sydney, he walked off the plane with some

---

[41] Joe and Lenore Pawsey Family Collection.

[42] In ESM 40.2, Rabi, Pawsey, Wild, we describe Paul Wild's trip to the US and his negotiation with Rabi about an offer of the Directorship of NRAO. He had changed his itinerary, postponing his departure from the US in order to accompany Pawsey and Lenore as they returned to Sydney.

Fig. 40.4 Pawsey in Princeton in late July 1962 (before 27 July); the cast supported the paralysed left arm. A crutch was required for walking due to the paralysed left leg. Photo by Ted Nicoll, Joe and Lenore Pawsey Family Collection

assistance, switching to a wheelchair for the transfer to the terminal. They were met by Stuart and Hastings, their sons, and a host of friends.[43]

During the first week in Australia, Pawsey visited his doctor on two occasions. Numerous letters of good wishes were received.[44] Colleagues visited, including Lindsay McCready and Brian Cooper with news of the "polarisation story"[45] and Fred White and Taffy Bowen. Pawsey went to the RPL laboratory to meet Paul Wild, Jim Roberts and Taffy Bowen. On 8 September 1962, his mother, Margaret Lade Pawsey, arrived from Melbourne. Five days later, Lenore wrote in her diary that Pawsey and his mother walked around the block in Vaucluse. The paralysis on the left side remained; in mid-August both his left arm and left foot were put in plaster casts to support the paralysis; he began spending most of his mornings in physical

---

[43] Met by families Pryor, McLean, Wild, Suzuki, Taffy Bowen, Jim Roberts. They went to tea at the Pryor's house with Lindsay McCready and the Christiansen family (Chris and Elspeth).

[44] A prominent example was from his former protégé, Joan Freeman, on 27 July. She wrote from the UK to both Joe and Lenore. She wrote Lenore with an ironic statement about good fortune: "... I don't know how much you believe in fate ... which may influence our lives, but I wonder if maybe your being in America just in time was the right thing so that just the right surgeon was available who had the skill to pull Joe through."

[45] The recent detection of Faraday rotation in the radio galaxy Centaurus A, published by Cooper and Price in *Nature*, 1962, vol 195, p. 1084 (Chap. 33).

therapy. He visited the RPL lab on some afternoons. Don Mathewson was a frequent driver for transportation to and from his home for lab visits and doctor visits.

Rabi remained concerned about Pawsey, writing to Wild on 20 August 1962 (ESM 40.2, Rabi, Pawsey, Wild) in which the NRAO Directorship offer to Wild continued to be discussed: "If character were all, Joe would now be a well man."[46] (Epigraph of this chapter). The next day Jerry Tape wrote Pawsey that he and Dave Heeschen would visit Australia in the near future. Pawsey replied on 5 September[47] 1961 in a dictated message:

> When we last met, we discussed the possibility of you [Tape], Rabi, Dave Heeschen coming out here to have a look at radio astronomy in Sydney. Since my return I have seen what has been going on and I think the case is much stronger than I had known then. In particular, there has been one discovery which I regard as one of the major break-throughs. You may remember when I wrote the report on future activities at Green Bank, I mentioned the possibility of observing linear polarisation and consequently getting information on these magnetic fields. The expected break-through is the discovery that certain of the radio sources which show linear polarisation show a very obvious Faraday rotation. [see Chap. 32] The plane of polarisation rotates with changing frequency in the precise manner which would be expected from passage of a wave though an ionised medium containing magnetic field. This is quite positive evidence for the existence of magnetic fields and it provides, for the first time in astronomy, an [estimate] of the strength of the magnetic field in space . . . With regard to my own plans, I cannot say any more than I could say in Princeton last. We just have to wait and see what sort of recovery I make.

On 25 September 1962, Pawsey wrote a revealing letter to Rabi. He mentioned that his left leg was now better than the left arm. The left hand was not much use at this time.

> It is very hard for me to properly assess my progress, but it is clear that the right thing for me to do is to continue with the present regime for a good many months before trying to take stock . . . I have walked for about a mile using a Canadian crutch, without undue fatigue . . . I seem to be normal mentally and I am going ahead with the editing,[48] a job which I undertook prior to my **disastrous** [our emphasis] visit to the US.

He was hoping for a visit from NRAO and AUI colleagues:

> . . . [B]ut I think as interim measure you [Rabi] and Dave Heeschen and probably Gerry [sic] Tape, ought to try and come out here and form your own assessment of what is going on and how it should affect US policy. I am a little reluctant to try to make any attempt to do this

---

[46]NAA C4660. On 17 August 1962, Wild wrote to Rabi that "Joe continues to convalesce—perhaps there is perceptible improvement. He is undergoing quite severe physiotherapeutic treatment daily, which he approves of whole-heartedly despite the discomfort. He puts in an appearance at the lab every other day and talks about radio astronomical problems with all his former zeal." On 28 August, Wild told Jan Oort that Pawsey "was mentally almost his normal self." Bowen was less optimistic on 23 August 1962 as he wrote Arthur Higgs: "[Pawsey] is still a very sick man—comes into the lab occasionally."

[47]C3830 Z3/1/XI, both 21 August and 5 September 1962.

[48]His major activity at the lab was editing the *Proceedings of the Institution of Radio Engineers Australia*—vol 24, no 2, February 1963. J.L. Pawsey as editor. (A short memorial text on page 94.) There were 22 articles besides the introduction by Pawsey, "Introduction to the Radio Astronomy issue", received by the journal 5 days after his death.

from this distance. **Science has to progress from the reaction of the individual to the circumstances of his day**. [our emphasis] But there are at least two items which I would like to bring to your attention.

The first was polarisation and Faraday rotation, already described by Pawsey in the letter to Tape. The second was the detection of radio sources at 20 or 6 cm with the new 300-ft or 140-ft telescopes, followed by radio source counts which could possibly lead to cosmological conclusions.

About 3 weeks later (19 October 1962), Dave Heeschen was appointed permanent Director of NRAO, a position he held until 1978. The VLA was planned and built under his leadership. A few days earlier (15 October 1962), Heeschen wrote Pawsey. Heeschen mentioned that he and Jerry Tape (the new AUI President) would be coming to Australia in late November to visit astronomy colleagues there.

In mid-September 1962, Lenore's diary states that Pawsey started having frequent spasms.[49] Some were severe, lasting 2 min or more. They continued every few days into October with simple twitches only on 2 days, 15 and 16 October 1962. The doctors were quite concerned about the continuing deterioration. Lenore Pawsey wrote in her diary on Monday 15 October that his walking was the "worst yet—hard to get out of chair". The following day she wrote: "mentally sluggish—started to wander in the lounge" as it was difficult for him to find the door of the dining room. His speech was confused: "What should my ambition be", "what shall I do now", "I can't find page one [out of eight]".

Lenore's diary has a record of a number of strikingly poignant statements made by her husband in this period: 20 October 1962,"You have turned out to be a ministering angel"; "You have done a hell of a good job with me"; 22 October (mainly in bed), "Thank you for all your kindness, I do appreciate it"; and on 24 October 1962 in response to her statement "We've had some wonderful times together", he replied, "We'll have some more, I hope" and "Don't worry Lenore, I'm completely satisfied with my choice of wife". (This is the last quote that appears in the diary.)

At some point during this time (mid- to late-October) Pawsey spoke to his older son Stuart with a poignant farewell:

> This above all: to thine ownself be true,
> And it must follow, as night the day,
> Thou canst not then be false to any man,
> Farewell: my blessing season this in thee![50]

At roughly the same time, Pawsey sent a letter to his daughter Margaret in Paris. Don McLean told Goss on 20 September 2008:

> He [Pawsey] wrote this letter to Margaret on one of these aerogram forms. It was all written down the right-hand side of the page; he left a huge blank margin down the left-hand side of

---

[49] Joe and Lenore Pawsey Family Collection. The first spasm had in fact occurred just upon arrival in Sydney on 31 July 1962, causing "severe arm and face" problems.

[50] From *Hamlet*, Lord Polonius to Laertes, Act 1, Scene 3. Quoted perfectly. Private communication October 2015 from Stuart Pawsey.

the page[51] which is most uncharacteristic . . . [The letter] basically said to her: "Lead your own life and forget about me". . . [McLean concluded] Joe realised that his condition was fatal.

On 26 October 1962, Pawsey was moved to Victoria Private Hospital in Potts Point, Sydney, a palliative institution. He had just finished editing the *Proceedings of the Institute of Radio Engineers Australia* Radio Astronomy Issue of February 1963 (Vol 24, no 2, after Pawsey's death). Wild wrote Rabi on 29 October with the news that Pawsey's deterioration had begun about 15 October ("mental processes are slow, but he enjoys company"[52]). On Saturday 2 November, Fred Hoyle came to the hospital in Potts Point to present the Hughes Medal of the Royal Society. Lenore, Stuart and Hastings Pawsey were present; the sad, solemn event produced strong memories for both sons.

On the previous day (25 October 1962) Pawsey's colleagues organised a well-wishes letter to Pawsey, likely initiated by Paul Wild. The total number of signatures was 31. See Fig. 40.5. The letter read:

Dear Dr Pawsey,

Most of us in your radio astronomy group have not seen as much of you in recent months as we would have liked. When we do meet there is usually so much work to discuss that other things that ought to be said get left unsaid. We know no words to express our sorrow that you should have been stricken so suddenly in the prime of your creative life, and this letter is to let you know the extent to which we appreciate the privilege of working in your team.

We appreciate not only your own contributions in radio astronomy, but your rare ability to dovetail the work of many individuals into a coherent and well-directed effort. We realise that by keeping your door open to us at all times, by listening patiently to new ideas in even their earliest and vaguest forms, by discussing the most minute details of papers, and by giving freely of your physical knowledge, experience and intuition you have been sacrificing most of your own research time to the success of the group as a whole. We would like to thank you for this.

We believe that although you can no longer play the same active role in leading the group, the Sydney radio astronomy effort (whether in Radiophysics or elsewhere) will continue to blossom, not merely because of the momentum you have given it over the last fifteen years, but because you have been responsible for giving a set of ordinary individuals the interest and drive to continue your work and the confidence to do well in the open field of international competition. Australian radio astronomy, like most unusual phenomena, can be traced in the main to a single cause, and we are well aware what that single cause is [that is, Pawsey himself].

With the deepest respect and affection of your colleagues,
SIGNATURES:

Paul Wild, Kevin Sheridan, Max Komesaroff, Steve Smerd, Chris Christiansen, John Bolton, Bruce Slee, S. Suzuki, Norman Labrum, Eric Hill, Jim Hindman, Alan Carter, Keith McAlister, Frank Gardner, Lindsay McCready, Brian Cooper, Alan Weiss, Brian Robinson, Peter Scheuer, Bernard Mills, Alec Little, Jim Roberts, Frank Kerr, Dick McGee, Dick

---

[51] Since he was paralysed on the left side, this event may represent the condition "hemispatial neglect".

[52] Paul Wild, 29 October 1962. NAA C4660/1.

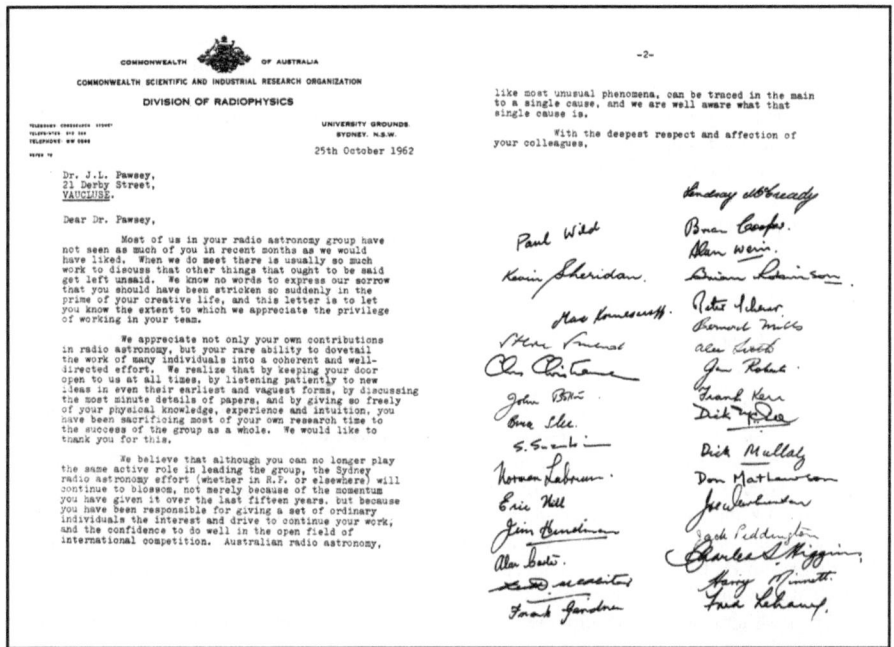

Fig. 40.5 The 31 signatures (top left) were Paul Wild, Kevin Sheridan, Max Komesaroff, Steve Smerd, Chris [W.N.] Christiansen, John Bolton, Bruce Slee, S. Suzuki, Norman Labrum, Eric Hill, Jim Hindman, Alan Carter, Keith McAlister, Frank Gardner, (top right) Lindsay McCready, Brian Cooper, Alan Weiss, Brian Robinson, Peter Scheuer, Bernard Mills, Alec Little, Jim Roberts, Frank Kerr, Dick McGee, Dick Mullay, Don Mathewson, Joe Warburton, Jack Piddington, Charles Higgins, Harry Minnett and Fred Lehany. A major omission was E.G. Bowen. Joe and Lenore Pawsey Family Collection

Mullaly, Don Mathewson, Joe Warburton, Jack Piddington, Charles S. Higgins, Harry Minnett, Fred Lehany.[53]

On 22 November, Thursday, a symposium was held at RPL discussing the last few months of Parkes 210-ft telescope research. The date was only 8 days before Pawsey's death.

A personal letter to Pawsey from Frank Kerr, the chair of the conference, was also sent on 25 October 1962:

---

[53] E.G. Bowen is missing from this list. Even though he was not a member of the radio astronomy group, his absence is surprising. Bowen and his wife attended the funeral on 2 December 1962. In contrast to many colleagues (e.g. Bok and numerous others), Bowen did not speak. Edward Bowen, their son, pointed out to Goss in 2016 that his mother, Vesta, was upset with Taffy by his refusal to join the speakers eulogising Pawsey.

Dear Joe,

We had a very successful symposium yesterday, going over the work of the last few months at Parkes. The presence of our three visitors (Dave Heeschen, Jerry Tape, and Fred Hoyle) added a lot to it.

We were very sorry you weren't able to be there, as you have contributed so much to all the work that was described. On behalf of everybody who was at the symposium, I would like to send you our very best wishes, and to tell you and Lenore that we are thinking of you very much.

Yours sincerely,

(F.J. Kerr)

The concluding remarks on 22 November 1962 by Kerr:

Portion of Concluding Remarks at Parkes Group Symposium.ei
November 22, 1962

Now I would like to talk on a rather different plane: I feel that we should not close without referring to the member of our group who is not here today.

In published papers, it is usual to make acknowledgements; this is not so easy for a spoken paper. I will try to remedy this on behalf of everyone who spoke today.

Most of the work described today was presented as from individuals, but we all acknowledge with pleasure that we are indebted to many other people in carrying out our work: to Dr Bowen, for his conception of the new telescope, and for taking the lion's share in providing it; to Harry Minnett, for his share in the design; to John Bolton, for taking the piece of hardware and developing it into a working system, and a well-run establishment; to John Shimmins, for keeping the telescope working; and so on.

But there is one other special acknowledgement that we would like to make: to Joe Pawsey, who has contributed to every paper presented today. Sometimes his contribution has been a direct one; for example, he first saw the great importance of polarisation work, and he would have carried it along if he had been able to. In other cases, his contribution has been indirect through his work in building up the group over the last fifteen years or so. In fact perhaps his greatest contribution has been that we all feel very strongly that we belong to a group.

I want to propose that we send Joe a message form [sic] this meeting, with our good wishes—our very best wishes—and telling him and his wife that we are thinking of them very much at this time. I take it that everyone is in agreement with this being done.

(carried by acclamation). F.J. Kerr

We will use Lenore's own words to recount the last few days of Joe Pawsey's life.[54] The last time he spoke was when Jerry Tape and Dave Heeschen visited him at Potts Point on 28 November,[55] Wednesday. She wrote to Tape:

---

[54] Letter from Lenore to Jerry Tape on 25 February 1963, found by Ken Kellermann in the Library of Congress Rabi collection.

[55] Joe and Lenore Pawsey Family Collection. Dave Heeschen had written Pawsey on 15 October 1962, announcing his visit in late November with Jerry Tape. He gave a summary of observations with the newly opened 300-ft transit telescope as well as a report on NRAO activities. He concluded: "We have heard only occasional second-hand reports of your own activities. I hope you are regaining your physical strength satisfactorily. I will try to write more frequently in the future, and keep you fully informed of all developments here." Tape and Heeschen visited Parkes and the radio astronomers in Sydney at RPL as well as the groups of Christiansen and Mills at Sydney University. After a short visit of 10 days, they returned to the US. To the AUI ExComm

The sudden burst of energy, as evidenced in the remark he made to you when you said you hoped some of the people from Green Bank would be over here again—[Joe said] "next time you must have some specific object in coming over", and you assured him that next time there were would be a definite research programme—was the last [words he spoke]. In fact, I don't think that he ever spoke to me again. I like to think that his last thoughts were of Green Bank to which he had been looking forward with so much pleasure. Thursday morning when I went in he looked awful and gave no sign of recognition, though he ate some lunch which I fed him, and I think he knew I was there. The boys and I went in again in the evening, but still no sign that he recognised us. The doctor expected him to live 2 or 3 more days, but he passed away 3:00 am that night [30 November 1962].

The funeral was Monday 3 December at 1400 at the Northern Suburbs Crematorium. Norman Lade (a first cousin) was the minister. The death certificate was signed by Dr Ian L. Thompson of Vaucluse.

Lenore's words to Tape on 25 February 1963 summarised the poignancy—the possibilities, now not to be:

> He naturally had a few qualms after having accepted the position, but he told me that after the few days he had in Green Bank in March, and meeting the staff, he was completely happy about things and felt that with the co-operation of people there, and AUI behind him, he could make a success of things ...

> If I could, I would do without Joe myself to give him back to the world of science. He had still so much to contribute. It was wonderful seeing you [Tape] and Dave [Heeschen] again and a great satisfaction to Joe to know that you at last visited Australia. It was comparatively easy being brave while he was alive. It is much harder now he is gone.

---

they reported on 14 December 1962 that they were impressed by the calibre of personnel in the three laboratories. The Parkes telescope was the "best movable telescope in the world today ... [T]he scientific programs represent a carefully coordinated effort on the part of the staff, and visitors are not encouraged unless they are prepared to become part of the general program." Bowen was a strong proponent of building a new larger version of Parkes with a diameter of 400–500 ft, "rather than large arrays with their complex electronic and data processing problems. Dr Bowen believes that the basic design [of the 210-foot can be applied to the larger instrument, even up to 600 feet]. Dr Heeschen does not agree with Dr Bowen's views on the value of arrays, which he considers necessary in obtaining high resolution." The AUI group were struck by the disadvantages of the large physical separation between Parkes and the laboratory in Sydney and also by the "overly authoritarian" nature of the administration of the Australian institutes compared to a similar US installation.

Chapter 41
Legacy

In conclusion, it is interesting to enquire if radio astronomy, this lusty child, has refunded anything to its parent, radio engineering. The answer is unequivocally yes. Firstly, over the past decade, the development which is probably of the greatest significance in radio engineering is the development of the low-noise receivers, the "maser" and "parametric amplifier". According to the pioneering papers on these subjects, the respective inventors were significantly stimulated by the knowledge that radio astronomy was in vital need of such receivers at the shorter wavelengths, particularly at 21 centimetres for the very weak hydrogen line.

Secondly, in meeting progressively more advanced requirements for aerials of high gain and high surface accuracy that are steerable and capable of tracking a source with high precision, notable advances in the art of aerial design have been made which are too recent to have yet found their way into more general radio engineering practice. As radio astronomy grows in stature and learns to rub shoulders with the astronomers it is thus beginning to repay some of the investment of specialized radio techniques that helped to bring it into being.

J L Pawsey, "Introduction to the Radio Astronomy Issue" *Proceedings of the Institute of Radio Engineers Australia*, February 1964

A scientist's legacy is modest: a set of publications, almost certainly doomed to become quickly outdated, or forgotten and lost, assumed to be largely irrelevant, with their number—as Pawsey's career shows—but a poor indicator of actual contribution to science. And this modest legacy is deeply valuable. As the editors of the Proc IRE Australia wrote:

Joseph Lade Pawsey died on November 30, 1962 after a prolonged illness. One of his last acts was to complete the editing of the present issue of *PROCEEDINGS of The Institution of Radio Engineers Australia,* and it gave him considerable satisfaction to know that the task was well and truly completed. This issue makes a fitting memorial to one who did so much to pioneer the subject of Radio Astronomy and helped put Australia in the forefront of this new branch of science.

Supplementary Information The online version contains supplementary material available at [https://doi.org/10.1007/978-3-031-07916-0_41].

W. M. Goss et al., *Joe Pawsey and the Founding of Australian Radio Astronomy*, Historical & Cultural Astronomy, https://doi.org/10.1007/978-3-031-07916-0_41

Many tributes flowed in for Pawsey after his death, and two significant obituaries were eventually published. But, as one of these Biographical Memoirs—written by Bernard Lovell, for the Royal Society (published 1964)—stated, "As an epilogue nothing could be more appropriate than the letter sent to Pawsey on 25 October 1962 (see Chap. 40 and Fig. 40.5), signed by the 31 members of his group in Sydney. It reveals the essence of the individual at work from those who were in a good position to judge.

Pawsey Memorial Funeral Service, Sydney 3 Dec 1962

Bart Bok was among the speakers at Pawsey's funeral service. He wrote to Lenore Pawsey the next day with a summary of his remarks. He emphasised the same intangible aspects of Pawsey's career as had his colleagues: Pawsey's guiding and mentoring young scientists, his critical assessment of the work of the new radio astronomers in the decades after WWII, and his vast scientific vision and ability to pose and answer fundamental questions.

Bok felt that Pawsey's contributions had brought respect to Australian science from throughout the world.[1]

> Dear Lenore,
>   I spoke first about Joe's contributions to science, his books and his scientific papers that will stand as lasting monuments to his activities, and made a brief reference to the many honours that Joe has received during his life. I then mentioned his intangible contributions, which appear nowhere in print as yet, but for which Joe will be remembered for many generations. First, I spoke of the manner in which Joe had inspired and encouraged young people. There are very few scientists in the world who will be able to look back upon a life in which they have helped produce so many distinguished scientists. The young men of Australia who are now the great names in radio astronomy, and who have helped place Australia at the top of the list in the field on a world-wide basis, all express great personal debts for the way in which Joe helped them get started and how he saw to it that their work came to fruition. Second, I mentioned that Joe was rated as the most profound critic among Australian radio astronomers. He possessed great scientific vision and perseverance in seeing to it that his ideas would be put to the test, and he had, more than anyone else, a remarkable power to simplify complex problems and present them in a new light. Two of Joe's former associates [likely, Christiansen and Mills] stressed that there was no one [else] in the scientific world from whom they would obtain more straightforward and unbiased answers to basic questions. He will be sorely missed not only by his friends and in the CSIRO, but also very much as a counsellor and advisor on national scientific problems, within the Academy and outside. He will be equally missed at international scientific meetings, where his quiet comments were always listened to with great respect. Joe's life as a scientist was a good one, for he opened doors in science and saw better than anyone else new directions, and he opened doors for his country and brought greater world-wide respect for Australian scientific achievement.

---

[1] Bok's letter from the Joe and Lenore Pawsey Family Collection.

Lastly, I spoke of Joe as a father and as a family man. It was always a pleasure to see him at home with you and with the children, and all of us have always thought of you as a closely knit and very happy family.[2]

Pawsey Lecture Series, Australian Institute of Physics, 1965–1999

Bok reiterated these sentiments in other tributes, including a public lecture in 1965 for the newly independent Australian Institute of Physics (AIP).

That the AIP existed at all, was in no small measure due to Pawsey himself. In the 1950s, Pawsey had been active in the Australian Branch of the UK Institute of Physics. Towards the end of the decade, there was increased sentiment to break away from the UK organisation, strongly supported by Pawsey. Pawsey became Branch president, 1960–1961. The final meeting of the Australian Branch was held on 21 August 1962; after this, the Australian Institute of Physics was born. Pawsey, of course, was gravely ill at the time.

So it was fitting that the new Australian Institute of Physics established the Pawsey Memorial Lecture in 1964–1965, a yearly event held at various Australian state capitols, recognising prominent scientists (see ESM 41.1, Pawsey Lecture Series).

On 25 February 1965, Bok gave the first lecture in this series, at the University of Sydney: "The Future of Galactic Research". He spoke of Pawsey's combination of profound understanding of electrical engineering with his newly acquired knowledge of astronomy.

The late J.L. Pawsey did as much as any scientist to initiate the work on Radio Astronomy in Australia and the country owes as much to him as to any one for having given Australia the undisputed place of leadership in the field in the world. It was tragic that he should have died at far too young an age, just when he was reaching his pinnacle performance. It is fitting that his colleagues should have established the Annual Pawsey Lecture to be delivered under the auspices of the Australian Institute of Physics.

Dr Pawsey was first of all a physicist in the fullest sense of the word. He was a man who knew how to apply to greatest advantage his profound knowledge of electronics and astronomy and use these in the development of the new science of Radio Astronomy. He was in a very special sense a great teacher and research director. It was he, more than anyone else, who guided the work of the remarkable group of young radio astronomers brought to Sydney and the CSIRO under the joint direction of Dr E.G. Bowen and himself. I need to mention here only a few: W.N. Christiansen, B.Y. Mills, J.P. Wild, J.G. Bolton, F.J. Kerr. There are at least a dozen more. Dr Pawsey was their critical friend and guided their work. Without asking for credit, he advised them in the early stages of the planning of their researches, followed their progress critically during the construction of equipment, communicated his ideas freely and read with care the first and subsequent drafts of their scientific papers. He brought to the Radiophysics Laboratory many of the younger radio astronomers

---

[2] In the Royal Society Memoir for Pawsey (*Biographical Memoirs Royal Soc of London*, vol 10. P 229, 1964), Lovell quoted portions of this letter, attributed to Bok, but with no indication that the text originated at the memorial service on 3 December 1962.

who now flourish at the CSIRO, at Sydney University and elsewhere. He retained his interest in people and in research during his tragic illness. During my last visit with him, a short time before his death, he and I talked at length about the new and wonderful studies [at the new Parkes 64 m radio telescope] on Faraday rotation, which he had helped to initiate. He was a good friend and a great man.

In ESM 41.1, Pawsey Lecture Series, we provide a list of those distinguished scientists who have given the Pawsey Lecture. The second Lecture was given by Jack Ratcliffe, "J.L. Pawsey's Research at Cambridge and Impact on Present Knowledge of the Small Scale Structure of the Ionosphere." Familiar names—including Wild, Woolley, Priestly, Bowen, Hanbury-Brown, Christiansen, Mills, and Bolton—filled the Lectures of the next decade; then younger colleagues, often from Mt Stromlo, followed. In 1990, the Lecture was given by one of the authors of this book, R D Ekers: "Revealing the Invisible Universe".

Obituary by Christiansen and Mills (1964): A Personal Assessment

This comprehensive description of Pawsey's life and legacy was written by his two closest colleagues, Christiansen and Mills (see also Chap. 33). The major achievements of the solar group at RPL were summarised by Christiansen and Mills, including the continuing ground breaking research of the group of Paul Wild and colleagues that developed in the 1950s.

As so many did in their recollections, Mills and Christiansen singled out Pawsey's leadership style for praise and memorialisation: how he could ask sharp questions in a friendly style, or give advice by asking questions;[3] how he liked simple answers and reacted with enthusiasm when new ideas appeared; how he stressed frankness with colleagues from outside RPL. (This aspect may have had its origin in his dislike of wartime secrecy). A number of Pawsey's favourite expressions appear in the text: "wildcat project" (or "long shot"), "inherent cussedness of nature" and "follow his nose":[4]

> Joseph Lade Pawsey died in Sydney on November 30, 1962. It is difficult indeed to over-estimate the value of his contribution to the recent development of the radio sciences and astronomy in Australia. Apart from his direct influence in the Radiophysics Division of the Commonwealth Scientific and Industrial Research Organisation where he founded and brilliantly led the radio astronomy group of the Division, his influence was felt in the field

---

[3] When Goss joined RPL as a post-doc in August 1967, he heard several "Pawsey stories" (perhaps apocryphal) from former senior colleagues who had earlier worked with Pawsey at RPL. Frequently the "Ohm's law" story was retold: At a seminar by a nervous junior colleague, Pawsey raised a question at the conclusion: "This was an excellent talk Mr X. However, your main conclusion may possibly be inconsistent with Ohm's law."

[4] See the discussion in Chap. 20 and ESM 20.1, Review of Recollections in which Pawsey's affinity for "wildcats" and "long shots" experiments was discussed.

of optical astronomy, in ionospheric research and in many applications of radiophysics techniques in other fields. Who of his associates can forget his painstaking and intellectually humble approach to the problems of a new field of discovery, his flashes of intuition, the depth of his physical understanding, his scientific honesty and his quiet but obstinate determination to see that the right decisions were made by scientific administrators?

... Apart from his integrity, the characteristics which most endeared Pawsey to his associates were his simplicity and enthusiasm. He insisted on treating any problem in its simplest terms, and was a master of the rapid "order of magnitude" calculation. This was one of the main factors in his success as a scientist and as the head of a scientific group ...

He entered with great enthusiasm into cooperative scientific activity. He was one of the most active members of a group of astronomers in the IAU concerned with redefining the positions of the galactic pole and the zero of galactic longitude. He produced a catalogue of reliably known discrete radio sources from data from various observatories, at a time when this was needed, and he published also a list of the radio observatories of all countries.

In the conclusion of the obituary text, Christiansen and Mills wrote:

At the age of fifty, Pawsey had already become the "Grand Old Man" of radio astronomy; he had pioneered a new branch of astronomy and built up in Sydney a scientific group which has made considerable contributions to science and had become well known and respected throughout the scientific world.

Immortalisation in Fiction by Fred Hoyle

In 1957, the eminent cosmologist Fred Hoyle published a work of science fiction, *The Black Cloud*. The story is said to occur in 1964, slightly in the future after publication of the novel. The plot centres on a cloud of gas that enters the solar system and causes disastrous climatic changes on Earth with immense mortality and suffering. As the behaviour of the cloud proves to be impossible to predict scientifically, scientists gradually realise that it is an intelligent alien superorganism.

As many reviewers have recognised, Hoyle based many of the characters on scientists he knew personally—including writing himself the main protagonist, Chris Kingsley, Professor of Astronomy, an "author surrogate".[5] Another character, Harry C. Leicester, an Australian scientist, is very clearly based on Joe Pawsey.

Leicester is asked to join the scientists in the UK as they prepare to deal with the catastrophe associated with the Black Cloud enveloping the earth. The prominent Cambridge University radio astronomer, John Marlborough, clearly based on Martin Ryle, also joins the group. Marlborough disappears from the book on page 136 (out of 226). Hoyle wrote about Ryle in an even-handed manner with no serious criticism.[6]

---

[5] On page 25 of the Valancourt edition of 2015, we read as Hoyle immodestly describes himself: "Chris Kingsley, Professor of Astronomy at the University of Cambridge, travelled by train ... for the meeting. It was unusual for him, the most theoretical of theoreticians, to be attending a meeting of amateur astronomers ...".

[6] Hoyle did include an off-hand reference to the Hoyle-Ryle conflicts over the Steady-state versus an evolving universe of the 1950s (see Chaps. 35 and 36). Kingsley (Hoyle) was asking whether he

Later in the book, Leicester's expertise on the properties of the ionosphere is decisive as the scientists make radio contact (at a wavelength of 1 cm!) with the CLOUD itself (on page 166).[7]

Leicester becomes one of the major players in the dramatic conclusion of the story, becoming one of closest partners of Kingsley. At the end of the book (last page but one), Leicester has a mysterious death. He is one of the few overseas scientists that attempt to return home from Britain after the world recovered from the Black Cloud's presence: "Against the advice [of the British government civil servant] he insisted on returning to his native Australia. He never reached Australia, being reported missing at sea."

Leicester is portrayed as having many valuable qualities, including a happy disposition and a dry sense of humour. All evidence suggests that Pawsey did not realise he was a character—a heroic one at that—in *The Black Cloud*.

In NRAO ONLINE.31 **The Black Cloud: Scientists in Science Fiction**, we present a text prepared by our colleague Tania Burchell, an astronomer, former colleague at NRAO and a science fiction writer. Tania presents a fascinating critique of the science fiction novel by Fred Hoyle.

Pawsey Medal[8]

In 1963, discussions had begun about another memorial for Pawsey, the Pawsey Medal. This was an initiative of Lenore Pawsey in February 1963. Details of this long drawn out process are described in ESM 41.2, Pawsey Medal. The process was drawn out because the Academy was reluctant to generate memorials in honour of all deceased scientists, stating that scientific fame was a matter for history to decide. But as we have seen—not least in the case of Ruby Payne Scott (and other scientists from social minority backgrounds)—"history" does not do this magically, like the "hidden hand" of the open market. History is *constructed*, partly by the loudness of supporting voices such as Pawsey's colleagues in 1962–1965, and by the existence

---

could trust Marlborough (Ryle, page 79): "Kingsley remembered his initial difficulties with the radio astronomy group [at Cambridge]." However, Kingsley did ask the Cambridge radio astronomers to carry out an almost impossible observation as they were to observe the Cloud at a declination of −30 deg in the 21 cm HI line to determine its velocity from the Doppler shift. The Cloud would be far in the south from the northern latitude of the UK, reaching only a maximum elevation of 8 degrees. The observation was also easily carried out by Leicester's group in Australia where the Cloud was almost at the zenith at transit.

[7]Using this communication channel, Hoyle has the Cloud provide a confirmatory statement about the validity of the Steady-state universe. Hoyle's summary quote about this event is described by Burchell in NRAO ONLINE.31.

[8]NAA C3830 Z8/31/A. Dr J.P. Wild Personal Files, Pawsey Award Arrangement for Funding and Establishment of the Award, 18 February 1963 to 20 December 1965 and 23 June 1971.

of such visible traces of contribution as a Medal; as well as by the (fairly chancy) interests of those who turn historian!

Fred White, indeed, responded to this very point in a letter from mid-August 1964 to E.M. Cherry, the President of the Australian Academy of Science. Two key sentences responded to the concerns of the Academy about the recognition of the recently deceased J.L. Pawsey. The entire correspondence is included in ESM 41.2, Pawsey Medal. Four key sentences are included here:

> We also believe that the recording of history is the responsibility especially of those who live contemporarily with the events of record. In the present case we are seeking to honour a man who left behind him evidence in the form of publications of only a fraction of his contribution to science. His major role was to stimulate by his imagination and his activity among his colleagues the growth of a completely new science [radio astronomy]. **Few men have played so great a role in the formation of a new science with such extreme unselfishness and disregard for personal prestige and gain.** [our emphasis]

The persistence of Paul Wild and Fred White finally paid off, and the Medal was announced in February 1965. Critical to the Academy's agreement to establish the Pawsey Medal was the recognition that Pawsey's major contribution was having stimulated the growth of a completely new science—radio astronomy. There were five conditions: (1) to be awarded for outstanding research in physics, "carried out largely in Australia"; (2) to be awarded to scientists not over 35 years in age, and (3) to be awarded at intervals, "not necessarily annually"; (4) the fund was to cover the cost of the medal and all travel expenses associated with the selection process; and (5) the Council of the Academy would reserve the right to reconsider the continuation of the award after a period of 25 years.[9] These conditions still apply today.

The first medal was awarded in 1967, to a theoretical ecologist, Robert M. May, later Baron May of Oxford. There followed 54 awardees from 1969 to 2020. (In 1998 and 1999 two Pawsey Medals were awarded each year.)[10] Sixteen of the 55 awardees are astronomers or astrophysicists, with five radio astronomers (Richard Manchester, Bryan Gaensler, Naomi McClure-Griffiths and Adam Deller of Swinburne University in 2020, and one of the authors, W. Miller Goss, in 1976). The 2001 Pawsey Medal was awarded to Nobel Laureate (in 2011) Brian Schmidt.

Pawsey has most recently been memorialised (somewhat ironically) in the Pawsey Supercomputing Centre in Western Australia. (Ruby Payne-Scott and Ron Bracewell have also been memorialised as CSIRO supercomputers.) The Centre is a joint venture between CSIRO and Western Australian government and Universities, with expected use by the planned Square Kilometre Array (not yet constructed at the time of writing) and current heavy use by astronomers. Thus Pawsey's name is now attached to both a humble, yet extensively useful, antenna engineering device—the

---

[9]Letter from L.G. Rees (Secretary of Physical Sciences Australian Academy of Science) to Fred White (Chairman of CSIRO) on 30 Sept 1964.

[10]The full list of awardees is included in the Australian Academy of Sciences web pages: https://www.science.org.au/pawsey-medal.

Pawsey stub—and to the most cutting edge tool in visionary "big" science and big data in Australia, just as the man himself had found his vision stretching from what was local, the Sun, to the Cosmos.

Chapter 42
Conclusion: J.L. Pawsey (1908–1962) and the Development of Radio Astronomy

> *Who of his associates can forget his painstaking and intellectually humble approach to the problems of a new field of discovery, his flashes of intuition, the depth of his physical understanding, his scientific honesty and his quiet but obstinate determination to see that the right decisions were made by scientific administrators?*
> *Christiansen and Mills (1964, p 137)*

What have we learned about Joe Pawsey, and how does a deeper understanding of his life and career provide a deeper understanding of the first 17 years of the field of radio astronomy? Along the way in assembling a detailed record of Pawsey's life and career, there have been many opportunities to develop an understanding of the complexities of science-in-action. What circumstances, what factors, influenced discoveries being made or missed? What analogies can be drawn between past and present radio astronomy research? What kind of social systems might best promote good science? In what ways do scientists contribute substantially beyond their direct intellectual contributions, for example to national prestige, or economic development, or the public communication of science? These questions have underpinned our analysis, and we return to them here.

We begin with a retrospective consideration of the man himself.

How does a lower middle-class boy from the country become a world leading scientist? Much must be attributed to being born in the right generation, and having, not merely access to education, but support to pursue it. But with these foundations established, Pawsey, like the majority of scientists, found his success rested on neither strategy nor plan, but on making successive good decisions that brought rich experiences.

He reflected on these questions himself, notably in two letters written to his mother on his birthday, in 1954 and in 1956. On 23 May 1954, in hospital in Potts Point, Sydney (recovering from varicose vein surgery), he looked back on his career:

Based on correspondence located in the Joe and Lenore Pawsey Family Collection.

© The Author(s) 2023
W. M. Goss et al., *Joe Pawsey and the Founding of Australian Radio Astronomy*,
Historical & Cultural Astronomy, https://doi.org/10.1007/978-3-031-07916-0_42

I look back with a comfortable degree of wonder that I made the grade that I did. I was nearly always second or third in a class or a group but when I moved into a more select group I still kept a high position. I have been very lucky in my career in that I always followed my nose and yet the way opened up ahead.

The 1851 [Exhibition fellowship for study at Cambridge] of course was my first great break. It gave me a real scientific chance. The move to EMI [television industry] was made because there was a reasonable scientific group and a reasonable salary. It was an excellent move and gave me most valuable experience.

The move to Sydney [late 1939] was stimulated by the idea of getting out of the bombing area but it led to good experience during the war and a first-class scientific opportunity after. It is not easy to guess what would have happened had I stayed at EMI. Certainly I should have got equally good wartime experience but I don't know where I should have gone after.

I think the moral of it all is not to look too far ahead but see that the path is a progressive one and then to what comes to hand as well as is feasible.

In 1956, he wrote from his home in Vaucluse, Sydney:

As you say I have achieved eminence in my profession and it is interesting to try to see why and how.

This letter included expressions of deep appreciation for his mother, with whom he had maintained such a long and detailed dialogue over the years. He reflected in this letter on the importance of her drive to support his education (the opportunity to which she herself had lacked access):

There is no doubt that the starting point in my scientific career was your single-minded insistence on my having a first-class education. Once properly started on that road I had sufficient ability to go to the PhD standard . . . But without your insistence I might well have wandered off that road. The goal then was to go forward in that walk of academic life in which I found I had most promise. This turned out to be a career in physics. Now I am where I am, the goal is the building up of the lab. Likewise, I wonder where this will lead.

In this letter, Pawsey reiterated the notion of "following his nose":

It is not that I have followed a clear vision from the farm to here but rather I have followed the path as it led on, with an urge to do well what there was to do. I am not brilliant but so far have had good judgement in picking the right things to do. My main strength seems to have been an ability to encourage others to do good work. If so, it is a very useful-attitude.

Pawsey expressed a similar assessment of his abilities in a letter dated 5 September 1960 to Lenard Huxley (soon to be Vice Chancellor of the Australian National University), as he contemplated his choices during the schism at RPL: "It seems to be that what **strength I have lies in my ability to stimulate and develop scientists at the research level. An exceptional proportion of those who work with me seem to reach top level** [our emphasis]."

Scientists who make this sort of contribution—the development of other scientists—rarely have their contributions recognised. Of course Pawsey was stimulating a very select group of former wartime radar developers and operators, primed with skills that maximised the chances of their achieving well. But it is equally true, as each reiterated, that his mentorship was indeed key to their eventual success.

Pawsey, a "straight-forward man of absolute honesty and integrity", as Sir Bernard Lovell (1964) termed him in the *Biographical Memoir* he wrote of Pawsey for the Royal Society, had a clear view of himself. He was not "brilliant", not a leading research scientist like his counterpart Sir Martin Ryle (Pawsey's role was indeed closer to that of his former PhD supervisor Jack Ratcliffe, whose support for Ryle enabled Ryle's own career to develop). But he had become someone whose contributions to a new area of science were both extensively diffused, and profound.

Joe Pawsey and the Founding of Australian Radio Astronomy

This book adds many details to existing histories of the foundational period, sometimes of how particular episodes unfolded, sometimes of how and why particular instruments were conceived, designed or constructed, sometimes of the people and personalities involved. For example, we have emphasised how it was the existence of the Radio Research Board (RRB), which supported basic radio research projects for 10 years despite the financial circumstances of the Great Depression, that provided the personnel and the skills that were needed for the wartime radar research program. Institutionally, the formation of the CSIR, and the inclusion of a scientist dedicated to the value of basic research among its leaders, was similarly a crucial factor underpinning RPL's turn to exploratory science after the war. At an intellectual level, it was only by understanding the specifics of Pawsey's PhD research and the ionospheric research of the early 1930s that we could make deep sense of the ways of thinking that eventually produced interferometry as a significant method in early radio astronomy. The ideas and techniques used to make sense of the communication fading due to interference of waves reflected from the ionosphere—an entity whose very existence was not proven until Pawsey was an undergraduate student—had a strong continuity with those that former radar scientists used to slowly characterise the unknown extraterrestrial phenomena they were now exploring.

We have also emphasised the significance of the early observations at Collaroy, which have tended to receive less attention than the dramatic results of sea-cliff interferometry that succeeded it at Dover Heights. Pawsey was the lead research scientist at Collaroy. The initial results obtained there provided sufficient success to venture further with what was then a highly speculative research direction, and provided the important "hot corona" result.

Along with Pawsey, we came to a much better understanding of other major figures from his career. In particular a comprehensive and previously unknown record of the ultimately tragic figure of brilliant theoretician David Martyn has been assembled. It is now possible to understand how personality traits, in conjunction with the deeply scarring events of his injudicious wartime *liaison*, left him marginalised in Australia for the remainder of his career, and produced the mental illness that would ultimately take his life (see NRAO ONLINE.7). The consequence was that Pawsey, and RPL, were deprived of a crucial theoretical insight in these first foundational decades.

Pawsey and the History of Ideas in Radio Astronomy

We set out to better understand both science in general (or at least the physical sciences) and the radio astronomy of this early period by examining them through the lens of Pawsey's life and career. Academic historians think of this as "locating" history. That is, instead of presenting a general story that tells how one discovery leads to another, as if this chain of discoveries could occur anywhere, a "located" history seeks to understand the importance of the particulars. This located history explored why it mattered that this period in science occurred at a particular time, in a particular place, and undertaken by particular people, each with strengths and weaknesses.

Paying attention to these details shows that, although the same scientific results would have been arrived at over time eventually, the *ways* in which discoveries and new insights occurred were strongly determined by contextual factors. The personalities and relationships of those involved, and the resources that were or were not available, shaped what was done, and what was possible. Various assumptions and mental models that scientists held (for example, that a discrete radio source must be a "star" or must come from within the galaxy), also had significant effects. This history demonstrates what every detailed history of science shows—namely, that science is indeed robust *over time*, but not *only*, or not directly, because of the use of the "scientific method". The prediction-experiment-confirmation model, i.e., the "scientific method", which is still omnipresent when applying for research grants and for applications for observer time on telescopes, is misleading as an account of how science "works" (Chalmers, 2013). Reconstructing the history of science shows how complicated the story of what was *actually* involved, is. The way science happens comprises an interplay of personalities, ideas, experiences, institutional constraints and affordances, and many other factors. In this book we followed Merton and discussed the role of serendipity, norms about appropriate scientific behaviour, priority disputes, and commitments to a scientific "ethos", particularly in relation to scientific internationalism. In some cases, such as in the disputes about how to interpret data from the source surveys of the 1950s, *epistemic values,* that is, values about knowledge, such as preferences for coherence, or empirical adequacy, or explanatory power, shaped how the scientists involved perceived and defended their work (Peels, 2018).

With these thoughts in mind, let us consider what social factors influenced the formative years of Australian radio astronomy. The lens of Pawsey's life adds this perspective to several episodes during this period. For example, social and scientific considerations were entangled in the events related to Pawsey's detection of the hot corona (see also Sullivan, 2009, p. 136). Here a scientific issue was at stake: the important distinction between a prediction (made by Martyn) that was confirmed, or an observation (made by Pawsey) that was explained, with the former a much stronger validation of a theory than the latter. But Martyn's miscommunication confounded this scientific distinction so badly that his actions appeared to Bowen and others as a blatant attempt to steal priority, and it required Pawsey's professionalism to salvage a positive outcome.

The entangling of social and scientific factors is also visible in the "discovery" of the galactic centre, to which we applied Hank van de Hulst's concept of "nanoherz" history. The discovery of the galactic centre extended far beyond the single moment in time in which John Bolton recognised the significance of the location of the peak signal that Dick McGee had observed using Bolton's rudimentary hole-in-the-ground antenna. McGee and Bolton's detection drew from an earlier detection by Piddington and Minnett and was championed by Pawsey through a publication process to which many contributed. The result was validated on the basis of quite different dynamical considerations using the hydrogen line observations made by other astronomers. Nor was even this the entire story—for it has continued to unfold—first with the recognition of the need for a new galactic coordinate system, and then, with increasingly accurate images and dynamical information, we have arrived at the 2020 Nobel prize in physics for the direct evidence that the centre of the galaxy is a super massive black hole!

Both epistemic values (such as differing preferences between theoretical coherence and the empirical adequacy of theory) and cognitive biases (such as interpreting new information to match pre-existing assumptions), were at play in some of the more turbulent interactions across Pawsey's 17 years in radio astronomy. For example, we can understand why there was continued acrimony during the long controversies over the interpretation of surveys in the 1950s, even after the scientific issues had been resolved! Some human factors were involved in this situation: Ryle could have prevented most, if not all, of the acrimony, had he only been willing to admit the errors in the 1 and 2C catalogues. Cognitive factors—particularly the well-understood heuristic of "anchoring" on an initial mental model of what the phenomenon is like, and then interpreting evidence to fit the anchor—were also evident. These factors, which under many circumstances have been shown to have *larger* effects in those with the highest levels of education and science literacy,[1] can be observed in many areas of science today.

Discoveries Made and Missed

The early period of radio astronomy was marked by a high rate of discovery, which then declined over time. Serendipitous discoveries were more frequent and played a more prominent role in this early period but have continued at a lower rate even up to the present time. The discovery of a new class of radio sources, the Fast Radio Bursts (FRBs), made with the 50-year-old Parkes Telescope in 2007, is an example. The high rate of discoveries in general, and serendipitous discoveries in particular, is normal in the early period of a new area of science, so we can use detailed historical

---

[1] See Neil et al. (1994) and Kahan et al. (2017).

studies such as these to extract some ideas about what conditions maximise opportunities for discovery, and how to value and allow for serendipity in science (Ekers, 2010). Some of the most important factors are the continual exploitation of new technology, the freedom of researchers to explore new ideas, and the time to do so.

In addition to considering how to best generate circumstances in which discoveries can occur, we can also use the details of Pawsey's career to consider how they come to be missed! A good example was that RPL missed making the first HI line detection in 1951. The factors then in play are just as relevant today. The chief factor was lack of time and personnel given the well-established existing projects. The scientific staff at RPL at the time were all firmly focussed on other projects of their own, and were reluctant to change direction to focus on the search for the HI line instead. Their reluctance can be partly attributed to the "sunk cost fallacy", that is, the tendency to continue a project simply because resources (especially time) have already been invested in it.[2]

At that time Pawsey did not forcibly redeploy any personnel to search for the HI line, although the swift work of Piddington, Christiansen and others to *confirm* the detection within a few weeks shows that earlier experiments with the existing technology would have been effective. His choices reflected his leadership style: he supported, rather than directed, researchers under most conditions. Is the flourishing that this leadership style enables to be valued, *even if* it occasionally has a cost in missed opportunities? Other factors were also at play. Pawsey's mental model of the HI line search, formed during his trip to the USA in 1947, was that the science was still confused and unclear.[3] "Anchored" on that first impression, he did not prioritise the search beyond suggesting it as a project for Mills. RPL's lack of access to the general astronomy community at that time meant that the significance of the detection, i.e., an update to his mental model, was not made apparent to him earlier.

In other cases, following the details of how early radio astronomy developed, revealed many circumstances where correct interpretations were arrived at for not entirely defensible reasons. There were also cases where very good reasoning led to what eventually turned out to be erroneous conclusions or lines of thought. The controversies concerning source surveys and their implications for the Steady-state and big bang models of the Universe in the 1950s contain many examples. Assumptions about what data was "normal" or atypical were involved: Ryle correctly interpreted the strongest radio source in the Northern Hemisphere, the quite distant Cygnus A, as typical of the other discrete radio sources, while Mills incorrectly assumed the strongest source in the Southern Hemisphere, the nearby Centaurus A, was typical. Bolton had noted that the Dover Heights survey (Bolton, Stanley and Slee 1954a) suggested an increasing density of sources at greater distance. But he dismissed this evidence, making a conceptual error about the role of the broad range

---

[2] https://en.wikipedia.org/wiki/Sunk_cost.

[3] Note that the HI line prediction by van der Hulst was in Dutch and in a Dutch publication that was not widely distributed.

of source luminosity. Mills's high-quality catalogue included strong evidence for evolution, but he made the wrong assumption about the average distances to the sources, misled by an unusual number of very nearby radio galaxies in the Southern Hemisphere. Ryle jumped to the (correct) evolution interpretation based on a flawed catalogue (and spurred on by the knowledge that this interpretation would annoy his colleague, the theoretical cosmologist Fred Hoyle).

We can also consider the construction of the Parkes radio telescope through this lens, undoubtedly the most important Australian development of the later 1950s. The Giant Radio Telescope (GRT) as it was then referred to, came about for primarily *non*scientific factors that were pushed forward by the entrepreneurial E.G. "Taffy" Bowen. Chief among these were the increase in Australia's prestige and importance in the international community and increased cooperation between Australia and the USA as part of the expansion of US influence through funding scientific projects in its allies. It fell to Pawsey to make the science case for the GRT, scientific projects that were very successful. But, as has been the case for almost all the great telescopes, it is the unanticipated science which has had the greatest impact. The lunar occultation leading to the discovery of quasars, the polarisation observations that confirmed the oblique rotator model for pulsars, the discovery of more pulsars than any other radio telescope on earth, and most recently the discovery of a new class of radio source, the Fast Radio Bursts. Much of the success of the GRT resulted from the factors we have discussed; the flexibility of the single dish combined with the freedom to explore, the continual development of new technology by astronomers with an intimate knowledge of the instrument.

The Conditions for Success

Beyond making discoveries, what can we learn from Pawsey's career to consider what factors support the success of a scientific research program? What enabled the Australians to create such an outstandingly successful research program in radio astronomy? Sullivan, Robertson and of course colleagues of Pawsey's, have all pointed to the context of wartime radar research. This circumstance resulted in a scientific team characterised by a strong work ethic and a can-do attitude that encouraged independent thinking and experimental trials. The team also had an unusual mix of physical insight and engineering skills, a product of their extensive experience in modifying and fine-tuning the antennas and various radio frequency electronic devices developed for radar (receiving amplifiers, recording devices, etc).

Pawsey himself exemplified the integration of physics and engineering skills typical of the first radio astronomers. Paul Wild wrote of him:

[O]n some days he would arrive unexpectedly at one's field station, usually at lunch time (accompanied by a type of sticky cake known as the lamington, which he found irresistible); or else infuriatingly near knock-off time. During all such visits one had to watch him like a hawk because he was a compulsive knob-twiddler. Some experimenters even claimed to have built into their equipment prominent functionless knobs as decoys, especially for Pawsey's benefit. (Wild, 1987)

Looking through the lens of Pawsey's life draws attention to the importance of understanding the materials and instruments as well as the observations and discoveries in the history of radio astronomy (Sullivan referred to this as "technoscience" (2009, p. 449)). For example, the Pawsey stub, and its application in the improved efficiency of radar antennas, was useful across a wide range of circumstances. The extent of its influence was analogous to that of Pawsey and Payne-Scott's recognition that interferometer observations could be understood as Fourier components, which led to the use of aperture synthesis, one of the most powerful tools in modern day radio astronomy (Yeang, 2013, Chapter 5).

Beyond the well-known context of radar research, a study of Pawsey's life reveals the "causes of the causes" (Marmot, 2015), the factors that influenced the development of radio astronomy at RPL. One such was the on-going issue of the impact of distance. The Australians entered the war in a continuing colonial relationship, dependent on Britain for most aspects of the radar research program and for the supply and/or training of research scientists. But Pawsey and Bowen emerged from the war with a strong American research network and a stronger sense of independence. But distance continued to matter. Most major scientific meetings took place overseas, and few leading scientists from the UK and USA found sufficient interest at RPL to overcome the financial and logistical disincentives and have then visited Sydney. So if the Australians wished to be part of an international community, they needed to travel. There was similarly a constant dilemma about whether or not to publish in local journals, which offered more opportunities for Australian scientists but were not widely read outside Australia.

As a result, Pawsey, Bowen and others at RPL remained strongly committed to the maintenance and growth of their networks outside Australia throughout their careers. In several chapters in this book, we set out the very packed itineraries that enabled Pawsey to build and maintain his networks, particularly in North America—a destination that helpfully matched his family connections. Pawsey's and Bowen's networks were mainly independent, with a few overlaps. Bowen's network was composed of science entrepreneurs and policymakers. Pawsey's was an extraordinary network of leading scientists across the range of subspecialities and research foci relevant to the growing field of radio astronomy. RPL depended on this for its intellectual resources.

The Australians also had certain advantages, some deriving from their comparative isolation. Post-war, the Sydney group had more financial resources than their British counterparts, helping them become established as the early leader in the field. Australia's small size and comparative distance from other research centres provided the advantage of more independence in ideas and research approaches. Without easy access to a range of experts, the Australians arguably developed broad skills and became innovators through necessity. The Mills Cross might be considered an example of such innovation. The high quality of the GRT, as a result of contracting a German firm (MAN), rather than a UK or USA firm, is another example of the unexpected benefits of distance.

The worst constraint imposed by distance was the lack of access to theorists, optical astronomers and experts from related disciplines, all of whom were far more readily accessible to University-associated groups in the UK and USA. These intellectual resources were brilliantly exploited at Cambridge, which saw early use of computing and uptake of ideas from crystallography to underpin the development of the aperture synthesis concept. While the role of computers in aperture synthesis was always recognised, what is now clear is the degree to which its exploitation separated developments in Cambridge and Sydney. In Australia, Pawsey and his colleagues, disconnected from code breakers and mathematicians, failed to envisage the relevance of punching holes in bits of paper-tape to feed the early experimental computing machines.

Teamwork, professional connections, and relationships also emerged as being of key importance to the institutional as well as intellectual conditions for success in radio astronomy. Institutionally, bringing together those with scientific capabilities and those talented at working with government and industry, was key to success at CSIRO from the beginning (Schedvin, 1987). Sir David Rivett's deep-seated value for the scientific ethos and his commitment to building it in Australia through the development of scientist-led research programs provided the key institutional support that would later enable radio astronomy to flourish However, the effective management of CSIR was as dependent on Rivett's industrial counterpart, Sir George Julius.

Post-war, a similar complementary mix of scientific and management flair characterised the leadership team of Pawsey and E.G. "Taffy" Bowen. Because their relationship concluded so poorly, it is easy to forget how much the two had in common: their shared wartime experience, their shared commitment to the development of young scientists, their shared purpose of increasing the stature of Australian science. Similarly, the difficult episodes between RPL and Ryle's group at the Cavendish Laboratory, Cambridge, and between Pawsey and Bolton at the time of Bolton's departure, make it easy to overlook the many years of pleasant, professional correspondence between these protagonists throughout the later 1950s. As mentioned, the international "networks" established by Bowen and Pawsey were essential components of the success of RPL in the 1945–1962 era.

It is important to avoid oversimplifying the portrayal of any historical figure. It would be a distortion to consider that all the scientific excellence fell on Pawsey's side, and all the entrepreneurship on Bowen's. It is also incorrect to see Pawsey as cautious and Bowen as a bold risk-taker, as Bowen and Bolton did. While Bowen and Bolton's frustration with Pawsey's cautiousness (for example over the GRT) was understandable, it was also true that Pawsey was an explicit supporter of what he termed "wildcat experiments". How do we reconcile Pawsey's conservative response in some circumstances with his enthusiasm to try new things? We suggest that he was often conservative in relation to physical interpretation, but not at all conservative about trying new experiments! He was also not in the least cautious (in contrast to Bowen, White or, famously, Martin Ryle) about sharing research results, or about other institutions, particularly the Australian National University, taking up radio astronomy.

Bowen, by contrast, showed technical risk aversion in areas where he was less conversant with the technology (for example, requiring Mills to build a prototype Cross initially), and managerial risk aversion about scientists managing projects. But he was much more comfortable with risk taking when it came to seeking funding for expensive new instruments on the basis of a relatively weak science case!

The differences that did exist between Bowen and Pawsey were an advantage at RPL for many years. But the advantages of difference come at a cost, and tensions were also longstanding between them. Together they achieved the GRT, and while the contributions of each were critical to the project in the end, their differences became too great. When their relationship deteriorated, other productive relationships, such as between Pawsey and Mills and Pawsey and Christiansen, became alliances in a distressing schism. It is true to say of that situation that Bowen's secret manoeuvring to gain Bolton as the scientific director of the GRT constituted distressing and unprofessional treatment of Pawsey.

The story of this schism is important beyond its impact on Pawsey personally. It represents a painful experience that many research leaders encounter in their mature careers when they discover that the approach that has worked so well for them in the past has reached its limits, or become unsuited to the institutional context as it changes. In this case, Pawsey's organisational model of small researcher-led projects oriented around researcher-developed instruments, was not well adapted to include big dishes. He had to grapple with the question of what kinds of projects and instruments and working arrangements should characterise the next generation of radio astronomy. We've argued that his own professional development, and his best chance of making a continued substantial contribution to radio astronomy, made his choice to accept the Directorship at NRAO the right decision.

We also suggest that Pawsey was likely to have been an effective Director at NRAO, at least initially. He would have brought to the then-struggling NRAO, the enormous asset of his high scientific credibility across his extraordinary North American network. At the time of Pawsey's appointment, NRAO was almost entirely staffed by astronomers from the Harvard school and not radio scientists, so Pawsey would have contributed one of his key strengths, his instrument-builder's knowledge and perspective. He had already embarked on a process of attracting instrument builders, such as Erickson and Weinreb. His calm professionalism may have been what was needed on a human level to establish the good relations and communications in an organisation marked at the time by a divisive management structure.

It is also true that in the longer term, Pawsey was unlikely to have shown the bold choices and real-world management skills that Dave Heeschen, who became NRAO Director after Pawsey's death, brought to the organisation. Pawsey's very expertise might have been the undoing of the ambitious VLA telescope project, with its baseline out to 30 km and an array of 36 (later 27) large dishes and 1 arc sec imaging capability to match optical telescopes at radio wavelengths. As Heeschen once remarked, the bold design of the VLA reflected inexperience: they had simply failed to realise that their proposal was "impossible"! Pawsey, of course, would have known all too well that arc second imaging through a turbulent ionosphere and atmosphere was impossible. The "self-calibration" technique that made it "possible" was only discovered after the construction of the VLA had started.

The VLA was opened in 1980—long after Pawsey is likely to have retired even had he lived. This instrument has become one of the most successful ground-based astronomy facilities ever built. It has provided exquisite radio source imaging capability based on the very Fourier Synthesis concept that Pawsey first articulated, but implemented in ways that Pawsey could never have imagined.

Conclusion

Joseph Lade Pawsey's values of service to science, of open internationalism in science, of intellectual integrity, and of work balanced with the priorities of family life, may be less visible than Nobel prizes and the attraction of some number of millions or billions of dollars in research grants. But they strike us as just as precious. In the same way, the many interpretations and concepts that scientists get wrong, are as necessary to the development of science, as those that turn out to be right. Reconsidering the early history of radio astronomy through the perspective provided by a close examination of his life, underscores the importance of cultivating a healthy social, institutional and intellectual ecosystem, for science to flourish. Such an ecosystem provides the capacity for big science projects, while maintaining niches for less conformist people or ideas. It can protect space and resources for risky, novel trials, and can consider how to leave room for serendipity. It allows for extensive conversation and debate, which can disrupt anchored models, or help people see and question existing assumptions. And it can connect ideas, skills and people from the system's most disparate parts, for the whole to flourish.

Correction to: Joe Pawsey and the Founding of Australian Radio Astronomy

Correction to:
W. M. Goss et al., *Joe Pawsey and the Founding of Australian Radio Astronomy*, **Historical & Cultural Astronomy,**
https://doi.org/10.1007/978-3-031-07916-0

This book was inadvertently published without updating the following corrections:

Abbreviations:

p. = page
l. = line
fn = footnote
col. = column

Corrections:

p. v, Frontispiece caption has been corrected to 'Joe Pawsey on the catwalk at the top of the linefeed support tower of the Vermilion River Radio Observatory, University of Illinois, September 1961.'
p.xxxvii, the text has been corrected to: and to colleagues Tom Sear, Peter Robertson, and Peter Hobbins
p. 33, fn 8, the text has been corrected to 'Tizard would lead the British radar research effort in the lead up, and during, World War II (Chap.9)'
p.86, l.3, the text has been corrected to 'Research Board in Australia, perhaps applying for a position'
p.111, 'White, A.l. Green' has been corrected to 'White A.L. Green'
p.117, l.1, first line: 'Harbour' has been corrected to 'Harbor'

The updated version of this book can be found at
https://doi.org/10.1007/978-3-031-07916-0

© The Author(s) 2024
W. M. Goss et al., *Joe Pawsey and the Founding of Australian Radio Astronomy*,
Historical & Cultural Astronomy, https://doi.org/10.1007/978-3-031-07916-0_43

p.119, 22nd line from top, 'fuze' has been corrected to 'fuse'

p.185, the text has been corrected to: lab and reachable by public buses. The station was called CA No 1 (Coastal Artil-

p.214 'Mark" L.W. Oliphan' has been corrected to 'Mark" L.E. Oliphant'

p.223, fn 5, 'Seeing' has been corrected to 'Seeding'

p.230, 'Austrasia' has been corrected to 'Australasia'

p.232, In Fig. 16.5 caption, 'Fig. 4' has been corrected to 'Fig. 16.4'

p.233, fn 22, '1955' has been corrected to '1945'

p.235, 'two' has been corrected to 'three'

p.257 'Hannes Alvén' has been corrected to 'Hannes Alfvén'

p.259 'Lindsey McCready' has been corrected to 'Lindsay McCready'

p.262, fn 55, the text has been corrected to: Charles Duguid, a well-known South Australian educator, a prominent advocate for Aboriginal

p.277, the text has been corrected to: Pawsey ended the 29 August 1951 letter on a positive note

p.290, the text has been corrected to: (ANU), located in the nation's capital city, Canberra. ANU, one of Australia's leading research universities...

p.299, the text has been corrected to: time. However, Greenstein also had discussions with van de Hulst

p.305, the text has been corrected to: Murray, Gardner, Piddington and Hindman), the groups of....

p. 313, In Fig. 21.4 caption, 'CSRIO' has been corrected to 'CSIRO'

314, In Fig 21.5 caption, the text has been corrected to: Fig. 21.5 Tour of Potts Hill, the 32-element 21 cm Grating Array. From right: Appleton, Bathalsar van der Pol (Phillips Laboratory), Fred White...

p.318, the text has been corrected to: Pawsey wrote with similar warmth of Hanbury Brown to Lovell...

p.319, the text has been corrected to: publication until one or two checks are applied." He requested that Hanbury Brown could

p.327, the text has been corrected to: California Institute of Technology, Caltech.

p. 345, In Fig. 23.3 caption, the text has been corrected to: cies. VIII. Discrete Sources at 100 Mc/s Between Declinations +50○ and −50○", Bolton, J.G.,

p.394, the text has been corrected to: spend 3–9 months in Sydney visiting RPL, specifically Maarten Schmidt and Gart

p.395, the text has been corrected to: complimentary copies. Hanbury Brown

p.402, the text has been corrected to: Alfvén, Herlofson, Kiepenheuer and Ginzburg...

418, footnote 32, the text has been corrected to: Full details provided in NRAO Online.44 additional note 1 "Wallis Disaffection with FFP - 1956 and 1957"

421, footnote 46, the text has been corrected to: Described in NRAO ONLINE.44 Additional Note 1.

p.429, the text has been corrected to: of Illinois in Urbana from 18 to 21 August 1957 where he attended a meeting of the American

442, footnote 8, the text has been corrected to: construction of the GRT.

446, footnote 15, the text has been corrected to: CSIRO in Melbourne.

p.451, the text has been corrected to: Bracewell, Brian Cooper, Marc Price, John Whiteoak and Frank Gardner

p.452, the text has been corrected to: uum observations with the Chris Cross. Frank Gardner

p.455, the text has been corrected to: Physical Sciences, ANU) about his version of the Super-Cross

457, footnote 10, the text has been corrected to: Also Bracewell to Pawsey, NAA C3830 Z3/1/Part 10.16 January 1960. Bracewell wrote,

p.459, the text has been corrected to: remainder with solar work). Jim Roberts and Don Mathewson were overseas,

461, footnote 20, the text has been corrected to: Based on an interview that Roslynn and Raymond...

p.462, the text has been corrected to: approved by University Senate, thus ending a 17-year career

p.484, the text has been corrected to: were five boxes: solar work (Wild), GRT (blank), receivers

p.487, the text has been corrected to: Catalogue of more than 1840 radio sources (at 408 MHz)

p.488, the text has been corrected to: star in the radio]. It is not a star. Measurements [by Jess Greenstein] on a high dispersion spectrum [from the 200-inch, was obtained by Allan Sandage] suggest [to Bolton] the lines are those of Neon [V], Argon [III] and Argon [IV] and that the red shift is 0.367

497, footnote 15, the text has been corrected to: of Cooper, Gardner, Robinson,...

498, footnote 16, the text has been corrected to: Parkes 30 Years of Radio Astronomy (1994), Goddard and Milne, CSIRO

p.501, the text has been corrected to: of the local Parkes and Goobang Shire citizens

p.513, the text has been corrected to: was expanding rapidly. For many years from this time Parkes dominated observations...

p.514, the text has been corrected to: Owens Valley as an interferometer.

p.521, the text has been corrected to: methods with the rejection of Hoyle, Bondi and Gold's "Steady-state" model of the cosmos

530, table, col 4, last line, the text has been corrected to: Schmidt et al (1963), Hazard et al (1963)

p.541, the text has been corrected to: type IV and Wild added type V. Different mechanisms were invoked

p.543, the text has been corrected to: was titled "Positions of three discrete sources of

p.547, the text has been corrected to: Here we again have Hoyle's principle of maximum

p.562, the text has been corrected to: similar to Mills: "I find a log N–log S slope of – 2.1 with a demarcation similar to

p.570, the text has been corrected to: which would have N(S) / S-1.5 in a static Euclidean Universe. He noted that the

p.575, the text has been corrected to: this should be a part of the preceding letter, thus: already done so and it will be most interesting to see what his results show . . . We look forward to seeing you and Paul Wild [at Jodrell Bank and the IAU].

p.581, the text has been corrected to: galaxy has been questioned by workers in Australia. A survey with the Mills-cross radio

599, footnote 17, the text has been corrected to: See also ESM 36.2 Peter Scheuer Source Count.

p.606, the text has been corrected to: in 2006). NEW PARAGRAPH: In NRAO ONLINE.35, there are a number of photographs from the 1958 Paris Symposium of Radio Astronomy, the Moscow IAU and a trip to the Crimean Astrophysical Observatory. Pawsey was accompanied by Professor Cecilia Payne-Gaposchkin of Harvard on the trip to the Crimea and additional sites in the southern USSR.

p.623, Fig. 37.5, Fig. 37.6, Fig. 37.7, the text has been corrected to: Fig. 37.5 Figure 1 from Ryle (1952) illustrating the principle of the phase switch, (a) connected in phase and (b) connected in anti-phase. Credit: Fig.1, "A new..., the text has been corrected to: Fig. 37.6 Figure 4 from Ryle (1952) showing total power output without phase switching. Credit: Fig. 4, "A new radio, the text has been corrected to: Fig. 37.7 Figure 9 from Ryle (1952) showing the interferometer output with phase switching. Credit: Fig. 9, "A new radio...

p.625, the text has been corrected to: sources (Hanbury Brown, 1974). Richard Twiss ...

p.626, the text has been corrected to: this Hanbury Brown–Twiss effect, was provided...

627, In Fig. 37.8 caption, the text has been corrected to: Scott, Alec Little and Chris

p.628, the text has been corrected to: as the basis for the Owens Valley Radio Observatory (OVRO) interferometer built

p.633, the text has been corrected to: Cambridge. Pearcey (1953) from CSIRO, Sydney, describes...

p.645, the text has been corrected to: In 1955 John Bolton from Sydney...

p.658, the text has been corrected to: David C. Hogg (of the "Hogg horn" antenna), J. R. Pierce and Arno Penzias.

p.660, the text has been corrected to: were compared with the 2C counts presented by Ryle...

p.661, the text has been corrected to: Candidate for President, who lost to Dwight D. Eisenhower in 1952 and 1956.

p.663, the text has been corrected to: Pawsey attended a dinner at Rudolf and Luise Minkowski's house...

p.665, the text has been corrected to: Evans, Paul Wild, J.F. Denisse, Monique Pick, John Firor, Einar Tandberg-Hanssen, Mrs... (3 commas)

p.666, the text has been corrected to: Danville, Illinois (see frontispiece), the first day...

684, footnote 11, the text has been corrected to: NRAO Archive, the same memo in two different locations https://www.nrao.edu/archives/items/show/9019 or https://www.nrao.edu/archives/items/show/16227

p.751, the text has been corrected to: Madsen, Sir John P. V. (1879-1969), Australian, Professor of Electrical Engineering, University of Sydney (1920-1948). Founder of the Radio Research Board (CSIR) in 1927. He played a major role in the establishment of the CSIR Division of Radiophysics as WWII began, creating the major development for Australian radar research. He influenced radar endeavours until 1945.

p.799, the text has been corrected to: 3C 273 Hazard, Jauncey, Goss, Herald. description of discovery of, 451, 481n, 490, 514, 516

p. 801, the text has been corrected to: Casey, R.G., 421, 438, 445, 447, 468, 489, 496, 500

p.810, the text has been corrected to: Pierce, J.R., 466, 645, 658

The correction chapters and the book have been updated.

Appendix A: Abbreviations

AAO	Anglo Australian Observatory late 1960s to 2010, now Australian Astronomical Observatory
AAS	Australian Academy of Science
AASW	Australian Association of Scientific Workers
AAT	Anglo-Australian Telescope
AC	Companion of the Order of Australia
AEI-Australia	Associated Electrical Industries, the holding company formed by the merger of Metrovick and British Thompson-Houston
AHQ	Air Force Headquarters
ALMA	Atacama Large Millimeter Array
AMF	Australian Military Forces (Australian Army)
ANCORS	Australian National Committee on Radio Science
ANGAU	Australia New Guinea Administrative Unit
ANU	Australia National University
ANZAAS	Australia New Zealand Association for the Advancement of Science
ARO	Algonquin Radio Observatory
ASLO	Australian Scientific Liaison Office
ASRLO	Australian Scientific Research Liaison Research Office (ASLO after 1949)
AST	Australian Synthesis Telescope (the proposal of 1970)
ASV	Air to Surface Vessel
AT	Australia Telescope, opened in 1988
ATCA	Australia Telescope Compact Array
ATNF	Australia Telescope National Facility
AURA	Associated Universities for Research in Astronomy
AW/LW	Aircraft warning-light weight
AWA	Amalgamated Wireless Australasia

W. M. Goss et al., *Joe Pawsey and the Founding of Australian Radio Astronomy*, Historical & Cultural Astronomy, https://doi.org/10.1007/978-3-031-07916-0

BIMA   Berkeley-Illinois-Maryland Association, a collaboration of the
                          Universities of California, Illinois, and Maryland that built and
                          operated the BIMA radio telescope array
Caltech   California Institute of Technology, Pasadena, CA
CARMA   Combined Array for Research in Millimeter-wave Astronomy
                          (Caltech)
CASW   Canadian Association of Scientific workers
CBE   Commander of the Most Excellent Order of the British Empire
CCIR   International Radio Consultative Committee
CERA   Commonwealth and Empire Radio for Civil Aviation
CHL   Chain Home Low flying
CMB   Cosmic Microwave Background
CME   Coronal mass ejections
CMP   Central meridian passage
COL   Chain home overseas low
CRAIA   CSIRO Radio Astronomy Image Archive https://www.atnf.csiro.
                          au/ImageArchive/index.html
CRPL   Central Radio Propagation Laboratory
CSIR   Council for Scientific and Industrial Research (Australia) 1926–
                          1949 then CSIRO
CSIRO   Commonwealth Scientific and Industrial Research Organisation
                          (Australia)
DAP   Department of Aircraft Production
DRSS   Directorate of Radio and Signal Supplies, Ministry of Munitions
DSIR   Department of Scientific and Industrial Research (UK)
EHT   Event Horizon Telescope, an international collaboration capturing
                          images of black holes using a virtual Earth-sized telescope
EMI   Electric and Musical Industries
EoR   Epoch of Reionisation
FFP   Freeman, Fox and Partners
FFT   Fast Fourier Transform
FRB   Fast Radio Burst
FST   Fleurs Synthesis Telescope
GAIA   Global Astrometric Interferometer for Astrophysics
GCI   Ground controlled Interception
GRT   Giant Radio Telescope—The working name for the Australian
                          project, after 1961 "Parkes telescope"
HI   Neutral hydrogen as observed with the 21 cm line
HII   "H-2" region, region of ionised hydrogen surrounding a hot star
HMV   His Master's Voice—The Gramophone Company Ltd
IAU   International Astronomical Union
ICSU   International Council for Science
IEEE   Institute of Electrical and Electronics Engineers
IGY   International Geophysical Year (IGY, 1 July 1957–31 December
                          1958)

ILS	Instrument Landing System
IRAM	Institut de Radioastronomie Millimétrique
LW/AW	Light Weight Air Warning
MAP	Ministry of Aircraft Production, US
MB	Mobile Base (radar)
MERLIN	Multi-Element Radio Linked Interferometer Network, an interferometer array of radio telescopes spread across England
MNRAS	Monthly Notices of the Royal Astronomical Society
MOST	Molonglo Observatory Synthesis Telescope
MPP	McCready, Pawsey, Payne-Scott paper of 1947
MRO	Molongolo Radio Observatory
MSH	Abbreviation for the Mills, Slee and Hill survey that catalogued over 2000 discrete radio sources between 1954 and 1957
MSO	Mount Stromlo Observatory, Australian National University, Canberra, ACT (in 1924 founded as the Commonwealth Solar Observatory, in about 1946 name changed to Commonwealth Observatory by Woolley)
MWA	Murchison Widefield Array in Western Australia
NAA	National Archives of Australia
NBS	National Bureau of Standards
NDRC	National Defence Research Committee
NRAO	National Radio Astronomy Observatory (USA)
NRC	National Research Council (Canada, Ottawa)
NRL	Naval Research Laboratory (USA, Washington DC)
NSF	National Science Foundation (USA)
NSWGR	New South Wales Government Railways (Australia)
OCD	European conference 1961
OECD	Organisation Economic Co-operation and Development, International organisation
OSRD	US Office of Scientific Research and Development
OVRO	Owens Valley Radio Observatory—Caltech
PMG	Postmaster General (Australia)
RAAF	Royal Australian Air Force
RAB	Radiophysics Advisory Board
RAN	Royal Australian Navy
RADAR	"Radio, Detection (or Direction Finding), Range" Invented by S.M. Tucker of the US Navy and adopted officially by the Navy in November 1940. Adopted by the British after 1 July 1943
RCM	Radar Counter Measures
RDF	Radio Direction Finding (later radar)
RIMU	Radio Installation and Maintenance Unit
RP and RPL	Radiophysics Laboratory (CSIRO Australia), also RP Radiophysics

RRB	Radio Research Board
RS	Radar station e.g. the Collaroy station north of Sydney was 54RS, that is the 54th radar station in the RAAF air warning system
SC	SuperCross proposal of Mills and Pawsey late 1950s, became the Molonglo Cross of the mid 1960sCross
ShD	Shore Defence radar
SMA	Submillimeter Array, an 8-element radio interferometer located atop Mauna Kea in Hawaii
SNR	Super Nova Remnant
SRLO	Scientific Research Liaison Office
STC	Standard Telephone and Cables—British radio, telephone, telegraph and telecommunications company with an Australian branch office
SVC	Slowly Varying Component of cm radio emission of the sun, originated by Denisse in 1949 "une composante lentement variable"
SWPA	South West Pacific Area, WWII, one of four major Allied command areas in the Pacific. North of Australia, mainly Australian and US forces
TRE	Telecommunications Research Establishment (UK), after 1942 the main centre of Royal Airforce research
UNSW	University of New South Wales (Sydney)
UWA	University of Western Australia (Perth)
URSI	Union Radio-Scientifique Internationale (International Union of Radio Science)
VLA	Very Large Array
VLBA	Very Long Baseline Array
VPI	Virginia Polytechnic Institute, Blacksburg, Virginia
WHC	Washington Hospital Center
WPC	Wave Propagation Committee
WSRT	Westerbork Synthesis Radio Telescope in the northeastern Netherlands
YAL	Young Australia League

Appendix B: Dramatis Personnae

Allen, Clabon "Cla" (1904–1987) Australian astronomer at Mt Stromlo, later director of the University of London Observatory and author of *Astrophysical Quantities*. In retirement, he was again at Mt Stromlo.

Appleton, Sir Edward Victor FRS (1892–1965) British space physicist, specialising in radio propagation; awarded the Nobel Prize in Physics for "his investigations of the physics of the upper atmosphere, especially for the discovery of the so-called Appleton layer."

Atkinson, Sally (1914–2012) Australian secretary to the Chief of RPL E.G. Bowen from 1946 to 1971. As Honorary Archivist at RPL from 1971 to 1992 she transferred over 60 m of files from RPL to the National Archives of Australia.

Baade, Walter (1893–1960) German American optical astronomer who introduced the concept of Population I and II stars. He worked at the Mt Wiison and Palomar Observatory in California and was involved in the identification of the first radio sources.

Baldwin, John Evan FRS (1931–2010) British Astronomer who worked at the Cavendish Astrophysics Group (formerly Mullard Radio Astronomy Observatory) from 1954.

Berkner, Lloyd V. (1905–1967) Prominent geophysicist who played a major role in US ionosphere programs starting in 1928. Member of the Byrd Antarctic expedition of 1928–1930, In 1933, he joined the Carnegie Institution of Washington, Department of Terrestrial Magnetism; active in the US Navy radar program, recruited by Vannevar Bush, returning to the Carnegie Institution as a consultant to the Department of Defence. In 1950, created a third International Polar year to be arranged 25 years (not the usual 50) after the previous events of the 1930s. In 1951, Berkner became the first full-time president of Associated Universities, Inc. (AUI), responsible for Brookhaven National Laboratory and eventually the National Radio Astronomy Observatory.

Bok, Bart J. (1906–1983) Dutch American astronomer whose primary research interest was the structure of the Milky Way; career was at Harvard, Mt Stromlo (Canberra, Australia) and the University of Arizona.

W. M. Goss et al., *Joe Pawsey and the Founding of Australian Radio Astronomy*, Historical & Cultural Astronomy, https://doi.org/10.1007/978-3-031-07916-0

Bok, Priscilla Fairfield (1896–1975). An active astronomer who collaborated with her husband Bart Bok at Harvard, Mt Stromlo and the University of Arizona. Phd 1921, University of California, Berkeley. Married 1929.

Bolton, John Gatenby FRS (1922–1993) British-Australian astronomer; his Cambridge study was interrupted for radar research in the UK Navy. Together with Gordon Stanley and Bruce Slee, he identified first known radio galaxies using a sea-cliff interferometer. He established Owens Valley Observatory in California in 1955 and was later the first Director of the Parkes Radio Telescope in Australia in 1961.

Bondi, Hermann FRS (1919–2005) British-Austrian mathematician and cosmologist best known for developing the steady state model of the universe with Fred Hoyle and Thomas Gold as an alternative to the Big Bang theory.

Bowen, Edward "Taffy" George FRS (1911–1991) British, born in Wales. He was a member of: the group headed by Sir Robert Watson-Watt who developed British radar in 1935; the Tizard Mission from the UK to the USA in 1940; the Radiation Laboratory, Massachusetts Institute of Technology, until 1943; the CSIR and later CSIRO Division of Radiophysics from 1944. He was the Chief of Division of Radiophysics from 1946 to 1971.

Bracewell, Ronald N. (1921–2007) Australian radio astronomer who designed and developed microwave radar equipment in the Radiophysics Laboratory of CSIR under Pawsey's direction during WWII. He had a Cambridge PhD in ionospheric research under Ratcliffe, then returned to CSIRO before moving to Stanford University in 1955. He wrote *The Fourier Transform and Its Applications*, McGraw-Hill, 1965. See *Four Pillars of Radio Astronomy*, Frater et al. (2017).

Builder, Geoffrey (1906–1960) Ionospheric physicist born in Western Australia. He did a PhD with Appleton, Kings College 1933, convincing Appleton of the advantages for investigating the ionosphere with the pulse-echo method (developed at the Carnegie Institution) over Appleton's own frequency-change method. He returned to Australia to the Radio Research Board 1933, then AWA, Australia Army radar in WWII. He was Senior Lecturer in physics at the University of Sydney 1947–1960.

Bush, Vannevar (1890–1974) Head of the U.S. Office of Scientific Research and Development (OSRD) during World War II. He helped initiate the Manhattan Project (the atomic bomb) and the Radiation Laboratory at MIT (radar development) during WWII. His institutional affiliations include the Massachusetts Institute of Technology, Tufts College and the Raytheon Corporation. He was the President of the Carnegie Institution of Washington from 1938 to 1955. He supported the funding of the GRT by the Carnegie Corporation of New York in the 1950s.

Chapman, Sydney FRS (1888–1970) British mathematician and geophysicist known for his work on the kinetic theory of gases, solar-terrestrial physics, and the Earth's ozone layer. From 1914 to 1919, he lectured at Cambridge, then held the Beyer Chair of Applied Mathematics at Manchester from 1919 to 1924. During the Second World War he was Deputy Scientific Advisor to the Army

Council. Following the war, he worked at Oxford until his retirement, then held various positions at the University of Alaska and University of Colorado.

Christiansen, W.N. "Chris" (1913–2007) Australian pioneer radio astronomer and electrical engineer. Christiansen built the first grating array for scanning the sun at the radio astronomy field station at Potts Hill, New South Wales. The Chris Cross Telescope at Badgerys Creek was named after him. From 1960, he was chairman of the electrical engineering department at the University of Sydney. Google doodle from 2013 https://www.google.com/doodles/wilbur-norman-christiansens-100th-birthday. See *Four Pillars of Radio Astronomy*, Frater et al. (2017).

Covington, Arthur Edwin (1913–2001) *Canadian* physicist who made the first *radio astronomy* measurements in *Canada*. He collected data on solar activity every day for more than 30 years at the National Research Council of Canada, Ottawa, Goth Hill Radio Observatory.

Day, George Arthur (1914–2004) Born in the UK. In 1936 as a member of the British Air Ministry worked on the early Chain Home radar set at Bawdsey, Surrey. Later in 1940, Flight Lieutenant in the Royal Air Force. February 1942, seconded to the RAAF and posted to Australia, working with Wing Commander George Pither. Worked with Harry Minnett at RPL and posted to Milne Bay (SE of Port Moresby, Papua, New Guinea) installing a COL Mark II radar set. Recruited by Bowen for RPL in 1946. Played a key role on site selection for the Parkes telescope. From 1961 to 1967, Station Manager for Parkes radio telescope, returned to radio astronomy group in Sydney in 1968.

Dollard, Charles (1907–1977) An educational executive and president of the Carnegie Corporation of New York from 1948 to 1954.

Edge, David O. (1932–2003), PhD in radio astronomy from the University of Cambridge 1959. Together with Michael J. Mulkay, he authored *Astronomy Transformed, The Emergence of Radio Astronomy in Britain*.

Erickson, William "Bill" C. (1930–2015) US radio astronomer. PhD 1956 University of Minnesota. Career at Carnegie Institute of Washington, Convair, Leiden University (Benelux Cross) and University of Maryland (founder of the Clark TPT radio telescope in California). A pioneer in low-frequency radio astronomy. Advisor in the planning for the NRAO Very Large Array and major player in the low-frequency use of the VLA (327 and 74 MHz) systems. Retired to Bunny Island in Tasmania in about 1985 where he constructed a broad-band, solar radio spectrometer that provided valuable data to the solar community up until about 2014. In 2005, the first recipient of the Grote Reber Medal.

Ewen, Harold Irving "Doc" (1922–2015) American. At Harvard University on 25 March 1951, he and Edward Purcell carried out the first detection of the hydrogen line at 21 cm. Later, he was President of the Ewen Knight Corporation and the Ewen Dae Corporation.

Faraday, Michael FRS (1791–1867) British. He established the concept of the electro-magnetic field and showed that magnetism could affect rays of light.

Fermi, Enrico FRS (1901–1954) Italian-born (later naturalised American) physicist and winner of the 1938 Nobel Prize in Physics. He was a professor in Rome at age

24, emigrating to the US in 1939 as professor of physics at Columbia. He played a major role in the Manhattan Project and became professor of physics at the University of Chicago in 1945.

Freeman, Joan (1919–1998) Australian nuclear physicist who started her career at CSIR doing wartime radar research. She received her PhD in physics at Cavendish, joining the Atomic Energy Research Establishment in the UK. Her autobiography is *A Passion for Physics—The Story of a Woman Physicist*. See ESM 9.6, Microwave Radar.

Freeman, Sir Ralph (1911–1998) British civil engineer. He followed his father (also Sir Ralph, 1880–1950, designer of the Sydney Harbour Bridge), joining Freeman, Fox & Partners, a firm of consulting engineers in 1939. FFP became the design team of the GRT in 1955 led by Sir Gilbert Roberts.

Ginzburg, Vitaly Lazarevich FRS (1916–2009) Nobel Prize in Physics, 1993. He was head of the Department of Theoretical Physics of the Lebedev Physical Institute of the Russian Academy of Sciences.

Giovanelli, Ronald Gordon (1915–1984) Australian physicist. He started optical solar physics at the CSIR in 1954 and originated the concept of magnetic reconnection in solar flares. He collaborated with the solar radio astronomy groups at RPL and founded the CSIRO optical observing facilities at Fleurs and later Culgoora.

Graham Smith, Sir Francis FRS (1923–) British radio astronomer. The 1951 publication in *Nature* (20 September) of precise radio position of Cygnus A by Smith led to the optical identification by Baade and Minkowski in 1954 with a 16.2 magnitude galaxy at a distance of 230 Mpc (redshift = 0.056). Director of the Royal Greenwich Observatory and later Jodrell Bank (University of Manchester). Thirteenth Astronomer Royal from 1982 to 1990.

Greenstein, Jesse (1909–2002) American astronomer, Professor of Astrophysics at Caltech. He was an early proponent of radio astronomy in the post WWII era, played a major role in establishment of radio astronomy at Caltech in the mid-1950s, and co-authored the first review paper in radio astronomy in 1947 with Grote Reber (*Observatory,* vol 47. p. 15)

Gum, Colin (1924–1960) Australian astronomer who catalogued emission nebulae in the southern sky at the Mount Stromlo Observatory using wide field photography. He died in a skiing accident in Switzerland in 1960 two years after joining Radiophysics.

Hanbury Brown, Robert FRS (1916–2002) British astronomer and physicist born in India. He made notable contributions to the development of radar during WWII. Post-war at Jodrell Bank, he imaged the northern sky at 1.9 m wavelength. He invented the intensity interferometer and resolved structure of Cygnus A. He was at Jodrell Bank and later the University of Sydney, at Narrabri NSW.

Hazard, Cyril (1928–) English radio astronomer. He held positions at Manchester, University of Sydney and University of Pittsburgh and was a collaborator of Hanbury Brown. In 1962 he observed the occultation of the radio source 3C273 to identify the first quasar.

Heaviside, Oliver FRS, (1850–1925) English self-taught electrical engineer, mathematician, and physicist who adapted complex numbers to the study of electrical circuits, invented mathematical techniques for the solution of differential equations (equivalent to Laplace transforms), reformulated Maxwell's field equations in terms of electric and magnetic forces and energy flux, and independently co-formulated vector analysis.

Heeschen, David Sutphin (1926–2012) Following the death of Pawsey in 1962, became Director of NRAO from 1962 to 1978 (acting director from 1961). Under his directorship, the Very Large Array (VLA) in New Mexico was funded and built. He is responsible for the NRAO's "Open Skies" policy whereby any competent scientist may propose for time on the NRAO telescopes. His philosophy for being a good director: *Hire good people leave them alone, don't take yourself too seriously and have fun!*

Helm, Australia "Austie" J. Australian farmer, the original owner of the GRT site in Gobang Shire, near Parkes, NSW.

Herlofson, Nicolai (1916–2004) Remembered for predicting synchrotron radiation from astronomical objects in 1950 together with Hannes Alvén. He was a pioneer in radar astronomy and contributed an understanding of cosmic rays and the role of magnetic fields in astrophysics.

Herschel, William FRS (1738–1822) British astronomer born in Germany who discovered infrared radiation in sunlight.

Hertz, Heinrich (1857–1894) German physicist who proved the existence of electromagnetic waves. The unit of frequency (cycles per second) is named "Hertz" or "Hz".

Hey, James Stanley FRS, (1909–2000) UK scientist who did radar research for UK army, WWII. With Army radars at 4–6 m wavelength, he discovered non-thermal radiation associated with a prominent solar flare on 27 and 28 February 1942 (published only 12 January 1946). After the war, Hey and his group discovered the first discrete radio source in the constellation of Cygnus. Afterwards, Hey continued radio astronomical observations at the Royal Radar Establishment at Malvern.

Hewish, Antony FRS, (1924–) English radio astronomer; working closely with Martin Ryle at the Mullard Radio Astronomy Observatory (MRAO), Cambridge, he won the Nobel Prize for Physics in 1974 for his role in the discovery of pulsars.

Higgs, Arthur J. (1904–1991) Born in 1904, Higgs graduated from Sydney University in 1926 with First Class Honours in physics. He then joined the staff of the Commonwealth Solar Observatory, in charge of the Radio Research Board's (of CSIR) cathode ray direction finding station at Mt Stromlo. From 1937 to 1941, he was engaged in ionospheric research. In 1941 he worked on radar. In July 1945 he became the Technical Secretary of Radiophysics until he retired from CSIRO on 3 January 1969.

Hill, Eric (1927–2016) Australian radio astronomer. He studied in Leiden in The Netherlands in 1950 under the supervision of Jan Oort, never completed his PhD. Returned to Australia in 1955 and worked at Radiophysics (RPL).

Högbom, Jan (1929–) Swedish astronomer who obtained a PhD in Cambridge, UK, with Ryle. He played a major role in the design of the Westerbork Synthesis Radio Telescope in the Netherlands, invented the CLEAN algorithm for processing images made with radio telescopes and with Christiansen, wrote the definitive book on radio telescope theory, *Radiotelescopes*, Cambridge University Press, 1969.

Hoyle, Fred FRS (1915–2001) British astronomer known for his proposal (along with Gold and Bondi) of the Steady-State theory of the universe. His major work was the theory of stellar nucleosynthesis at Cambridge University. He was an early advocate for the extragalactic origin of discrete radio sources.

Jansky, Karl Guthe (1905–1950) American physicist and engineer working at Bell Labs in New Jersey who first discovered radio waves arising from the Milky Way.

Jeffery, M. "Mike" (1926–1969) Engineer with Freeman Fox and Partners who played a major role in the Parkes telescope, Anglo-Australian Observatory 4 m and the Algonquin Radio Telescopes projects. See ESM 27.2 and NRAO ONLINE 45 and 47. He died in a ski accident in Australia in 1969.

Jelbart, Philip T. (1914–2013) Australia farmer and neighbour at the GRT site. He planted 5000 trees (windbreaks) at the GRT site and provided a photographic record of the pre-construction peg-event at the site of the GRT in July–August 1959.

Jennison, Roger Clifton (1922–2006) English radio astronomer at Jodrell Bank working with Hanbury Brown. Jennison made the discovery of the double nature of radio source Cygnus A with M.K. Das Gupta. He also made the first measurements of closure phase, recognising the great potential for this technique in radio interferometry. The method was used for the first VLBI measurement and a modified form of this approach ("Self-Calibration") is still used today at radio, optical and infrared wavelengths.

Keller, Geoffrey (1918–2007) Professor of Astronomy, Ohio State University (early 1950s to 1986); at the National Science Foundation 1958–1968, he was Program Director of Astronomy, 1957–1962.

Kerr, Frank John (1918–2000) Australian radio astronomer. In 1940 he joined the CSIR Radiophysics laboratory in Sydney, Australia, under the mentorship of Pawsey. In late 1951, Kerr used a specially built 36-foot transit telescope, the largest dish of its kind in Australia, for observations of the 21 cm hydrogen line for studies of the Milky Way and the Magellanic Clouds. He spent the end of his career at the University of Maryland.

Kiepenheuer, Karl Otto (1910–1975) German astronomer and astrophysicist; founded the Kiepenheuer Institute for Solar Physics. He recognised the importance of the synchrotron mechanism connecting cosmic rays and non-thermal radio emission.

Kragh, Helge Stjernholm (1944–) Danish physicist, philosopher and historian of science. He investigated the role of radio astronomy surveys on cosmology in *"Cosmology and Controversy—the Historical Development of Two Theories of the Universe"*, Princeton University Press, 1999.

Laby, Thomas Howell FRS (1880–1946) Australian Professor of Natural Philosophy at the University of Melbourne. He was J.L. Pawsey's mentor in the early 1930s.

Little, Alec (1925–1985) Australian radio astronomer who worked with Ruby Payne-Scott and Mills on instrument development and solar and cosmic radio observations. He later directed Molonglo Radio Observatory of the University of Sydney.

Lovell, Sir Bernard FRS (1913–2012) British, Professor at University of Manchester, Founder of the Jodrell Bank Observatory. During WWII, he was a prominent scientist in UK radar research. After the war, the Jodrell Bank Observatory became a pioneering research institution; the 250-foot telescope (now the Lovell Telescope) became operational in 1957. Lovell was a well-known public figure in the UK as a spokesman for science policy.

Madsen, Sir John P.V. (1879-1969), Australian, Professor of Electrical Engineering, University of Sydney (1920-1948). Founder of the Radio Research Board (CSIR) in 1927. He played a major role in the establishment of the CSIR Division of Radiophysics as WWII began, creating the major development for Australian radar research. He influenced radar endeavours until 1945.

Martyn, David Forbes FRS (1906–1970) Scottish physicist who came to Australia in 1930, leading CSIR ionospheric research. The first chief of CSIR's newly formed Radiophysics Laboratory in 1939; he left this position in October 1941 after a scandal. Post-war he was at Mt Stromlo as part of the CSIR and CSIRO Radio Research Board activities at the Commonwealth Observatory. In 1956 he moved to Camden, NSW, where the Upper Atmosphere Section of the CSIRO was specially created for him. He served as president of the Australian National Academy in 1969 until his death by suicide in 1970.

Mathewson, Don (1929–) Radiophysicist and astronomer known for discovering the Magellanic Stream in 1974; served as acting Director and then Director of Stromlo Observatory 1977–1986.

McCready, Lindsay (1910–1976) Australian radio astronomer and engineer who specialised in receiver systems. His career began at AWA, then he moved to RPL early in WWII. Post-war, he worked with Ruby Payne-Scott in 1945–1951, and later with Paul Wild on the development of solar dynamic spectrographs (frequency-time) at Penrith.

McGee, Richard "Dick" (1921–2012) Australian physicist whose work suggested radio source Sgr A as the Galactic Centre, leading to the redefining of the Galactic Coordinate system. He also pioneered work on HI in the Magellanic Clouds and molecules (eg OH) in the Milky Way.

Messel, Harry (1922–2015) Canadian-born Australian physicist and educator. He was at the University of Adelaide 1951–1952, and then was appointed Head of the School of Physics, University of Sydney in 1952. He recruited Hanbury Brown and Mills in the early 1960s.

Mills, Bernard Y. FRS (1920–2011) Recruited to RPL for radar research by Pawsey in late 1942. In 1948, he joined CSIR's newly formed radio astronomy group. The Mills Cross was planned in 1952–1953, becoming operational at Fleurs in 1954. Between 1954 and 1957, Bernie Mills, Eric Hill and Bruce Slee used the Mills Cross to carry out a survey of the southern sky and recorded more than 2000 sources of discrete radio emission (the MSH Catalogue). In 1960 he moved to the

University of Sydney, where he built the Molonglo Cross Telescope which became operational in 1967. See *Four Pillars of Radio Astronomy*, Frater et al. (2017).

Minkowski, Rudolph (1895–1976) German-born American astronomer, supernova expert. With Walter Baade, he made the first identification of the strong radio source Cygnus A with a faint galaxy in 1954 using the 200-in telescope at Palomar Observatory in California. He established close connections with RPL staff in the 1950s.

Minnett, Harry (1917–2003) Australian engineer who guided the design of the Parkes Radio Telescope (GRT), spending time in London working with FFP; later Chief of CSIRO Division of Radiophysics, 1978–1981.

Mulkay, Michael "Mike" Joseph (1936–) British sociologist of science. Together with David Edge, he authored *Astronomy Transformed, The Emergence of Radio Astronomy in Britain*. In the late 60s and early 70s, Mulkay used Kuhn's and Merton's work, both of which he felt had limitations, to formulate an approach that "opened the way for 'internalist' perspectives in the contemporary sociology of science".

Murray, John D. (1924–2019) Joined RPL in 1947 working with the solar group at Penrith and later Dapto. In 1953 he joined the HI group at Potts Hill and in 1955 worked at Murraybank. From 1961 to 1964 he worked in Leiden, returning to Australia in 1964 where he played a major role in HI work with the new Parkes telescope. In 1973, he played a major role with Don Mathewson in the discovery of the Magellanic stream with the 18 m Parkes dish. In the 1980s, he was instrumental in the construction of the electronics connecting the Parkes dish with the 64 m (now 70 m) Tidbinbilla dish—the Parkes Tidbinbilla Interferometer. He retired in 1989 after a 39-year career at RPL.

Nicoll, Frederick "Ted" H. (1908–2000) Lenore Pawsey's brother, from Battleford, Saskatchewan and J.L. Pawsey's classmate at Cambridge. After his PhD at the Cavendish, he worked for 5 years at EMI. In 1939 he moved to the US to work at RCA (Radio Corporation of America), moving to the RCA Laboratories in Princeton in 1942. Married Kate Neatby in 1934 in the UK. The Nicholl home was a home-base for the Pawsey family in the US for many visits in the years 1941–1962. He retired from RCA in 1973.

Oliphant, Sir Mark Laurence Elwin FRS (1901–2000) Australian physicist and humanitarian who played an important role in the first experimental demonstration of nuclear fusion and the development of nuclear weapons. The cavity magnetron was invented by Randell and Boot in Oliphant's Birmingham laboratory in 1940. After the war, Oliphant returned to Australia as the first director of the Research School of Physical Sciences and Engineering at the new Australian National University (ANU). In 1967, he was appointed Governor of South Australia.

Oort, Jan Hendrik FRS (1900–1992) Dutch astronomer; one of the discoverers of the rotation of the Milky Way in the late 1920s. In 1950 he discovered the "Oort Cloud" of comets in the solar system at distances of 100,000 times the earth-sun distance. He was the founder of radio astronomy in the Netherlands after WWII.

He proposed the Benelux Cross, which was later transformed into the Westerbork Synthesis Radio Telescope (WSRT).

Pawsey, Joseph Lade (14 May 1908–30 November 1962)

Pawsey, Greta Lenore Nicoll (1903–1974) married Joseph L. Pawsey on 7 September 1935.

Pawsey, Margaret Lenore (1937–1977) daughter of Lenore and J.L. Pawsey; married Donald J. McClean on 2 April 1960.

Pawsey, Stuart Frederick (1939–2020) son of Lenore and J.L. Pawsey; married Glenda Jean Powell on 14 December 1968.

Pawsey, Hastings Douglas (1945–) son of Lenore and J.L. Pawsey; married Elizabeth Russell Pain on 6 December 1967.

Payne-Scott, Ruby (1912–1981) Australian radar researcher at CSIR Division of Radiophysics during WWII. In March 1944, she carried out initial test observations of the Milky Way galactic plane at 11 cm wavelength, becoming the first female radio astronomer. Post-war, she was the discoverer of Type I and Type III solar bursts in Sydney, starting in October 1945. In March 1947, she was a co-discoverer of Type II bursts (along with Yabsley and Bolton). She and Alec Little carried out Michelson swept-lobe interferometer observations of solar bursts at Potts Hill in 1949–1951. She resigned from CSIRO in 1951 at the birth of her son, the prominent mathematician, Peter G. Hall FRS (1951–2016).

Penzias, Arno (1933–) German-born American physicist and radio astronomer. He was awarded the Nobel Prize in Physics 1978 with Robert W. Wilson for the discovery of the cosmic microwave background; he worked at Bell Labs.

Piddington, John "Jack" Hobart (1910–1997) Australian research physicist and radio scientist. He played an important role in pre-WWII radar research in the UK and Australia. In the immediate aftermath of Pearl Harbour, he played a major role in the first Air Warning Radar in Australia. He was an active participant in the first decade of radio astronomy starting 1945.

Purcell, Edward Mills FRS (1912–1997) American physicist; WWII cm radar research at the Radiation Lab in Cambridge, Mass. With his graduate student at Harvard, "Doc" Ewen was involved in detection of the 21 cm line of neutral hydrogen on 25 March 1951. In 1952 he shared the Nobel Prize in Physics with Felix Bloch for discovery of nuclear magnetic resonance in liquids and solids.

Rabi, Isidor Isaac (1898–1988) Physicist at Columbia University from 1929, becoming Professor of Physics in 1937. He was awarded the Nobel Prize in Physics in 1944 for his work in magnetic resonance. During WWII, he was Associate Director of the Radiation Lab at MIT. Rabi was a founding trustee of AUI in June 1946 and on the Executive Committee of AUI from 1946 to October 1959. He played a key role in the formation of the Brookhaven National Lab on Long Island in 1947 and in the establishment of the National Radio Astronomy Observatory in November 1956. Rabi was the AUI president from 21 April 1961 (replacing Haworth) to 19 October 1962 (replaced by Tape). During this key period, he appointed Pawsey on 17 December 1961 as the NRAO Director to take up the position on 1 October 1962 (not to occur due to Pawsey's illness and death on 30 November 1962).

Ratcliffe, John "Jack" Ashworth FRS (1902–1987) British radio physicist at the Cavendish Laboratory, University of Cambridge, UK. He pioneered research on the ionosphere including playing an important role in Appleton's ionospheric observations in 1924. During WWII he was a prominent radar scientist at the Telecommunications Research Establishment in the UK. Post-war, he helped initiate radio astronomy at Cambridge. He was Pawsey's thesis advisor in the early 1930s and Bracewell's in the late 1940s. From 1960 to 1966 he was Director of the Radio & Space Research Station at Slough.

Reber, Grote (1911–2002) American electrical engineer; the second radio astronomer following Karl Jansky. Reber built his own 31-foot (9 m) radio telescope—the first use of a parabolic dish for radio astronomy—conducting the first systematic radio sky survey at 160 MHz with a resolution (half power beamwidth) of 12.4°. He moved to Tasmania, Australia, in 1954 to work on low-frequency radio astronomy, remaining there until his death.

Rivett, David FRS (1885–1961) Australian chemist and science administrator, a major contributor to Australian science in the first half of the twentieth century. He served as Chief Executive Officer of the newly formed CSIR (Council for Scientific and Industrial Research) from 1927 to 1946, played a major role in the formation of the Division of Radiophysics in 1939 and the administration of the radar research programme during WWII. From 1946 to 1949, he chaired CSIR, retiring in 1949 as the new CSIRO (Commonwealth Scientific and Industrial Research Organisation) was formed in May 1949.

Roberts, Gilbert (1899–1978) British civil engineer and senior partner with the British firm Freeman, Fox & Partners. He was the major designer of the Parkes radio telescope.

Roberts, James "Jim" A. (1927–) Australian radio astronomer, a member of the Australian team led by John Bolton and Gordon Stanley that launched radio astronomy at Caltech, beginning in 1958. Later he had a prominent career at RPL in Sydney.

Roberts, Walter Orr (1915–1990) A prominent solar physicist, he made major contributions in the mid to late twentieth century. He and Pawsey were colleagues from 1947 to 1962, advising Pawsey about the possible NRAO Directorship. Roberts was the director of the High Altitude Observatory (HAO) in Climax, Colorado, from 1936 to 1961, becoming a professor of Astrogeophysics at the University of Colorado in 1956 until his death in 1990. He also directed the National Center for Atmospheric Research in Boulder, Colorado from 1960 to 1968.

Roderick, Jack W. (1913–1990) Canadian civil engineer who grew up in the UK, graduated University of Bristol and moved to University of Cambridge in 1944. In 1951, he became a Professor of Civil Engineering at the University of Sydney, retiring in 1978. He was an advisor to RPL in the 1950s for the GRT design. He visited London at the end of 1955 to present the design specifications to FFP. He also served as a member of the CSIRO RPL Technical Advisory Committee from July 1955 to June 1959.

Ross, Ian Clunies (1899–1959) Australian veterinary scientist who served as the Executive Officer of the CSIR from 1946 to 1949 and then Chairman of the CSIRO until his death in 1959.

Rowe, A.P. (1898–1976) UK radar pioneer and Australian university official. In the 1930s, Rowe played a major role in the development of radar. From 1935 he was secretary of the Committee for the Scientific Survey of Air Defence which was formed under the chairmanship of Henry Tizard to evaluate research in radio direction finding leading to the formation of the Chain Home system under the operational leadership of Robert Watson-Watt. In 1938–45 he was chief superintendent of the Telecommunications Research Establishment, the major UK research establishment for radar in WWII. Rowe moved to Australia in 1946 as an advisor to the Australian government, becoming the Vice-Chancellor of the University of Adelaide for 10 years in 1948.

Rutherford, Ernest, First Baron Rutherford of Nelson FRS (1871–1937) New Zealand-born British physicist known as the father of nuclear physics. Rutherford discovered the concept of radioactive half-life and the radioactive element radon, served as Director of the Cavendish lab from 1919 to 1937, and was awarded the Nobel Prize in Chemistry in 1908. He was one of J.L. Pawsey's advisors at the Cavendish Laboratory of the University of Cambridge 1931–1935.

Rydbeck, Olof (1911–1999) at Chalmers University of Technology in Gothenburg, Sweden. Radio astronomy pioneer who established the Onsala Radio Observatory in Sweden using WW2 Würzburg radar antennas.

Ryle, Martin FRS (1918–1984) British radio astronomer at Cambridge. For the development of radio astronomical aperture synthesis, he was awarded the Nobel Prize in Physics in 1974 (shared with Anthony Hewish). Ryle, Pawsey and Lovell were the first three radio astronomers to become Fellows of the Royal Society of London.

Sarkissian, John OAM (1962–) Operations Scientist at the CSIRO Parkes Radio Observatory since in 1996. Custodian for the history of the Parkes telescope and an expert in its space tracking activities, including the Apollo Moon landing mission.

Scheuer, Peter (1930–2001) German-British radio astronomer / theoretical astrophysicist. He earned his PhD in Cambridge with Martin Ryle in 1954. Pawsey initiated a RPL-Cavendish "Peace Treaty" with Ryle and Ratcliffe on 13 January 1959 with the suggestion that Scheuer come to RPL for an extended visit. Scheuer arrived in Sydney 28 January 1960, stayed at RPL for 3 years where he worked on radio link interferometry at 85.5 MHz with baselines of 6, 10, 17 and 32 km. He returned to Cambridge 1963, making major contributions in theoretical astrophysics, in particular the nature of extragalactic radio sources.

Schmidt, Maarten (1929–) Dutch astronomer, moved to the US in 1969 to work at Caltech. He optically identified 3C273 and obtained an optical spectrum using the 200-inch Hale Telescope on Mount Palomar. This spectrum showed that 3C 273 was receding at a rate of 47,000 km/s—the first quasar.

Shain, C. Alex (1922–1960) Australian; low-frequency research at RPL at the field stations at Hornsby and Fleurs.

Shklovskii, Iosif Samuilovich (1916–1985) Ukrainian, Soviet astronomer and astrophysicist. He specialised in theoretical astrophysics and radio astronomy, as well as the sun's corona, supernovae, and cosmic rays and their origins. He is noted especially for his suggestion that the radiation from the Crab Nebula was due to synchrotron radiation. His memoir, *Five Billion Vodka Bottles to the Moon: Tales of a Soviet Scientist,* was published posthumously in 1991.

Slee, O. Bruce (1924–2016) Australian radio astronomer. He made important contributions to galactic and extragalactic astronomy while working at CSIRO. He detected the sun in 1945 while a member of the Royal Australian Air Force using an aircraft warning radar antenna located near Darwin.

Smith, F. G., see Graham Smith, Sir Francis FRS

Southworth, George C. (1890–1972) American engineer at Bell Labs known for developing radio frequency wave guides. He carried out the first high-frequency observations of the sun during WWII, published in 1945.

Stanley, Gordon J. (1921–2001) Pioneer radio astronomer in Australia, New Zealand and the US. He moved from his native New Zealand during WWII to join the radar efforts at the CSIR Division of Radiophysics. After the war he worked with John Bolton and Bruce Slee at Dover Heights. With the former he was a key member of the 1948 Cosmic Noise Expedition to New Zealand which led to the optical identification of Taurus A, Virgo A and Centaurus A in 1949. In 1954 he moved to Caltech where he joined Bolton during the formation of the radio astronomy group at Owens Valley Radio Observatory. He was the observatory director from 1961 to 1975. Both Ekers and Goss worked with Stanley at OVRO during this period.

Struve, Otto FRS (1897–1963) Russian-born American astronomer. He directed Yerkes Observatory of the University of Chicago at Williams Bay, Wisconsin, was founding director of the McDonald Observatory near Fort Davis, Texas, and served as the first director of the National Radio Astronomy Observatory from 1959 to 1962.

Sullivan, Woodruff, T. III, (1944–) U.S. radio astronomer, known primarily for his work on the history of radio astronomy. He was a Professor of Astronomy at the University of Washington, Seattle. He is the author of *Cosmic Noise, A History of Early Radio Astronomy,* Cambridge University Press, 2009.

Swarup, Govind FRS (1929–2020) FRS Indian radio astronomer. Swarup worked at CSIRO in Radiophysics with Pawsey, W.N. "Chris" Christiansen, Paul Wild, Bernie Mills and John Bolton. He earned his PhD at Stanford University with Bracewell in late 1960. He founded the radio astronomy group at the Tata Institute of Fundamental Research in Bombay and was a leader in the design and implementation of the Giant Metrewave Radio Telescope.

Sweet, William H. (1910–2001) Neurosurgeon who cared for Pawsey in mid-1962 at Massachusetts General Hospital (MGH) in Boston. A former Rhodes scholar at Oxford, he became a physician in 1936 at Harvard and served as acting Chief of the Queen Elizabeth Hospital in Birmingham, England, Chief of the Neurosurgical Service at MGH, and professor of Surgery at Harvard Medical School. As a member of the Associated Universities, Inc board, he took an interest in the new National Radio Astronomy Observatory. In 1953, he was a co-inventor of the Positron Emission Tomography scan (PET).

Swenson, George W jnr (1922–2017) American engineer and radio telescope builder was a member of the astronomy and of the electrical engineering (EE) departments at the University of Illinois. He was an early contributor to the development of VLBI technology and was a member of the VLA design team.

Tape, Gerald "Jerry" Frederick (1915–2005) Radiation Lab (MIT) staff during WWII. In 1950 he joined the newly founded Brookhaven National Laboratories (AUI) on Long Island, was appointed AUI Vice-President in 1961, and presided as AUI President 19 October 1962 until 10 July 1963. He served on the Atomic Energy Commission from 1963 to 1969 and then returned to AUI as President on 1 May 1969, until the VLA opening 10 October 1980.

Tizard, Sir Henry FRS (1885–1959) chief scientific advisor to the UK Ministry of Defence. Tizard led what became known as the Tizard Mission to the United States. This introduced to the US the newly invented resonant-cavity magnetron (a major advance in radar technology, which in turn provided the basis for airborne interceptors using radar).

Townes, Charles H FRS (1915–2015) American physicist and Professor of physics at the University of California, Berkeley, for much of his career. For his creation of the maser, he shared the 1964 Nobel Prize in Physics with Basov and Prokhorov. Townes understood the role of masers and lasers in astronomy and made the first determination of the mass of the supermassive black hole at the centre of the Milky Way galaxy using molecular lines.

Tuve, Merle Antony (1901–1982) U.S. physicist and radio Astronomer at the Carnegie Institution of Washington (1966–1982). He made ionospheric observations using reflected radio pulses, and was an advisor on US radio astronomy policy 1955–1961.

van de Hulst, Hendrik Christoffel FRS (1918–2000) Dutch astronomer. During the last year of WWII, he predicted the existence of the 21 cm hyperfine line of atomic hydrogen. He was a leader in the formation of astronomical research from space in Europe.

Vonberg, Derek CBE (1921–2015) British electrical engineer, radio astronomer and medical research scientist. He studied at Imperial College then joined the Cavendish Laboratory in 1945 where he worked with Martin Ryle.

Wade, Campbell (1930–) American radio astronomer. He earned his PhD at Harvard in 1958, completed a postdoctoral Fellowship at CSIRO RPL from December 1957 to December 1959. Wade began work at NRAO in February 1960, working in Green Bank West Virginia, Charlottesville, Virginia, and Socorro, New Mexico. He was the first NRAO staff on the Plains of San Augustin (November 1965) and was VLA Director there from 1978 to 1980. He played a major role in choosing the NRAO VLBA sites from Hawaii to Saint Croix.

Wall, Jasper (1942–) Canadian radio astronomer. He worked in Australia with John Bolton (PhD Australian National University, 1970) and in Cambridge, UK, with Martin Ryle. He also worked at the Royal Greenwich Observatory 1979–1998 at Herstmonceux, ING La Palma, Cambridge, where he was Director from 1995 to

1998. He then went to the Department of Physics Oxford 1998–2002, and the University of British Columbia, Department of Physics after 2002.

Wallis, Barnes Neville FRS (1887–1979) British aeronautical designer and military engineer who served as chief of aeronautical research and development at the British Aircraft Corporation from 1945 to 1971. He invented the Master Equatorial coordinate conversion system used in the Parkes radio telescope. He is known for his invention of the "bouncing bomb" used by the Dambusters in WWII.

Warburton, J.A. (1924–2005) Australian; participated with Christiansen on construction and utilisation of the Solar Grating Array at Potts Hill before moving to work in Cloud Physics in 1957. In 1965, he moved to the Desert Research Institute of the University of Nevada in Reno, Nevada, where he ended his career in 1992 as the Executive Director of the Atmospheric Sciences Center. He specialised in weather modification studies.

Waterman, Alan Tower (1892–1967) Director of the National Science Foundation (1951–1963).

Watson-Watt, Sir Robert Alexander, FRS (1892–1973) British pioneer of radio direction finding and radar technology, leading to the development of a system that allowed WWII radio operators to quickly determine the location of an enemy radio. He played a major role in military efforts in 1930 to develop the "Chain Home", using radio signals to locate aircraft at long distances. Chain Home (a code name) was instrumental in the Battle of Britain (1940) as the RAAF defeated the German airforce.

Weaver, Harold F. (1917–2017) Professor of Astronomy at the University of California, Berkeley, and founder of the Radio Astronomy Laboratory in 1958. An expert on galactic structure. Founded the Hat Creek Observatory. Weaver and colleagues carried out major galactic 21 cm surveys of the Milky Way. A co-discoverer of the OH maser line in 1965. As an AUI trustee in 1961, he played a role in the selection of Pawsey as the Director of NRAO.

Weaver, Warren (1894–1978). Director of Rockefeller Foundation Natural Sciences 1932–1955, vice President Natural and Medical Sciences 1955–1959, vice-president of the Alfred P. Sloan Foundation 1959–1964. Weaver is credited for the invention of the term "Molecular Biology." During WWII, Weaver to set up the fire-control section of the National Defense Research Council. The system was used for directing the guns of aircraft against the enemy. He also established the Applied Mathematics Pane to provide assistance in military research. In the mid-1950s, he helped organise the two Rockefeller Foundation Grants to CSIR for the GRT construction (late 1955 and late 1959).

Westerhout, Gart (1927–2012) Dutch American radio astronomer who studied at Leiden University with J.H. Oort working on data from the newly completed 25 m Dwingeloo telescope. PhD in 1958. He played a major role in the galactic coordinate system of 1958 collaborating with Pawsey, Blaauw, Gum, Kerr (ESM 26.5). Moved to the University of Maryland in 1962 as Chairman of Astronomy Department. Became scientific director of the US Naval Observatory 1977–1993.

Westfold, Kevin C (1921–2001) Australian theoretical physicist. He worked in the CSIRO Radiophysics group in 1948, and later collaborated with John Bolton at

Caltech where he wrote the definitive paper on the polarisation of synchrotron radio emission. In his later career he became the Deputy Vice-Chancellor at Monash University in Australia.

White, Frederick William George FRS (1905–1994) New Zealand-born ionospheric physicist who became the second Chief of the CSIR Division of Radiophysics 1942–1944. He then served as a member of the CSIR Executive; in 1949 he became the Chief Executive Officer; in 1959 when Sir Ian Clunies Ross (Chairman from 1949 to 1959) died, he became Chairman, serving until 1970.

Wild, John Paul FRS (1923–2008) British-born Australian scientist who served in World War II as a radar officer in the Royal Navy, and then became a radio astronomer in Australia for CSIR, making discoveries based on radio observations of the sun and building and operating the world's first solar radio-spectrograph. Wild succeeded Taffy Bowen as Chief of the CSIRO Division of Radiophysics (1971–1978), was Chairman of CSIRO 1978–1985. See *Four Pillars of Radio Astronomy*, Frater et al. (2017).

Williamson, Ralph E. (1917–1982) American, Canadian astronomer who did his PhD with Chandrasekhar in 1943 at the University of Chicago working on stellar atmospheres. During WWII he was at Cornell. After the war he worked with Charles Seeger to start radio astronomy at Cornell. In 1946, he joined the astronomy department at the University of Toronto, becoming an Associate Professor in 1947. Pawsey tried unsuccessfully to recruit him to join RPL staff. In 1948 he wrote, "The Present Status of Microwave Astronomy," *Journal of the Royal Astronomical Society of Canada* 42 (1948). In 1953, Williamson left Toronto abruptly for Los Alamos National Labs in New Mexico where he worked on weapons physics and design.

Wills, H. "Howard" Arthur (1906–1989) Member of the Australian Aeronautical Research Laboratory (ARL), expert on fatigues tests. He organised a study of a GRT report from ARL in 1953 for a 250-foot GRT. He was a member of CSIRO's Technical Advisory Committee in 1955–1959 and Chief Defence Scientist 1968–1971.

Wilson, Robert W (1936–) American astronomer who discovered cosmic microwave background radiation (CMB) with Arno Penzias in 1964. The two shared the 1978 Nobel Prize in Physics for this discovery.

Wimperis, Harry Egerton (1876–1960) British aeronautical engineer best known for his role in setting up the Committee for the Scientific Survey of Air Defence under Henry Tizard, which led directly to the development and introduction of radar in the UK during WWII. He is also known for the development of the Drift Sight and Course Setting Bomb Sight during World War I, devices that revolutionised the art of bombing.

Woolley, Sir Richard van der Riet FRS (1906–1986) English astronomer who specialised in solar astronomy. In 1939 he was appointed director of the Commonwealth Solar Observatory in Canberra, Australia, served as Astronomer Royal and Director of the Royal Greenwich Observatory (1956–1971), and was Director of the new South African Astronomical Observatory (1972–1976).

Yabsley, Don E. (1923–2003) Astronomer with CSIRO; worked on radar during WWII; early solar research at Georges Heights, Sydney. He worked on the Parkes Radio Telescope in the 1960s perfecting methods of determining the accurate shape of the antenna. He played a major role in the design of the Australia Telescope in the 1980s.

Zwicky, Fritz (1898–1974) Swiss-American astronomer; an iconoclast scientist who inferred the existence of dark matter based on observations of the velocity dispersion of individual objects in clusters of galaxies. Based on his experience in observing supernovae with Walter Baade in the 1930s (the term "supernova" was coined by these two in 1934), Zwicky proposed that supernovae resulted from the transition of normal stars to neutron stars. He was a professor at Caltech.

Appendix C: Timeline of Key Events

1903   Greta Lenore Nicoll Pawsey born 10 July 1903

1908   Joseph Lade Pawsey born 14 May1908 in Ararat, Victoria

1919   Pawsey enters Camperdown Higher Elementary School

1921   Pawsey enters Wesley College in Melbourne in February

1924   Pawsey joins a Young Australia League (YAL) tour of Europe

1926   Pawsey starts at University of Melbourne

        Council of Scientific and Industrial Research (CSIR) established

1931   Pawsey awarded Masters degree by the University of Melbourne; studied atmospherics with Laby

        Pawsey awarded 1851 Exhibition Scholarship, which funds a PhD in the UK

        Pawsey admitted to a PhD at the University of Cambridge, 16 Oct 1931, supervised by Ratcliffe

1933   Jansky detects radio emission from the galaxy—beginning of radio astronomy

1934   Pawsey joins EMI in April, from academia to industry

1935   Pawsey marries Lenore Nicoll on 7 September 1935

        Pawsey awarded PhD at University of Cambridge

1939   WWII, radar technology developed

        Radiophysics Laboratory (RPL) established with D.F. Martyn, internationally prominent ionospheric theorist, as Chief of Radiophysics

        P.V. Madsen made Chair of Radiophysics Advisory Board

        Pawsey returns to Australia with a position in CSIR, Radiophysics

1940   Ella Kruse Horn (German citizen and suspected Nazi sympathiser) arrives in Australia; Christmas 1940 and early 1941, liaison with David Martyn

1941   Military Intelligence reports about Martyn and Ella Horn, Martyn not sent to the UK with Madsen

        Fred White arrives in Sydney from New Zealand to take over from Martyn

© The Author(s) 2023
W. M. Goss et al., *Joe Pawsey and the Founding of Australian Radio Astronomy*,
Historical & Cultural Astronomy, https://doi.org/10.1007/978-3-031-07916-0

Mid-year Ruby Payne-Scott joins RPL, moving from Amalgamated Wireless Australasia

1942   F.W.G. White formally takes over as Chief of Radiophysics from Martyn and remains until 1945

End of year, Bernard Y. Mills joins RPL

Hey (UK) detects solar radio emission but the detection is classified until the end of WWII

1943   R.N. Bracewell joins RPL to work with Pawsey

1944   van der Hulst predicts the HI line

Bowen joins CSIR

1945   J.N. Briton becomes Chief of Radiophysics until 1946

Norfolk Island Effect is an independent detection of radio emission from the sun

Pawsey becomes leader of the team investigating sources of "thermal and Cosmic" noise, marking the beginning of the Radio Astronomy group in Australia

The first successful experiment in Australia occurs at sunrise on 3 October 1945 at Collaroy Plateau using one of the RAAF radars when Pawsey detects a solar burst; hot million-degree corona required to explain the "quiescent" level

1946   Bowen takes over as Chief of Radiophysics until his retirement in 1971

Sunspots classified by Payne-Scott

John Bolton joins RPL

The first Australian paper on radio astronomy by Pawsey, McCready and Payne-Scott is published in *Nature*

Fourier synthesis paper submitted by Pawsey, Payne-Scott, McCready; theoretical basis for aperture synthesis imaging, not published until August 1947 (13 months delay)

Bracewell goes to England to complete his PhD

1947   J.P. Wild joins RPL

Radio bursts associated with optical sunspots by Pawsey, Payne-Scott, & McCready; first direct connection between radio and optical astronomy

Pawsey and his wife depart for the US by ship from Sydney

Pawsey visits: Caltech, Lenore's parents in Canada, Yerkes Observatory in Wisconsin, Washington DC, Princeton, New Jersey, MIT and Harvard, Cambridge, Mass., Christmas at Nicoll's home in Princeton

1948   Bolton finds six new discrete radio sources using Dover Heights sea-cliff interferometer

Pawsey's visits continue: Harvard visit to Bart Bok, Ottawa and Toronto (David Dunlap Observatory)

Pawseys depart for the UK by the *SS Queen Elizabeth*

In Europe, Pawsey visits Royal Astronomical Society London, the Cavendish Laboratory in Cambridge and Jodrell Bank in Manchester, conferences in Stockholm and Oslo

23 September Pawseys depart from London on *SS Orontes*, arriving in Sydney in October

W.N. Christiansen joins RPL. November partial solar eclipse observations mark entry of Mills and Christiansen to radio astronomy

Penrith field station established where Wild and McCready begin solar burst spectrum observations

Pawsey coins the name "Radio Astronomy"

1949 CSIR reorganised into Commonwealth Scientific and Industrial Research Organisation (CSIRO)

Bracewell returns from the UK to work at CSIRO

Jim Roberts goes to Cambridge, working with Hoyle for his PhD on radiation theory

Radio sources identified with optical counterparts by John Bolton, Taurus A with Crab nebula (snr), Centaurus A and Virgo A with NGC5128 and M87 (extragalactic nebula)

EDSAC 1, early UK computer, can do a 1-D Fourier Transform

CSIRAC early Australian computer operating

1950 Motion of solar bursts measured by Wild and Payne-Scott

Galactic radio emission explained as synchrotron radiation by Kiepenheuer

1951 HI detected at Harvard, based on prediction by van der Hulst

Synchrotron emission process to explain radio galaxies proposed by Russian physicist Ginzburg

Potts Hill array built by Christiansen for simultaneous observations with many antennas

Galactic centre radio source discovered at Potts Hill by Piddington and Minnett

Mills's 101 MHz survey at Badgery Creek

1952 International Union for Radio Science (URSI) meeting held in Sydney showcasing the achievements of Radiophysics

Mills proposes two classes of discrete radio sources: one Galactic and the other isotopically distributed, marking the beginning of controversy with Cambridge

Roberts completes PhD with Hoyle in Cambridge and returns to Australia

Bowen begins plans for GRT

Baade and Minkowski identify Cygnus A based on position by Smith, published 1954

1953 Mills Cross prototype tested at Potts Hill

Fleurs field station established and construction of Mills Cross begins

North-south array added to Potts Hill telescope

Crab nebula optical synchrotron emission explained, Shklovskii, leading to a synchrotron model for non-thermal radio sources

Mills accepts synchrotron radiation theory for galactic emission

Bolton and Slee, Dover Heights cliff interferometer survey: first evidence for evolution in radio source population

Bolton leaves Pawsey's radio astronomy group to join Bowen's cloud physics group

1954   Pawsey elected a Fellow of the Royal Society

Aerial smoothing paper published, Bracewell & Roberts, basic theorems of radio interferometry

British electronic computer, EDSAC1, used for radio astronomy aperture synthesis at the Cavendish Laboratory

Cambridge publishes the flawed 2C Catalogue

1955   *Radio Astronomy* published by Pawsey and Bracewell after a long delay, already partially out of date

CSIRAC moves from RPL to Melbourne, end of CSIR computer development

Christiansen calculates the first 2-D aperture synthesis image—the quiet sun.

Ryle's Halley Lecture announces 2C results. Source counts require evolving universe, controversy with Steady State cosmologists begins

Pawsey attends IAU Manchester symposium—synchrotron theory accepted, multiple papers putting the jigsaw puzzle together

IAU in Ireland

New galactic coordinates, Pawsey-Blaauw proposal at IAU

Beginning of the 85 MHz Mills Cross surveys; completed 1961

Bolton moves to Caltech and starts building the Owens Valley radio observatory in California

1956   Letters to *Scientific American* by Ryle & Mills; Sydney v Cambridge survey conflict

1957   M87 jet synchrotron emission theory by Burbidge completes the puzzle of the radio emission mechanism

Radio polarisation detected by Mayer as predicted for synchrotron radiation

Scheuer publishes a statistical method for analysing confusion limited surveys, the P(D) analysis

Pawsey in US and Canada from early August 1957 to late April 1958, a "try-out" visit to the US sponsored by the National Science Foundation as he is considered for a Directorship of NRAO by Associated Universities, Inc, at Greenbank, West Virginia

Pawsey builds a network of useful contacts through visits with multiple colleagues and observatories.

1958   IAU Paris symposium on Radio Astronomy; acrimonious debate on source counts and cosmology

Moscow IAU GA: new galactic coordinate system adopted (1958 revision)—Pawsey, Blaauw, Gum, Westerhout, Kerr, Oort, Rougoor published in 1960

1959   18 m Kennedy Dish added to E-W arm of Chris Cross to become the Fleurs
        Compound Interferometer
            Cambridge 3C Survey—observational survey results now converging
            First GRT construction contracts
1960   Mills moves to Department of Physics and Christiansen to Department of
        Electrical Engineering, University of Sydney
            Ryle & Hewish, Cambridge, publish the aperture synthesis method
            Bolton appointed Parkes Director, leaves Caltech at year end
            Bolton correspondence with Pawsey about redshift of 3C 48
1961   Bolton arrives back in Australia
            1 October Parkes 64 m Radio Telescope opens, also known as the Giant
        Radio Telescope (GRT)
            Pawsey offered NRAO directorship by AUI President I.I. Rabi same day
        as opening of GRT
            Pawsey at IAU in Berkeley, California, symposia in Santa Barbara,
        California, Cloud Croft, New Mexico, and visits to NRAO Green Bank
        and the Sugar Grove, West Virginia, 600-foot antenna site
            Confidential report to I.I. Rabi (AUI Director) on status of NRAO and
        personnel on 5 October 1961
1962   Parkes detects Faraday rotation in Centaurus A
            Pawsey leaves Radiophysics for NRAO
            In Green Bank in March Pawsey experiences paralysis due to
        glioblastoma
            Lenore Pawsey comes to the US for Pawsey's major surgery at
        Massachusetts General Hospital (June, Boston)
            Diagnosis implies that Pawsey cannot become NRAO director
            Paul Wild accompanies Pawsey and Lenore from NY back to Sydney
        (end July) after Pawsey's surgery in Boston
            Fred Hoyle presents Pawsey Hughes medal of the Royal Society in
        hospital, early November
            Pawsey dies 30 November 1962 in Australia
            David Heeschen appointed NRAO Director
            3C273 occultation at Parkes by Hazard & Schmidt, identification of the
        first quasar
            The Kennedy Dish moved from Fleurs to Parkes
1963   Schmidt measures redshift of 3C273—the first quasar
1964   CSIRAC shut down, end of Australia's role in the development of electronic
        computers
1967   First Pawsey Medal awarded to Sir Robert May by the Australian Academy
        of Science
            Paul Wild's radio heliograph makes 2-D images of solar bursts from the
        active sun
1974   Nobel prize shared by Ryle for aperture synthesis and Hewish for his role in
        the discovery of pulsars

Lenore Nicoll Pawsey dies 29 Nov

2013   Pawsey Supercomputer Centre named, Perth, Australia

2022   ***Joe Pawsey and the Founding of Radio Astronomy in Australia—Early Discoveries from the Sun to the Cosmos*** is published

Appendix D: Electronic Supplemental Material

Unlike the NRAO ONLINE material cited in the book, The Electronic Supplementary Material (ESM) of this book is part of its online edition. These are PDF documents cited in the text as, e.g., "ESM 1.1" and are linked to an online Appendix "Electronic Supplementary Material" of the respective chapter, e.g. https://link.springer.com/chapter/10.1007/978-3-031-07916-0_1#MOESM1, close to the bottom of the chapter webpage.

ESM 1.1 Additional details
ESM 6.1 Pawsey's letters
ESM 7.1 Helium Balloons
ESM 8.1 Helen Borland
ESM 8.2 Lenore Nicoll
ESM 9.1 Radar History
ESM 9.2 Radiophysics Laboratory 1940
ESM 9.3 Difficulties
ESM 9.4 Applied science
ESM 9.5 Light-Weight
ESM 9.6 Microwave Radar
ESM 9.7 Golden Year
ESM 9.8 Radar and Victory
ESM 10.1 Paul Wild
ESM 12.1 Collaroy
ESM 12.2 Marjorie Barnard
ESM 12.3 Distortions
ESM 12.4 Bowen, Collaroy
ESM 13.1 Historical Introduction
ESM 13.2 Fracas
ESM 14.1 Text of Pawsey
ESM 14.2 Final data
ESM 16.1 Proposed Coordination

© The Author(s) 2023   767
W. M. Goss et al., *Joe Pawsey and the Founding of Australian Radio Astronomy*,
Historical & Cultural Astronomy, https://doi.org/10.1007/978-3-031-07916-0

ESM 38.2 Controversy IAU
ESM 38.3 Pawsey Confidential
ESM 38.4 Struve and Appointment
ESM 38.5 Letters of congratulation
ESM 39.1 Pawsey—Rabi 1961
ESM 39.2 Sander Weinreb
ESM 39.3 William Erickson
ESM 39.4 Scheuer-Pawsey interactions
ESM 40.1 Correspondence, Lenore
ESM 40.2 Rabi, Pawsey, Wild
ESM 41.1 Pawsey Lecture Series
ESM 41.2 Pawsey Medal

Appendix E: NRAO ONLINE Supplementary Resources

References NRAO ONLINE.# are referring to sources available at https://science.nrao.edu/about/publications/pawsey.

Note that NRAO ONLINE 57 and 58 consist of additional related texts: NRAO ONLINE 57.1, 57.2, 57.3 and 57.4 and NRAO ONLINE 58.1 and 58.2.

NRAO ONLINE.1 Knuckey system, CSIRO Files
NRAO ONLINE.2 CenA, 1962, Nature publication
NRAO ONLINE.3 Schedvin, CSIRO History 1983
NRAO ONLINE.4 Martyn and lunar radar, 1930
NRAO ONLINE.5 Low level ionosphere, 1936
NRAO ONLINE.6 Martyn trip to UK, 1939
NRAO ONLINE.7 Martyn scandal 1940, breakdown 1954
NRAO ONLINE.8 Mills Cross vs Parkes Dish 1962
NRAO ONLINE.9 Darwin Radar Station 1942
NRAO ONLINE.10 Pither, RAAF History
NRAO ONLINE.11 Moran, threat from Air 1941–1942
NRAO ONLINE.12 Darwin Radar Failure 1942: Epilogue
NRAO ONLINE.13 Oliphant visit to Sydney RPL 1942
NRAOONLINE.14 High Frequency Radar Group, Melbourne 1942
NRAO ONLINE.15 Madsen Resignation as RAB Chair 1942
NRAO ONLINE.16 White Lecture 1943, Radar in Europe
NRAO ONLINE.17 Tizard visit to Australia 1943
NRAO ONLINE.18 Australian Group Radiation Lab 1944–45
NRAO ONLINE.19 Associated Press Brouhaha 1946
NRAO ONLINE.20 Solar Precis, RPL Post-war to 1963, Metre Wave
NRAO ONLINE.21 Eclipse Expedition Failure 1947, RPL and Cavendish
NRAO ONLINE.22 Pawsey Ionosphere Research 1947–1954
NRAO ONLINE.23 Solar Precis, RPL Post-war to 1963, Centimetre Wave
NRAO ONLINE.24 Martyn-Pawsey-Bowen Controversy over the Million Degree Corona 1946

© The Author(s) 2023
W. M. Goss et al., *Joe Pawsey and the Founding of Australian Radio Astronomy*,
Historical & Cultural Astronomy, https://doi.org/10.1007/978-3-031-07916-0

NRAO ONLINE.25 Cloud Physics 1947–48
NRAO ONLINE.26 Pawsey and Canada Connection
NRAO ONLINE.27 New Australia Journal Concerns 1948–59
NRAO ONLINE.28 IAU Commission Radio Astronomy 40 Report 1955
NRAO ONLINE.29 Pawsey Flinders Lecture 1957
NRAO ONLINE.30 Payne-Scott and Ryle Conflict, Type III Bursts
NRAO ONLINE.31 Burchell: Black Cloud Hoyle creates Pawsey as Harry Leicester, Australian Scientist
NRAO ONLINE.32 Menon, IAU 1961
NRAO ONLINE.33 Tata Institute, Pawsey and Indian Radio Astronomy
NRAO ONLINE.34 Helen Sim: URSI at 100 Years book, Australian radio astronomy
NRAO ONLINE.35 Pawsey images USSR 1958
NRAO ONLINE.36 Pawsey Post-war Public Policy
NRAO ONLINE.37 Mills 1953 Events–Baade, Prototype Cross and AAAS Boston
NRAO ONLINE.38 GRT 1951–1952
NRAO ONLINE.39 GRT 1953
NRAO ONLINE.40 GRT 1954
NRAO ONLINE.41 GRT 1955
NRAO ONLINE.42 GRT 1956
NRAO ONLINE.43 GRT 1957
NRAO ONLINE.44 GRT 1958
NRAO ONLINE.45 GRT 1959
NRAO ONLINE.46 GRT 1960
NRAO ONLINE.47 GRT 1961–1962
NRAO ONLINE.48 A Tale of Three Peg Events—Locating the Parkes Telescope, 1958–1959
NRAO ONLINE.49 Jim Roberts Autobiography *Have Gen Will Travel*
NRAO ONLINE.50 Helen Sim MSc thesis Rainmakers
NRAO ONLINE.51 Pawsey as Internationalist
NRAO ONLINE.52 Pawsey photos Solar Conference Sac Peak
NRAO ONLINE.53 Pawsey Bracewell textbook 1955
NRAO ONLINE.54 HI timeline
NRAO ONLINE.55 Duguid, Champion of Aboriginal Rights
NRAO ONLINE.56 TV Interview partial transcript, George Baker with Pawsey and G Giovanelli 1960 Nov 11 Horizons, ABC
NRAO ONLINE.57.1 Stromlo 1957 Conference New Galactic Coordinates
NRAO ONLINE.57.2 Groningen Meeting Nov 1957
NRAO ONLINE.57.3 Drake Galactic Centre Image 8 GHz 1959
NRAO ONLINE.57.4 Ovenden Critique, New Galactic Coordinates 1959
NRAO ONLINE.58.1 Struve: *Sky and Telescope*
NRAO ONLINE.58.2 Colin Gum Biographical Sketch
NRAO ONLINE.59 Radio Astronomy Becomes a New Discipline: J. L. Pawsey in North America and the United Kingdom, 1947–1948
NRAO ONLINE.60 Coordination of Solar Noise Research and Interferometer Techniques with Cambridge 1948–1949

References

(1952). Radar echoes from lightning discharges. *Nature* 169, 650.

Alexander, F. E. S. (1946). The Sun's radio energy. *Radio-Electronics, 1*(1), 16–17.

Alexander, T. B. (1945). *History of the development of the Australian LW/AW equipment.* RP 207/3, 11/1/45. CSIR.

Alfvén, H., & Herlofson, N. (1950). Cosmic radiation and radio stars. *Physical Review, 78*(5), 616.

Allen, C. W., & Gum, C. S. (1950). Survey of Galactic radio-noise at 200 Mc/s. *Australian Journal of Scientific Research A Physical Sciences, 3*, 224.

Allon, F. (2014). At home in the suburbs: Domesticity and nation in postwar Australia. *History Australia, 11*(1), 13–36.

Alsop, L. E., Giordmaine, J. A., Mayer, C. H., & Townes, C. H. (1958). Observations using a maser radiometer at 3-cm wave length. *The Astronomical Journal, 63*, 301.

Ambartzumian, V. A. (1958). *La structure et l'évolution de l'universe.* 11th Solvay conference report, pp. 241–249.

Ambartzumian, V. A., & Eddington, A. (1936). On the derivation of the frequently function of space velocities of the stars from the observed radial velocities. *Monthly Notices of the Royal Astronomical Society, 96*(3), 172–178.

Ankeny, R. A. (2019). *Repertoires as blueprints and frameworks for the doing of science.* https://www.inscits.org/

Ankeny, R. A., & Leonelli, S. (2016). Repertoires: A post-Kuhnian perspective on scientific change and collaborative research. *Studies in History and Philosophy of Science Part A, 60*, 18–28.

Appleton, E. V. (1924). Geophysical influences on the transmission of wireless waves. *Proceedings of the Physical Society of London, 37*(1), 16D.

Appleton, E. V. (1945). Departure of long-wave solar radiation from black-body intensity. *Nature, 156*, 534–535.

Appleton, E. V., & Barnett, M. A. (1925). Local reflection of wireless waves from the upper atmosphere. *Nature, 115*(2888), 333–334.

Appleton, E. V., & Piddington, J. H. (1938). The reflexion coefficients of ionospheric regions. *Proceedings of the Royal Society of London. Series A: Mathematical and Physical Sciences, 164*(919), 467–476.

Auerbach, J. (1999). *The Great Exhibition of 1851: A nation on display.* Yale University Press.

Australian Broadcasting Corporation. (1960, November 11). *Television programme HORIZONS, interview with Joseph Pawsey and Ron Giovanelli by Moderator George Baker.*

Australian Government Publishing Service. (1984). *Guidelines for the operation of National Research Facilities, a report to the Prime Minister by the Australian Science and Technology Council (ASTEC).* Australian Government Publishing Service.

© The Author(s) 2023

W. M. Goss et al., *Joe Pawsey and the Founding of Australian Radio Astronomy,*
Historical & Cultural Astronomy, https://doi.org/10.1007/978-3-031-07916-0

Baade, W., & Minkowski, R. (1954a). Identification of the radio sources in Cassiopeia, Cygnus A, and Puppis A. *The Astrophysical Journal, 119*, 206.

Baade, W., & Minkowski, R. (1954b). On the identification of radio sources. *The Astrophysical Journal, 119*, 215.

Baker, D. (1998). *Preacher, politician, patriot: A life of John Dunmore Lang*. Melbourne University Publishing.

Balick, B., & Brown, R. L. (1974). Intense sub-arcsecond structure in the Galactic center. *The Astrophysical Journal, 194*, 265–270.

Barnard, M. F. (1946). *Notes and narratives, 1940–1945, used in compiling "one single weapon: A history of radar"*. State Library of NSW archives, 1940–1946 MLMSS 887.

Bashford, A., & Macintyre, S. (Eds.). (2013). *The Cambridge history of Australia*. Cambridge University Press.

Baumbach, S. (1937). Strahlung, Ergiebigkeit Electronendichte der Sonnenkorona. *Astronomishe Nachrichten, 263*, 121.

Baxter, J. P., III. (1946). *Scientists against time*. Little, Brown.

Becklin, E. E., & Neugebauer, G. (1975). High-resolution maps of the Galactic center at 2.2 and 10 microns. *The Astrophysical Journal, 200*, L71–L74.

Beevers, C. A., & Lipson, H. (1936). A numerical method for two-dimensional Fourier synthesis. *Nature, 137*(3472), 825–826.

Beevers, C. A., & Lipson, H. (1985). A brief history of Fourier methods in crystal-structure determination. *Australian Journal of Physics, 38*(3), 263–272.

Bennett, A. S., & Simth, F. G. (1962). The preparation of the revised 3C catalogue of radio sources. *Monthly Notices of the Royal Astronomical Society, 125*(1), 75–86.

Bennett, B., & Hodge, J. (2011). Science and empire: Knowledge and networks of science across the British Empire, 1800–1970. *Palgrave Connect (Online Service)*.

Bertotti, B., Balbinot, R., Bergia, S., & Messina, A. (Eds.). (1990). *Modern cosmology in retrospect*. Cambridge University Press.

Bhathal, R. (2014). *Bowen, Edward George (1911–1991)*. Australian Dictionary of Biography, National Centre of Biography, Australian National University.

Bhathal, R., Sutherland, R., & Butcher, H. (2013). *Mt Stromlo observatory: From Bush observatory to the Nobel prize*. CSIRO.

Binnie, A. (2007). A short history of the Australian Institute of Physics: Part 2 from the formation of the Australian Branch of the Institute of Physics to the establishment of the Australian Institute of Physics. *Australian Physicist, 44*(4), 128–137.

Bird, A. Thomas Kuhn. In E. N. Zalta (Ed.), *The Stanford encyclopedia of philosophy* (Winter 2018 ed.) https://plato.stanford.edu/archives/win2018/entries/thomas-kuhn/

Blaauw, A. (1960). Optical determinations of the Galactic pole (paper IV). *Monthly Notices of the Royal Astronomical Society, 121*(2), 164–170.

Blaauw, A. (1994). *History of the IAU*. The birth and first half-century of the International Astronomical Union.

Blaauw, A., Gum, C. S., Pawsey, J. L., & Westerhout, G. (1959). Note: Definition of the new IAU system of Galactic co-ordinates. *The Astrophysical Journal, 130*, 702.

Blaauw, A., Gum, C. S., Pawsey, J. L., & Westerhout, G. (1960). The new IAU system of Galactic coordinates (1958 revision). *Monthly Notices of the Royal Astronomical Society, 121*(2), 123–131.

Blainey, G. (1982). *The tyranny of distance* (Rev. ed., pp. 1820–1870). Macmillan.

Bland-Hawthorn, J., & Robertson, P. (2014). Centre of the Galaxy–sixtieth anniversary of an Australian discovery. *Australian Physics, 51*, 194–199.

Blythe, J. H. (1957). A new type of pencil beam aerial for radio astronomy. *Monthly Notices of the Royal Astronomical Society, 117*(6), 644–651.

Boischot, A. (1958). Étude du rayonnement radioélectrique solaire sur 169 MHz é l'aide d'un grand interféromètre à réseau. *Annales d'Astrophysique, 21*, 273.

Boischot, A., & Denisse, J. (1957). Les emissions de type-iv et lorigine des rayons cosmiques associes aux eruptions chromospheriques. *Comptes Rendus Hebdomadaires des Séances de l'Académie des Sciences, 245*(25), 2194–2197.

Bok, B. J. (1955). Jodrell Bank symposium on radio astronomy. *Sky and Telescope, 15,* 21.

Bok, B. J. (1958). The Paris symposium on radio astronomy. *Sky and Telescope, 17,* 620.

Bok, B. J. (1961). Colin S. Gum. *Quarterly Journal of the Royal Astronomical Society, 2,* 37.

Bolton, J. G. (1948). Discrete sources of Galactic radio frequency noise. *Nature, 162*(4108), 141–142.

Bolton, J. G. (1953). Radio astronomy at URSI. *The Observatory, 73,* 23–26.

Bolton, J. G. (1960, September). *The discrete sources of cosmic radio emission.* URSI GA. Unpublished Caltech "Yellow Jacket" #6.

Bolton, J. G. (1982). Radio astronomy at Dover Heights. *Proceedings of the Astronomical Society of Australia, 4,* 349–358.

Bolton, J. G. (1990). The fortieth anniversary of extragalactic radio astronomy: Radiophysics in exile. *Publications of the Astronomical Society of Australia, 8*(4), 381–383.

Bolton, J. G., Slee, O. B., & Stanley, G. J. (1953). Galactic radiation at radio frequencies. VI. Low altitude scintillations of the discrete sources. *Australian Journal of Physics, 6*(4), 434–451.

Bolton, J. G., & Stanley, G. J. (1948a). Observations on the variable source of cosmic radio frequency radiation in the constellation of Cygnus. *Australian Journal of Scientific Research A Physical Sciences, 1,* 58.

Bolton, J. G., & Stanley, G. J. (1948b). Variable source of radio frequency radiation in the constellation of Cygnus. *Nature, 161*(4087), 312–313.

Bolton, J. G., Stanley, G. J., & Slee, O. B. (1949). Positions of three discrete sources of Galactic radio-frequency radiation. *Nature, 164*(4159), 101–102.

Bolton, J. G., Stanley, G. J., & Slee, O. B. (1954a). Galactic radiation at radio frequencies. VIII. Discrete sources at 100 Mc/s between declinations+ 50° and? 50°. *Australian Journal of Physics, 7*(1), 110–129.

Bolton, J. G., & Westfold, K. C. (1950). Structure of the Galaxy and the sense of rotation of spiral Nebulae. *Nature, 165*(4195), 487–488.

Bolton, J. G., Westfold, K. C., Stanley, G. J., & Slee, O. B. (1954b). Galactic radiations at radio frequencies. VII. Discrete sources with large angular widths. *Australian Journal of Physics, 7,* 96.

Bolton, J. G., & Wild, J. P. (1957). On the possibility of measuring interstellar magnetic fields by 21-CM Zeeman splitting. *The Astrophysical Journal, 125,* 296.

Bondi, H. (1960). *Cosmology.* Cambridge University Press. 1952, revised.

Born, M., & Wolf, E. (1959). *Principles of optics electromagnetic theory of propagation, interference and diffraction of light.* Pergamon Press.

Bowen, E. G. (1945). Radar in war. *Australian Journal of Science, 8,* 33.

Bowen, E. G. (1951). Report in 1951 May 11 meeting of the Royal Astronomical Society. *The Observatory, 71,* 97–99.

Bowen, E. G. (1956). A relation between meteor showers and the rainfall of November and December. *Tellus, 8*(3), 394–402.

Bowen, E. G. (1957). Relation between meteor showers and the rainfall of August, September, and October. *Australian Journal of Physics, 10*(3), 412–417.

Bowen, E. G. (1981). History of Australian astronomy: The pre-history of the Parkes 64-m telescope. *Publications of the Astronomical Society of Australia, 4*(2), 267–273.

Bowen, E. G. (1998). *Radar days.* CRC Press.

Boyd, R. L. F. (1951). *Proceedings of the conference on dynamics of ionized media held under the chairmanship and at the invitation of HSW Massey.* Department of Physics, University College.

Bracewell, R. N. (1952). Radio stars or radio Nebulae? *The Observatory, 72,* 27–29.

Bracewell, R. N. (Ed.). (1959). *Paris symposium on radio astronomy.* IAU symposium no. 9 and URSI symposium no. 1, 30 July–6 August, 1958. No. 9. Stanford University Press.

Bracewell, R. N. (1965). *The Fourier transform and its applications.* McGraw Hill.

Bracewell, R. N. (2002). The discovery of strong extragalactic polarization using the Parkes radio telescope. *Journal of Astronomical History and Heritage, 5*, 107–114.

Bracewell, R. N., Cooper, B. F. C., & Cousins, T. E. (1962). Polarization in the central component of Centaurus A. *Nature, 195*(4848), 1289–1290.

Bracewell, R. N., & Riddle, A. C. (1967). Inversion of fan-beam scans in radio astronomy. *The Astrophysical Journal, 150*, 427.

Bracewell, R. N., & Roberts, J. A. (1954). Aerial smoothing in radio astronomy. *Australian Journal of Physics, 7*(4), 615–640.

Bragg, W. L. (1929). The determination of parameters in crystal structures by means of Fourier series. *Proceedings of the Royal Society of London, 123*(792), 537–559. Series A, containing papers of a mathematical and physical character.

Branagan, D. F., & Holland, H. G. (Eds.). (1985). *Ever reaping something new: A science centenary*. University of Sydney.

Brannen, E., & Ferguson, H. I. (1956). The question of correlation between photons in coherent light rays. *Nature, 178*(4531), 481–482.

Brogan, A. H. (1990). *Committed to saving lives: A history of the Commonwealth Serum Laboratories*. Hyland House.

Brown, A. D., Kouri, N., & Hirst, W. (2012). Memory's malleability: Its role in shaping collective memory and social identity. *Frontiers in Psychology, 3*, 257.

Brown, L. (1999). *Technical and military imperatives: A radar history of World War 2*. CRC Press.

Brown, L. (2005). *Centennial history of the Carnegie Institution of Washington: Volume 2, the department of terrestrial magnetism*. Cambridge University Press.

Brown, Robert Hanbury – see Hanbury Brown.

Brown, R. L., & Johnston, K. (1983). The gas density and distribution within 2 parsecs of the Galactic Center. *The Astrophysical Journal, 268*, L85–L88.

Brown, R. L., Johnston, K. J., & Lo, K. Y. (1981). High resolution VLA observations of the Galactic center. *The Astrophysical Journal, 250*, 155–159.

Bryant, J. H. (1988). The first century of microwaves-1886 to 1986. *IEEE Transactions on Microwave Theory and Techniques, 36*(5), 830–858.

Buchwald, J. Z. (1985). *From Maxwell to microphysics: Aspects of electromagnetic theory in the last quarter of the nineteenth century*. University of Chicago Press.

Buckley-Moran, J. (1986). Australian scientists and the 'Cold War'. In *Intellectual suppression: Australian case histories, analysis and responses* (pp. 11–23). Angus & Robertson.

Budden, K. G. (1988). John Ashworth Ratcliffe. 12 December 1902–25 October 1987. *Biographical Memoirs of Fellows of the Royal Society, 34*, 671–711.

Bullock, E. M. (1999). Radar manufacture by the NSWGR during World War II. *Australian Railway Historical Society Bulletin, 50*(272).

Burbidge, G. R. (1956). On synchrotron radiation from Messier 87. *The Astrophysical Journal, 124*, 416.

Burbidge, G., & Hewitt, A. (1981). Telescopes for the 1980s. *Annual Reviews Monograph*.

Burgess, R. E. (1941). Noise in receiving aerial systems. *Proceedings of the Physical Society, 53*(3), 293.

Burgess, R. E. (1946). Fluctuation noise in a receiving aerial. *Proceedings of the Physical Society, 58*(3), 313.

Burke, B. F. (1954). Comments at Washington conference on radio astronomy. *Journal of Geophysical Research, 59*, 191.

Burke, B. F. (1957). Systematic distortion of the outer regions of the Galaxy. *AJ, 62*, 90.

Bush, V. (1995). *Science, the endless frontier*. Ayer Company Publishers.

Campbell, D. B. (2019). Radio astronomy at Cornell University: The early years, 1946 to 1962. *Journal of Astronomical History and Heritage, 22*(3), 503–520.

Campbell-Kelly, M. (2006). David John Wheeler 1927–2004. *Biographical Memoirs of Fellows of the Royal Society, 52*, 437–453.

Carpenter, M. (1957). *2. 21-cm. Observations in Sydney*. Symposium – International Astronomical Union, vol. 4, pp. 14–18.

Carty, B., & Griffen-Foley, B. (2011). *Australian radio history*. Bruce Carty.

Catterall, P. (2016). *Labour and the Free Churches, 1918–1939 radicalism, righteousness and religion*. Bloomsbury Academic.

CavMag. (2014, February). University of Cambridge, issue 11.

Cetina, K. K. (1991). *Merton's sociology of science: The first and the last sociology of science?*

Chalmers, A. F. (2013). *What is this thing called science?* (4th ed.). University of Queensland Press.

Chandrasekhar, S., & Fermi, E. (1953). Magnetic fields in spiral arms. *The Astrophysical Journal, 118*, 11.

Chowdhury, I. (2016). *Growing the tree of science: Homi Bhabha and the Tata Institute of Fundamental Research*. Oxford University Press.

Christiansen, W. N. (1989). An omission in radio astronomy histories. *Quarterly Journal of the Royal Astronomical Society, 30*, 357.

Christiansen, W. N., & Hindman, J. V. (1952). A preliminary survey of 1420 Mc/s. Line emission from Galactic hydrogen. *Australian Journal of Chemistry, 5*(3), 437–455.

Christiansen, W. N., & Högbom, J. A. (1969). *Radio telescopes*. Cambridge University Press.

Christiansen, W. N., Labrum, N. R., McAlister, K. R., & Mathewson, D. S. (1961). The crossed-grating interferometer: A new high-resolution radio telescope. *Proceedings of the IEE-Part B: Electronic and Communication Engineering, 108*(37), 48–58.

Christiansen, W. N., & Mathewson, D. S. (1958). Scanning the Sun with a highly directional array. *Proceedings of the IRE, 46*(1), 127–131.

Christiansen, W. N., Mathewson, D. S., & Pawsey, J. L. (1957a). Radio pictures of the Sun. *Nature, 180*(4593), 944–946.

Christiansen, W. N., & Mills, B. Y. (1964). Biographical memoir of Joseph Lade Pawsey. *Australian Physicist, 1*, 137–141.

Christiansen, W. N., & Warburton, J. A. (1955). The distribution of radio brightness over the solar disk at a wavelength of 21 centimetres. III. The quiet Sun? Two-dimensional observations. *Australian Journal of Physics, 8*(4), 474–486.

Christiansen, W. N., Warburton, J. A., & Davies, R. D. (1957b). The distribution of radio brightness over the solar disk at a wavelength of 21 centimetres. IV. The slowly varying component. *Australian Journal of Physics, 10*(4), 491–514.

Chubin, D. E. (1978). David O. Edge and Michael J. Mulkay, astronomy transformed: The emergence of radio astronomy in Britain (book review). *Technology and Culture, 19*(3), 580.

Clark, R. W., & Waterhouse, R. (1965). Tizard. *Physics Today, 18*(8), 49.

Cockburn, S., & Ellyard, D. (1981). *Oliphant, the life and times of Sir Mark Oliphant*. Axiom Books.

Colwell, R. C., & Friend, A. W. (1936a). The D region of the ionosphere. *Nature, 137*(3471), 782–782.

Colwell, R. C., & Friend, A. W. (1936b). The lower ionosphere. *Physical Review, 50*(7), 632.

Conklin, N. D. (2006). *Two paths to heaven's gate*. NRAO.

Cooley, J. W., & Tukey, J. W. (1965). An algorithm for the machine calculation of complex Fourier series. *Mathematics of Computation, 19*(90), 297–301.

Cooper, B. F. C., & Price, R. M. (1962). Faraday rotation effects associated with the radio source Centaurus A. *Nature, 195*(4846), 1084–1085.

Cork, E. C., & Pawsey, J. L. (1939). Long feeders for transmitting wide side-bands, with reference to the Alexandra Palace aerial-feeder system. *Journal of the Institution of Electrical Engineers, 84*(508), 448–467.

Cornwell, T. J., & Perley, R. A. (1991). *Radio interferometry: Theory, techniques, and applications*. Proceedings of the 131st IAU Colloquium, Socorro, NM, October 8–12, 1990, vol. 19.

Covington, A. E. (1983). Pearce/Williamson-an appreciation. *Journal of the Royal Astronomical Society of Canada, 77*, 97.

Covington, A. E. (1984). Beginnings of solar radio astronomy in Canada. In *The early years of radio astronomy: Reflections fifty years after Jansky's discovery* (pp. 317–334).

Crick, F. (1988). *What Mad Pursuit: A personal view of scientific discovery* (p. 138). Weidenfeld and Nicholson.

Crompton, R. W. (1991). Leonard George Holden Huxley 1902–1988. *Historical Records of Australian Science, 8*(4).

Davidson, T. W., & Martyn, D. F. (1964). A supposed dependence of meteor rates on lunar phase. *Journal of Geophysical Research, 69*(19), 3981–3987.

Davies, R. D., Graham-Smith, F., & Lyne, A. G. (2016). SIR ALFRED CHARLES BERNARD LOVELL OBE: 31 August 1913–6 August 2012. *Biographical Memoirs of Fellows of the Royal Society, 62*, 323–344.

Davis, L. (1989). Jessie street and war-time child care. *Lilith: A Feminist History Journal, 6*, 33.

de Solla Price, D. J. (1964). Automata and the origins of mechanism and mechanistic philosophy. *Technology and Culture*, 9–23.

Dellit, T. (2000). *Who were they: Royal Australian Air Force on Collaroy plateau in the second World War*. J.E. & D.M. Dellit.

Denisse, J. F. (1949). Relation entre les émissions radioélectriques solaires décimétriques et les taches du Soleil. *Comptes Rendus Hebdomadaires des Séances de l'Académie des Sciences, 228*(20), 1571–1572.

Dicke, R. H. (1946). The measurement of thermal radiation at microwave frequencies. *Review of Scientific Instruments, 17*(7), 268–275.

Domb, C. (2003). Fred Hoyle and Naval Radar 1941–5. *Astrophysics and Space Science, 285*(2), 293–302.

Dombrovsky, V. A. (1954). On the nature of the radiation from the Crab Nebula. *Doklady Akademii Nauk SSSR, 94*, 1021.

Dowden, R. L. (1957). Short-range echoes observed on ionospheric recorders. *Journal of Atmospheric and Terrestrial Physics, 11*(2), 111–117.

Drake, F. D. (1959). Radio resolution of the Galactic nucleus. *Sky and Telescope, 18*.

Drake, F., & Sobel, D. (1992). *Is anyone out there? The scientific search for extraterrestrial intelligence*. Delacorte Press.

Duguid, C. (1947). *The rocket range, Aborigines and war*. Rocket Range Protest Committee.

Duguid, C. (1972). *Doctor and the Aborigines*. Rigby.

DuShane, G. (1957). Radio astronomy at Green Bank. *Science, 126*(3280), 955–955.

Earl, J. A. (1961). Cloud-chamber observations of primary cosmic-ray electrons. *Physical Review Letters, 6*(3), 125.

Echterhoff, G., Higgins, E. T., & Groll, S. (2005). Audience-tuning effects on memory: The role of shared reality. *Journal of Personality and Social Psychology, 89*(3), 257.

Edge, D. O., & Mulkay, M. J. (1975). Fallstudien zu wissenschaftlichen Spezialgebieten. In *Wissenschaftssoziologie* (pp. 197–229). VS Verlag für Sozialwissenschaften.

Edge, D. O., & Mulkay, M. J. (1976). *Astronomy transformed. The emergence of radio astronomy in Britain*. Wiley.

Edge, D. O., Scheuer, P. A. G., & Shakeshaft, J. R. (1958). Evidence on the spatial distribution of radio sources derived from a survey at a frequency of 159 Mc/s. *Monthly Notices of the Royal Astronomical Society, 118*(2), 183–196.

Edge, D. O., Shakeshaft, J. R., McAdam, W. B., Baldwin, J. E., & Archer, S. (1959). A survey of radio sources at a frequency of 159 Mc/s. *Memoirs of the Royal Astronomical Society, 68*, 37–60.

Edlén, B., & Swings, P. (1942). Term analysis of the third spectrum of iron (Fe III). *Astrophysical Journal, 95*, 532.

Edmondson, F. K. (1955). Report from Dublin. *Sky and Telescope, 15*.

Egaña, A. A., & Anduaga, A. (2009). *Wireless and empire: Geopolitics, radio industry, and ionosphere in the British Empire, 1918–1939*. Oxford University Press.

Eickelkamp, U. (1999). *Don't ask for stories: The women from Ernabella and their art.* Aboriginal Studies Press.

Ekers, R. D. (1967). *The structure of Southern radio sources.* PhD thesis, Australian National University.

Ekers, J. A. (Ed.). (1969a). The Parkes catalogue of radio sources, declination zone +20 to -90. *Australian Journal of Physics,* (Suppl 7), 3–75.

Ekers, R. D. (1969b). Interferometric observations of the brightness distribution of Southern radio sources. *Australian Journal of Physics, Astrophysical Supplement, 6,* 3.

Ekers, R. D. (1970). Identification of Southern extragalactic radio sources. *Australian Journal of Physics, 23,* 217.

Ekers, R. D. (1993). *Achievements and challenges for Australian science* (Distinguished Lecture Series). Ian Clunies Ross Memorial Foundation.

Ekers, R. D. (2010). *Big and small.* Accelerating the rate of astronomical discovery – SPS5, IAU General Assembly, Rio de Janeiro, Brazil, August 11–14, 2009, Proceedings of science. arXiv:1004.4279E.

Ekers, R. D. (2014). Non-thermal radio astronomy. *Astroparticle Physics, 53,* 152–159.

Ekers, R. D., Goss, W. M., Schwarz, U. J., Downes, D., & Rogstad, D. H. (1975). A full synthesis map of Sgr A at 5 GHz. *Astronomy and Astrophysics, 43,* 159–166.

Ekers, R. D., Van Gorkom, J. H., Schwarz, U. J., & Goss, W. M. (1983). The radio structure of SGR A. *Astronomy and Astrophysics, 122,* 143–150.

Elder, F. R., Gurewitsch, A. M., Langmuir, R. V., & Pollock, H. C. (1947). Radiation from electrons in a synchrotron. *Physical Review, 71*(11), 829.

Eldridge, C. (2000). Electronic eyes for the allies: Anglo-American cooperation on radar development during World War II. *History and Technology, an International Journal, 17*(1), 1–20.

Ellsworth-Bowers, T. P., Glenn, J., Rosolowsky, E., Mairs, S., Evans, N. J., Battersby, C., . . . Bally, J. (2013). The bolocam Galactic plane survey. VIII. A mid-infrared kinematic distance discrimination method. *The Astrophysical Journal, 770*(1), 39.

Eshleman, V. R., Barthle, R. C., & Gallagher, P. B. (1960). Radar echoes from the Sun. *Science, 131*(3397), 329–332.

Evans, W. F. (1970). *History of the radiophysics advisory board 1939–1945.* CSIRO. 233 p.

Evans, W. F. (1973). *History of the radio research board, 1926–1945.* Commonwealth Scientific and Industrial Research Organization.

Ewen, H. I., & Purcell, E. M. (1951a). Observation of a line in the Galactic radio spectrum. *Nature, 168*(4270), 356.

Ewen, H. I., & Purcell, E. M. (1951b). Radiation from Galactic hydrogen at 1,420 Mc/sec. *Nature, 168*(356), 115–125.

Fainberg, J., & Stone, R. G. (1970). Type III solar radio burst storms observed at low frequencies. *Solar Physics, 15*(2), 433–445.

Farrell, R., & Hooker, C. (2012). The Simon–Kroes model of technical artifacts and the distinction between science and design. *Design Studies, 33*(5), 480–495.

Farrell, R., & Hooker, C. (2013). Design, science and wicked problems. *Design Studies, 34*(6), 681–705.

Fermi, E. (1949). On the origin of the cosmic radiation. *Physical Review, 75*(8), 1169.

Fine, C. (2017). *Testosterone rex: Unmaking the myths of our gendered minds.* Icon Books.

Fleming, A. (1937). Guglielmo Marconi and the development of radio-communication. *Journal of the Royal Society of Arts, 86*(4436), 41–64.

Fox, B. (1990). Forum: Mystery of the missing biography/A look at the life of Alan Blumlein. *New Scientist.*

Fox, K. (2018). *White, Sir Frederick William (Fred) (1905–1994).* Australian Dictionary of Biography, National Centre of Biography, Australian National University. https://adb.anu.edu.au/biography/white-sir-frederick-william-fred-1035/text35059

Frame, T. R., & Faulkner, D. (2003). *Stromlo: An Australian observatory.* Allen & Unwin.

Frater, R. H., Brooks, J. W., & Whiteoak, J. B. (1992). The Australia telescope-overview. *Journal of Electrical and Electronics Engineering, Australia, 12*(2), 103–112.

Frater, R. H., & Ekers, R. D. (2012). John Paul Wild 1923–2008. *Historical Records of Australian Science, 23*(2), 212–227.

Frater, R. H., Goss, W. M., & Wendt, H. W. (2013). Bernard Yarnton Mills AC FAA. 8 August 1920–25 April 2011. *Biographical Memoirs of Fellows of the Royal Society, 59*, 215–239.

Frater, R. H., Goss, W. M., & Wendt, H. W. (2017). *Four pillars of radio astronomy: Mills, Christiansen, Wild, Bracewell*. Springer.

Freeman, J. (1991). *A passion for physics*. CRC Press.

Friis, H. T. (1944). Noise figures of radio receivers. *Proceedings of the IRE, 32*(7), 419–422.

Gabites, J. (2000). *Miles Aylmer Fulton Barnett*. Dictionary of New Zealand biography. Ministry for culture and heritage.

Gaizauskas, V. (2008). Early years at the David Dunlap observatory. *THE R*, 224.

Galison, P. (1987). *How experiments end*. University of Chicago Press.

Galison, P., Hevly, B., & Weinberg, A. M. (1992). Big science: The growth of large-scale research. *Physics Today, 45*(11), 89.

Gardner, F. F., & Pawsey, J. L. (1953). Study of the ionospheric D-region using partial reflections. *Journal of Atmospheric and Terrestrial Physics, 3*(6), 321–344.

Gardner, F. F., & Whiteoak, J. B. (1963). Polarisation of radio sources and Faraday rotation effects in the Galaxy. *Nature, 197*(4873), 1162–1164.

Gascoigne, S. C. B. (1992). Bok, Woolley and Australian astronomy. *Historical Records of Australian Science, 9*(2), 119–126.

Gascoigne, S. C. B. (2012). *Woolley, Sir Richard Van Der Riet (1906–1986)*. In *Australian dictionary of biography* (Vol. 18). Manchester University Press.

Gattei, S. (2008). *Karl Popper's philosophy of science: Rationality without foundations* (Vol. 5). Routledge.

Gauss, G. F. (1805). *Fast Fourier transform* (unpublished).

Gillmor, C. S. (1976). The history of the term 'ionosphere'. *Nature, 262*(5567), 347–348.

Gillmor, C. S. (1986). Federal funding and knowledge growth in ionospheric physics, 1945–81. *Social Studies of Science, 16*(1), 105–133.

Gillmor, C. S. (1991). Ionospheric and radio physics in Australian science since the early days. In *International Science and National Scientific Identity* (pp. 181–204). Springer.

Ginzburg, V. L. (1946). On solar radiation in the radio spectrum. *Comptes Rendus (Doklady) de l'Académie de Sciences de l'URSS, 52*, 487.

Ginzburg, V. L. (1951). Cosmic rays as the source of Galactic radio emission. *Dokl Akad Nauk SSSR, 76*, 377. [In Russian].

Ginzburg, V. L. (1953). The origin of cosmic rays and radio astronomy. *Uspekhi Fizicheskikh Nauk, 51*, 343–392. [In Russian].

Ginzburg, V. L. (1959). Radio astronomy and the origin of cosmic rays. In *Symposium-International Astronomical Union* (Vol. 9, pp. 589–594). Cambridge University Press.

Ginzburg, V. L. (1982). Cosmic rays as the source of Galactic radio emission. In *Classics in Radio Astronomy* (pp. 93–99). Springer.

Glauber, R. J. (1963). Coherent and incoherent states of the radiation field. *Physical Review, 131*(6), 2766.

Goddard, B. R., Watkinson, A., & Mills, B. Y. (1960). An interferometer for the measurement of radio source sizes. *Australian Journal of Physics, 13*(4), 665–675.

Goddard, D. E. (Ed.). (1994). *Pioneering a new astronomy: Papers in memory of John G. Bolton*. CSIRO.

Goddard, D. E., & Haynes, R. (1994). Pioneering a new astronomy; John G. Bolton memorial symposium, Parkes, New South Wales, Australia, Dec. 9–10, 1993. *Australian Journal of Physics, 47*.

Goddard, D. E., & Milne, D. K. (Eds.). (1994). *Parkes: Thirty years of radio astronomy*. CSIRO.

Gordon, W. E. (2001). Henry G. Booker. In *Biographical Memoirs* (Vol. 79). National Academies Press.

Goss, W. M. (2013). *Making waves: The story of Ruby Payne-Scott: Australian pioneer radio astronomer*. Springer.

Goss, W. M. (2014). *Origins of radio astronomy at the Tata Institute of Fundamental Research and the role of JL Pawsey*. arXiv preprint arXiv:1408.3734.

Goss, W. M., Brown, R. L., & Lo, K. Y. (2003). The discovery of Sgr A. *Astronomische Nachrichten: Astronomical Notes, 324*(S1), 497–504.

Goss, W. M., & McGee, R. (1996). The discovery of the radio source Sagittarius A (Sgr A). *The Galactic Center, 102*, 369.

Goss, W. M., & McGee, R. (2009). *Under the radar: The first woman in radio astronomy: Ruby Payne-Scott* (Vol. 363). Springer.

Goss, W. M., Morris, D., & Sheshadri, G. (2016). *Radio astronomers at sea: Martin Ryle and V. Radhakrishnan correspondence, 1963–1966: The voyage of the 'Cygnus A' from the UK to Australia.*

Goss, W. M., Murray, J. D., & Radhakrishnan, V. (1970). The use of HI absorption to determine distance for 10 Galactic radio sources. *Publications of the Astronomical Society of Australia, 1*(7), 332–333.

Graham, A. W., Kenyon, K. H., Bull, L. J., Don, V. C. L., & Kuhlmann, K. (2020). History of astronomy in Australia: Big-impact astronomy from World War II until the lunar landing (1945–1969). *Galaxies, 9*, 24. https://doi.org/10.3390/galaxies9020024

Smith, G. (n.d.). *See Smith, F.G.*

Gredel, R. (Ed.). (1996). *The Galactic center: 4th ESO/CTIO Workshop, La Serena, Chile, March 10–15* (Vol. 102). Astronomical Society of the Pacific.

Greenstein, J. L., Henyey, L. G., & Keenan, P. C. (1946). Interstellar origin of cosmic radiation at radio-frequencies. *Nature, 157*(3998), 805–806.

Gregory, R. A., & Ferguson, A. (1941). Oliver Joseph Lodge. 1851–1940. *Obituary Notices of Fellows of the Royal Society, 3*(10), 551–574.

Griffiths, T. (1996). *Hunters and collectors: The antiquarian imagination in Australia*. Cambridge University Press.

Grimshaw, P. (1980). Women and the family in Australian history. In E. Windschuttle (Ed.), *Women, class and history: Feminist perspectives on Australia 1788 to 1978* (pp. 37–52). Fontana Books.

Guerlac, H. (1987). *Radar in World War II* (Vol. 8). Tomash Publishers.

Gum, C. S., Kerr, F. J., & Westerhout, G. (1960). A 21-cm determination of the principal plane of the Galaxy (paper II). *Monthly Notices of the Royal Astronomical Society, 121*(2), 132–149.

Gum, C. S., & Pawsey, J. L. (1960). Radio data relevant to the choice of a Galactic coordinate system. *Monthly Notices of the Royal Astronomical Society, 121*(2), 150–163.

Haddock, F. T., Mayer, C. H., & Sloanaker, R. M. (1954). Radio emission from the Orion Nebula and other sources at lambda 9.4 cm. *The Astrophysical Journal, 119*, 456.

Hagen, J. P., Haddock, F. T., & Reber, G. (1951). NRL Aleutian radio eclipse expedition. *Sky and Telescope, 10*, 111.

Hagen, J. P., & Hepburn, N. (1952). Solar outbursts at 8.5-mm. Wave-length. *Nature, 170*(4319), 244–245.

Halberstam, D. (1994). *The fifties*. Ballantine Books.

Hanbury Brown, R. (1953). A symposium on radio astronomy at Jodrell Bank. *The Observatory, 73*, 185–198.

Hanbury Brown, R. (1974). *The intensity interferometer*. Taylor & Francis.

Hanbury Brown, R. (1984). Measuring the sizes of stars. *Journal of Astrophysics and Astronomy, 5*(1), 19–30.

Hanbury Brown, R. (1991). *Boffin: A personal story of the early days of radar, radio astronomy and quantum optics*. CRC Press.

Hanbury Brown, R., & Hazard, C. (1951). Radio emission from the Andromeda Nebula. *Monthly Notices of the Royal Astronomical Society, 111*(4), 357–367.

Hanbury Brown, R., & Hazard, C. (1953). An extended radio-frequency source of extra-Galactic origin. *Nature, 172*(4387), 997–998.

Hanbury Brown, R., & Hazard, C. (1953b). A survey of 23 localized radio sources in the Northern Hemisphere. *Monthly Notices of the Royal Astronomical Society, 113*, 123.

Hanbury Brown, R., Jennison, R. C., & Gupta, M. D. (1952). Apparent angular sizes of discrete radio sources: Observations at Jodrell Bank, Manchester. *Nature, 170*(4338), 1061–1063.

Hanbury Brown, R., Minnett, H. C., & White, F. W. G. (1992). *Edward George Bowen*, 14 January 1911–12 August 1991.

Hanbury Brown, R., Palmer, H. P., & Thompson, A. R. (1955). XCVII. A rotating-lobe interferometer and its application to radio astronomy. *The London, Edinburgh, and Dublin Philosophical Magazine and Journal of Science, 46*(379), 857–866.

Hanbury Brown, R., & Twiss, R. Q. (1954). LXXIV. A new type of interferometer for use in radio astronomy. *The London, Edinburgh, and Dublin Philosophical Magazine and Journal of Science, 45*(366), 663–682.

Hargrave, P. J., & Ryle, M. (1974). Observations of Cygnus A with the 5-km radio telescope. *Monthly Notices of the Royal Astronomical Society, 166*(2), 305–327.

Harris, M. (2019). *Rocks, radio and radar: The extraordinary scientific, social and military life of Elizabeth Alexander* (Vol. 4). World Scientific.

Hartcup, G. (1970). "The challenge of war: Scientific and engineering contributions to World War Two.".

Harwit, M. (2019). *Cosmic discovery: The search, scope, and heritage of astronomy*. Cambridge University Press.

Hauptman, H. (1986). Direct methods and anomalous dispersion. *Chemica Scripta, 26*, 277–286.

Haynes, R., Haynes, R. D., Malin, D., & McGee, R. (1996). *Explorers of the Southern sky: A history of Australian astronomy*. Cambridge University Press.

Hazard, C., Jauncey, D., Goss, W. M., & Herald, D. (2018). The sequence of events that led to the 1963 publications in nature of 3C 273, the first Quasar and the first extragalactic radio jet. *Publications of the Astronomical Society of Australia, 35*.

Hazard, C., Mackey, M. B., & Shimmins, A. J. (1963). Investigation of the radio source 3C 273 by the method of lunar occultations. *Nature, 197*(4872), 1037–1039.

Hazard, C., & Walsh, D. (1959). *A comparison of an interferometer and total-power survey of discrete sources of radio-frequency radiation*. URSI Symp. 1, Paris symposium on radio astronomy 9, p. 477.

Hazard, C., & Walsh, D. (1959b). An experimental investigation of the effects of confusion in a survey of localized radio sources. *Monthly Notices of the Royal Astronomical Society, 119*(6), 648–656.

Heaviside, O. (1904). The radiation from an electron moving in an elliptic, or any other orbit. *Nature, 69*(1789), 342–343.

Heeschen, D. (1991). *Reminiscences of the early days of the VLA*. IAU Colloq. 131: Radio interferometry. Theory, techniques, and applications, vol. 19, p. 150.

Herbstreit, J. W., & Johler, J. R. (1948). Frequency variation of the intensity of cosmic radio noise. *Nature, 161*(4092), 515–516.

Hess, V. F. (1912). Observations in low level radiation during seven free balloon flights. *Physikalishce Zeitschrift, 13*, 1084–1091.

Hewish, A. (1958). The scattering of radio waves in the solar corona. *Monthly Notices of the Royal Astronomical Society, 118*(6), 534–546.

Hey, J. S. (1946). Solar radiations in the 4–6 metre radio wave-length band. *Nature, 157*(3976), 47–48.

Hey, J. S. (1949). Radio astronomy (Council report on the progress of astronomy). *Monthly Notices of the Royal Astronomical Society, 109*, 179.

Hey, J. S. (1973). *The evolution of radio astronomy*. Elek Science.

Hey, J. S., Parsons, S. J., & Phillips, J. W. (1946). Fluctuations in cosmic radiation at radio-frequencies. *Nature, 158*(4007), 234–234.

Hey, J. S., Parsons, S. J., & Phillips, J. W. (1948). Some characteristics of solar radio emissions. *Monthly Notices of the Royal Astronomical Society, 108*(5), 354–371.

Hodge, J. M. (2011). Science and empire: An overview of the historical scholarship. In B. M. Bennett & J. M. Hodge (Eds.), *Science and empire. Britain and the World*. Palgrave Macmillan.

Högbom, J. A. (1960). The structure and magnetic field of the solar corona. *Monthly Notices of the Royal Astronomical Society, 120*(6), 530–539.

Högbom, J. A. (1974). Aperture synthesis with a non-regular distribution of interferometer base-lines. *Astronomy & Astrophysics, Supplement Series, 15*, 417.

Högbom, J. A. (2003). Early work in imaging. *ASPC, 300*, 17.

Högbom, J. A., & Brouw, W. N. (1974). The synthesis radio telescope at Westerbork. Principles of operation, performance and data reduction. *Astronomy and Astrophysics, 33*, 289.

Holmes, K. (2016). Talking about mental illness: Life histories and mental health in modern Australia. *Australian Historical Studies, 47*(1), 25–40.

Home, R. W. (1983). Between classroom and industrial laboratory: The emergence of physics as a profession in Australia. *The Australian Physicist, 20*, 166.

Home, R. W. (1984). The problem of intellectual isolation in scientific life: WH Bragg and the Australian Scientific Community, 1886–1909. *Historical Records of Australian Science, 6*(1), 19–30.

Home, R. W. (Ed.). (1988a). *Australian science in the making*. Cambridge University Press.

Home, R. W. (1988b). *The physical sciences: String, sealing wax and self-sufficiency*. The Commonwealth of science: ANZAAS and the scientific enterprise in Australasia 1888–1988, pp. 147–165.

Home, R. W. (1990). *Threlfall, Sir Richard (1861–1932)*. In *Australian dictionary of biography* (Vol. 12).

Home, R. W. (1993). *Builder, Geoffrey (1906–1960)*. In *Australian Dictionary of Biography* (Vol. 13).

Home, R. W. (1993b). *Bailey, Victor Albert (1895–1964)*. Australian Dictionary of Biography, National Centre of Biography, Australian National University.

Home, R. W. (2000). *Martyn, David Forbes (1906–1970)*. Australian Dictionary of Biography, National Centre of Biography, Australian National University.

Home, R. W., & Needham, P. J. (1990). *Physics in Australia to 1945: Bibliography and biographical register*. Department of History and Philosophy of Science University of Melbourne and National Centre for Research and Development in Australian Studies Monash University.

Home, R. W., with the assistance of Needham, P. J. (1995). *Munro, George Hector*. Physics in Australia to 1945, Australian Science Archives Project.

Hooker, C. (2015). *Irresistible forces: Reflections on the history of women in Australian science*. The People & Environment Blog. Accessed September 1, 2015, from https://pateblog.nma.gov.au/2015/09/01/irresistible-forces/comment-page-1/

Hoyle, F. (1949). *Some recent researches in solar physics*. CUP Archive.

Hoyle, F. (1954). Generation of radio noise by cosmic sources. *Nature, 173*(4402), 483–484.

Hoyle, F. (1994). *Home is where the wind blows: Chapters from a cosmologist's life*. University Science Books.

Hull, A. W. (1921). The effect of a uniform magnetic field on the motion of electrons between coaxial cylinders. *Physical Review, 18*(1), 31.

Hyland, A. R., & Faulkner, D. J. (1989). From Sun to the Universe-the Woolley and BOK Directorships at Mount Stromlo. *Proceedings of the Astronomical Society of Australia, 8*.

Jaeger, J. (1943). Theory of the vertical field pattern for RDF [radio direction finding] stations. *RP, 174*.

Jansky, K. G. (1932). Directional studies of atmospherics at high frequencies. *Proceedings of the Institute of Radio Engineers, 20*(12), 1920–1932.

Jansky, K. G. (1933a). Electrical disturbances apparently of extraterrestrial origin. *Proceedings of the Institute of Radio Engineers, 21*(10), 1387–1398.

Jansky, K. G. (1933b). Radio waves from outside the solar system. *Nature, 132*(3323), 66–66.

Jarrell, R. A. (2014). Kevin Charles Westfold. In T. Hockey, V. Trimble, T. R. Williams, K. Bracher, R. A. Jarrell, J. D. Marché, J. Palmeri, & D. W. Green (Eds.), *Biographical encyclopedia of astronomers*. Springer.

Jarvie, I. C., Milford, K., & Miller, D. W. (Eds.). (2006). *Karl Popper: A centenary assessment Volume I* (Vol. 1). Ashgate Publishing.

Jennison, R. C. (1958). Phase closure. *MNRAS, 118*, 276.

Jennison, R. C. (1961). *Fourier transforms and convolutions for the experimentalist: RC Jennison* (p. 415). Pergamon Press.

Jennison, R. C. (1994). High resolution imaging forty years ago. In *Symposium-international astronomical union* (Vol. 158, pp. 337–341). Cambridge University Press.

Jennison, R. C., & Das Gupta, M. K. (1953). Fine structure of the extra-terrestrial radio source Cygnus I. *Nature, 172*(4387), 996–997.

Jones, C. (1995). *Something in the air: A history of radio in Australia*. Kangaroo Press.

Jones, R. V. (1978). *The wizard war: British scientific intelligence, 1939–1945*. Coward McCann.

Kahan, D., Jamieson, K. H., Landrum, A., & Winneg, K. (2017). Culturally antagonistic memes and the Zika virus: An experimental test. *Journal of Risk Research, 20*(1), 1–40.

Kahneman, D. (2011). *Thinking, fast and slow*. Macmillan.

Kai, K., Melrose, D. B., & Suzuki, S. (1985). In D. J. McLean & N. R. Labrum (Eds.), *Solar radio physics*. Cambridge University Press.

Kalberla, P. M. W., Mebold, U., & Reich, W. (1980). Time variable 21 cm lines and the stray radiation problem. *Astronomy and Astrophysics, 82*, 275–286.

Kanbur, R., & Venables, A. J. (Eds.). (2005). *Spatial inequality and development*. Oxford University Press.

Karim, M. T., & Mamajek, E. E. (2016). Revised geometric estimates of the north Galactic pole and the Sun's height above the Galactic midplane. *Monthly Notices of the Royal Astronomical Society*, stw2772.

Keller, G. (1961). Report of the advisory panel on radio telescopes. *ApJ, 134*, 927.

Kellermann, K. I. (1966). The radio source 1934–63. *Australian Journal of Physics, 19*, 195.

Kellermann, K. I. (2012). *Parkes @ 50 years young*. eprint arXiv:1210.0986.

Kellermann, K. I., Bouton, E. N., & Brandt, S. S. (2020). *Open skies: The national radio astronomy observatory and its impact on US radio astronomy* (p. 652). Springer.

Kellermann, K. I., & Sheets, B. (1984). *Serendipitous discoveries in radio astronomy*. Proceedings of a workshop held at the National Radio Astronomy Observatory, Green Bank, West Virginia, May 4–6, 1983: Workshop No. 7.

Kerensky, O. A. (1979). Gilbert Roberts. 18 February 1899–1 January 1978. *Biographical Memoirs of Fellows of the Royal Society, 25*, 477–503.

Kerin, S. (2011). *Doctor do-good: Charles Duguid and aboriginal advancement 1930s–1970s*. Australia Scholarly Publishing.

Kerr, F. J. (1948). Radio Superrefraction in the coastal regions of Australia. *Australian Journal of Scientific Research A Physical Sciences, 1*, 443.

Kerr, F. J. (1952). On the possibility of obtaining radar echoes from the Sun and planets. *Proceedings of the IRE, 40*(6), 660–666.

Kerr, F. J. (1957). A Magellanic effect on the Galaxy. *AJ, 62*, 93–93.

Kerr, F. J. (1971). *Colin Gum and the discovery of the Gum Nebula*. Conference paper, NASA. Goddard Space Flight Center, the Gum Nebula and related problems.

Kerr, F. J. (1984). Early days in radio and radar astronomy in Australia. In *The early years of radio astronomy-reflections fifty years after Jansky's discovery* (p. 133).

Kerr, F. J., Hindman, J. V., & Carpenter, M. S. (1956). Observations of the Southern Milky Way at 21 centimeters. *The Astronomical Journal, 61*, 7.

Kerr, F. J., Hindman, J. V., & Carpenter, M. S. (1957). The large-scale structure of the Galaxy. *Nature, 180*(4588), 677–679.

Kerr, F. J., Hindman, J., & Robinson, B. (1954a). HI in the large and small Magellanic Clouds. *The Australian Journal of Physics*.

Kerr, F. J., Hindman, J. V., & Robinson, B. J. (1954b). Observations of the 21 cm line from the Magellanic Clouds. *Australian Journal of Physics, 7*(2), 297–314.

Kerr, F. J., & Shain, C. A. (1951). Moon echoes and transmission through the ionosphere. *Proceedings of the IRE, 39*(3), 230–242.

Kerr, F. J., Shain, C. A., & Higgins, C. S. (1949). Moon echoes and penetration of the ionosphere. *Nature, 163*(4139), 310–313.

Kiepenheuer, K. O. (1950). Cosmic rays as the source of general Galactic radio emission. *Physical Review, 79*(4), 738.

Korchak, A. A., & Terletsky, Y. P. (1952). Electromagnetic radiation of cosmic-ray protons and Galactic radio radiation [In Russian]. *Zhurnal Eksperimental'noi i Teoreticheskoi Fiziki, 22*, 507–509.

Kragh, H. (1996). *Cosmology and controversy: The historical development of two theories of the Universe*. Princeton University Press.

Kraus, J. D., Ko, H. C., & Matt, S. (1954). Galactic and localized source observations at 250 megacycles per second. *The Astronomical Journal, 59*, 439–443.

Kraus, E. B., & Squires, P. (1947). Experiments on the stimulation of clouds to produce rain. *Nature, 159*(4041), 489–491.

Kuhn, T. S. (2012). *The structure of scientific revolutions*. University of Chicago Press.

Kuiper, G. P. (Ed.). (1953). *The Sun. The solar system* (Vol. 1). University of Chicago Press.

Kuiper, G. P., & Middlehurst, B. M. (Eds.). (1960). *Telescopes*. University of Chicago Press.

Laby, T. H., & Edge, A. B. B. (1931). *The principles & practice of geophysical prospecting: Being the report of the imperial geophysical experimental survey*. Cambridge University Press.

Lehany, F. J., & Yabsley, D. E. (1949). Solar radiation at 1200 Mc/s., 600 Mc/s., and 200 Mc/s. - *Australian Journal of Scientific Research A Physical Sciences, 2*, 48.

Lequeux, J. (1962). Mesures interférométriques à haute résolution du diamètre et de la structure des principales radio sources à 1420 MHz. *Annales d'Astrophysique, 25*, 221.

Lilley, A. E., & McClain, E. F. (1956). The hydrogen-line red shift of radio source Cygnus A. *Astrophysical Journal, 123*, 172.

Lipson, H., & Beevers, C. A. (1936). An improved numerical method of two-dimensional Fourier synthesis for crystals. *Proceedings of the Physical Society, 48*, 772–780.

Little, C. G., & Lovell, A. C. B. (1950). Origin of the fluctuations in the intensity of radio waves from Galactic sources: Jodrell Bank observations. *Nature, 165*(4194), 423–424.

Lockman, F. J., Ghigo, F. D., & Balser, D. S. (2007). *But it was fun: The first forty years of radio astronomy at Green Bank*.

Lockwood, D. (1984). *Australia's Pearl harbour: Darwin 1942*. Rigby.

Longair, M. S. (1966). On the interpretation of radio source counts. *Monthly Notices of the Royal Astronomical Society, 133*, 421.

Lorentz, H. A. (1909). *The theory of electrons and its application to the phenomena of light and radiant heat*. Dover.

Lovelace, N. (2012). History of the Queensland Country Women's Association: 'More than tea and scones'. *Queensland History Journal, 21*(9), 601.

Lovell, B. (1964). Joseph Lade Pawsey. 1908–1962. *Biographical Memoirs of Fellows of the Royal Society*, 229–243.

Lovell, B. (1968). *The story of Jodrell Bank*. Oxford University Press.

Lovell, B. (1973). *Out of the Zenith: Jodrell Bank: 1957–1970*. Oxford University Press.

Machin, K. E. (1951). Distribution of radiation across the solar disk at a frequency of 81.5 MC./S. *Nature, 167*(4257), 889–891.

MacKinnon, C., Simmons, E., & Mann, W. (2009). *The installation of 31 radar station, Darwin, 1942: An investigation by Colin MacKinnon June 1993 with a postscript and additional material by Ed Simmons*. Radar Returns, Williamtown, N.S.W.

MacLeod, R. (1980). On visiting the 'moving metropolis': Reflections on the architecture of imperial science. *Historical Records of Australian Science, 5*, 1–16.

MacLeod, R. (Ed.). (1988). *The commonwealth of science: ANZAAS and the scientific enterprise in Australasia, 1888–1988*. Oxford University Press.

MacLeod, R. (1999). The 'boffins' at Botany Bay: Radar at the university of Sydney 1939–1945. *Historical Records of Australian Science, 12*(4), 411.

MacLeod, R. (Ed.). (2000). *Nature and empire: Science and the colonial enterprise*. University of Chicago Press.

MacLeod, R. M., & Jarrell, R. A. (Eds.). (1994). *Dominions apart: Reflections on the culture of science and technology in Canada and Australia, 1850–1945* (Vol. 17, No. 1–2). Canadian Science and Technology Historical Association.

MacLeod, R. M., & Lewis, M. J. (1988). *Disease, medicine, and empire: Perspectives on Western medicine and the experience of European expansion*. Routledge.

Marine, G., Barr, S., Dirksen, E. M., & Ferry, W. H. (1961). Politics haunts astronomers at Berkeley. *Bulletin of the Atomic Scientists, 17*(8), 345–346.

Marmot, M. (2015). The health gap: The challenge of an unequal world. *The Lancet, 386*(10011), 2442–2444.

Martin, B., Baker, C. A., Manwell, C., & Pugh, C. (1986). Intellectual suppression: Australian case histories. In *Analysis and responses*. Angus & Robertson.

Martyn, D. F. (1946). Temperature radiation from the quiet Sun in the radio spectrum. *Nature, 158*(4018), 632–633.

Martyn, D. F. (1948). Solar radiation in the radio spectrum-I. Radiation from the quiet Sun. *Proceedings of the Royal Society of London. Series A: Mathematical and Physical Sciences, 193*(1032), 44–59.

Massey, H. S. W. (1971). David Forbes Martyn. 1906–1970. *Biographical Memoirs of Fellows of the Royal Society*, 497–510.

Massey, H. S. W., & Mohr, C. B. O. (1938, July). Anomalous scattering of α-particles and long-range nuclear forces. In *Mathematical proceedings of the Cambridge philosophical society* (Vol. 34, No. 3, pp. 498–501). Cambridge University Press.

Masterson, J. (1994). The Parkes Telescope—A 30-year photographic history. In D. E. Goddard & D. K. Milne (Eds.), *Parkes: Thirty years of Radio Astronomy* (p. 27). CSIRO Publications.

Mayer, C. H., McCullough, T. P., & Sloanaker, R. M. (1957). Evidence for polarized radio radiation from the Crab Nebula. *The Astrophysical Journal, 126*, 468.

Mayer, C. H., McCullough, T. P., & Sloanaker, R. M. (1962). Polarization of the radio emission of Taurus A, Cygnus A, and Centaurus A. *The Astronomical Journal, 67*, 581.

McCalman, J. (1993). *Journeyings: The biography of a middle-class generation 1920–1990*. Melbourne University Press.

McCann, D., & Thorne, P. (2000). The last of the first. In *CSIRAC: Australia's first computer*. University of Melbourne.

McClain, E. F. (1960). The 600 foot radio telescope. *Scientific American, 202*.

McCready, L. L., Pawsey, J. L., & Payne-Scott, R. (1947). Solar radiation at radio frequencies and its relation to sunspots. *Proceedings of the Royal Society of London. Series A: Mathematical and Physical Sciences, 190*(1022), 357–375.

McGee, R. X., & Bolton, J. G. (1954). Probable observation of the galactic nucleus at 400 Mc./s. *Nature, 173*(4412), 985–987.

McGee, R. X., Slee, O. B., & Stanley, G. J. (1955). Galactic survey at 400 Mc/s between declinations-17° and-49°. *Australian Journal of Physics, 8*, 347.

McLean, D. J., & Labrum, N. R. (Eds.). (1985). *Solar radiophysics. Studies of emission from the Sun at metre wavelengths*. Cambridge University Press.

McVittie, G. C. (1958). Model universes derived from counts of very distant radio sources. *Publications of the Astronomical Society of the Pacific, 70*, 152–159.

McVittie, G. C. (1959). *Remarks on cosmology.* URSI Symp. 1, Paris symposium on radio astronomy 9, p. 533.

Mellor, D. P. (1958). *The role of science and industry*, vol. 5, Australia in the War of 1939–45 Series Four, Australian War Memorial, Canberra.

Melrose, D. B., & Minnett, H. C. (1998). Jack Hobart Piddington 1910–1997. *Historical Records of Australian Science, 12*(2), 229.

Meredith, R. J. (1998). *Engineers' handbook of industrial microwave heating. No. 25.* Institution of Electrical Engineers.

Merton, R. K. (1942). A note on science and democracy. *J. Legal and Political Sociology, 1*, 115.

Merton, R. K. (1957). Priorities in scientific discovery: A chapter in the sociology of science. *American Sociological Review, 22*(6), 635–659.

Merton, R. K. (1968). The Matthew effect in science: The reward and communication systems of science are considered. *Science, 159*(3810), 56–63.

Merton, R. K., & Barber, E. (2004). *The travels and adventures of serendipity. A study in sociological semantics and the sociology of science.* Princeton University Press.

Meyer, M. (2010). Caring for weak ties—The natural history museum as a place of encounter between amateur and professional science. *Sociological Research Online, 15*(2), 133–146.

Michelson, A. A. (1890). On the application of interference methods to astronomical measurements. *Philosophical Magazine Series, 5*(30), 1–21.

Michelson, A. A., & Pease, F. G. (1921). Measurement of the diameter of alpha-orionis by the interferometer. *Astrophysical Journal, 53*, 249–259.

Mills archive University of Sydney. P154-series 8 file 2.

Mills, B. Y. (1952a). Apparent angular sizes of discrete radio sources: Observations at Sydney. *Nature, 170*(4338), 1063–1064.

Mills, B. Y. (1952b). The distribution of the discrete sources of cosmic radio radiation. *Australian Journal of Scientific Research A Physical Sciences, 5*, 266.

Mills, B. Y. (1953). The radio brightness distributions over four discrete sources of cosmic noise. *Australian Journal of Physics, 6*(4), 452–470.

Mills, B. Y. (1954). Abnormal galaxies as radio sources. *The Observatory, 74*, 248–249.

Mills, B. Y. (1955). The observation and interpretation of radio emission from some bright galaxies. *Australian Journal of Physics, 8*(3), 368–389.

Mills, B. Y. (1956). Letters. *Scientific American, 195*(6), 8.

Mills, B. Y. (1960). On the identification of extragalactic radio sources. *Australian Journal of Physics, 13*(3), 550–577.

Mills, B. Y. (1984). Radio sources and the log N-log S controversy. In *The early years of radio astronomy-reflections fifty years after Jansky's discovery* (p. 147).

Mills, B. (2006). An engineer becomes astronomer. *Annual Review of Astronomy and Astrophysics, 44*, 1–15.

Mills, B. Y., & Little, A. G. (1953). A high-resolution aerial system of a new type. *Australian Journal of Physics, 6*(3), 272–278.

Mills, B. Y., & Slee, O. B. (1957). A preliminary survey of radio sources in a limited region of the sky at a wavelength of 3–5 m. *Australian Journal of Physics, 10*(1), 162–194.

Mills, B. Y., Slee, O. B., & Hill, E. R. (1958). A catalogue of radio sources between declinations+ 10° and? 20°. *Australian Journal of Physics, 11*(3), 360–387.

Mills, B. Y., Slee, O. B., & Hill, E. R. (1960). A catalogue of radio sources between declinations - 20° and -50°. Australian Journal of Physics, 13, 676.

Mills, B. Y., Slee, O. B., & Hill, E. R. (1961). A catalogue of radio sources between declinations - 50° and -80°. Australian Journal of Physics, 14, 497–507.

Mills, B. Y., & Thomas, A. B. (1951). Observations of the sources of radio-frequency radiation in the constellation of Cygnus. *Australian Journal of Scientific Research A Physical Sciences, 4*, 158.

Minkowski archive University of California, Berkeley.

Minkowski, R. (1958). The problem of the identification of extragalactic radio sources. *Publications of the Astronomical Society of the Pacific, 70*, 143–151.

Minkowski, R. (1960). A new distant cluster of Galaxies. *The Astrophysical Journal, 132*, 908–910.

Minnett, H., Alexander, T. B., Bullock, E., Day, G., Fielder-Gill, W., Mills, B., & Richardson, R. C. (1998a). Light-weight air warning radar. *Historical Records of Australian Science, 12*(4), 457–467.

Minnett, H., Alexander, T. B., Cooper, F. C., & Porter, F. H. (1998b). Radar and the bombing of Darwin. *Historical Records of Australian Science, 12*(4), 429–455.

Minnett, H. C., & Labrum, N. R. (1950). Solar radiation at a wavelength of 3.18 centimetres. *Australian Journal of Scientific Research A Physical Sciences, 3*, 60.

Minnett, H. C., & Robertson, R. (1996). Sir Frederick William George White, CBE 26 May 1905–17 August 1994. *Biographical Memoirs of Fellows of the Royal Society, 42*, 497–521.

Mitra, S. N. (1986). Horizontal motion in ionospheric regions – A review. *IJRSP, 15*.

Mitroff, I. I. (1974). Norms and counter-norms in a select group of the Apollo moon scientists: A case study of the ambivalence of scientists. *American Sociological Review*, 579–595.

Mitton, S. (2011). *Fred Hoyle: A life in science*. Cambridge University Press.

Moran, M. (1980). *Threat to Australia from the air 1941–1943*. SR no. 4.

Morpurgo, J. E. (1972). *Barnes Wallis – A biography*. Longman.

Morton, D. C. (1985). The centre of our Galaxy: Is it a black hole. *Australian Physicist, 22*(8), 218–222.

Morton, P. (1989). *Fire across the desert: Woomera and the Anglo-Australian joint project 1946–1980*. Department of Defence Canberra (Australia).

Moskvitch, K. (2020). *Neutron stars: The quest to understand the zombies of the cosmos*. Harvard University Press.

Moyal, A. (1976). *Scientists in nineteenth century Australia: A documentary history*. Cassell.

Moyal, A. (1986). *A bright & savage land: Scientists in Colonial Australia*. Collins.

Moyal, A. (1994). *Portraits in science*. National Library of Australia.

Mulkay, M. J. (1975). Three models of scientific development. *The Sociological Review, 23*, 509–526.

Mulkay, M. J., & Edge, D. O. (1973). Cognitive, technical and social factors in the growth of radio astronomy. *Social Science Information, 12*(6), 25–61.

Muller, C. A., & Oort, J. H. (1951). Observation of a line in the Galactic radio spectrum: The interstellar hydrogen line at 1,420 Mc./sec., and an estimate of Galactic rotation. *Nature, 168*(4270), 357–358.

Muscio, W. T. (1984). *Australian radio: The technical story 1923–1983*. Kangaroo Press.

Myers, D. M. (1986). Madsen, Sir John Percival Vaissing (Vissing) (1879–1969). In *Australian dictionary of biography* (Vol. 10).

Nafe, J. E., Nelson, E. B., & Rabi, I. I. (1947). The hyperfine structure of atomic hydrogen and deuterium. *Physical Review, 71*(12), 914.

Nahin, P. J. (1987). *Oliver Heaviside: Sage in solitude*. IEEE Press.

Napier, P. J., Thompson, A. R., & Ekers, R. D. (1983). The very large array: Design and performance of a modern synthesis radio telescope. *Proceedings of the IEEE, 71*(11), 1295–1320.

Needell, A. A. (1987). Lloyd Berkner, Merle Tuve, and the Federal role in radio astronomy. *Osiris, 3*, 261–288.

Neff, S. G., Eilek, J. A., & Owen, F. N. (2015). The complex north transition region of Centaurus A: Radio structure. *The Astrophysical Journal, 802*(2), 87.

Neil, N., Malmfors, T., & Slovic, P. (1994). Intuitive toxicology: Expert and lay judgments of chemical risks. *Toxicologic Pathology, 22*(2), 198–201.

Newkirk, G., Jr. (1961). The solar corona in active regions and the thermal origin of the slowly varying component of solar radio radiation. *The Astrophysical Journal, 133*, 983.

Newton, H. W. (1955). The lineage of the great sunspots. *Vistas in Astronomy, 1*, 666–674.

Norman, A. (1961). Radiotelescope, deep probe into Universe. *Christian Science Monitor.*

Norris, R. P. (2016). Dawes review 5: Australian aboriginal astronomy and navigation. *Publications of the Astronomical Society of Australia, 33.*

Oatley, C. (2004). *Reginald Leslie Smith-Rose (1894–1980).* Oxford Dictionary of National Biography, Oxford University Press.

O'Brien, P. A. (1953). The distribution of radiation across the solar disk at metre wave-length. *Monthly Notices of the Royal Astronomical Society, 113*(5), 597–612.

O'Dea, C. P., & Baum, S. A. (1997). Constraints on radio source evolution from the compact steep spectrum and GHz peaked spectrum radio sources. *The Astronomical Journal, 113,* 148–161.

Ohlsson, J. (1932). Tables for the conversion of equatorial coordinates into Galactic coordinates. *Annals of the Observatory of Lund, 3,* 1–151.

Oort archive Leiden, University Library.

Oort, J. H., Kerr, F. J., & Westerhout, G. (1958). The Galactic system as a spiral Nebula (Council Note). *Monthly Notices of the Royal Astronomical Society, 118,* 379.

Oort, J. H., & Rougoor, G. W. (1960). The position of the Galactic centre (paper V). *Monthly Notices of the Royal Astronomical Society, 121*(2), 171–173.

Oort, J. H., & Walraven, T. (1956). Polarization and composition of the Crab Nebula. *Bulletin of the Astronomical Institutes of the Netherlands, 12,* 285.

Orchiston, W. (2012). The Parkes 18-m antenna: A brief historical evaluation. *Journal of Astronomical History and Heritage, 15*(2), 96–99.

Orchiston, W., & Mathewson, D. (2009). Chris Christiansen and the Chris Cross. *Journal of Astronomical History and Heritage, 12,* 11–32.

Osterbrock, D. E. (2002). *Walter Baade, dynamical astronomer at Goettingen, Hamburg, Mount Wilson, and Palomar observatories.* DDA 33: 10-03.

Otrupcek, R. E., & Wright, A. E. (1991). PKSCAT90-the Southern radio source database. *Proceedings of the Astronomical Society of Australia, 9,* 170.

Ovenden, M. W. (1959). Astronomy: Revision of the definition of Galactic coordinates. *Science Progress, 47*(187), 476–484.

Palmer, P., & Goss, W. M. (1996). *Nomenclature of the Galactic Center radio sources.* arXiv preprint astro-ph/9607153.

Parascandola, M. (2004). Skepticism, statistical methods, and the cigarette: A historical analysis of a methodological debate. *Perspectives in Biology and Medicine, 47*(2), 244–261.

Parkinson, A. l., & Australian Broadcasting Corporation. (2007). *Maralinga: Australia's nuclear waste cover-up.* ABC Books.

Pawsey, J. L. (1934). *Intensity variations of wireless waves.* PhD diss., Ph.D. thesis, Cambridge.

Pawsey, J. L. (1935). Further investigations of the amplitude variations of down-coming wireless waves. *Proceedings of the Cambridge Philosophical Society, 31.*

Pawsey, J. L. (1940). *Phase dependent elevation aerial, radiophysics laboratory technical report RP58/1.*

Pawsey, J. L. (1945). Atomic power and American work on the development of the atomic bomb. *Australian Journal of Science, 8,* 41–47.

Pawsey, J. L. (1946). Observation of million degree thermal radiation from the Sun at a wavelength of 1.5 metres. *Nature, 158*(4018), 633–634.

Pawsey, J. L. (1948). Geophysical discussion on solar radio noise. *Observatory, 68,* 178.

Pawsey, J. L. (1950). Solar radio-frequency radiation. *Proceedings of the IEE-Part III: Radio and Communication Engineering, 97*(49), 290–308.

Pawsey, J. L. (1953). Radio astronomy in Australia. *Journal of the Royal Astronomical Society of Canada, 47,* 137.

Pawsey, J. L. (1955). A catalogue of reliably known discrete sources of cosmic radio waves. *The Astrophysical Journal, 121,* 1.

Pawsey, J. L. (1955b). Radio star scintillations due to ionospheric focusing. *phio,* 172.

Pawsey, J. L. (1957). Preliminary statistics of discrete sources obtained with the "mills cross". In H. C. van der Hulst (Ed.), *IAU symposium no 4 radio astronomy* (p. 228).

Pawsey, J. L. (1958). Sydney investigations and very distant radio sources. *Publications of the Astronomical Society of the Pacific, 70*, 133–140.

Pawsey, J. L. (1960). Dr. C.S. Gum. *Nature, 186*(4729), 932–932.

Pawsey, J. L. (1961). Australian radio astronomy. *Australian Scientist*.

Pawsey, J. L. (1963). Introduction to the radio astronomy issue. *Proceedings of IRE (Aust.), 24*, 95–97.

Pawsey, J. L., & Bracewell, R. (1955). *Radio astronomy*. Clarendon Press.

Pawsey, J. L., & Harting, E. (1960). An attempt to detect linear polarization in the Galactic background radiation at 215 mc/s. *Australian Journal of Physics, 13*(4), 740–742.

Pawsey, J. L., & Hill, E. R. (1961). Cosmic radio waves and their interpretation. *Reports on Progress in Physics, 24*(1), 69.

Pawsey, J. L., McCready, L. L., & Gardner, F. F. (1951). Ionospheric thermal radiation at radio frequencies. *Journal of Atmospheric and Terrestrial Physics, 1*(5–6), 261–277.

Pawsey, J. L., Payne-Scott, R., & McCready, L. L. (1946). Radio-frequency energy from the Sun. *Nature, 157*(3980), 158–159.

Pawsey, J. L., & Smerd, S. F. (1953). Solar radio emission. In *The Sun* (p. 466). The University of Chicago Press.

Pawsey, J. L., Wark, W. J., & Fallon, R. (1930). Accurate measurement of the frequency of the carrier waves of Victorian broadcast stations. *Australasian Electrical Times*.

Pawsey, J. L., & Yabsley, D. E. (1949). Solar radio-frequency radiation of thermal origin. *Australian Journal of Scientific Research A Physical Sciences, 2*, 198.

Payne-Scott, R. (1949a). The noise-like character of solar radiation at metre wavelengths. *Australian Journal of Scientific Research A Physical Sciences, 2*, 228.

Payne-Scott, R. (1949b). Bursts of solar radiation at metre wavelengths. *Australian Journal of Scientific Research A Physical Sciences, 2*, 214.

Payne-Scott, R., & Little, A. G. (1951). The position and movement on the solar disk of sources of radiation at a frequency of 97 Mc/s. II. Noise storms. *Australian Journal of Scientific Research A Physical Sciences, 4*, 508.

Payne-Scott, R., Yabsley, D. E., & Bolton, J. G. (1947). Relative times of arrival of bursts of solar noise on different radio frequencies. *Nature, 160*(4060), 256–257.

Pedlar, A., Anantharamaiah, K. R., Ekers, R. D., Goss, W. M., & Van Gorkom, J. H. (1989). Radio studies of the Galactic center. I. The Sagittarius A complex. *Astrophysical Journal, 342*.

Pearcey, T. (1953). *Use of punched cards for Fourier synthesis*. Univ. Grounds, Radiophysics Lab. (Austral.).

Pecker, J. C. 1975. *Promotion of solar physics in Europe: Introduction to a debate*. CESRA-5, Committee of European solar radio astronomers, vol. 5, p. 125.

Peels, R. (2018). Epistemic values in the humanities and in the sciences. *History of Humanities, 3*(1), 89–111.

Penzias, A. A., & Wilson, R. (1965). A measurement of excess antenna temperature at 4080 Mc/s. *The Astrophysical Journal, 142*, 419–421.

Pickering, A. (1981). The hunting of the quark. *Isis, 72*, 216–236.

Piddington, J. H., & Hindman, J. V. (1949). Solar radiation at a wavelength of 10 centimetres including eclipse observations. *Australian Journal of Scientific Research A Physical Sciences, 2*, 524.

Piddington, J. H., & Minnett, H. C. (1949). Microwave thermal radiation from the moon. *Australian Journal of Scientific Research A Physical Sciences, 2*, 63.

Piddington, J. H., & Minnett, H. C. (1951a). Observations of Galactic radiation at frequencies of 1200 and 3000 Mc/s. *Australian Journal of Scientific Research A Physical Sciences, 4*, 459.

Piddington, J. H., & Minnett, H. C. (1951b). Solar radio-frequency emission from localized regions at very high temperatures. *Australian Journal of Scientific Research A Physical Sciences, 4*, 131.

Piddington, J. H., & Oliphant, M. L. (1971). David Forbes Martyn. *Historical Records of Australian Science, 2*(2), 47–60.

Piddington, J. H., & Trent, G. H. (1956). A survey of cosmic radio emission at 600 Mc/s. *Australian Journal of Physics, 9*(4), 481–493.

Pierce, G. W. (1910). *Principles of wireless telegraphy.* McGraw-Hill.

Pinch, T. (2015). Scientific controversies. In J. Wright (Ed.), *International encyclopedia of social and behavioral sciences* (pp. 281–286). Elsevier.

Pither, A. G. (1946). *An account of the development and use of radar in the Royal Australian Air Force.* Unpublished manuscript, Australian War Memorial, Canberra.

Popper, K. R. (1963). *Conjectures and refutations. The growth of scientific knowledge (essays and lectures).* Routledge & Kegan Paul.

Porter, F. H. (1988). *Adventures in radar: A story of the. secret war on Australia's Northern frontier.* Leader.

Potnis, V. R. (1960). In I. S. Shklovsky (Ed.), *Cosmic radio waves.* Cambridge, Harvard University Press. Translated by Richard B. Rodman and Carlos M. Varsavsky.

Powell, A. (2007). *The shadow's edge: Australia's Northern war.* Charles Darwin University Press.

Price, D. J. (1986). *Little science, big science. . . and beyond* (Vol. 480). Columbia University Press.

Price, M. (2012). *Parkes @ 50 years young.* eprint arXiv:1210.0986.

Pugsley, A., & Rowe, N. E. (1981). Barnes Neville Wallis. 26 September 1887–30 October 1979. *Biographical Memoirs of Fellows of the Royal Society, 27*, 603–627.

Purcell, E. M. (1956). The question of correlation between photons in coherent light rays. *Nature, 178*(4548), 1449–1450.

Puttock, M. J., & Minnett, H. C. (1966). Instrument for rapid measurement of surface deformations of a 210 ft radio telescope. *Proceedings of the Institution of Electrical Engineers, 113*(11), 1723–1730. IET Digital Library.

Radhakrishnan, V. (2006). Olof Rydbeck and early Swedish radio astronomy: A personal perspective. *Journal of Astronomical History and Heritage, 9*, 139–144.

Radhakrishnan, V., & Cooke, D. J. (1969). Magnetic poles and the polarization structure of pulsar radiation. *Astrophysical Letters, 3*, 225.

Radhakrishnan, V., Cooke, D. J., Komesaroff, M. M., & Morris, D. (1969). Evidence in support of a rotational model for the pulsar PSR 0833–45. *Nature, 221*(5179), 443–446.

Radhakrishnan, V., & Goss, W. M. (1972). The Parkes survey of 21-CENTIMETER absorption in discrete-source spectra. V. Note on the statistics of absorbing H I concentrations in the galactic disk. *The Astrophysical Journal Supplement Series, 24*, 161.

Ratcliffe, J. A. (1971). William Henry Eccles. 1875–1966. *Biographical Memoirs of Fellows of the Royal Society, 17*, 195.

Ratcliffe, J. A. (1974a). Experimental methods of ionospheric investigation 1925–1955. *Journal of Atmospheric and Terrestrial Physics, 36*(12), 2095–2103.

Ratcliffe, J. A. (1974b). The formation of the ionosphere. Ideas of the early years (1925–1955). *Journal of Atmospheric and Terrestrial Physics, 36*(12), 2167–2181.

Ratcliffe, J. A., & Pawsey, J. L. (1933). A study of the intensity variations of downcoming waves. In *Mathematical proceedings of the Cambridge Philosophical Society* (Vol. 29, p. 301). Cambridge University Press.

Reber, G. (1940). Cosmic static. *Proceedings of the IRE, 28*(2), 68–70.

Reber, G. (1944a). Cosmic static. *The Astrophysical Journal, 100*, 279.

Reber, G. (1944b, April 11). *Measurements of the noise level picked up by an S-band aerial, RP 209.*

Reber, G. (1946). Solar radiation at 480 Mc./sec. *Nature, 158*(4026), 945–945.

Reber, G. (1959). Radio interferometry at three kilometers altitude above the Pacific Ocean: Part I. Installation and ionosphere. *Journal of Geophysical Research, 64*(3), 287–293.

Reber, G., & Greenstein, J. L. (1947). Radio-frequency investigations of astronomical interest. *The Observatory, 67*, 15–26.

Reddish, V. C. (1962). Book review. *The Observatory, 81*, 207.

Reid, M. J., Readhead, A. C. S., Vermeulen, R. C., & Treuhaft, R. N. (1999). The proper motion of Sagittarius A*. I. First VLBA results. *The Astrophysical Journal, 524*(2), 816.

Rigden, J. S. (1987). *Rabi, scientist and citizen*. Basic Books.

Rivett, R. D. (1972). *David Rivett: Fighter for Australian science*. Camberwell.

Roberts, J. A. (1954). Radio astronomy. *Research, 7*, 388–389.

Roberts, J. A. (2002). *Have Gen will travel – Imperfect images from the life of a radio astronomer*, privately published by Jim Roberts.

Robertson, P. (1992). *Beyond Southern skies: Radio astronomy and the Parkes telescope*. Cambridge University Press.

Robertson, P. (2002). Smerd, Stefan Friedrich (1916–1978). In *Australian dictionary of biography* (Vol. 16). Manchester University Press.

Robertson, P. (2017). *Radio astronomer: John Bolton and a new window on the Universe*. NewSouth Publishing.

Robertson, P., & Bland-Hawthorn, J. (2014). Centre of the Galaxy–Sixtieth anniversary of an Australian discovery. *Australian Physics, 51*, 194–199.

Robertson, P., Cozens, G., Orchiston, W., Slee, B., & Wendt, H. (2010). Early Australian optical and radio observations of Centaurus A. *Publications of the Astronomical Society of Australia, 27*(4), 402–430.

Robinson, B. J. (1959). Experimental investigations of the ionospheric E-layer. *Reports on Progress in Physics, 22*(1), 241.

Robinson, B. J., Van Damme, K. J., & Koehler, J. A. (1963). Neutral hydrogen in the Virgo cluster. *Nature, 199*(4899), 1176–1177.

Rodgers, A. W., Campbell, C. T., & Whiteoak, J. B. (1960). A catalogue of H α-emission regions in the Southern Milky Way. *Monthly Notices of the Royal Astronomical Society, 121*(1), 103–110.

Roe, J. (2007). Barnard, Marjorie (Marjory) Faith (1897–1987). In *Australian dictionary of biography* (pp. 62–64). Melbourne University Press.

Roman, N. G. (Ed.). (1958). *Comparison of the large-scale structure of the Galactic system with that of other Stellar systems*. No. 5. CUP Archive.

Rowe, A. P. (1948). *One story of radar*. Cambridge University Press.

Ruse, M. (2012). Science and values: My debt to Ernan McMullin. *Zygon, 47*, 666–685.

Ryle, M. (1949). Evidence for the Stellar origin of cosmic rays. *Proceedings of the Physical Society. Section A, 62*(8), 491.

Ryle, M. (1952). A new radio interferometer and its application to the observation of weak radio stars. *Proceedings of the Royal Society of London. Series A: Mathematical and Physical Sciences, 211*(1106), 351–375.

Ryle, M. (1955). Halley lecture. "Radio stars and their cosmological significance". *The Observatory, 75*, 137–147.

Ryle, M. (1956). Radio galaxies. *Scientific American, 195*(3), 204–223.

Ryle, M. (1956b). Letters. *Scientific American, 195*(6), 10.

Ryle, M. (1957). The Mullard radio astronomy observatory, Cambridge. *Nature, 180*(4577), 110–112.

Ryle, M. (1958). Bakerian lecture. "The nature of the cosmic radio sources". *Proceedings of the Royal Society of London Series A, 248*, 289–308.

Ryle, M. (1975). Radio telescopes of large resolving power. *Reviews of Modern Physics, 47*(3), 557.

Ryle, M., & Clarke, R. W. (1961). An examination of the steady-state model in the light of some recent observations of radio sources. *Monthly Notices of the Royal Astronomical Society, 122*(4), 349–362.

Ryle, M., Elsmore, B., & Neville, A. C. (1965). High-resolution observations of the radio sources in Cygnus and Cassiopeia. *Nature, 205*, 1259–1262.

Ryle, M., & Hewish, A. (1960). The synthesis of large radio telescopes. *Monthly Notices of the Royal Astronomical Society, 120*(3), 220–230.

Ryle, M., & Neville, A. C. (1962). A radio survey of the North Polar region with a 4.5 minute of arc pencil-beam system. *Monthly Notices of the Royal Astronomical Society, 125*(1), 39–56.

Ryle, M., & Scheuer, P. A. G. (1955). The spatial distribution and the nature of radio stars. *Proceedings of the Royal Society of London Series A, 230*, 448–462.

Ryle, M., & Smith, F. G. (1948). A new intense source of radio-frequency radiation in the constellation of Cassiopeia. *Nature, 162*(4116), 462–463.

Ryle, M., Smith, F. G., & Elsmore, B. (1950). A preliminary survey of the radio stars in the Northern Hemisphere. *Monthly Notices of the Royal Astronomical Society, 110*(6), 508–523.

Ryle, M., & Vonberg, D. D. (1946). Solar radiation on 175 Mc./s. *Nature, 158*(4010), 339–340.

Ryle, M., & Vonberg, D. D. (1947). Relation between the intensity of solar radiation on 175 Mc./s. and 80 Mc./s. *Nature, 160*(4057), 157–159.

Ryle, M., & Vonberg, D. D. (1948). An investigation of radio-frequency radiation from the Sun. *Proceedings of the Royal Society of London. Series A: Mathematical and Physical Sciences, 193*(1032), 98–120.

Saha, M. N. (1946). Conditions of escape of radio-frequency energy from the Sun and the Stars. *Nature, 158*(4016), 549–549.

Sandage, A., & McVittie, G. C. (1962). *Problems of extra-Galactic research*. IAU symposium, vol. 15, p. 359.

Schacter, D. L. (2012). Constructive memory: Past and future. *Dialogues in Clinical Neuroscience, 14*(1), 7.

Schedvin, C. B. (1987). *Shaping science and industry: A history of Australia's council for scientific and industrial research 1926–49*. CSIRO.

Schedvin, C. B. (1988). *Rivett, Sir Albert Cherbury David (1885–1961)*. Australian Dictionary of Biography, National Centre of Biography, Australian National University.

Scheuer, P. A. G. (1954, September). PhD thesis Chapter 5, submitted, Cambridge.

Scheuer, P. A. G. (1957). A statistical method for analysing observations of faint radio stars. *Proceedings of the Cambridge Philosophical Society, 53*, 764–773.

Scheuer, P. A. G. (1984). The development of aperture synthesis at Cambridge. In *The Early Years of Radio Astronomy-Reflections fifty years after Jansky's discovery* (p. 249).

Scheuer, P. A. G., & Bertotti, R. (1990). Radio source counts. *Modern Cosmology in Retrospect, 331*.

Scheuer, P. A. G., Slee, O. B., & Fryar, C. F. (1963). Apparatus for investigating the angular structure of radio sources. *Proceedings of Institute of Radio Engineers (Australia), 24*, 185–190.

Schmeck, H. M., Jr. (1961, December 5). GIANT TELESCOPE PROBING UNIVERSE; radio device in Australia is passing its initial tests. *New York Times*, 23.

Schmidt, M. (1963). 3C 273: A star-like object with large red-shift. *Nature, 197*(4872), 1040–1040.

Schmidt, M. (1968). Space distribution and luminosity functions of quasi-Stellar radio sources. *The Astrophysical Journal, 151*, 393–409.

Schott, G. A. (1912). *Electromagnetic radiation and the mechanical reactions arising from it: Being an Adams Prize Essay in the University of Cambridge*. University Press.

Schulkin, M., Haddock, F. T., Decker, K. M., Mayar, C. H., & Hagen, J. P. (1948). Observation of a solar noise burst at 9500 Mc/s and a coincident solar flare. *Physical Review, 74*(7), 840.

Schwinger, J. (1949). On the classical radiation of accelerated electrons. *Physical Review, 75*(12), 1912.

Scott, P. F., & Ryle, M. (1961). The number-flux density relation for radio sources away from the Galactic plane. *Monthly Notices of the Royal Astronomical Society, 122*(5), 389–397.

Scott, P. F., Ryle, M., & Hewish, A. (1961). First results of radio star observations using the method of aperture synthesis. *Monthly Notices of the Royal Astronomical Society, 122*(1), 95–111.

Seeger, C. L., & Williamson, R. E. (1951). The pole of the Galaxy as determined from measurements at 205 Mc/sec. *The Astrophysical Journal, 113*, 21.

Senge, P. (1990). *The fifth discipline: The art & practice of learning organization*. Doubleday Currence.

Shain, C. A. (1951). Galactic radiation at 18.3 Mc/s. *Australian Journal of Scientific Research A Physical Sciences, 4*, 258.

Shakeshaft, J. R., Ryle, M., Baldwin, J. E., Elsmore, B., & Thomson, J. H. (1955). A survey of radio sources between declinations -38° and+ 83°. *Memoirs of the Royal Astronomical Society, 67,* 106.

Sheard, H. (2017). *All the little children: The story of Victoria's baby health centres.* MCHN.

Sheridan, K. V. (1963). Techniques for the investigation of solar radio bursts at metre wavelengths. *Proceedings of the Institution of Radio Engineers Australia, 24,* 174–184.

Sherratt, T., Griffiths, T., & Robin, L. (Eds.). (2005). *A change in the weather: Climate and culture in Australia.* National Museum of Australia Press.

Shimmins, A. J., Bolton, J. G., & Wall, J. V. (1968). Counts of radio sources at 2,700 MHz. *Nature, 217*(5131), 818–820.

Shklovskii, I. S. (1949). *Astronomicheskii Zhurnal, 26,* 10.

Shklovskii, I. S. (1953). On the nature of the optical radiation from the Crab Nebula. *Doklady Akademii Nauk SSSR, 90,* 983–986. [In Russian; English translation available in Lang and Gingerich, A source book in astronomy and astrophysics 1900–1975, Harvard University Press, 1979, 490–2].

Shklovskii, I. S. (1956). *Original Russian text; 1960, English translation, published by Harvard University Press and translated by Rodman and Varsavsky) page X of the English translation of Cosmic Radio Waves, from the translation of the 1956 preface.*

Shklovskii, I. S. (1957). On the nature of the emission from the Galaxy NGC4486. In H. C. van de Hulst (Ed.), *Radio astronomy, proceedings from 4th IAU symposium.* Cambridge University Press.

Sim, H. L. (1995). *The rise and fall of the rainmakers: A history of the CSIRO cloud-seeding experiments, 1947–1981.* MSciSoc thesis, University of New South Wales, Sydney.

Simmonds, E. (Ed.). (1992). More radar yarns. *Radar Returns.*

Simmonds, E., & Mann, W. (2009). AC/DRE A G Pither, CBE (Ret'd) (16/10/1908-2/7/1971). *Radar Returns, 14*(2), 5.

Simmonds, E., & Smith, N. (Eds.). (1991). *Radar yarns: Being memories and stories collected from RAAF personnel who served in ground-based radar during World War II, or a potpourri of people, places, problems and pleasantries* (Internet ed.). E.W. & E. Simmonds.

Simmonds, E., & Smith, N. (1995). *Echoes over the Pacific: An overview of allied air warning radar in the Pacific, from Pearl Harbor to the Philippines campaign.* E.W. & E. Simmonds.

Slee, B. (1994). Some memories of the Dover Heights field station, 1946–1954. *Australian Journal of Physics, 47*(5), 517–534.

Slottje, C., Kundu, M. R., & Gergely, T. (1980). Radio physics of the Sun. In *Proceedings of IAU Symposium* (Vol. 86, pp. 195–203).

Smart, J., & Quartly, M. (2015). *Respectable radicals: A history of the National Council of Women of Australia, 1896–2006.* Monash University Publishing.

Smerd, S. F. (1950a). Radio-frequency radiation from the quiet Sun. *Australian Journal of Scientific Research A Physical Sciences, 3,* 34.

Smerd, S. F. (1950b). A radio-frequency representation of the solar atmosphere. *Proceedings of the IEE-Part III: Radio and Communication Engineering, 97*(50), 447–452.

Smerd, S. F., & Westfold, K. C. (1949). LXXVII. The characteristics of radio-frequency radiation in an ionized gas, with applications to the transfer of radiation in the solar atmosphere. *The London, Edinburgh, and Dublin Philosophical Magazine and Journal of Science, 40*(307), 831–848.

Smith, E. J. (1949). Experiments in Seeding cumuliform cloud layers with dry ice. *Australian Journal of Scientific Research A Physical Sciences, 2,* 78.

Smith, F. G. (1950). Origin of the fluctuations in the intensity of radio waves from Galactic sources: Cambridge observations. *Nature, 165*(4194), 422–423.

Smith, F. G. (1951). An accurate determination of the positions of four radio stars. *Nature, 168*(4274), 555–555.

Smith, F. G. (1952a). Apparent angular sizes of discrete radio sources: Observations at Cambridge. *Nature, 170*(4338), 1065–1065.

Smith, F. G. (1952c). The determination of the position of a radio star. *Monthly Notices of the Royal Astronomical Society, 112*(5), 497–513.

Smith, F. G. (1952b). The measurement of the angular diameter of radio stars. *Proceedings of the Physical Society. Section B, 65*(12), 971.

Smith, F. G. (1960). *Radio astronomy*. Pelican Books.

Smith, F. G. (1986). Martin Ryle. *Biographical Memoirs of Fellows of the Royal Society, 32*, 497–524.

Smith, N., & Simmonds, E. (Eds.). (1991). *Radar yarns*. E.W. & E. Simmonds.

Smith-Rose, R. L. (2021, January 9). Aleksandr Popov. *Encyclopedia Britannica*. https://www.britannica.com/biography/Aleksandr-Popov-Russian-engineer

Smyth, H. D. W. (1945). *A general account of the development of methods of using atomic energy for military purposes under the auspices of the United States government, 1940–1945*. US Government Printing Office.

Southworth, G. C. (1945). Microwave radiation from the Sun. *Journal of the Franklin Institute, 239*(4), 285–297.

Southworth, G. C. (1956). Early history of radio astronomy. *The Scientific Monthly, 82*(2), 55–66.

Standage, T. (1998). *The Victorian internet: The remarkable story of the telegraph and the nineteenth century's on-line pioneers*. New York N.Y: Berkley Books.

Stanier, H. M. (1950). Distribution of radiation from the undisturbed Sun at a wave-length of 60 cm. *Nature, 165*(4192), 354–355.

Stanley, G. J. (1994). Recollections of John G Bolton at Dover Heights and Caltech. *Australian Journal of Physics, 47*(5), 507–516.

Stanley, G. J., & Slee, O. B. (1950). Galactic radiation at radio frequencies. II. The discrete sources. *Australian Journal of Scientific Research A Physical Sciences, 3*, 234.

Staveley-Smith, L., Wilson, W. E., Bird, T. S., Disney, M. J., Ekers, R. D., Freeman, K. C., ... Wright, A. E. (1996). The Parkes 21 CM multibeam receiver. *Publications of the Astronomical Society of Australia, 13*, 243–248.

Stewart, R. T. (1985). *Solar radiophysics*. CSIRO.

Strevens, M. (2006). The role of the Matthew effect in science. *Studies in History and Philosophy of Science Part A, 37*(2), 159–170.

Struve, O. (1958). Galactic co-ordinates. *Sky and Telescope, 17*.

Struve, W. (2006). 'Dedicated to the promotion of international understanding': A memorial for Kurt Offenburg at the State Library. *The La Trobe Journal, 78*, 56–71.

Sullivan, W. T., III. (1982). *Classics in radio astronomy*. Reidel Publishing.

Sullivan, W. T., III. (1984). *The early years of radio astronomy*. Cambridge University Press.

Sullivan, W. T., III. (1988). Frank Kerr and radio waves: From wartime radar to interstellar atoms. *The Outer Galaxy, 306*, 268.

Sullivan, W. T., III. (1990). The entry of radio astronomy into cosmology: Radio stars and Martin Ryle's 2C survey. *Modern Cosmology in Retrospect*, 309–330.

Sullivan, W. T., III. (2009). *Cosmic noise: A history of early radio astronomy*. Cambridge University Press.

Sullivan, W. T., III. *Papers*. Archives, National Radio Astronomy Observatory/Associated Universities.

Swarup, G. (2008). Reminiscences regarding Professor RN Christiansen. *Journal of Astronomical History and Heritage, 11*, 194–202.

Swarup, G., & Parthasarathy, R. (1955). Solar brightness distribution at a wavelength of 60 centimetres. I. The quiet Sun. *Australian Journal of Physics, 8*(4), 487–497.

Swarup, G., & Parthasarathy, R. (1958). Solar brightness distribution at a wavelength of 60 centimetres. II. Localized radio bright regions. *Australian Journal of Physics, 11*(3), 338–349.

Swords, S. S. (1986). *Technical history of the beginnings of RADAR*. IET.

Taylor, A. H. (1948). *Radio reminiscences: A half century. Introduction: The vocabulary of radio*. Naval Research Lab .

Taylor, P. (1980). *An end to silence: The building of the overland telegraph line from Adelaide to Darwin*. Methuen.

Thomas, B. M., & Robinson, B. (2005). Harry Clive Minnett 1917–2003. *Historical Records of Australian Science, 16*(2), 199–220.

Tonks, L., & Langmuir, I. (1929). Note on "oscillations in ionized gases". *Physical Review, 33*(6), 990.

Tuve, M. (1974). Early days of pulse radio at the Carnegie Institution. *Journal of Atmospheric and Terrestrial Physics, 36*, 2079–2083.

Unsöld, A. (1949). Origin of the radio frequency emission and cosmic radiation in the Milky Way. *Nature, 163*(4143), 489–491.

van de Hulst, H. C. (1945). The origin of radio waves from space (Herkomst der radiogolven uit het wereldruim). *Nerlandsch Tijdschrift voor Natuurkunde, 11*, 210.

van de Hulst, H. C. (1957). Radio astronomy. In *Proceedings from 4th IAU symposium*. Cambridge University Press.

Van der Kruit, P. C. (2019). *Jan Hendrik Oort: Master of the Galactic system* (Vol. 459). Springer.

Vashakidze, M. A. (1954). On the degree of polarisation of the radiation of nearby extragalactic Nebulae and of the Crab Nebula. *Astr. Circ, 147*(11).

Verschuur, G. L. (1968). Positive determination of an interstellar magnetic field by measurement of the Zeeman splitting of the 21-cm hydrogen line. *Physical Review Letters, 21*(11), 775.

Vickers- Armstrong (Aircraft_ Ltd. Morpurgo). (1972). *Vickers- Armstrong British patent application No. 29248/1955*. Improvements in telescope mountings.

von Hoerner, S. (1973). Radio source counts and cosmology. *The Astrophysical Journal, 186*, 741–766.

von Laue, M. (1914). Concerning the detection of X-ray interferences. *Nobel Lectures, Physics*.

Wall, J. (2016). *How a major 20th century discovery was lost in noise and confusion, a talk presented in celebrating the history of radio astronomy in Canada*. http://astroherzberg.org/radiohistory2016/

Wall, J. V., Shimmins, A. J., & Merkelijn, J. K. (1971). The Parkes 2700 MHz survey. Catalogues for the ±4 declination zone and for the selected regions. *Australian Journal of Physics, Astrophysical Supplement, 19*.

Watson, R. C., Jr. (2009). *Radar origins worldwide: History of its evolution in 13 nations through World War II*. Trafford Publishing.

Watson-Watt, R. (1957). *Three steps to victory*. Odhams Press.

Watson-Watt, R. A., Bainbridge-Bell, L. H., Wilkins, A. F., & Bowen, E. G. (1936). Return of radio waves from the middle atmosphere. *Nature, 137*(3473), 866.

Watson-Watt, R. A., Wilkins, A. F., & Bowen, E. G. (1937). The return of radio waves from the middle atmosphere—I. *Proceedings of the Royal Society of London. Series A: Mathematical and Physical Sciences, 161*(905), 181–196.

Weaver, H., Williams, D. R., Dieter, N. H., & Lum, W. T. (1965). Observations of a strong unidentified microwave line and of emission from the OH molecule. *Nature, 208*(5005), 29–31.

Webb, S. (1990). *From the watching of shadows: The origins of radiological tomography*. CRC Press.

Weinreb, S. (1962). An attempt to measure Zeeman splitting of the Galactic 21-CM hydrogen line. *The Astrophysical Journal, 136*, 1149–1152.

Weinreb, S., Barrett, A. H., Meeks, M. L., & Henry, J. C. (1963). Radio observations of OH in the interstellar medium. *Nature, 200*(4909), 829–831.

Weiss, A. A. (1963). The positions and movements of the sources of solar radio bursts of spectral type II. *Australian Journal of Physics, 16*(2), 240–271.

Wendt, H. (2011). Paul Wild and his investigation of the H-line. In *Highlighting the history of astronomy in the Asia-Pacific region* (pp. 543–546). Springer.

Wendt, H., & Orchiston, W. (2018). The contribution of the AN/TPS-3 radar antenna to Australian radio astronomy. *Journal of Astronomical History and Heritage, 21*(1), 65–80.

Wendt, H., Orchiston, W., & Slee, B. (2008). WN Christiansen and the initial Australian investigation of the 21 cm hydrogen line. *Journal of Astronomical History and Heritage, 11*, 185–193.

Westerhout, G. (1958). A survey of the continuous radiation from the Galactic system at a frequency of 1390 Mc/s. *Bulletin of the Astronomical Institutes of the Netherlands, 14*, 215.

Westfold, K. C. (1949). The wave equations for electromagnetic radiation in an ionized medium in a magnetic field. *Australian Journal of Scientific Research A Physical Sciences, 2*, 169.

Westfold, K. C. (1959). The polarisation of synchrotron radiation. *The Astrophysical Journal, 130*, 241.

Westfold, K. C. (1994). John Bolton: Some early memories. *Australian Journal of Physics, 47*(5), 535–540.

White, F. W. G. (1934). On the automatic registration of the amplitude of downcoming wireless waves part I. *Proceedings of the Physical Society, 46*, 805.

Wielebinski, R., & Shakeshaft, J. R. (1962). Faraday rotation of polarised Galactic radio emission. *Nature, 195*(4845), 982–983.

Wild, J. P. (1950). Observations of the spectrum of high-intensity solar radiation at metre wavelengths. II. Outbursts. *Australian Journal of Scientific Research A Physical Sciences, 3*, 399.

Wild, J. P. (1950b). Observations of the spectrum of high-intensity solar radiation at metre wavelengths. III. Isolated bursts. *Australian Journal of Scientific Research A Physical Sciences, 3*, 541.

Wild, J. P. (1951). Observations of the spectrum of high-intensity solar radiation at metre wavelengths. IV. Enhanced radiation. *Australian Journal of Scientific Research A Physical Sciences, 4*, 36.

Wild, J. P. (1952). The radio-frequency line spectrum of atomic hydrogen and its applications in astronomy. *The Astrophysical Journal, 115*, 206.

Wild, J. P. (1955). Outbursts of radio noise from the Sun. *VA, 1*(1), 573–584.

Wild, J. P. (1967). The culgoora radioheliograph. *Proceedings on Institution of Radio & Electronics Engineers Australia, 28*. 277, 279–291.

Wild, J. P. (1968). The exploration of the Sun by radio. *The Australian Physicist, 5*, 117.

Wild, J. P. (1985). In D. J. McLean & N. R. Labrum (Eds.), *The beginning in solar radiophysics in solar radiophysics: Studies of emission from the Sun at metre wavelengths* (pp. 3–17). Cambridge University Press.

Wild, J. P. (1987). The beginnings of radio astronomy in Australia. *Publications of the Astronomical Society of Australia, 7*(1), 95–102.

Wild, J. P. (1980). The SF Smerd memorial lecture: The Sun of Stefan Smerd. In *Symposium-international astronomical union* (Vol. 86, pp. 5–21). Cambridge University Press.

Wild, J. P., & McCready, L. L. (1950). Observations of the spectrum of high-intensity solar radiation at metre wavelengths. I. The apparatus and spectral types of solar burst observed. *Australian Journal of Scientific Research A Physical Sciences, 3*, 387.

Wild, J. P., Murray, J. D., & Rowe, W. C. (1953). Evidence of harmonics in the spectrum of a solar radio outburst. *Nature, 172*(4377), 533–534.

Wild, J. P., & Roberts, J. A. (1956). Regions of the ionosphere responsible for radio star scintillations. *Nature, 178*(4529), 377–378.

Wild, J. P., & Roberts, J. A. (1956b). The spectrum of radio-star scintillations and the nature of irregularities in the ionosphere. *Journal of Atmospheric and Terrestrial Physics, 8*(1-2), 55–75.

Wild, J. P., Roberts, J. A., & Murray, J. D. (1954). Radio evidence of the ejection of very fast particles from the Sun. *Nature, 173*(4403), 532–534.

Wild, J. P., & Sheridan, K. V. (1958). A swept-frequency interferometer for the study of high-intensity solar radiation at meter wavelengths. *Proceedings of the IRE, 46*(1), 160–171.

Wild, J. P., Sheridan, K. V., & Neylan, A. A. (1959a). An investigation of the speed of the solar disturbances responsible for type III radio bursts. *Australian Journal of Physics, 12*(4), 369–398.

Wild, J. P., Sheridan, K. V., & Trent, G. H. (1959b). The transverse motions of the sources of solar radio bursts. In *Symposium-International Astronomical Union* (Vol. 9, pp. 176–185). Cambridge University Press.

Wild, J. P., Smerd, S. F., & Weiss, A. A. (1963). Solar bursts. *Annual Review of Astronomy and Astrophysics, 1*, 291.

Wilkinson, P. N., Kellermann, K. I., Ekers, R. D., Cordes, J. M., & Lazio, T. J. W. (2004). The exploration of the unknown. *New Astronomy Reviews, 48*(11-12), 1551–1563.

Wilkinson, P., Kennewell, J. A., & Cole, D. (2018). The development of the Australian Space Forecast Centre (ASFC). *History of Geo- and Space Science, 9*(1), 53–63.

Williamson, R. E. (1948). The present status of microwave astronomy. *Journal of the Royal Astronomical Society of Canada, 42*, 9.

Willis, J. B., & Deane, J. F. (2006). Trevor pearcey and the first australian computer: A lost opportunity? *Historical Records of Australian Science, 17*(2), 209–225.

Wilson, D. B. (1971). The thought of late Victorian physicists: Oliver Lodge's Ethereal body. *Victorian Studies, 15*(1), 29–48.

Woolley, R. V. D. R. (1947). Galactic noise. *Monthly Notices of the Royal Astronomical Society, 107*, 308.

Wright, K. O. (1980). Obituary—Beals, Carlyle S. *Quarterly Journal of the Royal Astronomical Society, 21*, 212.

Yeang, C.-P. (2012). From mechanical objectivity to instrumentalizing theory: Inventing radio ionospheric sounders. *Historical Studies in the Natural Sciences, 42*(3), 190–234.

Yeang, C.-P. (2013). *Probing the sky with radio waves: From wireless technology to the development of atmospheric science.* University of Chicago Press.

Zachary, G. P. (2018). *Endless frontier: Vannevar Bush, engineer of the American century.* Simon and Schuster.

Zernike, F. (1938). The concept of degree of coherence and its application to optical problems. *Physica, 5*(8), 785–795.

Zernike, F. (1955). How I discovered phase contrast. *Science, 121*(3141), 345–349.

Zimmerman, D. (1996). *Top secret exchange: The Tizard mission and the scientific war.* McGill-Queen's Press-MQUP.

Index[1]

---

[1]Note, "f" following page numbers indicates figure or figure caption, *passim* indicates numerous, scattered mentions within page range.